PHYSICS

PRINCIPLES with APPLICATIONS

Fundamental Constants

Quantity	Symbol	Approximate Value	Current Best Value[†]
Speed of light in vacuum	c	3.00×10^8 m/s	2.99792458×10^8 m/s
Gravitational constant	G	6.67×10^{-11} N·m²/kg²	$6.6742(10) \times 10^{-11}$ N·m²/kg²
Avogadro's number	N_A	6.02×10^{23} mol⁻¹	$6.0221415(10) \times 10^{23}$ mol⁻¹
Gas constant	R	8.314 J/mol·K $= 1.99$ cal/mol·K $= 0.0821$ L·atm/mol·K	$8.314472(15)$ J/mol·K
Boltzmann's constant	k	1.38×10^{-23} J/K	$1.3806505(24) \times 10^{-23}$ J/K
Charge on electron	e	1.60×10^{-19} C	$1.60217653(14) \times 10^{-19}$ C
Stefan-Boltzmann constant	σ	5.67×10^{-8} W/m²·K⁴	$5.670400(40) \times 10^{-8}$ W/m²·K⁴
Permittivity of free space	$\epsilon_0 = \left(1/c^2\mu_0\right)$	8.85×10^{-12} C²/N·m²	$8.854187817\ldots \times 10^{-12}$ C²/N·m²
Permeability of free space	μ_0	$4\pi \times 10^{-7}$ T·m/A	$1.2566370614\ldots \times 10^{-6}$ T·m/A
Planck's constant	h	6.63×10^{-34} J·s	$6.6260693(11) \times 10^{-34}$ J·s
Electron rest mass	m_e	9.11×10^{-31} kg $= 0.000549$ u $= 0.511$ MeV/c^2	$9.1093826(16) \times 10^{-31}$ kg $= 5.4857990945(24) \times 10^{-4}$ u
Proton rest mass	m_p	1.6726×10^{-27} kg $= 1.00728$ u $= 938.3$ MeV/c^2	$1.67262171(29) \times 10^{-27}$ kg $= 1.00727646688(13)$ u
Neutron rest mass	m_n	1.6749×10^{-27} kg $= 1.008665$ u $= 939.6$ MeV/c^2	$1.67492728(29) \times 10^{-27}$ kg $= 1.00866491560(55)$ u
Atomic mass unit (1 u)		1.6605×10^{-27} kg $= 931.5$ MeV/c^2	$1.66053886(28) \times 10^{-27}$ kg $= 931.494043(80)$ MeV/c^2

[†] CODATA (12/03), Peter J. Mohr and Barry N. Taylor, National Institute of Standards and Technology. Numbers in parentheses indicate one-standard-deviation experimental uncertainties in final digits. Values without parentheses are exact (i.e., defined quantities).

Other Useful Data

Joule equivalent (1 cal)	4.186 J
Absolute zero (0 K)	-273.15°C
Acceleration due to gravity at Earth's surface (avg.)	9.80 m/s² ($= g$)
Speed of sound in air (20°C)	343 m/s
Density of air (dry)	1.29 kg/m³
Earth: Mass	5.98×10^{24} kg
Radius (mean)	6.38×10^3 km
Moon: Mass	7.35×10^{22} kg
Radius (mean)	1.74×10^3 km
Sun: Mass	1.99×10^{30} kg
Radius (mean)	6.96×10^5 km
Earth–Sun distance (mean)	149.6×10^6 km
Earth–Moon distance (mean)	384×10^3 km

The Greek Alphabet

Alpha	A	α	Nu	N	ν
Beta	B	β	Xi	Ξ	ξ
Gamma	Γ	γ	Omicron	O	o
Delta	Δ	δ	Pi	Π	π
Epsilon	E	ε	Rho	P	ρ
Zeta	Z	ζ	Sigma	Σ	σ
Eta	H	η	Tau	T	τ
Theta	Θ	θ	Upsilon	Y	υ
Iota	I	ι	Phi	Φ	ϕ, φ
Kappa	K	κ	Chi	X	χ
Lambda	Λ	λ	Psi	Ψ	ψ
Mu	M	μ	Omega	Ω	ω

Values of Some Numbers

$\pi = 3.1415927$	$\sqrt{2} = 1.4142136$	$\ln 2 = 0.6931472$	$\log_{10} e = 0.4342945$
$e = 2.7182818$	$\sqrt{3} = 1.7320508$	$\ln 10 = 2.3025851$	1 rad $= 57.2957795°$

Mathematical Signs and Symbols

$\propto$	is proportional to	$\leq$	is less than or equal to
$=$	is equal to	$\geq$	is greater than or equal to
$\approx$	is approximately equal to	Σ	sum of
$\neq$	is not equal to	$\bar{x}$	average value of x
$>$	is greater than	Δx	change in x
$\gg$	is much greater than	$\Delta x \to 0$	Δx approaches zero
$<$	is less than	$n!$	$n(n-1)(n-2)\ldots(1)$
$\ll$	is much less than		

Properties of Water

Density (4°C)	1.000 kg/m³
Heat of fusion (0°C)	333 kJ/kg (80 kcal/kg)
Heat of vaporization (100°C)	2260 kJ/kg (539 kcal/kg)
Specific heat (15°C)	4186 J/kg·C° (1.00 kcal/kg·C°)
Index of refraction	1.33

Unit Conversions (Equivalents)

Length

1 in. = 2.54 cm
1 cm = 0.3937 in.
1 ft = 30.48 cm
1 m = 39.37 in. = 3.281 ft
1 mi = 5280 ft = 1.609 km
1 km = 0.6214 mi
1 nautical mile (U.S.) = 1.151 mi = 6076 ft = 1.852 km
1 fermi = 1 femtometer (fm) = 10^{-15} m
1 angstrom (Å) = 10^{-10} m = 0.1 nm
1 light-year (ly) = 9.461×10^{15} m
1 parsec = 3.26 ly = 3.09×10^{16} m

Volume

1 liter (L) = 1000 mL = 1000 cm^3 = 1.0×10^{-3} m^3 =
   1.057 qt (U.S.) = 61.02 in.^{3}
1 gal (U.S.) = 4 qt (U.S.) = 231 in.^{3} = 3.785 L =
   0.8327 gal (British)
1 quart (U.S.) = 2 pints (U.S.) = 946 mL
1 pint (British) = 1.20 pints (U.S.) = 568 mL
1 m^3 = 35.31 ft^3

Speed

1 mi/h = 1.467 ft/s = 1.609 km/h = 0.447 m/s
1 km/h = 0.278 m/s = 0.621 mi/h
1 ft/s = 0.305 m/s = 0.682 mi/h
1 m/s = 3.281 ft/s = 3.600 km/h = 2.237 mi/h
1 knot = 1.151 mi/h = 0.5144 m/s

Angle

1 radian (rad) = 57.30° = 57°18′
1° = 0.01745 rad
1 rev/min (rpm) = 0.1047 rad/s

Time

1 day = 8.64×10^4 s
1 year = 3.156×10^7 s

Mass

1 atomic mass unit (u) = 1.6605×10^{-27} kg
1 kg = 0.0685 slug
[1 kg has a weight of 2.20 lb where g = 9.80 m/s^2.]

Force

1 lb = 4.45 N
1 N = 10^5 dyne = 0.225 lb

Energy and Work

1 J = 10^7 ergs = 0.738 ft·lb
1 ft·lb = 1.36 J = 1.29×10^{-3} Btu = 3.24×10^{-4} kcal
1 kcal = 4.186×10^3 J = 3.97 Btu
1 eV = 1.602×10^{-19} J
1 kWh = 3.60×10^6 J = 860 kcal

Power

1 W = 1 J/s = 0.738 ft·lb/s = 3.42 Btu/h
1 hp = 550 ft·lb/s = 746 W

Pressure

1 atm = 1.013 bar = 1.013×10^5 N/m^2
      = 14.7 lb/in.^{2} = 760 torr
1 lb/in.^{2} = 6.90×10^3 N/m^2
1 Pa = 1 N/m^2 = 1.45×10^{-4} lb/in.^{2}

SI Derived Units and Their Abbreviations

Quantity	Unit	Abbreviation	In Terms of Base Units[†]
Force	newton	N	kg·m/s^2
Energy and work	joule	J	kg·m^2/s^2
Power	watt	W	kg·m^2/s^3
Pressure	pascal	Pa	kg/(m·s^2)
Frequency	hertz	Hz	s^{-1}
Electric charge	coulomb	C	A·s
Electric potential	volt	V	kg·m^2/(A·s^3)
Electric resistance	ohm	Ω	kg·m^2/(A^2·s^3)
Capacitance	farad	F	A^2·s^4/(kg·m^2)
Magnetic field	tesla	T	kg/(A·s^2)
Magnetic flux	weber	Wb	kg·m^2/(A·s^2)
Inductance	henry	H	kg·m^2/(s^2·A^2)

[†] kg = kilogram (mass), m = meter (length), s = second (time), A = ampere (electric current).

Metric (SI) Multipliers

Prefix	Abbreviation	Value
yotta	Y	10^{24}
zeta	Z	10^{21}
exa	E	10^{18}
peta	P	10^{15}
tera	T	10^{12}
giga	G	10^9
mega	M	10^6
kilo	k	10^3
hecto	h	10^2
deka	da	10^1
deci	d	10^{-1}
centi	c	10^{-2}
milli	m	10^{-3}
micro	μ	10^{-6}
nano	n	10^{-9}
pico	p	10^{-12}
femto	f	10^{-15}
atto	a	10^{-18}
zepto	z	10^{-21}
yocto	y	10^{-24}

PHYSICS

PRINCIPLES WITH APPLICATIONS

SIXTH EDITION

Volume 1

DOUGLAS C. GIANCOLI

PEARSON
Prentice
Hall

Upper Saddle River, New Jersey 07458

Library of Congress Cataloging-in-Publication Data

Giancoli, Douglas C.
  Physics : principles with applications / Douglas C. Giancoli.-- 6th ed.
    p.   cm.
  Includes index.
  ISBN 0-13-035257-8 (vol.2 pbk. : alk. paper) — ISBN 0-13-060620-0 (full book vol 1 & 2. casebound. : alk. paper) —
  ISBN 0-13-035256-X (vol 1 reprint. pbk.. : alk. paper) — ISBN 0-13-184661-2 (Nasta edition : alk. paper) —
  ISBN 0-13-191183-X (International edition : alk. paper)

  1. Physics.   I. Title.

QC23.G399 2005
530—dc22   2004017226

Editor-in-Chief, Science: John Challice
Senior Acquisitions Editor: Erik Fahlgren
Senior Development Editor: Karen Karlin
Senior Production Editor: Susan Fisher
Production Editor: Chirag Thakkar
Vice President of Production and Manufacturing: David Riccardi
Executive Managing Editor: Kathleen Schiaparelli
Manufacturing Manager: Trudy Pisciotti
Manufacturing Buyer: Alan Fischer
Managing Editor, Audio and Visual Assets: Patricia Burns
AV Project Managers: Adam Velthaus and Connie Long
Assistant Managing Editor, Science Media: Nicole Bush
Associate Editor: Christian Botting
Media Editor: Michael J. Richards

Director of Creative Services: Paul Belfanti
Advertising and Promotions Manager: Elise Schneider
Creative Director: Carole Anson
Art Director: Maureen Eide
Illustration: Artworks
Marketing Manager: Mark Pfaltzgraff
Editor-in-Chief of Development: Carol Trueheart
Director, Image Research Center: Melinda Reo
Photo Research: Mary Teresa Giancoli and Jerry Marshall
Manager, Rights and Permissions: Cynthia Vincenti
Copy Editor: Jocelyn Phillips
Indexer: Steele/Katigbak
Editorial Assistant: Andrew Sobel
Composition: Emilcomp srl / Prepare Inc.

Cover Photo: K2, the world's second highest summit, 8611 m (see pp. 10–11) as seen from Concordia. It is said to be the most difficult of the world's highest peaks (over 8000 m—see Table 1–6 and Example 1–3), and was first climbed by Lino Lacedelli and Achille Compagnoni in 1954. (Art Wolfe/Getty Images, Inc.)

Printed in the United States of America
10 9 8 7 6 5 4 3 2

ISBN   0-13-035256-X

Pearson Education Ltd., *London*
Pearson Education Australia Pty., Limited, *Sydney*
Pearson Education Singapore, Pte. Ltd.
Pearson Education North Asia Ltd., *Hong Kong*
Pearson Education Canada, Ltd., *Toronto*
Pearson Educación de Mexico, S.A. de C.V.
Pearson Education—Japan, *Tokyo*
Pearson Education Malaysia, Pte. Ltd.

CONTENTS OF VOLUME 1

CONTENTS OF VOLUME 2

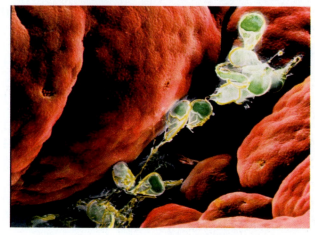

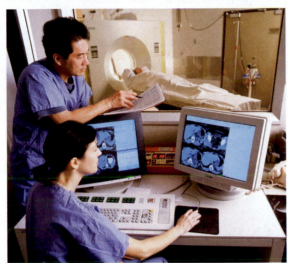

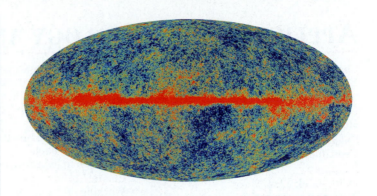

APPLICATIONS TO BIOLOGY AND MEDICINE

APPLICATIONS TO OTHER FIELDS AND EVERYDAY LIFE

PROBLEM SOLVING BOXES

PREFACE

See the World through Eyes that Know Physics

This book is written for students. It has been written to give students a thorough understanding of the basic concepts of physics in all its aspects, from mechanics to modern physics. It aims to explain physics in a readable and interesting manner that is accessible and clear, and to teach students by anticipating their needs and difficulties without oversimplifying. A second objective is to show students how useful physics is in their own lives and future professions by means of interesting applications. In addition, much effort has gone into techniques and approaches for solving problems.

This textbook is especially suited for students taking a one-year introductory course in physics that uses algebra and trigonometry but not calculus. Many of these students are majoring in biology or (pre)medicine, and others may be in architecture, technology, or the earth or environmental sciences. Many applications to these fields are intended to answer that common student query: "Why must I study physics?" The answer is that physics is fundamental to a full understanding of these fields, and here they can see how. Physics is all about us in the everyday world. It is the goal of this book to help students "see the world through eyes that know physics."

NEW ▶ Some of the new features in this sixth edition include (1) in-text Exercises for students to check their understanding; (2) new Approach paragraphs for worked-out Examples; (3) new Examples that step-by-step follow each Problem Solving Box; (4) new physics such as a rigorously updated Chapter 33 on cosmology and astrophysics to reflect the latest results in the recent "Cosmological Revolution"; and (5) new applications such as detailed physics-based descriptions of liquid crystal screens (LCD), digital cameras (with CCD), and expanded coverage of electrical safety and devices. These and other new aspects are highlighted below.

Physics and How to Understand It

I have avoided the common, dry, dogmatic approach of treating topics formally and abstractly first, and only later relating the material to the students' own experience. My approach is to recognize that physics is a description of reality and thus to start each topic with concrete observations and experiences that students can directly relate to. Then we move on to the generalizations and more formal treatment of the topic. Not only does this make the material more interesting and easier to understand, but it is closer to the way physics is actually practiced.

A major effort has been made to not throw too much at students reading the first few chapters. The basics have to be learned first; many aspects can come later, when the students are more prepared. If we don't overwhelm students with too much detail, especially at the start, maybe they can find physics interesting, fun, and helpful—and those who were afraid may lose their fear.

The *great laws of physics* are emphasized by giving them a tan-colored screen and a marginal note in capital letters enclosed in a rectangle. All important equations are given a number to distinguish them from less useful ones. To help make clear which equations are general and which are not, the limitations of important equations are given in brackets next to the equation, such as

$$x = x_0 + v_0 t + \tfrac{1}{2} a t^2. \qquad \text{[constant acceleration]}$$

Mathematics can be an obstacle to student understanding. I have aimed at including all steps in a derivation. Important mathematical tools, such as addition of vectors and trigonometry, are incorporated in the text where first needed, so they come with a context rather than in a scary introductory Chapter. Appendices contain a review of algebra and geometry (plus a few advanced topics: rotating reference frames, inertial forces, Coriolis effect; heat capacities of gases and equipartition of energy; Lorentz transformations). Système International (SI) units are used throughout. Other metric and British units are defined for informational purposes.

Chapter 1 is not a throwaway. It is fundamental to physics to realize that every measurement has an *uncertainty*, and how significant figures are used to reflect that. Converting units and being able to make rapid *estimates* are also basic. The cultural aspects at the start of Chapter 1 broaden a person's understanding of the world but do not have to be covered in class.

The many *applications* sometimes serve only as examples of physical principles. Others are treated in depth. They have been carefully chosen and integrated into the text so as not to interfere with the development of the physics, but rather to illuminate it. To make it easy to spot the applications, a Physics Applied marginal note is placed in the margin.

Color is used pedagogically to bring out the physics. Different types of vectors are given different colors (see the chart on page xxv). This book has been printed in 5 colors (5 passes through the presses) to provide better variety and definition for illustrating vectors and other concepts such as fields and rays. The photographs opening each Chapter, some of which have vectors superimposed on them, have been chosen so that the accompanying caption can be a sort of summary of the Chapter.

Some of the **new** aspects of physics and pedagogy in this sixth edition are:

Cosmological Revolution: The latest results in cosmology and astrophysics are presented with the generous help of top experts in the field. We give readers the latest results and interpretations from the present, ongoing "Golden Age of Cosmology."

Greater clarity: No topic, no paragraph in this book was overlooked in the search to improve the clarity of the presentation. Many changes and clarifications have been made, both small and not so small. One goal has been to eliminate phrases and sentences that may slow down the principle argument: keep to the essentials at first, give the elaborations later.

Vector notation, arrows: The symbols for vector quantities in the text and Figures now have a tiny arrow over them, so they are similar to what a professor writes by hand in lecture. The letters are still the traditional boldface: thus $\vec{\mathbf{v}}$ for velocity, $\vec{\mathbf{F}}$ for force.   ◄ N E W

Exercises within the text, for students to check their understanding. Answers are given at the end of the Chapter.   ◄ N E W

Step-by Step Examples, after a Problem Solving Box, as discussed on page xvii.   ◄ N E W

Conceptual Examples are not a new feature, but there are some new ones.

Examples modified: more math steps are spelled out, and many new Examples added: see page xvii.

Page layout: Complete Derivations. Even more than in the previous edition, serious attention has been paid to how each page is formatted. Great effort has been made to keep important derivations and arguments on facing pages. Students then don't have to turn back and forth. Throughout the book readers see before them, on two facing pages, an important slice of physics.

Subheads: Many of the Sections within a Chapter are now divided into subsections, thus breaking up the topics into more manageable "bites." They allow "pauses" for the students to rest or catch their breath.   ◄ N E W

NEW ▶ ***Marginal notes: Caution.*** Margin notes, in blue, point out main topics acting as a sort of outline and as an aid to find topics in review. They also point out applications and problem-solving hints. A new type, labeled CAUTION, points out possible misunderstandings discussed in the adjacent text.

Deletions. To keep the book from being too long, and also to reduce the burden on students in more advanced topics, many topics have been shortened or streamlined, and a few dropped.

New Physics Topics and Major Revisions

Here is a list of major changes or additions, but there are many others:

Symmetry used more, including for solving Problems

NEW ▶ Dimensional analysis, optional (Ch. 1)

More graphs in kinematics (Ch. 2)

Engine efficiency (Chs. 6, 15)

Work-energy principle, and conservation of energy: new subsection (Ch. 6); carried through in thermodynamics (Ch. 15) and electricity (Ch. 17)

NEW ▶ Force on tennis ball by racket (Ch. 7)

NEW ▶ Airplane wings, curve balls, sailboats, and other applications of Bernoulli's principle: improved and clarified with new material (Ch. 10)

Distinguish wave interference in space and in time (beats) (Ch. 11)

Doppler shift for light (Ch. 12 now, as well as Ch. 33)

NEW ▶ Giant star radius (Ch. 14)

First law of thermodynamics rewritten and extended, connected better to work-energy principle and energy conservation (Ch. 15)

Energy resources shortened (Ch. 15)

NEW ▶ SEER rating (Ch. 15)

NEW ▶ Separation of charge in nonconductors (Ch. 16)

NEW ▶ Gauss's law, optional (Ch. 16)

NEW ▶ Photocopiers and computer printers (Ch. 16)

Electric force and field directions emphasized more (Chs. 16, 17)

Electric potential related better to work, more detail (Ch. 17)

NEW ▶ Dielectric effect on capacitor with and without connection to voltage plus other details (Ch. 17)

NEW ▶ Parallel-plate capacitor derivation, optional (Ch. 17)

NEW ▶ Electric hazards, grounding, safety, current interrupters: expanded with much new material (Chs. 17, 18, 19 especially, 20, 21)

NEW ▶ Electric current, misconceptions discussed in Chapter 18

Superconductivity updated (Ch. 18)

Terminal voltage and emf reorganized, with more detail (Ch. 19)

Magnetic materials shortened (Ch. 20)

NEW ▶ Right-hand rules summarized in a Table (Ch. 20)

Faraday's and Lenz's laws expanded (Ch. 21)

AC circuits shortened (Ch. 21), displacement current downplayed (Ch. 22)

NEW ▶ Radiation pressure and momentum of EM waves (Ch. 22)

NEW ▶ Where to see yourself in a mirror; where you can actually *see* a lens image (Ch. 23)

NEW ▶ Liquid crystal displays (LCD) (Ch. 24)

NEW ▶ Physics behind digital cameras and CCD (Ch. 25)

NEW ▶ Seeing under water (Ch. 25)

Relativistic mass redone (Ch. 26)

NEW ▶ Revolutionary results in cosmology: flatness and age of universe, WMAP, SDSS, dark matter, and dark energy (Ch. 33)

NEW ▶ Specific heats of gases, equipartition of energy (Appendix)

Problem Solving, with New and Improved Approaches

Being able to solve problems is a valuable technique in general. Solving problems is also an effective way to understand the physics more deeply. Here are some of the ways this book uses to help students become effective problem solvers.

Problem Solving Boxes, about 20 of them, are found throughout the book (there is a list on p. xiii.). Each one outlines a step-by-step approach to solving problems in general, or specifically for the material being covered. The best students may find these "boxes" unnecessary (they can skip them), but many students may find it helpful to be reminded of the general approach and of steps they can take to get started. The general Problem Solving Box in Section 4–9 is placed there, after students have had some experience wrestling with problems, so they may be motivated to read it with close attention. Section 4–9 can be covered earlier if desired. Problem Solving Boxes are not intended to be a prescription, but rather a guide. Hence they sometimes follow the Examples to serve as a summary for future use.

Problem Solving Sections (such as Sections 2–6, 3–6, 4–7, 6–7, 8–6, and 13–8) are intended to provide extra drill in areas where solving problems is especially important.

Examples: Worked-out Examples, each with a title for easy reference, fall into four categories:

(1) The majority are regular worked-out Examples that serve as "practice problems." New ones have been added, a few old ones have been dropped, and many have been reworked to provide greater clarity, more math steps, more of "why we do it this way," and with the new Approach paragraph more discussion of the reasoning and approach. The aim is to "think aloud" with the students, leading them to develop insight. The level of the worked-out Examples for most topics increases gradually, with the more complicated ones being on a par with the most difficult Problems at the end of each Chapter. Many Examples provide relevant applications to various fields and to everyday life.

(2) *Step-by-step Examples:* After many of the Problem Solving Boxes, the next Example is done step-by-step following the steps of the preceding Box, just to show students how the Box can be used. Such solutions are long and can be redundant, so only one of each type is done in this manner.   ◀ N E W

(3) *Estimating Examples,* roughly 10% of the total, are intended to develop the skills for making order-of-magnitude estimates, even when the data are scarce, and even when you might never have guessed that any result was possible at all. See, for example, Section 1–7, Examples 1–6 to 1–9.

(4) *Conceptual Examples:* Each is a brief Socratic question intended to stimulate student response before reading the Response given.

APPROACH paragraph: Worked-out numerical Examples now all have a short introductory paragraph before the Solution, outlining an approach and the steps we can take to solve the given problem.   ◀ N E W

NOTE: Many Examples now have a brief "note" after the Solution, sometimes remarking on the Solution itself, sometimes mentioning an application, sometimes giving an alternate approach to solving the problem. These new Note paragraphs let the student know the Solution is finished, and now we mention a related issue(s).   ◀ N E W

Additional Examples: Some physics subjects require many different worked-out Examples to clarify the issues. But so many Examples in a row can be overwhelming to some students. In those places, a subhead "Additional Example(s)" is meant to suggest to students that they could skip these in a first reading. When students include them during a second reading of the Chapter, they can give power to solve a greater range of Problems.   ◀ N E W

Exercises within the text, after an Example or a derivation, which give students a chance to see if they have understood enough to answer a simple question or do a simple calculation. Answers are given at the bottom of the last page of each Chapter.   ◀ N E W

Problems at the end of each Chapter have been increased in quality and quantity. Some old ones have been replaced or rewritten to make them clearer, and/or have had their numerical values changed. Each Chapter contains a large group of Problems arranged by Section and graded according to (approximate) difficulty: level I Problems are simple, designed to give students confidence; level II are "normal" Problems, providing more of a challenge and often the combination of two different concepts; level III are the most complex and are intended as "extra credit" Problems that will challenge even superior students. The arrangement by Section number is to help the instructors choose which material they want to emphasize, and means that those Problems depend on material up to and including that Section: earlier material may also be relied upon.

General Problems are unranked and grouped together at the end of each Chapter, accounting for perhaps 30% of all Problems. These are not necessarily more difficult, but they may be more likely to call on material from earlier Chapters. They are useful for instructors who want to give students a few Problems without the clue as to what Section must be referred to or how hard they are.

Questions, also at the end of each Chapter, are conceptual. They help students to use and apply the principles and concepts, and thus deepen their understanding (or let them know they need to study more).

Assigning Problems

I suggest that instructors assign a significant number of the level I and level II Problems, as well as a small number of General Problems, and reserve level III Problems only as "extra credit" to stimulate the best students. Although most level I problems may seem easy, they help to build self-confidence—an important part of learning, especially in physics. Answers to odd-numbered Problems are given in the back of the book.

Organization

The general outline of this new edition retains a traditional order of topics: mechanics (Chapters 1 to 9); fluids, vibrations, waves, and sound (Chapters 10 to 12); kinetic theory and thermodynamics (Chapters 13 to 15); electricity and magnetism (Chapters 16 to 22); light (Chapters 23 to 25); and modern physics (Chapters 26 to 33). Nearly all topics customarily taught in introductory physics courses are included here.

The tradition of beginning with mechanics is sensible because it was developed first, historically, and because so much else in physics depends on it. Within mechanics, there are various ways to order topics, and this book allows for considerable flexibility. I prefer to cover statics after dynamics, partly because many students have trouble with the concept of force without motion. Furthermore, statics is a special case of dynamics—we study statics so that we can prevent structures from becoming dynamic (falling down). Nonetheless, statics (Chapter 9) could be covered earlier after a brief introduction to vectors. Another option is light, which I have placed after electricity and magnetism and EM waves. But light could be treated immediately after waves (Chapter 11). Special relativity (Chapter 26) could be treated along with mechanics, if desired—say, after Chapter 7.

Not every Chapter need be given equal weight. Whereas Chapter 4 or Chapter 21 might require $1\frac{1}{2}$ to 2 weeks of coverage, Chapter 12 or 22 may need only $\frac{1}{2}$ week or less. Because Chapter 11 covers standing waves, Chapter 12 could be left to the students to read on their own if little class time is available.

The book contains more material than can be covered in most one-year courses. Yet there is great flexibility in choice of topics. Sections marked with a star (*) are considered optional. They contain slightly more advanced physics material (perhaps material not usually covered in typical courses) and/or interesting applications. They contain no material needed in later Chapters, except perhaps in later optional Sections. Not all unstarred Sections must be covered; there remains considerable flexibility in the choice of material. For a brief course, all optional material could be dropped, as well as major parts of Chapters 10, 12, 19, 22, 28, 29, 32, and 33, and perhaps selected parts of Chapters 7, 8, 9, 15, 21, 24, 25, and 31. Topics not covered in class can be a resource to students for later study.

New Applications

Relevant applications of physics to biology and medicine, as well as to architecture, other fields, and everyday life, have always been a strong feature of this book, and continue to be. Applications are interesting in themselves, plus they answer the students' question, "Why must I study physics?" New applications have been added. Here are a few of the new ones (see list after Table of Contents, pages xii and xiii).

Digital cameras, charge coupled devices (CCD) (Ch. 25)

Liquid Crystal Displays (LCD) (Ch. 24)

Electric safety, hazards, and various types of current interrupters and circuit breakers (Chs. 17, 18, 19, 20, 21)

Photocopy machines (Ch. 16)

Inkjet and Laser printers (Ch. 16)

World's tallest peaks (unit conversion, Ch. 1)

Airport metal detectors (Ch. 21)

Capacitor uses (Ch. 17)

Underwater vision (Ch. 25)

SEER rating (Ch. 15)

Curve ball (Ch. 10)

Jump starting a car (Ch. 19)

RC circuits in pacemakers, turn signals, wipers (Ch. 19)

Digital voltmeters (Ch. 19)

◀ ALL
ARE
NEW

Thanks

Over 50 physics professors provided input and direct feedback on every aspect of the text: organization, content, figures, and suggestions for new Examples and Problems. The reviewers for this sixth edition are listed below. I owe each of them a debt of gratitude:

Zaven Altounian (McGill University)
David Amadio (Cypress Falls Senior High School)
Andrew Bacher (Indiana University)
Rama Bansil (Boston University)
Mitchell C. Begelman (University of Colorado)
Cornelius Bennhold (George Washington University)
Mike Berger (Indiana University)
George W. Brandenburg (Harvard University)
Robert Coakley (University of Southern Maine)
Renee D. Diehl (Penn State University)
Kathryn Dimiduk (University of New Mexico)
Leroy W. Dubeck (Temple University)
Andrew Duffy (Boston University)
John J. Dykla (Loyola University Chicago)
John Essick (Reed College)
David Faust (Mt. Hood Community College)
Gerald Feldman (George Washington University)
Frank A. Ferrone (Drexel University)
Alex Filippenko (University of California, Berkeley)
Richard Firestone (Lawrence Berkeley Lab)
Theodore Gotis (Oakton Community College)
J. Erik Hendrickson (University of Wisconsin, Eau Claire)
Laurent Hodges (Iowa State University)
Brian Houser (Eastern Washington University)
Brad Johnson (Western Washington University)
Randall S. Jones (Loyola College of Maryland)
Joseph A. Keane (St. Thomas Aquinas College)
Arthur Kosowsky (Rutgers University)
Amitabh Lath (Rutgers University)

Paul L. Lee (California State University, Northridge)
Jerome R. Long (Virginia Tech)
Mark Lucas (Ohio University)
Dan MacIsaac (Northern Arizona University)
William W. McNairy (Duke University)
Laszlo Mihaly (SUNY Stony Brook)
Peter J. Mohr (NIST)
Lisa K. Morris (Washington State University)
Paul Morris (Abilene Christian University)
Hon-Kie Ng (Florida State University)
Mark Oreglia (University of Chicago)
Lyman Page (Princeton University)
Bruce Partridge (Haverford College)
R. Daryl Pedigo (University of Washington)
Robert Pelcovits (Brown University)
Alan Pepper (Campbell School, Adelaide, Australia)
Kevin T. Pitts (University of Illinois)
Steven Pollock (University of Colorado, Boulder)
W. Steve Quon (Ventura College)
Michele Rallis (Ohio State University)
James J. Rhyne (University of Missouri, Columbia)
Paul L. Richards (University of California, Berkeley)
Dennis Rioux (University of Wisconsin, Oshkosh)
Robert Ross (University of Detroit, Mercy)
Roy S. Rubins (University of Texas, Arlington)
Wolfgang Rueckner (Harvard University Extension)
Randall J. Scalise (Southern Methodist University)
Arthur G. Schmidt (Northwestern University)
Cindy Schwarz (Vassar College)

Bartlett M. Sheinberg (Houston Community College)
J. L. Shinpaugh (East Carolina University)
Ross L. Spencer (Brigham Young University)
Mark Sprague (East Carolina University)
Michael G. Strauss (University of Oklahoma)
Chun Fu Su (Mississippi State University)
Ronald G. Taback (Youngstown State University)

Leo H. Takahashi (Pennsylvania State University, Beaver)
Raymond C. Turner (Clemson University)
Robert C. Webb (Texas A&M University)
Arthur Wiggins (Oakland Community College)
Stanley Wojcicki (Stanford University)
Edward L. Wright (University of California, Los Angeles)
Andrzej Zieminski (Indiana University)

I am grateful also to those other physicist reviewers of earlier editions:

David B. Aaron (South Dakota State University)
Narahari Achar (Memphis State University)
William T. Achor (Western Maryland College)
Arthur Alt (College of Great Falls)
John Anderson (University of Pittsburgh)
Subhash Antani (Edgewood College)
Atam P. Arya (West Virginia University)
Sirus Aryainejad (Eastern Illinois University)
Charles R. Bacon (Ferris State University)
Arthur Ballato (Brookhaven National Laboratory)
David E. Bannon (Chemeketa Community College)
Gene Barnes (California State University, Sacramento)
Isaac Bass
Jacob Becher (Old Dominion University)
Paul A. Bender (Washington State University)
Michael S. Berger (Indiana University)
Donald E. Bowen (Stephen F. Austin University)
Joseph Boyle (Miami-Dade Community College)
Peter Brancazio (Brooklyn College, CUNY)
Michael E. Browne (University of Idaho)
Michael Broyles (Collin County Community College)
Anthony Buffa (California Polytechnic State University)
David Bushnell (Northern Illinois University)
Neal M. Cason (University of Notre Dame)
H. R. Chandrasekhar (University of Missouri)
Ram D. Chaudhari (SUNY, Oswego)
K. Kelvin Cheng (Texas Tech University)
Lowell O. Christensen (American River College)
Mark W. Plano Clark (Doane College)
Irvine G. Clator (UNC, Wilmington)
Albert C. Claus (Loyola University of Chicago)
Scott Cohen (Portland State University)
Lawrence Coleman (University of California, Davis)
Lattie Collins (East Tennessee State University)
Sally Daniels (Oakland University)
Jack E. Denson (Mississippi State University)
Waren Deshotels (Marquette University)
Eric Dietz (California State University, Chico)
Frank Drake (University of California, Santa Cruz)
Paul Draper (University of Texas, Arlington)
Miles J. Dresser (Washington State University)
Ryan Droste (The College of Charleston)
F. Eugene Dunnam (University of Florida)
Len Feuerhelm (Oklahoma Christian University)
Donald Foster (Wichita State University)
Gregory E. Francis (Montana State University)
Philip Gash (California State University, Chico)
J. David Gavenda (University of Texas, Austin)
Simon George (California State University, Long Beach)
James Gerhart (University of Washington)
Bernard Gerstman (Florida International University)
Charles Glashausser (Rutgers University)
Grant W. Hart (Brigham Young University)
Hershel J. Hausman (Ohio State University)
Melissa Hill (Marquette University)
Mark Hillery (Hunter College)
Hans Hochheimer (Colorado State University)
Joseph M. Hoffman (Frostburg State University)

Peter Hoffman-Pinther (University of Houston, Downtown)
Alex Holloway (University of Nebraska, Omaha)
Fred W. Inman (Mankato State University)
M. Azad Islan (SUNY, Potsdam)
James P. Jacobs (University of Montana)
Larry D. Johnson (Northeast Louisiana University)
Gordon Jones (Mississippi State University)
Rex Joyner (Indiana Institute of Technology)
Sina David Kaviani (El Camino College)
Kirby W. Kemper (Florida State University)
Sanford Kern (Colorado State University)
James E. Kettler (Ohio University, Eastern Campus)
James R. Kirk (Edinboro University of Pennsylvania)
Alok Kuman (SUNY, Oswego)
Sung Kyu Kim (Macalester College)
Amer Lahamer (Berea College)
Clement Y. Lam (North Harris College)
David Lamp (Texas Tech University)
Peter Landry (McGill University)
Michael Lieber (University of Arkansas)
Bryan H. Long (Columbia State College)
Michael C. LoPresto (Henry Ford Community College)
James Madsen (University of Wisconsin, River Falls)
Ponn Mahes (Winthrop University)
Robert H. March (University of Wisconsin, Madison)
David Markowitz (University of Connecticut)
Daniel J. McLaughlin (University of Hartford)
E. R. Menzel (Texas Tech University)
Robert Messina
David Mills (College of the Redwoods)
George K. Miner (University of Dayton)
Victor Montemeyer (Middle Tennessee State University)
Marina Morrow (Lansing Community College)
Ed Nelson (University of Iowa)
Dennis Nemeschansky (USC)
Gregor Novak (Indiana University/Purdue University)
Roy J. Peterson (University of Colorado, Boulder)
Frederick M. Phelps (Central Michigan University)
Brian L. Pickering (Laney College)
T. A. K. Pillai (University of Wisconsin, La Crosse)
John Polo (Edinboro University of Pennsylvania)
Michael Ram (University of Buffalo)
John Reading (Texas A&M University)
David Reid (Eastern Michigan University)
Charles Richardson (University of Arkansas)
William Riley (Ohio State University)
Larry Rowan (University of North Carolina)
D. Lee Rutledge (Oklahoma State University)
Hajime Sakai (University of Massachusetts, Amherst)
Thomas Sayetta (East Carolina University)
Neil Schiller (Ocean County College)
Ann Schmiedekamp (Pennsylvania State University, Ogontz)
Juergen Schroeer (Illinois State University)
Mark Semon (Bates College)
James P. Sheerin (Eastern Michigan University)
Eric Sheldon (University of Massachusetts, Lowell)
K. Y. Shen (California State University, Long Beach)
Marc Sher (College of William and Mary)

Joseph Shinar (Iowa State University)
Thomas W. Sills (Wilbur Wright College)
Anthony A. Siluidi (Kent State University)
Michael A. Simon (Housatonic Community College)
Upindranath Singh (Embry-Riddle)
Michael I. Sobel (Brooklyn College)
Donald Sparks (Los Angeles Pierce College)
Thor F. Stromberg (New Mexico State University)
James F. Sullivan (University of Cincinnati)
Kenneth Swinney (Bevill State Community College)
Harold E. Taylor (Stockton State University)
John E. Teggins (Auburn University at Montgomery)
Colin Terry (Ventura College)
Michael Thoennessen (Michigan State University)
Kwok Yeung Tsang (Georgia Institute of Technology)

Jagdish K. Tuli (Brookhaven National Laboratory)
Paul Urone (CSU, Sacramento)
Linn D. Van Woerkom (Ohio State University)
S. L. Varghese (University of South Alabama)
Jearl Walker (Cleveland State University)
Robert A. Walking (University of Southern Maine)
Jai-Ching Wang (Alabama A&M University)
Thomas A. Weber (Iowa State University)
John C. Wells (Tennessee Technological)
Gareth Williams (San Jose State University)
Wendall S. Williams (Case Western Reserve University)
Jerry Wilson (Metropolitan State College at Denver)
Lowell Wood (University of Houston)
David Wright (Tidewater Community College)
Peter Zimmerman (Louisiana State University)

I owe special thanks to Profs. Bob Davis and J. Erik Hendrickson for much valuable input, and especially for working out all the Problems and producing the Solutions Manual with solutions to all Problems and Questions, as well as for providing the answers to odd-numbered Problems at the end of this book. Thanks as well to the team they managed (Profs. David Curott, Bryan Long, and Richard Louie) who also worked out all the Problems and Questions, each checking the others.

I am grateful to Profs. Robert Coakley, Lisa Morris, Kathryn Dimiduk, Robert Pelcovits, Raymond Turner, Cornelius Bennhold, Gerald Feldman, Alan Pepper, Michael Strauss, and Zaven Altounian, who inspired many of the Examples, Questions, Problems, and significant clarifications.

Chapter 33 on Cosmology and Astrophysics absorbed more time by far than any other Chapter because of the very recent, and ongoing, "revolutionary" results that I wanted to present. I was fortunate to receive generous input from some of the top experts in the field, to whom I owe a debt of gratitude: Paul Richards and Alex Filippenko (U.C. Berkeley), Lyman Page (Princeton and WMAP), Edward Wright (U.C.L.A. and WMAP), Mitchell Begelman (U. Colorado), Bruce Partridge (Haverford College), Arthur Kosowsky (Rutgers), and Michael Strauss (Princeton and SDSS).

I especially wish to thank Profs. Howard Shugart, Chris McKee, and many others at the University of California, Berkeley, Physics Department for helpful discussions, and for hospitality. Thanks also to Prof. Tito Arecchi and others at the Istituto Nazionale di Ottica, Florence, Italy.

Finally, I am most grateful to the many people at Prentice Hall with whom I worked on this project, especially Paul Corey, Erik Fahlgren, Andrew Sobel, Chirag Thakkar, John Challice, and above all to the highly professional and wonderfully dedicated Karen Karlin and Susan Fisher. The final responsibility for all errors lies with me. I welcome comments, corrections, and suggestions[†] as soon as possible to benefit students for the next reprint.

D.C.G.

[†] Please send to:
    email: physics_service@prenhall.com
    or by postal service: Physics Editor
                            Prentice Hall Inc.
                            One Lake Street
                            Upper Saddle River, NJ 07458

Available Supplements and Media

Supplements for the Student

Student Pocket Companion (0-13-035249-7)
by Biman Das (SUNY–Potsdam)
This 5" × 7" paperback book contains a summary of *Physics: Principles with Applications*, Sixth Edition, including key concepts, equations, and tips and hints.

Student Study Guide with Selected Solutions
(Volume I: 0-13-035239-X, Volume II: 0-13-146557-0)
by Joseph Boyle, (Miami-Dade Community College)
This study guide contains overviews, exercises, key phrases and terms, self-study exams, questions for review, and solutions to selected end-of-Chapter Problems for each Chapter of this textbook.

Mathematics for College Physics (0-13-141427-5)
by Biman Das (SUNY–Potsdam)
This text, for students who need help with the necessary mathematical tools, shows how mathematics is directly applied to physics, and discusses how to overcome math anxiety.

Ranking Task Exercises in Physics, Student Edition (0-13-144851-X)
by Thomas L. O'Kuma (Lee College), David P. Maloney (Indiana University–Purdue University at Fort Wayne), and Curtis J. Hieggelke (Joliet Junior College)
Ranking Tasks are an innovative type of conceptual exercise that ask students to make comparative judgments about variations on a particular physical situation. This supplement includes about 200 Ranking Task Exercises covering all of classical physics except optics.

PH GradeAssist Student Quick Start Guide (0-13-141926-9)
This student guide (with access code) contains information on how to register and use PH GradeAssist.

Interactive Physics Workbook, Second Edition (0-13-067108-8)
by Cindy Schwarz (Vassar College), John Ertel (Naval Academy), MSC.Software
This workbook and hybrid CD-ROM package is designed to help students visualize and work with specific physics problems by means of simulations created from Interactive Physics files. Forty problems of varying difficulty require students to make predictions, change variables, run, and visualize motion on the computer screen. The accompanying workbook/study guide provides instructions, a physics review, hints, and questions. The CD-ROM contains everything students need to run the simulations.

Physlet® Physics (0-13-101969-4)
by Wolfgang Christian and Mario Belloni (Davidson College)
This CD-ROM and text package has over 800 ready-to-run interactive Java applets which have been widely adopted by physics instructors. No web-server or Internet connection is required.

MCAT Physics Study Guide (0-13-627951-1)
by Joseph Boone (California Polytechnic State University–San Luis Obispo)
This MCAT study guide includes in-depth review, practice problems, and review questions.

Supplements for the Instructor

Test Item File (0-13-047311-1)
This test bank contains approximately 2800 multiple choice, true or false, short answer, and essay questions, of which about 25% are conceptual. All questions are ranked by level of difficulty and referenced to the corresponding text Section of this book. The Test Item File is also available in electronic format on the Instructor's Resource Center on CD-ROM.

Instructor's Solutions Manual
(Volume I: 0-13-035237-3, Volume II: 0-13-141545-X)
by Bob Davis (Taylor University) and J. Erik Hendrickson (University of Wisconsin-Eau Claire)
The Solutions Manual contains detailed worked solutions to every end-of-Chapter Problem in this textbook, as well as answers to the Questions. Electronic versions are available on the Instructor's Resource Center on CD-ROM for instructors with Microsoft Word or Word-compatible software.

Instructor's Resource Manual (0-13-035251-9)
by Katherine Whatley (University of North Carolina-Asheville)
This manual contains lecture outlines, notes, demonstration suggestions, suggested readings, and other teaching resources.

Instructor's Resource Center on CD-ROM (0-13-035246-2)
This two-CD set contains all text illustrations and tables in JPEG, Microsoft PowerPoint™, and Adobe PDF formats. Instructors can preview and sequence images, perform key-word searches, add lecture notes, and incorporate their own digital resources. Also contained is TestGenerator, an easy-to-use networkable program for creating tests ranging from short quizzes to long exams. Instructors can use the Question Editor to modify existing questions or problems, including algorithmic versions, or create new ones. The CDs also contain additional PowerPoint Lectures, plus electronic versions of the Instructor's Resource Manual, Instructor's Solutions Manual, and end-of-Chapter Questions and Problems for this book.

Transparency Pack (0-13-035245-4)
The pack includes approximately 400 full-color transparencies of images and Tables from this book.

"Physics You Can See" Video (0-205-12393-7)
This video contains eleven classic physics demonstrations, each 2 to 5 minutes long.

Course Management Systems

WebCT and Blackboard allow instructors to assign and grade homework online, manage their roster and grade book, and post course-related documents.

The content cartridges for WebCT and Blackboard are text-specific and include:

Just-in-Time Teaching tools: Warm-Ups, Puzzles, and Applications, by Gregor Novak and Andrew Gavrin (Indiana University—Purdue University, Indianapolis)

Ranking Task Exercises by Thomas L. O'Kuma (Lee College), David P. Maloney (Indiana University-Purdue University at Fort Wayne), and Curtis J. Hieggelke (Joliet Junior College)

Physlet® Problems by Wolfgang Christian and Mario Belloni (Davidson College)

Algorithmic Practice Problems by Carl Adler (East Carolina University)

MCAT Study Guide with questions from "Kaplan Test Prep and Admissions"

Companion Website (http://physics.prenhall.com/giancolippa)
This site contains practice problems, objectives, practice questions, destinations (links to related sites), and applications with links to related sites. Practice problems and questions are scored by computer, and results can be automatically e-mailed to the instructor.

Online Homework Systems

PH GradeAssist (www.prenhall.com/phga)
PH GradeAssist (PHGA) is Prentice Hall's online homework system. It includes content associated with Just-in-Time Teaching materials, Physlet Problems, conceptual and quantitative questions, and hundreds of end-of-Chapter Problems from this textbook. Many of the end-of-Chapter Problems have an algorithmically generated variant. Instructors can edit the questions, create new ones, and control important parameters such as how much a question is worth and when a student can take a quiz.

PH GradeAssist Instructor Quick Start Guide (0-13-141927-7)
This guide (with access code) helps instructors register for and use PH GradeAssist.

WebAssign (www.webassign.net)
WebAssign is a nationally hosted online homework system that allows instructors to create, post, collect, grade, and record assignments from a ready-to-use database of Problems and Questions from this textbook.

CAPA and LON-CAPA
Computer Assisted Personalized Approach (CAPA) is a locally hosted online homework system that allows instructors to create, post, collect, grade, and record assignments from a ready-to-use database of Problems and Questions from this textbook. The Learning Online Network with a Computer Assisted Personalized Approach (LON-CAPA) is an integrated system for online learning and assessment. It consists of a course management system, an individualized homework and automatic grading system, a data collection and data mining system, and a content delivery system that will provide gateways to and from NSF's National STEM Digital Library.

NOTES TO STUDENTS (AND INSTRUCTORS) ON THE FORMAT

1. Sections marked with a star (*) are considered optional. They can be omitted without interrupting the main flow of topics. No later material depends on them except possibly later starred Sections. They may be fun to read.

2. The customary conventions are used: symbols for quantities (such as m for mass) are italicized, whereas units (such as m for meter) are not italicized. Symbols for vectors are shown in boldface with a small arrow above: $\vec{\mathbf{F}}$.

3. Few equations are valid in all situations. Where practical, the limitations of important equations are stated in square brackets next to the equation. The equations that represent the great laws of physics are displayed with a tan background, as are a few other indispensable equations.

4. The number of significant figures (Section 1–4) should not be assumed to be greater or less than given: if a number is stated as (say) 6, with its units, it is meant to be 6 and not 6.0 or 6.00.

5. At the end of each Chapter is a set of Questions that students should attempt to answer (to themselves at least). These are followed by Problems which are ranked as level I, II, or III, according to estimated difficulty, with level I Problems being the easiest. Level II are normal Problems, and level III are for "extra credit." These ranked Problems are arranged by Section, but Problems for a given Section may depend on earlier material as well. There follows a group of General Problems, which are not arranged by Section nor ranked as to difficulty. Questions and Problems that relate to optional Sections are starred (*). Answers to odd-numbered Problems are given at the end of the book.

6. Being able to solve problems is a crucial part of learning physics, and provides a powerful means for understanding the concepts and principles. This book contains many aids to problem solving: (a) worked-out Examples and their solutions in the text (set off with a vertical blue line in the margin) which should be studied as an integral part of the text; (b) some of the worked-out Examples are Estimation Examples, which show how rough or approximate results can be obtained even if the given data are sparse (see Section 1–7); (c) special "Problem Solving Boxes" placed throughout the text to suggest a step-by-step approach to problem solving for a particular topic—but don't get the idea that every topic has its own "techniques," because the basics remain the same; some of these "Boxes" are followed by an Example that is solved by explicitly following the suggested steps; (d) special problem-solving Sections; (e) "Problem Solving" marginal notes (see point 9 below) which refer to hints for solving problems within the text; (f) Exercises within the text that you should work out immediately, and then check your response against the answer given at the bottom of the last page of that Chapter; (g) the Problems themselves at the end of each Chapter (point 5 above).

7. Conceptual Examples are conceptual rather than numerical. Each poses a question or two, which hopefully starts you to think and come up with a response. Give yourself a little time to come up with your own response before reading the Response given.

8. "Additional Examples" subheadings contain Examples that you could skip on a first reading, in case you are feeling overwhelmed. But a day or two later, when you read the Chapter a second time, try to work through these Examples too because they can give you more power in doing a wide range of Problems.

9. Margin notes: brief notes in the margin of almost every page are printed in blue and are of five types: (a) ordinary notes (the majority) that serve as a sort of outline of the text and can help you later locate important concepts and equations; (b) notes that refer to the great laws and principles of physics, and these are in capital letters and in a box for emphasis; (c) notes that refer to a problem-solving hint or technique treated in the text, and these say "Problem Solving"; (d) notes that refer to an application of physics in the text or an Example, and these say "Physics Applied"; (e) "Caution" notes that point out a possible misconception spelled out in the adjacent text.

10. This book is printed in full color—but not simply to make it more attractive. The color is used above all in the Figures, to give them greater clarity for our analysis. The Table on the next page is a summary of which colors are used for the different kinds of vectors, for field lines, and for other symbols and objects. These colors are used consistently throughout the book.

11. Math review, plus some additional topics, are found in Appendices. Useful data, conversion factors, and math formulas are found inside the front and back covers.

USE OF COLOR

Vectors

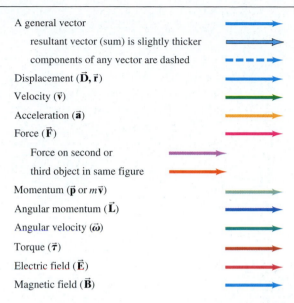

A general vector

   resultant vector (sum) is slightly thicker

   components of any vector are dashed

Displacement ($\vec{\mathbf{D}}$, $\vec{\mathbf{r}}$)

Velocity ($\vec{\mathbf{v}}$)

Acceleration ($\vec{\mathbf{a}}$)

Force ($\vec{\mathbf{F}}$)

   Force on second or

   third object in same figure

Momentum ($\vec{\mathbf{p}}$ or $m\vec{\mathbf{v}}$)

Angular momentum ($\vec{\mathbf{L}}$)

Angular velocity ($\vec{\boldsymbol{\omega}}$)

Torque ($\vec{\boldsymbol{\tau}}$)

Electric field ($\vec{\mathbf{E}}$)

Magnetic field ($\vec{\mathbf{B}}$)

Electricity and magnetism

Electric field lines

Equipotential lines

Magnetic field lines

Electric charge (+)   + or ● +

Electric charge (−)   − or ● −

Electric circuit symbols

Wire

Resistor

Capacitor

Inductor

Battery

Optics

Light rays

Object

Real image (dashed)

Virtual image (dashed and paler)

Other

Energy level (atom, etc.)

Measurement lines ⊢—1.0 m—⊣

Path of a moving object

Direction of motion or current

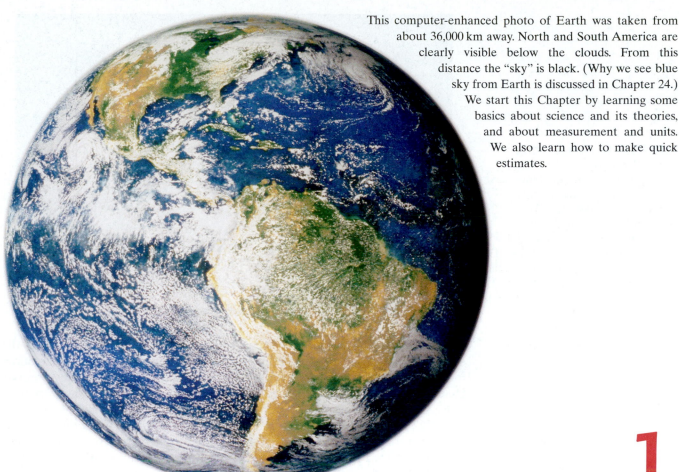

This computer-enhanced photo of Earth was taken from about 36,000 km away. North and South America are clearly visible below the clouds. From this distance the "sky" is black. (Why we see blue sky from Earth is discussed in Chapter 24.) We start this Chapter by learning some basics about science and its theories, and about measurement and units. We also learn how to make quick estimates.

CHAPTER **1**

Introduction, Measurement, Estimating

Physics is the most basic of the sciences. It deals with the behavior and structure of matter. The field of physics is usually divided into *classical physics* which includes motion, fluids, heat, sound, light, electricity, and magnetism; and *modern physics* which includes the topics of relativity, atomic structure, condensed matter, nuclear physics, elementary particles, and cosmology and astrophysics. We will cover all these topics in this book, beginning with motion (or mechanics, as it is often called). But before we begin on the physics itself, we take a brief look at how this overall activity called "science," including physics, is actually practiced.

1–1 The Nature of Science

The principal aim of all sciences, including physics, is generally considered to be the search for order in our observations of the world around us. Many people think that science is a mechanical process of collecting facts and devising theories. But it is not so simple. Science is a creative activity that in many respects resembles other creative activities of the human mind.

FIGURE 1–1 Aristotle is the central figure (dressed in blue) at the top of the stairs (the figure next to him is Plato) in this famous Renaissance portrayal of *The School of Athens*, painted by Raphael around 1510. Also in this painting, considered one of the great masterpieces in art, are Euclid (drawing a circle at the lower right), Ptolemy (extreme right with globe), Pythagoras, Socrates, and Diogenes.

Observation and experiment

One important aspect of science is **observation** of events, which includes the design and carrying out of experiments. But observation requires imagination, for scientists can never include everything in a description of what they observe. Hence, scientists must make judgments about what is relevant in their observations and experiments. Consider, for example, how two great minds, Aristotle (384–322 B.C.; Fig. 1–1) and Galileo (1564–1642; Fig. 2–17), interpreted motion along a horizontal surface. Aristotle noted that objects given an initial push along the ground (or on a tabletop) always slow down and stop. Consequently, Aristotle argued that the natural state of an object is at rest. Galileo, in his reexamination of horizontal motion in the early 1600s, imagined that if friction could be eliminated, an object given an initial push along a horizontal surface would continue to move indefinitely without stopping. He concluded that for an object to

Motion is as natural as rest

be in motion was just as natural as for it to be at rest. By inventing a new approach, Galileo founded our modern view of motion (Chapters 2, 3, and 4). Galileo made this intellectual leap conceptually, without actually eliminating friction.

Theories

Observation, with careful experimentation and measurement, is one side of the scientific process. The other side is the invention or creation of **theories** to explain and order the observations. Theories are never derived directly from observations. Observations may help inspire a theory, and theories are accepted or rejected based on observation and experiment.

Theories are inspirations that come from the minds of human beings. For example, the idea that matter is made up of atoms (the atomic theory) was not arrived at by direct observation of atoms—we can't see atoms directly. Rather, the idea sprang from creative minds. The theory of relativity, the electromagnetic theory of light, and Newton's law of universal gravitation were likewise the result of human imagination.

The great theories of science may be compared, as creative achievements, with great works of art or literature. But how does science differ from these other

Testing a theory

creative activities? One important difference is that science requires **testing** of its ideas or theories to see if their predictions are borne out by experiment. But theories are not "proved" by testing. First of all, no measuring instrument is perfect, so exact confirmation cannot be possible. Furthermore, it is not possible to test a theory for every possible set of circumstances. Hence a theory can never be absolutely "proved." Indeed, the history of science tells us that long-held theories are sometimes replaced by new ones.

Theory acceptance

A new theory is accepted by scientists in some cases because its predictions are quantitatively in better agreement with experiment than those of the older

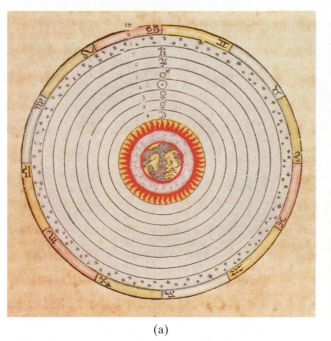

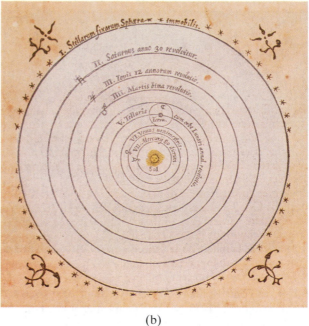

(a)   (b)

FIGURE 1–2 (a) Ptolemy's geocentric view of the universe. Note at the center the four elements of the ancients: Earth, water, air (clouds around the Earth), and fire; then the circles, with symbols, for the Moon, Mercury, Venus, Sun, Mars, Jupiter, Saturn, the fixed stars, and the signs of the zodiac. (b) An early representation of Copernicus's heliocentric view of the universe with the Sun at the center. (See Chapter 5.)

theory. But in many cases, a new theory is accepted only if it explains a greater *range* of phenomena than does the older one. Copernicus's Sun-centered theory of the universe (Fig. 1–2b), for example, was originally no more accurate than Ptolemy's Earth-centered theory (Fig. 1–2a) for predicting the motion of heavenly bodies (Sun, Moon, planets). But Copernicus's theory had consequences that Ptolemy's did not, such as predicting the moonlike phases of Venus. A simpler and richer theory, one which unifies and explains a greater variety of phenomena, is more useful and beautiful to a scientist. And this aspect, as well as quantitative agreement, plays a major role in the acceptance of a theory.

An important aspect of any theory is how well it can quantitatively predict phenomena, and from this point of view a new theory may often seem to be only a minor advance over the old one. For example, Einstein's theory of relativity gives predictions that differ very little from the older theories of Galileo and Newton in nearly all everyday situations. Its predictions are better mainly in the extreme case of very high speeds close to the speed of light. But quantitative prediction is not the only important outcome of a theory. Our view of the world is affected as well. As a result of Einstein's theory of relativity, for example, our concepts of space and time have been completely altered, and we have come to see mass and energy as a single entity (via the famous equation $E = mc^2$).

FIGURE 1–3 Studies on the forces in structures by Leonardo da Vinci (1452–1519).

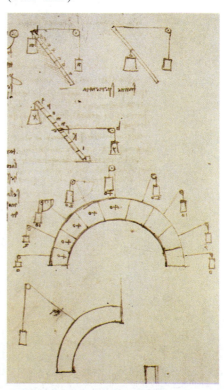

1–2 Physics and its Relation to Other Fields

For a long time science was more or less a united whole known as natural philosophy. Not until a century or two ago did the distinctions between physics and chemistry and even the life sciences become prominent. Indeed, the sharp distinction we now see between the arts and the sciences is itself but a few centuries old. It is no wonder then that the development of physics has both influenced and been influenced by other fields. For example, the notebooks (Fig. 1–3) of Leonardo da Vinci, the great Renaissance artist, researcher, and engineer, contain the first references to the forces acting within a structure, a subject we consider as physics today; but then, as now, it has great relevance to architecture and building.

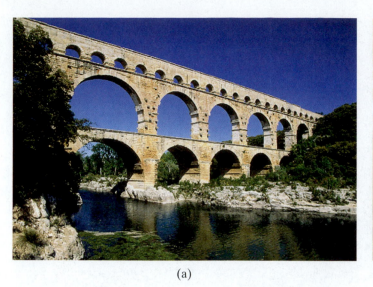

(a)   (b)

FIGURE 1–4 (a) This Roman aqueduct was built 2000 years ago and still stands. (b) Collapse of the Hartford Civic Center in 1978, just two years after it was built.

Early work in electricity that led to the discovery of the electric battery and electric current was done by an eighteenth-century physiologist, Luigi Galvani (1737–1798). He noticed the twitching of frogs' legs in response to an electric spark and later that the muscles twitched when in contact with two dissimilar metals (Chapter 18). At first this phenomenon was known as "animal electricity," but it shortly became clear that electric current itself could exist in the absence of an animal.

Physics applies to many fields

Physics is used in many fields. A zoologist, for example, may find physics useful in understanding how prairie dogs and other animals can live underground without suffocating. A physical therapist will do a more effective job if aware of the principles of center of gravity and the action of forces within the human body. A knowledge of the operating principles of optical and electronic equipment is helpful in a variety of fields. Life scientists and architects alike will be interested in the nature of heat loss and gain in human beings and the resulting comfort or discomfort. Architects themselves may not have to calculate, for example, the dimensions of the pipes in a heating system or the forces involved in a given structure to determine if it will remain standing (Fig. 1–4). But architects must know the principles behind these analyses in order to make realistic designs and to communicate effectively with engineering consultants and other specialists. From the aesthetic or psychological point of view, too, architects must be aware of the forces involved in a structure—for instability, even if only illusory, can be discomforting to those who must live or work in the structure.

The list of ways in which physics relates to other fields is extensive. In the Chapters that follow we will discuss many such applications as we carry out our principal aim of explaining basic physics.

1–3 Models, Theories, and Laws

Models

When scientists are trying to understand a particular set of phenomena, they often make use of a **model**. A model, in the scientific sense, is a kind of analogy or mental image of the phenomena in terms of something else we are already familiar with. One example is the wave model of light. We cannot see waves of light as we can water waves. But it is valuable to think of light as if it were made up of waves, because experiments indicate that light behaves in many respects as water waves do.

The purpose of a model is to give us an approximate mental or visual picture—something to hold onto—when we cannot see or understand what actually is happening. Models often give us a deeper understanding: the analogy to a known system (for instance, water waves in the above example) can suggest new experiments to perform and can provide ideas about what other related phenomena might occur.

You may wonder what the difference is between a theory and a model. Usually, a model is relatively simple and provides a structural similarity to the phenomena being studied. A **theory** is broader, more detailed, and can give quantitatively testable predictions, often with great precision. It is important, however, not to confuse a model or a theory with the real system or the phenomena themselves.

Theories (vs. models)

Scientists give the title **law** to certain concise but general statements about how nature behaves (that energy is conserved, for example). Sometimes the statement takes the form of a relationship or equation between quantities (such as Newton's second law, $F = ma$).

Laws

and

To be called a law, a statement must be found experimentally valid over a wide range of observed phenomena. For less general statements, the term **principle** is often used (such as Archimedes' principle).

principles

Scientific laws are different from political laws in that the latter are *prescriptive*: they tell us how we ought to behave. Scientific laws are *descriptive*: they do not say how nature *should* behave, but rather are meant to describe how nature *does* behave. As with theories, laws cannot be tested in the infinite variety of cases possible. So we cannot be sure that any law is absolutely true. We use the term "law" when its validity has been tested over a wide range of cases, and when any limitations and the range of validity are clearly understood.

Scientists normally do their work as if the accepted laws and theories were true. But they are obliged to keep an open mind in case new information should alter the validity of any given law or theory.

1–4 Measurement and Uncertainty; Significant Figures

In the quest to understand the world around us, scientists try to work out relationships among physical quantities that can be measured.

Uncertainty

Accurate, precise measurements are an important part of physics. But no measurement is absolutely precise. There is an uncertainty associated with every measurement. Among the most important sources of uncertainty, other than blunders, are the limited accuracy of every measuring instrument and the inability to read an instrument beyond some fraction of the smallest division shown. For example, if you were to use a centimeter ruler to measure the width of a board (Fig. 1–5), the result could be claimed to be precise to about 0.1 cm (1 mm), the smallest division on the ruler, although half of this value might be a valid claim as well. The reason for this is that it is difficult for the observer to estimate between the smallest divisions. Furthermore, the ruler itself may not have been manufactured to an accuracy any better than this.[†]

Every measurement has an uncertainty

FIGURE 1–5 Measuring the width of a board with a centimeter ruler. Accuracy is about ± 1 mm.

[†]There is a technical difference between "precision" and "accuracy." **Precision** in a strict sense refers to the repeatability of the measurement using a given instrument. For example, if you measure the width of a board many times, getting results like 8.81 cm, 8.85 cm, 8.78 cm, 8.82 cm (estimating between the 0.1-cm marks as best as possible each time), you could say the measurements give a *precision* a bit better than 0.1 cm. **Accuracy** refers to how close a measurement is to the true value. For example, if the ruler shown in Fig. 1–5 was manufactured with a 2% error, the accuracy of its measurement of the board's width (about 8.8 cm) would be about 2% of 8.8 cm, or about ± 0.2 cm. Estimated uncertainty is meant to take both accuracy and precision into account.

When giving the result of a measurement, it is important to state the **estimated uncertainty** in the measurement. For example, the width of a board might be written as 8.8 ± 0.1 cm. The ± 0.1 cm ("plus or minus 0.1 cm") represents the estimated uncertainty in the measurement, so that the actual width most likely lies between 8.7 and 8.9 cm. The **percent uncertainty** is simply the ratio of the uncertainty to the measured value, multiplied by 100. For example, if the measurement is 8.8 and the uncertainty about 0.1 cm, the percent uncertainty is

Stating the uncertainty

$$\frac{0.1}{8.8} \times 100\% \approx 1\%$$

where ≈ means "is roughly equal to."

Assumed uncertainty

Often the uncertainty in a measured value is not specified explicitly. In such cases, the uncertainty is generally assumed to be one or a few units in the last digit specified. For example, if a length is given as 8.8 cm, the uncertainty is assumed to be about 0.1 cm or 0.2 cm. It is important in this case that you do not write 8.80 cm, for this implies an uncertainty on the order of 0.01 cm; it assumes that the length is probably between 8.79 cm and 8.81 cm, when actually you believe it is between 8.7 and 8.9 cm.

CONCEPTUAL EXAMPLE 1–1 **Is the diamond yours?** A friend asks to borrow your precious diamond for a day to show her family. You are a bit worried, so you carefully have your diamond weighed on a scale which reads 8.17 grams. The scale's accuracy is claimed to be ± 0.05 gram. The next day you weigh the returned diamond again, getting 8.09 grams. Is this your diamond?

RESPONSE The scale readings are measurements and do not necessarily give the "true" value of the mass. Each measurement could have been high or low by up to 0.05 gram or so. The actual mass of your diamond lies most likely between 8.12 grams and 8.22 grams. The actual mass of the returned diamond is most likely between 8.04 grams and 8.14 grams. These two ranges overlap, so there is not a strong reason to doubt that the returned diamond is yours, at least based on the scale readings.

Significant Figures

Which digits are significant?

The number of reliably known digits in a number is called the number of **significant figures**. Thus there are four significant figures in the number 23.21 cm and two in the number 0.062 cm (the zeros in the latter are merely place holders that show where the decimal point goes). The number of significant figures may not always be clear. Take, for example, the number 80. Are there one or two significant figures? If we say it is *about* 80 km between two cities, there is only one significant figure (the 8) since the zero is merely a place holder. If it is *exactly* 80 km within an accuracy of 1 or 2 km, then the 80 has two significant figures.[†] If it is precisely 80 km, to within ± 0.1 km, then we write 80.0 km.

When making measurements, or when doing calculations, you should avoid the temptation to keep more digits in the final answer than is justified. For example, to calculate the area of a rectangle 11.3 cm by 6.8 cm, the result of multiplication would be 76.84 cm². But this answer is clearly not accurate to 0.01 cm², since (using the outer limits of the assumed uncertainty for each measurement) the result could be between 11.2 cm × 6.7 cm = 75.04 cm² and 11.4 cm × 6.9 cm = 78.66 cm². At best, we can quote the answer as 77 cm², which implies an uncertainty of about 1 or 2 cm². The other two digits (in the number 76.84 cm²) must be dropped since they are not significant. As a rough general rule (i.e., in the absence of a detailed consideration of uncertainties), we can say that *the final result of a multiplication or division should have only as many digits as the number with the least number of significant figures used in the calculation.* In our example, 6.8 cm has the least number of significant figures, namely two. Thus the result 76.84 cm² needs to be rounded off to 77 cm².

➡ **PROBLEM SOLVING**

Number of significant figures in final result should be same as least significant input value

---

[†] If the 80 has two significant figures, some people prefer to write it 80., with a decimal point. This is not usually done, so the number of significant figures in 80 can be ambiguous unless something is said about it such as "about" (meaning 80 ± 10), or "very nearly" or "precisely" (meaning 80 ± 1).

EXERCISE A The area of a rectangle 4.5 cm by 3.25 cm is correctly given by (a) 14.625 cm²; (b) 14.63 cm²; (c) 14.6 cm²; (d) 15 cm².

When adding or subtracting numbers, the final result is no more accurate than the least accurate number used. For example, the result of subtracting 0.57 from 3.6 is 3.0 (and not 3.03).

Keep in mind when you use a calculator that all the digits it produces may not be significant. When you divide 2.0 by 3.0, the proper answer is 0.67, and not some such thing as 0.666666666. Digits should not be quoted in a result, unless they are truly significant figures. However, to obtain the most accurate result, you should normally *keep one or more extra significant figures throughout a calculation, and round off only in the final result.* (With a calculator, you can keep all its digits in intermediate results.) Note also that calculators sometimes give too few significant figures. For example, when you multiply 2.5 × 3.2, a calculator may give the answer as simply 8. But the answer is good to two significant figures, so the proper answer is 8.0. See Fig. 1–6.

EXERCISE B Do 0.00324 and 0.00056 have the same number of significant figures?

Be careful not to confuse significant figures with the number of decimal places.

EXERCISE C For each of the following numbers, state the number of significant figures and the number of decimal places: (a) 1.23; (b) 0.123; (c) 0.0123.

CONCEPTUAL EXAMPLE 1–2 | **Significant figures.** Using a protractor (Fig. 1–7), you measure an angle to be 30°. (a) How many significant figures should you quote in this measurement? (b) Use a calculator to find the cosine of the angle you measured.

RESPONSE (a) If you look at a protractor, you will see that the precision with which you can measure an angle is about one degree (certainly not 0.1°). So you can quote two significant figures, namely, 30° (not 30.0°). (b) If you enter cos 30° in your calculator, you will get a number like 0.866025403. However, the angle you entered is known only to two significant figures, so its cosine is correctly given by 0.87; i.e., you must round your answer to two significant figures.

NOTE We discuss trigonometric functions like cosine in Chapter 3.

Scientific Notation

We commonly write numbers in "powers of ten," or "scientific" notation—for instance 36,900 as 3.69×10^4, or 0.0021 as 2.1×10^{-3}. One advantage of scientific notation (discussed in Appendix A) is that it allows the number of significant figures to be clearly expressed. For example, it is not clear whether 36,900 has three, four, or five significant figures. With powers of ten notation the ambiguity can be avoided: if the number is known to an accuracy of three significant figures, we write 3.69×10^4, but if it is known to four, we write 3.690×10^4.

* Percent Error

The significant figures rule is only approximate, and in some cases may underestimate the precision of the answer. Suppose for example we divide 97 by 92:

$$\frac{97}{92} = 1.05 \approx 1.1.$$

Both 97 and 92 have two significant figures, so the rule says to give the answer as 1.1. Yet the numbers 97 and 92 both imply an uncertainty of ± 1 if no other uncertainty is stated. Now 92 ± 1 and 97 ± 1 both imply an accuracy of about 1% ($1/92 \approx 0.01 = 1\%$). But the final result to two significant figures is 1.1, with an implied uncertainty of ± 0.1, which is an uncertainty of $0.1/1.1 \approx 0.1 \approx 10\%$. In this case it is better to give the answer as 1.05 (which is three significant figures). Why? Because 1.05 implies an uncertainty of ± 0.01 which is $0.01/1.05 \approx 0.01 \approx 1\%$, just like the uncertainty in the original numbers 92 and 97.

SUGGESTION: Use the significant figures rule, but consider the % uncertainty too, and add an extra digit if it gives a more realistic estimate of uncertainty.

⚠️ **CAUTION**

Calculators err with significant figures

➡️ **PROBLEM SOLVING**

Report only the proper number of significant figures in the final result. Keep extra digits during the calculation.

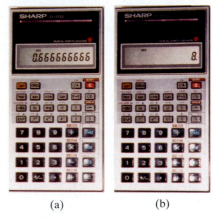

(a)   (b)

FIGURE 1–6 These two calculators show the wrong number of significant figures. In (a), 2.0 was divided by 3.0. The correct final result would be 0.67. In (b), 2.5 was multiplied by 3.2. The correct result is 8.0.

FIGURE 1–7 Example 1–2. A protractor used to measure an angle.

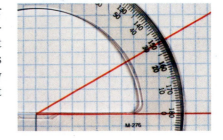

1–5 Units, Standards, and the SI System

The measurement of any quantity is made relative to a particular standard or **unit**, and this unit must be specified along with the numerical value of the quantity. For example, we can measure length in units such as inches, feet, or miles, or in the metric system in centimeters, meters, or kilometers. To specify that the length of a particular object is 18.6 is meaningless. The unit *must* be given; for clearly, 18.6 meters is very different from 18.6 inches or 18.6 millimeters.

For any unit we use, such as the meter for distance or the second for time, we need to define a **standard** which defines exactly how long one meter or one second is. It is important that standards be chosen that are readily reproducible so that anyone needing to make a very accurate measurement can refer to the standard in the laboratory.

Length

Standard of length (meter)

The first truly international standard was the **meter** (abbreviated m) established as the standard of **length** by the French Academy of Sciences in the 1790s. The standard meter was originally chosen to be one ten-millionth of the distance from the Earth's equator to either pole,[†] and a platinum rod to represent this length was made. (One meter is, very roughly, the distance from the tip of your nose to the tip of your finger, with arm and hand stretched out to the side.) In 1889, the meter was defined more precisely as the distance between two finely engraved marks on a particular bar of platinum–iridium alloy. In 1960, to provide greater precision and reproducibility, the meter was redefined as 1,650,763.73 wavelengths of a particular orange light emitted by the gas krypton-86. In 1983 the meter was again redefined, this time in terms of the speed of light (whose best measured value in terms of the older definition of the meter was 299,792,458 m/s, with an uncertainty of 1 m/s). The new definition reads: "The meter is the length of path traveled by light in vacuum during a time interval of 1/299,792,458 of a second." [‡]

British units of length (inch, foot, mile) are now defined in terms of the meter. The inch (in.) is defined as precisely 2.54 centimeters (cm; 1 cm = 0.01 m). Other conversion factors are given in the Table on the inside of the front cover of this book. Table 1–1 presents some typical lengths, from very small to very large, rounded off to the nearest power of ten. See also Fig. 1–8. [Note that the abbreviation for inches (in.) is the only one with a period, to distinguish it from the word "in".]

FIGURE 1–8 Some lengths: (a) viruses (about 10^{-7} m long) attacking a cell; (b) Mt. Everest's height is on the order of 10^4 m (8850 m, to be precise).

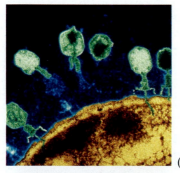
(a)

(b)

TABLE 1–1 Some Typical Lengths or Distances (order of magnitude)

Length (or Distance)	Meters (approximate)
Neutron or proton (radius)	10^{-15} m
Atom	10^{-10} m
Virus [see Fig. 1–8a]	10^{-7} m
Sheet of paper (thickness)	10^{-4} m
Finger width	10^{-2} m
Football field length	10^{2} m
Height of Mt. Everest [see Fig. 1–8b]	10^{4} m
Earth diameter	10^{7} m
Earth to Sun	10^{11} m
Earth to nearest star	10^{16} m
Earth to nearest galaxy	10^{22} m
Earth to farthest galaxy visible	10^{26} m

[†]Modern measurements of the Earth's circumference reveal that the intended length is off by about one-fiftieth of 1%. Not bad!

[‡]The new definition of the meter has the effect of giving the speed of light the exact value of 299,792,458 m/s.

TABLE 1–2 Some Typical Time Intervals	
Time Interval	Seconds (approximate)
Lifetime of very unstable subatomic particle	10^{-23} s
Lifetime of radioactive elements	10^{-22} s to 10^{28} s
Lifetime of muon	10^{-6} s
Time between human heartbeats	10^{0} s $(=1\,\text{s})$
One day	10^{5} s
One year	3×10^{7} s
Human life span	2×10^{9} s
Length of recorded history	10^{11} s
Humans on Earth	10^{14} s
Life on Earth	10^{17} s
Age of Universe	10^{18} s

TABLE 1–3 Some Masses	
Object	Kilograms (approximate)
Electron	10^{-30} kg
Proton, neutron	10^{-27} kg
DNA molecule	10^{-17} kg
Bacterium	10^{-15} kg
Mosquito	10^{-5} kg
Plum	10^{-1} kg
Human	10^{2} kg
Ship	10^{8} kg
Earth	6×10^{24} kg
Sun	2×10^{30} kg
Galaxy	10^{41} kg

Time

The standard unit of **time** is the **second** (s). For many years, the second was defined as 1/86,400 of a mean solar day. The standard second is now defined more precisely in terms of the frequency of radiation emitted by cesium atoms when they pass between two particular states. [Specifically, one second is defined as the time required for 9,192,631,770 periods of this radiation.] There are, by definition, 60 s in one minute (min) and 60 minutes in one hour (h). Table 1–2 presents a range of measured time intervals, rounded off to the nearest power of ten.

Mass

The standard unit of **mass** is the **kilogram** (kg). The standard mass is a particular platinum–iridium cylinder, kept at the International Bureau of Weights and Measures near Paris, France, whose mass is defined as exactly 1 kg. A range of masses is presented in Table 1–3. [For practical purposes, 1 kg weighs about 2.2 pounds on Earth.]

When dealing with atoms and molecules, we usually use the **unified atomic mass unit** (u). In terms of the kilogram,

$$1\,\text{u} = 1.6605 \times 10^{-27}\,\text{kg}.$$

The definitions of other standard units for other quantities will be given as we encounter them in later Chapters.

Unit Prefixes

In the metric system, the larger and smaller units are defined in multiples of 10 from the standard unit, and this makes calculation particularly easy. Thus 1 kilometer (km) is 1000 m, 1 centimeter is $\frac{1}{100}$ m, 1 millimeter (mm) is $\frac{1}{1000}$ m or $\frac{1}{10}$ cm, and so on. The prefixes "centi-," "kilo-," and others are listed in Table 1–4 and can be applied not only to units of length, but to units of volume, mass, or any other metric unit. For example, a centiliter (cL) is $\frac{1}{100}$ liter (L), and a kilogram (kg) is 1000 grams (g).

Systems of Units

When dealing with the laws and equations of physics it is very important to use a consistent set of units. Several systems of units have been in use over the years. Today the most important is the **Système International** (French for International System), which is abbreviated SI. In SI units, the standard of length is the meter, the standard for time is the second, and the standard for mass is the kilogram. This system used to be called the MKS (meter-kilogram-second) system.

A second metric system is the **cgs system**, in which the centimeter, gram, and second are the standard units of length, mass, and time, as abbreviated in the title. The **British engineering system** takes as its standards the foot for length, the pound for force, and the second for time.

TABLE 1–4 Metric (SI) Prefixes		
Prefix	Abbreviation	Value
yotta	Y	10^{24}
zetta	Z	10^{21}
exa	E	10^{18}
peta	P	10^{15}
tera	T	10^{12}
giga	G	10^{9}
mega	M	10^{6}
kilo	k	10^{3}
hecto	h	10^{2}
deka	da	10^{1}
deci	d	10^{-1}
centi	c	10^{-2}
milli	m	10^{-3}
micro[†]	μ	10^{-6}
nano	n	10^{-9}
pico	p	10^{-12}
femto	f	10^{-15}
atto	a	10^{-18}
zepto	z	10^{-21}
yocto	y	10^{-24}

[†] μ is the Greek letter "mu."

➡ **PROBLEM SOLVING**
Always use a consistent set of units

SI units

TABLE 1–5 SI Base Quantities and Units

Quantity	Unit	Unit Abbreviation
Length	meter	m
Time	second	s
Mass	kilogram	kg
Electric current	ampere	A
Temperature	kelvin	K
Amount of substance	mole	mol
Luminous intensity	candela	cd

SI units are the principal ones used today in scientific work. We will therefore use SI units almost exclusively in this book, although we will give the cgs and British units for various quantities when introduced.

Base vs. Derived Quantities

Physical quantities can be divided into two categories: *base quantities* and *derived quantities*. The corresponding units for these quantities are called *base units* and *derived units*. A **base quantity** must be defined in terms of a standard. Scientists, in the interest of simplicity, want the smallest number of base quantities possible consistent with a full description of the physical world. This number turns out to be seven, and those used in the SI are given in Table 1–5. All other quantities can be defined in terms of these seven base quantities,† and hence are referred to as **derived quantities**. An example of a derived quantity is speed, which is defined as distance divided by the time it takes to travel that distance. A Table inside the front cover lists many derived quantities and their units in terms of base units. To define any quantity, whether base or derived, we can specify a rule or procedure, and this is called an **operational definition**.

1–6 Converting Units

Any quantity we measure, such as a length, a speed, or an electric current, consists of a number *and* a unit. Often we are given a quantity in one set of units, but we want it expressed in another set of units. For example, suppose we measure that a table is 21.5 inches wide, and we want to express this in centimeters. We must use a **conversion factor**, which in this case is

$$1 \text{ in.} = 2.54 \text{ cm}$$

or, written another way,

$$1 = 2.54 \text{ cm/in.}$$

Since multiplying by one does not change anything, the width of our table, in cm, is

$$21.5 \text{ inches} = (21.5 \text{ in.}) \times \left(2.54 \frac{\text{cm}}{\text{in.}}\right) = 54.6 \text{ cm.}$$

Note how the units (inches in this case) cancelled out. A Table containing many unit conversions is found inside the front cover of this book. Let's take some Examples.

FIGURE 1–9 The world's second highest peak, K2, whose summit is considered the most difficult of the 8,000-ers. K2 is seen here from the north (China). Our cover shows K2 from the south (Pakistan). Example 1–3.

PHYSICS APPLIED
The world's tallest peaks

EXAMPLE 1–3 **The 8000-m peaks.** The fourteen tallest peaks in the world (Fig. 1–9 and Table 1–6) are referred to as "eight-thousanders," meaning their summits are over 8000 m above sea level. What is the elevation, in feet, of an elevation of 8000 m?

APPROACH We need simply to convert meters to feet, and we can start with the conversion factor 1 in. = 2.54 cm, which is exact. That is, 1 in. = 2.5400 cm to any number of significant figures.

SOLUTION One foot is 12 in., so we can write

$$1 \text{ ft} = (12 \text{ in.})\left(2.54 \frac{\text{cm}}{\text{in.}}\right) = 30.48 \text{ cm} = 0.3048 \text{ m.}$$

The units cancel (colored slashes), and our result is exact. We can rewrite this

---

†The only exceptions are for angle (radians—see Chapter 8) and solid angle (steradian). No general agreement has been reached as to whether these are base or derived quantities.

equation to find the number of feet in 1 meter:

$$1 \text{ m} = \frac{1 \text{ ft}}{0.3048} = 3.28084 \text{ ft}.$$

We multiply this equation by 8,000.0 (to have five significant figures):

$$8,000.0 \text{ m} = (8,000.0 \text{ m})\left(3.28084 \frac{\text{ft}}{\text{m}}\right) = 26,247 \text{ ft}.$$

An elevation of 8000 m is 26,247 ft above sea level.

NOTE We could have done the conversion all in one line:

$$8000 \text{ m} = (8000 \text{ m})\left(\frac{100 \text{ cm}}{1 \text{ m}}\right)\left(\frac{1 \text{ in.}}{2.54 \text{ cm}}\right)\left(\frac{1 \text{ ft}}{12 \text{ in.}}\right) = 26,247 \text{ ft}.$$

The key is to multiply conversion factors, each equal to one (= 1.0000), and to make sure the units cancel.

EXERCISE D There are only 14 eight-thousand-meter peaks in the world (see Example 1–3) and their names and elevations are given in Table 1–6. They are all in the Himalaya mountain range in India, Pakistan, Tibet, and China. Determine the elevation of the world's three highest peaks in feet.

TABLE 1–6 The 8000-m Peaks	
Peak	**Height (m)**
Mt. Everest	8850
K2	8611
Kangchenjunga	8586
Lhotse	8516
Makalu	8462
Cho Oyu	8201
Dhaulagiri	8167
Manaslu	8156
Nanga Parbat	8125
Annapurna	8091
Gasherbrum I	8068
Broad Peak	8047
Gasherbrum II	8035
Shisha Pangma	8013

EXAMPLE 1–4 **Area of a semiconductor chip.** A silicon chip has an area of 1.25 square inches. Express this in square centimeters.

APPROACH We use the same conversion factor, 1 in. = 2.54 cm, but this time we have to use it twice.

SOLUTION Because 1 in. = 2.54 cm, then 1 in.^{2} = (2.54 cm)^2 = 6.45 cm^2. So

$$1.25 \text{ in.}^2 = (1.25 \text{ in.}^2)\left(2.54 \frac{\text{cm}}{\text{in.}}\right)^2 = (1.25 \text{ in.}^2)\left(6.45 \frac{\text{cm}^2}{\text{in.}^2}\right) = 8.06 \text{ cm}^2.$$

EXAMPLE 1–5 **Speeds.** Where the posted speed limit is 55 miles per hour (mi/h or mph), what is this speed (a) in meters per second (m/s) and (b) in kilometers per hour (km/h)?

APPROACH We again use the conversion factor 1 in. = 2.54 cm, and we recall that there are 5280 ft in a mile and 12 inches in a foot; also, one hour contains (60 min/h) × (60 s/min) = 3600 s/h.

SOLUTION (a) We can write 1 mile as

$$1 \text{ mi} = (5280 \text{ ft})\left(12 \frac{\text{in.}}{\text{ft}}\right)\left(2.54 \frac{\text{cm}}{\text{in.}}\right)\left(\frac{1 \text{ m}}{100 \text{ cm}}\right) = 1609 \text{ m}.$$

Note that each conversion factor is equal to one. We also know that 1 hour contains 3600 s, so

Conversion factors = 1

$$55 \frac{\text{mi}}{\text{h}} = \left(55 \frac{\text{mi}}{\text{h}}\right)\left(1609 \frac{\text{m}}{\text{mi}}\right)\left(\frac{1 \text{ h}}{3600 \text{ s}}\right) = 25 \frac{\text{m}}{\text{s}},$$

where we rounded off to two significant figures.
(b) Now we use 1 mi = 1609 m = 1.609 km; then

$$55 \frac{\text{mi}}{\text{h}} = \left(55 \frac{\text{mi}}{\text{h}}\right)\left(1.609 \frac{\text{km}}{\text{mi}}\right) = 88 \frac{\text{km}}{\text{h}}.$$

NOTE These unit conversions are very handy. You can always look them up in the Table inside the front cover.

EXERCISE E Would a driver traveling at 15 m/s in a 35 mi/h zone be exceeding the speed limit?

When changing units, you can avoid making an error in the use of conversion factors by checking that units cancel out properly. For example, in our conversion of 1 mi to 1609 m in Example 1–5(a), if we had incorrectly used the factor $\left(\frac{100 \text{ cm}}{1 \text{ m}}\right)$ instead of $\left(\frac{1 \text{ m}}{100 \text{ cm}}\right)$, the meter units would not have cancelled out; we would not have ended up with meters.

➡ **PROBLEM SOLVING**

Unit conversion is wrong if units do not cancel

1–7 Order of Magnitude: Rapid Estimating

We are sometimes interested only in an approximate value for a quantity. This might be because an accurate calculation would take more time than it is worth or would require additional data that are not available. In other cases, we may want to make a rough estimate in order to check an accurate calculation made on a calculator, to make sure that no blunders were made when the numbers were entered.

➡ **PROBLEM SOLVING**

How to make a rough estimate

A rough estimate is made by rounding off all numbers to one significant figure and its power of 10, and after the calculation is made, again only one significant figure is kept. Such an estimate is called an **order-of-magnitude estimate** and can be accurate within a factor of 10, and often better. In fact, the phrase "order of magnitude" is sometimes used to refer simply to the power of 10.

To give you some idea of how useful and powerful rough estimates can be, let us do a few "worked-out Examples."

🚶 **PHYSICS APPLIED**

Estimating the volume (or mass) of a lake; see also Fig. 1–10

EXAMPLE 1–6 | **ESTIMATE** | **Volume of a lake.** Estimate how much water there is in a particular lake, Fig. 1–10a, which is roughly circular, about 1 km across, and you guess it has an average depth of about 10 m.

APPROACH No lake is a perfect circle, nor can lakes be expected to have a perfectly flat bottom. We are only estimating here. To estimate the volume, we can use a simple model of the lake as a cylinder: we multiply the average depth of the lake times its roughly circular surface area, as if the lake were a cylinder (Fig. 1–10b).

SOLUTION The volume V of a cylinder is the product of its height h times the area of its base: $V = h\pi r^2$, where r is the radius of the circular base.[†] The radius r is $\frac{1}{2}$ km = 500 m, so the volume is approximately

$$V = h\pi r^2 \approx (10\,\text{m}) \times (3) \times (5 \times 10^2\,\text{m})^2 \approx 8 \times 10^6\,\text{m}^3 \approx 10^7\,\text{m}^3,$$

where π was rounded off to 3. So the volume is on the order of $10^7\,\text{m}^3$, ten million cubic meters. Because of all the estimates that went into this calculation, the order-of-magnitude estimate $(10^7\,\text{m}^3)$ is probably better to quote than the $8 \times 10^6\,\text{m}^3$ figure.

[†]Formulas like this for volume, area, etc., are found inside the back cover of this book.

(a)

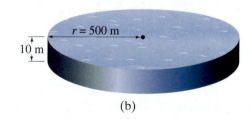

(b)

FIGURE 1–10 Example 1–6. (a) How much water is in this lake? (Photo is of one of the Rae Lakes in the Sierra Nevada of California.) (b) Model of the lake as a cylinder. [We could go one step further and estimate the mass or weight of this lake. We will see later that water has a density of 1000 kg/m³, so this lake has a mass of about $(10^3\,\text{kg/m}^3)(10^7\,\text{m}^3) \approx 10^{10}\,\text{kg}$, which is about 10 billion kg or 10 million metric tons. (A metric ton is 1000 kg, about 2200 lbs, slightly larger than a British ton, 2000 lbs.)]

NOTE To express our result in U.S. gallons, we see in the Table on the inside front cover that 1 liter = 10^{-3} m^3 ≈ $\frac{1}{4}$ gallon. Hence, the lake contains $(10^7\,\text{m}^3)(1\,\text{gallon}/4 \times 10^{-3}\,\text{m}^3) \approx 2 \times 10^9$ gallons of water.

EXAMPLE 1–7 | **ESTIMATE** | **Thickness of a page.** Estimate the thickness of a page of this book.

➡ P R O B L E M   S O L V I N G
Use symmetry when possible

APPROACH At first you might think that a special measuring device, a micrometer (Fig. 1–11), is needed to measure the thickness of one page since an ordinary ruler clearly won't do. But we can use a trick or, to put it in physics terms, make use of a *symmetry*: we can make the reasonable assumption that all the pages of this book are equal in thickness.

SOLUTION We can use a ruler to measure hundreds of pages at once. If you measure the thickness of the first 500 pages of this book (page 1 to page 500), you might get something like 1.5 cm. Note that 500 pages counted front and back is 250 separate pieces of paper. So one page must have a thickness of about

$$\frac{1.5\,\text{cm}}{250\,\text{pages}} \approx 6 \times 10^{-3}\,\text{cm} = 6 \times 10^{-2}\,\text{mm},$$

or less than a tenth of a millimeter (0.1 mm).

EXAMPLE 1–8 | **ESTIMATE** | **Total number of heartbeats.** Estimate the total number of beats a typical human heart makes in a lifetime.

APPROACH A typical resting heart rate is 70 beats/min. But during exercise it can be a lot higher. A reasonable average might be 80 beats/min.

SOLUTION If an average person lives 70 years ≈ 2×10^9 s (see Table 1–2),

$$\left(80\,\frac{\text{beats}}{\text{min}}\right)\left(\frac{1\,\text{min}}{60\,\text{s}}\right)(2 \times 10^9\,\text{s}) \approx 3 \times 10^9,$$

or 3 trillion.

FIGURE 1–11 Example 1–7. A micrometer, which is used for measuring small thicknesses.

Now let's take a simple Example of how a diagram can be useful for making an estimate. It cannot be emphasized enough how important it is to draw a diagram when trying to solve a physics problem.

FIGURE 1–12 Example 1–9. Diagrams are really useful!

EXAMPLE 1–9 | **ESTIMATE** | **Height by triangulation.** Estimate the height of the building shown in Fig. 1–12, by "triangulation," with the help of a bus-stop pole and a friend.

APPROACH By standing your friend next to the pole, you estimate the height of the pole to be 3 m. You next step away from the pole until the top of the pole is in line with the top of the building, Fig. 1–12a. You are 5 ft 6 in. tall, so your eyes are about 1.5 m above the ground. Your friend is taller, and when she stretches out her arms, one hand touches you, and the other touches the pole, so you estimate that distance as 2 m (Fig. 1–12a). You then pace off the distance from the pole to the base of the building with big, 1-m-long, steps, and you get a total of 16 steps or 16 m.

SOLUTION Now you draw, to scale, the diagram shown in Fig. 1–12b using these measurements. You can measure, right on the diagram, the last side of the triangle to be about $x = 13$ m. Alternatively, you can use similar triangles to obtain the height x:

$$\frac{1.5\,\text{m}}{2\,\text{m}} = \frac{x}{18\,\text{m}}, \quad \text{so} \quad x \approx 13\tfrac{1}{2}\,\text{m}.$$

Finally you add in your eye height of 1.5 m above the ground to get your final result: the building is about 15 m tall.

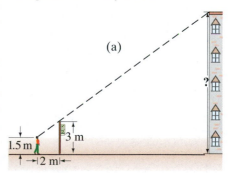

(a)

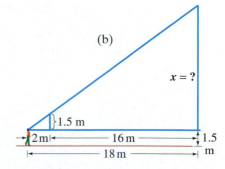

(b)

Another technique for estimating, this one made famous by Enrico Fermi to his physics students, is to estimate the number of piano tuners in a city, say, Chicago or San Francisco. To get a rough order-of-magnitude estimate of the number of piano tuners today in San Francisco, a city of about 700,000 inhabitants, we can proceed by estimating the number of functioning pianos, how often each piano is tuned, and how many pianos each tuner can tune. To estimate the number of pianos in San Francisco, we note that certainly not everyone has a piano. A guess of 1 family in 3 having a piano would correspond to 1 piano per 12 persons, assuming an average family of 4 persons. As an order of magnitude, let's say 1 piano per 10 people. This is certainly more reasonable than 1 per 100 people, or 1 per every person, so let's proceed with the estimate that 1 person in 10 has a piano, or about 70,000 pianos in San Francisco. Now a piano tuner needs an hour or two to tune a piano. So let's estimate that a tuner can tune 4 or 5 pianos a day. A piano ought to be tuned every 6 months or a year—let's say once each year. A piano tuner tuning 4 pianos a day, 5 days a week, 50 weeks a year can tune about 1000 pianos a year. So San Francisco, with its (very) roughly 70,000 pianos, needs about 70 piano tuners. This is, of course, only a rough estimate.[†] It tells us that there must be many more than 10 piano tuners, and surely not as many as 1000. If you were estimating the number of car mechanics, on the other hand, your estimate would be rather different!

→ **PROBLEM SOLVING**

Estimating how many piano tuners there are in a city

* 1–8 Dimensions and Dimensional Analysis[‡]

When we speak of the **dimensions** of a quantity, we are referring to the type of units or base quantities that make it up. The dimensions of area, for example, are always length squared, abbreviated $[L^2]$, using square brackets; the units can be square meters, square feet, cm^2, and so on. Velocity, on the other hand, can be measured in units of km/h, m/s, or mi/h, but the dimensions are always a length $[L]$ divided by a time $[T]$; that is, $[L/T]$.

The formula for a quantity may be different in different cases, but the dimensions remain the same. For example, the area of a triangle of base b and height h is $A = \frac{1}{2}bh$, whereas the area of a circle of radius r is $A = \pi r^2$. The formulas are different in the two cases, but the dimensions of area in both cases are the same: $[L^2]$.

When we specify the dimensions of a quantity, we usually do so in terms of base quantities, not derived quantities. For example, force, which we will see later has the same units as mass $[M]$ times acceleration $[L/T^2]$, has dimensions of $[ML/T^2]$.

Dimensional analysis

Dimensions can be used as a help in working out relationships, and such a procedure is referred to as **dimensional analysis**.[§] One useful technique is the use of dimensions to check if a relationship is *incorrect*. A simple rule applies here: we add or subtract quantities only if they have the same dimensions (we don't add centimeters and hours). This implies that the quantities on each side of an equals sign must have the same dimensions. (In numerical calculations, the units must also be the same on both sides of an equation.)

For example, suppose you derived the equation $v = v_0 + \frac{1}{2}at^2$, where v is the speed of an object after a time t, v_0 is the object's initial speed, and the object undergoes an acceleration a. Let's do a dimensional check to see if this equation is correct; note that numerical factors, like the $\frac{1}{2}$ here, do not affect

[†] A check of the San Francisco Yellow Pages (done after this calculation) reveals about 50 listings. Each of these listings may employ more than one tuner, but on the other hand, each may also do repairs as well as tuning. In any case, our estimate is reasonable.

[‡] Some Sections of this book, such as this one, may be considered *optional* at the discretion of the instructor. See the Preface for more details.

[§] The techniques described in the next few paragraphs may seem more meaningful after you have studied a few Chapters of this book. Reading this Section now will give you an overview of the subject, and you can then return to it later as needed.

dimensional checks. We write a dimensional equation as follows, remembering that the dimensions of speed are $[L/T]$ and (as we shall see in Chapter 2) the dimensions of acceleration are $[L/T^2]$:

$$\left[\frac{L}{T}\right] \stackrel{?}{=} \left[\frac{L}{T}\right] + \left[\frac{L}{T^2}\right][T^2]$$

$$\stackrel{?}{=} \left[\frac{L}{T}\right] + [L].$$

The dimensions are incorrect: on the right side, we have the sum of quantities whose dimensions are not the same. Thus we conclude that an error was made in the derivation of the original equation.

If such a dimensional check does come out correct, it does not prove that the equation is correct. For example, a dimensionless numerical factor (such as $\frac{1}{2}$ or 2π) could be wrong. Thus a dimensional check can only tell you when a relationship is wrong. It can't tell you if it is completely right.

Dimensional analysis can also be used as a quick check on an equation you are not sure about. For example, suppose that you can't remember whether the equation for the period T (the time to make one back-and-forth swing) of a simple pendulum of length l is $T = 2\pi\sqrt{l/g}$ or $T = 2\pi\sqrt{g/l}$, where g is the acceleration due to gravity and, like all accelerations, has dimensions $[L/T^2]$. (Do not worry about these formulas—the correct one will be derived in Chapter 11; what we are concerned about here is a person's forgetting whether it contains l/g or g/l.) A dimensional check shows that the former (l/g) is correct:

$$[T] = \sqrt{\frac{[L]}{[L/T^2]}} = \sqrt{[T^2]} = [T],$$

whereas the latter (g/l) is not:

$$[T] \neq \sqrt{\frac{[L/T^2]}{[L]}} = \sqrt{\frac{1}{[T^2]}} = \frac{1}{[T]}.$$

Note that the constant 2π has no dimensions and so can't be checked using dimensions.

Summary

[The Summary that appears at the end of each Chapter in this book gives a brief overview of the main ideas of the Chapter. The Summary *cannot* serve to give an understanding of the material, which can be accomplished only by a detailed reading of the Chapter.]

Physics, like other sciences, is a creative endeavor. It is not simply a collection of facts. Important **theories** are created with the idea of explaining **observations**. To be accepted, theories are "tested" by comparing their predictions with the results of actual experiments. Note that, in general, a theory cannot be "proved" in an absolute sense.

Scientists often devise models of physical phenomena. A **model** is a kind of picture or analogy that helps to describe the phenomena in terms of something we already know. A **theory**, often developed from a model, is usually deeper and more complex than a simple model.

A scientific **law** is a concise statement, often expressed in the form of an equation, which quantitatively describes a wide range of phenomena.

Measurements play a crucial role in physics, but can never be perfectly precise. It is important to specify the **uncertainty** of a measurement either by stating it directly using the $\pm$ notation, and/or by keeping only the correct number of **significant figures**.

Physical quantities are always specified relative to a particular standard or **unit**, and the unit used should always be stated. The commonly accepted set of units today is the **Système International** (SI), in which the standard units of length, mass, and time are the **meter**, **kilogram**, and **second**.

When converting units, check all **conversion factors** for correct cancellation of units.

Making rough, **order-of-magnitude estimates** is a very useful technique in science as well as in everyday life.

[*The **dimensions** of a quantity refer to the combination of base quantities that comprise it. Velocity, for example, has dimensions of [length/time] or $[L/T]$. Working with only the dimensions of the various quantities in a given relationship (this technique is called **dimensional analysis**) make it possible to check a relationship for correct form.]

Questions

1. What are the merits and drawbacks of using a person's foot as a standard? Consider both (a) a particular person's foot, and (b) any person's foot. Keep in mind that it is advantagous that fundamental standards be accessible (easy to compare to), invariable (do not change), indestructible, and reproducible.

2. When traveling a highway in the mountains, you may see elevation signs that read "914 m (3000 ft)." Critics of the metric system claim that such numbers show the metric system is more complicated. How would you alter such signs to be more consistent with a switch to the metric system?

3. Why is it incorrect to think that the more digits you represent in your answer, the more accurate it is?

4. What is wrong with this road sign:
   Memphis 7 mi (11.263 km)?

5. For an answer to be complete, the units need to be specified. Why?

6. Discuss how the notion of symmetry could be used to estimate the number of marbles in a 1-liter jar.

7. You measure the radius of a wheel to be 4.16 cm. If you multiply by 2 to get the diameter, should you write the result as 8 cm or as 8.32 cm? Justify your answer.

8. Express the sine of 30.0° with the correct number of significant figures.

9. A recipe for a soufflé specifies that the measured ingredients must be exact, or the soufflé will not rise. The recipe calls for 6 large eggs. The size of "large" eggs can vary by 10%, according to the USDA specifications. What does this tell you about how exactly you need to measure the other ingredients?

10. List assumptions useful to estimate the number of car mechanics in (a) San Francisco, (b) your hometown, and then make the estimates.

Problems

[The Problems at the end of each Chapter are ranked I, II, or III according to estimated difficulty, with (I) Problems being easiest. Level (III) Problems are meant mainly as a challenge for the best students, for "extra credit." The Problems are arranged by Sections, meaning that the reader should have read up to and including that Section, but not only that Section—Problems often depend on earlier material. Each Chapter also has a group of General Problems that are not arranged by Section and not ranked.]

1–4 Measurement, Uncertainty, Significant Figures

(*Note:* In Problems, assume a number like 6.4 is accurate to ±0.1; and 950 is ±10 unless 950 is said to be "precisely" or "very nearly" 950, in which case assume 950 ± 1.)

1. (I) The age of the universe is thought to be about 14 billion years. Assuming two significant figures, write this in powers of ten in (a) years, (b) seconds.

2. (I) How many significant figures do each of the following numbers have: (a) 214, (b) 81.60, (c) 7.03, (d) 0.03, (e) 0.0086, (f) 3236, and (g) 8700?

3. (I) Write the following numbers in powers of ten notation: (a) 1.156, (b) 21.8, (c) 0.0068, (d) 27.635, (e) 0.219, and (f) 444.

4. (I) Write out the following numbers in full with the correct number of zeros: (a) 8.69×10^4, (b) 9.1×10^3, (c) 8.8×10^{-1}, (d) 4.76×10^2, and (e) 3.62×10^{-5}.

5. (II) What, approximately, is the percent uncertainty for the measurement given as $1.57 \, \text{m}^2$?

6. (II) What is the percent uncertainty in the measurement $3.76 \pm 0.25 \, \text{m}$?

7. (II) Time intervals measured with a stopwatch typically have an uncertainty of about 0.2 s, due to human reaction time at the start and stop moments. What is the percent uncertainty of a handtimed measurement of (a) 5 s, (b) 50 s, (c) 5 min?

8. (II) Add $(9.2 \times 10^3 \, \text{s}) + (8.3 \times 10^4 \, \text{s}) + (0.008 \times 10^6 \, \text{s})$.

9. (II) Multiply $2.079 \times 10^2 \, \text{m}$ by 0.082×10^{-1}, taking into account significant figures.

10. (III) What is the area, and its approximate uncertainty, of a circle of radius $3.8 \times 10^4 \, \text{cm}$?

11. (III) What, roughly, is the percent uncertainty in the volume of a spherical beach ball whose radius is $r = 2.86 \pm 0.09 \, \text{m}$?

1–5 and 1–6 Units, Standards, SI, Converting Units

12. (I) Write the following as full (decimal) numbers with standard units: (a) 286.6 mm, (b) 85 μV, (c) 760 mg, (d) 60.0 ps, (e) 22.5 fm, (f) 2.50 gigavolts.

13. (I) Express the following using the prefixes of Table 1–4: (a) 1×10^6 volts, (b) 2×10^{-6} meters, (c) 6×10^3 days, (d) 18×10^2 bucks, and (e) 8×10^{-9} pieces.

14. (I) Determine your own height in meters, and your mass in kg.

15. (I) The Sun, on average, is 93 million miles from Earth. How many meters is this? Express (a) using powers of ten, and (b) using a metric prefix.

16. (II) What is the conversion factor between (a) ft^2 and yd^2, (b) m^2 and ft^2?

17. (II) An airplane travels at 950 km/h. How long does it take to travel 1.00 km?

18. (II) A typical atom has a diameter of about 1.0×10^{-10} m. (a) What is this in inches? (b) Approximately how many atoms are there along a 1.0-cm line?

19. (II) Express the following sum with the correct number of significant figures: $1.80 \, \text{m} + 142.5 \, \text{cm} + 5.34 \times 10^5 \, \mu\text{m}$.

20. (II) Determine the conversion factor between (a) km/h and mi/h, (b) m/s and ft/s, and (c) km/h and m/s.

21. (II) How much longer (percentage) is a one-mile race than a 1500-m race ("the metric mile")?

22. (II) A *light-year* is the distance light travels in one year (at speed $= 2.998 \times 10^8 \, \text{m/s}$). (a) How many meters are there in 1.00 light-year? (b) An astronomical unit (AU) is the average distance from the Sun to Earth, 1.50×10^8 km. How many AU are there in 1.00 light-year? (c) What is the speed of light in AU/h?

23. (III) The diameter of the Moon is 3480 km. (a) What is the surface area of the Moon? (b) How many times larger is the surface area of the Earth?

1–7 Order of Magnitude Estimating

(*Note:* Remember that for rough estimates, only round numbers are needed both as input to calculations and as final results.)

24. (I) Estimate the order of magnitude (power of ten) of: (a) 2800, (b) 86.30×10^2, (c) 0.0076, and (d) 15.0×10^8.

25. (II) Estimate how many books can be shelved in a college library with 3500 square meters of floor space. Assume 8 shelves high, having books on both sides, with corridors 1.5 m wide. Assume books are about the size of this one, on average.

26. (II) Estimate how many hours it would take a runner to run (at 10 km/h) across the United States from New York to California.

27. (II) Estimate how long it would take one person to mow a football field using an ordinary home lawn mower (Fig. 1–13). Assume the mower moves with a 1 km/h speed, and has a 0.5 m width.

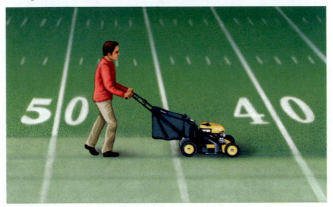

FIGURE 1–13 Problem 27.

28. (II) Estimate the number of liters of water a human drinks in a lifetime.

29. (II) Make a rough estimate of the volume of your body (in cm³).

30. (II) Make a rough estimate, for a typical suburban house, of the % of its outside wall area that consists of window area.

31. (III) The rubber worn from tires mostly enters the atmosphere as particulate pollution. Estimate how much rubber (in kg) is put into the air in the United States every year. To get started, a good estimate for a tire tread's depth is 1 cm when new, and the density of rubber is about 1200 kg/m³.

* 1–8 Dimensions

* 32. (II) The speed, v, of an object is given by the equation $v = At^3 - Bt$, where t refers to time. What are the dimensions of A and B?

* 33. (II) Three students derive the following equations in which x refers to distance traveled, v the speed, a the acceleration (m/s²), and t the time, and the subscript $(_0)$ means a quantity at time $t = 0$: (a) $x = vt^2 + 2at$, (b) $x = v_0 t + \frac{1}{2}at^2$, and (c) $x = v_0 t + 2at^2$. Which of these could possibly be correct according to a dimensional check?

General Problems

34. Global positioning satellites (GPS) can be used to determine positions with great accuracy. The system works by determining the distance between the observer and each of several satellites orbiting Earth. If one of the satellites is at a distance of 20,000 km from you, what percent accuracy in the distance is required if we desire a 2-meter uncertainty? How many significant figures do we need to have in the distance?

35. Computer chips (Fig. 1–14) are etched on circular silicon wafers of thickness 0.60 mm that are sliced from a solid cylindrical silicon crystal of length 30 cm. If each wafer can hold 100 chips, what is the maximum number of chips that can be produced from one entire cylinder?

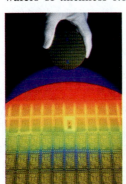

FIGURE 1–14 Problem 35. The wafer held by the hand (above) is shown below, enlarged and illuminated by colored light. Visible are rows of integrated circuits (chips).

36. (a) How many seconds are there in 1.00 year? (b) How many nanoseconds are there in 1.00 year? (c) How many years are there in 1.00 second?

37. A typical adult human lung contains about 300 million tiny cavities called alveoli. Estimate the average diameter of a single alveolus.

38. One hectare is defined as 10^4 m². One acre is 4×10^4 ft². How many acres are in one hectare?

39. Use Table 1–3 to estimate the total number of protons or neutrons in (a) a bacterium, (b) a DNA molecule, (c) the human body, (d) our Galaxy.

40. Estimate the number of gallons of gasoline consumed by the total of all automobile drivers in the United States, per year.

41. Estimate the number of gumballs in the machine of Fig. 1–15.

FIGURE 1–15
Problem 41. Estimate the number of gumballs in the machine.

42. An average family of four uses roughly 1200 liters (about 300 gallons) of water per day. (One liter = 1000 cm³.) How much depth would a lake lose per year if it uniformly covered an area of 50 square kilometers and supplied a local town with a population of 40,000 people? Consider only population uses, and neglect evaporation and so on.

43. How big is a ton? That is, what is the volume of something that weighs a ton? To be specific, estimate the diameter of a 1-ton rock, but first make a wild guess: will it be 1 ft across, 3 ft, or the size of a car? [*Hint*: Rock has mass per volume about 3 times that of water, which is 1 kg per liter (10^3 cm³) or 62 lb per cubic foot.]

44. A heavy rainstorm dumps 1.0 cm of rain on a city 5 km wide and 8 km long in a 2-h period. How many metric tons $\left(1 \text{ metric ton} = 10^3 \text{ kg}\right)$ of water fell on the city? [1 cm³ of water has a mass of 1 gram = 10^{-3} kg.] How many gallons of water was this?

45. Hold a pencil in front of your eye at a position where its blunt end just blocks out the Moon (Fig. 1–16). Make appropriate measurements to estimate the diameter of the Moon, given that the Earth–Moon distance is 3.8×10^5 km.

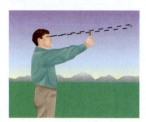

FIGURE 1–16 Problem 45. How big is the Moon?

46. Estimate how many days it would take to walk around the world, assuming 10 h walking per day at 4 km/h.

47. Noah's ark was ordered to be 300 cubits long, 50 cubits wide, and 30 cubits high. The cubit was a unit of measure equal to the length of a human forearm, elbow to the tip of the longest finger. Express the dimensions of Noah's ark in meters, and estimate its volume (m³).

48. One liter (1000 cm³) of oil is spilled onto a smooth lake. If the oil spreads out uniformly until it makes an oil slick just one molecule thick, with adjacent molecules just touching, estimate the diameter of the oil slick. Assume the oil molecules have a diameter of 2×10^{-10} m.

49. Jean camps beside a wide river and wonders how wide it is. She spots a large rock on the bank directly across from her. She then walks upstream until she judges that the angle between her and the rock, which she can still see clearly, is now at an angle of 30° downstream (Fig. 1–17). Jean measures her stride to be about one yard long. The distance back to her camp is 120 strides. About how far across, both in yards and in meters, is the river?

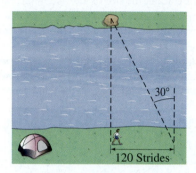

FIGURE 1–17
Problem 49.

120 Strides

50. A watch manufacturer claims that its watches gain or lose no more than 8 seconds in a year. How accurate is this watch, expressed as a percentage?

51. The diameter of the Moon is 3480 km. What is the volume of the Moon? How many Moons would be needed to create a volume equal to that of Earth?

52. An angstrom (symbol Å) is a unit of length, defined as 10^{-10} m, which is on the order of the diameter of an atom. (*a*) How many nanometers are in 1.0 angstrom? (*b*) How many femtometers or fermis (the common unit of length in nuclear physics) are in 1.0 angstrom? (*c*) How many angstroms are in 1.0 meter? (*d*) How many angstroms are in 1.0 light-year (see Problem 22)?

53. Determine the percent uncertainty in θ, and in $\sin\theta$, when (*a*) $\theta = 15.0° \pm 0.5°$, (*b*) $\theta = 75.0° \pm 0.5°$.

54. If you began walking along one of Earth's lines of longitude and walked until you had changed latitude by 1 minute of arc (there are 60 minutes per degree), how far would you have walked (in miles)? This distance is called a "nautical mile."

Answers to Exercises

A: (d).

B: No: 3, 2.

C: All three have three significant figures, although the number of decimal places is (*a*) 2, (*b*) 3, (*c*) 4.

D: Mt. Everest, 29,035 ft; K2, 28,251 ft; Kangchenjunga, 28,169 ft.

E: No: 15 m/s ≈ 34 mi/h.

A high-speed car has released a parachute to reduce its speed quickly. The directions of the car's velocity and acceleration are shown by the green ($\vec{v}$) and gold ($\vec{a}$) arrows. Motion is described using the concepts of velocity and acceleration. We see here that the acceleration $\vec{a}$ can sometimes be in the opposite direction from the velocity $\vec{v}$. We will also examine in detail motion with constant acceleration, including the vertical motion of objects falling under gravity.

CHAPTER 2

Describing Motion: Kinematics in One Dimension

The motion of objects—baseballs, automobiles, joggers, and even the Sun and Moon—is an obvious part of everyday life. It was not until the sixteenth and seventeenth centuries that our modern understanding of motion was established. Many individuals contributed to this understanding, particularly Galileo Galilei (1564–1642) and Isaac Newton (1642–1727).

The study of the motion of objects, and the related concepts of force and energy, form the field called **mechanics**. Mechanics is customarily divided into two parts: **kinematics**, which is the description of how objects move, and **dynamics**, which deals with force and why objects move as they do. This Chapter and the next deal with kinematics.

For now we only discuss objects that move without rotating (Fig. 2–1a). Such motion is called **translational motion**. In this Chapter we will be concerned with describing an object that moves along a straight-line path, which is one-dimensional translational motion. In Chapter 3 we will describe translational motion in two (or three) dimensions along paths that are not straight. (We discuss rotation, as in Fig. 2–1b, in Chapter 8.)

We will often use the concept, or *model*, of an idealized **particle** which is considered to be a mathematical point and to have no spatial extent (no size). A particle can undergo only translational motion. The particle model is useful in many real situations where we are interested only in translational motion and the object's size is not so significant. For example, we might consider a billiard ball, or even a spacecraft traveling toward the Moon, as a particle for many purposes.

(a)   (b)

FIGURE 2–1 The pinecone in (a) undergoes pure translation as it falls, whereas in (b) it is rotating as well as translating.

2–1 Reference Frames and Displacement

Any measurement of position, distance, or speed must be made with respect to a **reference frame**, or **frame of reference**. For example, while you are on a train traveling at 80 km/h, suppose a person walks past you toward the front of the train at a speed of, say, 5 km/h (Fig. 2–2). This 5 km/h is the person's speed with respect to the train as frame of reference. With respect to the ground, that person is moving at a speed of 80 km/h + 5 km/h = 85 km/h. It is always important to specify the frame of reference when stating a speed. In everyday life, we usually mean "with respect to the Earth" without even thinking about it, but the reference frame must be specified whenever there might be confusion.

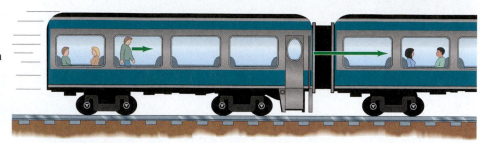

FIGURE 2–2 A person walks toward the front of a train at 5 km/h. The train is moving 80 km/h with respect to the ground, so the walking person's speed, relative to the ground, is 85 km/h.

When specifying the motion of an object, it is important to specify not only the speed but also the direction of motion. Often we can specify a direction by using north, east, south, and west, and by "up" and "down." In physics, we often draw a set of **coordinate axes**, as shown in Fig. 2–3, to represent a frame of reference. We can always place the origin 0, and the directions of the x and y axes, as we like for convenience. The x and y axes are always perpendicular to each other. Objects positioned to the right of the origin of coordinates (0) on the x axis have an x coordinate which we usually choose to be positive; objects to the left of 0 then have a negative x coordinate. The position along the y axis is usually considered positive when above 0, and negative when below 0, although the reverse convention can be used if convenient. Any point on the plane can be specified by giving its x and y coordinates. In three dimensions, a z axis perpendicular to the x and y axes is added.

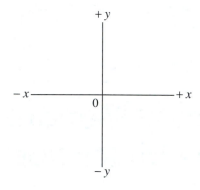

FIGURE 2–3 Standard set of xy coordinate axes.

For one-dimensional motion, we often choose the x axis as the line along which the motion takes place. Then the **position** of an object at any moment is given by its x coordinate. If the motion is vertical, as for a dropped object, we usually use the y axis.

We need to make a distinction between the *distance* an object has traveled and its **displacement**, which is defined as the *change in position* of the object. That is, *displacement is how far the object is from its starting point*. To see the distinction between total distance and displacement, imagine a person walking 70 m to the east and then turning around and walking back (west) a distance of 30 m (see Fig. 2–4). The total *distance* traveled is 100 m, but the *displacement* is only 40 m since the person is now only 40 m from the starting point.

Displacement

FIGURE 2–4 A person walks 70 m east, then 30 m west. The total distance traveled is 100 m (path is shown dashed in black); but the displacement, shown as a blue arrow, is 40 m to the east.

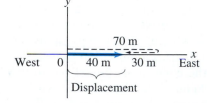

Displacement is a quantity that has both magnitude and direction. Such quantities are called **vectors**, and are represented by arrows in diagrams. For example, in Fig. 2–4, the blue arrow represents the displacement whose magnitude is 40 m and whose direction is to the right (east).

We will deal with vectors more fully in Chapter 3. For now, we deal only with motion in one dimension, along a line. In this case, vectors which point in one direction will have a positive sign, whereas vectors that point in the opposite direction will have a negative sign, along with their magnitude.

Consider the motion of an object over a particular time interval. Suppose that at some initial time, call it t_1, the object is on the x axis at the position x_1 in the coordinate system shown in Fig. 2–5. At some later time, t_2, suppose the object has moved to position x_2. The displacement of our object is $x_2 - x_1$, and is represented by the arrow pointing to the right in Fig. 2–5. It is convenient to write

$$\Delta x = x_2 - x_1,$$

where the symbol Δ (Greek letter delta) means "change in." Then Δx means "the change in x," or "change in position," which is the displacement. Note that the "change in" any quantity means the final value of that quantity, minus the initial value.

Suppose $x_1 = 10.0 \text{ m}$ and $x_2 = 30.0 \text{ m}$. Then

$$\Delta x = x_2 - x_1 = 30.0 \text{ m} - 10.0 \text{ m} = 20.0 \text{ m},$$

so the displacement is 20.0 m in the positive direction, as in Fig. 2–5.

Now consider an object moving to the left as shown in Fig. 2–6. Here the object, say, a person, starts at $x_1 = 30.0 \text{ m}$ and walks to the left to the point $x_2 = 10.0 \text{ m}$. In this case

$$\Delta x = x_2 - x_1 = 10.0 \text{ m} - 30.0 \text{ m} = -20.0 \text{ m},$$

and the blue arrow representing the vector displacement points to the left. The displacement is 20.0 m in the negative direction. This example illustrates that for one-dimensional motion along the x axis, a vector pointing to the right has a positive sign, whereas a vector pointing to the left has a negative sign.

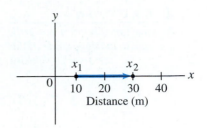
FIGURE 2–5 The arrow represents the displacement $x_2 - x_1$. Distances are in meters.

Δ means final value minus initial value

FIGURE 2–6 For the displacement $\Delta x = x_2 - x_1 = 10.0 \text{ m} - 30.0 \text{ m}$, the displacement vector points to the left.

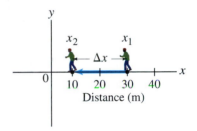

2–2 Average Velocity

Consider a racing sprinter, a galloping horse, a speeding Ferrari, or a rocket shot off into space. The most obvious aspect of their motion is how fast they are moving, which brings us to the idea of speed and velocity.

The term "speed" refers to how far an object travels in a given time interval, regardless of direction. If a car travels 240 kilometers (km) in 3 hours (h), we say its average speed was 80 km/h. In general, the **average speed** of an object is defined as *the total distance traveled along its path divided by the time it takes to travel this distance*:

$$\text{average speed} = \frac{\text{distance traveled}}{\text{time elapsed}}. \qquad \textbf{(2–1)}$$

Average speed

The terms "velocity" and "speed" are often used interchangeably in ordinary language. But in physics we make a distinction between the two. Speed is simply a positive number, with units. **Velocity**, on the other hand, is used to signify both the *magnitude* (numerical value) of how fast an object is moving and also the *direction* in which it is moving. (Velocity is therefore a vector.) There is a second difference between speed and velocity: namely, the **average velocity** is defined in terms of *displacement*, rather than total distance traveled:

Velocity

$$\text{average velocity} = \frac{\text{displacement}}{\text{time elapsed}} = \frac{\text{final position} - \text{initial position}}{\text{time elapsed}}.$$

Average velocity

Average speed and average velocity have the same magnitude when the motion is all in one direction. In other cases, they may differ: recall the walk we described earlier, in Fig. 2–4, where a person walked 70 m east and then 30 m west. The total distance traveled was 70 m + 30 m = 100 m, but the displacement was 40 m. Suppose this walk took 70 s to complete. Then the average speed was:

$$\frac{\text{distance}}{\text{time elapsed}} = \frac{100 \text{ m}}{70 \text{ s}} = 1.4 \text{ m/s.}$$

The magnitude of the average velocity, on the other hand, was:

$$\frac{\text{displacement}}{\text{time elapsed}} = \frac{40 \text{ m}}{70 \text{ s}} = 0.57 \text{ m/s.}$$

This difference between the speed and the magnitude of the velocity can occur when we calculate *average* values.

To discuss one-dimensional motion of an object in general, suppose that at some moment in time, call it t_1, the object is on the x axis at position x_1 in a coordinate system, and at some later time, t_2, suppose it is at position x_2. The elapsed time is $t_2 - t_1$; during this time interval the displacement of our object is $\Delta x = x_2 - x_1$. Then the average velocity, defined as *the displacement divided by the elapsed time*, can be written

Average velocity

$$\bar{v} = \frac{x_2 - x_1}{t_2 - t_1} = \frac{\Delta x}{\Delta t}, \tag{2-2}$$

where v stands for velocity and the bar (¯) over the v is a standard symbol meaning "average."

The **elapsed time**, or **time interval**, $t_2 - t_1$, is the time that has passed during our chosen period of observation.

For the usual case of the $+x$ axis to the right, note that if x_2 is less than x_1, the object is moving to the left, and then $\Delta x = x_2 - x_1$ is less than zero. The sign of the displacement, and thus of the average velocity, indicates the direction: the average velocity is positive for an object moving to the right along the $+x$ axis and negative when the object moves to the left. The direction of the average velocity is always the same as the direction of the displacement.

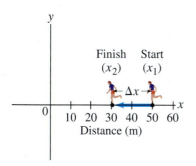

FIGURE 2–7 Example 2–1. A person runs from $x_1 = 50.0$ m to $x_2 = 30.5$ m. The displacement is −19.5 m.

EXAMPLE 2–1 **Runner's average velocity.** The position of a runner as a function of time is plotted as moving along the x axis of a coordinate system. During a 3.00-s time interval, the runner's position changes from $x_1 = 50.0$ m to $x_2 = 30.5$ m, as shown in Fig. 2–7. What was the runner's average velocity?

APPROACH We want to find the average velocity, which is the displacement divided by the elapsed time.

SOLUTION The displacement is $\Delta x = x_2 - x_1 = 30.5$ m − 50.0 m = −19.5 m. The elapsed time, or time interval, is $\Delta t = 3.00$ s. The average velocity is

$$\bar{v} = \frac{\Delta x}{\Delta t} = \frac{-19.5 \text{ m}}{3.00 \text{ s}} = -6.50 \text{ m/s.}$$

The displacement and average velocity are negative, which tells us that the runner is moving to the left along the x axis, as indicated by the arrow in Fig. 2–7. Thus we can say that the runner's average velocity is 6.50 m/s to the left.

EXAMPLE 2–2 **Distance a cyclist travels.** How far can a cyclist travel in 2.5 h along a straight road if her average velocity is 18 km/h?

APPROACH We are given the average velocity and the time interval (= 2.5 h). We want to find the distance traveled, so we solve Eq. 2–2 for Δx.

SOLUTION We rewrite Eq. 2–2 as $\Delta x = \bar{v} \Delta t$, and find

$$\Delta x = \bar{v} \Delta t = (18 \text{ km/h})(2.5 \text{ h}) = 45 \text{ km.}$$

2–3 Instantaneous Velocity

If you drive a car 150 km along a straight road in one direction for 2.0 h, the magnitude of your average velocity is 75 km/h. It is unlikely, though, that you were moving at precisely 75 km/h at every instant. To deal with this situation we need the concept of *instantaneous velocity*, which is the velocity at any instant of time. (Its magnitude is the number, with units, indicated by a speedometer; Fig. 2–8.) More precisely, the **instantaneous velocity** at any moment is defined as *the average velocity during an infinitesimally short time interval.* That is, starting with Eq. 2–2,

$$\bar{v} = \frac{\Delta x}{\Delta t},$$

we define instantaneous velocity as the average velocity as we let Δt become extremely small, approaching zero. We can write the definition of instantaneous velocity, v, for one-dimensional motion as

$$v = \lim_{\Delta t \to 0} \frac{\Delta x}{\Delta t}. \tag{2–3}$$

FIGURE 2–8 Car speedometer showing mi/h in white, and km/h in orange.

Instantaneous velocity

The notation $\lim_{\Delta t \to 0}$ means the ratio $\Delta x / \Delta t$ is to be evaluated in the limit of Δt approaching zero.

For instantaneous velocity we use the symbol v, whereas for average velocity we use $\bar{v}$, with a bar. In the rest of this book, when we use the term "velocity," it will refer to instantaneous velocity. When we want to speak of the average velocity, we will make this clear by including the word "average."

Note that the *instantaneous* speed always equals the magnitude of the instantaneous velocity. Why? Because the distance and the magnitude of the displacement become the same when they become infinitesimally small.

If an object moves at a uniform (that is, constant) velocity during a particular time interval, then its instantaneous velocity at any instant is the same as its average velocity (see Fig. 2–9a). But in many situations this is not the case. For example, a car may start from rest, speed up to 50 km/h, remain at that velocity for a time, then slow down to 20 km/h in a traffic jam, and finally stop at its destination after traveling a total of 15 km in 30 min. This trip is plotted on the graph of Fig. 2–9b. Also shown on the graph is the average velocity (dashed line), which is $\bar{v} = \Delta x / \Delta t = 15 \text{ km}/0.50 \text{ h} = 30 \text{ km/h}$.

FIGURE 2–9 Velocity of a car as a function of time: (a) at constant velocity; (b) with varying velocity.

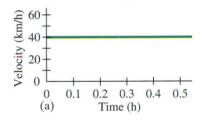

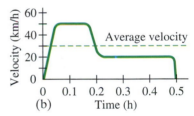

2–4 Acceleration

An object whose velocity is changing is said to be accelerating. For instance, a car whose velocity increases in magnitude from zero to 80 km/h is accelerating. Acceleration specifies how rapidly the velocity of an object is changing.

Average acceleration is defined as the change in velocity divided by the time taken to make this change:

$$\text{average acceleration} = \frac{\text{change of velocity}}{\text{time elapsed}}.$$

In symbols, the average acceleration, $\bar{a}$, during a time interval $\Delta t = t_2 - t_1$ over which the velocity changes by $\Delta v = v_2 - v_1$, is defined as

$$\bar{a} = \frac{v_2 - v_1}{t_2 - t_1} = \frac{\Delta v}{\Delta t}. \tag{2–4}$$

Average acceleration

Acceleration is also a vector, but for one-dimensional motion, we need only use a plus or minus sign to indicate direction relative to a chosen coordinate system.

The **instantaneous acceleration**, a, can be defined in analogy to instantaneous velocity, for any specific instant:

$$a = \lim_{\Delta t \to 0} \frac{\Delta v}{\Delta t}. \tag{2–5}$$

Here Δv is the very small change in velocity during the very short time interval Δt.

EXAMPLE 2–3 **Average acceleration.** A car accelerates along a straight road from rest to 75 km/h in 5.0 s, Fig. 2–10. What is the magnitude of its average acceleration?

APPROACH Average acceleration is the change in velocity divided by elapsed time, 5.0 s. The car starts from rest, so $v_1 = 0$. The final velocity is $v_2 = 75$ km/h.

SOLUTION From Eq. 2–4, the average acceleration is

$$\bar{a} = \frac{v_2 - v_1}{t_2 - t_1} = \frac{75 \text{ km/h} - 0 \text{ km/h}}{5.0 \text{ s}} = 15 \frac{\text{km/h}}{\text{s}}.$$

This is read as "fifteen kilometers per hour per second" and means that, on average, the velocity changed by 15 km/h during each second. That is, assuming the acceleration was constant, during the first second the car's velocity increased from zero to 15 km/h. During the next second its velocity increased by another 15 km/h, reaching a velocity of 30 km/h at $t = 2.0$ s, and so on. See Fig. 2–10.

NOTE Our result contains two different time units: hours and seconds. We usually prefer to use only seconds. To do so we can change km/h to m/s (see Section 1–6, and Example 1–5):

$$75 \text{ km/h} = \left(75 \frac{\cancel{\text{km}}}{\cancel{\text{h}}}\right)\left(\frac{1000 \text{ m}}{1 \cancel{\text{km}}}\right)\left(\frac{1 \cancel{\text{h}}}{3600 \text{ s}}\right) = 21 \text{ m/s}.$$

Then

$$\bar{a} = \frac{21 \text{ m/s} - 0.0 \text{ m/s}}{5.0 \text{ s}} = 4.2 \frac{\text{m/s}}{\text{s}} = 4.2 \frac{\text{m}}{\text{s}^2}.$$

We almost always write the units for acceleration as m/s² (meters per second squared), as we just did, instead of m/s/s. This is possible because:

$$\frac{\text{m/s}}{\text{s}} = \frac{\text{m}}{\text{s} \cdot \text{s}} = \frac{\text{m}}{\text{s}^2}.$$

According to the calculation in Example 2–3, the velocity changed on the average by 4.2 m/s during each second, for a total change of 21 m/s over the 5.0 s.

FIGURE 2–10 Example 2–3. The car is shown at the start with $v_1 = 0$ at $t_1 = 0$. The car is shown three more times, at $t = 1.0$ s, $t = 2.0$ s, and at the end of our time interval, $t_2 = 5.0$ s. We assume the acceleration is constant and equals 15 km/h/s. The green arrows represent the velocity vectors; the length of each arrow represents the magnitude of the velocity at that moment. The acceleration vector is the orange arrow. Distances are not to scale.

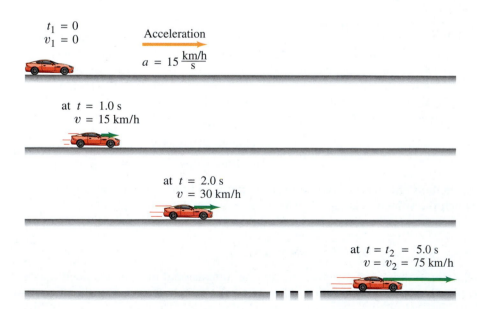

Note that *acceleration tells us how quickly the velocity changes*, whereas *velocity tells us how quickly the position changes.*

! **CAUTION**

Distinguish velocity from acceleration

CONCEPTUAL EXAMPLE 2–4 **Velocity and acceleration.** (*a*) If the velocity of an object is zero, does it mean that the acceleration is zero? (*b*) If the acceleration is zero, does it mean that the velocity is zero? Think of some examples.

! **CAUTION**

If v or a is zero, is the other zero too?

RESPONSE A zero velocity does not necessarily mean that the acceleration is zero, nor does a zero acceleration mean that the velocity is zero. (*a*) For example, when you put your foot on the gas pedal of your car which is at rest, the velocity starts from zero but the acceleration is not zero since the velocity of the car changes. (How else could your car start forward if its velocity weren't changing—that is, accelerating?) (*b*) As you cruise along a straight highway at a constant velocity of 100 km/h, your acceleration is zero: $a = 0$, $v \neq 0$.

EXERCISE A A car is advertised to go from zero to 60 mi/h in 6.0 s. What does this say about the car: (*a*) it is fast (high speed); or (*b*) it accelerates well?

EXAMPLE 2–5 **Car slowing down.** An automobile is moving to the right along a straight highway, which we choose to be the positive x axis (Fig. 2–11). Then the driver puts on the brakes. If the initial velocity (when the driver hits the brakes) is $v_1 = 15.0$ m/s, and it takes 5.0 s to slow down to $v_2 = 5.0$ m/s, what was the car's average acceleration?

APPROACH We are given the initial and final velocities and the elapsed time, so we can calculate $\bar{a}$ using Eq. 2–4.

SOLUTION We use Eq. 2–4 and call the initial time $t_1 = 0$; then $t_2 = 5.0$ s. (Note that our choice of $t_1 = 0$ doesn't affect the calculation of $\bar{a}$ because only $\Delta t = t_2 - t_1$ appears in Eq. 2–4.) Then

$$\bar{a} = \frac{5.0 \text{ m/s} - 15.0 \text{ m/s}}{5.0 \text{ s}} = -2.0 \text{ m/s}^2.$$

The negative sign appears because the final velocity is less than the initial velocity. In this case the direction of the acceleration is to the left (in the negative x direction)—even though the velocity is always pointing to the right. We say that the acceleration is 2.0 m/s² to the left, and it is shown in Fig. 2–11 as an orange arrow.

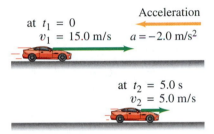

FIGURE 2–11 Example 2–5, showing the position of the car at times t_1 and t_2, as well as the car's velocity represented by the green arrows. The acceleration vector (orange) points to the left as the car slows down while moving to the right.

Deceleration

When an object is slowing down, we sometimes say it is **decelerating**. But be careful: deceleration does *not* mean that the acceleration is necessarily negative. For an object moving to the right along the positive x axis and slowing down (as in Fig. 2–11), the acceleration *is* negative. But the same car moving to the left (decreasing x), and slowing down, has positive acceleration that points to the right, as shown in Fig. 2–12. We have a deceleration whenever the magnitude of the velocity is decreasing, and then the velocity and acceleration point in opposite directions.

! **CAUTION**

Deceleration means the magnitude of the velocity is decreasing; it does not *necessarily mean a is negative*

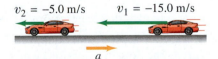

$v_2 = -5.0$ m/s $v_1 = -15.0$ m/s

a

FIGURE 2–12 The car of Example 2–5, now moving to the *left* and decelerating. The acceleration is

$$a = \frac{v_2 - v_1}{\Delta t} = \frac{-5.0 \text{ m/s} - (-15.0 \text{ m/s})}{5.0 \text{ s}} = \frac{-5.0 \text{ m/s} + 15.0 \text{ m/s}}{5.0 \text{ s}} = +2.0 \text{ m/s}.$$

EXERCISE B A car moves along the x axis. What is the sign of the car's acceleration if it is moving in the positive x direction with (*a*) increasing speed or (*b*) decreasing speed? What is the sign of the acceleration if the car moves in the negative direction with (*c*) increasing speed or (*d*) decreasing speed?

2-5 Motion at Constant Acceleration

Let a = constant

Many practical situations occur in which the acceleration is constant or nearly constant. We now examine this situation when the magnitude of the acceleration is constant and the motion is in a straight line. In this case, the instantaneous and average accelerations are equal.

We now use our definitions of velocity and acceleration to derive a set of extremely useful equations that relate x, v, a, and t when a is constant, allowing us to determine any one of these variables if we know the others.

To simplify our notation, let us take the initial time in any discussion to be zero, and we call it t_0: $t_1 = t_0 = 0$. (This is effectively starting a stopwatch at t_0.)

x (at $t = 0$) $= x_0$
v (at $t = 0$) $= v_0$
t = elapsed time

We can then let $t_2 = t$ be the elapsed time. The initial position (x_1) and the initial velocity (v_1) of an object will now be represented by x_0 and v_0, since they represent x and v at $t = 0$. At time t the position and velocity will be called x and v (rather than x_2 and v_2). The average velocity during the time interval $t - t_0$ will be (Eq. 2–2)

$$\bar{v} = \frac{x - x_0}{t - t_0} = \frac{x - x_0}{t}$$

since we chose $t_0 = 0$. The acceleration, assumed constant in time, is (Eq. 2–4)

$$a = \frac{v - v_0}{t}.$$

A common problem is to determine the velocity of an object after any elapsed time t, when we are given the object's constant acceleration. We can solve such problems by solving for v in the last equation to obtain:

v related to a and t
(a = constant, t = elapsed time)

$$v = v_0 + at. \qquad \text{[constant acceleration]} \quad \textbf{(2–6)}$$

For example, it may be known that the acceleration of a particular motorcycle is 4.0 m/s^2, and we wish to determine how fast it will be going after an elapsed time $t = 6.0 \text{ s}$ when it starts from rest $(v_0 = 0 \text{ at } t_0 = 0)$. At $t = 6.0 \text{ s}$, the velocity will be $v = at = (4.0 \text{ m/s}^2)(6.0 \text{ s}) = 24 \text{ m/s}$.

Next, let us see how to calculate the position of an object after a time t when it is undergoing constant acceleration. The definition of average velocity (Eq. 2–2) is $\bar{v} = (x - x_0)/t$, which we can rewrite as

$$x = x_0 + \bar{v}t. \qquad \textbf{(2–7)}$$

Because the velocity increases at a uniform rate, the average velocity, $\bar{v}$, will be midway between the initial and final velocities:

⚠ **CAUTION**

Average velocity, but only if
a = constant

$$\bar{v} = \frac{v_0 + v}{2}. \qquad \text{[constant acceleration]} \quad \textbf{(2–8)}$$

(Careful: Eq. 2–8 is not necessarily valid if the acceleration is not constant.) We combine the last two Equations with Eq. 2–6 and find

$$x = x_0 + \bar{v}t = x_0 + \left(\frac{v_0 + v}{2}\right)t$$

$$= x_0 + \left(\frac{v_0 + v_0 + at}{2}\right)t$$

or

x related to a and t
(a = constant)

$$x = x_0 + v_0 t + \tfrac{1}{2}at^2. \qquad \text{[constant acceleration]} \quad \textbf{(2–9)}$$

Equations 2–6, 2–8, and 2–9 are three of the four most useful equations for motion at constant acceleration. We now derive the fourth equation, which is useful in situations where the time t is not known. We begin with Eq. 2–7 and substitute in Eq. 2–8:

$$x = x_0 + \bar{v}t = x_0 + \left(\frac{v + v_0}{2}\right)t.$$

Next we solve Eq. 2–6 for t, obtaining

$$t = \frac{v - v_0}{a},$$

and substituting this into the previous equation we have

$$x = x_0 + \left(\frac{v + v_0}{2}\right)\left(\frac{v - v_0}{a}\right) = x_0 + \frac{v^2 - v_0^2}{2a}.$$

We solve this for v^2 and obtain

$$v^2 = v_0^2 + 2a(x - x_0), \qquad \text{[constant acceleration]} \quad \textbf{(2–10)}$$

v related to a and x
(a = constant)

which is the useful equation we sought.

We now have four equations relating position, velocity, acceleration, and time, when the acceleration a is constant. We collect these kinematic equations here in one place for future reference (the tan background screen emphasizes their usefulness):

$$v = v_0 + at \qquad \text{[a = constant]} \quad \textbf{(2–11a)}$$
$$x = x_0 + v_0 t + \tfrac{1}{2}at^2 \qquad \text{[a = constant]} \quad \textbf{(2–11b)}$$
$$v^2 = v_0^2 + 2a(x - x_0) \qquad \text{[a = constant]} \quad \textbf{(2–11c)}$$
$$\bar{v} = \frac{v + v_0}{2}. \qquad \text{[a = constant]} \quad \textbf{(2–11d)}$$

Kinematic equations

for constant acceleration

(we'll use them a lot)

These useful equations are not valid unless a is a constant. In many cases we can set $x_0 = 0$, and this simplifies the above equations a bit. Note that x represents position, not distance, that $x - x_0$ is the displacement, and that t is the elapsed time.

EXAMPLE 2–6 **Runway design.** You are designing an airport for small planes. One kind of airplane that might use this airfield must reach a speed before takeoff of at least 27.8 m/s (100 km/h), and can accelerate at 2.00 m/s². (*a*) If the runway is 150 m long, can this airplane reach the required speed for take off? (*b*) If not, what minimum length must the runway have?

APPROACH The plane's acceleration is given as constant $(a = 2.00 \text{ m/s}^2)$, so we can use the kinematic equations for constant acceleration. In (*a*), we are given that the plane can travel a distance of 150 m. The plane starts from rest, so $v_0 = 0$ and we take $x_0 = 0$. We want to find its velocity, to determine if it will be at least 27.8 m/s. We want to find v when we are given:

Known	Wanted
$x_0 = 0$	v
$v_0 = 0$	
$x = 150$ m	
$a = 2.00$ m/s²	

SOLUTION (*a*) Of the above four equations, Eq. 2–11c will give us v when we know $v_0, a, x,$ and x_0:

$$v^2 = v_0^2 + 2a(x - x_0)$$
$$= 0 + 2(2.0 \text{ m/s}^2)(150 \text{ m}) = 600 \text{ m}^2/\text{s}^2$$
$$v = \sqrt{600 \text{ m}^2/\text{s}^2} = 24.5 \text{ m/s}.$$

This runway length is *not* sufficient.

(*b*) Now we want to find the minimum length of runway, $x - x_0$, given $v = 27.8$ m/s and $a = 2.00$ m/s². So we again use Eq. 2–11c, but rewritten as

$$(x - x_0) = \frac{v^2 - v_0^2}{2a} = \frac{(27.8 \text{ m/s})^2 - 0}{2(2.0 \text{ m/s}^2)} = 193 \text{ m}.$$

A 200-m runway is more appropriate for this plane.

2–6 Solving Problems

Before doing more worked-out Examples, let us look at how to approach problem solving. First, it is important to note that physics is *not* a collection of equations to be memorized. (In fact, rather than memorizing the very useful Eqs. 2–11, it is better to understand how to derive them from the definitions of velocity and acceleration as we did above.) Simply searching for an equation that might work can lead you to a wrong result and will surely not help you understand physics. A better approach is to use the following (rough) procedure, which we put in a special "Box." (Other such Problem Solving Boxes, as an aid, will be found throughout the book.)

PROBLEM SOLVING

1. Read and **reread** the whole problem carefully before trying to solve it.
2. Decide what **object** (or objects) you are going to study, and for what **time interval**. You can often choose the initial time to be $t = 0$.
3. **Draw** a **diagram** or picture of the situation, with coordinate axes wherever applicable. [You can place the origin of coordinates and the axes wherever you like to make your calculations easier. You also choose which direction is positive and which is negative. Usually we choose the x axis to the right as positive.]
4. Write down what quantities are "**known**" or "given," and then what you *want* to know. Consider quantities both at the beginning and at the end of the chosen time interval. You may need to "translate" stated language into physical terms, such as "starts from rest" means $v_0 = 0$.
5. Think about which **principles of physics** apply in this problem. Use common sense and your own experiences. Then plan an approach.
6. Consider which **equations** (and/or definitions) relate the quantities involved. Before using them, be sure their **range of validity** includes your problem (for example, Eqs. 2–11 are valid only when the acceleration is constant). If you find an applicable

equation that involves only known quantities and one desired unknown, **solve** the equation algebraically for the unknown. In many instances several sequential calculations, or a combination of equations, may be needed. It is often preferable to solve algebraically for the desired unknown before putting in numerical values.

7. Carry out the **calculation** if it is a numerical problem. Keep one or two extra digits during the calculations, but round off the final answer(s) to the correct number of significant figures (Section 1–4).
8. Think carefully about the result you obtain: Is it **reasonable**? Does it make sense according to your own intuition and experience? A good check is to do a rough **estimate** using only powers of ten, as discussed in Section 1–7. Often it is preferable to do a rough estimate at the *start* of a numerical problem because it can help you focus your attention on finding a path toward a solution.
9. A very important aspect of doing problems is keeping track of **units**. An equals sign implies the units on each side must be the same, just as the numbers must. If the units do not balance, a mistake has no doubt been made. This can serve as a **check** on your solution (but it only tells you if you're wrong, not if you're right). And: always use a consistent set of units.

➡ **PROBLEM SOLVING**

"Starting from rest" means
$v = 0$ *at* $t = 0$ *[i.e., $v_0 = 0$]*

FIGURE 2–13 Example 2–7.

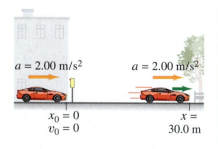

$a = 2.00 \text{ m/s}^2$ $a = 2.00 \text{ m/s}^2$

$x_0 = 0$ $x =$
$v_0 = 0$ 30.0 m

EXAMPLE 2–7 **Acceleration of a car.** How long does it take a car to cross a 30.0-m-wide intersection after the light turns green, if the car accelerates from rest at a constant 2.00 m/s²?

APPROACH We follow the Problem Solving Box, step by step.
SOLUTION
1. **Reread** the problem. Be sure you understand what it asks for (here, a time period).
2. The **object** under study is the car. We need to choose the **time interval** during which we look at the car's motion: we choose $t = 0$, the initial time, to be the moment the car starts to accelerate from rest ($v_0 = 0$); the time t is the instant the car has traveled the full 30.0-m width of the intersection.
3. **Draw** a **diagram**: the situation is shown in Fig. 2–13, where the car is shown moving along the positive x axis. We choose $x_0 = 0$ at the front bumper of the car before it starts to move.

4. The "**knowns**" and the "wanted" are shown in the Table in the margin, and we choose $x_0 = 0$. Note that "starting from rest" means $v = 0$ at $t = 0$; that is, $v_0 = 0$.

5. The **physics**: the motion takes place at constant acceleration, so we can use the kinematic equations, Eqs. 2–11.

6. **Equations**: we want to find the time, given the distance and acceleration; Eq. 2–11b is perfect since the only unknown quantity is t. Setting $v_0 = 0$ and $x_0 = 0$ in Eq. 2–11b $\left(x = x_0 + v_0 t + \frac{1}{2} at^2\right)$, we can solve for t:

$$x = \tfrac{1}{2} at^2,$$
$$t^2 = \frac{2x}{a},$$

so

$$t = \sqrt{\frac{2x}{a}}.$$

Known	Wanted
$x_0 = 0$	t
$x = 30.0 \text{ m}$	
$a = 2.00 \text{ m/s}^2$	
$v_0 = 0$	

7. The **calculation**:

$$t = \sqrt{\frac{2x}{a}} = \sqrt{\frac{2(30.0 \text{ m})}{2.00 \text{ m/s}^2}} = 5.48 \text{ s}.$$

This is our answer. Note that the units come out correctly.

8. We can check the **reasonableness** of the answer by calculating the final velocity $v = at = (2.00 \text{ m/s}^2)(5.48 \text{ s}) = 10.96 \text{ m/s}$, and then finding $x = x_0 + \bar{v}t = 0 + \frac{1}{2}(10.96 \text{ m/s} + 0)(5.48 \text{ s}) = 30.0 \text{ m}$, which is our given distance.

➡ **PROBLEM SOLVING**
Check your answer

9. We checked the **units**, and they came out perfectly (seconds).

NOTE In steps 6 and 7, when we took the square root, we should have written $t = \pm\sqrt{2x/a} = \pm 5.48 \text{ s}$. Mathematically there are two solutions. But the second solution, $t = -5.48 \text{ s}$, is a time *before* our chosen time interval and makes no sense physically. We say it is "unphysical" and ignore it.

We explicitly followed the steps of the Problem Solving Box in Example 2–7. In upcoming Examples, we will use our usual "approach" and "solution" to avoid being wordy.

EXAMPLE 2–8 **ESTIMATE** **Air bags.** Suppose you want to design an air-bag system that can protect the driver in a head-on collision at a speed of 100 km/h (60 mph). Estimate how fast the air bag must inflate (Fig. 2–14) to effectively protect the driver. How does the use of a seat belt help the driver?

APPROACH We assume the acceleration is roughly constant, so we can use Eqs. 2–11. Both Eqs. 2–11a and 2–11b contain t, our desired unknown. They both contain a, so we must first find a, which we can do using Eq. 2–11c if we know the distance x over which the car crumples. A rough estimate might be about 1 meter. We choose the time interval to start at the instant of impact with the car moving at $v_0 = 100$ km/h, and to end when the car comes to rest ($v = 0$) after traveling 1 m.

SOLUTION We convert the given initial speed to SI units: 100 km/h = 100×10^3 m/3600 s = 28 m/s. We then find the acceleration from Eq. 2–11c:

$$a = -\frac{v_0^2}{2x} = -\frac{(28 \text{ m/s})^2}{2.0 \text{ m}} = -390 \text{ m/s}^2.$$

This enormous acceleration takes place in a time given by (Eq. 2–11a):

$$t = \frac{v - v_0}{a} = \frac{0 - 28 \text{ m/s}}{-390 \text{ m/s}^2} = 0.07 \text{ s}.$$

To be effective, the air bag would need to inflate faster than this.

What does the air bag do? It spreads the force over a large area of the chest (to avoid puncture of the chest by the steering wheel). The seat belt keeps the person in a stable position against the expanding air bag.

⊛ **PHYSICS APPLIED**
Car safety—air bags

FIGURE 2–14 An air bag deploying on impact. Example 2–8.

FIGURE 2–15 Example 2–9: stopping distance for a braking car.

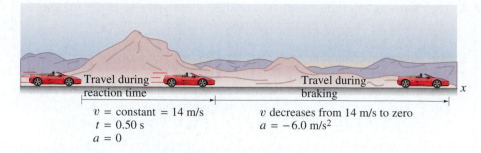

Travel during reaction time
v = constant = 14 m/s
t = 0.50 s
a = 0

Travel during braking
v decreases from 14 m/s to zero
a = −6.0 m/s²

PHYSICS APPLIED
Braking distances

EXAMPLE 2–9 **ESTIMATE** **Braking distances.** Estimate the minimum stopping distance for a car, which is important for traffic safety and traffic design. The problem is best dealt with in two parts, two separate time intervals. (1) The first time interval begins when the driver decides to hit the brakes, and ends when the foot touches the brake pedal. This is the "reaction time" during which the speed is constant, so $a = 0$. (2) The second time interval is the actual braking period when the vehicle slows down ($a \neq 0$) and comes to a stop. The stopping distance depends on the reaction time of the driver, the initial speed of the car (the final speed is zero), and the acceleration of the car. For a dry road and good tires, good brakes can decelerate a car at a rate of about 5 m/s² to 8 m/s². Calculate the total stopping distance for an initial velocity of 50 km/h (14 m/s ≈ 31 mi/h) and assume the acceleration of the car is −6.0 m/s² (the minus sign appears because the velocity is taken to be in the positive x direction and its magnitude is decreasing). Reaction time for normal drivers varies from perhaps 0.3 s to about 1.0 s; take it to be 0.50 s.

APPROACH During the "reaction time," part (1), the car moves at constant speed of 14 m/s, so $a = 0$. Once the brakes are applied, part (2), the acceleration is $a = -6.0$ m/s² and is constant over this time interval. For both parts a is constant, so we can use Eqs. 2–11.

SOLUTION Part (1). We take $x_0 = 0$ for the first part of the problem, in which the car travels at a constant speed of 14 m/s during the time interval when the driver is reacting (0.50 s). See Fig. 2–15 and the Table in the margin. To find x, the position of the car at $t = 0.50$ s (when the brakes are applied), we cannot use Eq. 2–11c because x is multiplied by a, which is zero. But Eq. 2–11b works:

$$x = v_0 t + 0 = (14 \text{ m/s})(0.50 \text{ s}) = 7.0 \text{ m}.$$

Thus the car travels 7.0 m during the driver's reaction time, until the moment the brakes are applied. We will use this result as input to part (2).

Part (2). Now we consider the second time interval, during which the brakes are applied and the car is brought to rest. We have an initial position $x_0 = 7.0$ m (result of part (1)), and other variables are shown in the Table in the margin. Equation 2–11a doesn't contain x; Eq. 2–11b contains x but also the unknown t. Equation 2–11c, $v^2 - v_0^2 = 2a(x - x_0)$, is what we want; after setting $x_0 = 7.0$ m, we solve for x, the final position of the car (when it stops):

$$x = x_0 + \frac{v^2 - v_0^2}{2a}$$

$$= 7.0 \text{ m} + \frac{0 - (14 \text{ m/s})^2}{2(-6.0 \text{ m/s}^2)} = 7.0 \text{ m} + \frac{-196 \text{ m}^2/\text{s}^2}{-12 \text{ m/s}^2}$$

$$= 7.0 \text{ m} + 16 \text{ m} = 23 \text{ m}.$$

The car traveled 7.0 m while the driver was reacting and another 16 m during the braking period before coming to a stop. The total distance traveled was then 23 m. Figure 2–16 shows a graph of v vs. t: v is constant from $t = 0$ to $t = 0.50$ s and decreases linearly, to zero, after $t = 0.50$ s.

NOTE From the equation above for x, we see that the stopping distance after you hit the brakes $(= x - x_0)$ increases with the *square* of the initial speed, not just linearly with speed. If you are traveling twice as fast, it takes four times the distance to stop.

Part 1: Reaction time

Known	Wanted
$t = 0.50$ s	x
$v_0 = 14$ m/s	
$v = 14$ m/s	
$a = 0$	
$x_0 = 0$	

Part 2: Braking

Known	Wanted
$x_0 = 7.0$ m	x
$v_0 = 14$ m/s	
$v = 0$	
$a = -6.0$ m/s²	

FIGURE 2–16 Example 2–9. Graph of v vs. t.

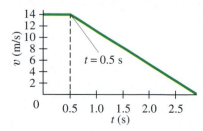

$t = 0.5$ s

The analysis of motion we have been discussing in this Chapter is basically algebraic. It is sometimes helpful to use a graphical interpretation as well; see the optional Section 2–8.

2–7 Falling Objects

One of the most common examples of uniformly accelerated motion is that of an object allowed to fall freely near the Earth's surface. That a falling object is accelerating may not be obvious at first. And beware of thinking, as was widely believed until the time of Galileo (Fig. 2–17), that heavier objects fall faster than lighter objects and that the speed of fall is proportional to how heavy the object is.

Galileo's analysis of falling objects made use of his new and creative technique of imagining what would happen in idealized (simplified) cases. For free fall, he postulated that all objects would fall with the *same constant acceleration* in the absence of air or other resistance. He showed that this postulate predicts that for an object falling from rest, the distance traveled will be proportional to the square of the time (Fig. 2–18); that is, $d \propto t^2$. We can see this from Eq. 2–11b, but Galileo was the first to derive this mathematical relation. [Among Galileo's great contributions to science was to establish such mathematical relations, and to insist on specific experimental consequences that could be quantitatively checked, such as $d \propto t^2$.]

To support his claim that falling objects increase in speed as they fall, Galileo made use of a clever argument: a heavy stone dropped from a height of 2 m will drive a stake into the ground much further than will the same stone dropped from a height of only 0.2 m. Clearly, the stone must be moving faster in the former case.

As we saw, Galileo also claimed that *all* objects, light or heavy, fall with the *same* acceleration, at least in the absence of air. If you hold a piece of paper horizontally in one hand and a heavier object—say, a baseball—in the other, and release them at the same time as in Fig. 2–19a, the heavier object will reach the ground first. But if you repeat the experiment, this time crumpling the paper into a small wad (see Fig. 2–19b), you will find that the two objects reach the floor at nearly the same time.

Galileo was sure that air acts as a resistance to very light objects that have a large surface area. But in many circumstances this air resistance is negligible. In a chamber from which the air has been removed, even light objects like a feather or a horizontally held piece of paper will fall with the same acceleration as any other object (see Fig. 2–20). Such a demonstration in vacuum was not possible in Galileo's time, which makes Galileo's achievement all the greater. Galileo is often called the "father of modern science," not only for the content of his science (astronomical discoveries, inertia, free fall), but also for his style or approach to science (idealization and simplification, mathematization of theory, theories that have testable consequences, experiments to test theoretical predictions).

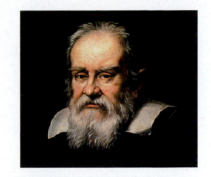

FIGURE 2–17 Galileo Galilei (1564–1642).

⚠ **CAUTION**

The speed of a falling object is NOT proportional to its mass or weight

FIGURE 2–18 Multiflash photograph of a falling apple, at equal time intervals. The apple falls farther during each successive interval, which means it is accelerating.

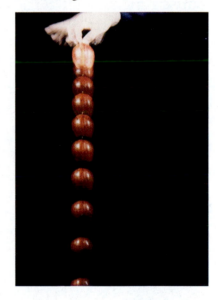

(a)   (b)

FIGURE 2–19 (a) A ball and a light piece of paper are dropped at the same time. (b) Repeated, with the paper wadded up.

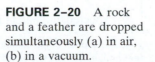

FIGURE 2–20 A rock and a feather are dropped simultaneously (a) in air, (b) in a vacuum.

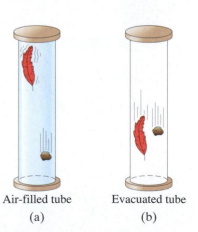

Air-filled tube   Evacuated tube

(a)   (b)

Galileo's specific contribution to our understanding of the motion of falling objects can be summarized as follows:

Galileo's hypothesis: free fall is at constant acceleration g

at a given location on the Earth and in the absence of air resistance, all objects fall with the same constant acceleration.

We call this acceleration the **acceleration due to gravity** on the Earth, and we give it the symbol g. Its magnitude is approximately

Acceleration due to gravity

$$g = 9.80 \, \text{m/s}^2. \qquad \text{[at surface of Earth]}$$

In British units g is about $32 \, \text{ft/s}^2$. Actually, g varies slightly according to latitude and elevation, but these variations are so small that we will ignore them for most purposes. The effects of air resistance are often small, and we will neglect them for the most part. However, air resistance will be noticeable even on a reasonably heavy object if the velocity becomes large.[†] Acceleration due to gravity is a vector, as is any acceleration, and its direction is toward the center of the Earth.

When dealing with freely falling objects we can make use of Eqs. 2–11, where for a we use the value of g given above. Also, since the motion is vertical we will substitute y in place of x, and y_0 in place of x_0. We take $y_0 = 0$ unless otherwise specified. *It is arbitrary whether we choose y to be positive in the upward direction or in the downward direction; but we must be consistent about it throughout a problem's solution.*

➡ **PROBLEM SOLVING**

You choose y to be positive either up or down

"Drop" means $v_0 = 0$

FIGURE 2–21 Example 2–10. (a) An object dropped from a tower falls with progressively greater speed and covers greater distance with each successive second. (See also Fig. 2–18.) (b) Graph of y vs. t.

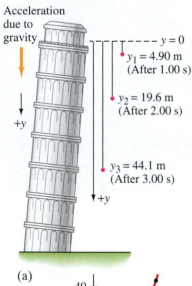

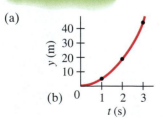

EXAMPLE 2–10 Falling from a tower. Suppose that a ball is dropped $(v_0 = 0)$ from a tower 70.0 m high. How far will the ball have fallen after a time $t_1 = 1.00 \, \text{s}$, $t_2 = 2.00 \, \text{s}$, and $t_3 = 3.00 \, \text{s}$?

APPROACH Let us take y as positive downward. We neglect any air resistance. Thus the acceleration is $a = g = +9.80 \, \text{m/s}^2$, which is positive because we have chosen downward as positive. We set $v_0 = 0$ and $y_0 = 0$. We want to find the position y of the ball after three different time intervals. Equation 2–11b, with x replaced by y, relates the given quantities (t, a, and v_0) to the unknown y.

SOLUTION We set $t = t_1 = 1.00 \, \text{s}$ in Eq. 2–11b:

$$y_1 = v_0 t_1 + \tfrac{1}{2} a t_1^2$$
$$= 0 + \tfrac{1}{2} a t_1^2 = \tfrac{1}{2}(9.80 \, \text{m/s}^2)(1.00 \, \text{s})^2 = 4.90 \, \text{m}.$$

The ball has fallen a distance of 4.90 m during the time interval $t = 0$ to $t_1 = 1.00 \, \text{s}$. Similarly, after 2.00 s ($= t_2$), the ball's position is

$$y_2 = \tfrac{1}{2} a t_2^2 = \tfrac{1}{2}(9.80 \, \text{m/s}^2)(2.00 \, \text{s})^2 = 19.6 \, \text{m}.$$

Finally, after 3.00 s ($= t_3$), the ball's position is (see Fig. 2–21)

$$y_3 = \tfrac{1}{2} a t_3^2 = \tfrac{1}{2}(9.80 \, \text{m/s}^2)(3.00 \, \text{s})^2 = 44.1 \, \text{m}.$$

NOTE Whenever we say "dropped," we mean $v_0 = 0$.

EXAMPLE 2–11 Thrown down from a tower. Suppose the ball in Example 2–10 is *thrown* downward with an initial velocity of 3.00 m/s, instead of being dropped. (*a*) What then would be its position after 1.00 s and 2.00 s? (*b*) What would its speed be after 1.00 s and 2.00 s? Compare with the speeds of a dropped ball.

APPROACH We can approach this in the same way as in Example 2–10. Again we use Eq. 2–11b, but now v_0 is not zero, it is $v_0 = 3.00 \, \text{m/s}$.

SOLUTION (*a*) At $t = 1.00 \, \text{s}$, the position of the ball as given by Eq. 2–11b is

$$y = v_0 t + \tfrac{1}{2} a t^2 = (3.00 \, \text{m/s})(1.00 \, \text{s}) + \tfrac{1}{2}(9.80 \, \text{m/s}^2)(1.00 \, \text{s})^2 = 7.90 \, \text{m}.$$

At $t = 2.00 \, \text{s}$, (time interval $t = 0$ to $t = 2.00 \, \text{s}$), the position is

$$y = v_0 t + \tfrac{1}{2} a t^2 = (3.00 \, \text{m/s})(2.00 \, \text{s}) + \tfrac{1}{2}(9.80 \, \text{m/s}^2)(2.00 \, \text{s})^2 = 25.6 \, \text{m}.$$

As expected, the ball falls farther each second than if it were dropped with $v_0 = 0$.

[†]The speed of an object falling in air (or other fluid) does not increase indefinitely. If the object falls far enough, it will reach a maximum velocity called the **terminal velocity** due to air resistance.

(*b*) The velocity is obtained from Eq. 2–11a:

$$v = v_0 + at$$
$$= 3.00 \text{ m/s} + (9.80 \text{ m/s}^2)(1.00 \text{ s}) = 12.8 \text{ m/s} \qquad [\text{at } t_1 = 1.00 \text{ s}]$$
$$= 3.00 \text{ m/s} + (9.80 \text{ m/s}^2)(2.00 \text{ s}) = 22.6 \text{ m/s}. \qquad [\text{at } t_2 = 2.00 \text{ s}]$$

In Example 2–10, when the ball was dropped $(v_0 = 0)$, the first term (v_0) in these equations was zero, so

$$v = 0 + at$$
$$= (9.80 \text{ m/s}^2)(1.00 \text{ s}) = 9.80 \text{ m/s} \qquad [\text{at } t_1 = 1.00 \text{ s}]$$
$$= (9.80 \text{ m/s}^2)(2.00 \text{ s}) = 19.6 \text{ m/s}. \qquad [\text{at } t_2 = 2.00 \text{ s}]$$

NOTE For both Examples 2–10 and 2–11, the speed increases linearly in time by 9.80 m/s during each second. But the speed of the downwardly thrown ball at any moment is always 3.00 m/s (its initial speed) higher than that of a dropped ball.

EXAMPLE 2–12 **Ball thrown upward, I.** A person throws a ball *upward* into the air with an initial velocity of 15.0 m/s. Calculate (*a*) how high it goes, and (*b*) how long the ball is in the air before it comes back to his hand.

APPROACH We are not concerned here with the throwing action, but only with the motion of the ball *after* it leaves the thrower's hand (Fig. 2–22) and until it comes back to his hand again. Let us choose y to be positive in the upward direction and negative in the downward direction. (This is a different convention from that used in Examples 2–10 and 2–11, and so illustrates our options.) The acceleration due to gravity will have a negative sign, $a = -g = -9.80 \text{ m/s}^2$. As the ball rises, its speed decreases until it reaches the highest point (B in Fig. 2–22), where its speed is zero for an instant; then it descends, with increasing speed.

SOLUTION (*a*) We consider the time interval from when the ball leaves the thrower's hand until the ball reaches the highest point. To determine the maximum height, we calculate the position of the ball when its velocity equals zero ($v = 0$ at the highest point). At $t = 0$ (point A in Fig. 2–22) we have $y_0 = 0$, $v_0 = 15.0 \text{ m/s}$, and $a = -9.80 \text{ m/s}^2$. At time t (maximum height), $v = 0$, $a = -9.80 \text{ m/s}^2$, and we wish to find y. We use Eq. 2–11c, replacing x with y: $v^2 = v_0^2 + 2ay$. We solve this equation for y:

$$y = \frac{v^2 - v_0^2}{2a} = \frac{0 - (15.0 \text{ m/s})^2}{2(-9.80 \text{ m/s}^2)} = 11.5 \text{ m}.$$

The ball reaches a height of 11.5 m above the hand.

(*b*) Now we need to choose a different time interval to calculate how long the ball is in the air before it returns to his hand. We could do this calculation in two parts by first determining the time required for the ball to reach its highest point, and then determining the time it takes to fall back down. However, it is simpler to consider the time interval for the entire motion from A to B to C (Fig. 2–22) in one step and use Eq. 2–11b. We can do this because y (or x) represents position or displacement, and not the total distance traveled. Thus, at both points A and C, $y = 0$. We use Eq. 2–11b with $a = -9.80 \text{ m/s}^2$ and find

$$y = v_0 t + \tfrac{1}{2} a t^2$$
$$0 = (15.0 \text{ m/s}) t + \tfrac{1}{2} (-9.80 \text{ m/s}^2) t^2.$$

This equation is readily factored (we factor out one t):

$$(15.0 \text{ m/s} - 4.90 \text{ m/s}^2 \, t) \, t = 0.$$

There are two solutions:

$$t = 0 \qquad \text{and} \qquad t = \frac{15.0 \text{ m/s}}{4.90 \text{ m/s}^2} = 3.06 \text{ s}.$$

The first solution ($t = 0$) corresponds to the initial point (A) in Fig. 2–22, when the ball was first thrown from $y = 0$. The second solution, $t = 3.06 \text{ s}$, corresponds to point C, when the ball has returned to $y = 0$. Thus the ball is in the air 3.06 s.

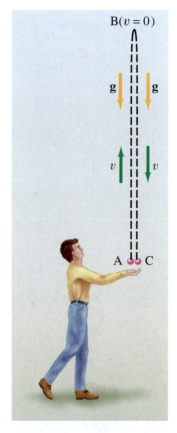

FIGURE 2–22 An object thrown into the air leaves the thrower's hand at A, reaches its maximum height at B, and returns to the original position at C. Examples 2–12, 2–13, 2–14, and 2–15.

We did not consider the throwing action in this Example. Why? Because during the throw, the thrower's hand is touching the ball and accelerating the ball at a rate unknown to us—the acceleration is *not* g. We consider only the time when the ball is in the air and the acceleration is equal to g.

Every quadratic equation (where the variable is squared) mathematically produces two solutions. In physics, sometimes only one solution corresponds to the real situation, as in Example 2–7, in which case we ignore the "unphysical" solution. But in Example 2–12, both solutions to our equation in t^2 are physically meaningful: $t = 0$ and $t = 3.06$ s.

CONCEPTUAL EXAMPLE 2–13 **Two possible misconceptions.** Give examples to show the error in these two common misconceptions: (1) that acceleration and velocity are always in the same direction, and (2) that an object thrown upward has zero acceleration at the highest point (B in Fig. 2–22).

RESPONSE Both are wrong. (1) Velocity and acceleration are *not* necessarily in the same direction. When the ball in Example 2–12 is moving upward, its velocity is positive (upward), whereas the acceleration is negative (downward). (2) At the highest point (B in Fig. 2–22), the ball has zero velocity for an instant. Is the acceleration also zero at this point? No. The velocity near the top of the arc points upward, then becomes zero (for zero time) at the highest point, and then points downward. Gravity does not stop acting, so $a = -g = -9.80 \, \text{m/s}^2$ even there. Thinking that $a = 0$ at point B would lead to the conclusion that upon reaching point B, the ball would stay there: if the acceleration (= rate of change of velocity) were zero, the velocity would stay zero at the highest point, and the ball would stay up there without falling. In sum, the acceleration of gravity always points down toward the Earth, even when the object is moving up.

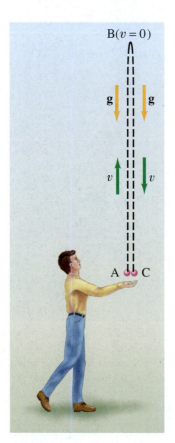

FIGURE 2–22 (Repeated for Examples 2–13, 2–14, and 2–15.)

EXAMPLE 2–14 **Ball thrown upward, II.** Let us consider again the ball thrown upward of Example 2–12, and make more calculations. Calculate (a) how much time it takes for the ball to reach the maximum height (point B in Fig. 2–22), and (b) the velocity of the ball when it returns to the thrower's hand (point C).

APPROACH Again we assume the acceleration is constant, so Eqs. 2–11 are valid. We have the height of 11.5 m from Example 2–12. Again we take y as positive upward.

SOLUTION (a) We consider the time interval between the throw ($t = 0$, $v_0 = 15.0 \, \text{m/s}$) and the top of the path ($y = +11.5 \, \text{m}$, $v = 0$, and we want to find t). The acceleration is constant at $a = -g = -9.80 \, \text{m/s}^2$. Both Eqs. 2–11a and 2–11b contain the time t with other quantities known. Let us use Eq. 2–11a with $a = -9.80 \, \text{m/s}^2$, $v_0 = 15.0 \, \text{m/s}$, and $v = 0$:

$$v = v_0 + at;$$

setting $v = 0$ and solving for t gives

$$t = -\frac{v_0}{a} = -\frac{15.0 \, \text{m/s}}{-9.80 \, \text{m/s}^2} = 1.53 \, \text{s}.$$

This is just half the time it takes the ball to go up and fall back to its original position [3.06 s, calculated in part (b) of Example 2–12]. Thus it takes the same time to reach the maximum height as to fall back to the starting point.

(b) Now we consider the time interval from the throw ($t = 0$, $v_0 = 15.0 \, \text{m/s}$) until the ball's return to the hand, which occurs at $t = 3.06$ s (as calculated in Example 2–12), and we want to find v when $t = 3.06$ s:

$$v = v_0 + at = 15.0 \, \text{m/s} - (9.80 \, \text{m/s}^2)(3.06 \, \text{s}) = -15.0 \, \text{m/s}.$$

NOTE The ball has the same magnitude of velocity when it returns to the starting point as it did initially, but in the opposite direction (this is the meaning of the negative sign). Thus, as we gathered from part (a), the motion is symmetrical about the maximum height.

EXERCISE C Two balls are thrown from a cliff. One is thrown directly up, the other directly down. Both balls have the same initial speed, and both hit the ground below the cliff. Which ball hits the ground at the greater speed: (a) the ball thrown upward, (b) the ball thrown downward, or (c) both the same? Ignore air resistance. [*Hint*: See the result of Example 2–14, part (b).]

The acceleration of objects such as rockets and fast airplanes is often given as a multiple of $g = 9.80 \text{ m/s}^2$. For example, a plane pulling out of a dive and undergoing 3.00 g's would have an acceleration of $(3.00)(9.80 \text{ m/s}^2) = 29.4 \text{ m/s}^2$.

Acceleration expressed in g's

EXERCISE D If a car is said to accelerate at 0.50 g, what is its acceleration in m/s^2?

Additional Example—Using the Quadratic Formula

EXAMPLE 2–15 Ball thrown upward, III. For the ball in Example 2–14, calculate at what time t the ball passes a point 8.00 m above the person's hand.

APPROACH We choose the time interval from the throw ($t = 0$, $v_0 = 15.0 \text{ m/s}$) until the time t (to be determined) when the ball is at position $y = 8.00 \text{ m}$, using Eq. 2–11b.

SOLUTION We want t, given $y = 8.00 \text{ m}$, $y_0 = 0$, $v_0 = 15.0 \text{ m/s}$, and $a = -9.80 \text{ m/s}^2$. We use Eq. 2–11b:

$$y = y_0 + v_0 t + \tfrac{1}{2} a t^2$$

$$8.00 \text{ m} = 0 + (15.0 \text{ m/s}) t + \tfrac{1}{2}(-9.80 \text{ m/s}^2) t^2.$$

To solve any quadratic equation of the form $at^2 + bt + c = 0$, where a, b, and c are constants (a is *not* acceleration here), we use the **quadratic formula** (see Appendix A–4):

➡ **PROBLEM SOLVING**
Using the quadratic formula

$$t = \frac{-b \pm \sqrt{b^2 - 4ac}}{2a}.$$

We rewrite our y equation just above in standard form, $at^2 + bt + c = 0$:

$$(4.90 \text{ m/s}^2) t^2 - (15.0 \text{ m/s}) t + (8.00 \text{ m}) = 0.$$

So the coefficient a is 4.90 m/s^2, b is -15.0 m/s, and c is 8.00 m. Putting these into the quadratic formula, we obtain

$$t = \frac{15.0 \text{ m/s} \pm \sqrt{(15.0 \text{ m/s})^2 - 4(4.90 \text{ m/s}^2)(8.00 \text{ m})}}{2(4.90 \text{ m/s}^2)},$$

which gives us $t = 0.69 \text{ s}$ and $t = 2.37 \text{ s}$. Are both solutions valid? Yes, because the ball passes $y = 8.00 \text{ m}$ when it goes up ($t = 0.69 \text{ s}$) and again when it comes down ($t = 2.37 \text{ s}$).

For some people, graphs can be a help in understanding. Figure 2–23 shows graphs of y vs. t and v vs. t for the ball thrown upward in Fig. 2–22, incorporating the results of Examples 2–12, 2–14, and 2–15. We shall discuss some useful properties of graphs in the next Section.

We will use the word "vertical" a lot in this book. What does it mean? (Try to respond before reading on.) Vertical is defined as the line along which an object falls. Or, if you put a small sphere on the end of a string and let it hang, the string represents a vertical line (sometimes called a *plumb line*).

EXERCISE E What does *horizontal* mean?

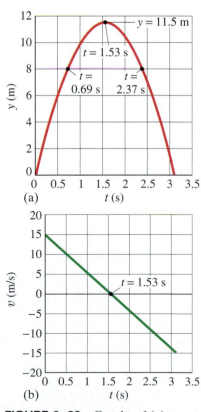

FIGURE 2–23 Graphs of (a) y vs. t, (b) v vs. t for a ball thrown upward, Examples 2–12, 2–14, and 2–15.

*2–8 Graphical Analysis of Linear Motion[†]

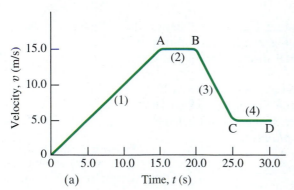

FIGURE 2–24 Graph of position vs. time for an object moving at a uniform velocity of 11 m/s.

Velocity = slope of x vs. t graph

Slope of a curve

Figure 2–9 showed the graph of the velocity of a car versus time for two cases of linear motion: (a) constant velocity, and (b) a particular case in which the magnitude of the velocity varied. It is also useful to graph, or "plot," the position x (or y) as a function of time, as we did in Fig. 2–23a. The time t is considered the independent variable and is measured along the horizontal axis. The position, x, the dependent variable, is measured along the vertical axis.

Let us make a graph of x vs. t, and make the choice that at $t = 0$, the position is $x_0 = 0$. First we consider a car moving at a constant velocity of 40 km/h, which is equivalent to 11 m/s. Equation 2–11b tells us $x = vt$, and we see that x increases by 11 m every second. Thus, the position increases linearly in time, so the graph of x vs. t is a straight line, as shown in Fig. 2–24. Each point on this straight line tells us the car's position at a particular time. For example, at $t = 3.0\,s$, the position is 33 m, and at $t = 4.0\,s$, $x = 44\,m$, as indicated by the dashed lines. The small (shaded) triangle on the graph indicates the **slope** of the straight line, which is defined as the change in the dependent variable (Δx) divided by the corresponding change in the independent variable (Δt):

$$\text{slope} = \frac{\Delta x}{\Delta t}.$$

We see, using the definition of average velocity (Eq. 2–2), that the *slope of the x vs. t graph is equal to the velocity*. And, as can be seen from the small triangle on the graph, $\Delta x/\Delta t = (11\,m)/(1.0\,s) = 11\,m/s$, which is the given velocity.

The slope of the x vs. t graph is everywhere the same if the velocity is constant, as in Fig. 2–24. But if the velocity changes, as in Fig. 2–25a, the slope of the x vs. t graph also varies. Consider, for example, a car that (1) accelerates uniformly from rest to 15 m/s in 15 s, after which (2) it remains at a constant velocity of 15 m/s for the next 5.0 s; (3) during the following 5.0 s, the car slows down uniformly to 5.0 m/s, and then (4) remains at this constant velocity. This velocity as a function of time is shown in the graph of Fig. 2–25a. To construct the x vs. t graph, we can use Eq. 2–11b ($x = x_0 + v_0 t + \frac{1}{2}at^2$) with constant acceleration for the interval $t = 0$ to $t = 15\,s$ and for $t = 20\,s$ to $t = 25\,s$; for the constant velocity period $t = 15\,s$ to $t = 20\,s$, and after $t = 25\,s$, we set $a = 0$. The result is the x vs. t graph of Fig. 2–25b.

From the origin to point A, the x vs. t graph (Fig. 2–25b) is not a straight line, but is curved. The **slope** of a curve at any point is defined as the *slope of the tangent to the curve at that point*. (The *tangent* is a straight line drawn so it touches the curve only at that one point, but does not pass across or through the curve.) For example, the tangent to the x vs. t curve at the time $t = 10.0\,s$ is drawn on the graph of Fig. 2–25b. A triangle is drawn with Δt chosen to be 4.0 s;

[†]Some Sections of this book, such as this one, may be considered *optional* at the discretion of the instructor. See the Preface for more details.

FIGURE 2–25 (a) Velocity vs. time and (b) displacement vs. time for an object with variable velocity. (See text.)

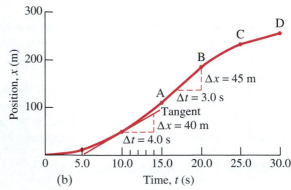

Δx can be measured off the graph for this chosen Δt and is found to be 40 m. Thus, the slope of the curve at $t = 10.0$ s, which equals the instantaneous velocity at that instant, is $v = \Delta x/\Delta t = 40\text{ m}/4.0\text{ s} = 10\text{ m/s}$.

In the region between A and B (Fig. 2–25b) the x vs. t graph is a straight line because the slope (equal to the velocity) is constant. The slope can be measured using the triangle shown for the time interval between $t = 17$ s and $t = 20$ s, where the increase in x is 45 m: $\Delta x/\Delta t = 45\text{ m}/3.0\text{ s} = 15\text{ m/s}$.

The slope of an x vs. t graph at any point is $\Delta x/\Delta t$ and thus equals the velocity of the object being described at that moment. Similarly, the slope at any point of a v vs. t graph is $\Delta v/\Delta t$ and so (by Eq. 2–4) equals the acceleration at that moment.

Suppose we were given the x vs. t graph of Fig. 2–25b. We could measure the slopes at a number of points and plot these slopes as a function of time. Since the slope equals the velocity, we could thus reconstruct the v vs. t graph! In other words, given the graph of x vs. t, we can determine the velocity as a function of time using graphical methods, instead of using equations. This technique is particularly useful when the acceleration is not constant, for then Eqs. 2–11 cannot be used.

If, instead, we are given the v vs. t graph, as in Fig. 2–25a, we can determine the position, x, as a function of time using a graphical procedure, which we illustrate by applying it to the v vs. t graph of Fig. 2–25a. We divide the total time interval into subintervals, as shown in Fig. 2–26a, where only six are shown (by dashed vertical lines). In each interval, a *horizontal* dashed line is drawn to indicate the average velocity during that time interval. For example, in the first interval, the velocity increases at a constant rate from zero to 5.0 m/s, so $\bar{v} = 2.5$ m/s; and in the fourth interval the velocity is a constant 15 m/s, so $\bar{v} = 15$ m/s (no horizontal dashed line is shown in Fig. 2–26a since it coincides with the curve itself). The displacement (change in position) during any subinterval is $\Delta x = \bar{v}\,\Delta t$. Thus the displacement during each subinterval equals the product of $\bar{v}$ and Δt, which is just the *area of the rectangle* (height × base = $\bar{v} \times \Delta t$), shown shaded in rose, for that interval. The total displacement after 25 s, say, will be the sum of the areas of the first five rectangles.

If the velocity varies a great deal, it may be difficult to estimate $\bar{v}$ from the graph. To reduce this difficulty, we can choose to divide the time interval into many more—but narrower—subintervals of time, making each Δt smaller as shown in Fig. 2–26b. More intervals give a better approximation. Ideally, we could let Δt approach zero; this leads to the techniques of integral calculus, which we don't discuss here. The result, in any case, is that *the total displacement between any two times is equal to the area under the v vs. t graph between these two times.*

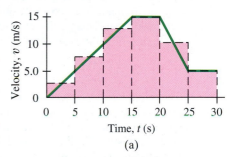

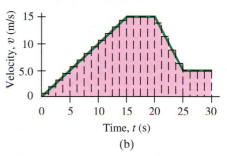

FIGURE 2–26 Determining the displacement from the graph of v vs. t is done by calculating areas.

Displacement = area under v vs. t graph

EXAMPLE 2–16 **Displacement using v vs. t graph.** A space probe accelerates uniformly from 50 m/s at $t = 0$ to 150 m/s at $t = 10$ s. How far did it move between $t = 2.0$ s and $t = 6.0$ s?

APPROACH A graph of v vs. t can be drawn as shown in Fig. 2–27. We need to calculate the area of the shaded region, which is a trapezoid. The area will be the average of the heights (in units of velocity) times the width (which is 4.0 s).
SOLUTION The acceleration is $a = (150\text{ m/s} - 50\text{ m/s})/10\text{ s} = 10\text{ m/s}^2$. Using Eq. 2–11a, or Fig. 2–27, at $t = 2.0$ s, $v = 70$ m/s; and at $t = 6.0$ s, $v = 110$ m/s. Thus the area, ($\bar{v} \times \Delta t$), which equals Δx, is

$$\Delta x = \left(\frac{70\text{ m/s} + 110\text{ m/s}}{2}\right)(4.0\text{ s}) = 360\text{ m}.$$

NOTE For this case of constant acceleration, we could use Eqs. 2–11 and we would get the same result.

FIGURE 2–27 Example 2–16. The shaded area represents the displacement during the time interval $t = 2.0$ s to $t = 6.0$ s.

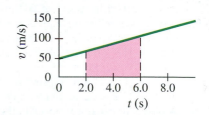

In cases where the acceleration is not constant, the area can be obtained by counting squares on graph paper.

Summary

[The Summary that appears at the end of each Chapter in this book gives a brief overview of the main ideas of the Chapter. The Summary *cannot* serve to give an understanding of the material, which can be accomplished only by a detailed reading of the Chapter.]

Kinematics deals with the description of how objects move. The description of the motion of any object must always be given relative to some particular **reference frame**.

The **displacement** of an object is the change in position of the object.

Average speed is the distance traveled divided by the elapsed time or time interval, Δt, the time period over which we choose to make our observations. An object's **average velocity** over a particular time interval Δt is its displacement Δx during that time interval, divided by Δt:

$$\bar{v} = \frac{\Delta x}{\Delta t}. \tag{2-2}$$

The **instantaneous velocity**, whose magnitude is the same as the *instantaneous speed*, is defined as the average velocity taken over an infinitesimally short time interval.

Acceleration is the change of velocity per unit time. An object's **average acceleration** over a time interval Δt is

$$\bar{a} = \frac{\Delta v}{\Delta t}, \tag{2-4}$$

where Δv is the change of velocity during the time interval Δt. **Instantaneous acceleration** is the average acceleration taken over an infinitesimally short time interval.

If an object has position x_0 and velocity v_0 at time $t = 0$ and moves in a straight line with constant acceleration, the velocity v and position x at a later time t are related to the acceleration a, the initial position x_0, and the initial velocity v_0 by Eqs. 2–11:

$$v = v_0 + at, \qquad\qquad x = x_0 + v_0 t + \tfrac{1}{2}at^2,$$
$$v^2 = v_0^2 + 2a(x - x_0), \qquad \bar{v} = \frac{v + v_0}{2}. \tag{2-11}$$

Objects that move vertically near the surface of the Earth, either falling or having been projected vertically up or down, move with the constant downward **acceleration due to gravity**, whose magnitude is $g = 9.80 \text{ m/s}^2$ if air resistance can be ignored. We can apply Eqs. 2–11 for constant acceleration to objects that move up or down freely near the Earth's surface.

[*The slope of a curve at any point on a graph is the slope of the tangent to the curve at that point. If the graph is x vs. t, the slope is $\Delta x / \Delta t$ and equals the velocity at that point. The area under a v vs. t graph equals the displacement between any two chosen times.]

Questions

1. Does a car speedometer measure speed, velocity, or both?
2. Can an object have a varying speed if its velocity is constant? If yes, give examples.
3. When an object moves with constant velocity, does its average velocity during any time interval differ from its instantaneous velocity at any instant?
4. In drag racing, is it possible for the car with the greatest speed crossing the finish line to lose the race? Explain.
5. If one object has a greater speed than a second object, does the first necessarily have a greater acceleration? Explain, using examples.
6. Compare the acceleration of a motorcycle that accelerates from 80 km/h to 90 km/h with the acceleration of a bicycle that accelerates from rest to 10 km/h in the same time.
7. Can an object have a northward velocity and a southward acceleration? Explain.
8. Can the velocity of an object be negative when its acceleration is positive? What about vice versa?
9. Give an example where both the velocity and acceleration are negative.
10. Two cars emerge side by side from a tunnel. Car A is traveling with a speed of 60 km/h and has an acceleration of 40 km/h/min. Car B has a speed of 40 km/h and has an acceleration of 60 km/h/min. Which car is passing the other as they come out of the tunnel? Explain your reasoning.
11. Can an object be increasing in speed as its acceleration decreases? If so, give an example. If not, explain.
12. A baseball player hits a foul ball straight up into the air. It leaves the bat with a speed of 120 km/h. In the absence of air resistance, how fast will the ball be traveling when the catcher catches it?
13. As a freely falling object speeds up, what is happening to its acceleration due to gravity—does it increase, decrease, or stay the same?
14. How would you estimate the maximum height you could throw a ball vertically upward? How would you estimate the maximum speed you could give it?
15. You travel from point A to point B in a car moving at a constant speed of 70 km/h. Then you travel the same distance from point B to another point C, moving at a constant speed of 90 km/h. Is your average speed for the entire trip from A to C 80 km/h? Explain why or why not.
16. In a lecture demonstration, a 3.0-m-long vertical string with ten bolts tied to it at equal intervals is dropped from the ceiling of the lecture hall. The string falls on a tin plate, and the class hears the clink of each bolt as it hits the plate. The sounds will not occur at equal time intervals. Why? Will the time between clinks increase or decrease near the end of the fall? How could the bolts be tied so that the clinks occur at equal intervals?
17. Which one of these motions is *not* at constant acceleration: a rock falling from a cliff, an elevator moving from the second floor to the fifth floor making stops along the way, a dish resting on a table?
18. An object that is thrown vertically upward will return to its original position with the same speed as it had initially if air resistance is negligible. If air resistance is appreciable, will this result be altered, and if so, how? [*Hint:* The acceleration due to air resistance is always in a direction opposite to the motion.]
19. Can an object have zero velocity and nonzero acceleration at the same time? Give examples.
20. Can an object have zero acceleration and nonzero velocity at the same time? Give examples.

*21. Describe in words the motion plotted in Fig. 2–28 in terms of v, a, etc. [*Hint*: First try to duplicate the motion plotted by walking or moving your hand.]

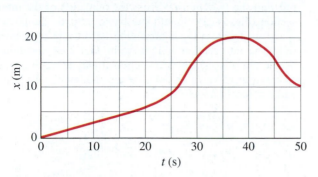

FIGURE 2–28 Question 21, Problems 50, 51, and 55.

*22. Describe in words the motion of the object graphed in Fig. 2–29.

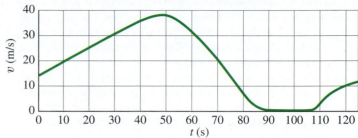

FIGURE 2–29 Question 22, Problems 49 and 54.

Problems

[The Problems at the end of each Chapter are ranked I, II, or III according to estimated difficulty, with (I) Problems being easiest. Level III are meant as challenges for the best students. The Problems are arranged by Section, meaning that the reader should have read up to and including that Section, but not only that Section—Problems often depend on earlier material. Finally, there is a set of unranked "General Problems" not arranged by Section number.]

2–1 to 2–3 Speed and Velocity

1. (I) What must be your car's average speed in order to travel 235 km in 3.25 h?

2. (I) A bird can fly 25 km/h. How long does it take to fly 15 km?

3. (I) If you are driving 110 km/h along a straight road and you look to the side for 2.0 s, how far do you travel during this inattentive period?

4. (I) Convert 35 mi/h to (a) km/h, (b) m/s, and (c) ft/s.

5. (I) A rolling ball moves from $x_1 = 3.4$ cm to $x_2 = -4.2$ cm during the time from $t_1 = 3.0$ s to $t_2 = 6.1$ s. What is its average velocity?

6. (II) A particle at $t_1 = -2.0$ s is at $x_1 = 3.4$ cm and at $t_2 = 4.5$ s is at $x_2 = 8.5$ cm. What is its average velocity? Can you calculate its average speed from these data?

7. (II) You are driving home from school steadily at 95 km/h for 130 km. It then begins to rain and you slow to 65 km/h. You arrive home after driving 3 hours and 20 minutes. (a) How far is your hometown from school? (b) What was your average speed?

8. (II) According to a rule-of-thumb, every five seconds between a lightning flash and the following thunder gives the distance to the flash in miles. Assuming that the flash of light arrives in essentially no time at all, estimate the speed of sound in m/s from this rule.

9. (II) A person jogs eight complete laps around a quarter-mile track in a total time of 12.5 min. Calculate (a) the average speed and (b) the average velocity, in m/s.

10. (II) A horse canters away from its trainer in a straight line, moving 116 m away in 14.0 s. It then turns abruptly and gallops halfway back in 4.8 s. Calculate (a) its average speed and (b) its average velocity for the entire trip, using "away from the trainer" as the positive direction.

11. (II) Two locomotives approach each other on parallel tracks. Each has a speed of 95 km/h with respect to the ground. If they are initially 8.5 km apart, how long will it be before they reach each other? (See Fig. 2–30).

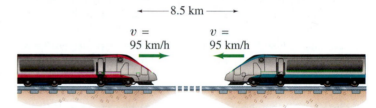

FIGURE 2–30 Problem 11.

12. (II) A car traveling 88 km/h is 110 m behind a truck traveling 75 km/h. How long will it take the car to reach the truck?

13. (II) An airplane travels 3100 km at a speed of 790 km/h, and then encounters a tailwind that boosts its speed to 990 km/h for the next 2800 km. What was the total time for the trip? What was the average speed of the plane for this trip? [*Hint*: Think carefully before using Eq. 2–11d.]

14. (II) Calculate the average speed and average velocity of a complete round-trip in which the outgoing 250 km is covered at 95 km/h, followed by a 1.0-hour lunch break, and the return 250 km is covered at 55 km/h.

15. (III) A bowling ball traveling with constant speed hits the pins at the end of a bowling lane 16.5 m long. The bowler hears the sound of the ball hitting the pins 2.50 s after the ball is released from his hands. What is the speed of the ball? The speed of sound is 340 m/s.

2–4 Acceleration

16. (I) A sports car accelerates from rest to 95 km/h in 6.2 s. What is its average acceleration in m/s²?

17. (I) A sprinter accelerates from rest to 10.0 m/s in 1.35 s. What is her acceleration (a) in m/s², and (b) in km/h²?

18. (II) At highway speeds, a particular automobile is capable of an acceleration of about 1.6 m/s^2. At this rate, how long does it take to accelerate from 80 km/h to 110 km/h?

19. (II) A sports car moving at constant speed travels 110 m in 5.0 s. If it then brakes and comes to a stop in 4.0 s, what is its acceleration in m/s^2? Express the answer in terms of "g's," where $1.00 g = 9.80 \text{ m/s}^2$.

20. (III) The position of a racing car, which starts from rest at $t = 0$ and moves in a straight line, is given as a function of time in the following Table. Estimate (a) its velocity and (b) its acceleration as a function of time. Display each in a Table and on a graph.

t (s)	0	0.25	0.50	0.75	1.00	1.50	2.00	2.50
x (m)	0	0.11	0.46	1.06	1.94	4.62	8.55	13.79

t (s)	3.00	3.50	4.00	4.50	5.00	5.50	6.00
x (m)	20.36	28.31	37.65	48.37	60.30	73.26	87.16

2–5 and 2–6 Motion at Constant Acceleration

21. (I) A car accelerates from 13 m/s to 25 m/s in 6.0 s. What was its acceleration? How far did it travel in this time? Assume constant acceleration.

22. (I) A car slows down from 23 m/s to rest in a distance of 85 m. What was its acceleration, assumed constant?

23. (I) A light plane must reach a speed of 33 m/s for takeoff. How long a runway is needed if the (constant) acceleration is 3.0 m/s^2?

24. (II) A world-class sprinter can burst out of the blocks to essentially top speed (of about 11.5 m/s) in the first 15.0 m of the race. What is the average acceleration of this sprinter, and how long does it take her to reach that speed?

25. (II) A car slows down uniformly from a speed of 21.0 m/s to rest in 6.00 s. How far did it travel in that time?

26. (II) In coming to a stop, a car leaves skid marks 92 m long on the highway. Assuming a deceleration of 7.00 m/s^2, estimate the speed of the car just before braking.

27. (II) A car traveling 85 km/h strikes a tree. The front end of the car compresses and the driver comes to rest after traveling 0.80 m. What was the average acceleration of the driver during the collision? Express the answer in terms of "g's," where $1.00 g = 9.80 \text{ m/s}^2$.

28. (II) Determine the stopping distances for a car with an initial speed of 95 km/h and human reaction time of 1.0 s, for an acceleration (a) $a = -4.0 \text{ m/s}^2$; (b) $a = -8.0 \text{ m/s}^2$.

29. (III) Show that the equation for the stopping distance of a car is $d_S = v_0 t_R - v_0^2/(2a)$, where v_0 is the initial speed of the car, t_R is the driver's reaction time, and a is the constant acceleration (and is negative).

30. (III) A car is behind a truck going 25 m/s on the highway. The car's driver looks for an opportunity to pass, guessing that his car can accelerate at 1.0 m/s^2. He gauges that he has to cover the 20-m length of the truck, plus 10 m clear room at the rear of the truck and 10 m more at the front of it. In the oncoming lane, he sees a car approaching, probably also traveling at 25 m/s. He estimates that the car is about 400 m away. Should he attempt the pass? Give details.

31. (III) A runner hopes to complete the 10,000-m run in less than 30.0 min. After exactly 27.0 min, there are still 1100 m to go. The runner must then accelerate at 0.20 m/s^2 for how many seconds in order to achieve the desired time?

32. (III) A person driving her car at 45 km/h approaches an intersection just as the traffic light turns yellow. She knows that the yellow light lasts only 2.0 s before turning red, and she is 28 m away from the near side of the intersection (Fig. 2–31). Should she try to stop, or should she speed up to cross the intersection before the light turns red? The intersection is 15 m wide. Her car's maximum deceleration is -5.8 m/s^2, whereas it can accelerate from 45 km/h to 65 km/h in 6.0 s. Ignore the length of her car and her reaction time.

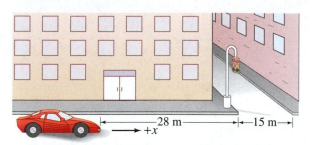

FIGURE 2–31 Problem 32.

2–7 Falling Objects [neglect air resistance]

33. (I) A stone is dropped from the top of a cliff. It hits the ground below after 3.25 s. How high is the cliff?

34. (I) If a car rolls gently ($v_0 = 0$) off a vertical cliff, how long does it take it to reach 85 km/h?

35. (I) Estimate (a) how long it took King Kong to fall straight down from the top of the Empire State Building (380 m high), and (b) his velocity just before "landing"?

36. (II) A baseball is hit nearly straight up into the air with a speed of 22 m/s. (a) How high does it go? (b) How long is it in the air?

37. (II) A ballplayer catches a ball 3.0 s after throwing it vertically upward. With what speed did he throw it, and what height did it reach?

38. (II) An object starts from rest and falls under the influence of gravity. Draw graphs of (a) its speed and (b) the distance it has fallen, as a function of time from $t = 0$ to $t = 5.00 \text{ s}$. Ignore air resistance.

39. (II) A helicopter is ascending vertically with a speed of 5.20 m/s. At a height of 125 m above the Earth, a package is dropped from a window. How much time does it take for the package to reach the ground? [*Hint*: The package's initial speed equals the helicopter's.]

40. (II) For an object falling freely from rest, show that the distance traveled during each successive second increases in the ratio of successive odd integers (1, 3, 5, etc.). This was first shown by Galileo. See Figs. 2–18 and 2–21.

41. (II) If air resistance is neglected, show (algebraically) that a ball thrown vertically upward with a speed v_0 will have the same speed, v_0, when it comes back down to the starting point.

42. (II) A stone is thrown vertically upward with a speed of 18.0 m/s. (a) How fast is it moving when it reaches a height of 11.0 m? (b) How long is required to reach this height? (c) Why are there two answers to (b)?

43. (III) Estimate the time between each photoflash of the apple in Fig. 2–18 (or number of photoflashes per second). Assume the apple is about 10 cm in diameter. [*Hint*: Use two apple positions, but not the unclear ones at the top.]

44. (III) A falling stone takes 0.28 s to travel past a window 2.2 m tall (Fig. 2–32). From what height above the top of the window did the stone fall?

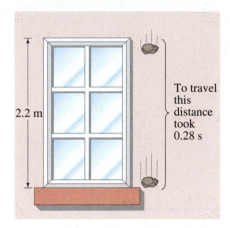

2.2 m

To travel this distance took 0.28 s

FIGURE 2–32
Problem 44.

45. (III) A rock is dropped from a sea cliff, and the sound of it striking the ocean is heard 3.2 s later. If the speed of sound is 340 m/s, how high is the cliff?

46. (III) Suppose you adjust your garden hose nozzle for a hard stream of water. You point the nozzle vertically upward at a height of 1.5 m above the ground (Fig. 2–33). When you quickly move the nozzle away from the vertical, you hear the water striking the ground next to you for another 2.0 s. What is the water speed as it leaves the nozzle?

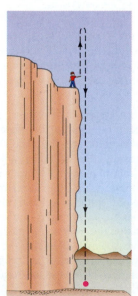

1.5 m

FIGURE 2–33
Problem 46.

47. (III) A stone is thrown vertically upward with a speed of 12.0 m/s from the edge of a cliff 70.0 m high (Fig. 2–34). (a) How much later does it reach the bottom of the cliff? (b) What is its speed just before hitting? (c) What total distance did it travel?

FIGURE 2–34
Problem 47.

48. (III) A baseball is seen to pass upward by a window 28 m above the street with a vertical speed of 13 m/s. If the ball was thrown from the street, (a) what was its initial speed, (b) what altitude does it reach, (c) when was it thrown, and (d) when does it reach the street again?

* **2–8 Graphical Analysis**

* **49.** (I) Figure 2–29 shows the velocity of a train as a function of time. (a) At what time was its velocity greatest? (b) During what periods, if any, was the velocity constant? (c) During what periods, if any, was the acceleration constant? (d) When was the magnitude of the acceleration greatest?

* **50.** (II) The position of a rabbit along a straight tunnel as a function of time is plotted in Fig. 2–28. What is its instantaneous velocity (a) at $t = 10.0$ s and (b) at $t = 30.0$ s? What is its average velocity (c) between $t = 0$ and $t = 5.0$ s, (d) between $t = 25.0$ s and $t = 30.0$ s, and (e) between $t = 40.0$ s and $t = 50.0$ s?

* **51.** (II) In Fig. 2–28, (a) during what time periods, if any, is the velocity constant? (b) At what time is the velocity greatest? (c) At what time, if any, is the velocity zero? (d) Does the object move in one direction or in both directions during the time shown?

* **52.** (II) A certain type of automobile can accelerate approximately as shown in the velocity–time graph of Fig. 2–35. (The short flat spots in the curve represent shifting of the gears.) (a) Estimate the average acceleration of the car in second gear and in fourth gear. (b) Estimate how far the car traveled while in fourth gear.

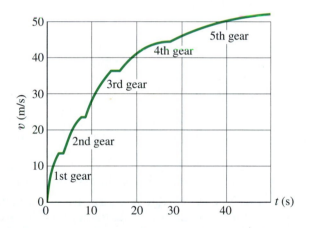

FIGURE 2–35 Problems 52 and 53. The velocity of an automobile as a function of time, starting from a dead stop. The jumps in the curve represent gear shifts.

* **53.** (II) Estimate the average acceleration of the car in the previous Problem (Fig. 2–35) when it is in (a) first, (b) third, and (c) fifth gear. (d) What is its average acceleration through the first four gears?

* **54.** (II) In Fig. 2–29, estimate the distance the object traveled during (a) the first minute, and (b) the second minute.

* **55.** (II) Construct the v vs. t graph for the object whose displacement as a function of time is given by Fig. 2–28.

* **56.** (II) Figure 2–36 is a position versus time graph for the motion of an object along the x axis. Consider the time interval from A to B. (*a*) Is the object moving in the positive or negative direction? (*b*) Is the object speeding up or slowing down? (*c*) Is the acceleration of the object positive or negative? Now consider the time interval from D to E. (*d*) Is the object moving in the positive or negative direction? (*e*) Is the object speeding up or slowing down? (*f*) Is the acceleration of the object positive or negative? (*g*) Finally, answer these same three questions for the time interval from C to D.

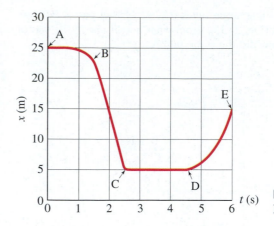

FIGURE 2–36
Problem 56.

General Problems

57. A person jumps from a fourth-story window 15.0 m above a firefighter's safety net. The survivor stretches the net 1.0 m before coming to rest, Fig. 2–37. (*a*) What was the average deceleration experienced by the survivor when she was slowed to rest by the net? (*b*) What would you do to make it "safer" (that is, to generate a smaller deceleration): would you stiffen or loosen the net? Explain.

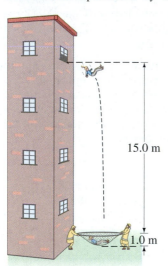

FIGURE 2–37
Problem 57.

58. The acceleration due to gravity on the Moon is about one-sixth what it is on Earth. If an object is thrown vertically upward on the Moon, how many times higher will it go than it would on Earth, assuming the same initial velocity?

59. A person who is properly constrained by an over-the-shoulder seat belt has a good chance of surviving a car collision if the deceleration does not exceed about 30 "g's" $(1.0\ g = 9.8\ \text{m/s}^2)$. Assuming uniform deceleration of this value, calculate the distance over which the front end of the car must be designed to collapse if a crash brings the car to rest from 100 km/h.

60. Agent Bond is standing on a bridge, 12 m above the road below, and his pursuers are getting too close for comfort. He spots a flatbed truck approaching at 25 m/s, which he measures by knowing that the telephone poles the truck is passing are 25 m apart in this country. The bed of the truck is 1.5 m above the road, and Bond quickly calculates how many poles away the truck should be when he jumps down from the bridge onto the truck to make his getaway. How many poles is it?

61. Suppose a car manufacturer tested its cars for front-end collisions by hauling them up on a crane and dropping them from a certain height. (*a*) Show that the speed just before a car hits the ground, after falling from rest a vertical distance H, is given by $\sqrt{2gH}$. What height corresponds to a collision at (*b*) 60 km/h? (*c*) 100 km/h?

62. Every year the Earth travels about 10^9 km as it orbits the Sun. What is Earth's average speed in km/h?

63. A 95-m-long train begins uniform acceleration from rest. The front of the train has a speed of 25 m/s when it passes a railway worker who is standing 180 m from where the front of the train started. What will be the speed of the last car as it passes the worker? (See Fig. 2–38.)

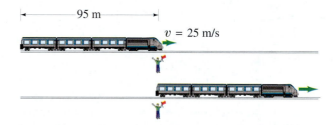

FIGURE 2–38 Problem 63.

64. A person jumps off a diving board 4.0 m above the water's surface into a deep pool. The person's downward motion stops 2.0 m below the surface of the water. Estimate the average deceleration of the person while under the water.

65. In the design of a rapid transit system, it is necessary to balance the average speed of a train against the distance between stops. The more stops there are, the slower the train's average speed. To get an idea of this problem, calculate the time it takes a train to make a 9.0-km trip in two situations: (*a*) the stations at which the trains must stop are 1.8 km apart (a total of 6 stations, including those at the ends); and (*b*) the stations are 3.0 km apart (4 stations total). Assume that at each station the train accelerates at a rate of 1.1 m/s^2 until it reaches 90 km/h, then stays at this speed until its brakes are applied for arrival at the next station, at which time it decelerates at -2.0 m/s^2. Assume it stops at each intermediate station for 20 s.

66. Pelicans tuck their wings and free fall straight down when diving for fish. Suppose a pelican starts its dive from a height of 16.0 m and cannot change its path once committed. If it takes a fish 0.20 s to perform evasive action, at what minimum height must it spot the pelican to escape? Assume the fish is at the surface of the water.

67. In putting, the force with which a golfer strikes a ball is planned so that the ball will stop within some small distance of the cup, say, 1.0 m long or short, in case the putt is missed. Accomplishing this from an uphill lie (that is, putting downhill, see Fig. 2–39) is more difficult than from a downhill lie. To see why, assume that on a particular green the ball decelerates constantly at 2.0 m/s² going downhill, and constantly at 3.0 m/s² going uphill. Suppose we have an uphill lie 7.0 m from the cup. Calculate the allowable range of initial velocities we may impart to the ball so that it stops in the range 1.0 m short to 1.0 m long of the cup. Do the same for a downhill lie 7.0 m from the cup. What in your results suggests that the downhill putt is more difficult?

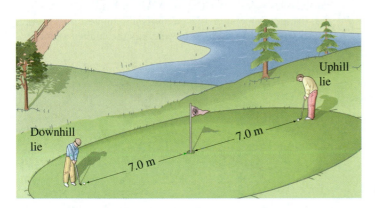

FIGURE 2–39 Problem 67. Golf on Wednesday morning.

68. A fugitive tries to hop on a freight train traveling at a constant speed of 6.0 m/s. Just as an empty box car passes him, the fugitive starts from rest and accelerates at $a = 4.0$ m/s² to his maximum speed of 8.0 m/s. (a) How long does it take him to catch up to the empty box car? (b) What is the distance traveled to reach the box car?

69. A stone is dropped from the roof of a high building. A second stone is dropped 1.50 s later. How far apart are the stones when the second one has reached a speed of 12.0 m/s?

70. A race car driver must average 200.0 km/h over the course of a time trial lasting ten laps. If the first nine laps were done at 198.0 km/h, what average speed must be maintained for the last lap?

71. A bicyclist in the Tour de France crests a mountain pass as he moves at 18 km/h. At the bottom, 4.0 km farther, his speed is 75 km/h. What was his average acceleration (in m/s²) while riding down the mountain?

72. Two children are playing on two trampolines. The first child can bounce up one-and-a-half times higher than the second child. The initial speed up of the second child is 5.0 m/s. (a) Find the maximum height the second child reaches. (b) What is the initial speed of the first child? (c) How long was the first child in the air?

73. An automobile traveling 95 km/h overtakes a 1.10-km-long train traveling in the same direction on a track parallel to the road. If the train's speed is 75 km/h, how long does it take the car to pass it, and how far will the car have traveled in this time? See Fig. 2–40. What are the results if the car and train are traveling in opposite directions?

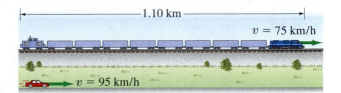

FIGURE 2–40 Problem 73.

74. A baseball pitcher throws a baseball with a speed of 44 m/s. In throwing the baseball, the pitcher accelerates the ball through a displacement of about 3.5 m, from behind the body to the point where it is released (Fig. 2–41). Estimate the average acceleration of the ball during the throwing motion.

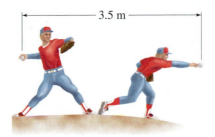

FIGURE 2–41 Problem 74.

75. A rocket rises vertically, from rest, with an acceleration of 3.2 m/s² until it runs out of fuel at an altitude of 1200 m. After this point, its acceleration is that of gravity, downward. (a) What is the velocity of the rocket when it runs out of fuel? (b) How long does it take to reach this point? (c) What maximum altitude does the rocket reach? (d) How much time (total) does it take to reach maximum altitude? (e) With what velocity does the rocket strike the Earth? (f) How long (total) is it in the air?

76. Consider the street pattern shown in Fig. 2–42. Each intersection has a traffic signal, and the speed limit is 50 km/h. Suppose you are driving from the west at the speed limit. When you are 10 m from the first intersection, all the lights turn green. The lights are green for 13 s each. (a) Calculate the time needed to reach the third stoplight. Can you make it through all three lights without stopping? (b) Another car was stopped at the first light when all the lights turned green. It can accelerate at the rate of 2.0 m/s² to the speed limit. Can the second car make it through all three lights without stopping?

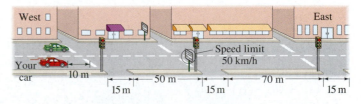

FIGURE 2–42 Problem 76.

77. A police car at rest, passed by a speeder traveling at a constant 120 km/h, takes off in hot pursuit. The police officer catches up to the speeder in 750 m, maintaining a constant acceleration. (a) Qualitatively plot the position vs. time graph for both cars from the police car's start to the catch-up point. Calculate (b) how long it took the police officer to overtake the speeder, (c) the required police car acceleration, and (d) the speed of the police car at the overtaking point.

78. A stone is dropped from the roof of a building; 2.00 s after that, a second stone is thrown straight down with an initial speed of 25.0 m/s, and the two stones land at the same time. (a) How long did it take the first stone to reach the ground? (b) How high is the building? (c) What are the speeds of the two stones just before they hit the ground?

79. Two stones are thrown vertically up at the same time. The first stone is thrown with an initial velocity of 11.0 m/s from a 12th-floor balcony of a building and hits the ground after 4.5 s. With what initial velocity should the second stone be thrown from a 4th-floor balcony so that it hits the ground at the same time as the first stone? Make simple assumptions, like equal-height floors.

80. If there were no air resistance, how long would it take a free-falling parachutist to fall from a plane at 3200 m to an altitude of 350 m, where she will pull her ripcord? What would her speed be at 350 m? (In reality, the air resistance will restrict her speed to perhaps 150 km/h.)

81. A fast-food restaurant uses a conveyor belt to send the burgers through a grilling machine. If the grilling machine is 1.1 m long and the burgers require 2.5 min to cook, how fast must the conveyor belt travel? If the burgers are spaced 15 cm apart, what is the rate of burger production (in burgers/min)?

82. Bill can throw a ball vertically at a speed 1.5 times faster than Joe can. How many times higher will Bill's ball go than Joe's?

83. You stand at the top of a cliff while your friend stands on the ground below you. You drop a ball from rest and see that it takes 1.2 s for the ball to hit the ground below. Your friend then picks up the ball and throws it up to you, such that it just comes to rest in your hand. What is the speed with which your friend threw the ball?

84. Two students are asked to find the height of a particular building using a barometer. Instead of using the barometer as an altitude-measuring device, they take it to the roof of the building and drop it off, timing its fall. One student reports a fall time of 2.0 s, and the other, 2.3 s. How much difference does the 0.3 s make for the estimates of the building's height?

* 85. Figure 2–43 shows the position vs. time graph for two bicycles, A and B. (a) Is there any instant at which the two bicycles have the same velocity? (b) Which bicycle has the larger acceleration? (c) At which instant(s) are the bicycles passing each other? Which bicycle is passing the other? (d) Which bicycle has the highest instantaneous velocity? (e) Which bicycle has the higher average velocity?

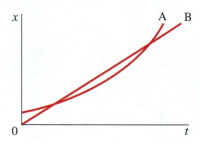

FIGURE 2–43 Problem 85.

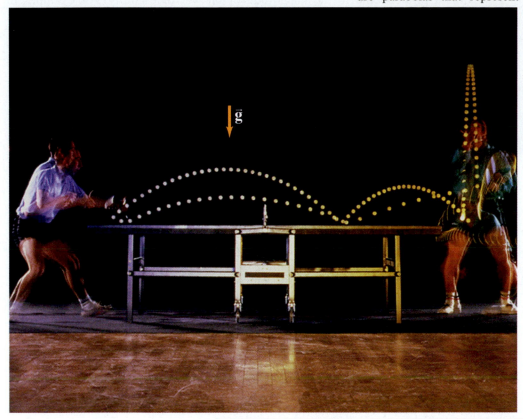

This multiflash photograph of a ping pong ball shows examples of motion in two dimensions. The arcs of the ping pong ball are parabolas that represent "projectile motion." Galileo analyzed projectile motion into its horizontal and vertical components; the gold arrow represents the downward acceleration of gravity, $\vec{g}$. We will discuss how to manipulate vectors and how to add them. Besides analyzing projectile motion, we will also see how to work with relative velocity.

CHAPTER 3

Kinematics in Two Dimensions; Vectors

In Chapter 2 we dealt with motion along a straight line. We now consider the description of the motion of objects that move in paths in two (or three) dimensions. In particular, we discuss an important type of motion known as *projectile motion*: objects projected outward near the surface of the Earth, such as struck baseballs and golf balls, kicked footballs, and other projectiles. Before beginning our discussion of motion in two dimensions, we first need to present a new tool—vectors—and how to add them.

3–1 Vectors and Scalars

We mentioned in Chapter 2 that the term *velocity* refers not only to how fast something is moving but also to its direction. A quantity such as velocity, which has *direction* as well as *magnitude*, is a **vector** quantity. Other quantities that are also vectors are displacement, force, and momentum. However, many quantities have no direction associated with them, such as mass, time, and temperature. They are specified completely by a number and units. Such quantities are called **scalar** quantities.

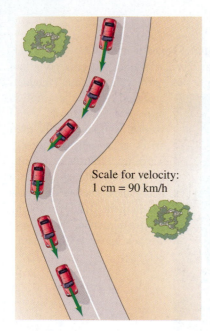

FIGURE 3–1 Car traveling on a road. The green arrows represent the velocity vector at each position.

Scale for velocity:
1 cm = 90 km/h

FIGURE 3–2 Combining vectors in one dimension.

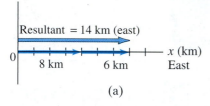

Resultant = 14 km (east)

8 km   6 km   East

x (km)

(a)

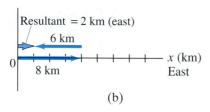

Resultant = 2 km (east)

6 km

8 km   East

x (km)

(b)

Drawing a diagram of a particular physical situation is always helpful in physics, and this is especially true when dealing with vectors. On a diagram, each vector is represented by an arrow. The arrow is always drawn so that it points in the direction of the vector quantity it represents. The length of the arrow is drawn proportional to the magnitude of the vector quantity. For example, in Fig. 3–1, green arrows have been drawn representing the velocity of a car at various places as it rounds a curve. The magnitude of the velocity at each point can be read off Fig. 3–1 by measuring the length of the corresponding arrow and using the scale shown (1 cm = 90 km/h).

When we write the symbol for a vector, we will always use boldface type, with a tiny arrow over the symbol. Thus for velocity we write $\vec{v}$. If we are concerned only with the magnitude of the vector, we will write simply v, in italics, as we do for other symbols.

3–2 Addition of Vectors—Graphical Methods

Because vectors are quantities that have direction as well as magnitude, they must be added in a special way. In this Chapter, we will deal mainly with displacement vectors, for which we now use the symbol $\vec{D}$, and velocity vectors, $\vec{v}$. But the results will apply for other vectors we encounter later.

We use simple arithmetic for adding scalars. Simple arithmetic can also be used for adding vectors if they are in the same direction. For example, if a person walks 8 km east one day, and 6 km east the next day, the person will be 8 km + 6 km = 14 km east of the point of origin. We say that the *net* or *resultant* displacement is 14 km to the east (Fig. 3–2a). If, on the other hand, the person walks 8 km east on the first day, and 6 km west (in the reverse direction) on the second day, then the person will end up 2 km from the origin (Fig. 3–2b), so the resultant displacement is 2 km to the east. In this case, the resultant displacement is obtained by subtraction: 8 km − 6 km = 2 km.

But simple arithmetic cannot be used if the two vectors are not along the same line. For example, suppose a person walks 10.0 km east and then walks 5.0 km north. These displacements can be represented on a graph in which the positive y axis points north and the positive x axis points east, Fig. 3–3. On this graph, we draw an arrow, labeled $\vec{D}_1$, to represent the displacement vector of the 10.0-km displacement to the east. Then we draw a second arrow, $\vec{D}_2$, to represent the 5.0-km displacement to the north. Both vectors are drawn to scale, as in Fig. 3–3.

After taking this walk, the person is now 10.0 km east and 5.0 km north of the point of origin. The **resultant displacement** is represented by the arrow labeled $\vec{D}_R$ in Fig. 3–3. Using a ruler and a protractor, you can measure on this diagram that the person is 11.2 km from the origin at an angle $\theta = 27°$ north of east. In other words, the resultant displacement vector has a magnitude of 11.2 km and makes an angle $\theta = 27°$ with the positive x axis. The magnitude (length) of $\vec{D}_R$ can also be obtained using the theorem of Pythagoras in this case, since D_1, D_2, and D_R form a right triangle with D_R as the hypotenuse. Thus

$$D_R = \sqrt{D_1^2 + D_2^2} = \sqrt{(10.0 \text{ km})^2 + (5.0 \text{ km})^2} = \sqrt{125 \text{ km}^2} = 11.2 \text{ km}.$$

You can use the Pythagorean theorem, of course, only when the vectors are *perpendicular* to each other.

FIGURE 3–3 A person walks 10.0 km east and then 5.0 km north. These two displacements are represented by the vectors $\vec{D}_1$ and $\vec{D}_2$, which are shown as arrows. The resultant displacement vector, $\vec{D}_R$, which is the vector sum of $\vec{D}_1$ and $\vec{D}_2$, is also shown. Measurement on the graph with ruler and protractor shows that $\vec{D}_R$ has a magnitude of 11.2 km and points at an angle $\theta = 27°$ north of east.

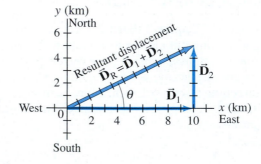

The resultant displacement vector, $\vec{\mathbf{D}}_R$, is the sum of the vectors $\vec{\mathbf{D}}_1$ and $\vec{\mathbf{D}}_2$. That is,

$$\vec{\mathbf{D}}_R = \vec{\mathbf{D}}_1 + \vec{\mathbf{D}}_2.$$ *Vector equation*

This is a *vector* equation. An important feature of adding two vectors that are not along the same line is that the magnitude of the resultant vector is not equal to the sum of the magnitudes of the two separate vectors, but is smaller than their sum:

$$D_R < D_1 + D_2.$$ [vectors not along the same line]

In our example (Fig. 3–3), $D_R = 11.2 \text{ km}$, whereas $D_1 + D_2$ equals 15 km. Note also that we cannot set $\vec{\mathbf{D}}_R$ equal to 11.2 km, because we have a vector equation and 11.2 km is only a part of the resultant vector, its magnitude. We could write something like this, though: $\vec{\mathbf{D}}_R = \vec{\mathbf{D}}_1 + \vec{\mathbf{D}}_2 = (11.2 \text{ km}, 27° \text{ N of E})$.

EXERCISE A Under what conditions can the magnitude of the resultant vector above be $D_R = D_1 + D_2$?

Figure 3–3 illustrates the general rules for graphically adding two vectors together, no matter what angles they make, to get their sum. The rules are as follows:

1. On a diagram, draw one of the vectors—call it $\vec{\mathbf{D}}_1$—to scale.
2. Next draw the second vector, $\vec{\mathbf{D}}_2$, to scale, placing its tail at the tip of the first vector and being sure its direction is correct.   *Tail-to-tip method of adding vectors*
3. The arrow drawn from the tail of the first vector to the tip of the second vector represents the *sum*, or **resultant**, of the two vectors.

The length of the resultant vector represents its magnitude. Note that vectors can be translated parallel to themselves (maintaining the same length and angle) to accomplish these manipulations. The length of the resultant can be measured with a ruler and compared to the scale. Angles can be measured with a protractor. This method is known as the **tail-to-tip method of adding vectors**.

It is not important in which order the vectors are added. For example, a displacement of 5.0 km north, to which is added a displacement of 10.0 km east, yields a resultant of 11.2 km and angle $\theta = 27°$ (see Fig. 3–4), the same as when they were added in reverse order (Fig. 3–3). That is,

$$\vec{\mathbf{V}}_1 + \vec{\mathbf{V}}_2 = \vec{\mathbf{V}}_2 + \vec{\mathbf{V}}_1.$$

The tail-to-tip method of adding vectors can be extended to three or more vectors. The resultant is drawn from the tail of the first vector to the tip of the last one added. An example is shown in Fig. 3–5; the three vectors could represent displacements (northeast, south, west) or perhaps three forces. Check for yourself that you get the same resultant no matter in which order you add the three vectors.

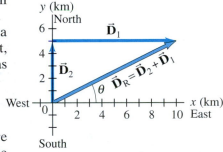

FIGURE 3–4 If the vectors are added in reverse order, the resultant is the same. (Compare to Fig. 3–3.)

FIGURE 3–5 The resultant of three vectors: $\vec{\mathbf{V}}_R = \vec{\mathbf{V}}_1 + \vec{\mathbf{V}}_2 + \vec{\mathbf{V}}_3$.

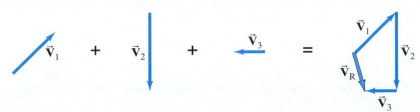

*Parallelogram method of
adding vectors*

A second way to add two vectors is the **parallelogram method**. It is fully equivalent to the tail-to-tip method. In this method, the two vectors are drawn starting from a common origin, and a parallelogram is constructed using these two vectors as adjacent sides as shown in Fig. 3–6b. The resultant is the diagonal drawn from the common origin. In Fig. 3–6a, the tail-to-tip method is shown, and it is clear that both methods yield the same result.

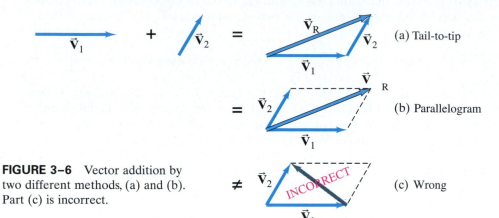

FIGURE 3–6 Vector addition by two different methods, (a) and (b). Part (c) is incorrect.

⚠️ **CAUTION**

Be sure to use the correct diagonal on parallelogram to get the resultant

It is a common error to draw the sum vector as the diagonal running between the tips of the two vectors, as in Fig. 3–6c. *This is incorrect*: it does not represent the sum of the two vectors. (In fact, it represents their difference, $\vec{V}_2 - \vec{V}_1$, as we will see in the next Section.)

CONCEPTUAL EXAMPLE 3–1 **Range of vector lengths.** Suppose two vectors each have length 3.0 units. What is the range of possible lengths for the vector representing the sum of the two?

RESPONSE The sum can take on any value from 6.0 (= 3.0 + 3.0) where the vectors point in the same direction, to 0 (= 3.0 − 3.0) when the vectors are antiparallel.

EXERCISE B If the two vectors of Conceptual Example 3–1 are perpendicular to each other, what is the resultant vector length?

3–3 Subtraction of Vectors, and Multiplication of a Vector by a Scalar

Given a vector $\vec{V}$, we define the *negative* of this vector $(-\vec{V})$ to be a vector with the same magnitude as $\vec{V}$ but opposite in direction, Fig. 3–7. Note, however, that no vector is ever negative in the sense of its magnitude: the magnitude of every vector is positive. Rather, a minus sign tells us about its direction.

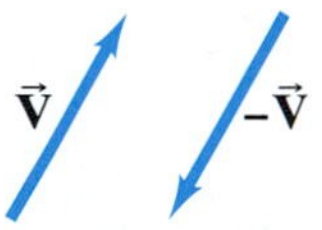

We can now define the subtraction of one vector from another: the difference between two vectors $\vec{V}_2 - \vec{V}_1$ is defined as

$$\vec{V}_2 - \vec{V}_1 = \vec{V}_2 + (-\vec{V}_1).$$

That is, the difference between two vectors is equal to the sum of the first plus the negative of the second. Thus our rules for addition of vectors can be applied as shown in Fig. 3–8 using the tail-to-tip method.

FIGURE 3–7 The negative of a vector is a vector having the same length but opposite direction.

FIGURE 3–8 Subtracting two vectors: $\vec{V}_2 - \vec{V}_1$.

A vector $\vec{\mathbf{V}}$ can be multiplied by a scalar c. We define their product so that $c\vec{\mathbf{V}}$ has the same direction as $\vec{\mathbf{V}}$ and has magnitude cV. That is, multiplication of a vector by a positive scalar c changes the magnitude of the vector by a factor c but doesn't alter the direction. If c is a negative scalar, the magnitude of the product $c\vec{\mathbf{V}}$ is still cV (without the minus sign), but the direction is precisely opposite to that of $\vec{\mathbf{V}}$. See Fig. 3–9.

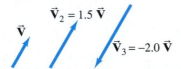

FIGURE 3–9 Multiplying a vector $\vec{\mathbf{V}}$ by a scalar c gives a vector whose magnitude is c times greater and in the same direction as $\vec{\mathbf{V}}$ (or opposite direction if c is negative).

3–4 Adding Vectors by Components

Adding vectors graphically using a ruler and protractor is often not sufficiently accurate and is not useful for vectors in three dimensions. We discuss now a more powerful and precise method for adding vectors. But do not forget graphical methods—they are always useful for visualizing, for checking your math, and thus for getting the correct result.

Consider first a vector $\vec{\mathbf{V}}$ that lies in a particular plane. It can be expressed as the sum of two other vectors, called the **components** of the original vector. The components are usually chosen to be along two perpendicular directions. The process of finding the components is known as **resolving the vector into its components**. An example is shown in Fig. 3–10; the vector $\vec{\mathbf{V}}$ could be a displacement vector that points at an angle $\theta = 30°$ north of east, where we have chosen the positive x axis to be to the east and the positive y axis north. This vector $\vec{\mathbf{V}}$ is resolved into its x and y components by drawing dashed lines out from the tip (A) of the vector (lines AB and AC) making them perpendicular to the x and y axes. Then the lines 0B and 0C represent the x and y components of $\vec{\mathbf{V}}$, respectively, as shown in Fig. 3–10b. These **vector components** are written $\vec{\mathbf{V}}_x$ and $\vec{\mathbf{V}}_y$. We generally show vector components as arrows, like vectors, but dashed. The *scalar components*, V_x and V_y, are numbers, with units, that are given a positive or negative sign depending on whether they point along the positive or negative x or y axis. As can be seen in Fig. 3–10, $\vec{\mathbf{V}}_x + \vec{\mathbf{V}}_y = \vec{\mathbf{V}}$ by the parallelogram method of adding vectors.

Resolving a vector into components

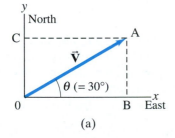

(a)

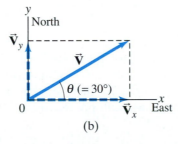

(b)

FIGURE 3–10 Resolving a vector $\vec{\mathbf{V}}$ into its components along an arbitrarily chosen set of x and y axes. The components, once found, themselves represent the vector. That is, the components contain as much information as the vector itself.

Space is made up of three dimensions, and sometimes it is necessary to resolve a vector into components along three mutually perpendicular directions. In rectangular coordinates the components are $\vec{\mathbf{V}}_x$, $\vec{\mathbf{V}}_y$, and $\vec{\mathbf{V}}_z$. Resolution of a vector in three dimensions is merely an extension of the above technique. We will mainly be concerned with situations in which the vectors are in a plane and two components are all that are necessary.

To add vectors using the method of components, we need to use the trigonometric functions sine, cosine, and tangent, which we now review.

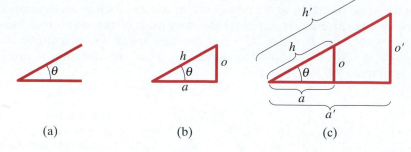

FIGURE 3–11 Starting with an angle θ as in (a), we can construct right triangles of different sizes, (b) and (c), but the ratio of the lengths of the sides does not depend on the size of the triangle.

(a)   (b)   (c)

Given any angle θ, as in Fig. 3–11a, a right triangle can be constructed by drawing a line perpendicular to either of its sides, as in Fig. 3–11b. The longest side of a right triangle, opposite the right angle, is called the hypotenuse, which we label h. The side opposite the angle θ is labeled o, and the side adjacent is labeled a. We let h, o, and a represent the lengths of these sides, respectively. We now define the three trigonometric functions, sine, cosine, and tangent (abbreviated sin, cos, tan), in terms of the right triangle, as follows:

Trigonometric functions defined

$$\sin\theta = \frac{\text{side opposite}}{\text{hypotenuse}} = \frac{o}{h}$$

$$\cos\theta = \frac{\text{side adjacent}}{\text{hypotenuse}} = \frac{a}{h} \qquad \textbf{(3–1)}$$

$$\tan\theta = \frac{\text{side opposite}}{\text{side adjacent}} = \frac{o}{a}.$$

If we make the triangle bigger, but keep the same angles, then the ratio of the length of one side to the other, or of one side to the hypotenuse, remains the same. That is, in Fig. 3–11c we have: $a/h = a'/h'$; $o/h = o'/h'$; and $o/a = o'/a'$. Thus the values of sine, cosine, and tangent do not depend on how big the triangle is. They depend only on the size of the angle. The values of sine, cosine, and tangent for different angles can be found using a scientific calculator, or from the Table in Appendix A.

A useful trigonometric identity is

$$\sin^2\theta + \cos^2\theta = 1 \qquad \textbf{(3–2)}$$

which follows from the Pythagorean theorem ($o^2 + a^2 = h^2$ in Fig. 3–11). That is:

$$\sin^2\theta + \cos^2\theta = \frac{o^2}{h^2} + \frac{a^2}{h^2} = \frac{o^2 + a^2}{h^2} = \frac{h^2}{h^2} = 1.$$

(See also Appendix A for other details on trigonometric functions and identities.)

The use of trigonometric functions for finding the components of a vector is illustrated in Fig. 3–12, where a vector and its two components are thought of as making up a right triangle. We then see that the sine, cosine, and tangent are as given in the Figure. If we multiply the definition of $\sin\theta = V_y/V$ by V on both sides, we get

Components

$$V_y = V\sin\theta. \qquad \textbf{(3–3a)}$$

of a

Similarly, from the definition of $\cos\theta$, we obtain

vector

$$V_x = V\cos\theta. \qquad \textbf{(3–3b)}$$

Note that θ is chosen (by convention) to be the angle that the vector makes with the positive x axis.

Using Eqs. 3–3, we can calculate V_x and V_y for any vector, such as that illustrated in Fig. 3–10 or Fig. 3–12. Suppose $\vec{V}$ represents a displacement of 500 m

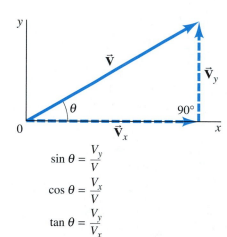

$$\sin\theta = \frac{V_y}{V}$$

$$\cos\theta = \frac{V_x}{V}$$

$$\tan\theta = \frac{V_y}{V_x}$$

$$V^2 = V_x^2 + V_y^2$$

FIGURE 3–12 Finding the components of a vector using trigonometric functions.

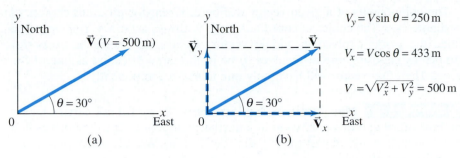

FIGURE 3–13 (a) Vector $\vec{V}$ represents a displacement of 500 m at a 30° angle north of east. (b) The components of $\vec{V}$ are $\vec{V}_x$ and $\vec{V}_y$, whose magnitudes are given on the right.

in a direction 30° north of east, as shown in Fig. 3–13. Then $V = 500$ m. From a calculator or Tables, $\sin 30° = 0.500$ and $\cos 30° = 0.866$. Then

$$V_x = V \cos \theta = (500 \text{ m})(0.866) = 433 \text{ m (east)},$$
$$V_y = V \sin \theta = (500 \text{ m})(0.500) = 250 \text{ m (north)}.$$

There are two ways to specify a vector in a given coordinate system:

1. We can give its components, V_x and V_y.
2. We can give its magnitude V and the angle θ it makes with the positive x axis.

Two ways to specify a vector

We can shift from one description to the other using Eqs. 3–3, and, for the reverse, by using the theorem of Pythagoras† and the definition of tangent:

$$V = \sqrt{V_x^2 + V_y^2} \qquad (3\text{–}4a)$$

$$\tan \theta = \frac{V_y}{V_x} \qquad (3\text{–}4b)$$

Components related to magnitude and direction

as can be seen in Fig. 3–12.

We can now discuss how to add vectors using components. The first step is to resolve each vector into its components. Next we can see, using Fig. 3–14, that the addition of any two vectors $\vec{V}_1$ and $\vec{V}_2$ to give a resultant, $\vec{V} = \vec{V}_1 + \vec{V}_2$, implies that

$$V_x = V_{1x} + V_{2x}$$
$$V_y = V_{1y} + V_{2y}. \qquad (3\text{–}5)$$

Adding vectors analytically (by components)

That is, the sum of the x components equals the x component of the resultant, and similarly for y. That this is valid can be verified by a careful examination of Fig. 3–14. But note that we add all the x components together to get the x component of the resultant; and we add all the y components together to get the y component of the resultant. We do *not* add x components to y components.

If the magnitude and direction of the resultant vector are desired, they can be obtained using Eqs. 3–4.

†In three dimensions, the theorem of Pythagoras becomes $V = \sqrt{V_x^2 + V_y^2 + V_z^2}$, where V_z is the component along the third, or z, axis.

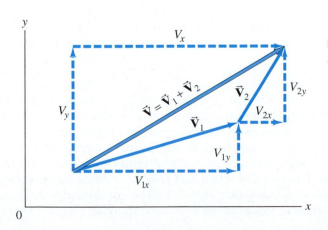

FIGURE 3–14 The components of $\vec{V} = \vec{V}_1 + \vec{V}_2$ are $V_x = V_{1x} + V_{2x}$ and $V_y = V_{1y} + V_{2y}$.

The components of a given vector will be different for different choices of coordinate axes. The choice of coordinate axes is always arbitrary. You can often reduce the work involved in adding vectors by a good choice of axes—for example, by choosing one of the axes to be in the same direction as one of the vectors. Then that vector will have only one nonzero component.

Choice of axes can simplify effort needed

EXAMPLE 3–2 **Mail carrier's displacement.** A rural mail carrier leaves the post office and drives 22.0 km in a northerly direction. She then drives in a direction 60.0° south of east for 47.0 km (Fig. 3–15a). What is her displacement from the post office?

APPROACH We resolve each vector into its x and y components. We add the x components together, and then the y components together, giving us the x and y components of the resultant. We choose the positive x axis to be east and the positive y axis to be north, since those are the compass directions used on most maps.

SOLUTION Resolve each displacement vector into its components, as shown in Fig. 3–15b. Since $\vec{D}_1$ has magnitude 22.0 km and points north, it has only a y component:

$$D_{1x} = 0, \qquad D_{1y} = 22.0 \text{ km}.$$

$\vec{D}_2$ has both x and y components:

$$D_{2x} = +(47.0 \text{ km})(\cos 60°) = +(47.0 \text{ km})(0.500) = +23.5 \text{ km}$$
$$D_{2y} = -(47.0 \text{ km})(\sin 60°) = -(47.0 \text{ km})(0.866) = -40.7 \text{ km}.$$

Notice that D_{2y} is negative because this vector component points along the negative y axis. The resultant vector, $\vec{D}$, has components:

$$D_x = D_{1x} + D_{2x} = \quad 0 \text{ km} + \quad 23.5 \text{ km} = +23.5 \text{ km}$$
$$D_y = D_{1y} + D_{2y} = 22.0 \text{ km} + (-40.7 \text{ km}) = -18.7 \text{ km}.$$

This specifies the resultant vector completely:

$$D_x = 23.5 \text{ km}, \qquad D_y = -18.7 \text{ km}.$$

We can also specify the resultant vector by giving its magnitude and angle using Eqs. 3–4:

$$D = \sqrt{D_x^2 + D_y^2} = \sqrt{(23.5 \text{ km})^2 + (-18.7 \text{ km})^2} = 30.0 \text{ km}$$
$$\tan \theta = \frac{D_y}{D_x} = \frac{-18.7 \text{ km}}{23.5 \text{ km}} = -0.796.$$

A calculator with an INV TAN, an ARC TAN, or a TAN^{-1} key gives $\theta = \tan^{-1}(-0.796) = -38.5°$. The negative sign means $\theta = 38.5°$ below the x axis, Fig. 3–15c. So, the resultant displacement is 30.0 km directed at 38.5° in a southeasterly direction.

NOTE Always be attentive about the quadrant in which the resultant vector lies. An electronic calculator does not fully give this information, but a good diagram does.

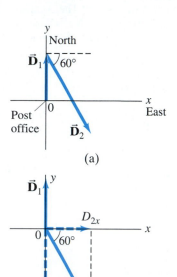

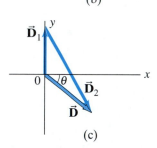

FIGURE 3–15 Example 3–2. (a) The two displacement vectors, $\vec{D}_1$ and $\vec{D}_2$. (b) $\vec{D}_2$ is resolved into its components. (c) $\vec{D}_1$ and $\vec{D}_2$ are added graphically to obtain the resultant $\vec{D}$. The component method of adding the vectors is explained in the Example.

➡ **PROBLEM SOLVING**

Identify the correct quadrant by drawing a careful diagram

The signs of trigonometric functions depend on which "quadrant" the angle falls in: for example, the tangent is positive in the first and third quadrants (from 0° to 90°, and 180° to 270°), but negative in the second and fourth quadrants; see Appendix A–7. The best way to keep track of angles, and to check any vector result, is always to draw a vector diagram. A vector diagram gives you something tangible to look at when analyzing a problem, and provides a check on the results.

The following Problem Solving Box should not be considered a prescription. Rather it is a summary of things to do to get you thinking and involved in the problem at hand.

Here is a brief summary of how to add two or more vectors using components:

1. **Draw a diagram**, adding the vectors graphically by either the parallelogram or tail-to-tip method.

2. **Choose x and y axes.** Choose them in a way, if possible, that will make your work easier. (For example, choose one axis along the direction of one of the vectors so that vector will have only one component.)

3. **Resolve each vector** into its x and y components, showing each component along its appropriate (x or y) axis as a (dashed) arrow.

4. **Calculate each component** (when not given) using sines and cosines. If θ_1 is the angle that vector $\vec{\mathbf{V}}_1$ makes with the positive x axis, then:
$$V_{1x} = V_1 \cos\theta_1, \qquad V_{1y} = V_1 \sin\theta_1.$$

Pay careful attention to **signs**: any component that points along the negative x or y axis gets a $-$ sign.

5. **Add** the x **components** together to get the x component of the resultant. Ditto for y:
$$V_x = V_{1x} + V_{2x} + \text{any others}$$
$$V_y = V_{1y} + V_{2y} + \text{any others}.$$

This is the answer: the components of the resultant vector. Check signs to see if they fit the quadrant shown in your diagram (point 1 above).

6. If you want to know the **magnitude and direction** of the resultant vector, use Eqs. 3–4:
$$V = \sqrt{V_x^2 + V_y^2}, \qquad \tan\theta = \frac{V_y}{V_x}.$$

The vector diagram you already drew helps to obtain the correct position (quadrant) of the angle θ.

EXAMPLE 3–3 **Three short trips.** An airplane trip involves three legs, with two stopovers, as shown in Fig. 3–16a. The first leg is due east for 620 km; the second leg is southeast (45°) for 440 km; and the third leg is at 53° south of west, for 550 km, as shown. What is the plane's total displacement?

APPROACH We follow the steps in the above Problem Solving Box.

SOLUTION

1. **Draw a diagram** such as Fig. 3–16a, where $\vec{\mathbf{D}}_1$, $\vec{\mathbf{D}}_2$, and $\vec{\mathbf{D}}_3$ represent the three legs of the trip, and $\vec{\mathbf{D}}_R$ is the plane's total displacement.

2. **Choose axes**: Axes are also shown in Fig. 3–16a.

3. **Resolve components**: It is imperative to draw a good figure. The components are drawn in Fig. 3–16b. Instead of drawing all the vectors starting from a common origin, as we did in Fig. 3–15b, here we draw them "tail-to-tip" style, which is just as valid and may make it easier to see.

4. **Calculate the components**:
$$\vec{\mathbf{D}}_1: D_{1x} = +D_1 \cos 0° = D_1 = 620\,\text{km}$$
$$\phantom{\vec{\mathbf{D}}_1:} D_{1y} = +D_1 \sin 0° = 0\,\text{km}$$
$$\vec{\mathbf{D}}_2: D_{2x} = +D_2 \cos 45° = +(440\,\text{km})(0.707) = +311\,\text{km}$$
$$\phantom{\vec{\mathbf{D}}_2:} D_{2y} = -D_2 \sin 45° = -(440\,\text{km})(0.707) = -311\,\text{km}$$
$$\vec{\mathbf{D}}_3: D_{3x} = -D_3 \cos 53° = -(550\,\text{km})(0.602) = -331\,\text{km}$$
$$\phantom{\vec{\mathbf{D}}_3:} D_{3y} = -D_3 \sin 53° = -(550\,\text{km})(0.799) = -439\,\text{km}.$$

We have given a minus sign to each component that in Fig. 3–16b points in the $-x$ or $-y$ direction. The components are shown in the Table in the margin.

5. **Add the components**: We add the x components together, and we add the y components together to obtain the x and y components of the resultant:
$$D_x = D_{1x} + D_{2x} + D_{3x} = 620\,\text{km} + 311\,\text{km} - 331\,\text{km} = 600\,\text{km}$$
$$D_y = D_{1y} + D_{2y} + D_{3y} = \phantom{620\,\text{km} + }0\,\text{km} - 311\,\text{km} - 439\,\text{km} = -750\,\text{km}.$$

The x and y components are 600 km and -750 km, and point respectively to the east and south. This is one way to give the answer.

6. **Magnitude and direction**: We can also give the answer as
$$D_R = \sqrt{D_x^2 + D_y^2} = \sqrt{(600)^2 + (-750)^2}\,\text{km} = 960\,\text{km}$$
$$\tan\theta = \frac{D_y}{D_x} = \frac{-750\,\text{km}}{600\,\text{km}} = -1.25, \qquad \text{so } \theta = -51°.$$

Thus, the total displacement has magnitude 960 km and points 51° below the x axis (south of east), as was shown in our original sketch, Fig. 3–16a.

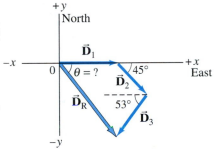

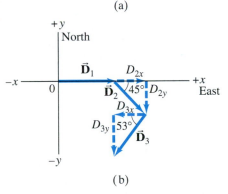

FIGURE 3–16 Example 3–3.

Vector	Components	
	x (km)	y (km)
$\vec{\mathbf{D}}_1$	620	0
$\vec{\mathbf{D}}_2$	311	-311
$\vec{\mathbf{D}}_3$	-331	-439
$\vec{\mathbf{D}}_R$	600	-750

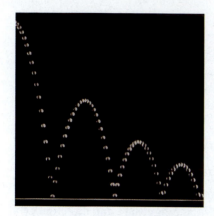

FIGURE 3–17 This strobe photograph of a ball making a series of bounces shows the characteristic "parabolic" path of projectile motion.

3–5 Projectile Motion

In Chapter 2, we studied the motion of objects in one dimension in terms of displacement, velocity, and acceleration, including purely vertical motion of falling bodies undergoing acceleration due to gravity. Now we examine the more general motion of objects moving through the air in two dimensions near the Earth's surface, such as a golf ball, a thrown or batted baseball, kicked footballs, and speeding bullets. These are all examples of **projectile motion** (see Fig. 3–17), which we can describe as taking place in two dimensions. Although air resistance is often important, in many cases its effect can be ignored, and we will ignore it in the following analysis. We will not be concerned now with the process by which the object is thrown or projected. We consider only its motion *after* it has been projected, and *before* it lands or is caught—that is, we analyze our projected object only when it is moving freely through the air under the action of gravity alone. Then the acceleration of the object is that due to gravity, which acts downward with magnitude $g = 9.80 \text{ m/s}^2$, and we assume it is constant.[†]

Horizontal and vertical motion analyzed separately

Galileo was the first to describe projectile motion accurately. He showed that it could be understood by analyzing the horizontal and vertical components of the motion separately. For convenience, we assume that the motion begins at time $t = 0$ at the origin of an xy coordinate system (so $x_0 = y_0 = 0$).

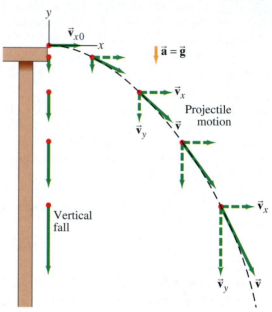

FIGURE 3–18 Projectile motion of a small ball projected horizontally. The dashed black line represents the path of the object. The velocity vector $\vec{v}$ at each point is in the direction of motion and thus is tangent to the path. The velocity vectors are green arrows, and velocity components are dashed. (A vertically falling object starting at the same point is shown at the left for comparison; v_y is the same for the falling object and the projectile.)

$\vec{v}$ is tangent to the path

Let us look at a (tiny) ball rolling off the end of a horizontal table with an initial velocity in the horizontal (x) direction, v_{x0}. See Fig. 3–18, where an object falling vertically is also shown for comparison. The velocity vector $\vec{v}$ at each instant points in the direction of the ball's motion at that instant and is always tangent to the path. Following Galileo's ideas, we treat the horizontal and vertical components of the velocity, v_x and v_y, separately, and we can apply the kinematic equations (Eqs. 2–11a through 2–11c) to the x and y components of the motion.

Vertical motion ($a_y = \text{constant} = -g$)

First we examine the vertical (y) component of the motion. At the instant the ball leaves the table's top ($t = 0$), it has only an x component of velocity. Once the ball leaves the table (at $t = 0$), it experiences a vertically downward acceleration g, the acceleration due to gravity. Thus v_y is initially zero ($v_{y0} = 0$) but increases continually in the downward direction (until the ball hits the ground). Let us take y to be positive upward. Then $a_y = -g$, and from Eq. 2–11a we can write $v_y = -gt$ since we set $v_{y0} = 0$. The vertical displacement is given by $y = -\frac{1}{2}gt^2$.

[†]This restricts us to objects whose distance traveled and maximum height above the Earth are small compared to the Earth's radius (6400 km).

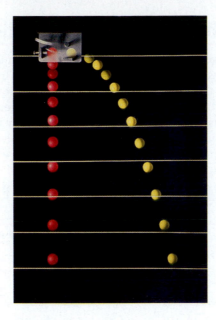

FIGURE 3–19 Multiple-exposure photograph showing positions of two balls at equal time intervals. One ball was dropped from rest at the same time the other was projected horizontally outward. The vertical position of each ball is seen to be the same.

In the horizontal direction, on the other hand, there is no acceleration (we are ignoring air resistance). So the horizontal component of velocity, v_x, remains constant, equal to its initial value, v_{x0}, and thus has the same magnitude at each point on the path. The horizontal displacement is then given by $x = v_{x0} t$. The two vector components, $\vec{v}_x$ and $\vec{v}_y$, can be added vectorially at any instant to obtain the velocity $\vec{v}$ at that time (that is, for each point on the path), as shown in Fig. 3–18.

Horizontal motion
$(a_x = 0, v_x = constant)$

One result of this analysis, which Galileo himself predicted, is that *an object projected horizontally will reach the ground in the same time as an object dropped vertically*. This is because the vertical motions are the same in both cases, as shown in Fig. 3–18. Figure 3–19 is a multiple-exposure photograph of an experiment that confirms this.

EXERCISE C Two balls having different speeds roll off the edge of a horizontal table at the same time. Which hits the floor sooner, the faster ball or the slower one?

If an object is projected at an upward angle, as in Fig. 3–20, the analysis is similar, except that now there is an initial vertical component of velocity, v_{y0}. Because of the downward acceleration of gravity, v_y gradually decreases with time until the object reaches the highest point on its path, at which point $v_y = 0$. Subsequently the object moves downward (Fig. 3–20) and v_y increases in the downward direction, as shown (that is, becoming more negative). As before, v_x remains constant.

Object projected upward

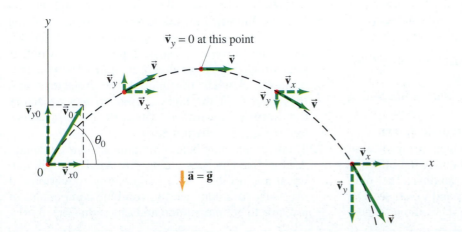

FIGURE 3–20 Path of a projectile fired with initial velocity $\vec{v}_0$ at angle θ to the horizontal. Path is shown in black, the velocity vectors are green arrows, and velocity components are dashed.

3-6 Solving Problems Involving Projectile Motion

We now work through several Examples of projectile motion quantitatively. We use the kinematic equations (2–11a through 2–11c) separately for the vertical and horizontal components of the motion. These equations are shown separately for the x and y components of the motion in Table 3–1, for the general case of two-dimensional motion at constant acceleration. Note that x and y are the respective displacements, that v_x and v_y are the components of the velocity, and that a_x and a_y are the components of the acceleration, each of which is constant. The subscript 0 means "at $t = 0$."

TABLE 3–1 General Kinematic Equations for Constant Acceleration in Two Dimensions

x component (horizontal)		y component (vertical)
$v_x = v_{x0} + a_x t$	(Eq. 2–11a)	$v_y = v_{y0} + a_y t$
$x = x_0 + v_{x0} t + \frac{1}{2} a_x t^2$	(Eq. 2–11b)	$y = y_0 + v_{y0} t + \frac{1}{2} a_y t^2$
$v_x^2 = v_{x0}^2 + 2a_x(x - x_0)$	(Eq. 2–11c)	$v_y^2 = v_{y0}^2 + 2a_y(y - y_0)$

We can simplify these equations for the case of projectile motion because we can set $a_x = 0$. See Table 3–2, which assumes y is positive upward, so $a_y = -g = -9.80 \text{ m/s}^2$. Note that if θ is chosen relative to the $+x$ axis, as in Fig. 3–20, then

$$v_{x0} = v_0 \cos \theta, \quad \text{and} \quad v_{y0} = v_0 \sin \theta.$$

➡ **PROBLEM SOLVING**

Choice of time interval

In doing Problems involving projectile motion, we must consider a time interval for which our chosen object is in the air, influenced only by gravity. We do not consider the throwing (or projecting) process, nor the time after the object lands or is caught, because then other influences act on the object, and we can no longer set $\vec{a} = \vec{g}$.

TABLE 3–2 Kinematic Equations for Projectile Motion
(*y* positive upward; $a_x = 0$, $a_y = -g = -9.80 \text{ m/s}^2$)

Horizontal Motion ($a_x = 0$, $v_x = $ constant)		Vertical Motion† ($a_y = -g = $ constant)
$v_x = v_{x0}$	(Eq. 2–11a)	$v_y = v_{y0} - gt$
$x = x_0 + v_{x0} t$	(Eq. 2–11b)	$y = y_0 + v_{y0} t - \frac{1}{2}gt^2$
	(Eq. 2–11c)	$v_y^2 = v_{y0}^2 - 2g(y - y_0)$

†If y is taken positive downward, the minus $(-)$ signs in front of g become $+$ signs.

PROBLEM SOLVING Projectile Motion

Our approach to solving problems in Section 2–6 also applies here. Solving problems involving projectile motion can require creativity, and cannot be done just by following some rules. Certainly you must avoid just plugging numbers into equations that seem to "work."

1. As always, **read** carefully; **choose** the object (or objects) you are going to analyze.

2. **Draw** a careful **diagram** showing what is happening to the object.

3. **Choose** an origin and an *xy* **coordinate system**.

4. Decide on the **time interval**, which for projectile motion can only include motion under the effect of gravity alone, not throwing or landing. The time interval must be the same for the *x* and *y* analyses. The *x* and *y* motions are connected by the common time.

5. **Examine** the horizontal (x) and vertical (y) **motions** separately. If you are given the initial velocity, you may want to resolve it into its *x* and *y* components.

6. List the **known** and **unknown** quantities, choosing $a_x = 0$ and $a_y = -g$ or $+g$, where $g = 9.80 \text{ m/s}^2$, and using the $+$ or $-$ sign, depending on whether you choose *y* positive down or up. Remember that v_x never changes throughout the trajectory, and that $v_y = 0$ at the highest point of any trajectory that returns downward. The velocity just before landing is generally not zero.

7. Think for a minute before jumping into the equations. A little planning goes a long way. **Apply** the relevant **equations** (Table 3–2), combining equations if necessary. You may need to combine components of a vector to get magnitude and direction (Eqs. 3–4).

EXAMPLE 3–4 **Driving off a cliff.** A movie stunt driver on a motorcycle speeds horizontally off a 50.0-m-high cliff. How fast must the motorcycle leave the cliff top to land on level ground below, 90.0 m from the base of the cliff where the cameras are? Ignore air resistance.

APPROACH We explicitly follow the steps of the Problem Solving Box.

SOLUTION

1. and 2. **Read, choose the object, and draw a diagram.** Our object is the motor-cycle and driver, taken as a single unit. The diagram is shown in Fig. 3–21.

3. **Choose a coordinate system.** We choose the y direction to be positive upward, with the top of the cliff as $y_0 = 0$. The x direction is horizontal with $x_0 = 0$ at the point where the motorcycle leaves the cliff.

4. **Choose a time interval.** We choose our time interval to begin ($t = 0$) just as the motorcycle leaves the cliff top at position $x_0 = 0$, $y_0 = 0$; our time interval ends just before the motorcycle hits the ground below.

5. **Examine x and y motions.** In the horizontal (x) direction, the acceleration $a_x = 0$, so the velocity is constant. The value of x when the motorcycle reaches the ground is $x = +90.0$ m. In the vertical direction, the accelera-tion is the acceleration due to gravity, $a_y = -g = -9.80 \text{ m/s}^2$. The value of y when the motorcycle reaches the ground is $y = -50.0$ m. The initial velocity is horizontal and is our unknown, v_{x0}; the initial vertical velocity is zero, $v_{y0} = 0$.

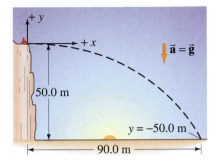

FIGURE 3–21 Example 3–4.

6. **List knowns and unknowns.** See the Table in the margin. Note that in addition to not knowing the initial horizontal velocity v_{x0} (which stays constant until landing), we also do not know the time t when the motorcycle reaches the ground.

Known	Unknown
$x_0 = y_0 = 0$	v_{x0}
$x = 90.0$ m	t
$y = -50.0$ m	
$a_x = 0$	
$a_y = -g = -9.80 \text{ m/s}^2$	
$v_{y0} = 0$	

7. **Apply relevant equations.** The motorcycle maintains constant v_x as long as it is in the air. The time it stays in the air is determined by the y motion—when it hits the ground. So we first find the time using the y motion, and then use this time value in the x equations. To find out how long it takes the motorcycle to reach the ground below, we use Eq. 2–11b (Table 3–2) for the vertical (y) direction with $y_0 = 0$ and $v_{y0} = 0$:

$$y = y_0 + v_{y0}t + \tfrac{1}{2}a_y t^2$$
$$= 0 + 0 + \tfrac{1}{2}(-g)t^2$$

or

$$y = -\tfrac{1}{2}gt^2.$$

We solve for t and set $y = -50.0$ m:

$$t = \sqrt{\frac{2y}{-g}} = \sqrt{\frac{2(-50.0 \text{ m})}{-9.80 \text{ m/s}^2}} = 3.19 \text{ s}.$$

To calculate the initial velocity, v_{x0}, we again use Eq. 2–11b, but this time for the horizontal (x) direction, with $a_x = 0$ and $x_0 = 0$:

$$x = x_0 + v_{x0}t + \tfrac{1}{2}a_x t^2$$
$$= 0 + v_{x0}t + 0$$

or

$$x = v_{x0}t.$$

Then

$$v_{x0} = \frac{x}{t} = \frac{90.0 \text{ m}}{3.19 \text{ s}} = 28.2 \text{ m/s},$$

which is about 100 km/h (roughly 60 mi/h).

NOTE In the time interval of the projectile motion, the only acceleration is g in the negative y direction. The acceleration in the x direction is zero.

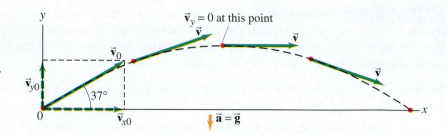

FIGURE 3–22 Example 3–5.

$\vec{v}_y = 0$ at this point

$\vec{a} = \vec{g}$

37°

PHYSICS APPLIED

Sports

EXAMPLE 3–5 **A kicked football.** A football is kicked at an angle $\theta_0 = 37.0°$ with a velocity of 20.0 m/s, as shown in Fig. 3–22. Calculate (*a*) the maximum height, (*b*) the time of travel before the football hits the ground, (*c*) how far away it hits the ground, (*d*) the velocity vector at the maximum height, and (*e*) the acceleration vector at maximum height. Assume the ball leaves the foot at ground level, and ignore air resistance and rotation of the ball.

APPROACH This may seem difficult at first because there are so many questions. But we can deal with them one at a time. We take the y direction as positive upward, and treat the x and y motions separately. The total time in the air is again determined by the y motion. The x motion occurs at constant velocity. The y component of velocity varies, being positive (upward) initially, decreasing to zero at the highest point, and then becoming negative as the football falls.

SOLUTION We resolve the initial velocity into its components (Fig. 3–22):

$$v_{x0} = v_0 \cos 37.0° = (20.0 \text{ m/s})(0.799) = 16.0 \text{ m/s}$$
$$v_{y0} = v_0 \sin 37.0° = (20.0 \text{ m/s})(0.602) = 12.0 \text{ m/s}.$$

(*a*) We consider a time interval that begins just after the football loses contact with the foot until it reaches its maximum height. During this time interval, the acceleration is g downward. At the maximum height, the velocity is horizontal (Fig. 3–22), so $v_y = 0$; and this occurs at a time given by $v_y = v_{y0} - gt$ with $v_y = 0$ (see Eq. 2–11a in Table 3–2). Thus

$$t = \frac{v_{y0}}{g} = \frac{(12.0 \text{ m/s})}{(9.80 \text{ m/s}^2)} = 1.22 \text{ s}.$$

From Eq. 2–11b, with $y_0 = 0$, we have

$$y = v_{y0}t - \tfrac{1}{2}gt^2$$
$$= (12.0 \text{ m/s})(1.22 \text{ s}) - \tfrac{1}{2}(9.80 \text{ m/s}^2)(1.22 \text{ s})^2 = 7.35 \text{ m}.$$

Alternatively, we could have used Eq. 2–11c, solved for y, and found

$$y = \frac{v_{y0}^2 - v_y^2}{2g} = \frac{(12.0 \text{ m/s})^2 - (0 \text{ m/s})^2}{2(9.80 \text{ m/s}^2)} = 7.35 \text{ m}.$$

The maximum height is 7.35 m.

(*b*) To find the time it takes for the ball to return to the ground, we consider a different time interval, starting at the moment the ball leaves the foot ($t = 0, y_0 = 0$) and ending just before the ball touches the ground ($y = 0$ again). We can use Eq. 2–11b with $y_0 = 0$ and also set $y = 0$ (ground level):

$$y = y_0 + v_{y0}t - \tfrac{1}{2}gt^2$$
$$0 = 0 + (12.0 \text{ m/s})t - \tfrac{1}{2}(9.80 \text{ m/s}^2)t^2.$$

This equation can be easily factored:

$$\left[\tfrac{1}{2}(9.80 \text{ m/s}^2)t - 12.0 \text{ m/s}\right]t = 0.$$

There are two solutions, $t = 0$ (which corresponds to the initial point, y_0), and

$$t = \frac{2(12.0 \text{ m/s})}{(9.80 \text{ m/s}^2)} = 2.45 \text{ s},$$

which is the total travel time of the football.

NOTE The time $t = 2.45\,\text{s}$ for the whole trip is double the time to reach the highest point, calculated in (a). That is, the time to go up equals the time to come back down to the same level, but only in the absence of air resistance.

Time up = time down

(c) The total distance traveled in the x direction is found by applying Eq. 2–11b with $x_0 = 0$, $a_x = 0$, $v_{x0} = 16.0\,\text{m/s}$:

$$x = v_{x0}t = (16.0\,\text{m/s})(2.45\,\text{s}) = 39.2\,\text{m}.$$

(d) At the highest point, there is no vertical component to the velocity. There is only the horizontal component (which remains constant throughout the flight), so $v = v_{x0} = v_0 \cos 37.0° = 16.0\,\text{m/s}$.

(e) The acceleration vector is the same at the highest point as it is throughout the flight, which is $9.80\,\text{m/s}^2$ downward.

NOTE We treated the football as if it were a particle, ignoring its rotation. We also ignored air resistance, which is considerable on a rotating football, so our results are not very accurate.

EXERCISE D Two balls are thrown in the air at different angles, but each reaches the same height. Which ball remains in the air longer: the one thrown at the steeper angle or the one thrown at a shallower angle?

CONCEPTUAL EXAMPLE 3–6 **Where does the apple land?** A child sits upright in a wagon which is moving to the right at constant speed as shown in Fig. 3–23. The child extends her hand and throws an apple straight upward (from her own point of view, Fig. 3–23a), while the wagon continues to travel forward at constant speed. If air resistance is neglected, will the apple land (a) behind the wagon, (b) in the wagon, or (c) in front of the wagon?

RESPONSE The child throws the apple straight up from her own reference frame with initial velocity $\vec{v}_{y0}$ (Fig. 3–23a). But when viewed by someone on the ground, the apple also has an initial horizontal component of velocity equal to the speed of the wagon, $\vec{v}_{x0}$. Thus, to a person on the ground, the apple will follow the path of a projectile as shown in Fig. 3–23b. The apple experiences no horizontal acceleration, so $\vec{v}_{x0}$ will stay constant and equal to the speed of the wagon. As the apple follows its arc, the wagon will be directly under the apple at all times because they have the same horizontal velocity. When the apple comes down, it will drop right into the outstretched hand of the child. The answer is (b).

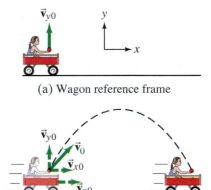

(a) Wagon reference frame

(b) Ground reference frame

FIGURE 3–23 Example 3–6.

CONCEPTUAL EXAMPLE 3–7 **The wrong strategy.** A boy on a small hill aims his water-balloon slingshot horizontally, straight at a second boy hanging from a tree branch a distance d away, Fig. 3–24. At the instant the water balloon is released, the second boy lets go and falls from the tree, hoping to avoid being hit. Show that he made the wrong move. (He hadn't studied physics yet.) Ignore air resistance.

RESPONSE Both the water balloon and the boy in the tree start falling at the same instant, and in a time t they each fall the same vertical distance $y = \frac{1}{2}gt^2$, much like Fig. 3–19. In the time it takes the water balloon to travel the horizontal distance d, the balloon will have the same y position as the falling boy. Splat. If the boy had stayed in the tree, he would have avoided the humiliation.

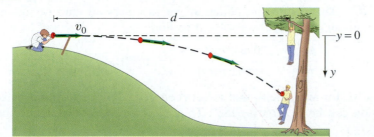

FIGURE 3–24 Example 3–7.

EXERCISE E A package is dropped from a plane flying at constant velocity parallel to the ground. If air resistance is ignored, the package will (a) fall behind the plane, (b) remain directly below the plane until hitting the ground, (c) move ahead of the plane, or (d) it depends on the speed of the plane.

Horizontal range of a projectile

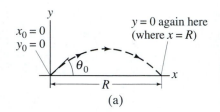

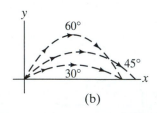

FIGURE 3–25 Example 3–8. (a) The range R of a projectile; (b) there are generally two angles θ_0 that will give the same range. Can you show that if one angle is θ_{01}, the other is $\theta_{02} = 90° - \theta_{01}$?

EXAMPLE 3–8 **Level horizontal range.** (a) Derive a formula for the horizontal range R of a projectile in terms of its initial velocity v_0 and angle θ_0. The horizontal *range* is defined as the horizontal distance the projectile travels before returning to its original height (which is typically the ground); that is, y (final) $= y_0$. See Fig. 3–25a. (b) Suppose one of Napoleon's cannons had a muzzle velocity, v_0, of 60.0 m/s. At what angle should it have been aimed (ignore air resistance) to strike a target 320 m away?

APPROACH The situation is the same as in Example 3–5, except we are not now given numbers in (a). We will algebraically manipulate equations to obtain our result.

SOLUTION (a) We set $x_0 = 0$ and $y_0 = 0$ at $t = 0$. After the projectile travels a horizontal distance R, it returns to the same level, $y = 0$, the final point. We choose our time interval to start ($t = 0$) just after the projectile is fired and to end when it returns to the same vertical height. To find a general expression for R, we set both $y = 0$ and $y_0 = 0$ in Eq. 2–11b for the vertical motion, and obtain

$$y = y_0 + v_{y0}t + \tfrac{1}{2}a_y t^2$$

so

$$0 = 0 + v_{y0}t - \tfrac{1}{2}gt^2.$$

We solve for t, which gives two solutions: $t = 0$ and $t = 2v_{y0}/g$. The first solution corresponds to the initial instant of projection and the second is the time when the projectile returns to $y = 0$. Then the range, R, will be equal to x at the moment t has this value, which we put into Eq. 2–11b for the *horizontal* motion ($x = v_{x0}t$, with $x_0 = 0$). Thus we have:

$$R = x = v_{x0}t = v_{x0}\left(\frac{2v_{y0}}{g}\right) = \frac{2v_{x0}v_{y0}}{g} = \frac{2v_0^2 \sin\theta_0 \cos\theta_0}{g} \qquad [y = y_0]$$

where we have written $v_{x0} = v_0 \cos\theta_0$ and $v_{y0} = v_0 \sin\theta_0$. This is the result we sought. It can be rewritten, using the trigonometric identity $2\sin\theta\cos\theta = \sin 2\theta$ (Appendix A or inside the rear cover):

Level range formula
$$[y \, (final) = y_0]$$

$$R = \frac{v_0^2 \sin 2\theta_0}{g}. \qquad [y = y_0]$$

We see that the maximum range, for a given initial velocity v_0, is obtained when $\sin 2\theta$ takes on its maximum value of 1.0, which occurs for $2\theta_0 = 90°$; so

$$\theta_0 = 45° \text{ for maximum range, and } R_{max} = v_0^2/g.$$

[When air resistance is important, the range is less for a given v_0, and the maximum range is obtained at an angle smaller than 45°.]

NOTE The maximum range increases by the square of v_0, so doubling the muzzle velocity of a cannon increases its maximum range by a factor of 4.

(b) We put $R = 320$ m into the equation we just derived, and (assuming, unrealistically, no air resistance) we solve it to find

$$\sin 2\theta_0 = \frac{Rg}{v_0^2} = \frac{(320 \text{ m})(9.80 \text{ m/s}^2)}{(60.0 \text{ m/s})^2} = 0.871.$$

We want to solve for an angle θ_0 that is between 0° and 90°, which means $2\theta_0$ in this equation can be as large as 180°. Thus, $2\theta_0 = 60.6°$ is a solution, but

$2\theta_0 = 180° - 60.6° = 119.4°$ is also a solution (see Appendix A–7). In general we will have two solutions (see Fig. 3–25b), which in the present case are given by

$$\theta_0 = 30.3° \quad \text{or} \quad 59.7°.$$

Either angle gives the same range. Only when $\sin 2\theta_0 = 1$ (so $\theta_0 = 45°$) is there a single solution (that is, both solutions are the same).

Additional Example: slightly more Complicated, but Fun

EXAMPLE 3–9 **A punt.** Suppose the football in Example 3–5 was a punt and left the punter's foot at a height of 1.00 m above the ground. How far did the football travel before hitting the ground? Set $x_0 = 0$, $y_0 = 0$.

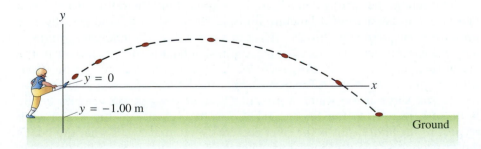 PHYSICS APPLIED
Sports

APPROACH The x and y motions are again treated separately. But we cannot use the range formula from Example 3–8 because it is valid only if y (final) $= y_0$, which is not the case here. Now we have $y_0 = 0$, and the football hits the ground where $y = -1.00$ m (see Fig. 3–26). We choose our time interval to start when the ball leaves his foot ($t = 0$, $y_0 = 0$, $x_0 = 0$) and end just before the ball hits the ground ($y = -1.00$ m). We can get x from Eq. 2–11b, $x = v_{x0}t$, since we know that $v_{x0} = 16.0$ m/s from Example 3–5. But first we must find t, the time at which the ball hits the ground, which we obtain from the y motion.

PROBLEM SOLVING
Do not use any formula unless you are sure its range of validity fits the problem. The range formula does not apply here because $y \neq y_0$

SOLUTION With $y = -1.00$ m and $v_{y0} = 12.0$ m/s (see Example 3–5), we use the equation

$$y = y_0 + v_{y0}t - \tfrac{1}{2}gt^2,$$

and obtain

$$-1.00\,\text{m} = 0 + (12.0\,\text{m/s})t - (4.90\,\text{m/s}^2)t^2.$$

We rearrange this equation into standard form so we can use the quadratic formula (Appendix A–4; also Example 2–15):

$$(4.90\,\text{m/s}^2)t^2 - (12.0\,\text{m/s})t - (1.00\,\text{m}) = 0.$$

Using the quadratic formula gives

$$t = \frac{12.0\,\text{m/s} \pm \sqrt{(12.0\,\text{m/s})^2 - 4(4.90\,\text{m/s}^2)(-1.00\,\text{m})}}{2(4.90\,\text{m/s}^2)}$$

$$= 2.53\,\text{s} \quad \text{or} \quad -0.081\,\text{s}.$$

The second solution would correspond to a time prior to the kick, so it doesn't apply. With $t = 2.53$ s for the time at which the ball touches the ground, the horizontal distance the ball traveled is (using $v_{x0} = 16.0$ m/s from Example 3–5):

$$x = v_{x0}t = (16.0\,\text{m/s})(2.53\,\text{s}) = 40.5\,\text{m}.$$

Our assumption in Example 3–5 that the ball leaves the foot at ground level results in an underestimate of about 1.3 m in the distance traveled.

FIGURE 3–26 Example 3–9: the football leaves the punter's foot at $y = 0$, and reaches the ground where $y = -1.00$ m.

(a)

(b)

(c)

FIGURE 3–27 Examples of projectile motion—sparks (small hot glowing pieces of metal), water, and fireworks. All exhibit the parabolic path characteristic of projectile motion, although the effects of air resistance can be seen to alter the path of some trajectories.

➡ **PROBLEM SOLVING**

Subscripts for adding velocities: first subscript for the object; second subscript for the reference frame

* 3–7 Projectile Motion Is Parabolic

We now show that the path followed by any projectile is a parabola, if we ignore air resistance and assume that $\vec{g}$ is constant. To show this, we need to find y as a function of x by eliminating t between the two equations for horizontal and vertical motion (Eq. 2–11b), and we set $x_0 = y_0 = 0$:

$$x = v_{x0}t$$
$$y = v_{y0}t - \tfrac{1}{2}gt^2.$$

From the first equation, we have $t = x/v_{x0}$, and we substitute this into the second one to obtain

$$y = \left(\frac{v_{y0}}{v_{x0}}\right)x - \left(\frac{g}{2v_{x0}^2}\right)x^2.$$

If we write $v_{x0} = v_0 \cos\theta_0$ and $v_{y0} = v_0 \sin\theta_0$, we can also write

$$y = (\tan\theta_0)x - \left(\frac{g}{2v_0^2 \cos^2\theta_0}\right)x^2.$$

In either case, we see that y as a function of x has the form

$$y = Ax - Bx^2,$$

where A and B are constants for any specific projectile motion. This is the well-known equation for a parabola. See Figs. 3–17 and 3–27.

The idea that projectile motion is parabolic was, in Galileo's day, at the forefront of physics research. Today we discuss it in Chapter 3 of introductory physics!

* 3–8 Relative Velocity

We now consider how observations made in different reference frames are related to each other. For example, consider two trains approaching one another, each with a constant speed of 80 km/h with respect to the Earth. Observers on the Earth beside the tracks will measure 80 km/h for the speed of each train. Observers on either of the trains (a different reference frame) will measure a speed of 160 km/h for the other train approaching them.

Similarly, when one car traveling 90 km/h passes a second car traveling in the same direction at 75 km/h, the first car has a speed relative to the second car of 90 km/h − 75 km/h = 15 km/h.

When the velocities are along the same line, simple addition or subtraction is sufficient to obtain the relative velocity. But if they are not along the same line, we must use vector addition. We emphasize, as mentioned in Section 2–1, that when specifying a velocity, it is important to specify what the reference frame is.

When determining relative velocity, it is easy to make a mistake by adding or subtracting the wrong velocities. It is important, therefore, to draw a diagram and use a careful labeling process. Each velocity is labeled by *two subscripts: the first refers to the object, the second to the reference frame in which it has this velocity.* For example, suppose a boat is to cross a river to the opposite side, as shown in Fig. 3–28. We let $\vec{v}_{BW}$ be the velocity of the **B**oat with respect to the **W**ater. (This is also what the boat's velocity would be relative to the shore if the water were still.) Similarly, $\vec{v}_{BS}$ is the velocity of the **B**oat with respect to the **S**hore, and $\vec{v}_{WS}$ is the velocity of the **W**ater with respect to the **S**hore (this is the river current). Note that $\vec{v}_{BW}$ is what the boat's motor produces (against the water), whereas $\vec{v}_{BS}$ is equal to $\vec{v}_{BW}$ plus the effect of the current, $\vec{v}_{WS}$. Therefore, the velocity of the boat relative to the shore is

(see vector diagram, Fig. 3–28)

$$\vec{v}_{BS} = \vec{v}_{BW} + \vec{v}_{WS}.$$ (3–6)   *Follow the subscripts*

By writing the subscripts using this convention, we see that the inner subscripts (the two W's) on the right-hand side of Eq. 3–6 are the same, whereas the outer subscripts on the right of Eq. 3–6 (the B and the S) are the same as the two subscripts for the sum vector on the left, $\mathbf{v}_{\rightarrow BS}$. By following this convention (first subscript for the object, second for the reference frame), one can write down the correct equation relating velocities in different reference frames.[†] Equation 3–6 is valid in general and can be extended to three or more velocities. For example, if a fisherman on the boat walks with a velocity $\vec{v}_{FB}$ relative to the boat, his velocity relative to the shore is $\vec{v}_{FS} = \vec{v}_{FB} + \vec{v}_{BW} + \vec{v}_{WS}$. The equations involving relative velocity will be correct when adjacent inner subscripts are identical and when the outermost ones correspond exactly to the two on the velocity on the left of the equation. But this works only with plus signs (on the right), not minus signs.

It is often useful to remember that for any two objects or reference frames, A and B, the velocity of A relative to B has the same magnitude, but opposite direction, as the velocity of B relative to A:

$$\vec{v}_{BA} = -\vec{v}_{AB}.$$ (3–7)

For example, if a train is traveling 100 km/h relative to the Earth in a certain direction, objects on the Earth (such as trees) appear to an observer on the train to be traveling 100 km/h in the opposite direction.

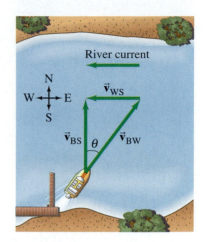

FIGURE 3–28 To move directly across the river, the boat must head upstream at an angle θ. Velocity vectors are shown as green arrows:

$\vec{v}_{BS}$ = velocity of **B**oat with respect to the **S**hore,

$\vec{v}_{BW}$ = velocity of **B**oat with respect to the **W**ater,

$\vec{v}_{WS}$ = velocity of **W**ater with respect to the **S**hore (river current).

CONCEPTUAL EXAMPLE 3–10   **Crossing a river.** A man in a small motor boat is trying to cross a river that flows due west with a strong current. The man starts on the south bank and is trying to reach the north bank directly north from his starting point. Should he (*a*) head due north, (*b*) head due west, (*c*) head in a northwesterly direction, (*d*) head in a northeasterly direction?

RESPONSE If the man heads straight across the river, the current will drag the boat downstream (westward). To overcome the river's westward current, the boat must acquire an eastward component of velocity as well as a northward component. Thus the boat must (*d*) head in a northeasterly direction (see Fig. 3–28). The actual angle depends on the strength of the current and how fast the boat moves relative to the water. If the current is weak and the motor is strong, then the boat can head almost, but not quite, due north.

EXAMPLE 3–11   **Heading upstream.** A boat's speed in still water is $v_{BW} = 1.85$ m/s. If the boat is to travel directly across a river whose current has speed $v_{WS} = 1.20$ m/s, at what upstream angle must the boat head? (See Fig. 3–29.)

APPROACH We reason as in Example 3–10, and use subscripts as in Eq. 3–6. Figure 3–29 has been drawn with $\vec{v}_{BS}$, the velocity of the **B**oat relative to the **S**hore, pointing directly across the river since this is how the boat is supposed to move. (Note that $\vec{v}_{BS} = \vec{v}_{BW} + \vec{v}_{WS}$.) To accomplish this, the boat needs to head upstream to offset the current pulling it downstream.

SOLUTION Vector $\vec{v}_{BW}$ points upstream at an angle θ as shown. From the diagram,

$$\sin \theta = \frac{v_{WS}}{v_{BW}} = \frac{1.20 \text{ m/s}}{1.85 \text{ m/s}} = 0.6486.$$

Thus $\theta = 40.4°$, so the boat must head upstream at a $40.4°$ angle.

FIGURE 3–29 Example 3–11.

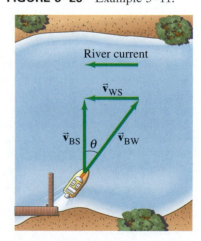

[†]We thus would know by inspection that (for example) the equation $\vec{v}_{BW} = \vec{v}_{BS} + \vec{v}_{WS}$ is wrong: the inner subscripts are not the same, and the outer ones on the right are not the same as the subscripts on the left.

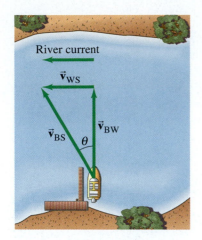

River current

$\vec{v}_{WS}$

$\vec{v}_{BW}$

$\vec{v}_{BS}$ θ

FIGURE 3–30 Example 3–12. A boat heading directly across a river whose current moves at 1.20 m/s.

EXAMPLE 3–12 **Heading across the river.** The same boat $(v_{BW} = 1.85 \text{ m/s})$ now heads directly across the river whose current is still 1.20 m/s. (a) What is the velocity (magnitude and direction) of the boat relative to the shore? (b) If the river is 110 m wide, how long will it take to cross and how far downstream will the boat be then?

APPROACH The boat now heads directly across the river and is pulled downstream by the current, as shown in Fig. 3–30. The boat's velocity with respect to the shore, $\vec{v}_{BS}$, is the sum of its velocity with respect to the water, $\vec{v}_{BW}$, plus the velocity of the water with respect to the shore, $\vec{v}_{WS}$:

$$\vec{v}_{BS} = \vec{v}_{BW} + \vec{v}_{WS},$$

just as before.

SOLUTION (a) Since $\vec{v}_{BW}$ is perpendicular to $\vec{v}_{WS}$, we can get v_{BS} using the theorem of Pythagoras:

$$v_{BS} = \sqrt{v_{BW}^2 + v_{WS}^2} = \sqrt{(1.85 \text{ m/s})^2 + (1.20 \text{ m/s})^2} = 2.21 \text{ m/s}.$$

We can obtain the angle (note how θ is defined in the diagram) from:

$$\tan \theta = v_{WS}/v_{BW} = (1.20 \text{ m/s})/(1.85 \text{ m/s}) = 0.6486.$$

A calculator with an INV TAN, an ARC TAN, or a TAN^{-1} key gives $\theta = \tan^{-1}(0.6486) = 33.0°$. Note that this angle is not equal to the angle calculated in Example 3–11.

(b) The travel time for the boat is determined by the time it takes to cross the river. Given the river's width $D = 110 \text{ m}$, we can use the velocity component in the direction of D, $v_{BW} = D/t$. Solving for t, we get $t = 110 \text{ m}/1.85 \text{ m/s} = 60 \text{ s}$. The boat will have been carried downstream, in this time, a distance

$$d = v_{WS}t = (1.20 \text{ m/s})(60 \text{ s}) = 72 \text{ m}.$$

NOTE There is no acceleration in this Example, so the motion involves only constant velocities (of the boat or of the river).

Summary

A quantity such as velocity, that has both a magnitude and a direction, is called a **vector**. A quantity such as mass, that has only a magnitude, is called a **scalar**.

Addition of vectors can be done graphically by placing the tail of each successive arrow at the tip of the previous one. The sum, or **resultant vector**, is the arrow drawn from the tail of the first vector to the tip of the last vector. Two vectors can also be added using the parallelogram method.

Vectors can be added more accurately by adding their **components** along chosen axes with the aid of trigonometric functions. A vector of magnitude V making an angle θ with the x axis has components

$$V_x = V \cos \theta, \qquad V_y = V \sin \theta. \qquad \textbf{(3–3)}$$

Given the components, we can find a vector's magnitude and direction from

$$V = \sqrt{V_x^2 + V_y^2}, \qquad \tan \theta = \frac{V_y}{V_x}. \qquad \textbf{(3–4)}$$

Projectile motion is the motion of an object in an arc near the Earth's surface under the effect of gravity alone. It can be analyzed as two separate motions if air resistance can be ignored. The horizontal component of motion is at constant velocity, whereas the vertical component is at constant acceleration, $\vec{g}$, just as for a body falling vertically under the action of gravity.

[*The velocity of an object relative to one frame of reference can be found by vector addition if its velocity relative to a second frame of reference, and the **relative velocity** of the two reference frames, are known.]

Questions

1. One car travels due east at 40 km/h, and a second car travels north at 40 km/h. Are their velocities equal? Explain.

2. Can you give several examples of an object's motion in which a great distance is traveled but the displacement is zero?

3. Can the displacement vector for a particle moving in two dimensions ever be longer than the length of path traveled by the particle over the same time interval? Can it ever be less? Discuss.

4. During baseball practice, a batter hits a very high fly ball and then runs in a straight line and catches it. Which had the greater displacement, the batter or the ball?

5. If $\vec{V} = \vec{V}_1 + \vec{V}_2$, is V necessarily greater than V_1 and/or V_2? Discuss.

6. Two vectors have length $V_1 = 3.5$ km and $V_2 = 4.0$ km. What are the maximum and minimum magnitudes of their vector sum?

7. Can two vectors of unequal magnitude add up to give the zero vector? Can *three* unequal vectors? Under what conditions?

8. Can the magnitude of a vector ever (*a*) be equal to one of its components, or (*b*) be less than one of its components?

9. Can a particle with constant speed be accelerating? What if it has constant velocity?

10. A child wishes to determine the speed a slingshot imparts to a rock. How can this be done using only a meter stick, a rock, and the slingshot?

11. It was reported in World War I that a pilot flying at an altitude of 2 km caught in his bare hands a bullet fired at the plane! Using the fact that a bullet slows down considerably due to air resistance, explain how this incident occurred.

12. At some amusement parks, to get on a moving "car" the riders first hop onto a moving walkway and then onto the cars themselves. Why is this done?

13. If you are riding on a train that speeds past another train moving in the same direction on an adjacent track, it appears that the other train is moving backward. Why?

14. If you stand motionless under an umbrella in a rainstorm where the drops fall vertically, you remain relatively dry. However, if you start running, the rain begins to hit your legs even if they remain under the umbrella. Why?

15. A person sitting in an enclosed train car, moving at constant velocity, throws a ball straight up into the air in her reference frame. (*a*) Where does the ball land? What is your answer if the car (*b*) accelerates, (*c*) decelerates, (*d*) rounds a curve, (*e*) moves with constant velocity but is open to the air?

16. Two rowers, who can row at the same speed in still water, set off across a river at the same time. One heads straight across and is pulled downstream somewhat by the current. The other one heads upstream at an angle so as to arrive at a point opposite the starting point. Which rower reaches the opposite side first?

17. How do you think a baseball player "judges" the flight of a fly ball? Which equation in this Chapter becomes part of the player's intuition?

18. In archery, should the arrow be aimed directly at the target? How should your angle of aim depend on the distance to the target?

19. A projectile is launched at an angle of 30° to the horizontal with a speed of 30 m/s. How does the horizontal component of its velocity 1.0 s after launch compare with its horizontal component of velocity 2.0 s after launch?

20. Two cannonballs, A and B, are fired from the ground with identical initial speeds, but with θ_A larger than θ_B. (*a*) Which cannonball reaches a higher elevation? (*b*) Which stays longer in the air? (*c*) Which travels farther?

Problems

3–2 to 3–4 Vector Addition

1. (I) A car is driven 215 km west and then 85 km southwest. What is the displacement of the car from the point of origin (magnitude and direction)? Draw a diagram.

2. (I) A delivery truck travels 18 blocks north, 10 blocks east, and 16 blocks south. What is its final displacement from the origin? Assume the blocks are equal length.

3. (I) Show that the vector labeled "incorrect" in Fig. 3–6c is actually the difference of the two vectors. Is it $\vec{V}_2 - \vec{V}_1$, or $\vec{V}_1 - \vec{V}_2$?

4. (I) If $V_x = 6.80$ units and $V_y = -7.40$ units, determine the magnitude and direction of $\vec{V}$.

5. (II) Graphically determine the resultant of the following three vector displacements: (1) 34 m, 25° north of east; (2) 48 m, 33° east of north; and (3) 22 m, 56° west of south.

6. (II) The components of a vector $\vec{V}$ can be written (V_x, V_y, V_z). What are the components and length of a vector which is the sum of the two vectors, $\vec{V}_1$ and $\vec{V}_2$, whose components are (8.0, −3.7, 0.0) and (3.9, −8.1, −4.4)?

7. (II) $\vec{V}$ is a vector 14.3 units in magnitude and points at an angle of 34.8° above the negative x axis. (*a*) Sketch this vector. (*b*) Find V_x and V_y. (*c*) Use V_x and V_y to obtain (again) the magnitude and direction of $\vec{V}$. [*Note*: Part (*c*) is a good way to check if you've resolved your vector correctly.]

8. (II) Vector $\vec{V}_1$ is 6.6 units long and points along the negative x axis. Vector $\vec{V}_2$ is 8.5 units long and points at +45° to the positive x axis. (*a*) What are the x and y components of each vector? (*b*) Determine the sum $\vec{V}_1 + \vec{V}_2$ (magnitude and angle).

9. (II) An airplane is traveling 735 km/h in a direction 41.5° west of north (Fig. 3–31). (a) Find the components of the velocity vector in the northerly and westerly directions. (b) How far north and how far west has the plane traveled after 3.00 h?

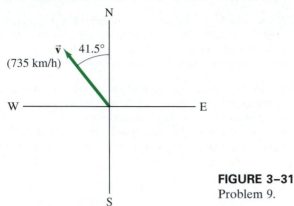

FIGURE 3–31
Problem 9.

10. (II) Three vectors are shown in Fig. 3–32. Their magnitudes are given in arbitrary units. Determine the sum of the three vectors. Give the resultant in terms of (a) components, (b) magnitude and angle with the x axis.

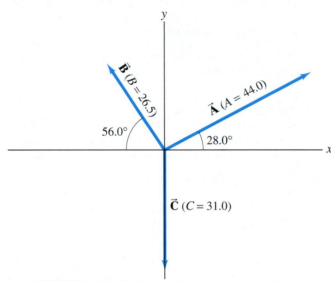

FIGURE 3–32   Problems 10, 11, 12, 13, and 14. Vector magnitudes are given in arbitrary units.

11. (II) Determine the vector $\vec{\mathbf{A}} - \vec{\mathbf{C}}$, given the vectors $\vec{\mathbf{A}}$ and $\vec{\mathbf{C}}$ in Fig. 3–32.

12. (II) (a) Given the vectors $\vec{\mathbf{A}}$ and $\vec{\mathbf{B}}$ shown in Fig. 3–32, determine $\vec{\mathbf{B}} - \vec{\mathbf{A}}$. (b) Determine $\vec{\mathbf{A}} - \vec{\mathbf{B}}$ without using your answer in (a). Then compare your results and see if they are opposite.

13. (II) For the vectors given in Fig. 3–32, determine (a) $\vec{\mathbf{A}} - \vec{\mathbf{B}} + \vec{\mathbf{C}}$, (b) $\vec{\mathbf{A}} + \vec{\mathbf{B}} - \vec{\mathbf{C}}$, and (c) $\vec{\mathbf{C}} - \vec{\mathbf{A}} - \vec{\mathbf{B}}$.

14. (II) For the vectors shown in Fig. 3–32, determine (a) $\vec{\mathbf{B}} - 2\vec{\mathbf{A}}$, (b) $2\vec{\mathbf{A}} - 3\vec{\mathbf{B}} + 2\vec{\mathbf{C}}$.

15. (II) The summit of a mountain, 2450 m above base camp, is measured on a map to be 4580 m horizontally from the camp in a direction 32.4° west of north. What are the components of the displacement vector from camp to summit? What is its magnitude? Choose the x axis east, y axis north, and z axis up.

16. (II) You are given a vector in the xy plane that has a magnitude of 70.0 units and a y component of -55.0 units. What are the two possibilities for its x component?

3–5 and 3–6 Projectile Motion (neglect air resistance)

17. (I) A tiger leaps horizontally from a 6.5-m-high rock with a speed of 3.5 m/s. How far from the base of the rock will she land?

18. (I) A diver running 1.8 m/s dives out horizontally from the edge of a vertical cliff and 3.0 s later reaches the water below. How high was the cliff, and how far from its base did the diver hit the water?

19. (II) A fire hose held near the ground shoots water at a speed of 6.8 m/s. At what angle(s) should the nozzle point in order that the water land 2.0 m away (Fig. 3–33)? Why are there two different angles? Sketch the two trajectories.

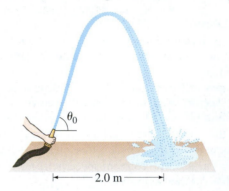

FIGURE 3–33   Problem 19.

20. (II) Romeo is chucking pebbles gently up to Juliet's window, and he wants the pebbles to hit the window with only a horizontal component of velocity. He is standing at the edge of a rose garden 4.5 m below her window and 5.0 m from the base of the wall (Fig. 3–34). How fast are the pebbles going when they hit her window?

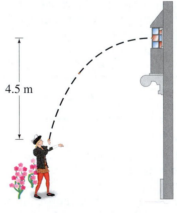

FIGURE 3–34
Problem 20.

21. (II) A ball is thrown horizontally from the roof of a building 45.0 m tall and lands 24.0 m from the base. What was the ball's initial speed?

22. (II) A football is kicked at ground level with a speed of 18.0 m/s at an angle of 35.0° to the horizontal. How much later does it hit the ground?

23. (II) A ball thrown horizontally at 22.2 m/s from the roof of a building lands 36.0 m from the base of the building. How tall is the building?

24. (II) An athlete executing a long jump leaves the ground at a 28.0° angle and travels 7.80 m. (a) What was the takeoff speed? (b) If this speed were increased by just 5.0%, how much longer would the jump be?

25. (II) Determine how much farther a person can jump on the Moon as compared to the Earth if the takeoff speed and angle are the same. The acceleration due to gravity on the Moon is one-sixth what it is on Earth.

26. (II) A hunter aims directly at a target (on the same level) 75.0 m away. (a) If the bullet leaves the gun at a speed of 180 m/s, by how much will it miss the target? (b) At what angle should the gun be aimed so as to hit the target?

27. (II) The pilot of an airplane traveling 180 km/h wants to drop supplies to flood victims isolated on a patch of land 160 m below. The supplies should be dropped how many seconds before the plane is directly overhead?

28. (II) Show that the speed with which a projectile leaves the ground is equal to its speed just before it strikes the ground at the end of its journey, assuming the firing level equals the landing level.

29. (II) Suppose the kick in Example 3–5 is attempted 36.0 m from the goalposts, whose crossbar is 3.00 m above the ground. If the football is directed correctly between the goalposts, will it pass over the bar and be a field goal? Show why or why not.

30. (II) A projectile is fired with an initial speed of 65.2 m/s at an angle of 34.5° above the horizontal on a long flat firing range. Determine (a) the maximum height reached by the projectile, (b) the total time in the air, (c) the total horizontal distance covered (that is, the range), and (d) the velocity of the projectile 1.50 s after firing.

31. (II) A projectile is shot from the edge of a cliff 125 m above ground level with an initial speed of 65.0 m/s at an angle of 37.0° with the horizontal, as shown in Fig. 3–35. (a) Determine the time taken by the projectile to hit point P at ground level. (b) Determine the range X of the projectile as measured from the base of the cliff. At the instant just before the projectile hits point P, find (c) the horizontal and the vertical components of its velocity, (d) the magnitude of the velocity, and (e) the angle made by the velocity vector with the horizontal. (f) Find the maximum height above the cliff top reached by the projectile.

32. (II) A shotputter throws the shot with an initial speed of 15.5 m/s at a 34.0° angle to the horizontal. Calculate the horizontal distance traveled by the shot if it leaves the athlete's hand at a height of 2.20 m above the ground.

33. (II) At what projection angle will the range of a projectile equal its maximum height?

34. (III) Revisit Conceptual Example 3–7, and assume that the boy with the slingshot is *below* the boy in the tree (Fig. 3–36), and so aims *upward*, directly at the boy in the tree. Show that again the boy in the tree makes the wrong move by letting go at the moment the water balloon is shot.

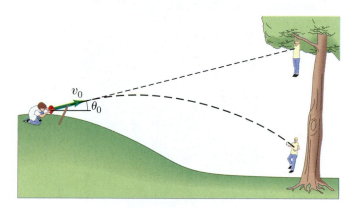

FIGURE 3–36 Problem 34.

35. (III) A rescue plane wants to drop supplies to isolated mountain climbers on a rocky ridge 235 m below. If the plane is traveling horizontally with a speed of 250 km/h (69.4 m/s), (a) how far in advance of the recipients (horizontal distance) must the goods be dropped (Fig. 3–37a)? (b) Suppose, instead, that the plane releases the supplies a horizontal distance of 425 m in advance of the mountain climbers. What vertical velocity (up or down) should the supplies be given so that they arrive precisely at the climbers' position (Fig. 3–37b)? (c) With what speed do the supplies land in the latter case?

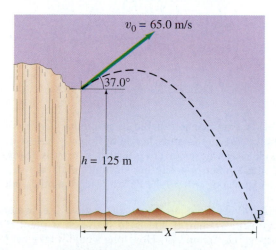

FIGURE 3–35 Problem 31.

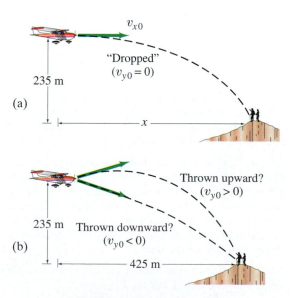

FIGURE 3–37 Problem 35.

* **36.** (I) A person going for a morning jog on the deck of a cruise ship is running toward the bow (front) of the ship at 2.2 m/s while the ship is moving ahead at 7.5 m/s. What is the velocity of the jogger relative to the water? Later, the jogger is moving toward the stern (rear) of the ship. What is the jogger's velocity relative to the water now?

* **37.** (II) Huck Finn walks at a speed of 0.60 m/s across his raft (that is, he walks perpendicular to the raft's motion relative to the shore). The raft is traveling down the Mississippi River at a speed of 1.70 m/s relative to the river bank (Fig. 3–38). What is Huck's velocity (speed and direction) relative to the river bank?

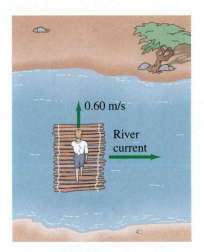

FIGURE 3–38 Problem 37.

* **38.** (II) You are driving south on a highway at 25 m/s (approximately 55 mi/h) in a snowstorm. When you last stopped, you noticed that the snow was coming down vertically, but it is passing the windows of the moving car at an angle of 30° to the horizontal. Estimate the speed of the snowflakes relative to the car and relative to the ground.

* **39.** (II) A boat can travel 2.30 m/s in still water. (a) If the boat points its prow directly across a stream whose current is 1.20 m/s, what is the velocity (magnitude and direction) of the boat relative to the shore? (b) What will be the position of the boat, relative to its point of origin, after 3.00 s? (See Fig. 3–30.)

* **40.** (II) Two planes approach each other head-on. Each has a speed of 785 km/h, and they spot each other when they are initially 11.0 km apart. How much time do the pilots have to take evasive action?

* **41.** (II) An airplane is heading due south at a speed of 600 km/h. If a wind begins blowing from the southwest at a speed of 100 km/h (average), calculate: (a) the velocity (magnitude and direction) of the plane relative to the ground, and (b) how far from its intended position will it be after 10 min if the pilot takes no corrective action. [*Hint*: First draw a diagram.]

* **42.** (II) In what direction should the pilot aim the plane in Problem 41 so that it will fly due south?

* **43.** (II) Determine the speed of the boat with respect to the shore in Example 3–11.

* **44.** (II) A passenger on a boat moving at 1.50 m/s on a still lake walks up a flight of stairs at a speed of 0.50 m/s (Fig. 3–39). The stairs are angled at 45° pointing in the direction of motion as shown. What is the velocity of the passenger relative to the water?

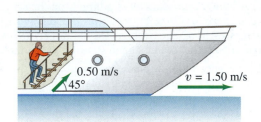

FIGURE 3–39 Problem 44.

* **45.** (II) A motorboat whose speed in still water is 2.60 m/s must aim upstream at an angle of 28.5° (with respect to a line perpendicular to the shore) in order to travel directly across the stream. (a) What is the speed of the current? (b) What is the resultant speed of the boat with respect to the shore? (See Fig. 3–28.)

* **46.** (II) A boat, whose speed in still water is 1.70 m/s, must cross a 260-m-wide river and arrive at a point 110 m upstream from where it starts (Fig. 3–40). To do so, the pilot must head the boat at a 45° upstream angle. What is the speed of the river's current?

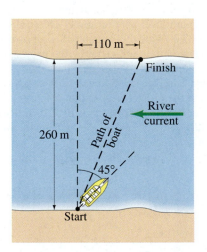

FIGURE 3–40 Problem 46.

* **47.** (II) A swimmer is capable of swimming 0.45 m/s in still water. (a) If she aims her body directly across a 75-m-wide river whose current is 0.40 m/s, how far downstream (from a point opposite her starting point) will she land? (b) How long will it take her to reach the other side?

* **48.** (II) (a) At what upstream angle must the swimmer in Problem 47 aim, if she is to arrive at a point directly across the stream? (b) How long would it take her?

* **49.** (III) An airplane whose air speed is 620 km/h is supposed to fly in a straight path 35.0° north of east. But a steady 95 km/h wind is blowing from the north. In what direction should the plane head?

* **50.** (III) An unmarked police car, traveling a constant 95 km/h, is passed by a speeder traveling 145 km/h. Precisely 1.00 s after the speeder passes, the policeman steps on the accelerator. If the police car's acceleration is 2.00 m/s^2, how much time elapses after the police car is passed until it overtakes the speeder (assumed moving at constant speed)?

* **51.** (III) Assume in Problem 50 that the speeder's speed is not known. If the police car accelerates uniformly as given above, and overtakes the speeder after 7.00 s, what was the speeder's speed?

* **52.** (III) Two cars approach a street corner at right angles to each other (Fig. 3–41). Car 1 travels at a speed relative to Earth $v_{1E} = 35$ km/h, and car 2 at $v_{2E} = 55$ km/h. What is the relative velocity of car 1 as seen by car 2? What is the velocity of car 2 relative to car 1?

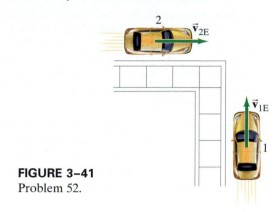

FIGURE 3–41
Problem 52.

General Problems

53. William Tell must split the apple atop his son's head from a distance of 27 m. When William aims directly at the apple, the arrow is horizontal. At what angle must he aim it to hit the apple if the arrow travels at a speed of 35 m/s?

54. A plumber steps out of his truck, walks 50 m east and 25 m south, and then takes an elevator 10 m down into the subbasement of a building where a bad leak is occurring. What is the displacement of the plumber relative to his truck? Give your answer in components, and also give the magnitude and angles with the x axis in the vertical and horizontal planes. Assume x is east, y is north, and z is up.

55. On mountainous downhill roads, escape routes are sometimes placed to the side of the road for trucks whose brakes might fail. Assuming a constant upward slope of 32°, calculate the horizontal and vertical components of the acceleration of a truck that slowed from 120 km/h to rest in 6.0 s. See Fig. 3–42.

56. What is the y component of a vector (in the xy plane) whose magnitude is 88.5 and whose x component is 75.4? What is the direction of this vector (angle it makes with the x axis)?

57. Raindrops make an angle θ with the vertical when viewed through a moving train window (Fig. 3–43). If the speed of the train is v_T, what is the speed of the raindrops in the reference frame of the Earth in which they are assumed to fall vertically?

FIGURE 3–43 Problem 57.

58. A light plane is headed due south with a speed of 155 km/h relative to still air. After 1.00 hour, the pilot notices that they have covered only 125 km and their direction is not south but southeast (45.0°). What is the wind velocity?

59. A car moving at 95 km/h passes a 1.00-km-long train traveling in the same direction on a track that is parallel to the road. If the speed of the train is 75 km/h, how long does it take the car to pass the train, and how far will the car have traveled in this time? What are the results if the car and train are instead traveling in opposite directions?

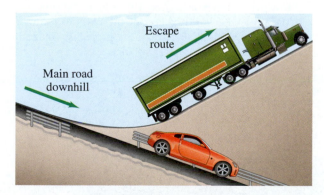

FIGURE 3–42 Problem 55.

60. An Olympic long jumper is capable of jumping 8.0 m. Assuming his horizontal speed is 9.1 m/s as he leaves the ground, how long is he in the air and how high does he go? Assume that he lands standing upright—that is, the same way he left the ground.

61. *Apollo* astronauts took a "nine iron" to the Moon and hit a golf ball about 180 m! Assuming that the swing, launch angle, and so on, were the same as on Earth where the same astronaut could hit it only 35 m, estimate the acceleration due to gravity on the surface of the Moon. (Neglect air resistance in both cases, but on the Moon there is none!)

62. When Babe Ruth hit a homer over the 7.5-m-high right-field fence 95 m from home plate, roughly what was the minimum speed of the ball when it left the bat? Assume the ball was hit 1.0 m above the ground and its path initially made a 38° angle with the ground.

63. The cliff divers of Acapulco push off horizontally from rock platforms about 35 m above the water, but they must clear rocky outcrops at water level that extend out into the water 5.0 m from the base of the cliff directly under their launch point. See Fig. 3–44. What minimum pushoff speed is necessary to clear the rocks? How long are they in the air?

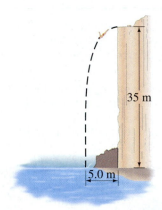

FIGURE 3–44 Problem 63.

64. At serve, a tennis player aims to hit the ball horizontally. What minimum speed is required for the ball to clear the 0.90-m-high net about 15.0 m from the server if the ball is "launched" from a height of 2.50 m? Where will the ball land if it just clears the net (and will it be "good" in the sense that it lands within 7.0 m of the net)? How long will it be in the air? See Fig. 3–45.

FIGURE 3–45 Problem 64.

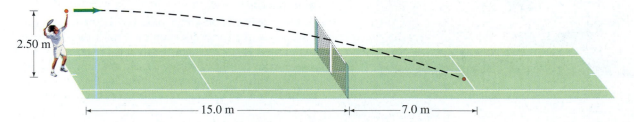

65. Spymaster Paul, flying a constant 215 km/h horizontally in a low-flying helicopter, wants to drop secret documents into his contact's open car which is traveling 155 km/h on a level highway 78.0 m below. At what angle (to the horizontal) should the car be in his sights when the packet is released (Fig. 3–46)?

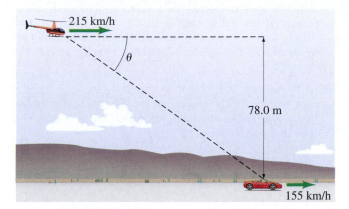

FIGURE 3–46 Problem 65.

66. The speed of a boat in still water is v. The boat is to make a round trip in a river whose current travels at speed u. Derive a formula for the time needed to make a round trip of total distance D if the boat makes the round trip by moving (*a*) upstream and back downstream, (*b*) directly across the river and back. We must assume $u < v$; why?

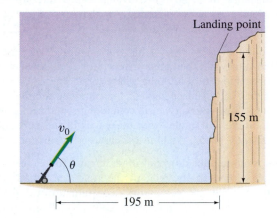

FIGURE 3–47 Problem 67.

67. A projectile is launched from ground level to the top of a cliff which is 195 m away and 155 m high (see Fig. 3–47). If the projectile lands on top of the cliff 7.6 s after it is fired, find the initial velocity of the projectile (magnitude and direction). Neglect air resistance.

68. (a) A skier is accelerating down a 30.0° hill at 1.80 m/s² (Fig. 3–48). What is the vertical component of her acceleration? (b) How long will it take her to reach the bottom of the hill, assuming she starts from rest and accelerates uniformly, if the elevation change is 335 m?

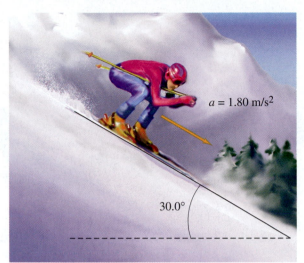

$a = 1.80 \text{ m/s}^2$

30.0°

FIGURE 3–48 Problem 68.

69. A basketball leaves a player's hands at a height of 2.10 m above the floor. The basket is 2.60 m above the floor. The player likes to shoot the ball at a 38.0° angle. If the shot is made from a horizontal distance of 11.00 m and must be accurate to ± 0.22 m (horizontally), what is the range of initial speeds allowed to make the basket?

70. A high diver leaves the end of a 5.0-m-high diving board and strikes the water 1.3 s later, 3.0 m beyond the end of the board. Considering the diver as a particle, determine (a) her initial velocity, $\vec{v}_0$, (b) the maximum height reached, and (c) the velocity $\vec{v}_f$ with which she enters the water.

71. A stunt driver wants to make his car jump over eight cars parked side by side below a horizontal ramp (Fig. 3–49). (a) With what minimum speed must he drive off the horizontal ramp? The vertical height of the ramp is 1.5 m above the cars, and the horizontal distance he must clear is 20 m. (b) If the ramp is now tilted upward, so that "takeoff angle" is 10° above the horizontal, what is the new minimum speed?

20 m

1.5 m

Must clear this point!

FIGURE 3–49 Problem 71.

72. A batter hits a fly ball which leaves the bat 0.90 m above the ground at an angle of 61° with an initial speed of 28 m/s heading toward centerfield. Ignore air resistance. (a) How far from home plate would the ball land if not caught? (b) The ball is caught by the centerfielder who, starting at a distance of 105 m from home plate, runs straight toward home plate at a constant speed and makes the catch at ground level. Find his speed.

73. At $t = 0$ a batter hits a baseball with an initial speed of 32 m/s at a 55° angle to the horizontal. An outfielder is 85 m from the batter at $t = 0$, and, as seen from home plate, the line of sight to the outfielder makes a horizontal angle of 22° with the plane in which the ball moves (see Fig. 3–50). What speed and direction must the fielder take in order to catch the ball at the same height from which it was struck? Give angle with respect to the outfielder's line of sight to home plate.

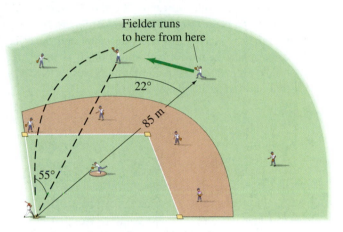

Fielder runs
to here from here

22°

85 m

55°

FIGURE 3–50 Problem 73.

74. A ball is shot from the top of a building with an initial velocity of 18 m/s at an angle $\theta = 42°$ above the horizontal. (a) What are the x and y components of the initial velocity? (b) If a nearby building is the same height and 55 m away, how far below the top of the building will the ball strike the nearby building?

75. You buy a plastic dart gun, and being a clever physics student you decide to do a quick calculation to find its maximum horizontal range. You shoot the gun straight up, and it takes 4.0 s for the dart to land back at the barrel. What is the maximum horizontal range of your gun?

Answers to Exercises

A: When the two vectors D_1 and D_2 point in the same direction.

B: $3\sqrt{2} = 4.24$.

C: They hit at the same time.

D: Both balls reach the same height; therefore they are in the air for the same length of time.

E: (b).

This airplane is taking off. It is accelerating, increasing in speed rapidly. To do so, a force must be exerted on it according to Newton's second law, $\Sigma\vec{F} = m\vec{a}$. What exerts this force? The two jet engines of this plane exert a strong force on the gases they push out toward the rear of the plane (labeled $\vec{F}_{GP}$). According to Newton's third law, these ejected gases exert an equal and opposite force on the airplane in the forward direction. It is these "reaction" forces exerted on the **p**lane by the **g**ases, labeled $\vec{F}_{PG}$, that accelerate the plane forward.

CHAPTER 4

Dynamics: Newton's Laws of Motion

We have discussed how motion is described in terms of velocity and acceleration. Now we deal with the question of *why* objects move as they do: What makes an object at rest begin to move? What causes an object to accelerate or decelerate? What is involved when an object moves in a circle? We can answer in each case that a force is required. In this Chapter, we will investigate the connection between force and motion, which is the subject called **dynamics**.

We begin with intuitive ideas of what a force is, and then discuss Newton's three laws of motion. We next look at several types of force, including friction and the force of gravity. We then apply Newton's laws to real problems.

4–1 Force

Intuitively, we experience **force** as any kind of a push or a pull on an object. When you push a stalled car or a grocery cart (Fig. 4–1), you are exerting a force on it. When a motor lifts an elevator, or a hammer hits a nail, or the wind blows the leaves of a tree, a force is being exerted. We say that an object falls because of the *force of gravity*.

FIGURE 4–1 A force exerted on a grocery cart—in this case exerted by a child.

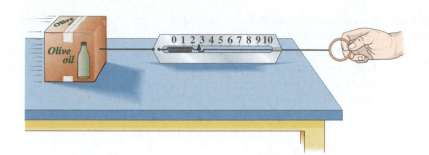

FIGURE 4–2 A spring scale used to measure a force.

If an object is at rest, to start it moving requires force—that is, a force is needed to accelerate an object from zero velocity to a nonzero velocity. For an object already moving, if you want to change its velocity—either in direction or in magnitude—again a force is required. In other words, to accelerate an object, a force is required.

One way to measure the magnitude (or strength) of a force is to use a spring scale (Fig. 4–2). Normally, such a spring scale is used to find the weight of an object; by weight we mean the force of gravity acting on the object (Section 4–6). The spring scale, once calibrated, can be used to measure other kinds of forces as well, such as the pulling force shown in Fig. 4–2.

Measuring force

A force exerted in different directions has a different effect. Clearly, force has direction as well as magnitude, and is indeed a vector that follows the rules of vector addition discussed in Chapter 3. We can represent any force on a diagram by an arrow, just as we did with velocity. The direction of the arrow is the direction of the push or pull, and its length is drawn proportional to the magnitude of the force.

4–2 Newton's First Law of Motion

What is the relationship between force and motion? Aristotle (384–322 B.C.) believed that a force was required to keep an object moving along a horizontal plane. To Aristotle, the natural state of an object was at rest, and a force was believed necessary to keep an object in motion. Furthermore, Aristotle argued, the greater the force on the object, the greater its speed.

Aristotle

vs.

Some 2000 years later, Galileo disagreed: He maintained that it is just as natural for an object to be in motion with a constant velocity as it is for it to be at rest.

Galileo

To understand Galileo's idea, consider the following observations involving motion along a horizontal plane. To push an object with a rough surface along a tabletop at constant speed requires a certain amount of force. To push an equally heavy object with a very smooth surface across the table at the same speed will require less force. If a layer of oil or other lubricant is placed between the surface of the object and the table, then almost no force is required to move the object. Notice that in each successive step, less force is required. As the next step, we imagine that the object does not rub against the table at all— or there is a perfect lubricant between the object and the table—and theorize that once started, the object would move across the table at constant speed with *no* force applied. A steel ball bearing rolling on a hard horizontal surface approaches this situation. So does a puck on an air table, in which a thin layer of air reduces friction almost to zero.

It was Galileo's genius to imagine such an idealized world—in this case, one where there is no friction—and to see that it could lead to a more accurate and richer understanding of the real world. This idealization led him to his remarkable conclusion that if no force is applied to a moving object, it will continue to move with constant speed in a straight line. An object slows down only if a force is exerted on it. Galileo thus interpreted friction as a force akin to ordinary pushes and pulls.

Friction as a force

To push an object across a table at constant speed requires a force from your hand that can balance out the force of friction (Fig. 4–3). When the object moves at constant speed, your pushing force is equal in magnitude to the friction force, but these two forces are in opposite directions, so the *net* force on the object (the vector sum of the two forces) is zero. This is consistent with Galileo's viewpoint, for the object moves with constant speed when no net force is exerted on it.

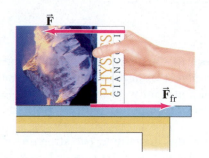

FIGURE 4–3 $\vec{\mathbf{F}}$ represents the force applied by the person and $\vec{\mathbf{F}}_{fr}$ represents the force of friction.

Upon this foundation laid by Galileo, Isaac Newton (Fig. 4–4) built his great theory of motion. Newton's analysis of motion is summarized in his famous "three laws of motion." In his great work, the *Principia* (published in 1687), Newton readily acknowledged his debt to Galileo. In fact, **Newton's first law of motion** is close to Galileo's conclusions. It states that

Every object continues in its state of rest, or of uniform velocity in a straight line, as long as no net force acts on it.

The tendency of an object to maintain its state of rest or of uniform motion in a straight line is called **inertia**. As a result, Newton's first law is often called the **law of inertia**.

Inertia

FIGURE 4–4
Isaac Newton (1642–1727).

Inertial reference frames

| CONCEPTUAL EXAMPLE 4–1 | **Newton's first law.** A school bus comes to a sudden stop, and all of the backpacks on the floor start to slide forward. What force causes them to do that?

RESPONSE It isn't "force" that does it. The backpacks continue their state of motion, maintaining their velocity (friction may slow them down), as the velocity of the bus decreases.

Inertial Reference Frames

Newton's first law does not hold in every reference frame. For example, if your reference frame is fixed in an accelerating car, an object such as a cup resting on the dashboard may begin to move toward you (it stayed at rest as long as the car's velocity remained constant). The cup accelerated toward you, but neither you nor anything else exerted a force on it in that direction. Similarly, in the reference frame of the bus in Example 4–1, there was no force pushing the backpacks forward. In accelerating reference frames, Newton's first law does not hold. Reference frames in which Newton's first law does hold are called **inertial reference frames** (the law of inertia is valid in them). For most purposes, we can usually assume that reference frames fixed on the Earth are inertial frames. (This is not precisely true, due to the Earth's rotation, but usually it is close enough.) Any reference frame that moves with constant velocity (say, a car or an airplane) relative to an inertial frame is also an inertial reference frame. Reference frames where the law of inertia does *not* hold, such as the accelerating reference frames discussed above, are called **noninertial** reference frames. How can we be sure a reference frame is inertial or not? By checking to see if Newton's first law holds. Thus Newton's first law serves as the definition of inertial reference frames.

4–3 Mass

Newton's second law, which we come to in the next Section, makes use of the concept of mass. Newton used the term *mass* as a synonym for *quantity of matter*. This intuitive notion of the mass of an object is not very precise because the concept "quantity of matter" is not very well defined. More precisely, we can say that **mass** is a *measure of the inertia* of an object. The more mass an object has, the greater the force needed to give it a particular acceleration. It is harder to start it moving from rest, or to stop it when it is moving, or to change its velocity sideways out of a straight-line path. A truck has much more inertia than a baseball moving at the same speed, and it requires a much greater force to change the truck's velocity at the same rate as the ball's. The truck therefore has much more mass.

Mass as inertia

To quantify the concept of mass, we must define a standard. In SI units, the unit of mass is the **kilogram** (kg) as we discussed in Chapter 1, Section 1–5.

The terms *mass* and *weight* are often confused with one another, but it is important to distinguish between them. Mass is a property of an object itself (a measure of an object's inertia, or its "quantity of matter"). Weight, on the other hand, is a force, the pull of gravity acting on an object. To see the difference, suppose we take an object to the Moon. The object will weigh only about one-sixth as much as it did on Earth, since the force of gravity is weaker. But its mass will be the same. It will have the same amount of matter as on Earth, and will have just as much inertia—for in the absence of friction, it will be just as hard to start it moving on the Moon as on Earth, or to stop it once it is moving. (More on weight in Section 4–6.)

⚠ CAUTION
Distinguish mass from weight

4–4 Newton's Second Law of Motion

Newton's first law states that if no net force is acting on an object at rest, the object remains at rest; or if the object is moving, it continues moving with constant speed in a straight line. But what happens if a net force *is* exerted on an object? Newton perceived that the object's velocity will change (Fig. 4–5). A net force exerted on an object may make its velocity increase. Or, if the net force is in a direction opposite to the motion, the force will reduce the object's velocity. If the net force acts sideways on a moving object, the *direction* of the object's velocity changes (and the magnitude may as well). Since a change in velocity is an acceleration (Section 2–4), we can say that *a net force causes acceleration*.

What precisely is the relationship between acceleration and force? Everyday experience can suggest an answer. Consider the force required to push a cart when friction is small enough to ignore. (If there is friction, consider the *net* force, which is the force you exert minus the force of friction.) Now if you push with a gentle but constant force for a certain period of time, you will make the cart accelerate from rest up to some speed, say 3 km/h. If you push with twice the force, the cart will reach 3 km/h in half the time. The acceleration will be twice as great. If you triple the force, the acceleration is tripled, and so on. Thus, the acceleration of an object is directly proportional[†] to the net applied force. But the acceleration depends on the mass of the object as well. If you push an empty grocery cart with the same force as you push one that is filled with groceries, you will find that the full cart accelerates more slowly. The greater the mass, the less the acceleration for the same net force. The mathematical relation, as Newton argued, is that the acceleration of an object is inversely proportional to its mass. These relationships are found to hold in general and can be summarized as follows:

FIGURE 4–5 The bobsled accelerates because the team exerts a force.

> **The acceleration of an object is directly proportional to the net force acting on it, and is inversely proportional to its mass. The direction of the acceleration is in the direction of the net force acting on the object.**

This is **Newton's second law of motion**.

NEWTON'S SECOND LAW OF MOTION

[†] A review of proportionality is given in Appendix A, at the back of this book.

Newton's second law can be written as an equation:

$$\vec{a} = \frac{\Sigma \vec{F}}{m},$$

Net force

where $\vec{a}$ stands for acceleration, m for the mass, and $\Sigma \vec{F}$ for the *net force* on the object. The symbol Σ (Greek "sigma") stands for "sum of"; $\vec{F}$ stands for force, so $\Sigma \vec{F}$ means the *vector sum of all forces* acting on the object, which we define as the **net force**.

We rearrange this equation to obtain the familiar statement of Newton's second law:

NEWTON'S SECOND LAW OF MOTION

$$\Sigma \vec{F} = m\vec{a}. \qquad (4\text{–}1)$$

Newton's second law relates the description of motion to the cause of motion, force. It is one of the most fundamental relationships in physics. From Newton's second law we can make a more precise definition of **force** as *an action capable of accelerating an object.*

Force defined

Every force $\vec{F}$ is a vector, with magnitude and direction. Equation 4–1 is a vector equation valid in any inertial reference frame. It can be written in component form in rectangular coordinates as

$$\Sigma F_x = ma_x, \qquad \Sigma F_y = ma_y, \qquad \Sigma F_z = ma_z.$$

If the motion is all along a line (one-dimensional), we can leave out the subscripts and simply write $\Sigma F = ma$.

Unit of force: the newton

In SI units, with the mass in kilograms, the unit of force is called the **newton** (N). One newton, then, is the force required to impart an acceleration of $1\,\text{m/s}^2$ to a mass of 1 kg. Thus $1\,\text{N} = 1\,\text{kg·m/s}^2$.

In cgs units, the unit of mass is the gram (g) as mentioned earlier.[†] The unit of force is the *dyne*, which is defined as the net force needed to impart an acceleration of $1\,\text{cm/s}^2$ to a mass of 1 g. Thus $1\,\text{dyne} = 1\,\text{g·cm/s}^2$. It is easy to show that $1\,\text{dyne} = 10^{-5}\,\text{N}$.

In the British system, the unit of force is the *pound* (abbreviated lb), where $1\,\text{lb} = 4.44822\,\text{N} \approx 4.45\,\text{N}$. The unit of mass is the *slug*, which is defined as that mass which will undergo an acceleration of $1\,\text{ft/s}^2$ when a force of 1 lb is applied to it. Thus $1\,\text{lb} = 1\,\text{slug·ft/s}^2$. Table 4–1 summarizes the units in the different systems.

➥ **PROBLEM SOLVING**
Use a consistent set of units

It is very important that only one set of units be used in a given calculation or problem, with the SI being preferred. If the force is given in, say, newtons, and the mass in grams, then before attempting to solve for the acceleration in SI units, we must change the mass to kilograms. For example, if the force is given as 2.0 N along the x axis and the mass is 500 g, we change the latter to 0.50 kg, and the acceleration will then automatically come out in m/s^2 when Newton's second law is used (we set $1\,\text{N} = 1\,\text{kg·m/s}^2$):

$$a_x = \frac{\Sigma F_x}{m} = \frac{2.0\,\text{N}}{0.50\,\text{kg}} = \frac{2.0\,\text{kg·m/s}^2}{0.50\,\text{kg}} = 4.0\,\text{m/s}^2.$$

TABLE 4–1
Units for Mass and Force

System	Mass	Force
SI	kilogram (kg)	newton (N) $(= \text{kg·m/s}^2)$
cgs	gram (g)	dyne $(= \text{g·cm/s}^2)$
British	slug	pound (lb)

Conversion factors: $1\,\text{dyne} = 10^{-5}\,\text{N}$; $1\,\text{lb} \approx 4.45\,\text{N}$.

EXAMPLE 4–2 **ESTIMATE** **Force to accelerate a fast car.** Estimate the net force needed to accelerate (*a*) a 1000-kg car at $\frac{1}{2}g$; (*b*) a 200-g apple at the same rate.

APPROACH We can use Newton's second law to find the net force needed for each object, because we are given the mass and the acceleration. This is an estimate (the $\frac{1}{2}$ is not said to be precise) so we round off to one significant figure.

[†]Be careful not to confuse g for gram with g for the acceleration due to gravity. The latter is always italicized (or boldface when a vector).

SOLUTION (a) The car's acceleration is $a = \frac{1}{2}g = \frac{1}{2}(9.8\,\text{m/s}^2) \approx 5\,\text{m/s}^2$. We use Newton's second law to get the net force needed to achieve this acceleration:

$$\Sigma F = ma \approx (1000\,\text{kg})(5\,\text{m/s}^2) = 5000\,\text{N}.$$

(If you are used to British units, to get an idea of what a 5000-N force is, you can divide by 4.45 N/lb and get a force of about 1000 lb.)

(b) For the apple, $m = 200\,\text{g} = 0.200\,\text{kg}$, so

$$\Sigma F = ma \approx (0.200\,\text{kg})(5\,\text{m/s}^2) = 1\,\text{N}.$$

EXAMPLE 4–3 **Force to stop a car.** What average net force is required to bring a 1500-kg car to rest from a speed of 100 km/h within a distance of 55 m?

APPROACH We can use Newton's second law, $\Sigma F = ma$, to determine the force if we know the mass and acceleration of the car. We are given the mass, but we will have to calculate the acceleration a. We assume the acceleration is constant, so we can use the kinematic equations, Eqs. 2–11, to calculate it.

$v_0 = 100$ km/h $v = 0$

$x = 0$ $x = 55\,\text{m}$ x (m)

FIGURE 4–6 Example 4–3.

SOLUTION We assume the motion is along the $+x$ axis (Fig. 4–6). We are given the initial velocity $v_0 = 100\,\text{km/h} = 28\,\text{m/s}$ (Section 1–6), the final velocity $v = 0$, and the distance traveled $x - x_0 = 55\,\text{m}$. From Eq. 2–11c, we have

$$v^2 = v_0^2 + 2a(x - x_0),$$

so

$$a = \frac{v^2 - v_0^2}{2(x - x_0)} = \frac{0 - (28\,\text{m/s})^2}{2(55\,\text{m})} = -7.1\,\text{m/s}^2.$$

The net force required is then

$$\Sigma F = ma = (1500\,\text{kg})(-7.1\,\text{m/s}^2) = -1.1 \times 10^4\,\text{N}.$$

The force must be exerted in the direction *opposite* to the initial velocity, which is what the negative sign means.

NOTE When we assume the acceleration is constant, even though it may not be precisely true, we are determining an "average" acceleration and we obtain an "average" net force (or vice versa).

Newton's second law, like the first law, is valid only in inertial reference frames (Section 4–2). In the noninertial reference frame of an accelerating car, for example, a cup on the dashboard starts sliding—it accelerates—even though the net force on it is zero; thus $\Sigma \vec{\mathbf{F}} = m\vec{\mathbf{a}}$ doesn't work in such an accelerating reference frame.

4–5 Newton's Third Law of Motion

Newton's second law of motion describes quantitatively how forces affect motion. But where, we may ask, do forces come from? Observations suggest that a force applied to any object is always applied *by another object*. A horse pulls a wagon, a person pushes a grocery cart, a hammer pushes on a nail, a magnet attracts a paper clip. In each of these examples, a force is exerted *on* one object, and that force is exerted *by* another object. For example, the force exerted *on* the nail is exerted *by* the hammer.

A force is exerted on an object and is exerted by another object

But Newton realized that things are not so one-sided. True, the hammer exerts a force on the nail (Fig. 4–7). But the nail evidently exerts a force back on the hammer as well, for the hammer's speed is rapidly reduced to zero upon contact. Only a strong force could cause such a rapid deceleration of the hammer. Thus, said Newton, the two objects must be treated on an equal basis. The hammer exerts a force on the nail, and the nail exerts a force back on the hammer. This is the essence of **Newton's third law of motion**:

NEWTON'S THIRD LAW OF MOTION

> **Whenever one object exerts a force on a second object, the second exerts an equal force in the opposite direction on the first.**

This law is sometimes paraphrased as "to every action there is an equal and opposite reaction." This is perfectly valid. But to avoid confusion, it is very important to remember that the "action" force and the "reaction" force are acting on *different* objects.

⚠ C A U T I O N

Action and reaction forces act on different objects

FIGURE 4–7 A hammer striking a nail. The hammer exerts a force on the nail and the nail exerts a force back on the hammer. The latter force decelerates the hammer and brings it to rest.

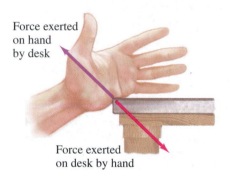

Force exerted on hand by desk

Force exerted on desk by hand

FIGURE 4–8 If your hand pushes against the edge of a desk (the force vector is shown in red), the desk pushes back against your hand (this force vector is shown in a different color, violet, to remind us that this force acts on a different object).

As evidence for the validity of Newton's third law, look at your hand when you push against the edge of a desk, Fig. 4–8. Your hand's shape is distorted, clear evidence that a force is being exerted on it. You can *see* the edge of the desk pressing into your hand. You can even *feel* the desk exerting a force on your hand; it hurts! The harder you push against the desk, the harder the desk pushes back on your hand. (You only feel forces exerted *on* you; when you exert a force on another object, what you feel is that object pushing back on you.)

As another demonstration of Newton's third law, consider the ice skater in Fig. 4–9. There is very little friction between her skates and the ice, so she will move freely if a force is exerted on her. She pushes against the wall; and then *she* starts moving backward. The force she exerts on the wall cannot make *her* start moving, for that force acts on the wall. Something had to exert a force *on her* to start her moving, and that force could only have been exerted by the wall. The force with which the wall pushes on her is, by Newton's third law, equal and opposite to the force she exerts on the wall.

When a person throws a package out of a boat (initially at rest), the boat starts moving in the opposite direction. The person exerts a force on the package. The package exerts an equal and opposite force back on the person, and this force propels the person (and the boat) backward slightly.

Rocket propulsion also is explained using Newton's third law (Fig. 4–10). A common misconception is that rockets accelerate because the gases rushing out the back of the engine push against the ground or the atmosphere. Not true. What happens, instead, is that a rocket exerts a strong force on the gases, expelling them; and the gases exert an equal and opposite force *on the rocket*. It is this latter force that propels the rocket forward—the force exerted *on* the rocket *by* the gases. Thus, a space vehicle is maneuvered in empty space by firing its rockets in the direction opposite to that in which it needs to accelerate. When the rocket pushes on the gases in one direction, the gases push back on the rocket in the opposite direction.

FIGURE 4–9 An example of Newton's third law: when an ice skater pushes against the wall, the wall pushes back and this force causes her to accelerate away.

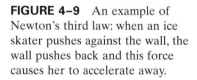

Force on skater   Force on wall

Rocket acceleration

FIGURE 4–10 Another example of Newton's third law: the launch of a rocket. The rocket engine pushes the gases downward, and the gases exert an equal and opposite force upward on the rocket, accelerating it upward. (A rocket does *not* accelerate as a result of its propelling gases pushing against the ground.)

Consider how we walk. A person begins walking by pushing with the foot backward against the ground. The ground then exerts an equal and opposite force forward on the person (Fig. 4–11), and it is this force, *on* the person, that moves the person forward. (If you doubt this, try walking normally where there is no friction, such as on very smooth slippery ice.) In a similar way, a bird flies forward by exerting a backward force on the air, but it is the air pushing forward on the bird's wings that propels the bird forward.

How we can walk

FIGURE 4–11 We can walk forward because, when one foot pushes backward against the ground, the ground pushes forward on that foot. The two forces shown *act on different objects.*

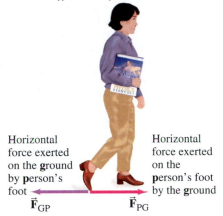

Horizontal force exerted on the ground by person's foot

Horizontal force exerted on the person's foot by the ground

$\vec{F}_{GP}$ $\vec{F}_{PG}$

Inanimate objects can exert a force (due to elasticity)

| **CONCEPTUAL EXAMPLE 4–4** | **What exerts the force on a car?** What makes a car go forward? |

RESPONSE A common answer is that the engine makes the car move forward. But it is not so simple. The engine makes the wheels go around. But if the tires are on slick ice or deep mud, they just spin. Friction is needed. On *solid* ground, the tires push backward against the ground because of friction. By Newton's third law, the ground pushes on the tires in the opposite direction, accelerating the car forward.

We tend to associate forces with active objects such as humans, animals, engines, or a moving object like a hammer. It is often difficult to see how an inanimate object at rest, such as a wall or a desk, or the wall of an ice rink (Fig. 4–9), can exert a force. The explanation is that every material, no matter how hard, is elastic (springy), at least to some degree. A stretched rubber band can exert a force on a wad of paper and accelerate it to fly across the room. Other materials may not stretch as readily as rubber, but they do stretch or compress when a force is applied to them. And just as a stretched rubber band exerts a force, so does a stretched (or compressed) wall, desk, or car fender.

From the examples discussed above, we can see how important it is to remember *on* what object a given force is exerted and *by* what object that force is exerted. A force influences the motion of an object only when it is applied *on* that object. A force exerted *by* an object does not influence that same object; it only influences the other object *on* which it is exerted. Thus, to avoid confusion, the two prepositions *on* and *by* must always be used—and used with care.

One way to keep clear which force acts on which object is to use double subscripts. For example, the force exerted on the **P**erson by the **G**round as the person walks in Fig. 4–11 can be labeled $\vec{F}_{PG}$. And the force exerted on the ground by the person is $\vec{F}_{GP}$. By Newton's third law

$$\vec{F}_{GP} = -\vec{F}_{PG}. \qquad (4\text{–}2)$$

NEWTON'S THIRD LAW OF MOTION

$\vec{F}_{GP}$ and $\vec{F}_{PG}$ have the same magnitude (Newton's third law), and the minus sign reminds us that these two forces are in opposite directions.

Note carefully that the two forces shown in Fig. 4–11 act on different objects—hence we used slightly different colors for the vector arrows representing these forces. These two forces would never appear together in a sum of forces in Newton's second law, $\Sigma\vec{F} = m\vec{a}$. Why not? Because they act on different objects: $\vec{a}$ is the acceleration of one particular object, and $\Sigma\vec{F}$ must include *only* the forces on that *one* object.

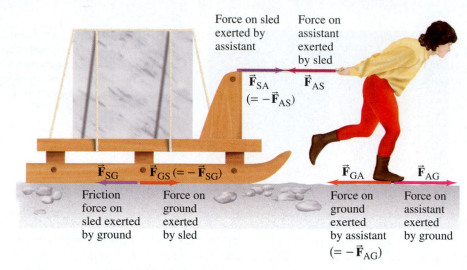

Force on sled exerted by assistant
Force on assistant exerted by sled

$\vec{F}_{SA}$ ($= -\vec{F}_{AS}$) $\vec{F}_{AS}$

$\vec{F}_{SG}$ $\vec{F}_{GS} (= -\vec{F}_{SG})$ $\vec{F}_{GA}$ $\vec{F}_{AG}$

Friction force on sled exerted by ground
Force on ground exerted by sled
Force on ground exerted by assistant ($= -\vec{F}_{AG}$)
Force on assistant exerted by ground

FIGURE 4–12 Example 4–5, showing only horizontal forces. Seventy-year-old Michelangelo has selected a fine block of marble for his next sculpture. Shown here is his assistant pulling it on a sled away from the quarry. Forces on the assistant are shown as red (magenta) arrows. Forces on the sled are purple arrows. Forces acting on the ground are orange arrows. Action–reaction forces that are equal and opposite are labeled by the same subscripts but reversed (such as $\vec{F}_{GA}$ and $\vec{F}_{AG}$) and are of different colors because they act on different objects.

➡ **PROBLEM SOLVING**

A study of Newton's second and third laws

Force on assistant exerted by sled

$\vec{F}_{AS}$

$\vec{F}_{AG}$

Force on assistant exerted by ground

FIGURE 4–13 Example 4–5. The horizontal forces on the assistant.

CONCEPTUAL EXAMPLE 4–5 **Third law clarification.** Michelangelo's assistant has been assigned the task of moving a block of marble using a sled (Fig. 4–12). He says to his boss, "When I exert a forward force on the sled, the sled exerts an equal and opposite force backward. So how can I ever start it moving? No matter how hard I pull, the backward reaction force always equals my forward force, so the net force must be zero. I'll never be able to move this load." Is this a case of a little knowledge being dangerous? Explain.

RESPONSE Yes. Although it is true that the action and reaction forces are equal in magnitude, the assistant has forgotten that they are exerted on different objects. The forward ("action") force is exerted by the assistant on the sled (Fig. 4–12), whereas the backward "reaction" force is exerted by the sled on the assistant. To determine if the *assistant* moves or not, we must consider only the forces *on the assistant* and then apply $\Sigma\vec{F} = m\vec{a}$, where $\Sigma\vec{F}$ is the net force *on the assistant*, $\vec{a}$ is the acceleration of the assistant, and m is the assistant's mass. There are two forces on the assistant that affect his forward motion; they are shown as bright red (magenta) arrows in Figs. 4–12 and 4–13: they are (1) the horizontal force $\vec{F}_{AG}$ exerted on the assistant by the ground (the harder he pushes backward against the ground, the harder the ground pushes forward on him— Newton's third law), and (2) the force $\vec{F}_{AS}$ exerted on the assistant by the sled, pulling backward on him; see Fig. 4–13. If he pushes hard enough on the ground, the force on him exerted by the ground, $\vec{F}_{AG}$, will be larger than the sled pulling back, $\vec{F}_{AS}$, and the assistant accelerates forward (Newton's second law). The sled, on the other hand, accelerates forward when the force on *it* exerted by the assistant is greater than the frictional force exerted backward on it by the ground (that is, when $\vec{F}_{SA}$ has greater magnitude than $\vec{F}_{SG}$ in Fig. 4–12).

Using double subscripts to clarify Newton's third law can become cumbersome, and we won't usually use them in this way. Nevertheless, if there is any confusion in your mind about a given force, go ahead and use them to identify *on* what object and *by* what object the force is exerted. We will usually use a single subscript referring to what exerts the force on the object being discussed.

EXERCISE A A massive truck collides head-on with a small sports car. (*a*) Which vehicle experiences the greater force of impact? (*b*) Which experiences the greater acceleration? (*c*) Which of Newton's laws is useful to obtain the correct answer?

4–6 Weight—the Force of Gravity; and the Normal Force

As we saw in Chapter 2, Galileo claimed that all objects dropped near the surface of the Earth will fall with the same acceleration, $\vec{g}$, if air resistance can be neglected. The force that causes this acceleration is called the *force of gravity* or *gravitational force*. What exerts the gravitational force on an object? It is the Earth, as we will discuss in Chapter 5, and the force acts vertically[†] downward, toward the center of the Earth. Let us apply Newton's second law to an object of mass m falling due to gravity; for the acceleration, $\vec{\mathbf{a}}$, we use the downward acceleration due to gravity, $\vec{\mathbf{g}}$. Thus, the **gravitational force** on an object, $\vec{\mathbf{F}}_G$, can be written as

$$\vec{\mathbf{F}}_G = m\vec{\mathbf{g}}. \qquad (4\text{–}3)$$

Weight = gravitational force

The direction of this force is down toward the center of the Earth. The magnitude of the force of gravity on an object is commonly called the object's **weight**.

In SI units, $g = 9.80\ \text{m/s}^2 = 9.80\ \text{N/kg}$,[‡] so the weight of a 1.00-kg mass on Earth is $1.00\ \text{kg} \times 9.80\ \text{m/s}^2 = 9.80\ \text{N}$. We will mainly be concerned with the weight of objects on Earth, but we note that on the Moon, on other planets, or in space, the weight of a given mass will be different than it is on Earth. For example, on the Moon the acceleration due to gravity is about one-sixth what it is on Earth, and a 1.0-kg mass weighs only 1.7 N. Although we will not use British units, we note that for practical purposes on the Earth, a mass of 1 kg weighs about 2.2 lb. (On the Moon, 1 kg weighs only about 0.4 lb.)

CAUTION

Mass vs. weight

The force of gravity acts on an object when it is falling. When an object is at rest on the Earth, the gravitational force on it does not disappear, as we know if we weigh it on a spring scale. The same force, given by Eq. 4–3, continues to act. Why, then, doesn't the object move? From Newton's second law, the net force on an object that remains at rest is zero. There must be another force on the object to balance the gravitational force. For an object resting on a table, the table exerts this upward force; see Fig. 4–14a. The table is compressed slightly beneath the object, and due to its elasticity, it pushes up on the object as shown. The force exerted by the table is often called a **contact force**, since it occurs when two objects are in contact. (The force of your hand pushing on a cart is also a contact force.) When a contact force acts *perpendicular* to the common surface of contact, it is referred to as the **normal force** ("normal" means perpendicular); hence it is labeled $\vec{\mathbf{F}}_N$ in Fig. 4–14a.

Contact force

Normal force

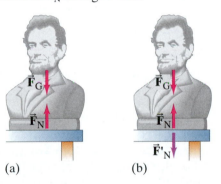

(a)   (b)

FIGURE 4–14 (a) The net force on an object at rest is zero according to Newton's second law. Therefore the downward force of gravity $(\vec{\mathbf{F}}_G)$ on an object must be balanced by an upward force (the normal force $\vec{\mathbf{F}}_N$) exerted by the table in this case. (b) $\vec{\mathbf{F}}'_N$ is the force exerted on the table by the statue and is the reaction force to $\vec{\mathbf{F}}_N$ per Newton's third law. ($\vec{\mathbf{F}}'_N$ is shown in a different color to remind us it acts on a different object.) The reaction to $\vec{\mathbf{F}}_G$ is not shown.

The two forces shown in Fig. 4–14a are both acting on the statue, which remains at rest, so the vector sum of these two forces must be zero (Newton's second law). Hence $\vec{\mathbf{F}}_G$ and $\vec{\mathbf{F}}_N$ must be of equal magnitude and in opposite directions. But they are *not* the equal and opposite forces spoken of in Newton's third law. The action and reaction forces of Newton's third law act on *different objects*, whereas the two forces shown in Fig. 4–14a act on the *same* object. For each of the forces shown in Fig. 4–14a, we can ask, "What is the reaction force?" The upward force, $\vec{\mathbf{F}}_N$, on the statue is exerted by the table. The reaction to this force is a force exerted by the statue downward on the table. It is shown in Fig. 4–14b, where it is labeled $\vec{\mathbf{F}}'_N$. This force, $\vec{\mathbf{F}}'_N$, exerted on the table by the statue, is the reaction force to $\vec{\mathbf{F}}_N$ in accord with Newton's third law. What about the other force on the statue, the force of gravity $\vec{\mathbf{F}}_G$ exerted by the Earth? Can you guess what the reaction is to this force? We will see in Chapter 5 that the reaction force is also a gravitational force, exerted on the Earth by the statue.

CAUTION

*Weight and normal force are **not** action–reaction pairs*

[†]The concept of "vertical" is tied to gravity. The best definition of *vertical* is that it is the direction in which objects fall. A surface that is "horizontal," on the other hand, is a surface on which a round object won't start rolling: gravity has no effect. Horizontal is perpendicular to vertical.

[‡]Since $1\ \text{N} = 1\ \text{kg}\cdot\text{m/s}^2$ (Section 4–4), $1\ \text{m/s}^2 = 1\ \text{N/kg}$.

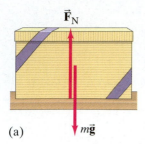

(a)

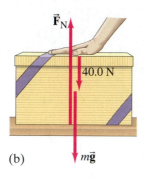

(b)

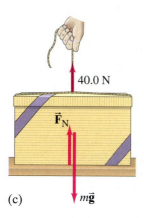

(c)

FIGURE 4–15 Example 4–6.
(a) A 10-kg gift box is at rest on
a table. (b) A person pushes down
on the box with a force of 40.0 N.
(c) A person pulls upward on the
box with a force of 40.0 N. The forces
are all assumed to act along a line;
they are shown slightly displaced in
order to be distinguishable. Only
forces acting on the box are shown.

⚠ **CAUTION**

*The normal force is not
necessarily equal to the weight*

⚠ **CAUTION**

The normal force, $\vec{F}_N$*, is not
necessarily vertical*

EXAMPLE 4–6 **Weight, normal force, and a box.** A friend has given you
a special gift, a box of mass 10.0 kg with a mystery surprise inside. The box is
resting on the smooth (frictionless) horizontal surface of a table (Fig. 4–15a).
(a) Determine the weight of the box and the normal force exerted on it by the
table. (b) Now your friend pushes down on the box with a force of 40.0 N, as in
Fig. 4–15b. Again determine the normal force exerted on the box by the table.
(c) If your friend pulls upward on the box with a force of 40.0 N (Fig. 4–15c),
what now is the normal force exerted on the box by the table?

APPROACH The box is at rest on the table, so the net force on the box in
each case is zero (Newton's second law). The weight of the box equals mg in
all three cases.

SOLUTION (a) The weight of the box is $mg = (10.0\,\text{kg})(9.80\,\text{m/s}^2) = 98.0\,\text{N}$,
and this force acts downward. The only other force on the box is the normal
force exerted upward on it by the table, as shown in Fig. 4–15a. We chose the
upward direction as the positive y direction; then the net force ΣF_y on the box
is $\Sigma F_y = F_N - mg$. The box is at rest, so the net force on it must be zero
(Newton's second law, $\Sigma F_y = ma_y$, and $a_y = 0$). Thus

$$\Sigma F_y = F_N - mg = 0,$$

and we have in this case

$$F_N = mg.$$

The normal force on the box, exerted by the table, is 98.0 N upward, and has
magnitude equal to the box's weight.

(b) Your friend is pushing down on the box with a force of 40.0 N. So instead
of only two forces acting on the box, now there are three forces acting on the
box, as shown in Fig. 4–15b. The weight of the box is still $mg = 98.0\,\text{N}$. The
net force is $\Sigma F_y = F_N - mg - 40.0\,\text{N}$, and is equal to zero because the box
remains at rest. Thus, since $a = 0$, Newton's second law gives

$$\Sigma F_y = F_N - mg - 40.0\,\text{N} = 0.$$

We solve this equation for the normal force:

$$F_N = mg + 40.0\,\text{N} = 98.0\,\text{N} + 40.0\,\text{N} = 138.0\,\text{N},$$

which is greater than in (a). The table pushes back with more force when a person
pushes down on the box. The normal force is not always equal to the weight!

(c) The box's weight is still 98.0 N and acts downward. The force exerted by
your friend and the normal force both act upward (positive direction), as
shown in Fig. 4–15c. The box doesn't move since your friend's upward force is
less than the weight. The net force, again set to zero in Newton's second law
because $a = 0$, is

$$\Sigma F_y = F_N - mg + 40.0\,\text{N} = 0,$$

so

$$F_N = mg - 40.0\,\text{N} = 98.0\,\text{N} - 40.0\,\text{N} = 58.0\,\text{N}.$$

The table does not push against the full weight of the box because of the
upward pull exerted by your friend.

NOTE The weight of the box ($= mg$) does not change as a result of your
friend's push or pull. Only the normal force is affected.

Recall that the normal force is elastic in origin (the table in Fig. 4–15 sags
slightly under the weight of the box). The normal force in Example 4–6 is
vertical, perpendicular to the horizontal table. The normal force is not always
vertical, however. When you push against a vertical wall, for example, the
normal force with which the wall pushes back on you is horizontal. For an
object on a plane inclined at an angle to the horizontal, such as a skier or car on
a hill, the normal force acts perpendicular to the plane and so is not vertical.

EXAMPLE 4–7 **Accelerating the box.** What happens when a person pulls upward on the box in Example 4–6 (*c*) with a force equal to, or greater than, the box's weight, say $F_P = 100.0\,\text{N}$ rather than the 40.0 N shown in Fig. 4–15c?

APPROACH We can start just as in Example 4–6, but be ready for a surprise.

SOLUTION The net force on the box is

$$\Sigma F_y = F_N - mg + F_P$$
$$= F_N - 98.0\,\text{N} + 100.0\,\text{N},$$

and if we set this equal to zero (thinking the acceleration might be zero), we would get $F_N = -2.0\,\text{N}$. This is nonsense, since the negative sign implies F_N points downward, and the table surely cannot *pull* down on the box (unless there's glue on the table). The least F_N can be is zero, which it will be in this case. What really happens here is that the box accelerates upward because the net force is not zero. The net force (setting the normal force $F_N = 0$) is

$$\Sigma F_y = F_P - mg = 100.0\,\text{N} - 98.0\,\text{N}$$
$$= 2.0\,\text{N}$$

upward. See Fig. 4–16. We apply Newton's second law and see that the box moves upward with an acceleration

$$a_y = \frac{\Sigma F_y}{m} = \frac{2.0\,\text{N}}{10.0\,\text{kg}}$$
$$= 0.20\,\text{m/s}^2.$$

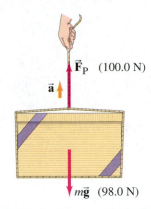

FIGURE 4–16 Example 4–7. The box accelerates upward because $F_P > mg$.

An Additional Example

EXAMPLE 4–8 **Apparent weight loss.** A 65-kg woman descends in an elevator that briefly accelerates at 0.20*g* downward when leaving a floor. She stands on a scale that reads in kg. (*a*) During this acceleration, what is her weight and what does the scale read? (*b*) What does the scale read when the elevator descends at a constant speed of 2.0 m/s?

APPROACH Figure 4–17 shows all the forces that act on the woman (and *only* those that act on her). The direction of the acceleration is downward, which we take as positive.

SOLUTION (*a*) From Newton's second law,

$$\Sigma F = ma$$
$$mg - F_N = m\,(0.20g).$$

We solve for F_N:

$$F_N = mg - 0.20mg = 0.80mg,$$

and it acts upward. The normal force $\vec{F}_N$ is the force the scale exerts on the person, and is equal and opposite to the force she exerts on the scale: $F'_N = 0.80mg$ downward. Her weight (force of gravity on her) is still $mg = (65\,\text{kg})(9.8\,\text{m/s}^2) = 640\,\text{N}$. But the scale, needing to exert a force of only $0.80mg$, will give a reading of $0.80m = 52\,\text{kg}$.
(*b*) Now there is no acceleration, $a = 0$, so by Newton's second law, $mg - F_N = 0$ and $F_N = mg$. The scale reads her true mass of 65 kg.

NOTE The scale in (*a*) may give a reading of 52 kg (as an "apparent mass"), but her mass doesn't change as a result of the acceleration: it stays at 65 kg.

FIGURE 4–17 Example 4–8.

4–7 Solving Problems with Newton's Laws: Free-Body Diagrams

Newton's second law tells us that the acceleration of an object is proportional to the *net force* acting on the object. The **net force**, as mentioned earlier, is the *vector sum* of all forces acting on the object. Indeed, extensive experiments have shown that forces do add together as vectors precisely according to the rules we developed in Chapter 3. For example, in Fig. 4–18, two forces of equal magnitude (100 N each) are shown acting on an object at right angles to each other. Intuitively, we can see that the object will start moving at a 45° angle and thus the net force acts at a 45° angle. This is just what the rules of vector addition give. From the theorem of Pythagoras, the magnitude of the resultant force is $F_R = \sqrt{(100\,\text{N})^2 + (100\,\text{N})^2} = 141\,\text{N}$.

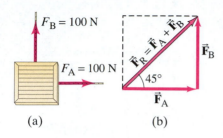

FIGURE 4–18 (a) Two forces, $\vec{F}_A$ and $\vec{F}_B$, exerted by workers A and B, act on a crate. (b) The sum, or resultant, of $\vec{F}_A$ and $\vec{F}_B$ is $\vec{F}_R$.

FIGURE 4–19 Example 4–9: Two force vectors act on a boat.

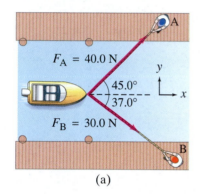

(a)

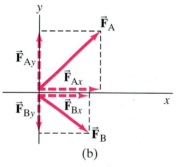

(b)

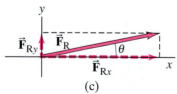

(c)

EXAMPLE 4–9 Adding force vectors. Calculate the sum of the two forces exerted on the boat by workers A and B in Fig. 4–19a.

APPROACH We add force vectors like any other vectors as described in Chapter 3. The first step is to choose an xy coordinate system, as in Fig. 4–19a, and then resolve vectors into their components.

SOLUTION The two force vectors are shown resolved into components in Fig. 4–19b. We add the forces using the method of components. The components of $\vec{F}_A$ are

$$F_{Ax} = F_A \cos 45.0° = (40.0\,\text{N})(0.707) = 28.3\,\text{N},$$
$$F_{Ay} = F_A \sin 45.0° = (40.0\,\text{N})(0.707) = 28.3\,\text{N}.$$

The components of $\vec{F}_B$ are

$$F_{Bx} = +F_B \cos 37.0° = +(30.0\,\text{N})(0.799) = +24.0\,\text{N},$$
$$F_{By} = -F_B \sin 37.0° = -(30.0\,\text{N})(0.602) = -18.1\,\text{N}.$$

F_{By} is negative because it points along the negative y axis. The components of the resultant force are (see Fig. 4–19c)

$$F_{Rx} = F_{Ax} + F_{Bx} = 28.3\,\text{N} + 24.0\,\text{N} = 52.3\,\text{N},$$
$$F_{Ry} = F_{Ay} + F_{By} = 28.3\,\text{N} - 18.1\,\text{N} = 10.2\,\text{N}.$$

To find the magnitude of the resultant force, we use the Pythagorean theorem:

$$F_R = \sqrt{F_{Rx}^2 + F_{Ry}^2} = \sqrt{(52.3)^2 + (10.2)^2}\,\text{N} = 53.3\,\text{N}.$$

The only remaining question is the angle θ that the net force $\vec{F}_R$ makes with the x axis. We use:

$$\tan\theta = \frac{F_{Ry}}{F_{Rx}} = \frac{10.2\,\text{N}}{52.3\,\text{N}} = 0.195,$$

and $\tan^{-1}(0.195) = 11.0°$. The net force on the boat has magnitude 53.3 N and acts at an 11.0° angle to the x axis.

➡ **PROBLEM SOLVING**

Free-body diagram

Identifying every force

When solving problems involving Newton's laws and force, it is very important to draw a diagram showing all the forces acting *on* each object involved. Such a diagram is called a **free-body diagram**, or **force diagram**: choose one object, and draw an arrow to represent each force acting on it. Include *every* force acting on that object. Do not show forces that the chosen object exerts on *other* objects. To help you identify each and every force that is exerted on your chosen object, ask yourself what other objects could exert a force on it. If your problem involves more than one object, a separate free-body diagram is needed for each object.

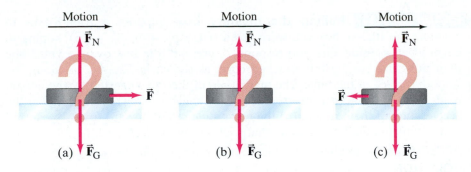

FIGURE 4–20 Example 4–10. Which is the correct free-body diagram for a hockey puck sliding across frictionless ice?

CONCEPTUAL EXAMPLE 4–10 **The hockey puck.** A hockey puck is sliding at constant velocity across a flat horizontal ice surface that is assumed to be frictionless. Which of the sketches in Fig. 4–20 is the correct free-body diagram for this puck? What would your answer be if the puck slowed down?

RESPONSE Did you choose (*a*)? If so, can you answer the question: what exerts the horizontal force labeled $\vec{F}$ on the puck? If you say that it is the force needed to maintain the motion, ask yourself: what exerts this force? Remember that another object must exert any force—and there simply isn't any possibility here. Therefore, (*a*) is wrong. Besides, the force $\vec{F}$ in Fig. 4–20a would give rise to an acceleration by Newton's second law. It is (*b*) that is correct, as long as there is no friction. No net force acts on the puck, and the puck slides at constant velocity across the ice.

In the real world, where even smooth ice exerts at least a tiny friction force, then (*c*) is the correct answer. The tiny friction force is in the direction opposite to the motion, and the puck's velocity decreases, even if very slowly.

Here now is a brief summary of how to approach solving problems involving Newton's laws.

PROBLEM SOLVING **Newton's Laws; Free-Body Diagrams**

1. **Draw a sketch** of the situation.

2. Consider only one object (at a time), and draw a **free-body diagram** for that object, showing *all* the forces acting *on* that object. Include any unknown forces that you have to solve for. Do not show any forces that the chosen object exerts on other objects. Draw the arrow for each force vector reasonably accurately for direction and magnitude. Label each force, including forces you must solve for, as to its source (gravity, person, friction, and so on).

   If several objects are involved, draw a free-body diagram for each object *separately*, showing all the forces acting *on that object* (and *only* forces acting on that object). For each (and every) force, you must be clear about: *on* what object that force

acts, and *by* what object that force is exerted. Only forces acting *on* a given object can be included in $\Sigma\vec{F} = m\vec{a}$ for that object.

3. Newton's second law involves vectors, and it is usually important to **resolve vectors** into components. **Choose** *x* and *y* **axes** in a way that simplifies the calculation. For example, it often saves work if you choose one coordinate axis to be in the direction of the acceleration.

4. For each object, **apply Newton's second law** to the *x* and *y* components separately. That is, the *x* component of the net force on that object is related to the *x* component of that object's acceleration: $\Sigma F_x = ma_x$, and similarly for the *y* direction.

5. **Solve** the equation or equations for the unknown(s).

This Problem Solving Box should not be considered a prescription. Rather it is a summary of things to do that will start you thinking and getting involved in the problem at hand.

When we are concerned only about translational motion, all the forces on a given object can be drawn as acting at the center of the object, thus treating the object as a point particle. However, for problems involving rotation or statics, the place *where* each force acts is also important, as we shall see in Chapters 8 and 9.

Force arrow placement on diagrams

In the Examples that follow, we assume that all surfaces are very smooth so that friction can be ignored. (Friction, and Examples using it, are discussed in Section 4–8.)

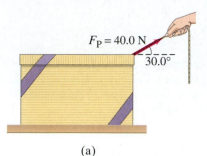

$F_P = 40.0$ N

$30.0°$

(a)

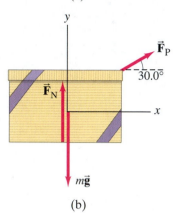

$\vec{F}_P$

$30.0°$

$\vec{F}_N$

$m\vec{g}$

(b)

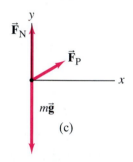

$\vec{F}_N$

$\vec{F}_P$

$m\vec{g}$

(c)

FIGURE 4–21 (a) Pulling the box, Example 4–11; (b) is the free-body diagram for the box, and (c) is the free-body diagram considering all the forces to act at a point (translational motion only, which is what we have here).

EXAMPLE 4–11 **Pulling the mystery box.** Suppose a friend asks to examine the 10.0-kg box you were given (Example 4–6, Fig. 4–15), hoping to guess what is inside; and you respond, "Sure, pull the box over to you." She then pulls the box by the attached cord, as shown in Fig. 4–21a, along the smooth surface of the table. The magnitude of the force exerted by the person is $F_P = 40.0$ N, and it is exerted at a 30.0° angle as shown. Calculate (a) the acceleration of the box, and (b) the magnitude of the upward force F_N exerted by the table on the box. Assume that friction can be neglected.

APPROACH We follow the Problem Solving Box on the previous page.

SOLUTION

1. **Draw a sketch**: The situation is shown in Fig. 4–21a; it shows the box and the force applied by the person, F_P.

2. **Free-body diagram**: Figure 4–21b shows the free-body diagram of the box. To draw it correctly, we show *all* the forces acting on the box and *only* the forces acting on the box. They are: the force of gravity $m\vec{g}$; the normal force exerted by the table $\vec{F}_N$; and the force exerted by the person $\vec{F}_P$. We are interested only in translational motion, so we can show the three forces acting at a point, Fig. 4–21c.

3. **Choose axes and resolve vectors**: We expect the motion to be horizontal, so we choose the x axis horizontal and the y axis vertical. The pull of 40.0 N has components

$$F_{Px} = (40.0 \text{ N})(\cos 30.0°) = (40.0 \text{ N})(0.866) = 34.6 \text{ N},$$
$$F_{Py} = (40.0 \text{ N})(\sin 30.0°) = (40.0 \text{ N})(0.500) = 20.0 \text{ N}.$$

In the horizontal (x) direction, $\vec{F}_N$ and $m\vec{g}$ have zero components. Thus the horizontal component of the net force is F_{Px}.

4. (a) **Apply Newton's second law** to determine the x component of the acceleration:

$$F_{Px} = ma_x.$$

5. (a) **Solve**:

$$a_x = \frac{F_{Px}}{m} = \frac{(34.6 \text{ N})}{(10.0 \text{ kg})} = 3.46 \text{ m/s}^2.$$

The acceleration of the box is 3.46 m/s² to the right.

(b) Next we want to find F_N.

4. (b) **Apply Newton's second law** to the vertical (y) direction, with upward as positive:

$$\Sigma F_y = ma_y$$
$$F_N - mg + F_{Py} = ma_y.$$

5. (b) **Solve**: We have $mg = (10.0 \text{ kg})(9.80 \text{ m/s}^2) = 98.0$ N and, from point 3 above, $F_{Py} = 20.0$ N. Furthermore, since $F_{Py} < mg$, the box does not move vertically, so $a_y = 0$. Thus

$$F_N - 98.0 \text{ N} + 20.0 \text{ N} = 0,$$

so

$$F_N = 78.0 \text{ N}.$$

NOTE F_N is less than mg: the table does not push against the full weight of the box because part of the pull exerted by the person is in the upward direction.

Tension in a Flexible Cord

When a flexible cord pulls on an object, the cord is said to be under **tension**, and the force it exerts on the object is the tension F_T. If the cord has negligible mass, the force exerted at one end is transmitted undiminished to each adjacent piece of cord along the entire length to the other end. Why? Because $\Sigma \vec{F} = m\vec{a} = 0$ for the cord if the cord's mass m is zero (or negligible) no matter what $\vec{a}$ is. Hence the forces pulling on the cord at its two ends must add up to zero (F_T and $-F_T$). Note that flexible cords and strings can only pull. They can't push because they bend.

FIGURE 4–22 Example 4–12. (a) Two boxes, A and B, are connected by a cord. A person pulls horizontally on box A with force $F_P = 40.0$ N. (b) Free-body diagram for box A. (c) Free-body diagram for box B.

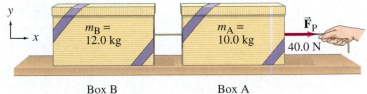

Box B   Box A

(a)

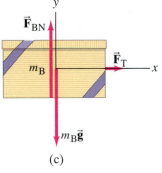

(b)

Our next Example involves two boxes connected by a cord. We can refer to this group of objects as a system. A *system* is any group of one or more objects we choose to consider and study.

EXAMPLE 4–12 **Two boxes connected by a cord.** Two boxes, A and B, are connected by a lightweight cord and are resting on a smooth (frictionless) table. The boxes have masses of 12.0 kg and 10.0 kg. A horizontal force F_P of 40.0 N is applied to the 10.0-kg box, as shown in Fig. 4–22a. Find (a) the acceleration of each box, and (b) the tension in the cord connecting the boxes.

APPROACH We streamline our approach by not listing each step. We have two boxes so we need to draw a free-body diagram for each box. To draw them correctly, we must consider the forces on *each* box by itself, so that Newton's second law can be applied to each. The person exerts a force F_P on box A. Box A exerts a force F_T on the connecting cord, and the cord exerts an opposite but equal magnitude force F_T back on box A (Newton's third law). These two horizontal forces on box A are shown in Fig. 4–22b, along with the force of gravity $m_A \vec{g}$ downward and the normal force $\vec{F}_{AN}$ exerted upward by the table. The cord is light, so we neglect its mass. The tension at each end of the cord is thus the same. Hence the cord exerts a force F_T on the second box. Figure 4–22c shows the forces on box B, which are $\vec{F}_T$, $m_B \vec{g}$, and the normal force $\vec{F}_{BN}$. There will be only horizontal motion. We take the positive x axis to the right.

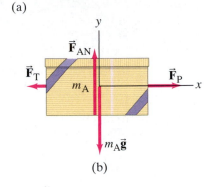

(c)

SOLUTION (a) We apply $\Sigma F_x = m a_x$ to box A:

$$\Sigma F_x = F_P - F_T = m_A a_A. \qquad \text{[box A]}$$

For box B, the only horizontal force is F_T, so

$$\Sigma F_x = F_T = m_B a_B. \qquad \text{[box B]}$$

The boxes are connected, and if the cord remains taut and doesn't stretch, then the two boxes will have the same acceleration a. Thus $a_A = a_B = a$. We are given $m_A = 10.0$ kg and $m_B = 12.0$ kg. We can add the two equations above to eliminate an unknown (F_T) and obtain

$$(m_A + m_B)a = F_P - F_T + F_T = F_P$$

or

$$a = \frac{F_P}{m_A + m_B} = \frac{40.0 \text{ N}}{22.0 \text{ kg}} = 1.82 \text{ m/s}^2.$$

This is what we sought.

Alternate Solution We would have obtained the same result had we considered a single system, of mass $m_A + m_B$, acted on by a net horizontal force equal to F_P. (The tension forces F_T would then be considered internal to the system as a whole, and summed together would make zero contribution to the net force on the *whole* system.)

(b) From the equation above for box B ($F_T = m_B a_B$), the tension in the cord is

$$F_T = m_B a = (12.0 \text{ kg})(1.82 \text{ m/s}^2) = 21.8 \text{ N}.$$

Thus, F_T is less than $F_P(= 40.0 \text{ N})$, as we expect, since F_T acts to accelerate only m_B.

NOTE It might be tempting to say that the force the person exerts, F_P, acts not only on box A but also on box B. It doesn't. F_P acts only on box A. It affects box B via the tension in the cord, F_T, which acts on box B and accelerates it.

➡ **PROBLEM SOLVING**

An alternate analysis

⚠ **CAUTION**

For any object, use only the forces on that object in calculating $\Sigma F = ma$

Additional Examples

Here are some more worked-out Examples to give you practice in solving a wide range of Problems.

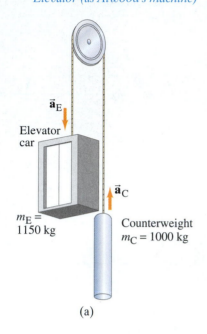

(a)

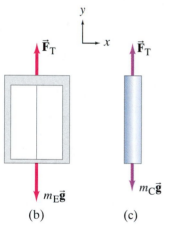

(b)   (c)

FIGURE 4–23 Example 4–13. (a) Atwood's machine in the form of an elevator–counterweight system. (b) and (c) Free-body diagrams for the two objects.

➡ **PROBLEM SOLVING**

Check your result by seeing if it works in situations where the answer is easily guessed

EXAMPLE 4–13 Elevator and counterweight (Atwood's machine). A system of two objects suspended over a pulley by a flexible cable, as shown in Fig. 4–23a, is sometimes referred to as an *Atwood's machine*. Consider the real-life application of an elevator (m_E) and its counterweight (m_C). To minimize the work done by the motor to raise and lower the elevator safely, m_E and m_C are similar in mass. We leave the motor out of the system for this calculation, and assume that the cable's mass is negligible and that the mass of the pulley, as well as any friction, is small and ignorable. These assumptions ensure that the tension F_T in the cable has the same magnitude on both sides of the pulley. Let the mass of the counterweight be $m_C = 1000$ kg. Assume the mass of the empty elevator is 850 kg, and its mass when carrying four passengers is $m_E = 1150$ kg. For the latter case $(m_E = 1150$ kg$)$, calculate (a) the acceleration of the elevator and (b) the tension in the cable.

APPROACH Again we have two objects, and we will need to apply Newton's second law to each of them separately. Each mass has two forces acting on it: gravity downward and the cable tension pulling upward, $\vec{F}_T$. Figures 4–23b and c show the free-body diagrams for the elevator (m_E) and for the counterweight (m_C). The elevator, being the heavier, will accelerate downward, whereas the counterweight will accelerate upward. The magnitudes of their accelerations will be equal (we assume the cable doesn't stretch). For the counterweight, $m_C g = (1000$ kg$)(9.80$ m/s$^2) = 9800$ N, so F_T must be greater than 9800 N (in order that m_C will accelerate upward). For the elevator, $m_E g = (1150$ kg$)(9.80$ m/s$^2) = 11,300$ N, which must have greater magnitude than F_T so that m_E accelerates downward. Thus our calculation must give F_T between 9800 N and 11,300 N.

SOLUTION (a) To find F_T as well as the acceleration a, we apply Newton's second law, $\Sigma F = ma$, to each object. We take upward as the positive y direction for both objects. With this choice of axes, $a_C = a$ because m_C accelerates upward, and $a_E = -a$ because m_E accelerates downward. Thus

$$F_T - m_E g = m_E a_E = -m_E a$$
$$F_T - m_C g = m_C a_C = +m_C a.$$

We can subtract the first equation from the second to get

$$(m_E - m_C)g = (m_E + m_C)a,$$

where a is now the only unknown. We solve this for a:

$$a = \frac{m_E - m_C}{m_E + m_C} g = \frac{1150 \text{ kg} - 1000 \text{ kg}}{1150 \text{ kg} + 1000 \text{ kg}} g = 0.070 \, g = 0.68 \text{ m/s}^2.$$

The elevator (m_E) accelerates downward (and the counterweight m_C upward) at $a = 0.070g = 0.68$ m/s^2.

(b) The tension in the cable F_T can be obtained from either of the two $\Sigma F = ma$ equations, setting $a = 0.070g = 0.68$ m/s^2:

$$F_T = m_E g - m_E a = m_E(g - a)$$
$$= 1150 \text{ kg} (9.80 \text{ m/s}^2 - 0.68 \text{ m/s}^2) = 10,500 \text{ N},$$

$$F_T = m_C g + m_C a = m_C(g + a)$$
$$= 1000 \text{ kg} (9.80 \text{ m/s}^2 + 0.68 \text{ m/s}^2) = 10,500 \text{ N},$$

which are consistent. As predicted, our result lies between 9800 N and 11,300 N.

NOTE We can check our equation for the acceleration a in this Example by noting that if the masses were equal $(m_E = m_C)$, then our equation above for a would give $a = 0$, as we should expect. Also, if one of the masses is zero (say, $m_C = 0$), then the other mass $(m_E \neq 0)$ would be predicted by our equation to accelerate at $a = g$, again as expected.

CONCEPTUAL EXAMPLE 4–14 | **The advantage of a pulley.** A mover is trying to lift a piano (slowly) up to a second-story apartment (Fig. 4–24). He is using a rope looped over two pulleys as shown. What force must he exert on the rope to slowly lift the piano's 2000-N weight?

RESPONSE The magnitude of the tension force F_T within the rope is the same at any point along the rope if we assume we can ignore its mass. First notice the forces acting on the lower pulley at the piano. The weight of the piano pulls down on the pulley via a short cable. The tension in the rope, looped through this pulley, pulls up *twice*, once on each side of the pulley. Let us apply Newton's second law to the pulley–piano combination (of mass m):

$$2F_T - mg = ma.$$

To move the piano with constant speed (set $a = 0$ in this equation) thus requires a tension in the rope, and hence a pull on the rope, of $F_T = mg/2$. The mover can exert a force equal to half the piano's weight. We say the pulley has given a **mechanical advantage** of 2, since without the pulley the mover would have to exert twice the force.

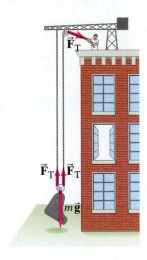

FIGURE 4–24 Example 4–14.

EXAMPLE 4–15 | **Getting the car out of the mud.** Finding her car stuck in the mud, a bright graduate of a good physics course ties a strong rope to the back bumper of the car, and the other end to a boulder, as shown in Fig. 4–25a. She pushes at the midpoint of the rope with her maximum effort, which she estimates to be a force $F_P \approx 300 \, \text{N}$. The car just begins to budge with the rope at an angle θ (see the Figure), which she estimates to be 5°. With what force is the rope pulling on the car? Neglect the mass of the rope.

APPROACH First, note that the tension in a rope is always along the rope. Any component perpendicular to the rope would cause the rope to bend or buckle (as it does here where $\vec{F}_P$ acts)—in other words, a rope can support a tension force only along its length. Let $\vec{F}_{BR}$ and $\vec{F}_{CR}$ be the forces on the boulder and on the car, exerted via the tension in the rope, as shown in Fig. 4–25a. Let us choose to look at the forces on the tiny section of rope where she pushes. The free-body diagram is shown in Fig. 4–25b, which shows $\vec{F}_P$ as well as the tensions in the rope (note that we have used Newton's third law: $\vec{F}_{RB} = -\vec{F}_{BR}$, $\vec{F}_{RC} = -\vec{F}_{CR}$). At the moment the car budges, the acceleration is still essentially zero, so $\vec{a} = 0$.

SOLUTION For the x component of $\Sigma\vec{F} = m\vec{a} = 0$ on that small section of rope (Fig. 4–25b), we have

$$\Sigma F_x = F_{RB} \cos \theta - F_{RC} \cos \theta = 0.$$

Hence $F_{RB} = F_{RC}$, and these forces represent the magnitude of the tension in the rope, call it F_T; then we can write $F_T = F_{RB} = F_{RC}$. In the y direction, the forces acting are F_P, and the components of F_{RB} and F_{RC} that point in the negative y direction (each equal to $F_T \sin \theta$). So for the y component of $\Sigma\vec{F} = m\vec{a}$, we have

$$\Sigma F_y = F_P - 2F_T \sin \theta = 0.$$

We solve this for F_T, and insert $\theta = 5°$ and $F_P \approx 300 \, \text{N}$, which were given:

$$F_T = \frac{F_P}{2 \sin \theta} \approx \frac{300 \, \text{N}}{2 \sin 5°} \approx 1700 \, \text{N}.$$

When our physics graduate exerted a force of 300 N on the rope, the force produced on the car was 1700 N. She was able to magnify her effort almost six times using this technique!

NOTE Notice the symmetry of the problem, which ensures that $F_{RB} = F_{RC}$.

NOTE Compare Figs. 4–25a and b. Notice that we cannot write down Newton's second law using Fig. 4–25a because the force vectors are not acting on the same object. It is only by choosing a tiny section of rope as our object, and using Newton's third law (in this case, the boulder and the car pulling back on the rope with forces F_{RB} and F_{RC}), that all forces apply to the same object.

How to get out of the mud

FIGURE 4–25 Example 4–15. (a) Getting a car out of the mud, showing the forces on the boulder, on the car, and exerted by the person. (b) The free-body diagram: forces on a small segment of rope.

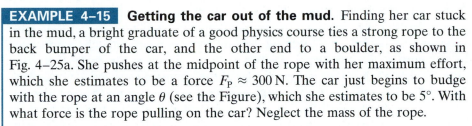

(a)

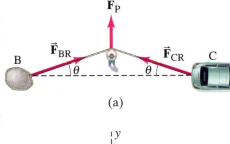

(b)

➡ **PROBLEM SOLVING**

Use any symmetry present to simplify a problem

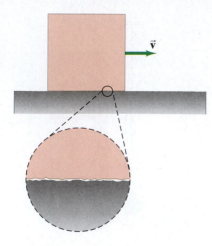

FIGURE 4–26 An object moving to the right on a table or floor. The two surfaces in contact are rough, at least on a microscopic scale.

Kinetic friction

$\vec{\mathbf{F}}_{fr} \perp \vec{\mathbf{F}}_N$

FIGURE 4–27 When an object is pulled by an applied force $(\vec{\mathbf{F}}_A)$ along a surface, the force of friction $\vec{\mathbf{F}}_{fr}$ opposes the motion. The magnitude of $\vec{\mathbf{F}}_{fr}$ is proportional to the magnitude of the normal force (F_N).

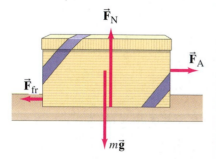

4–8 Problems Involving Friction, Inclines

Friction

Until now we have ignored friction, but it must be taken into account in most practical situations. Friction exists between two solid surfaces because even the smoothest looking surface is quite rough on a microscopic scale, Fig. 4–26. When we try to slide an object across another surface, these microscopic bumps impede the motion. Exactly what is happening at the microscopic level is not yet fully understood. It is thought that the atoms on a bump of one surface may come so close to the atoms of the other surface that attractive electric forces between the atoms can "bond" as a tiny weld between the two surfaces. Sliding an object across a surface is often jerky, perhaps due to the making and breaking of these bonds. Even when a round object rolls across a surface, there is still some friction, called *rolling friction*, although it is generally much less than when an object slides across a surface. We focus now on sliding friction, which is usually called **kinetic friction** (*kinetic* is from the Greek for "moving").

When an object slides along a rough surface, the force of kinetic friction acts opposite to the direction of the object's velocity. The magnitude of the force of kinetic friction depends on the nature of the two sliding surfaces. For given surfaces, experiment shows that the friction force is approximately proportional to the *normal force* between the two surfaces, which is the force that either object exerts on the other, perpendicular to their common surface of contact (see Fig. 4–27). The force of friction between hard surfaces in many cases depends very little on the total surface area of contact; that is, the friction force on this book is roughly the same whether it is being slid on its wide face or on its spine, assuming the surfaces have the same smoothness. We consider a simple model of friction in which we make this assumption that the friction force is independent of area. Then we write the proportionality between the friction force F_{fr} and the normal force F_N as an equation by inserting a constant of proportionality, μ_k:

$$F_{fr} = \mu_k F_N.$$

This relation is not a fundamental law; it is an experimental relation between the magnitude of the friction force F_{fr}, which acts parallel to the two surfaces, and the magnitude of the normal force F_N, which acts perpendicular to the surfaces. It is *not* a vector equation since the two forces have directions perpendicular to one another. The term μ_k is called the *coefficient of kinetic friction*, and its value depends on the nature of the two surfaces. Measured values for a variety of surfaces are given in Table 4–2. These are only approximate, however, since μ depends on whether the surfaces are wet or dry, on how much they have been sanded or rubbed, if any burrs remain, and other such factors. But μ_k is roughly independent of the sliding speed, as well as the area in contact.

TABLE 4–2 Coefficients of Friction[†]

Surfaces	Coefficient of Static Friction, μ_s	Coefficient of Kinetic Friction, μ_k
Wood on wood	0.4	0.2
Ice on ice	0.1	0.03
Metal on metal (lubricated)	0.15	0.07
Steel on steel (unlubricated)	0.7	0.6
Rubber on dry concrete	1.0	0.8
Rubber on wet concrete	0.7	0.5
Rubber on other solid surfaces	1–4	1
Teflon® on Teflon in air	0.04	0.04
Teflon on steel in air	0.04	0.04
Lubricated ball bearings	<0.01	<0.01
Synovial joints (in human limbs)	0.01	0.01

[†] Values are approximate and intended only as a guide.

What we have been discussing up to now is *kinetic friction*, when one object slides over another. There is also **static friction**, which refers to a force parallel to the two surfaces that can arise even when they are not sliding. Suppose an object such as a desk is resting on a horizontal floor. If no horizontal force is exerted on the desk, there also is no friction force. But now suppose you try to push the desk, and it doesn't move. You are exerting a horizontal force, but the desk isn't moving, so there must be another force on the desk keeping it from moving (the net force is zero on an object that doesn't move). This is the force of *static friction* exerted by the floor on the desk. If you push with a greater force without moving the desk, the force of static friction also has increased. If you push hard enough, the desk will eventually start to move, and kinetic friction takes over. At this point, you have exceeded the maximum force of static friction, which is given by $F_{fr}(\text{max}) = \mu_s F_N$, where μ_s is the *coefficient of static friction* (Table 4–2). Since the force of static friction can vary from zero to this maximum value, we write

$$F_{fr} \leq \mu_s F_N.$$

Static friction

You may have noticed that it is often easier to keep a heavy object sliding than it is to start it sliding in the first place. This is consistent with μ_s generally being greater than μ_k (see Table 4–2).

EXAMPLE 4–16 **Friction: static and kinetic.**

Our 10.0-kg mystery box rests on a horizontal floor. The coefficient of static friction is $\mu_s = 0.40$ and the coefficient of kinetic friction is $\mu_k = 0.30$. Determine the force of friction, F_{fr}, acting on the box if a horizontal external applied force F_A is exerted on it of magnitude: (*a*) 0, (*b*) 10 N, (*c*) 20 N, (*d*) 38 N, and (*e*) 40 N.

APPROACH We don't know, right off, if we are dealing with static friction or kinetic friction, nor if the box remains at rest or accelerates. We need to draw a free-body diagram, and then determine in each case whether or not the box will move, by using Newton's second law. The forces on the box are gravity $m\vec{g}$, the normal force exerted by the floor $\vec{F}_N$, the horizontal applied force $\vec{F}_A$, and the friction force $\vec{F}_{fr}$, as shown in Fig. 4–27.

SOLUTION The free-body diagram of the box is shown in Fig. 4–27. In the vertical direction there is no motion, so Newton's second law in the vertical direction gives $\Sigma F_y = ma_y = 0$, which tells us $F_N - mg = 0$. Hence the normal force is

$$F_N = mg = (10.0\,\text{kg})(9.8\,\text{m/s}^2) = 98\,\text{N}.$$

(*a*) Since no external force F_A is applied in this first case, the box doesn't move, and $F_{fr} = 0$.

(*b*) The force of static friction will oppose any applied force up to a maximum of

$$\mu_s F_N = (0.40)(98\,\text{N}) = 39\,\text{N}.$$

When the applied force is $F_A = 10\,\text{N}$, the box will not move. Since $\Sigma F_x = F_A - F_{fr} = 0$, then $F_{fr} = 10\,\text{N}$.

(*c*) An applied force of 20 N is also not sufficient to move the box. Thus $F_{fr} = 20\,\text{N}$ to balance the applied force.

(*d*) The applied force of 38 N is still not quite large enough to move the box; so the friction force has now increased to 38 N to keep the box at rest.

(*e*) A force of 40 N will start the box moving since it exceeds the maximum force of static friction, $\mu_s F_N = (0.40)(98\,\text{N}) = 39\,\text{N}$. Instead of static friction, we now have kinetic friction, and its magnitude is

$$F_{fr} = \mu_k F_N = (0.30)(98\,\text{N}) = 29\,\text{N}.$$

There is now a net (horizontal) force on the box of magnitude $F = 40\,\text{N} - 29\,\text{N} = 11\,\text{N}$, so the box will accelerate at a rate

$$a_x = \frac{\Sigma F}{m} = \frac{11\,\text{N}}{10\,\text{kg}} = 1.1\,\text{m/s}^2$$

as long as the applied force is 40 N. Figure 4–28 shows a graph that summarizes this Example.

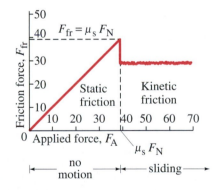

FIGURE 4–28 Example 4–16. Magnitude of the force of friction as a function of the external force applied to an object initially at rest. As the applied force is increased in magnitude, the force of static friction increases linearly to just match it, until the applied force equals $\mu_s F_N$. If the applied force increases further, the object will begin to move, and the friction force drops to a roughly constant value characteristic of kinetic friction.

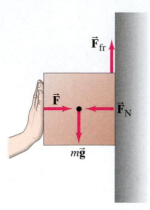

FIGURE 4–29 Example 4–17.

CONCEPTUAL EXAMPLE 4–17 **A box against a wall.** You can hold a box against a rough wall (Fig. 4–29) and prevent it from slipping down by pressing hard horizontally. How does the application of a horizontal force keep an object from moving vertically?

RESPONSE This won't work well if the wall is slippery. You need friction. Even then, if you don't press hard enough, the box will slip. The horizontal force you apply produces a normal force on the box exerted by the wall. The force of gravity mg, acting downward on the box, can now be balanced by an upward friction force whose magnitude is proportional to the normal force. The harder you push, the greater F_N is and the greater F_{fr} can be. If you don't press hard enough, then $mg > \mu_s F_N$ and the box begins to slide down.

Additional Examples

Here are some more worked-out Examples that can help you for solving Problems.

CONCEPTUAL EXAMPLE 4–18 **To push or to pull a sled?** Your little sister wants a ride on her sled. If you are on flat ground, will you exert less force if you push her or pull her? See Figs. 4–30a and b. Assume the same angle θ in each case.

RESPONSE Let us draw free-body diagrams for the sled–sister combination, as shown in Figs. 4–30c and d. They show, for the two cases, the forces exerted by you, $\vec{F}$ (an unknown), by the snow, $\vec{F}_N$ and $\vec{F}_{fr}$, and gravity $m\vec{g}$. (a) If you push her, and $\theta > 0$, there is a vertically downward component to your force. Hence the normal force upward exerted by the ground (Fig. 4–30c) will be larger than mg (where m is the mass of sister plus sled). (b) If you pull her, your force has a vertically upward component, so the normal force F_N will be less than mg, Fig. 4–30d. Because the friction force is proportional to the normal force, F_{fr} will be less if you pull her. So you exert less force if you pull her.

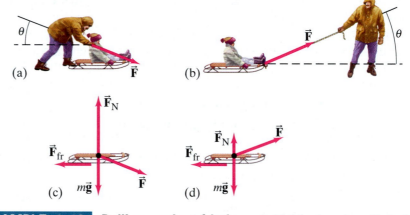

FIGURE 4–30 Example 4–18.

EXAMPLE 4–19 **Pulling against friction.** A 10.0-kg box is pulled along a horizontal surface by a force F_P of 40.0 N applied at a 30.0° angle. This is like Example 4–11 except now there is friction, and we assume a coefficient of kinetic friction of 0.30. Calculate the acceleration.

APPROACH The free-body diagram is like that in Fig. 4–21, but with one more force, that of friction; see Fig. 4–31.

SOLUTION The calculation for the vertical (y) direction is just the same as in Example 4–11, where we saw that $F_{Py} = 20.0$ N, $F_{Px} = 34.6$ N, and the normal force is $F_N = 78.0$ N. Now we apply Newton's second law for the horizontal (x) direction (positive to the right), and include the friction force:

$$F_{Px} - F_{fr} = ma_x.$$

The friction force is kinetic as long as $F_{fr} = \mu_k F_N$ is less than $F_{Px}(=34.6 \text{ N})$, which it is:

$$F_{fr} = \mu_k F_N = (0.30)(78.0 \text{ N}) = 23.4 \text{ N}.$$

FIGURE 4–31 Example 4–19.

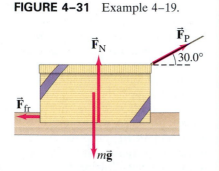

Hence the box does accelerate:

$$a_x = \frac{F_{Px} - F_{fr}}{m} = \frac{34.6\,\text{N} - 23.4\,\text{N}}{10.0\,\text{kg}} = 1.1\,\text{m/s}^2.$$

In the absence of friction, as we saw in Example 4–11, the acceleration would be much greater than this.

NOTE Our final answer has only two significant figures because our least significant input value $(\mu_k = 0.30)$ has two.

EXERCISE B If $\mu_k F_N$ were greater than F_{Px}, what would you conclude?

EXAMPLE 4–20 **Two boxes and a pulley.** In Fig. 4–32a, two boxes are connected by a cord running over a pulley. The coefficient of kinetic friction between box A and the table is 0.20. We ignore the mass of the cord and pulley and any friction in the pulley, which means we can assume that a force applied to one end of the cord will have the same magnitude at the other end. We wish to find the acceleration, a, of the system, which will have the same magnitude for both boxes assuming the cord doesn't stretch. As box B moves down, box A moves to the right.

APPROACH We need a free-body diagram for each box, Figs. 4–32b and c, so we can apply Newton's second law to each. The forces on box A are the pulling force of the cord F_T, gravity $m_A g$, the normal force exerted by the table F_N, and a friction force exerted by the table F_{fr}; the forces on box B are gravity $m_B g$, and the cord pulling up, F_T.

SOLUTION Box A does not move vertically, so Newton's second law tells us the normal force just balances the weight,

$$F_N = m_A g = (5.0\,\text{kg})(9.8\,\text{m/s}^2) = 49\,\text{N}.$$

In the horizontal direction, there are two forces on box A (Fig. 4–32b): F_T, the tension in the cord (whose value we don't know), and the force of friction

$$F_{fr} = \mu_k F_N = (0.20)(49\,\text{N}) = 9.8\,\text{N}.$$

The horizontal acceleration is what we wish to find; we use Newton's second law in the x direction, $\Sigma F_{Ax} = m_A a_x$, which becomes (taking the positive direction to the right and setting $a_{Ax} = a$):

$$\Sigma F_{Ax} = F_T - F_{fr} = m_A a. \qquad \text{[box A]}$$

Next consider box B. The force of gravity $m_B g = (2.0\,\text{kg})(9.8\,\text{m/s}^2) = 19.6\,\text{N}$ pulls downward; and the cord pulls upward with a force F_T. So we can write Newton's second law for box B (taking the downward direction as positive):

$$\Sigma F_{By} = m_B g - F_T = m_B a. \qquad \text{[box B]}$$

[Notice that if $a \neq 0$, then F_T is not equal to $m_B g$.]

We have two unknowns, a and F_T, and we also have two equations. We solve the box A equation for F_T:

$$F_T = F_{fr} + m_A a,$$

and substitute this into the box B equation:

$$m_B g - F_{fr} - m_A a = m_B a.$$

Now we solve for a and put in numerical values:

$$a = \frac{m_B g - F_{fr}}{m_A + m_B} = \frac{19.6\,\text{N} - 9.8\,\text{N}}{5.0\,\text{kg} + 2.0\,\text{kg}} = 1.4\,\text{m/s}^2,$$

which is the acceleration of box A to the right, and of box B down.

If we wish, we can calculate F_T using the first equation:

$$F_T = F_{fr} + m_A a = 9.8\,\text{N} + (5.0\,\text{kg})(1.4\,\text{m/s}^2) = 17\,\text{N}.$$

NOTE Box B is not in free fall. It does not fall at $a = g$ because an additional force, F_T, is acting upward on it.

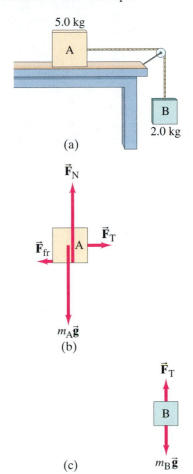

FIGURE 4–32 Example 4–20.

5.0 kg

A

(a)

B

2.0 kg

$\vec{F}_N$

$\vec{F}_{fr}$ A $\vec{F}_T$

$m_A \vec{g}$

(b)

$\vec{F}_T$

B

(c) $m_B \vec{g}$

⚠ **CAUTION**

Tension in a cord supporting a falling object may not equal object's weight

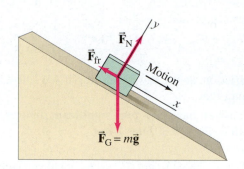

FIGURE 4–33 Forces on an object sliding down an incline.

Inclines

Now we consider what happens when an object slides down an incline, such as a hill or ramp. Such problems are interesting because gravity is the accelerating force, yet the acceleration is not vertical. Solving problems is usually easier if we choose the xy coordinate system so the x axis points along the incline and the y axis is perpendicular to the incline, as shown in Fig. 4–33. Note also that the normal force is not vertical, but is perpendicular to the sloping surface of the plane in Fig. 4–33.

→ **PROBLEM SOLVING**

Good choice of coordinate system simplifies the calculation

EXERCISE C Is the gravitational force always perpendicular to an inclined plane? Is it always vertical?

EXERCISE D Is the normal force always perpendicular to an inclined plane? Is it always vertical?

PHYSICS APPLIED

Skiing

EXAMPLE 4–21 **The skier.** The skier in Fig. 4–34 has just begun descending the 30° slope. Assuming the coefficient of kinetic friction is 0.10, calculate (a) her acceleration and (b) the speed she will reach after 4.0 s.

APPROACH We choose the x axis along the slope, positive pointing downslope in the direction of the skier's motion. The y axis is perpendicular to the surface as shown. The forces acting on the skier are gravity, $\vec{F}_G = m\vec{g}$, which points vertically downward (*not* perpendicular to the slope), and the two forces exerted on her skis by the snow—the normal force perpendicular to the snowy slope (*not* vertical), and the friction force parallel to the surface. These three forces are shown acting at one point in Fig. 4–34b, for convenience, and is our free-body diagram for the skier.

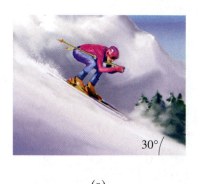

(a)

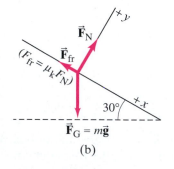

(b)

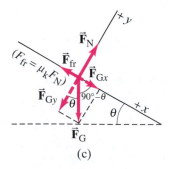

(c)

FIGURE 4–34 Example 4–21. A skier descending a slope; $\vec{F}_G = m\vec{g}$ is the force of gravity (weight) on the skier.

SOLUTION We have to resolve only one vector into components, the weight $\vec{\mathbf{F}}_G$, and its components are shown as dashed lines in Fig. 4–34c. To be general, we use θ rather than 30° for now. We use the definitions of sine ("side opposite") and cosine ("side adjacent") to obtain the components:

$$F_{Gx} = mg \sin \theta,$$

$$F_{Gy} = -mg \cos \theta.$$

where F_{Gy} is in the negative y direction.

(*a*) To calculate the skier's acceleration down the hill, a_x, we apply Newton's second law to the x direction:

$$\Sigma F_x = ma_x$$

$$mg \sin \theta - \mu_k F_N = ma_x$$

where the two forces are the x component of the gravity force ($+x$ direction) and the friction force ($-x$ direction). We want to find the value of a_x, but we don't yet know F_N in the last equation. Let's see if we can get F_N from the y component of Newton's second law:

$$\Sigma F_y = ma_y$$

$$F_N - mg \cos \theta = ma_y = 0$$

where we set $a_y = 0$ because there is no motion in the y direction (perpendicular to the slope). Thus we can solve for F_N:

$$F_N = mg \cos \theta$$

and we can substitute this into our equation above for ma_x:

$$mg \sin \theta - \mu_k (mg \cos \theta) = ma_x.$$

There is an m in each term which can be canceled out. Thus (setting $\theta = 30°$ and $\mu_k = 0.10$):

$$a_x = g \sin 30° - \mu_k g \cos 30°$$

$$= 0.50g - (0.10)(0.866)g = 0.41g.$$

The skier's acceleration is 0.41 times the acceleration of gravity, which in numbers is $a = (0.41)(9.8 \text{ m/s}^2) = 4.0 \text{ m/s}^2$. It is interesting that the mass canceled out here, and so we have the useful conclusion that *the acceleration doesn't depend on the mass*. That such a cancellation sometimes occurs, and thus may give a useful conclusion as well as saving calculation, is a big advantage of working with the algebraic equations and putting in the numbers only at the end.

(*b*) The speed after 4.0 s is found, since the acceleration is constant, by using Eq. 2–11a:

$$v = v_0 + at$$

$$= 0 + (4.0 \text{ m/s}^2)(4.0 \text{ s}) = 16 \text{ m/s},$$

where we assumed a start from rest.

➡ **PROBLEM SOLVING**

It is often helpful to put in numbers only at the end

In problems involving a slope or "inclined plane," it is common to make an error in the direction of the normal force or in the direction of gravity. The normal force is *not* vertical in Example 4–21. It is perpendicular to the slope or plane. And gravity is *not* perpendicular to the slope or plane—gravity acts vertically downward toward the center of the Earth.

⚠ **CAUTION**

Directions of gravity and the normal force

A basic part of a physics course is solving problems effectively. The approach discussed here, though emphasizing Newton's laws, can be applied generally for other topics discussed throughout this book.

PROBLEM SOLVING In General

1. **Read** and reread written problems carefully. A common error is to skip a word or two when reading, which can completely change the meaning of a problem.

2. **Draw** an accurate picture or diagram of the situation. (This is probably the most overlooked, yet most crucial, part of solving a problem.) Use arrows to represent vectors such as velocity or force, and label the vectors with appropriate symbols. When dealing with forces and applying Newton's laws, make sure to include all forces on a given object, including unknown ones, and make clear what forces act on what object (otherwise you may make an error in determining the *net force* on a particular object). A separate **free-body diagram** needs to be drawn for each object involved, and it must show *all* the forces acting on a given object (and only on that object). Do not show forces that act on other objects.

3. Choose a convenient xy **coordinate system** (one that makes your calculations easier, such as one axis in the direction of the acceleration). Vectors are to be resolved into components along the coordinate axes. When using Newton's second law, apply $\Sigma \vec{F} = m\vec{a}$ separately to x and y components, remembering that x direction forces are related to a_x, and similarly for y. If more than one object is involved, you can choose different (convenient) coordinate systems for each.

4. List the knowns and the unknowns (what you are trying to determine), and decide what you need in order to find the unknowns. For problems in the present Chapter, we use Newton's laws. More generally, it may help to see if one or more **relationships** (or **equations**) relate the unknowns to the knowns. But be sure each relationship is applicable in the given case. It is very important

to know the limitations of each formula or relationship—when it is valid and when not. In this book, the more general equations have been given numbers, but even these can have a limited range of validity (often stated in brackets to the right of the equation).

5. Try to solve the problem approximately, to see if it is doable (to check if enough information has been given) and reasonable. Use your intuition, and make **rough calculations**—see "Order of Magnitude Estimating" in Section 1–7. A rough calculation, or a reasonable guess about what the range of final answers might be, is very useful. And a rough calculation can be checked against the final answer to catch errors in calculation, such as in a decimal point or the powers of 10.

6. **Solve** the problem, which may include algebraic manipulation of equations and/or numerical calculations. Recall the mathematical rule that you need as many independent equations as you have unknowns; if you have three unknowns, for example, then you need three independent equations. It is usually best to work out the algebra symbolically before putting in the numbers. Why? Because (*a*) you can then solve a whole class of similar problems with different numerical values; (*b*) you can check your result for cases already understood (say, $\theta = 0°$ or $90°$); (*c*) there may be cancellations or other simplifications; (*d*) there is usually less chance for numerical error; and (*e*) you may gain better insight into the problem.

7. Be sure to keep track of **units**, for they can serve as a check (they must balance on both sides of any equation).

8. Again consider if your answer is **reasonable**. The use of dimensional analysis, described in Section 1–8, can also serve as a check for many problems.

Summary

Newton's three laws of motion are the basic classical laws describing motion.

Newton's first law (the **law of inertia**) states that if the net force on an object is zero, an object originally at rest remains at rest, and an object in motion remains in motion in a straight line with constant velocity.

Newton's second law states that the acceleration of an object is directly proportional to the net force acting on it, and inversely proportional to its mass:

$$\Sigma \vec{F} = m\vec{a}. \qquad (4\text{–}1)$$

Newton's second law is one of the most important and fundamental laws in classical physics.

Newton's third law states that whenever one object exerts a force on a second object, the second object always exerts a force on the first object which is equal in magnitude but opposite in direction:

$$\vec{F}_{AB} = -\vec{F}_{BA} \qquad (4\text{--}2)$$

where $\vec{F}_{BA}$ is the force on object B exerted by object A.

The tendency of an object to resist a change in its motion is called **inertia**. **Mass** is a measure of the inertia of an object.

Weight refers to the **gravitational force** on an object, and is equal to the product of the object's mass m and the acceleration of gravity $\vec{g}$:

$$\vec{F}_G = m\vec{g}. \qquad (4\text{--}3)$$

Force, which is a vector, can be considered as a push or pull; or, from Newton's second law, force can be defined as an action capable of giving rise to acceleration. The **net force** on an object is the vector sum of all forces acting on it.

When two objects slide over one another, the force of friction that each object exerts on the other can be written approximately as $F_{fr} = \mu_k F_N$, where F_N is the **normal force** (the force each object exerts on the other perpendicular to their contact surfaces), and μ_k is the coefficient of **kinetic friction**. If the objects are at rest relative to each other, then F_{fr} is just large enough to hold them at rest and satisfies the inequality $F_{fr} < \mu_s F_N$, where μ_s is the coefficient of **static friction**.

For solving problems involving the forces on one or more objects, it is essential to draw a **free-body diagram** for each object, showing all the forces acting on only that object. Newton's second law can be applied to the vector components for each object.

Questions

1. Why does a child in a wagon seem to fall backward when you give the wagon a sharp pull forward?

2. A box rests on the (frictionless) bed of a truck. The truck driver starts the truck and accelerates forward. The box immediately starts to slide toward the rear of the truck bed. Discuss the motion of the box, in terms of Newton's laws, as seen (a) by Mary standing on the ground beside the truck, and (b) by Chris who is riding on the truck (Fig. 4–35).

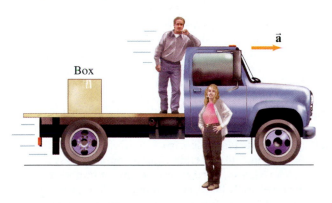

FIGURE 4–35 Question 2.

3. If the acceleration of an object is zero, are no forces acting on it? Explain.

4. Only one force acts on an object. Can the object have zero acceleration? Can it have zero velocity? Explain.

5. When a golf ball is dropped to the pavement, it bounces back up. (a) Is a force needed to make it bounce back up? (b) If so, what exerts the force?

6. If you walk along a log floating on a lake, why does the log move in the opposite direction?

7. Why might your foot hurt if you kick a heavy desk or a wall?

8. When you are running and want to stop quickly, you must decelerate quickly. (a) What is the origin of the force that causes you to stop? (b) Estimate (using your own experience) the maximum rate of deceleration of a person running at top speed to come to rest.

9. A stone hangs by a fine thread from the ceiling, and a section of the same thread dangles from the bottom of the stone (Fig. 4–36). If a person gives a sharp pull on the dangling thread, where is the thread likely to break: below the stone or above it? What if the person gives a slow and steady pull? Explain your answers.

FIGURE 4–36 Question 9.

10. The force of gravity on a 2-kg rock is twice as great as that on a 1-kg rock. Why then doesn't the heavier rock fall faster?

11. Would a spring scale carried to the Moon give accurate results if the scale had been calibrated (a) in pounds, or (b) in kilograms?

12. You pull a box with a constant force across a frictionless table using an attached rope held horizontally. If you now pull the rope with the same force at an angle to the horizontal (with the box remaining flat on the table), does the acceleration of the box (a) remain the same, (b) increase, or (c) decrease? Explain.

13. When an object falls freely under the influence of gravity there is a net force mg exerted on it by the Earth. Yet by Newton's third law the object exerts an equal and opposite force on the Earth. Why doesn't the Earth move?

14. Compare the effort (or force) needed to lift a 10-kg object when you are on the Moon with the force needed to lift it on Earth. Compare the force needed to throw a 2-kg object horizontally with a given speed on the Moon and on Earth.

15. According to Newton's third law, each team in a tug of war (Fig. 4–37) pulls with equal force on the other team. What, then, determines which team will win?

FIGURE 4–37 Question 15. A tug of war. Describe the forces on each of the teams and on the rope.

16. A person exerts an upward force of 40 N to hold a bag of groceries. Describe the "reaction" force (Newton's third law) by stating (*a*) its magnitude, (*b*) its direction, (*c*) *on* what object it is exerted, and (*d*) *by* what object it is exerted.

17. When you stand still on the ground, how large a force does the ground exert on you? Why doesn't this force make you rise up into the air?

18. Whiplash sometimes results from an automobile accident when the victim's car is struck violently from the rear. Explain why the head of the victim seems to be thrown backward in this situation. Is it really?

19. A heavy crate rests on the bed of a flatbed truck. When the truck accelerates, the crate remains where it is on the truck, so it, too, accelerates. What force causes the crate to accelerate?

20. A block is given a push so that it slides up a ramp. After the block reaches its highest point, it slides back down but the magnitude of its acceleration is less on the descent than on the ascent. Why?

21. What would your bathroom scale read if you weighed yourself on an inclined plane? Assume the mechanism functions properly, even at an angle.

Problems

4–4 to 4–6 Newton's Laws, Gravitational Force, Normal Force

1. (I) What force is needed to accelerate a child on a sled (total mass = 60.0 kg) at 1.25 m/s^2?

2. (I) A net force of 265 N accelerates a bike and rider at 2.30 m/s^2. What is the mass of the bike and rider together?

3. (I) How much tension must a rope withstand if it is used to accelerate a 960-kg car horizontally along a frictionless surface at 1.20 m/s^2?

4. (I) What is the weight of a 76-kg astronaut (*a*) on Earth, (*b*) on the Moon ($g = 1.7 \text{ m/s}^2$), (*c*) on Mars ($g = 3.7 \text{ m/s}^2$), (*d*) in outer space traveling with constant velocity?

5. (II) A 20.0-kg box rests on a table. (*a*) What is the weight of the box and the normal force acting on it? (*b*) A 10.0-kg box is placed on top of the 20.0-kg box, as shown in Fig. 4–38. Determine the normal force that the table exerts on the 20.0-kg box and the normal force that the 20.0-kg box exerts on the 10.0-kg box.

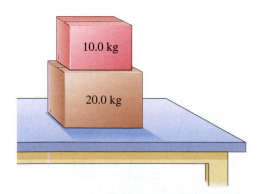

FIGURE 4–38 Problem 5.

6. (II) What average force is required to stop an 1100-kg car in 8.0 s if the car is traveling at 95 km/h?

7. (II) What average force is needed to accelerate a 7.00-gram pellet from rest to 125 m/s over a distance of 0.800 m along the barrel of a rifle?

8. (II) A fisherman yanks a fish vertically out of the water with an acceleration of 2.5 m/s^2 using very light fishing line that has a breaking strength of 22 N. The fisherman unfortunately loses the fish as the line snaps. What can you say about the mass of the fish?

9. (II) A 0.140-kg baseball traveling 35.0 m/s strikes the catcher's mitt, which, in bringing the ball to rest, recoils backward 11.0 cm. What was the average force applied by the ball on the glove?

10. (II) How much tension must a rope withstand if it is used to accelerate a 1200-kg car vertically upward at 0.80 m/s^2?

11. (II) A particular race car can cover a quarter-mile track (402 m) in 6.40 s starting from a standstill. Assuming the acceleration is constant, how many "g's" does the driver experience? If the combined mass of the driver and race car is 485 kg, what horizontal force must the road exert on the tires?

12. (II) A 12.0-kg bucket is lowered vertically by a rope in which there is 163 N of tension at a given instant. What is the acceleration of the bucket? Is it up or down?

13. (II) An elevator (mass 4850 kg) is to be designed so that the maximum acceleration is 0.0680*g*. What are the maximum and minimum forces the motor should exert on the supporting cable?

14. (II) A 75-kg petty thief wants to escape from a third-story jail window. Unfortunately, a makeshift rope made of sheets tied together can support a mass of only 58 kg. How might the thief use this "rope" to escape? Give a quantitative answer.

15. (II) A person stands on a bathroom scale in a motionless elevator. When the elevator begins to move, the scale briefly reads only 0.75 of the person's regular weight. Calculate the acceleration of the elevator, and find the direction of acceleration.

16. (II) The cable supporting a 2125-kg elevator has a maximum strength of 21,750 N. What maximum upward acceleration can it give the elevator without breaking?

17. (II) (a) What is the acceleration of two falling sky divers (mass 132 kg including parachute) when the upward force of air resistance is equal to one-fourth of their weight? (b) After popping open the parachute, the divers descend leisurely to the ground at constant speed. What now is the force of air resistance on the sky divers and their parachute? See Fig. 4–39.

FIGURE 4–39 Problem 17.

18. (III) A person jumps from the roof of a house 3.9-m high. When he strikes the ground below, he bends his knees so that his torso decelerates over an approximate distance of 0.70 m. If the mass of his torso (excluding legs) is 42 kg, find (a) his velocity just before his feet strike the ground, and (b) the average force exerted on his torso by his legs during deceleration.

4–7 Newton's Laws and Vectors

19. (I) A box weighing 77.0 N rests on a table. A rope tied to the box runs vertically upward over a pulley and a weight is hung from the other end (Fig. 4–40). Determine the force that the table exerts on the box if the weight hanging on the other side of the pulley weighs (a) 30.0 N, (b) 60.0 N, and (c) 90.0 N.

FIGURE 4–40
Problem 19.

20. (I) Draw the free-body diagram for a basketball player (a) just before leaving the ground on a jump, and (b) while in the air. See Fig. 4–41.

FIGURE 4–41
Problem 20.

21. (I) Sketch the free-body diagram of a baseball (a) at the moment it is hit by the bat, and again (b) after it has left the bat and is flying toward the outfield.

22. (I) A 650-N force acts in a northwesterly direction. A second 650-N force must be exerted in what direction so that the resultant of the two forces points westward? Illustrate your answer with a vector diagram.

23. (II) Arlene is to walk across a "high wire" strung horizontally between two buildings 10.0 m apart. The sag in the rope when she is at the midpoint is 10.0°, as shown in Fig. 4–42. If her mass is 50.0 kg, what is the tension in the rope at this point?

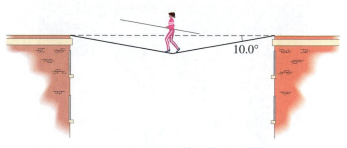

FIGURE 4–42 Problem 23.

24. (II) The two forces $\vec{F}_1$ and $\vec{F}_2$ shown in Fig. 4–43a and b (looking down) act on a 27.0-kg object on a frictionless tabletop. If $F_1 = 10.2$ N and $F_2 = 16.0$ N, find the net force on the object and its acceleration for (a) and (b).

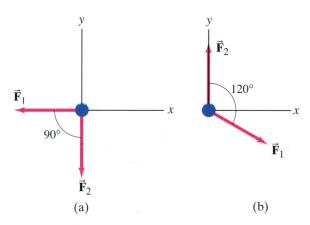

(a)   (b)

FIGURE 4–43 Problem 24.

25. (II) One 3.2-kg paint bucket is hanging by a massless cord from another 3.2-kg paint bucket, also hanging by a massless cord, as shown in Fig. 4–44. (a) If the buckets are at rest, what is the tension in each cord? (b) If the two buckets are pulled upward with an acceleration of 1.60 m/s² by the upper cord, calculate the tension in each cord.

FIGURE 4–44
Problem 25.

26. (II) A person pushes a 14.0-kg lawn mower at constant speed with a force of $F = 88.0\,\text{N}$ directed along the handle, which is at an angle of $45.0°$ to the horizontal (Fig. 4–45). (*a*) Draw the free-body diagram showing all forces acting on the mower. Calculate (*b*) the horizontal friction force on the mower, then (*c*) the normal force exerted vertically upward on the mower by the ground. (*d*) What force must the person exert on the lawn mower to accelerate it from rest to 1.5 m/s in 2.5 seconds, assuming the same friction force?

FIGURE 4–45 Problem 26.

27. (II) Two snowcats tow a housing unit to a new location at McMurdo Base, Antarctica, as shown in Fig. 4–46. The sum of the forces $\vec{\mathbf{F}}_A$ and $\vec{\mathbf{F}}_B$ exerted on the unit by the horizontal cables is parallel to the line L, and $F_A = 4500\,\text{N}$. Determine F_B and the magnitude of $\vec{\mathbf{F}}_A + \vec{\mathbf{F}}_B$.

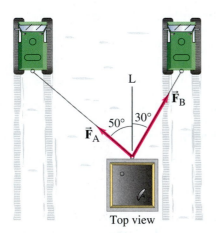

FIGURE 4–46
Problem 27.

28. (II) A train locomotive is pulling two cars of the same mass behind it, Fig. 4–47. Determine the ratio of the tension in the coupling between the locomotive and the first car (F_{T1}), to that between the first car and the second car (F_{T2}), for any nonzero acceleration of the train.

FIGURE 4–47 Problem 28.

29. (II) A window washer pulls herself upward using the bucket–pulley apparatus shown in Fig. 4–48. (*a*) How hard must she pull downward to raise herself slowly at constant speed? (*b*) If she increases this force by 15%, what will her acceleration be? The mass of the person plus the bucket is 65 kg.

FIGURE 4–48
Problem 29.

30. (II) At the instant a race began, a 65-kg sprinter exerted a force of 720 N on the starting block at a $22°$ angle with respect to the ground. (*a*) What was the horizontal acceleration of the sprinter? (*b*) If the force was exerted for 0.32 s, with what speed did the sprinter leave the starting block?

31. (II) Figure 4–49 shows a block (mass m_A) on a smooth horizontal surface, connected by a thin cord that passes over a pulley to a second block (m_B), which hangs vertically. (*a*) Draw a free-body diagram for each block, showing the force of gravity on each, the force (tension) exerted by the cord, and any normal force. (*b*) Apply Newton's second law to find formulas for the acceleration of the system and for the tension in the cord. Ignore friction and the masses of the pulley and cord.

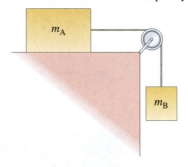

FIGURE 4–49
Problem 31. Mass m_A rests on smooth horizontal surface, m_B hangs vertically.

32. (II) A pair of fuzzy dice is hanging by a string from your rearview mirror. While you are accelerating from a stoplight to 28 m/s in 6.0 s, what angle θ does the string make with the vertical? See Fig. 4–50.

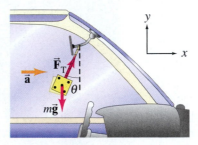

FIGURE 4–50
Problem 32.

33. (III) Three blocks on a frictionless horizontal surface are in contact with each other, as shown in Fig. 4–51. A force $\vec{F}$ is applied to block A (mass m_A). (a) Draw a free-body diagram for each block. Determine (b) the acceleration of the system (in terms of m_A, m_B, and m_C), (c) the net force on each block, and (d) the force of contact that each block exerts on its neighbor. (e) If $m_A = m_B = m_C = 12.0$ kg and $F = 96.0$ N, give numerical answers to (b), (c), and (d). Do your answers make sense intuitively?

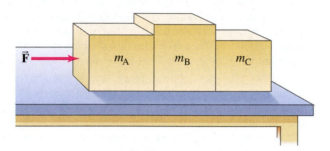

FIGURE 4–51 Problem 33.

34. (III) The two masses shown in Fig. 4–52 are each initially 1.80 m above the ground, and the massless frictionless pulley is 4.8 m above the ground. What maximum height does the lighter object reach after the system is released? [*Hint*: First determine the acceleration of the lighter mass and then its velocity at the moment the heavier one hits the ground. This is its "launch" speed. Assume it doesn't hit the pulley.]

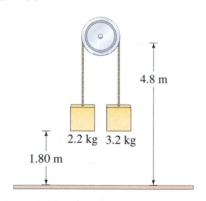

FIGURE 4–52 Problem 34.

35. (III) Suppose two boxes on a frictionless table are connected by a heavy cord of mass 1.0 kg. Calculate the acceleration of each box and the tension at each end of the cord, using the free-body diagrams shown in Fig. 4–53. Assume $F_P = 40.0$ N, and ignore sagging of the cord. Compare your results to Example 4–12 and Fig. 4–22.

4–8 Newton's Laws with Friction; Inclines

36. (I) If the coefficient of kinetic friction between a 35-kg crate and the floor is 0.30, what horizontal force is required to move the crate at a steady speed across the floor? What horizontal force is required if μ_k is zero?

37. (I) A force of 48.0 N is required to start a 5.0-kg box moving across a horizontal concrete floor. (a) What is the coefficient of static friction between the box and the floor? (b) If the 48.0-N force continues, the box accelerates at 0.70 m/s^2. What is the coefficient of kinetic friction?

38. (I) Suppose that you are standing on a train accelerating at $0.20g$. What minimum coefficient of static friction must exist between your feet and the floor if you are not to slide?

39. (I) What is the maximum acceleration a car can undergo if the coefficient of static friction between the tires and the ground is 0.80?

40. (II) The coefficient of static friction between hard rubber and normal street pavement is about 0.8. On how steep a hill (maximum angle) can you leave a car parked?

41. (II) A 15.0-kg box is released on a 32° incline and accelerates down the incline at 0.30 m/s^2. Find the friction force impeding its motion. What is the coefficient of kinetic friction?

42. (II) A car can decelerate at -4.80 m/s^2 without skidding when coming to rest on a level road. What would its deceleration be if the road were inclined at 13° uphill? Assume the same static friction coefficient.

43. (II) (a) A box sits at rest on a rough 30° inclined plane. Draw the free-body diagram, showing all the forces acting on the box. (b) How would the diagram change if the box were sliding down the plane? (c) How would it change if the box were sliding up the plane after an initial shove?

44. (II) Drag-race tires in contact with an asphalt surface have a very high coefficient of static friction. Assuming a constant acceleration and no slipping of tires, estimate the coefficient of static friction needed for a drag racer to cover 1.0 km in 12 s, starting from rest.

45. (II) The coefficient of kinetic friction for a 22-kg bobsled on a track is 0.10. What force is required to push it down a 6.0° incline and achieve a speed of 60 km/h at the end of 75 m?

46. (II) For the system of Fig. 4–32 (Example 4–20) how large a mass would box A have to have to prevent any motion from occurring? Assume $\mu_s = 0.30$.

47. (II) A box is given a push so that it slides across the floor. How far will it go, given that the coefficient of kinetic friction is 0.20 and the push imparts an initial speed of 4.0 m/s?

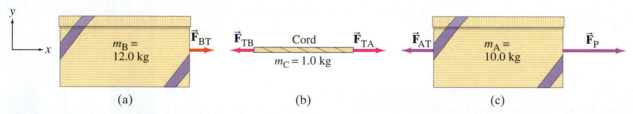

FIGURE 4–53 Problem 35. Free-body diagrams for two boxes on a table connected by a heavy cord, and being pulled to the right as in Fig. 4–22a. Vertical forces, $\vec{F}_N$ and $\vec{F}_G$, are not shown.

48. (II) Two crates, of mass 75 kg and 110 kg, are in contact and at rest on a horizontal surface (Fig. 4–54). A 620-N force is exerted on the 75-kg crate. If the coefficient of kinetic friction is 0.15, calculate (a) the acceleration of the system, and (b) the force that each crate exerts on the other. (c) Repeat with the crates reversed.

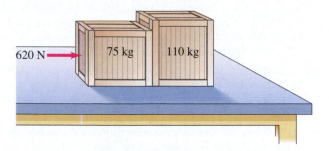

620 N   75 kg   110 kg

FIGURE 4–54 Problem 48.

49. (II) A flatbed truck is carrying a heavy crate. The coefficient of static friction between the crate and the bed of the truck is 0.75. What is the maximum rate at which the driver can decelerate and still avoid having the crate slide against the cab of the truck?

50. (II) On an icy day, you worry about parking your car in your driveway, which has an incline of 12°. Your neighbor's driveway has an incline of 9.0°, and the driveway across the street is at 6.0°. The coefficient of static friction between tire rubber and ice is 0.15. Which driveway(s) will be safe to park in?

51. (II) A child slides down a slide with a 28° incline, and at the bottom her speed is precisely half what it would have been if the slide had been frictionless. Calculate the coefficient of kinetic friction between the slide and the child.

52. (II) The carton shown in Fig. 4–55 lies on a plane tilted at an angle $\theta = 22.0°$ to the horizontal, with $\mu_k = 0.12$. (a) Determine the acceleration of the carton as it slides down the plane. (b) If the carton starts from rest 9.30 m up the plane from its base, what will be the carton's speed when it reaches the bottom of the incline?

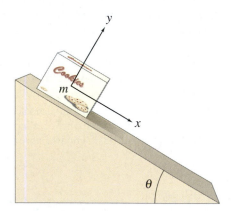

FIGURE 4–55 Carton on inclined plane.
Problems 52 and 53.

53. (II) A carton is given an initial speed of 3.0 m/s up the 22.0° plane shown in Fig. 4–55. (a) How far up the plane will it go? (b) How much time elapses before it returns to its starting point? Ignore friction.

54. (II) A roller coaster reaches the top of the steepest hill with a speed of 6.0 km/h. It then descends the hill, which is at an average angle of 45° and is 45.0 m long. Estimate its speed when it reaches the bottom. Assume $\mu_k = 0.18$.

55. (II) An 18.0-kg box is released on a 37.0° incline and accelerates down the incline at 0.270 m/s². Find the friction force impeding its motion. How large is the coefficient of kinetic friction?

56. (II) A small box is held in place against a rough wall by someone pushing on it with a force directed upward at 28° above the horizontal. The coefficients of static and kinetic friction between the box and wall are 0.40 and 0.30, respectively. The box slides down unless the applied force has magnitude 13 N. What is the mass of the box?

57. (II) Piles of snow on slippery roofs can become dangerous projectiles as they melt. Consider a chunk of snow at the ridge of a roof with a pitch of 30°. (a) What is the minimum value of the coefficient of static friction that will keep the snow from sliding down? (b) As the snow begins to melt, the coefficient of static friction decreases and the snow eventually slips. Assuming that the distance from the chunk to the edge of the roof is 5.0 m and the coefficient of kinetic friction is 0.20, calculate the speed of the snow chunk when it slides off the roof. (c) If the edge of the roof is 10.0 m above ground, what is the speed of the snow when it hits the ground?

58. (III) (a) Show that the minimum stopping distance for an automobile traveling at speed v is equal to $v^2/2\mu_s g$, where μ_s is the coefficient of static friction between the tires and the road, and g is the acceleration of gravity. (b) What is this distance for a 1200-kg car traveling 95 km/h if $\mu_s = 0.75$?

59. (III) A coffee cup on the dashboard of a car slides forward on the dash when the driver decelerates from 45 km/h to rest in 3.5 s or less, but not if he decelerates in a longer time. What is the coefficient of static friction between the cup and the dash?

60. (III) A small block of mass m is given an initial speed v_0 up a ramp inclined at angle θ to the horizontal. It travels a distance d up the ramp and comes to rest. Determine a formula for the coefficient of kinetic friction between block and ramp.

61. (III) The 75-kg climber in Fig. 4–56 is supported in the "chimney" by the friction forces exerted on his shoes and back. The static coefficients of friction between his shoes and the wall, and between his back and the wall, are 0.80 and 0.60, respectively. What is the minimum normal force he must exert? Assume the walls are vertical and that friction forces are both at a maximum. Ignore his grip on the rope.

FIGURE 4–56
Problem 61.

62. (III) Boxes are moved on a conveyor belt from where they are filled to the packing station 11.0 m away. The belt is initially stationary and must finish with zero speed. The most rapid transit is accomplished if the belt accelerates for half the distance, then decelerates for the final half of the trip. If the coefficient of static friction between a box and the belt is 0.60, what is the minimum transit time for each box?

63. (III) A block (mass m_1) lying on a frictionless inclined plane is connected to a mass m_2 by a massless cord passing over a pulley, as shown in Fig. 4–57. (a) Determine a formula for the acceleration of the system of the two blocks in terms of m_1, m_2, θ and g. (b) What conditions apply to masses m_1 and m_2 for the acceleration to be in one direction (say, m_1 down the plane), or in the opposite direction?

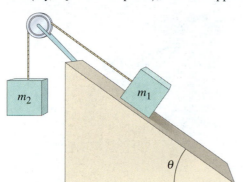

FIGURE 4–57
Problems 63 and 64.

64. (III) (a) Suppose the coefficient of kinetic friction between m_1 and the plane in Fig. 4–57 is $\mu_k = 0.15$, and that $m_1 = m_2 = 2.7$ kg. As m_2 moves down, determine the magnitude of the acceleration of m_1 and m_2, given $\theta = 25°$. (b) What smallest value of μ_k will keep this system from accelerating?

65. (III) A bicyclist of mass 65 kg (including the bicycle) can coast down a 6.0° hill at a steady speed of 6.0 km/h because of air resistance. How much force must be applied to climb the hill at the same speed and same air resistance?

General Problems

66. According to a simplified model of a mammalian heart, at each pulse approximately 20 g of blood is accelerated from 0.25 m/s to 0.35 m/s during a period of 0.10 s. What is the magnitude of the force exerted by the heart muscle?

67. A person has a reasonable chance of surviving an automobile crash if the deceleration is no more than 30 "g's." Calculate the force on a 70-kg person undergoing this acceleration. What distance is traveled if the person is brought to rest at this rate from 100 km/h?

68. (a) If the horizontal acceleration produced by an earthquake is a, and if an object is going to "hold its place" on the ground, show that the coefficient of static friction with the ground must be at least $\mu_s = a/g$. (b) The famous Loma Prieta earthquake that stopped the 1989 World Series produced ground accelerations of up to 4.0 m/s² in the San Francisco Bay Area. Would a chair have started to slide on a linoleum floor with coefficient of static friction 0.25?

69. An 1150-kg car pulls a 450-kg trailer. The car exerts a horizontal force of 3.8×10^3 N against the ground in order to accelerate. What force does the car exert on the trailer? Assume an effective friction coefficient of 0.15 for the trailer.

70. Police investigators, examining the scene of an accident involving two cars, measure 72-m-long skid marks of one of the cars, which nearly came to a stop before colliding. The coefficient of kinetic friction between rubber and the pavement is about 0.80. Estimate the initial speed of that car assuming a level road.

71. A car starts rolling down a 1-in-4 hill (1-in-4 means that for each 4 m traveled along the road, the elevation change is 1 m). How fast is it going when it reaches the bottom after traveling 55 m? (a) Ignore friction. (b) Assume an effective coefficient of friction equal to 0.10.

72. A 2.0-kg purse is dropped from the top of the Leaning Tower of Pisa and falls 55 m before reaching the ground with a speed of 29 m/s. What was the average force of air resistance?

73. A cyclist is coasting at a steady speed of 12 m/s but enters a muddy stretch where the effective coefficient of friction is 0.60. Will the cyclist emerge from the muddy stretch without having to pedal if the mud lasts for 11 m? If so, what will be the speed upon emerging?

74. A city planner is working on the redesign of a hilly portion of a city. An important consideration is how steep the roads can be so that even low-powered cars can get up the hills without slowing down. A particular small car, with a mass of 1100 kg, can accelerate on a level road from rest to 21 m/s (75 km/h) in 14.0 s. Using these data, calculate the maximum steepness of a hill.

75. Francesca, who likes physics experiments, dangles her watch from a thin piece of string while the jetliner she is in takes off from JFK Airport (Fig. 4–58). She notices that the string makes an angle of 25° with respect to the vertical as the aircraft accelerates for takeoff, which takes about 18 s. Estimate the takeoff speed of the aircraft.

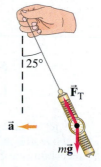

FIGURE 4–58
Problem 75.

76. A 28.0-kg block is connected to an empty 1.35-kg bucket by a cord running over a frictionless pulley (Fig. 4–59). The coefficient of static friction between the table and the block is 0.450 and the coefficient of kinetic friction between the table and the block is 0.320. Sand is gradually added to the bucket until the system just begins to move. (a) Calculate the mass of sand added to the bucket. (b) Calculate the acceleration of the system.

28.0 kg

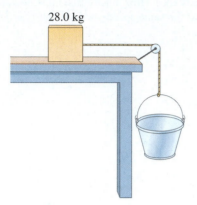

FIGURE 4–59 Problem 76.

77. In the design of a supermarket, there are to be several ramps connecting different parts of the store. Customers will have to push grocery carts up the ramps and it is obviously desirable that this not be too difficult. The engineer has done a survey and found that almost no one complains if the force directed up the ramp is no more than 20 N. Ignoring friction, at what maximum angle θ should the ramps be built, assuming a full 30-kg grocery cart?

78. (a) What minimum force F is needed to lift the piano (mass M) using the pulley apparatus shown in Fig. 4–60? (b) Determine the tension in each section of rope: F_{T1}, F_{T2}, F_{T3}, and F_{T4}.

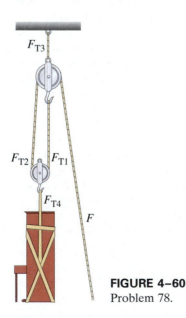

FIGURE 4–60
Problem 78.

79. A jet aircraft is accelerating at 3.5 m/s² at an angle of 45° above the horizontal. What is the total force that the cockpit seat exerts on the 75-kg pilot?

80. In the design process for a child-restraint chair, an engineer considers the following set of conditions: A 12-kg child is riding in the chair, which is securely fastened to the seat of an automobile (Fig. 4–61). Assume the automobile is involved in a head-on collision with another vehicle. The initial speed v_0 of the car is 45 km/h, and this speed is reduced to zero during the collision time of 0.20 s. Assume a constant car deceleration during the collision and estimate the net horizontal force F that the straps of the restraint chair must exert on the child in order to keep her fixed to the chair. Treat the child as a particle and state any additional assumptions made during your analysis.

FIGURE 4–61 Problem 80.

81. A 7650-kg helicopter accelerates upward at 0.80 m/s² while lifting a 1250-kg frame at a construction site, Fig. 4–62. (a) What is the lift force exerted by the air on the helicopter rotors? (b) What is the tension in the cable (ignore its mass) that connects the frame to the helicopter? (c) What force does the cable exert on the helicopter?

FIGURE 4–62 Problem 81.

82. A super high-speed 12-car Italian train has a mass of 660 metric tons (660,000 kg). It can exert a maximum force of 400 kN horizontally against the tracks, whereas at maximum velocity (300 km/h), it exerts a force of about 150 kN. Calculate (a) its maximum acceleration, and (b) estimate the force of air resistance at top speed.

83. A 65-kg ice skater coasts with no effort for 75 m until she stops. If the coefficient of kinetic friction between her skates and the ice is $\mu_k = 0.10$, how fast was she moving at the start of her coast?

84. Two rock climbers, Bill and Karen, use safety ropes of similar length. Karen's rope is more elastic, called a *dynamic rope* by climbers. Bill has a *static rope*, not recommended for safety purposes in pro climbing. Karen falls freely about 2.0 m and then the rope stops her over a distance of 1.0 m (Fig. 4–63). (*a*) Estimate, assuming that the force is constant, how large a force she will feel from the rope. (Express the result in multiples of her weight.) (*b*) In a similar fall, Bill's rope stretches by 30 cm only. How many times his weight will the rope pull on him? Which climber is more likely to be hurt?

FIGURE 4–63
Problem 84.

85. A fisherman in a boat is using a "10-lb test" fishing line. This means that the line can exert a force of 45 N without breaking (1 lb = 4.45 N). (*a*) How heavy a fish can the fisherman land if he pulls the fish up vertically at constant speed? (*b*) If he accelerates the fish upward at 2.0 m/s², what maximum weight fish can he land? (*c*) Is it possible to land a 15-lb trout on 10-lb test line? Why or why not?

86. An elevator in a tall building is allowed to reach a maximum speed of 3.5 m/s going down. What must the tension be in the cable to stop this elevator over a distance of 2.6 m if the elevator has a mass of 1300 kg including occupants?

87. Two boxes, $m_1 = 1.0$ kg with a coefficient of kinetic friction of 0.10, and $m_2 = 2.0$ kg with a coefficient of 0.20, are placed on a plane inclined at $\theta = 30°$. (*a*) What acceleration does each box experience? (*b*) If a taut string is connected to the boxes (Fig. 4–64), with m_2 initially farther down the slope, what is the acceleration of each box? (*c*) If the initial configuration is reversed with m_1 starting lower with a taut string, what is the acceleration of each box?

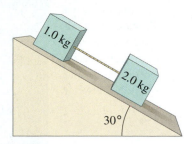

FIGURE 4–64 Problem 87.

88. A 75.0-kg person stands on a scale in an elevator. What does the scale read (in N and in kg) when the elevator is (*a*) at rest, (*b*) ascending at a constant speed of 3.0 m/s, (*c*) falling at 3.0 m/s, (*d*) accelerating upward at 3.0 m/s², (*e*) accelerating downward at 3.0 m/s²?

89. Three mountain climbers who are roped together are ascending an icefield inclined at 21.0° to the horizontal. The last climber slips, pulling the second climber off his feet. The first climber is able to hold them both. If each climber has a mass of 75 kg, calculate the tension in each of the two sections of rope between the three climbers. Ignore friction between the ice and the fallen climbers.

Answers to Exercises

A: (*a*) The same; (*b*) the sports car; (*c*) third law for part (a), second law for part (b).

B: The force applied by the person is insufficient to keep the box moving.

C: No; yes.

D: Yes; no.

The astronauts in the upper left of this photo are working on the space shuttle. As they orbit the Earth—at a rather high speed—they experience apparent weightlessness. The Moon, in the background, also is orbiting the Earth at high speed. Both the Moon and the space shuttle move in nearly circular orbits, and each undergoes a centripetal acceleration. What keeps the Moon and the space shuttle (and its astronauts) from moving off in a straight line away from Earth? It is the force of gravity. Newton's law of universal gravitation states that all objects attract all other objects with a force proportional to their masses and inversely proportional to the square of the distance between them.

CHAPTER **5**

Circular Motion; Gravitation

FIGURE 5–1 A small object moving in a circle, showing how the velocity changes. At each point, the instantaneous velocity is in a direction tangent to the circular path.

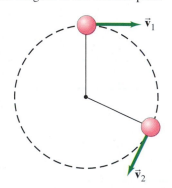

An object moves in a straight line if the net force on it acts in the direction of motion, or the net force is zero. If the net force acts at an angle to the direction of motion at any moment, then the object moves in a curved path. An example of the latter is projectile motion, which we discussed in Chapter 3. Another important case is that of an object moving in a circle, such as a ball at the end of a string revolving around one's head, or the nearly circular motion of the Moon about the Earth.

In this Chapter, we study the circular motion of objects, and how Newton's laws of motion apply. We also discuss how Newton conceived of another great law by applying the concepts of circular motion to the motion of the Moon and the planets. This is the law of universal gravitation, which was the capstone of Newton's analysis of the physical world.

5–1 Kinematics of Uniform Circular Motion

An object that moves in a circle at constant speed v is said to experience **uniform circular motion**. The *magnitude* of the velocity remains constant in this case, but the *direction* of the velocity continuously changes as the object moves around the circle (Fig. 5–1). Because acceleration is defined as the rate of change of velocity, a change in direction of velocity constitutes an acceleration, just as a change in magnitude of velocity does. Thus, an object revolving in a circle is continuously accelerating, even when the speed remains constant ($v_1 = v_2 = v$). We now investigate this acceleration quantitatively.

Acceleration is defined as

$$\vec{a} = \frac{\vec{v}_2 - \vec{v}_1}{\Delta t} = \frac{\Delta \vec{v}}{\Delta t},$$

where $\Delta \vec{v}$ is the change in velocity during the short time interval Δt. We will eventually consider the situation in which Δt approaches zero and thus obtain the instantaneous acceleration. But for purposes of making a clear drawing, Fig. 5–2, we consider a nonzero time interval. During the time interval Δt, the particle in Fig. 5–2a moves from point A to point B, covering a distance Δl *along the arc* which subtends an angle $\Delta \theta$. The change in the velocity vector is $\vec{v}_2 - \vec{v}_1 = \Delta \vec{v}$, and is shown in Fig. 5–2b.

If we let Δt be very small (approaching zero), then Δl and $\Delta \theta$ are also very small, and $\vec{v}_2$ will be almost parallel to $\vec{v}_1$; $\Delta \vec{v}$ will be essentially perpendicular to them (Fig. 5–2c). Thus $\Delta \vec{v}$ points toward the center of the circle. Since $\vec{a}$, by definition, is in the same direction as $\Delta \vec{v}$, it too must point toward the center of the circle. Therefore, this acceleration is called **centripetal acceleration** ("center-pointing" acceleration) or **radial acceleration** (since it is directed along the radius, toward the center of the circle), and we denote it by $\vec{a}_R$.

We next determine the magnitude of the centripetal (radial) acceleration, a_R. Because CA in Fig. 5–2a is perpendicular to $\vec{v}_1$, and CB is perpendicular to $\vec{v}_2$, it follows that the angle $\Delta \theta$, defined as the angle between CA and CB, is also the angle between $\vec{v}_1$ and $\vec{v}_2$. Hence the vectors $\vec{v}_1$, $\vec{v}_2$, and $\Delta \vec{v}$ in Fig. 5–2b form a triangle that is geometrically similar[†] to triangle CAB in Fig. 5–2a. If we take $\Delta \theta$ to be very small (letting Δt be very small) and setting $v = v_1 = v_2$ because the magnitude of the velocity is assumed not to change, we can write

$$\frac{\Delta v}{v} \approx \frac{\Delta l}{r}.$$

This is an exact equality when Δt approaches zero, for then the arc length Δl equals the cord length AB. We want to find the instantaneous acceleration, so we let Δt approach zero, write the above expression as an equality, and then solve for Δv:

$$\Delta v = \frac{v}{r} \Delta l.$$

To get the centripetal acceleration, a_R, we divide Δv by Δt:

$$a_R = \frac{\Delta v}{\Delta t} = \frac{v}{r} \frac{\Delta l}{\Delta t}.$$

But $\Delta l / \Delta t$ is just the linear speed, v, of the object, so

$$a_R = \frac{v^2}{r}. \tag{5–1}$$

Centripetal (radial) acceleration

Equation 5–1 is valid even when v is not constant.

To summarize, *an object moving in a circle of radius r at constant speed v has an acceleration whose direction is toward the center of the circle and whose magnitude is $a_R = v^2/r$.* It is not surprising that this acceleration depends on v and r. The greater the speed v, the faster the velocity changes direction; and the larger the radius, the less rapidly the velocity changes direction.

[†] Appendix A contains a review of geometry.

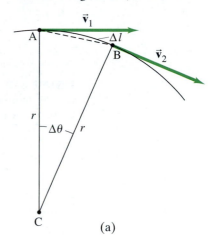
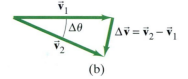
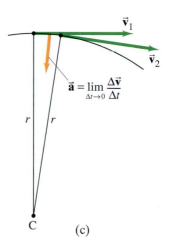

FIGURE 5–2 Determining the change in velocity, $\Delta \vec{v}$, for a particle moving in a circle. The length Δl is the distance along the arc, from A to B.

CAUTION

In uniform circular motion, the speed is constant, but the acceleration is not zero

The acceleration vector points toward the center of the circle. But the velocity vector always points in the direction of motion, which is tangential to the circle. Thus the velocity and acceleration vectors are perpendicular to each other at every point in the path for uniform circular motion (Fig. 5–3). This is another example that illustrates the error in thinking that acceleration and velocity are always in the same direction. For an object falling vertically, $\vec{a}$ and $\vec{v}$ are indeed parallel. But in circular motion, $\vec{a}$ and $\vec{v}$ are perpendicular, not parallel (nor were they parallel in projectile motion, Section 3–5).

CAUTION

The direction of motion ($\vec{v}$) and the acceleration ($\vec{a}$) are not in the same direction; instead, $\vec{a} \perp \vec{v}$

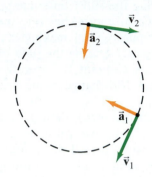

FIGURE 5–3 For uniform circular motion, $\vec{a}$ is always perpendicular to $\vec{v}$.

Period and frequency

Circular motion is often described in terms of the **frequency** f, the number of revolutions per second. The **period** T of an object revolving in a circle is the time required for one complete revolution. Period and frequency are related by

$$T = \frac{1}{f}. \tag{5–2}$$

For example, if an object revolves at a frequency of 3 rev/s, then each revolution takes $\frac{1}{3}$ s. For an object revolving in a circle (of circumference $2\pi r$) at constant speed v, we can write

$$v = \frac{2\pi r}{T},$$

since in one revolution the object travels one circumference.

EXAMPLE 5–1 **Acceleration of a revolving ball.** A 150-g ball at the end of a string is revolving uniformly in a horizontal circle of radius 0.600 m, as in Fig. 5–1 or 5–3. The ball makes 2.00 revolutions in a second. What is its centripetal acceleration?

APPROACH The centripetal acceleration is $a_R = v^2/r$. We are given r, and we can find the speed of the ball, v, from the given radius and frequency.

SOLUTION If the ball makes two complete revolutions per second, then the ball travels in a complete circle in a time interval equal to 0.500 s, which is its period T. The distance traveled in this time is the circumference of the circle, $2\pi r$, where r is the radius of the circle. Therefore, the ball has speed

$$v = \frac{2\pi r}{T} = \frac{2(3.14)(0.600 \text{ m})}{(0.500 \text{ s})} = 7.54 \text{ m/s}.$$

The centripetal acceleration[†] is

$$a_R = \frac{v^2}{r} = \frac{(7.54 \text{ m/s})^2}{(0.600 \text{ m})} = 94.7 \text{ m/s}^2.$$

EXERCISE A If the string is doubled in length to 1.20 m but all else stays the same, by what factor will the centripetal acceleration change?

[†]Differences in the final digit can depend on whether you keep all digits in your calculator for v (which gives $a_R = 94.7 \text{ m/s}^2$), or if you use $v = 7.54 \text{ m/s}$ in which case you get $a_R = 94.8 \text{ m/s}^2$. Both results are valid since our assumed accuracy is about $\pm 0.1 \text{ m/s}$ (see Section 1–4).

EXAMPLE 5–2 **Moon's centripetal acceleration.** The Moon's nearly circular orbit about the Earth has a radius of about 384,000 km and a period T of 27.3 days. Determine the acceleration of the Moon toward the Earth.

APPROACH Again we need to find the velocity v in order to find a_R. We will need to convert to SI units to get v in m/s.

SOLUTION In one orbit around the Earth, the Moon travels a distance $2\pi r$, where $r = 3.84 \times 10^8$ m is the radius of its circular path. The time required for one complete orbit is the Moon's period of 27.3 d. The speed of the Moon in its orbit about the Earth is $v = 2\pi r/T$. The period T in seconds is $T = (27.3 \text{ d})(24.0 \text{ h/d})(3600 \text{ s/h}) = 2.36 \times 10^6$ s. Therefore,

$$a_R = \frac{v^2}{r} = \frac{(2\pi r)^2}{T^2 r} = \frac{4\pi^2 r}{T^2} = \frac{4\pi^2 (3.84 \times 10^8 \text{ m})}{(2.36 \times 10^6 \text{ s})^2}$$

$$= 0.00272 \text{ m/s}^2 = 2.72 \times 10^{-3} \text{ m/s}^2.$$

We can write this acceleration in terms of $g = 9.80 \text{ m/s}^2$ (the acceleration of gravity at the Earth's surface) as

$$a = 2.72 \times 10^{-3} \text{ m/s}^2 \left(\frac{g}{9.80 \text{ m/s}^2}\right) = 2.78 \times 10^{-4} g.$$

NOTE The centripetal acceleration of the Moon, $a = 2.78 \times 10^{-4} g$, is *not* the acceleration of gravity for objects at the Moon's surface due to the Moon's gravity. Rather, it is the acceleration due to the *Earth's* gravity for any object (such as the Moon) that is 384,000 km from the Earth. Notice how small this acceleration is compared to the acceleration of objects near the Earth's surface.

> ⚠ **CAUTION**
>
> *Distinguish Moon's gravity on objects at its surface, from Earth's gravity acting on Moon (this Example)*

5–2 Dynamics of Uniform Circular Motion

According to Newton's second law ($\Sigma\vec{F} = m\vec{a}$), an object that is accelerating must have a net force acting on it. An object moving in a circle, such as a ball on the end of a string, must therefore have a force applied to it to keep it moving in that circle. That is, a net force is necessary to give it centripetal acceleration. The magnitude of the required force can be calculated using Newton's second law for the radial component, $\Sigma F_R = m a_R$, where a_R is the centripetal acceleration, $a_R = v^2/r$, and ΣF_R is the total (or net) force in the radial direction:

$$\Sigma F_R = m a_R = m \frac{v^2}{r}. \qquad \text{[circular motion]} \quad \textbf{(5–3)}$$

For uniform circular motion ($v = $ constant), the acceleration is a_R, which is directed toward the center of the circle at any moment. Thus the *net force too must be directed toward the center of the circle* (Fig. 5–4). A net force is necessary because otherwise, if no net force were exerted on the object, it would not move in a circle but in a straight line, as Newton's first law tells us. The direction of the net force is continually changing so that it is always directed toward the center of the circle. This force is sometimes called a centripetal ("pointing toward the center") force. But be aware that "centripetal force" does not indicate some new kind of force. The term merely describes the *direction* of the net force needed to provide a circular path: the net force is directed toward the circle's center. The force *must be applied by other objects*. For example, to swing a ball in a circle on the end of a string, you pull on the string and the string exerts the force on the ball. (Try it.)

Force is needed to provide centripetal acceleration

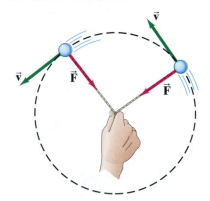

FIGURE 5–4 A force is required to keep an object moving in a circle. If the speed is constant, the force is directed toward the circle's center.

> ⚠ **CAUTION**
>
> *Centripetal force is not a new kind of force*
> *(Every force must be exerted by an object)*

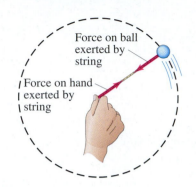

FIGURE 5–5 Swinging a ball on the end of a string.

FIGURE 5–6 If centrifugal force existed, the revolving ball would fly outward as in (a) when released. In fact, it flies off tangentially as in (b). For example, in (c) sparks fly in straight lines tangentially from the edge of a rotating grinding wheel.

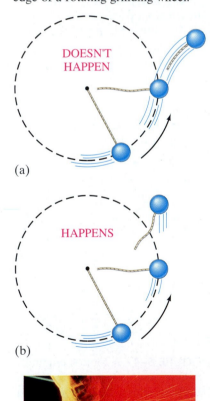

(a)

(b)

(c)

There is a common misconception that an object moving in a circle has an outward force acting on it, a so-called centrifugal ("center-fleeing") force. This is incorrect: *there is no outward force* on the revolving object. Consider, for example, a person swinging a ball on the end of a string around her head (Fig. 5–5). If you have ever done this yourself, you know that you feel a force pulling outward on your hand. The misconception arises when this pull is interpreted as an outward "centrifugal" force pulling on the ball that is transmitted along the string to your hand. This is not what is happening at all. To keep the ball moving in a circle, you pull *inwardly* on the string, and the string exerts this force on the ball. The ball exerts an equal and opposite force on the string (Newton's third law), and *this* is the outward force your hand feels (see Fig. 5–5).

The force *on the ball* is the one exerted *inwardly* on it by you, via the string. To see even more convincing evidence that a "centrifugal force" does not act on the ball, consider what happens when you let go of the string. If a centrifugal force were acting, the ball would fly outward, as shown in Fig. 5–6a. But it doesn't; the ball flies off tangentially (Fig. 5–6b), in the direction of the velocity it had at the moment it was released, because the inward force no longer acts. Try it and see!

EXAMPLE 5–3 **ESTIMATE** **Force on revolving ball (horizontal).** Estimate the force a person must exert on a string attached to a 0.150-kg ball to make the ball revolve in a horizontal circle of radius 0.600 m. The ball makes 2.00 revolutions per second ($T = 0.500$ s), as in Example 5–1.

APPROACH First we need to draw the free-body diagram for the ball. The forces acting on the ball are the force of gravity, $m\vec{\mathbf{g}}$ downward, and the tension force $\vec{\mathbf{F}}_T$ that the string exerts toward the hand at the center (which occurs because the person exerts that same force on the string). The free-body diagram for the ball is as shown in Fig. 5–7. The ball's weight complicates matters and makes it impossible to revolve a ball with the cord perfectly horizontal. We assume the weight is small, and put $\phi \approx 0$ in Fig. 5–7. Thus $\vec{\mathbf{F}}_T$ will act nearly horizontally and, in any case, provides the force necessary to give the ball its centripetal acceleration.

SOLUTION We apply Newton's second law to the radial direction, which we assume is horizontal:

$$(\Sigma F)_R = ma_R,$$

where $a_R = v^2/r$ and $v = 2\pi r/T = 2\pi(0.600 \text{ m})/(0.500 \text{ s}) = 7.54 \text{ m/s}$. Thus

$$F_T = m\frac{v^2}{r} = (0.150 \text{ kg})\frac{(7.54 \text{ m/s})^2}{(0.600 \text{ m})} \approx 14 \text{ N}.$$

NOTE We keep only two significant figures in the answer because $mg = (0.150 \text{ kg})(9.80 \text{ m/s}^2) = 1.5 \text{ N}$, being about $\frac{1}{10}$ of our result, is small but not so small as to justify stating a more precise answer since we ignored the effect of mg.

NOTE To include the effect of $m\vec{\mathbf{g}}$, resolve $\vec{\mathbf{F}}_T$ in Fig. 5–7 into components, and set the horizontal component of $\vec{\mathbf{F}}_T$ equal to mv^2/r and its vertical component equal to mg.

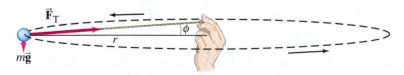

FIGURE 5–7 Example 5–3.

EXAMPLE 5–4 **Revolving ball (vertical circle).** A 0.150-kg ball on the end of a 1.10-m-long cord (negligible mass) is swung in a *vertical* circle. (*a*) Determine the minimum speed the ball must have at the top of its arc so that the ball continues moving in a circle. (*b*) Calculate the tension in the cord at the bottom of the arc, assuming the ball is moving at twice the speed of part (*a*).

APPROACH The ball moves in a vertical circle and is *not* undergoing uniform circular motion. The radius is assumed constant, but the speed v changes because of gravity. Nonetheless, Eq. 5–1 is valid at each point along the circle, and we use it at points 1 and 2. The free-body diagram is shown in Fig. 5–8 for both positions 1 and 2.

SOLUTION (*a*) At the top (point 1), two forces act on the ball: $m\vec{\mathbf{g}}$, the force of gravity, and $\vec{\mathbf{F}}_{T1}$, the tension force the cord exerts at point 1. Both act downward, and their vector sum acts to give the ball its centripetal acceleration a_R. We apply Newton's second law, for the vertical direction, choosing downward as positive since the acceleration is downward (toward the center):

$$(\Sigma F)_R = ma_R$$

$$F_{T1} + mg = m\frac{v_1^2}{r}. \qquad \text{[at top]}$$

From this equation we can see that the tension force F_{T1} at point 1 will get larger if v_1 (ball's speed at top of circle) is made larger, as expected. But we are asked for the *minimum* speed to keep the ball moving in a circle. The cord will remain taut as long as there is tension in it. But if the tension disappears (because v_1 is too small) the cord can go limp, and the ball will fall out of its circular path. Thus, the minimum speed will occur if $F_{T1} = 0$, for which we have

$$mg = m\frac{v_1^2}{r}. \qquad \text{[minimum speed at top]}$$

We solve for v_1:

$$v_1 = \sqrt{gr} = \sqrt{(9.80\,\text{m/s}^2)(1.10\,\text{m})} = 3.28\,\text{m/s}.$$

This is the minimum speed at the top of the circle if the ball is to continue moving in a circular path.

(*b*) When the ball is at the bottom of the circle (point 2 in Fig. 5–8), the cord exerts its tension force F_{T2} upward, whereas the force of gravity, $m\vec{\mathbf{g}}$, still acts downward. So we apply Newton's second law, this time choosing upward as positive since the acceleration is upward (toward the center):

$$(\Sigma F)_R = ma_R$$

$$F_{T2} - mg = m\frac{v_2^2}{r}. \qquad \text{[at bottom]}$$

The speed v_2 is given as twice that in (*a*), namely 6.56 m/s. We solve for F_{T2}:

$$F_{T2} = m\frac{v_2^2}{r} + mg$$

$$= (0.150\,\text{kg})\frac{(6.56\,\text{m/s})^2}{(1.10\,\text{m})} + (0.150\,\text{kg})(9.80\,\text{m/s}^2) = 7.34\,\text{N}.$$

EXERCISE B In a tumble dryer, the speed of the drum should be just large enough so that the clothes are carried nearly to the top of the drum and then fall away, rather than being pressed against the drum for the whole revolution. Determine whether this speed will be different for heavier wet clothes than for lighter dry clothes.

EXERCISE C A rider on a Ferris wheel moves in a vertical circle of radius r at constant speed v (Fig. 5–9). Is the normal force that the seat exerts on the rider at the top of the wheel (*a*) less than, (*b*) more than, or (*c*) the same as, the force the seat exerts at the bottom of the wheel?

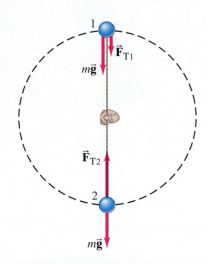

FIGURE 5–8 Example 5–4. Free-body diagrams for positions 1 and 2.

Cord tension and gravity together provide centripetal acceleration

Gravity provides centripetal acceleration

String tension and gravity acting in opposite directions provide centripetal acceleration

FIGURE 5–9 Exercise C.

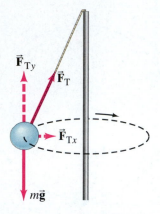

FIGURE 5–10   Example 5–5.

CONCEPTUAL EXAMPLE 5–5 **Tetherball.** The game of tetherball is played with a ball tied to a pole with a string. After the ball is struck, it revolves around the pole as shown in Fig. 5–10. In what direction is the acceleration of the ball, and what force causes the acceleration?

RESPONSE If the ball revolves in a horizontal plane as shown, then the acceleration points horizontally toward the center of the ball's circular path (not toward the top of the pole). The force responsible for the acceleration may not be obvious at first, since there seems to be no force pointing directly horizontally. But it is the *net* force (the sum of $m\vec{g}$ and $\vec{F}_T$ here) that must point in the direction of the acceleration. The vertical component of the string tension, F_{Ty}, balances the ball's weight, $m\vec{g}$. The horizontal component of the string tension, F_{Tx}, is the force that produces the centripetal acceleration toward the center.

PROBLEM SOLVING | Uniform Circular Motion

1. **Draw a free-body diagram**, showing all the forces acting on each object under consideration. Be sure you can identify the source of each force (tension in a cord, Earth's gravity, friction, normal force, and so on). Don't put in something that doesn't belong (like a centrifugal force).

2. **Determine** which of the forces, or which of their components, act to provide the centripetal acceleration—that is, all the **forces or components that act** radially, toward or away from the center of the circular path. The sum of these forces (or components) provides the centripetal acceleration, $a_R = v^2/r$.

3. **Choose a convenient coordinate system**, preferably with one axis along the acceleration direction.

4. **Apply Newton's second law** to the radial component:

$$(\Sigma F)_R = ma_R = m\frac{v^2}{r}. \qquad \text{[radial direction]}$$

5–3 Highway Curves, Banked and Unbanked

PHYSICS APPLIED
Driving around a curve

An example of circular dynamics occurs when an automobile rounds a curve, say to the left. In such a situation, you may feel that you are thrust outward toward the right side door. But there is no mysterious centrifugal force pulling on you. What is happening is that you tend to move in a straight line, whereas the car has begun to follow a curved path. To make you go in the curved path, the seat (friction) or the door of the car (direct contact) exerts a force on you (Fig. 5–11). The car also must have a force exerted on it toward the center of the curve if it is to move in that curve. On a flat road, this force is supplied by friction between the tires and the pavement.

FIGURE 5–11   The road exerts an inward force (friction against the tires) on a car to make it move in a circle. The car exerts an inward force on the passenger.

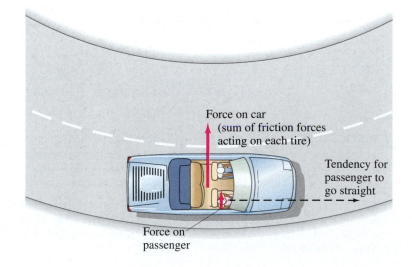

Force on car
(sum of friction forces
acting on each tire)

Tendency for
passenger to
go straight

Force on
passenger

If the wheels and tires of the car are rolling normally without slipping or sliding, the bottom of the tire is at rest against the road at each instant; so the friction force the road exerts on the tires is static friction. But if the static friction force is not great enough, as under icy conditions, sufficient friction force cannot be applied and the car will skid out of a circular path into a more nearly straight path. See Fig. 5–12. Once a car skids or slides, the friction force becomes kinetic friction, which is less than static friction.

FIGURE 5–12 Race car heading into a curve. From the tire marks we see that most cars experienced a sufficient friction force to give them the needed centripetal acceleration for rounding the curve safely. But, we also see tire tracks of cars on which there was not sufficient force—and which followed more nearly straight-line paths.

EXAMPLE 5–6 **Skidding on a curve.** A 1000-kg car rounds a curve on a flat road of radius 50 m at a speed of 50 km/h (14 m/s). Will the car follow the curve, or will it skid? Assume: (a) the pavement is dry and the coefficient of static friction is $\mu_s = 0.60$; (b) the pavement is icy and $\mu_s = 0.25$.

APPROACH The forces on the car are gravity mg downward, the normal force F_N exerted upward by the road, and a horizontal friction force due to the road. They are shown in Fig. 5–13, which is the free-body diagram for the car. The car will follow the curve if the maximum static friction force is greater than the mass times the centripetal acceleration.

SOLUTION In the vertical direction there is no acceleration. Newton's second law tells us that the normal force F_N on the car is equal to the weight mg since the road is flat:

$$F_N = mg = (1000 \text{ kg})(9.8 \text{ m/s}^2) = 9800 \text{ N}.$$

In the horizontal direction the only force is friction, and we must compare it to the force needed to produce the centripetal acceleration to see if it is sufficient. The net horizontal force required to keep the car moving in a circle around the curve is

$$(\Sigma F)_R = ma_R = m\frac{v^2}{r} = (1000 \text{ kg})\frac{(14 \text{ m/s})^2}{(50 \text{ m})} = 3900 \text{ N}.$$

Now we compute the maximum total static friction force (the sum of the friction forces acting on each of the four tires) to see if it can be large enough to provide a safe centripetal acceleration. For (a), $\mu_s = 0.60$, and the maximum friction force attainable (recall from Section 4–8 that $F_{fr} \le \mu_s F_N$) is

$$(F_{fr})_{max} = \mu_s F_N = (0.60)(9800 \text{ N}) = 5900 \text{ N}.$$

Since a force of only 3900 N is needed, and that is, in fact, how much will be exerted by the road as a static friction force, the car can follow the curve. But in (b) the maximum static friction force possible is

$$(F_{fr})_{max} = \mu_s F_N = (0.25)(9800 \text{ N}) = 2500 \text{ N}.$$

The car will skid because the ground cannot exert sufficient force (3900 N is needed) to keep it moving in a curve of radius 50 m at a speed of 50 km/h.

FIGURE 5–13 Example 5–6. Forces on a car rounding a curve on a flat road. (a) Front view, (b) top view.

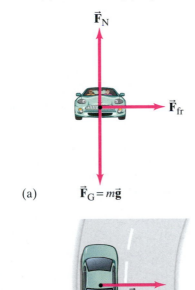

(a)

(b)

The possibility of skidding is worse if the wheels lock (stop rotating) when the brakes are applied too hard. When the tires are rolling, static friction exists. But if the wheels lock (stop rotating), the tires slide and the friction force, which is now kinetic friction, is less. More importantly, the *direction* of the friction force changes suddenly if the wheels lock. Static friction can point perpendicular to the velocity, as in Fig. 5–13b, but if the car slides, kinetic friction points *opposite* to the velocity. The force no longer points toward the center of the circle, and the car cannot continue in a curved path (see Fig. 5–12). Even worse, if the road is wet or icy, locking of the wheels occurs with less force on the brake pedal since there is less road friction to keep the wheels turning rather than sliding. Antilock brakes (ABS) are designed to limit brake pressure just before the point where sliding would occur, by means of delicate sensors and a fast computer.

PHYSICS APPLIED

Antilock brakes

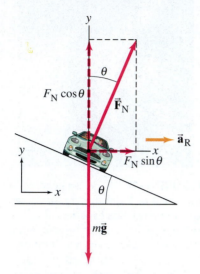

FIGURE 5–14 Normal force on a car rounding a banked curve, resolved into its horizontal and vertical components. The centripetal acceleration is horizontal (*not* parallel to the sloping road). The friction force on the tires, not shown, could point up or down along the slope, depending on the car's speed. The friction force will be zero for one particular speed.

CAUTION
F_N is not always equal to mg

Horizontal component of normal force acts to provide centripetal acceleration (friction is desired to be zero— otherwise it too would contribute)

Banking angle (friction not needed)

The banking of curves can reduce the chance of skidding. The normal force exerted by a banked road, acting perpendicular to the road, will have a component toward the center of the circle (Fig. 5–14), thus reducing the reliance on friction. For a given banking angle θ, there will be one speed for which no friction at all is required. This will be the case when the horizontal component of the normal force toward the center of the curve, $F_N \sin \theta$ (see Fig. 5–14), is just equal to the force required to give a vehicle its centripetal acceleration—that is, when

$$F_N \sin \theta = m \frac{v^2}{r}. \qquad \text{[no friction required]}$$

The banking angle of a road, θ, is chosen so that this condition holds for a particular speed, called the "design speed."

EXAMPLE 5–7 Banking angle. (*a*) For a car traveling with speed v around a curve of radius r, determine a formula for the angle at which a road should be banked so that no friction is required. (*b*) What is this angle for an expressway off-ramp curve of radius 50 m at a design speed of 50 km/h?

APPROACH Even though the road is banked, the car is still moving along a horizontal circle, so the centripetal acceleration needs to be horizontal. We choose our x and y axes as horizontal and vertical so that a_R, which is horizontal, is along the x axis. The forces on the car are the Earth's gravity mg downward, and the normal force F_N exerted by the road perpendicular to its surface. See Fig. 5–14, where the components of F_N are also shown. We don't need to consider the friction of the road because we are designing a road to be banked so as to eliminate dependence on friction.

SOLUTION (*a*) For the horizontal direction, $\Sigma F_R = ma_R$ gives

$$F_N \sin \theta = \frac{mv^2}{r}.$$

Since there is no vertical motion, the y component of the acceleration is zero, so $\Sigma F_y = ma_y$ gives us

$$F_N \cos \theta - mg = 0.$$

Thus,

$$F_N = \frac{mg}{\cos \theta}.$$

[Note in this case that $F_N \geq mg$ since $\cos \theta \leq 1$.]
We substitute this relation for F_N into the equation for the horizontal motion,

$$F_N \sin \theta = m \frac{v^2}{r},$$

and obtain

$$\frac{mg}{\cos \theta} \sin \theta = m \frac{v^2}{r}$$

or

$$mg \tan \theta = m \frac{v^2}{r},$$

so

$$\tan \theta = \frac{v^2}{rg}.$$

This is the formula for the banking angle θ: no friction needed at speed v.
(*b*) For $r = 50$ m and $v = 50$ km/h (or 14 m/s),

$$\tan \theta = \frac{(14 \text{ m/s})^2}{(50 \text{ m})(9.8 \text{ m/s}^2)} = 0.40,$$

so $\theta = 22°$.

EXERCISE D To negotiate an unbanked curve at a *faster* speed, a driver puts a couple of sand bags in his van aiming to increase the force of friction between the tires and the road. Will the sand bags help?

EXERCISE E Can a heavy truck and a small car travel safely at the same speed around an icy, banked-curve road?

*5–4 Nonuniform Circular Motion

Circular motion at constant speed occurs when the net force on an object is exerted toward the center of the circle. If the net force is not directed toward the center but is at an angle, as shown in Fig. 5–15a, the force has two components. The component directed toward the center of the circle, F_R, gives rise to the centripetal acceleration, a_R, and keeps the object moving in a circle. The component tangent to the circle, F_{tan}, acts to increase (or decrease) the speed, and thus gives rise to a component of the acceleration tangent to the circle, a_{tan}. When the speed of the object is changing, a tangential component of force is acting.

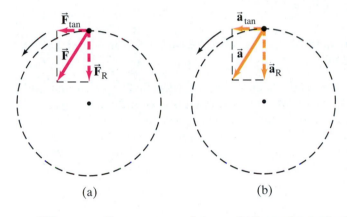

(a)   (b)

FIGURE 5–15 The speed of an object moving in a circle changes if the force on it has a tangential component, F_{tan}. Part (a) shows the force $\vec{F}$ and its vector components; part (b) shows the acceleration vector and its vector components.

When you first start revolving a ball on the end of a string around your head, you must give it tangential acceleration. You do this by pulling on the string with your hand displaced from the center of the circle. In athletics, a hammer thrower accelerates the hammer tangentially in a similar way so that it reaches a high speed before release.

The tangential component of the acceleration, a_{tan}, is equal to the rate of change of the *magnitude* of the object's velocity:

$$a_{tan} = \frac{\Delta v}{\Delta t}.$$

The radial (centripetal) acceleration arises from the change in *direction* of the velocity and, as we have seen (Eq. 5–1), is given by

$$a_R = \frac{v^2}{r}.$$

The tangential acceleration always points in a direction tangent to the circle, and is in the direction of motion (parallel to $\vec{v}$, which is always tangent to the circle) if the speed is increasing, as shown in Fig. 5–15b. If the speed is decreasing, $\vec{a}_{tan}$ points antiparallel to $\vec{v}$. In either case, $\vec{a}_{tan}$ and $\vec{a}_R$ are always perpendicular to each other; and *their directions change* continually as the object moves along its circular path. The total vector acceleration $\vec{a}$ is the sum of these two:

$$\vec{a} = \vec{a}_{tan} + \vec{a}_R.$$

Since $\vec{a}_R$ and $\vec{a}_{tan}$ are always perpendicular to each other, the magnitude of $\vec{a}$ at any moment is

$$a = \sqrt{a_{tan}^2 + a_R^2}.$$

EXAMPLE 5–8 **Two components of acceleration.** A race car starts from rest in the pit area and accelerates at a uniform rate to a speed of 35 m/s in 11 s, moving on a circular track of radius 500 m. Assuming constant tangential acceleration, find (a) the tangential acceleration, and (b) the radial acceleration, at the instant when the speed is $v = 15$ m/s.

APPROACH The tangential acceleration relates to the change in speed of the car, and can be calculated as $a_{tan} = \Delta v/\Delta t$. The centripetal acceleration relates to the change in the *direction* of the velocity vector and is calculated using $a_R = v^2/r$.

SOLUTION (a) During the 11-s time interval, we assume the tangential acceleration a_{tan} is constant. Its magnitude is

$$a_{tan} = \frac{\Delta v}{\Delta t} = \frac{(35 \text{ m/s} - 0 \text{ m/s})}{11 \text{ s}} = 3.2 \text{ m/s}^2.$$

(b) When $v = 15$ m/s, the centripetal acceleration is

$$a_R = \frac{v^2}{r} = \frac{(15 \text{ m/s})^2}{(500 \text{ m})} = 0.45 \text{ m/s}^2.$$

EXERCISE F When the speed of the race car in Example 5–8 is 30 m/s, how are (a) a_{tan} and (b) a_R changed?

These concepts can be used for an object moving along any curved path, such as that shown in Fig. 5–16. We can treat any portion of the curve as an arc of a circle with a radius of curvature r. The velocity at any point is always tangent to the path. The acceleration can be written, in general, as a vector sum of two components: the tangential component $a_{tan} = \Delta v/\Delta t$, and the radial (centripetal) component $a_R = v^2/r$.

FIGURE 5–16 Object following a curved path (solid line). At point P the path has a radius of curvature r. The object has velocity $\vec{v}$, tangential acceleration $\vec{a}_{tan}$ (the object is increasing in speed), and radial (centripetal) acceleration $\vec{a}_R$ (magnitude $a_R = v^2/r$) which points toward the center of curvature C.

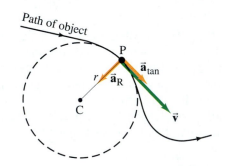

Path of object

*5-5 Centrifugation

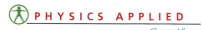

A useful device that nicely illustrates circular motion is the centrifuge, or the very high speed ultracentrifuge. These devices are used to sediment materials quickly or to separate materials. Test tubes are held in the centrifuge rotor, which is accelerated to very high rotational speeds: see Fig. 5–17, where one test tube is shown in two positions as the rotor turns. The small green dot represents a small particle, perhaps a macromolecule, in a fluid-filled test tube. When the tube is at position A and the rotor is turning, the particle has a tendency to move in a straight line in the direction of the dashed arrow. But the fluid, resisting the motion of the particles, exerts a centripetal force that keeps the particles moving nearly in a circle. Usually, the resistance of the fluid (a liquid, a gas, or a gel, depending on the application) does not quite equal mv^2/r, and the particles eventually reach the bottom of the tube. The purpose of a centrifuge is to provide an "effective gravity" much larger than normal gravity because of the high rotational speeds, thus causing more rapid sedimentation.

EXAMPLE 5–9 **Ultracentrifuge.** The rotor of an ultracentrifuge rotates at 50,000 rpm (revolutions per minute). The top of a 4.00-cm-long test tube (Fig. 5–17) is 6.00 cm from the rotation axis and is perpendicular to it. The bottom of the tube is 10.00 cm from the axis of rotation. Calculate the centripetal acceleration, in "g's," at the top and the bottom of the tube.

APPROACH We can calculate the centripetal acceleration from $a_R = v^2/r$. We divide by $g = 9.80 \text{ m/s}^2$ to find a_R in g's.

SOLUTION At the top of the tube, a particle revolves in a circle of circumference $2\pi r$, which is a distance

$$2\pi r = (2\pi)(0.0600 \text{ m}) = 0.377 \text{ m per revolution.}$$

It makes 5.00×10^4 such revolutions each minute, or, dividing by 60 s/min, 833 rev/s. The time to make one revolution, the period T, is

$$T = \frac{1}{(833 \text{ rev/s})} = 1.20 \times 10^{-3} \text{ s/rev.}$$

The speed of the particle is then

$$v = \frac{2\pi r}{T} = \left(\frac{0.377 \text{ m/rev}}{1.20 \times 10^{-3} \text{ s/rev}} \right) = 3.14 \times 10^2 \text{ m/s.}$$

The centripetal acceleration is

$$a_R = \frac{v^2}{r} = \frac{(3.14 \times 10^2 \text{ m/s})^2}{0.0600 \text{ m}} = 1.64 \times 10^6 \text{ m/s}^2,$$

which, dividing by $g = 9.80 \text{ m/s}^2$, is 1.67×10^5 g's.

At the bottom of the tube ($r = 0.1000$ m), the speed is

$$v = \frac{2\pi r}{T} = \frac{(2\pi)(0.1000 \text{ m})}{1.20 \times 10^{-3} \text{ s/rev}} = 523.6 \text{ m/s.}$$

Then

$$a_R = \frac{v^2}{r} = \frac{(523.6 \text{ m/s})^2}{(0.1000 \text{ m})} = 2.74 \times 10^6 \text{ m/s}^2$$

$$= 2.80 \times 10^5 \text{ g's,}$$

or 280,000 g's.

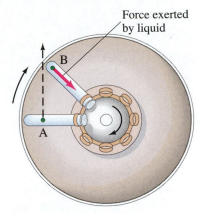

Force exerted by liquid

FIGURE 5–17 Two positions of a rotating test tube in a centrifuge (top view). At A, the green dot represents a macromolecule or other particle being sedimented. It would tend to follow the dashed line, heading toward the bottom of the tube, but the fluid resists this motion by exerting a force on the particle as shown at point B.

5–6 Newton's Law of Universal Gravitation

Besides developing the three laws of motion, Sir Isaac Newton also examined the motion of the planets and the Moon. In particular, he wondered about the nature of the force that must act to keep the Moon in its nearly circular orbit around the Earth.

Newton was also thinking about the problem of gravity. Since falling objects accelerate, Newton had concluded that they must have a force exerted on them, a force we call the force of gravity. Whenever an object has a force exerted *on* it, that force is exerted *by* some other object. But what *exerts* the force of gravity? Every object on the surface of the Earth feels the force of gravity, and no matter where the object is, the force is directed toward the center of the Earth (Fig. 5–18). Newton concluded that it must be the Earth itself that exerts the gravitational force on objects at its surface.

According to legend, Newton noticed an apple drop from a tree. He is said to have been struck with a sudden inspiration: If gravity acts at the tops of trees, and even at the tops of mountains, then perhaps it acts all the way to the Moon!

FIGURE 5–18 Anywhere on Earth, whether in Alaska, Peru, or Australia, the force of gravity acts downward toward the Earth's center.

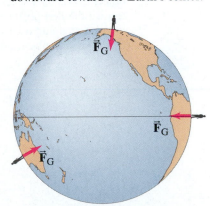

With this idea that it is the Earth's gravity that holds the Moon in its orbit, Newton developed his great theory of gravitation. But there was controversy at the time. Many thinkers had trouble accepting the idea of a force "acting at a distance." Typical forces act through contact—your hand pushes a cart and pulls a wagon, a bat hits a ball, and so on. But gravity acts without contact, said Newton: the Earth exerts a force on a falling apple and on the Moon, even though there is no contact, and the two objects may even be very far apart.

Newton set about determining the magnitude of the gravitational force that the Earth exerts on the Moon as compared to the gravitational force on objects at the Earth's surface. The centripetal acceleration of the Moon, as we calculated in Example 5–2, is $a_R = 0.00272 \text{ m/s}^2$. In terms of the acceleration of gravity at the Earth's surface, $g = 9.80 \text{ m/s}^2$,

The Moon's acceleration toward Earth

$$a_R = \frac{0.00272 \text{ m/s}^2}{9.8 \text{ m/s}^2} \approx \frac{1}{3600} g.$$

That is, the acceleration of the Moon toward the Earth is about $\frac{1}{3600}$ as great as the acceleration of objects at the Earth's surface. The Moon is 384,000 km from the Earth, which is about 60 times the Earth's radius of 6380 km. That is, the Moon is 60 times farther from the Earth's center than are objects at the Earth's surface. But $60 \times 60 = 60^2 = 3600$. Again that number 3600. Newton concluded that the gravitational force exerted by the Earth on any object decreases with the square of its distance r from the Earth's center:

$$\text{force of gravity} \propto \frac{1}{r^2}.$$

The Moon is 60 Earth radii away, so it feels a gravitational force only $\frac{1}{60^2} = \frac{1}{3600}$ times as strong as an equal mass would at the Earth's surface.

Newton realized that the force of gravity on an object depends not only on distance but also on the object's mass. In fact, it is directly proportional to its mass, as we have seen. According to Newton's third law, when the Earth exerts its gravitational force on any object, such as the Moon, that object exerts an equal and opposite force on the Earth (Fig. 5–19). Because of this symmetry, Newton reasoned, the magnitude of the force of gravity must be proportional to *both* the masses. Thus

$$F \propto \frac{m_E m_{Obj}}{r^2},$$

where m_E is the mass of the Earth, m_{Obj} the mass of the other object, and r the distance from the Earth's center to the center of the other object.

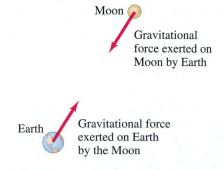

FIGURE 5–19 The gravitational force one object exerts on a second object is directed toward the first object, and (by Newton's third law) is equal and opposite to the force exerted by the second object on the first.

Moon

Gravitational force exerted on Moon by Earth

Earth

Gravitational force exerted on Earth by the Moon

Newton went a step further in his analysis of gravity. In his examination of the orbits of the planets, he concluded that the force required to hold the planets in their orbits around the Sun seems to diminish as the inverse square of their distance from the Sun. This led him to believe that it is also the gravitational force that acts between the Sun and each of the planets to keep them in their orbits. And if gravity acts between these objects, why not between all objects?

Thus he proposed his **law of universal gravitation**, which we can state as follows:

> **Every particle in the universe attracts every other particle with a force that is proportional to the product of their masses and inversely proportional to the square of the distance between them. This force acts along the line joining the two particles.**

NEWTON'S

LAW

OF

UNIVERSAL

GRAVITATION

The magnitude of the gravitational force can be written as

$$F = G\frac{m_1 m_2}{r^2}, \qquad (5\text{--}4)$$

where m_1 and m_2 are the masses of the two particles, r is the distance between them, and G is a universal constant which must be measured experimentally and has the same numerical value for all objects.

The value of G must be very small, since we are not aware of any force of attraction between ordinary-sized objects, such as between two baseballs. The force between two ordinary objects was first measured by Henry Cavendish in 1798, over 100 years after Newton published his law. To detect and measure the incredibly small force between ordinary objects, he used an apparatus like that shown in Fig. 5–20. Cavendish confirmed Newton's hypothesis that two objects attract one another, and that Eq. 5–4 accurately describes this force. In addition, because Cavendish could measure F, m_1, m_2, and r accurately, he was able to determine the value of the constant G as well. The accepted value today is

$$G = 6.67 \times 10^{-11}\,\text{N}\cdot\text{m}^2/\text{kg}^2.$$

[Strictly speaking, Eq. 5–4 gives the magnitude of the gravitational force that one particle exerts on a second particle that is a distance r away. For an extended object (that is, not a point), we must consider how to measure the distance r. This is often best done using integral calculus, which Newton himself invented. Newton showed that for two uniform spheres, Eq. 5–4 gives the correct force where r is the distance between their centers. When extended objects are small compared to the distance between them (as for the Earth–Sun system), little inaccuracy results from considering them as point particles.]

EXAMPLE 5–10 ESTIMATE Can you attract another person gravitationally? A 50-kg person and a 75-kg person are sitting on a bench. Estimate the magnitude of the gravitational force each exerts on the other.

APPROACH This is an estimate: we let the distance between the people be $\frac{1}{2}$ m, and round off G to $10^{-10}\,\text{N}\cdot\text{m}^2/\text{kg}^2$.

SOLUTION We use Eq. 5–4:

$$F = G\frac{m_1 m_2}{r^2} \approx \frac{(10^{-10}\,\text{N}\cdot\text{m}^2/\text{kg}^2)(50\,\text{kg})(75\,\text{kg})}{(0.5\,\text{m})^2} \approx 10^{-6}\,\text{N},$$

which is unnoticeably small unless very delicate instruments are used.

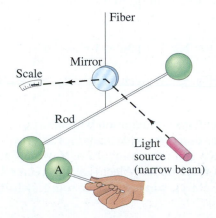

Fiber

Mirror

Scale

Rod

Light source
(narrow beam)

A

FIGURE 5–20 Schematic diagram of Cavendish's apparatus. Two spheres are attached to a lightweight horizontal rod, which is suspended at its center by a thin fiber. When a third sphere labeled A is brought close to one of the suspended spheres, the gravitational force causes the latter to move, and this twists the fiber slightly. The tiny movement is magnified by the use of a narrow light beam directed at a mirror mounted on the fiber. The beam reflects onto a scale. Previous determination of how large a force will twist the fiber a given amount then allows one to determine the magnitude of the gravitational force between two objects.

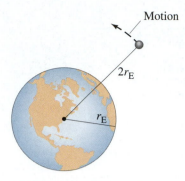

Motion

$2r_E$

r_E

FIGURE 5–21 Example 5–11.

EXAMPLE 5–11 **Spacecraft at $2r_E$.** What is the force of gravity acting on a 2000-kg spacecraft when it orbits two Earth radii from the Earth's center (that is, a distance $r_E = 6380$ km above the Earth's surface, Fig. 5–21)? The mass of the Earth is $M_E = 5.98 \times 10^{24}$ kg.

APPROACH We could plug all the numbers into Eq. 5–4, but there is a simpler approach. The spacecraft is twice as far from the Earth's center as when it is at the surface of the Earth. Therefore, since the force of gravity decreases as the square of the distance (and $\frac{1}{2^2} = \frac{1}{4}$), the force of gravity on the satellite will be only one-fourth its weight at the Earth's surface.

SOLUTION At the surface of the Earth, $F_G = mg$. At a distance from the Earth's center of $2r_E$, F_G is $\frac{1}{4}$ as great:

$$F_G = \tfrac{1}{4}mg = \tfrac{1}{4}(2000\text{ kg})(9.80\text{ m/s}^2)$$
$$= 4900\text{ N}.$$

FIGURE 5–22 Example 5–12. Orientation of Sun (S), Earth (E), and Moon (M) at right angles to each other (not to scale).

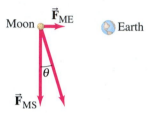

Moon $\vec{F}_{ME}$ Earth

θ

$\vec{F}_{MS}$

Sun

EXAMPLE 5–12 **Force on the Moon.** Find the net force on the Moon $(m_M = 7.35 \times 10^{22}\text{ kg})$ due to the gravitational attraction of both the Earth $(m_E = 5.98 \times 10^{24}\text{ kg})$ and the Sun $(m_S = 1.99 \times 10^{30}\text{ kg})$, assuming they are at right angles to each other as in Fig. 5–22.

APPROACH The forces on our object, the Moon, are the gravitational force exerted on the Moon by the Earth F_{ME} and that exerted by the Sun F_{MS}, as shown in the free-body diagram of Fig. 5–22. We use the law of universal gravitation to find the magnitude of each force, and then add the two forces as vectors.

SOLUTION The Earth is 3.84×10^5 km $= 3.84 \times 10^8$ m from the Moon, so F_{ME} (the gravitational force on the Moon due to the Earth) is

$$F_{ME} = \frac{(6.67 \times 10^{-11}\text{ N}\cdot\text{m}^2/\text{kg}^2)(7.35 \times 10^{22}\text{ kg})(5.98 \times 10^{24}\text{ kg})}{(3.84 \times 10^8\text{ m})^2}$$
$$= 1.99 \times 10^{20}\text{ N}.$$

The Sun is 1.50×10^8 km from the Earth and the Moon, so F_{MS} (the gravitational force on the Moon due to the Sun) is

$$F_{MS} = \frac{(6.67 \times 10^{-11}\text{ N}\cdot\text{m}^2/\text{kg}^2)(7.35 \times 10^{22}\text{ kg})(1.99 \times 10^{30}\text{ kg})}{(1.50 \times 10^{11}\text{ m})^2}$$
$$= 4.34 \times 10^{20}\text{ N}.$$

The two forces act at right angles in the case we are considering (Fig. 5–22), so we can apply the Pythagorean theorem to find the magnitude of the total force:

$$F = \sqrt{(1.99 \times 10^{20}\text{ N})^2 + (4.34 \times 10^{20}\text{ N})^2} = 4.77 \times 10^{20}\text{ N}.$$

The force acts at an angle θ (Fig. 5–22) given by $\theta = \tan^{-1}(1.99/4.34) = 24.6°$.

⚠ **CAUTION**

Distinguish between Newton's second law and the law of universal gravitation

Don't confuse the law of universal gravitation with Newton's second law of motion, $\Sigma\vec{F} = m\vec{a}$. The former describes a particular force, gravity, and how its strength varies with the distance and masses involved. Newton's second law, on the other hand, relates the net force on an object (i.e., the vector sum of all the different forces acting on the object, whatever their sources) to the mass and acceleration of that object.

5-7 Gravity Near the Earth's Surface; Geophysical Applications

When Eq. 5–4 is applied to the gravitational force between the Earth and an object at its surface, m_1 becomes the mass of the Earth m_E, m_2 becomes the mass of the object m, and r becomes the distance of the object from the Earth's center,[†] which is the radius of the Earth r_E. This force of gravity due to the Earth is the weight of the object, which we have been writing as mg. Thus,

$$mg = G\frac{mm_E}{r_E^2}.$$

We can solve this for g, the acceleration of gravity at the Earth's surface:

$$g = G\frac{m_E}{r_E^2}. \qquad (5-5)$$

g in terms of G

Thus, the acceleration of gravity at the surface of the Earth, g, is determined by m_E and r_E. (Don't confuse G with g; they are very different quantities, but are related by Eq. 5–5.)

! **CAUTION**
Distinguish G from g

Until G was measured, the mass of the Earth was not known. But once G was measured, Eq. 5–5 could be used to calculate the Earth's mass, and Cavendish was the first to do so. Since $g = 9.80 \text{ m/s}^2$ and the radius of the Earth is $r_E = 6.38 \times 10^6 \text{ m}$, then, from Eq. 5–5, we obtain

$$m_E = \frac{gr_E^2}{G} = \frac{(9.80 \text{ m/s}^2)(6.38 \times 10^6 \text{ m})^2}{6.67 \times 10^{-11} \text{ N} \cdot \text{m}^2/\text{kg}^2} = 5.98 \times 10^{24} \text{ kg}$$

Mass of the Earth

for the mass of the Earth.

Equation 5–5 can be applied to other planets, where g, m, and r would refer to that planet.

EXAMPLE 5–13 ESTIMATE **Gravity on Everest.** Estimate the effective value of g on the top of Mt. Everest, 8850 m (29,035 ft) above sea level. That is, what is the acceleration due to gravity of objects allowed to fall freely at this altitude?

APPROACH The force of gravity (and the acceleration due to gravity g) depends on the distance from the center of the Earth, so there will be an effective value g' on top of Mt. Everest which will be smaller than g at sea level. We assume the Earth is a uniform sphere (a reasonable "estimate").

SOLUTION We use Eq. 5–5, with r_E replaced by $r = 6380 \text{ km} + 8.9 \text{ km} = 6389 \text{ km} = 6.389 \times 10^6 \text{ m}$:

$$g = G\frac{m_E}{r^2} = \frac{(6.67 \times 10^{-11} \text{ N} \cdot \text{m}^2/\text{kg}^2)(5.98 \times 10^{24} \text{ kg})}{(6.389 \times 10^6 \text{ m})^2} = 9.77 \text{ m/s}^2,$$

which is a reduction of about 3 parts in a thousand (0.3%).

NOTE This is an estimate because, among other things, we ignored the mass accumulated under the mountaintop.

Note that Eq. 5–5 does not give precise values for g at different locations because the Earth is not a perfect sphere. The Earth not only has mountains and valleys, and bulges at the equator, but also its mass is not distributed precisely uniformly (see Table 5–1). The Earth's rotation also affects the value of g. However, for most practical purposes, when an object is near the Earth's surface, we will simply use $g = 9.80 \text{ m/s}^2$ and write the weight of an object as mg.

[†]That the distance is measured from the Earth's center does not imply that the force of gravity somehow emanates from that one point. Rather, all parts of the Earth attract gravitationally, but the net effect is a force acting toward the Earth's center.

TABLE 5–1
Acceleration Due to Gravity at Various Locations on Earth

Location	Elevation (m)	g (m/s²)
New York	0	9.803
San Francisco	0	9.800
Denver	1650	9.796
Pikes Peak	4300	9.789
Sydney, Australia	0	9.798
Equator	0	9.780
North Pole (calculated)	0	9.832

The value of g can vary locally on the Earth's surface because of the presence of irregularities and rocks of different densities. Such variations in g, known as "gravity anomalies," are very small—on the order of 1 part per 10^6 or 10^7 in the value of g. But they can be measured by "gravimeters" which detect variations in g to 1 part in 10^9. Geophysicists use such measurements as part of their investigations into the structure of the Earth's crust, and in mineral and oil exploration. Mineral deposits, for example, often have a greater density than does surrounding material. Because of the greater mass in a given volume, g can have a slightly greater value on top of such a deposit than at its flanks. "Salt domes," under which petroleum is often found, have a lower than average density; searches for a slight reduction in the value of g in certain locales have led to the discovery of oil.

5–8 Satellites and "Weightlessness"

Satellite Motion

Artificial satellites circling the Earth are now commonplace (Fig. 5–23). A satellite is put into orbit by accelerating it to a sufficiently high tangential speed with the use of rockets, as shown in Fig. 5–24. If the speed is too high, the spacecraft will not be confined by the Earth's gravity and will escape, never to return. If the speed is too low, it will return to Earth. Satellites are usually put into circular (or nearly circular) orbits, because such orbits require the least takeoff speed.

FIGURE 5–23   A satellite circling the Earth.

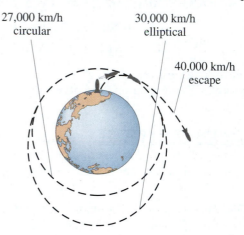

FIGURE 5–24   Artificial satellites launched at different speeds.

It is sometimes asked: "What keeps a satellite up?" The answer is: its high speed. If a satellite stopped moving, it would fall directly to Earth. But at the very high speed a satellite has, it would quickly fly out into space (Fig. 5–25) if it weren't for the gravitational force of the Earth pulling it into orbit. In fact, a satellite *is* falling (accelerating toward Earth), but its high tangential speed keeps it from hitting Earth.

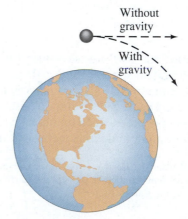

FIGURE 5–25   A moving satellite "falls" out of a straight-line path toward the Earth.

EXAMPLE 5–14 **Geosynchronous satellite.** A *geosynchronous* satellite is one that stays above the same point on the Earth, which is possible only if it is above a point on the equator. Such satellites are used for TV and radio transmission, for weather forecasting, and as communication relays. Determine (*a*) the height above the Earth's surface such a satellite must orbit, and (*b*) such a satellite's speed. (*c*) Compare to the speed of a satellite orbiting 200 km above Earth's surface.

PHYSICS APPLIED
Geosynchronous satellites

APPROACH To remain above the same point on Earth as the Earth rotates, the satellite must have a period of one day. We can apply Newton's second law, $F = ma$, where $a = v^2/r$ if we assume the orbit is circular.

SOLUTION (*a*) The only force on the satellite is the force of universal gravitation. So Eq. 5–4 gives us the force F, which we insert into Newton's second law:

$$F = ma$$

$$G\frac{m_{Sat}\,m_E}{r^2} = m_{Sat}\frac{v^2}{r}. \qquad \text{[satellite equation]}$$

This equation has two unknowns, r and v. But the satellite revolves around the Earth with the same period that the Earth rotates on its axis, namely once in 24 hours. Thus the speed of the satellite must be

$$v = \frac{2\pi r}{T},$$

where $T = 1$ day $= (24\,\text{h})(3600\,\text{s/h}) = 86{,}400\,\text{s}$. We substitute this into the "satellite equation" above and obtain (after canceling m_{Sat} on both sides)

$$G\frac{m_E}{r^2} = \frac{(2\pi r)^2}{rT^2}.$$

After cancelling an r, we can solve for r^3:

$$r^3 = \frac{Gm_E T^2}{4\pi^2} = \frac{(6.67 \times 10^{-11}\,\text{N}\cdot\text{m}^2/\text{kg}^2)(5.98 \times 10^{24}\,\text{kg})(86{,}400\,\text{s})^2}{4\pi^2}$$

$$= 7.54 \times 10^{22}\,\text{m}^3.$$

Taking the cube root, we get $r = 4.23 \times 10^7\,\text{m}$, or 42,300 km from the Earth's center. We subtract the Earth's radius of 6380 km to find that a geosynchronous satellite must orbit about 36,000 km (about 6 r_E) above the Earth's surface.
(*b*) We solve for v in the satellite equation given in part (*a*):

$$v = \sqrt{\frac{Gm_E}{r}} = \sqrt{\frac{(6.67 \times 10^{-11}\,\text{N}\cdot\text{m}^2/\text{kg}^2)(5.98 \times 10^{24}\,\text{kg})}{(4.23 \times 10^7\,\text{m})}} = 3070\,\text{m/s}.$$

We get the same result if we use $v = 2\pi r/T$.
(*c*) The equation in part (*b*) for v shows $v \propto \sqrt{1/r}$. So for $r = r_E + h = 6380\,\text{km} + 200\,\text{km} = 6580\,\text{km}$, we get

$$v' = v\sqrt{\frac{r}{r'}} = (3070\,\text{m/s})\sqrt{\frac{(42{,}300\,\text{km})}{(6580\,\text{km})}} = 7780\,\text{m/s}.$$

NOTE The center of a satellite orbit is always at the center of the Earth; so it is not possible to have a satellite orbiting above a fixed point on the Earth at any latitude other than 0°.

EXERCISE G Two satellites orbit the Earth in circular orbits of the same radius. One satellite is twice as massive as the other. Which of the following statements is true about the speeds of these satellites? (*a*) The heavier satellite moves twice as fast as the lighter one. (*b*) The two satellites have the same speed. (*c*) The lighter satellite moves twice as fast as the heavier one. (*d*) The heavier satellite moves four times as fast as the lighter one.

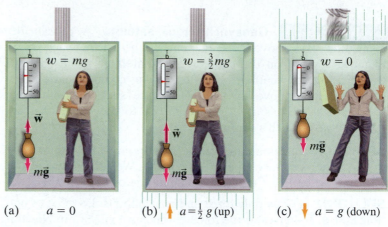

FIGURE 5–26 (a) An object in an elevator at rest exerts a force on a spring scale equal to its weight. (b) In an elevator accelerating upward at $\frac{1}{2}g$, the object's apparent weight is $1\frac{1}{2}$ times larger than its true weight. (c) In a freely falling elevator, the object experiences "weightlessness": the scale reads zero.

(a) $a = 0$ (b) $a = \frac{1}{2}g$ (up) (c) $a = g$ (down)

Weightlessness

"Weightlessness" in a falling elevator

People and other objects in a satellite circling the Earth are said to experience apparent weightlessness. Let us first look at a simpler case, that of a falling elevator. In Fig. 5–26a, an elevator is at rest with a bag hanging from a spring scale. The scale reading indicates the downward force exerted on it by the bag. This force, exerted *on* the scale, is equal and opposite to the force exerted *by* the scale upward on the bag, and we call its magnitude w. Two forces act on the bag: the downward gravitational force and the upward force exerted by the scale (Newton's third law) equal to w. Because the bag is not accelerating, when we apply $\Sigma F = ma$ to the bag in Fig. 5–26a we obtain

$$w - mg = 0,$$

where mg is the weight of the bag. Thus, $w = mg$, and since the scale indicates the force w exerted on it by the bag, it registers a force equal to the weight of the bag, as we expect.

If, now, the elevator has an acceleration, a, then applying $\Sigma F = ma$ to the bag, we have

$$w - mg = ma.$$

Solving for w, we have

$$w = mg + ma. \qquad [a \text{ is } + \text{ upward}]$$

We have chosen the positive direction up. Thus, if the acceleration a is up, a is positive; and the scale, which measures w, will read more than mg. We call w the *apparent weight* of the bag, which in this case would be greater than its actual weight (mg). If the elevator accelerates downward, a will be negative and w, the apparent weight, will be less than mg. The direction of the velocity $\vec{v}$ doesn't matter. Only the direction of the acceleration $\vec{a}$ influences the scale reading.

Suppose, for example, the elevator's acceleration is $\frac{1}{2}g$ upward; then we find

$$w = mg + m(\tfrac{1}{2}g) = \tfrac{3}{2}mg.$$

That is, the scale reads $1\frac{1}{2}$ times the actual weight of the bag (Fig. 5–26b). The apparent weight of the bag is $1\frac{1}{2}$ times its real weight. The same is true of the person: her apparent weight (equal to the normal force exerted on her by the elevator floor) is $1\frac{1}{2}$ times her real weight. We can say that she is experiencing $1\frac{1}{2}g$'s, just as astronauts experience so many g's at a rocket's launch.

If, instead, the elevator's acceleration is $a = -\frac{1}{2}g$ (downward), then $w = mg - \frac{1}{2}mg = \frac{1}{2}mg$. That is, the scale reads half the actual weight. If the elevator is in *free fall* (for example, if the cables break), then $a = -g$ and $w = mg - mg = 0$. The scale reads zero. See Fig. 5–26c. The bag appears weightless. If the person in the elevator accelerating at $-g$ let go of a pencil, say, it would not fall to the floor. True, the pencil would be falling with acceleration g. But so would the floor of the elevator and the person. The pencil would hover right in front of the person. This phenomenon is called *apparent weightlessness* because in the reference frame of the person, objects don't fall or seem to have weight—yet gravity does not disappear. It is still acting on the object, whose

FIGURE 5–27 Experiencing weightlessness on Earth.

(a)

(b)

(c)

weight is still *mg*. The objects seem weightless only because the elevator is in free fall, and there is no contact force to make us feel the weight.

The "weightlessness" experienced by people in a satellite orbit close to the Earth is the same apparent weightlessness experienced in a freely falling elevator. It may seem strange, at first, to think of a satellite as freely falling. But a satellite is indeed falling toward the Earth, as was shown in Fig. 5–25. The force of gravity causes it to "fall" out of its natural straight-line path. The acceleration of the satellite must be the acceleration due to gravity at that point, since the only force acting on it is gravity. Thus, although the force of gravity acts on objects within the satellite, the objects experience an apparent weightlessness because they, and the satellite, are all accelerating as in free fall.

Figure 5–27 shows some examples of "free fall," or apparent weightlessness, experienced by people on Earth for brief moments.

A different situation occurs when a spacecraft is out in space far from the Earth, the Moon, and other attracting objects. The force of gravity due to the Earth and other heavenly bodies will then be quite small because of the distances involved, and people in such a spacecraft will experience real weightlessness.

"Weightlessness" in a satellite

5–9 Kepler's Laws and Newton's Synthesis

More than a half century before Newton proposed his three laws of motion and his law of universal gravitation, the German astronomer Johannes Kepler (1571–1630) had worked out a detailed description of the motion of the planets about the Sun: three empirical findings that we now refer to as **Kepler's laws of planetary motion**. They are summarized as follows, with additional explanation in Figs. 5–28 and 5–29.

Kepler's first law: The path of each planet about the Sun is an ellipse with the Sun at one focus (Fig. 5–28).

Kepler's second law: Each planet moves so that an imaginary line drawn from the Sun to the planet sweeps out equal areas in equal periods of time (Fig. 5–29).

Kepler's third law: The ratio of the squares of the periods T of any two planets revolving about the Sun is equal to the ratio of the cubes of their mean distances s from the Sun: $(T_1/T_2)^2 = (s_1/s_2)^3$. [Actually, s is the semimajor axis, defined as half the long (major) axis of the orbit, as shown in Fig. 5–28. We can also call it the mean distance of the planet from the Sun.] Present-day data are given in Table 5–2: see the last column.

Kepler arrived at his laws through careful analysis of experimental data. Fifty years later, Newton was able to show that Kepler's laws could be derived mathematically from the law of universal gravitation and the laws of motion. Newton also showed that for any reasonable form for the gravitational force law, only one that depends on the inverse square of the distance is fully consistent with Kepler's laws. He thus used Kepler's laws as evidence in favor of his law of universal gravitation, Eq. 5–4.

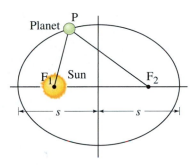

FIGURE 5–28 (a) *Kepler's first law*. An ellipse is a closed curve such that the sum of the distances from any point P on the curve to two fixed points (called the foci, F_1 and F_2) remains constant. That is, the sum of the distances, $F_1 P + F_2 P$, is the same for all points on the curve. A circle is a special case of an ellipse in which the two foci coincide, at the center of the circle.

FIGURE 5–29 *Kepler's second law*. The two shaded regions have equal areas. The planet moves from point 1 to point 2 in the same time as it takes to move from point 3 to point 4. Planets move fastest in that part of their orbit where they are closest to the Sun. Exaggerated scale.

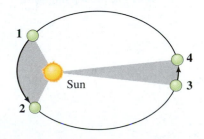

TABLE 5–2 Planetary Data Applied to Kepler's Third Law

Planet	Mean Distance from Sun, s (10^6 km)	Period, T (Earth years)	s^3/T^2 (10^{24} km^3/y^2)
Mercury	57.9	0.241	3.34
Venus	108.2	0.615	3.35
Earth	149.6	1.0	3.35
Mars	227.9	1.88	3.35
Jupiter	778.3	11.86	3.35
Saturn	1427	29.5	3.34
Uranus	2870	84.0	3.35
Neptune	4497	165	3.34
Pluto	5900	248	3.34

Derivation of
Kepler's third law

We will derive Kepler's third law for the special case of a circular orbit. (Most planetary orbits are close to a circle.) First, we write Newton's second law of motion, $\Sigma F = ma$. For F we use the gravitational force (Eq. 5–4) between the Sun and a planet of mass m_1, and for a the centripetal acceleration, v^2/r. We assume the mass of the Sun, M_S, is much greater than the mass of its planets. Then

$$\Sigma F = ma$$
$$G\frac{m_1 M_S}{r_1^2} = m_1 \frac{v_1^2}{r_1}.$$

Here r_1 is the distance of one planet from the Sun, and v_1 is its average speed in orbit; M_S is the mass of the Sun, since it is the gravitational attraction of the Sun that keeps each planet in its orbit. The period T_1 of the planet is the time required for one complete orbit, a distance equal to the circumference of its orbit, $2\pi r_1$. Thus

$$v_1 = \frac{2\pi r_1}{T_1}.$$

We substitute this formula for v_1 into the equation above:

$$G\frac{m_1 M_S}{r_1^2} = m_1 \frac{4\pi^2 r_1}{T_1^2}.$$

We rearrange this to get

$$\frac{T_1^2}{r_1^3} = \frac{4\pi^2}{GM_S}. \qquad (5\text{–}6a)$$

We derived this for planet 1 (say, Mars). The same derivation would apply for a second planet (say, Saturn) orbiting the Sun,

$$\frac{T_2^2}{r_2^3} = \frac{4\pi^2}{GM_S},$$

where T_2 and r_2 are the period and orbit radius, respectively, for the second planet. Since the right sides of the two previous equations are equal, we have $T_1^2/r_1^3 = T_2^2/r_2^3$ or, rearranging,

Kepler's third law

$$\left(\frac{T_1}{T_2}\right)^2 = \left(\frac{r_1}{r_2}\right)^3, \qquad (5\text{–}6b)$$

which is Kepler's third law.

The derivations of Eqs. 5–6a and 5–6b (Kepler's third law) compared two planets revolving around the Sun; but they are general enough to be applied to other systems. For example, we could apply Eq. 5–6a to our Moon revolving around Earth (then M_S would be M_E, the mass of the Earth). Or we could apply Eq. 5–6b to compare two moons revolving around Jupiter. But Kepler's third law applies only to objects orbiting the same attracting center. Do not use Eq. 5–6b to compare, say, the Moon's orbit around the Earth to the orbit of Mars around the Sun because they depend on different attracting centers.

 **CAUTION**
Compare orbits of objects only around the same center

In the following Examples, we assume the orbits are circles, although it is not quite true in general.

EXAMPLE 5–15 **Where is Mars?** Mars' period (its "year") was noted by Kepler to be about 687 days (Earth days), which is (687 d/365 d) = 1.88 yr. Determine the distance of Mars from the Sun using the Earth as a reference.

APPROACH We know the periods of Earth and Mars, and the distance from the Sun to Earth. We can use Kepler's third law to obtain the distance from the Sun to Mars.

SOLUTION The period of the Earth is $T_E = 1$ yr, and the distance of Earth from the Sun is $r_{ES} = 1.50 \times 10^{11}$ m. From Kepler's third law (Eq. 5–6b):

$$\frac{r_{MS}}{r_{ES}} = \left(\frac{T_M}{T_E}\right)^{\frac{2}{3}} = \left(\frac{1.88 \text{ yr}}{1 \text{ yr}}\right)^{\frac{2}{3}} = 1.52.$$

So Mars is 1.52 times the Earth's distance from the Sun, or 2.28×10^{11} m.

EXAMPLE 5–16 **The Sun's mass determined.** Determine the mass of the Sun given the Earth's distance from the Sun as $r_{ES} = 1.5 \times 10^{11}$ m.

PHYSICS APPLIED
Determining the Sun's mass

APPROACH Equation 5–6a, relates the mass of the Sun M_S to the period and distance of any planet. We use the Earth.

SOLUTION The Earth's period is $T_E = 1$ yr $= (365\frac{1}{4}\text{d})(24 \text{ h/d})(3600 \text{ s/h}) = 3.16 \times 10^7$ s. We solve Eq. 5–6a for M_S:

$$M_S = \frac{4\pi^2 r_{ES}^3}{GT_E^2} = \frac{4\pi^2(1.5 \times 10^{11} \text{ m})^3}{(6.67 \times 10^{-11} \text{ N}\cdot\text{m}^2/\text{kg}^2)(3.16 \times 10^7 \text{ s})^2} = 2.0 \times 10^{30} \text{ kg}.$$

Accurate measurements on the orbits of the planets indicated that they did not precisely follow Kepler's laws. For example, slight deviations from perfectly elliptical orbits were observed. Newton was aware that this was to be expected because any planet would be attracted gravitationally not only by the Sun but also (to a much lesser extent) by the other planets. Such deviations, or **perturbations**, in the orbit of Saturn were a hint that helped Newton formulate the law of universal gravitation, that all objects attract gravitationally. Observation of other perturbations later led to the discovery of Neptune and Pluto. Deviations in the orbit of Uranus, for example, could not all be accounted for by perturbations due to the other known planets. Careful calculation in the nineteenth century indicated that these deviations could be accounted for if another planet existed farther out in the solar system. The position of this planet was predicted from the deviations in the orbit of Uranus, and telescopes focused on that region of the sky quickly found it; the new planet was called Neptune. Similar but much smaller perturbations of Neptune's orbit led to the discovery of Pluto in 1930.

Perturbations and discovery of planets

Starting in the mid–1990s, planets revolving about distant stars (Fig. 5–30) were inferred from the regular "wobble" of each star due to the gravitational attraction of the revolving planet(s).

Planets around other stars

The development by Newton of the law of universal gravitation and the three laws of motion was a major intellectual achievement: with these laws, he was able to describe the motion of objects on Earth and in the heavens. The motions of heavenly bodies and objects on Earth were seen to follow the same laws. For this reason, and also because Newton integrated the results of earlier scientists into his system, we sometimes speak of *Newton's synthesis.*

Newton's synthesis

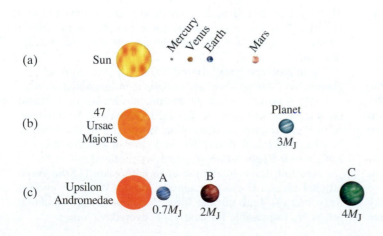

FIGURE 5–30 Our solar system (a) is compared to recently discovered planets orbiting (b) the star 47 Ursae Majoris and (c) the star Upsilon Andromedae with at least three planets. M_J is the mass of Jupiter. (Sizes not to scale.)

The laws formulated by Newton are referred to as **causal laws**. By **causality** we mean the idea that one occurrence can cause another. When a rock strikes a window, we infer that the rock *causes* the window to break. This idea of "cause and effect" relates to Newton's laws: the acceleration of an object was seen to be *caused* by the net force acting on it.

5–10 Types of Forces in Nature

We have already discussed that Newton's law of universal gravitation, Eq. 5–4, describes how a particular type of force—gravity—depends on the masses of the objects involved and the distance between them. Newton's second law, $\Sigma \vec{F} = m\vec{a}$, on the other hand, tells how an object will accelerate due to *any* type of force. But what are the types of forces that occur in nature besides gravity?

In the twentieth century, physicists came to recognize four fundamental forces in nature: (1) the gravitational force; (2) the electromagnetic force (we shall see later that electric and magnetic forces are intimately related); (3) the strong nuclear force; and (4) the weak nuclear force. In this Chapter, we discussed the gravitational force in detail. The nature of the electromagnetic force will be discussed in Chapters 16 to 22. The strong and weak nuclear forces, which are discussed in Chapters 30 to 32, operate at the level of the atomic nucleus; although they manifest themselves in such phenomena as radioactivity and nuclear energy, they are much less obvious in our daily lives.

Electroweak and GUT

Physicists have been working on theories that would unify these four forces—that is, to consider some or all of these forces as different manifestations of the same basic force. So far, the electromagnetic and weak nuclear forces have been theoretically united to form *electroweak* theory, in which the electromagnetic and weak forces are seen as two different manifestations of a single *electroweak force*. Attempts to further unify the forces, such as in *grand unified theories* (GUT), are hot research topics today.

Everyday forces are gravity and electromagnetic

But where do everyday forces fit into this scheme? Ordinary forces, other than gravity, such as pushes, pulls, and other contact forces like the normal force and friction, are today considered to be due to the electromagnetic force acting at the atomic level. For example, the force your fingers exert on a pencil is the result of electrical repulsion between the outer electrons of the atoms of your finger and those of the pencil.

Summary

An object moving in a circle of radius r with constant speed v is said to be in **uniform circular motion**. It has a **centripetal acceleration** a_R that is directed radially toward the center of the circle (also called **radial acceleration**), and has magnitude

$$a_R = \frac{v^2}{r}. \tag{5–1}$$

The direction of the velocity vector and that of the acceleration $\vec{a}_R$ are continually changing in direction, but are perpendicular to each other at each moment.

A force is needed to keep a particle revolving in a circle, and the direction of this force is toward the center of the circle. This force may be due to gravity, to tension in a cord, to a component of the normal force, to another type of force, or to a combination of forces.

[*When the speed of circular motion is not constant, the acceleration has two components, tangential as well as centripetal.]

Newton's **law of universal gravitation** states that every particle in the universe attracts every other particle with a force proportional to the product of their masses and inversely proportional to the square of the distance between them:

$$F = G\frac{m_1 m_2}{r^2}. \tag{5–4}$$

The direction of this force is along the line joining the two particles. It is this gravitational force that keeps the Moon revolving around the Earth, and the planets revolving around the Sun.

Satellites revolving around the Earth are acted on by gravity, but "stay up" because of their high tangential speed.

[*Newton's three laws of motion, plus his law of universal gravitation, constituted a wide-ranging theory of the universe. With them, motion of objects on Earth and in the heavens could be accurately described. And they provided a theoretical base for **Kepler's laws** of planetary motion.]

The four fundamental forces in nature are (1) the gravitational force, (2) electromagnetic force, (3) strong nuclear force, and (4) weak nuclear force. The first two fundamental forces are responsible for nearly all "everyday" forces.

Questions

1. Sometimes people say that water is removed from clothes in a spin dryer by centrifugal force throwing the water outward. What is wrong with this statement?

2. Will the acceleration of a car be the same when the car travels around a sharp curve at a constant 60 km/h as when it travels around a gentle curve at the same speed? Explain.

3. Suppose a car moves at constant speed along a hilly road. Where does the car exert the greatest and least forces on the road: (a) at the top of a hill, (b) at a dip between two hills, (c) on a level stretch near the bottom of a hill?

4. Describe all the forces acting on a child riding a horse on a merry-go-round. Which of these forces provides the centripetal acceleration of the child?

5. A bucket of water can be whirled in a vertical circle without the water spilling out, even at the top of the circle when the bucket is upside down. Explain.

6. How many "accelerators" do you have in your car? There are at least three controls in the car which can be used to cause the car to accelerate. What are they? What accelerations do they produce?

7. A child on a sled comes flying over the crest of a small hill, as shown in Fig. 5–31. His sled does not leave the ground (he does not achieve "air"), but he feels the normal force between his chest and the sled decrease as he goes over the hill. Explain this decrease using Newton's second law.

FIGURE 5–31 Question 7.

8. Why do bicycle riders lean inward when rounding a curve at high speed?

9. Why do airplanes bank when they turn? How would you compute the banking angle given its speed and radius of the turn?

10. A girl is whirling a ball on a string around her head in a horizontal plane. She wants to let go at precisely the right time so that the ball will hit a target on the other side of the yard. When should she let go of the string?

11. Does an apple exert a gravitational force on the Earth? If so, how large a force? Consider an apple (a) attached to a tree, and (b) falling.

12. If the Earth's mass were double what it is, in what ways would the Moon's orbit be different?

13. Which pulls harder gravitationally, the Earth on the Moon, or the Moon on the Earth? Which accelerates more?

14. The Sun's gravitational pull on the Earth is much larger than the Moon's. Yet the Moon's is mainly responsible for the tides. Explain. [*Hint*: Consider the difference in gravitational pull from one side of the Earth to the other.]

15. Will an object weigh more at the equator or at the poles? What two effects are at work? Do they oppose each other?

16. The gravitational force on the Moon due to the Earth is only about half the force on the Moon due to the Sun. Why isn't the Moon pulled away from the Earth?

17. Is the centripetal acceleration of Mars in its orbit around the Sun larger or smaller than the centripetal acceleration of the Earth?

18. Would it require less speed to launch a satellite (a) toward the east or (b) toward the west? Consider the Earth's rotation direction.

19. When will your apparent weight be the greatest, as measured by a scale in a moving elevator: when the elevator (a) accelerates downward, (b) accelerates upward, (c) is in free fall, (d) moves upward at constant speed? In which case would your weight be the least? When would it be the same as when you are on the ground?

20. What keeps a satellite up in its orbit around the Earth?

21. Astronauts who spend long periods in outer space could be adversely affected by weightlessness. One way to simulate gravity is to shape the spaceship like a cylindrical shell that rotates, with the astronauts walking on the inside surface (Fig. 5–32). Explain how this simulates gravity. Consider (a) how objects fall, (b) the force we feel on our feet, and (c) any other aspects of gravity you can think of.

FIGURE 5–32 Question 21 and Problem 45.

22. Explain how a runner experiences "free fall" or "apparent weightlessness" between steps.

* 23. The Earth moves faster in its orbit around the Sun in January than in July. Is the Earth closer to the Sun in January, or in July? Explain. [*Note*: This is not much of a factor in producing the seasons—the main factor is the tilt of the Earth's axis relative to the plane of its orbit.]

* 24. The mass of Pluto was not known until it was discovered to have a moon. Explain how this discovery enabled an estimate of Pluto's mass.

Problems

5–1 to 5–3 Uniform Circular Motion; Highway Curves

1. (I) A child sitting 1.10 m from the center of a merry-go-round moves with a speed of 1.25 m/s. Calculate (*a*) the centripetal acceleration of the child, and (*b*) the net horizontal force exerted on the child (mass = 25.0 kg).

2. (I) A jet plane traveling 1890 km/h (525 m/s) pulls out of a dive by moving in an arc of radius 6.00 km. What is the plane's acceleration in *g*'s?

3. (I) Calculate the centripetal acceleration of the Earth in its orbit around the Sun, and the net force exerted on the Earth. What exerts this force on the Earth? Assume that the Earth's orbit is a circle of radius 1.50×10^{11} m. [*Hint*: see the Tables inside the front cover of this book.]

4. (I) A horizontal force of 210 N is exerted on a 2.0-kg discus as it rotates uniformly in a horizontal circle (at arm's length) of radius 0.90 m. Calculate the speed of the discus.

5. (II) Suppose the space shuttle is in orbit 400 km from the Earth's surface, and circles the Earth about once every 90 minutes. Find the centripetal acceleration of the space shuttle in its orbit. Express your answer in terms of *g*, the gravitational acceleration at the Earth's surface.

6. (II) What is the magnitude of the acceleration of a speck of clay on the edge of a potter's wheel turning at 45 rpm (revolutions per minute) if the wheel's diameter is 32 cm?

7. (II) A ball on the end of a string is revolved at a uniform rate in a vertical circle of radius 72.0 cm, as shown in Fig. 5–33. If its speed is 4.00 m/s and its mass is 0.300 kg, calculate the tension in the string when the ball is (*a*) at the top of its path, and (*b*) at the bottom of its path.

8. (II) A 0.45-kg ball, attached to the end of a horizontal cord, is rotated in a circle of radius 1.3 m on a frictionless horizontal surface. If the cord will break when the tension in it exceeds 75 N, what is the maximum speed the ball can have?

9. (II) What is the maximum speed with which a 1050-kg car can round a turn of radius 77 m on a flat road if the coefficient of static friction between tires and road is 0.80? Is this result independent of the mass of the car?

10. (II) How large must the coefficient of static friction be between the tires and the road if a car is to round a level curve of radius 85 m at a speed of 95 km/h?

11. (II) A device for training astronauts and jet fighter pilots is designed to rotate a trainee in a horizontal circle of radius 12.0 m. If the force felt by the trainee on her back is 7.85 times her own weight, how fast is she rotating? Express your answer in both m/s and rev/s.

12. (II) A coin is placed 11.0 cm from the axis of a rotating turntable of variable speed. When the speed of the turntable is slowly increased, the coin remains fixed on the turntable until a rate of 36 rpm is reached and the coin slides off. What is the coefficient of static friction between the coin and the turntable?

13. (II) At what minimum speed must a roller coaster be traveling when upside down at the top of a circle (Fig. 5–34) so that the passengers will not fall out? Assume a radius of curvature of 7.4 m.

FIGURE 5–34 Problem 13.

14. (II) A sports car of mass 950 kg (including the driver) crosses the rounded top of a hill (radius = 95 m) at 22 m/s. Determine (*a*) the normal force exerted by the road on the car, (*b*) the normal force exerted by the car on the 72-kg driver, and (*c*) the car speed at which the normal force on the driver equals zero.

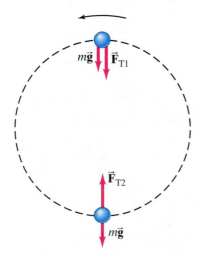

FIGURE 5–33 Problem 7.

15. (II) How many revolutions per minute would a 15-m-diameter Ferris wheel need to make for the passengers to feel "weightless" at the topmost point?

16. (II) A bucket of mass 2.00 kg is whirled in a vertical circle of radius 1.10 m. At the lowest point of its motion the tension in the rope supporting the bucket is 25.0 N. (a) Find the speed of the bucket. (b) How fast must the bucket move at the top of the circle so that the rope does not go slack?

17. (II) How fast (in rpm) must a centrifuge rotate if a particle 9.00 cm from the axis of rotation is to experience an acceleration of 115,000 g's?

18. (II) In a "Rotor-ride" at a carnival, people are rotated in a cylindrically walled "room." (See Fig. 5–35.) The room radius is 4.6 m, and the rotation frequency is 0.50 revolutions per second when the floor drops out. What is the minimum coefficient of static friction so that the people will not slip down? People on this ride say they were "pressed against the wall." Is there really an outward force pressing them against the wall? If so, what is its source? If not, what is the proper description of their situation (besides "scary")? [Hint: First draw the free-body diagram for a person.]

FIGURE 5–35   Problem 18.

19. (II) A flat puck (mass M) is rotated in a circle on a frictionless air-hockey tabletop, and is held in this orbit by a light cord connected to a dangling block (mass m) through a central hole as shown in Fig. 5–36. Show that the speed of the puck is given by

$$v = \sqrt{\frac{mgR}{M}}.$$

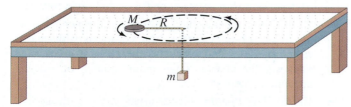

FIGURE 5–36   Problem 19.

20. (II) Redo Example 5–3, precisely this time, by not ignoring the weight of the ball which revolves on a string 0.600 m long. In particular, find the magnitude of $\vec{\mathbf{F}}_T$, and the angle it makes with the horizontal. [Hint: Set the horizontal component of $\vec{\mathbf{F}}_T$ equal to ma_R; also, since there is no vertical motion, what can you say about the vertical component of $\vec{\mathbf{F}}_T$?]

21. (III) If a curve with a radius of 88 m is perfectly banked for a car traveling 75 km/h, what must be the coefficient of static friction for a car not to skid when traveling 95 km/h?

22. (III) A 1200-kg car rounds a curve of radius 67 m banked at an angle of 12°. If the car is traveling at 95 km/h, will a friction force be required? If so, how much and in what direction?

23. (III) Two blocks, of masses m_1 and m_2, are connected to each other and to a central post by cords as shown in Fig. 5–37. They rotate about the post at a frequency f (revolutions per second) on a frictionless horizontal surface at distances r_1 and r_2 from the post. Derive an algebraic expression for the tension in each segment of the cord.

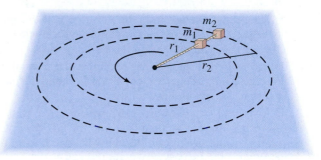

FIGURE 5–37   Problem 23.

24. (III) A pilot performs an evasive maneuver by diving vertically at 310 m/s. If he can withstand an acceleration of 9.0 g's without blacking out, at what altitude must he begin to pull out of the dive to avoid crashing into the sea?

* 5–4 Nonuniform Circular Motion

* 25. (I) Determine the tangential and centripetal components of the net force exerted on the car (by the ground) in Example 5–8 when its speed is 15 m/s. The car's mass is 1100 kg.

* 26. (II) A car at the Indianapolis 500 accelerates uniformly from the pit area, going from rest to 320 km/h in a semicircular arc with a radius of 220 m. Determine the tangential and radial acceleration of the car when it is halfway through the turn, assuming constant tangential acceleration. If the curve were flat, what would the coefficient of static friction have to be between the tires and the road to provide this acceleration with no slipping or skidding?

* 27. (III) A particle revolves in a horizontal circle of radius 2.90 m. At a particular instant, its acceleration is 1.05 m/s², in a direction that makes an angle of 32.0° to its direction of motion. Determine its speed (a) at this moment, and (b) 2.00 s later, assuming constant tangential acceleration.

5–6 and 5–7 Law of Universal Gravitation

28. (I) Calculate the force of Earth's gravity on a spacecraft 12,800 km (2 Earth radii) above the Earth's surface if its mass is 1350 kg.

29. (I) At the surface of a certain planet, the gravitational acceleration g has a magnitude of 12.0 m/s². A 21.0-kg brass ball is transported to this planet. What is (a) the mass of the brass ball on the Earth and on the planet, and (b) the weight of the brass ball on the Earth and on the planet?

30. (II) Calculate the acceleration due to gravity on the Moon. The Moon's radius is 1.74×10^6 m and its mass is 7.35×10^{22} kg.

31. (II) A hypothetical planet has a radius 1.5 times that of Earth, but has the same mass. What is the acceleration due to gravity near its surface?

32. (II) A hypothetical planet has a mass 1.66 times that of Earth, but the same radius. What is g near its surface?

33. (II) Two objects attract each other gravitationally with a force of 2.5×10^{-10} N when they are 0.25 m apart. Their total mass is 4.0 kg. Find their individual masses.

34. (II) Calculate the effective value of g, the acceleration of gravity, at (a) 3200 m, and (b) 3200 km, above the Earth's surface.

35. (II) What is the distance from the Earth's center to a point outside the Earth where the gravitational acceleration due to the Earth is $\frac{1}{10}$ of its value at the Earth's surface?

36. (II) A certain neutron star has five times the mass of our Sun packed into a sphere about 10 km in radius. Estimate the surface gravity on this monster.

37. (II) A typical white-dwarf star, which once was an average star like our Sun but is now in the last stage of its evolution, is the size of our Moon but has the mass of our Sun. What is the surface gravity on this star?

38. (II) You are explaining why astronauts feel weightless while orbiting in the space shuttle. Your friends respond that they thought gravity was just a lot weaker up there. Convince them and yourself that it isn't so by calculating the acceleration of gravity 250 km above the Earth's surface in terms of g.

39. (II) Four 9.5-kg spheres are located at the corners of a square of side 0.60 m. Calculate the magnitude and direction of the total gravitational force exerted on one sphere by the other three.

40. (II) Every few hundred years most of the planets line up on the same side of the Sun. Calculate the total force on the Earth due to Venus, Jupiter, and Saturn, assuming all four planets are in a line (Fig. 5–38). The masses are $M_V = 0.815 M_E$, $M_J = 318 M_E$, $M_S = 95.1 M_E$, and their mean distances from the Sun are 108, 150, 778, and 1430 million km, respectively. What fraction of the Sun's force on the Earth is this?

FIGURE 5–38 Problem 40. (Not to scale.)

41. (II) Given that the acceleration of gravity at the surface of Mars is 0.38 of what it is on Earth, and that Mars' radius is 3400 km, determine the mass of Mars.

42. (III) Determine the mass of the Sun using the known value for the period of the Earth and its distance from the Sun. [Note: Compare your answer to that obtained using Kepler's laws, Example 5–16.]

5–8 Satellites; Weightlessness

43. (I) Calculate the speed of a satellite moving in a stable circular orbit about the Earth at a height of 3600 km.

44. (I) The space shuttle releases a satellite into a circular orbit 650 km above the Earth. How fast must the shuttle be moving (relative to Earth) when the release occurs?

45. (II) At what rate must a cylindrical spaceship rotate if occupants are to experience simulated gravity of 0.60 g? Assume the spaceship's diameter is 32 m, and give your answer as the time needed for one revolution. (See Question 21, Fig 5–32.)

46. (II) Determine the time it takes for a satellite to orbit the Earth in a circular "near-Earth" orbit. A "near-Earth" orbit is one at a height above the surface of the Earth which is very small compared to the radius of the Earth. Does your result depend on the mass of the satellite?

47. (II) At what horizontal velocity would a satellite have to be launched from the top of Mt. Everest to be placed in a circular orbit around the Earth?

48. (II) During an *Apollo* lunar landing mission, the command module continued to orbit the Moon at an altitude of about 100 km. How long did it take to go around the Moon once?

49. (II) The rings of Saturn are composed of chunks of ice that orbit the planet. The inner radius of the rings is 73,000 km, while the outer radius is 170,000 km. Find the period of an orbiting chunk of ice at the inner radius and the period of a chunk at the outer radius. Compare your numbers with Saturn's mean rotation period of 10 hours and 39 minutes. The mass of Saturn is 5.7×10^{26} kg.

50. (II) A Ferris wheel 24.0 m in diameter rotates once every 15.5 s (see Fig. 5–9). What is the ratio of a person's apparent weight to her real weight (a) at the top, and (b) at the bottom?

51. (II) What is the apparent weight of a 75-kg astronaut 4200 km from the center of the Earth's Moon in a space vehicle (a) moving at constant velocity, and (b) accelerating toward the Moon at 2.9 m/s²? State the "direction" in each case.

52. (II) Suppose that a binary-star system consists of two stars of equal mass. They are observed to be separated by 360 million km and take 5.7 Earth years to orbit about a point midway between them. What is the mass of each?

53. (II) What will a spring scale read for the weight of a 55-kg woman in an elevator that moves (a) upward with constant speed of 6.0 m/s, (b) downward with constant speed of 6.0 m/s, (c) upward with acceleration of 0.33 g, (d) downward with acceleration 0.33 g, and (e) in free fall?

54. (II) A 17.0-kg monkey hangs from a cord suspended from the ceiling of an elevator. The cord can withstand a tension of 220 N and breaks as the elevator accelerates. What was the elevator's minimum acceleration (magnitude and direction)?

55. (III) (a) Show that if a satellite orbits very near the surface of a planet with period T, the density (mass/volume) of the planet is $\rho = m/V = 3\pi/GT^2$. (b) Estimate the density of the Earth, given that a satellite near the surface orbits with a period of about 85 min.

*5–9 Kepler's Laws

* 56. (I) Use Kepler's laws and the period of the Moon (27.4 d) to determine the period of an artificial satellite orbiting very near the Earth's surface.

* 57. (I) The asteroid Icarus, though only a few hundred meters across, orbits the Sun like the planets. Its period is 410 d. What is its mean distance from the Sun?

* 58. (I) Neptune is an average distance of 4.5×10^9 km from the Sun. Estimate the length of the Neptunian year given that the Earth is 1.50×10^8 km from the Sun on the average.

* 59. (II) Halley's comet orbits the Sun roughly once every 76 years. It comes very close to the surface of the Sun on its closest approach (Fig. 5–39). Estimate the greatest distance of the comet from the Sun. Is it still "in" the Solar System? What planet's orbit is nearest when it is out there? [*Hint:* The mean distance s in Kepler's third law is half the sum of the nearest and farthest distance from the Sun.]

Halley's comet
Sun

FIGURE 5–39
Problem 59.

* 60. (II) Our Sun rotates about the center of the Galaxy $(M_G \approx 4 \times 10^{41}$ kg$)$ at a distance of about 3×10^4 light-years $(1 \text{ ly} = 3 \times 10^8 \text{ m/s} \times 3.16 \times 10^7 \text{ s/y} \times 1 \text{ y})$. What is the period of our orbital motion about the center of the Galaxy?

* 61. (II) Table 5–3 gives the mass, period, and mean distance for the four largest moons of Jupiter (those discovered by Galileo in 1609). (*a*) Determine the mass of Jupiter using the data for Io. (*b*) Determine the mass of Jupiter using data for each of the other three moons. Are the results consistent?

TABLE 5–3 Principal Moons of Jupiter

Moon	Mass (kg)	Period (Earth days)	Mean distance from Jupiter (km)
Io	8.9×10^{22}	1.77	422×10^3
Europa	4.9×10^{22}	3.55	671×10^3
Ganymede	15×10^{22}	7.16	1070×10^3
Callisto	11×10^{22}	16.7	1883×10^3

* 62. (II) Determine the mass of the Earth from the known period and distance of the Moon.

* 63. (II) Determine the mean distance from Jupiter for each of Jupiter's moons, using Kepler's third law. Use the distance of Io and the periods given in Table 5–3. Compare to the values in the Table.

* 64. (II) The asteroid belt between Mars and Jupiter consists of many fragments (which some space scientists think came from a planet that once orbited the Sun but was destroyed). (*a*) If the center of mass of the asteroid belt (where the planet would have been) is about three times farther from the Sun than the Earth is, how long would it have taken this hypothetical planet to orbit the Sun? (*b*) Can we use these data to deduce the mass of this planet?

* 65. (III) A science-fiction tale describes an artificial "planet" in the form of a band completely encircling a sun (Fig. 5–40). The inhabitants live on the inside surface (where it is always noon). Imagine that this sun is exactly like our own, that the distance to the band is the same as the Earth–Sun distance (to make the climate temperate), and that the ring rotates quickly enough to produce an apparent gravity of g as on Earth. What will be the period of revolution, this planet's year, in Earth days?

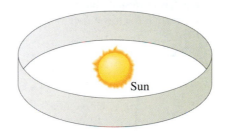
Sun

FIGURE 5–40
Problem 65.

General Problems

66. Tarzan plans to cross a gorge by swinging in an arc from a hanging vine (Fig. 5–41). If his arms are capable of exerting a force of 1400 N on the vine, what is the maximum speed he can tolerate at the lowest point of his swing? His mass is 80 kg, and the vine is 5.5 m long.

FIGURE 5–41
Problem 66.

67. How far above the Earth's surface will the acceleration of gravity be half what it is on the surface?

68. On an ice rink, two skaters of equal mass grab hands and spin in a mutual circle once every 2.5 s. If we assume their arms are each 0.80 m long and their individual masses are 60.0 kg, how hard are they pulling on one another?

69. Because the Earth rotates once per day, the apparent acceleration of gravity at the equator is slightly less than it would be if the Earth didn't rotate. Estimate the magnitude of this effect. What fraction of g is this?

70. At what distance from the Earth will a spacecraft traveling directly from the Earth to the Moon experience zero net force because the Earth and Moon pull with equal and opposite forces?

71. You know your mass is 65 kg, but when you stand on a bathroom scale in an elevator, it says your mass is 82 kg. What is the acceleration of the elevator, and in which direction?

72. A projected space station consists of a circular tube that will rotate about its center (like a tubular bicycle tire) (Fig. 5–42). The circle formed by the tube has a diameter of about 1.1 km. What must be the rotation speed (revolutions per day) if an effect equal to gravity at the surface of the Earth (1.0 g) is to be felt?

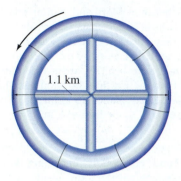

FIGURE 5–42
Problem 72.

73. A jet pilot takes his aircraft in a vertical loop (Fig. 5–43). (a) If the jet is moving at a speed of 1300 km/h at the lowest point of the loop, determine the minimum radius of the circle so that the centripetal acceleration at the lowest point does not exceed 6.0 g's. (b) Calculate the 78-kg pilot's effective weight (the force with which the seat pushes up on him) at the bottom of the circle, and (c) at the top of the circle (assume the same speed).

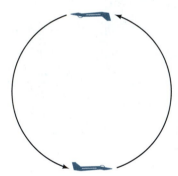

FIGURE 5–43
Problem 73.

74. Derive a formula for the mass of a planet in terms of its radius r, the acceleration due to gravity at its surface g_P, and the gravitational constant G.

75. A plumb bob (a mass m hanging on a string) is deflected from the vertical by an angle θ due to a massive mountain nearby (Fig. 5–44). (a) Find an approximate formula for θ in terms of the mass of the mountain, m_M, the distance to its center, D_M, and the radius and mass of the Earth. (b) Make a rough estimate of the mass of Mt. Everest, assuming it has the shape of a cone 4000 m high and base of diameter 4000 m. Assume its mass per unit volume is 3000 kg per m³. (c) Estimate the angle θ of the plumb bob if it is 5 km from the center of Mt. Everest.

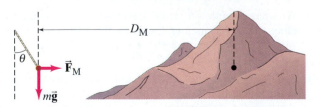

FIGURE 5–44 Problem 75.

76. A curve of radius 67 m is banked for a design speed of 95 km/h. If the coefficient of static friction is 0.30 (wet pavement), at what range of speeds can a car safely handle the curve?

77. How long would a day be if the Earth were rotating so fast that objects at the equator were apparently weightless?

78. Two equal-mass stars maintain a constant distance apart of 8.0×10^{10} m and rotate about a point midway between them at a rate of one revolution every 12.6 yr. (a) Why don't the two stars crash into one another due to the gravitational force between them? (b) What must be the mass of each star?

79. A train traveling at a constant speed rounds a curve of radius 235 m. A lamp suspended from the ceiling swings out to an angle of 17.5° throughout the curve. What is the speed of the train?

80. Jupiter is about 320 times as massive as the Earth. Thus, it has been claimed that a person would be crushed by the force of gravity on a planet the size of Jupiter since people can't survive more than a few g's. Calculate the number of g's a person would experience at the equator of such a planet. Use the following data for Jupiter: mass = 1.9×10^{27} kg, equatorial radius = 7.1×10^4 km, rotation period = 9 hr 55 min. Take the centripetal acceleration into account.

81. Astronomers using the Hubble Space Telescope deduced the presence of an extremely massive core in the distant galaxy M87, so dense that it could be a black hole (from which no light escapes). They did this by measuring the speed of gas clouds orbiting the core to be 780 km/s at a distance of 60 light-years $(5.7 \times 10^{17}$ m) from the core. Deduce the mass of the core, and compare it to the mass of our Sun.

82. A car maintains a constant speed v as it traverses the hill and valley shown in Fig. 5–45. Both the hill and valley have a radius of curvature R. (a) How do the normal forces acting on the car at A, B, and C compare? (Which is largest? Smallest?) Explain. (b) Where would the driver feel heaviest? Lightest? Explain. (c) How fast can the car go without losing contact with the road at A?

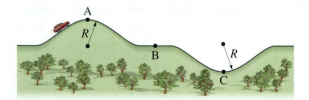

FIGURE 5–45   Problem 82.

83. The Navstar Global Positioning System (GPS) utilizes a group of 24 satellites orbiting the Earth. Using "triangulation" and signals transmitted by these satellites, the position of a receiver on the Earth can be determined to within an accuracy of a few centimeters. The satellite orbits are distributed evenly around the Earth, with four satellites in each of six orbits, allowing continuous navigational "fixes." The satellites orbit at an altitude of approximately 11,000 nautical miles [1 nautical mile = 1.852 km = 6076 ft]. (a) Determine the speed of each satellite. (b) Determine the period of each satellite.

84. The *Near Earth Asteroid Rendezvous* (*NEAR*), after traveling 2.1 billion km, is meant to orbit the asteroid Eros at a height of about 15 km. Eros is roughly 40 km × 6 km × 6 km. Assume Eros has a density (mass/volume) of about $2.3 \times 10^3 \, \text{kg/m}^3$. (*a*) What will be the period of *NEAR* as it orbits Eros? (*b*) If Eros were a sphere with the same mass and density, what would its radius be? (*c*) What would *g* be at the surface of this spherical Eros?

85. You are an astronaut in the space shuttle pursuing a satellite in need of repair. You are in a circular orbit of the same radius as the satellite (400 km above the Earth), but 25 km behind it. (*a*) How long will it take to overtake the satellite if you reduce your orbital radius by 1.0 km? (*b*) By how much must you reduce your orbital radius to catch up in 7.0 hours?

* 86. The comet Hale-Bopp has a period of 3000 years. (*a*) What is its mean distance from the Sun? (*b*) At its closest approach, the comet is about 1 A.U. from the Sun (1 A.U. = distance from Earth to the Sun). What is the farthest distance? (*c*) What is the ratio of the speed at the closest point to the speed at the farthest point? [*Hint*: Use Kepler's second law and estimate areas by a triangle (as in Fig. 5–29, but smaller distance travelled; see also Hint for Problem 59.]

87. Estimate what the value of *G* would need to be if you could actually "feel" yourself gravitationally attracted to someone near you. Make reasonable assumptions, like $F \approx 1 \, \text{N}$.

* 88. The Sun rotates around the center of the Milky Way Galaxy (Fig. 5–46) at a distance of about 30,000 light-years from the center $\left(1 \, \text{ly} = 9.5 \times 10^{15} \, \text{m}\right)$. If it takes about 200 million years to make one rotation, estimate the mass of our Galaxy. Assume that the mass distribution of our Galaxy is concentrated mostly in a central uniform sphere. If all the stars had about the mass of our Sun $\left(2 \times 10^{30} \, \text{kg}\right)$, how many stars would there be in our Galaxy?

89. Four 1.0-kg masses are located at the corners of a square 0.50 m on each side. Find the magnitude and direction of the gravitational force on a fifth 1.0-kg mass placed at the midpoint of the bottom side of the square.

90. A satellite of mass 5500 kg orbits the Earth (mass = $6.0 \times 10^{24} \, \text{kg}$) and has a period of 6200 s. Find (*a*) the magnitude of the Earth's gravitational force on the satellite, (*b*) the altitude of the satellite.

91. What is the acceleration experienced by the tip of the 1.5-cm-long sweep second hand on your wrist watch?

92. While fishing, you get bored and start to swing a sinker weight around in a circle below you on a 0.25-m piece of fishing line. The weight makes a complete circle every 0.50 s. What is the angle that the fishing line makes with the vertical? [*Hint*: See Fig. 5–10.]

93. A circular curve of radius *R* in a new highway is designed so that a car traveling at speed v_0 can negotiate the turn safely on glare ice (zero friction). If a car travels too slowly, then it will slip toward the center of the circle. If it travels too fast, then it will slip away from the center of the circle. If the coefficient of static friction increases, a car can stay on the road while traveling at any speed within a range from v_{min} to v_{max}. Derive formulas for v_{min} and v_{max} as functions of μ_s, v_0, and *R*.

94. Amtrak's high speed train, the *Acela*, utilizes tilt of the cars when negotiating curves. The angle of tilt is adjusted so that the main force exerted on the passengers, to provide the centripetal acceleration, is the normal force. The passengers experience less friction force against the seat, thus feeling more comfortable. Consider an *Acela* train that rounds a curve with a radius of 620 m at a speed of 160 km/h (approximately 100 mi/h). (*a*) Calculate the friction force needed on a train passenger of mass 75 kg if the track is not banked and the train does not tilt. (*b*) Calculate the friction force on the passenger if the train tilts to its maximum tilt of 8.0° toward the center of the curve.

Sun

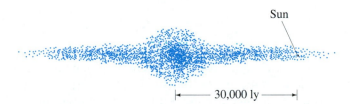

|← 30,000 ly →|

FIGURE 5–46 Problem 88. Edge-on view of our Galaxy.

Answers to Exercises

A: A factor of two (doubles).
B: Speed is independent of the mass of the clothes.
C: (*a*).
D: No.

E: Yes.
F: (*a*) No change; (*b*) four times larger.
G: (*b*).

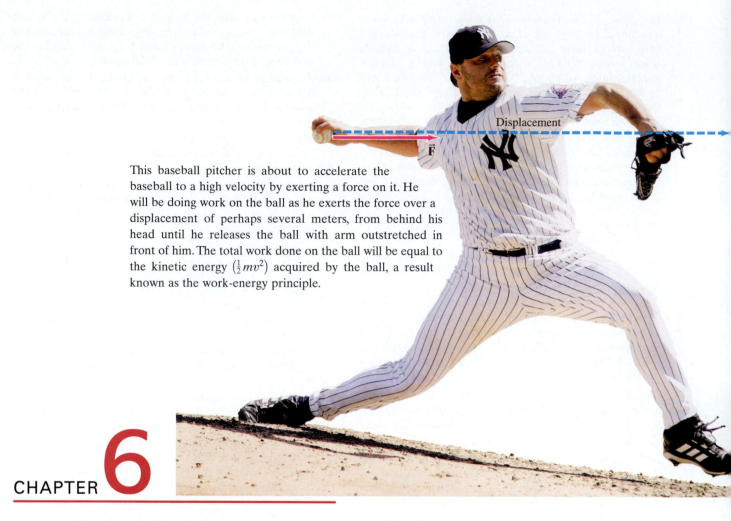

This baseball pitcher is about to accelerate the baseball to a high velocity by exerting a force on it. He will be doing work on the ball as he exerts the force over a displacement of perhaps several meters, from behind his head until he releases the ball with arm outstretched in front of him. The total work done on the ball will be equal to the kinetic energy $\left(\frac{1}{2}mv^2\right)$ acquired by the ball, a result known as the work-energy principle.

6

Work and Energy

Until now we have been studying the translational motion of an object in terms of Newton's three laws of motion. In this analysis, *force* has played a central role as the quantity determining the motion. In this Chapter and the next, we discuss an alternative analysis of the translational motion of objects in terms of the quantities *energy* and *momentum*. The significance of energy and momentum is that they are *conserved*. That is, in quite general circumstances they remain constant. That conserved quantities exist gives us not only a deeper insight into the nature of the world, but also gives us another way to approach solving practical problems.

The conservation laws of energy and momentum are especially valuable in dealing with systems of many objects, in which a detailed consideration of the forces involved would be difficult or impossible. These laws are applicable to a wide range of phenomena, including the atomic and subatomic worlds, where Newton's laws do not apply.

This Chapter is devoted to the very important concepts of *work* and *energy*. These two quantities are scalars and so have no direction associated with them, which often makes them easier to work with than vector quantities.

6–1 Work Done by a Constant Force

The word *work* has a variety of meanings in everyday language. But in physics, work is given a very specific meaning to describe what is accomplished when a force acts on an object, and the object moves through a distance. Specifically, the **work** done on an object by a constant force (constant in both magnitude and direction) is defined to be *the product of the magnitude of the displacement times the component of the force parallel to the displacement*. In equation form, we can write

$$W = F_{\parallel} d,$$

where $F_{\parallel}$ is the component of the constant force $\vec{F}$ parallel to the displacement $\vec{d}$. We can also write

$$W = Fd \cos \theta, \tag{6–1}$$

where F is the magnitude of the constant force, d is the magnitude of the displacement of the object, and θ is the angle between the directions of the force and the displacement (Fig. 6–1). The $\cos \theta$ factor appears in Eq. 6–1 because $F \cos \theta \, (= F_{\parallel})$ is the component of $\vec{F}$ that is parallel to $\vec{d}$. Work is a scalar quantity—it has only magnitude, which can be positive or negative.

Let us first consider the case in which the motion and the force are in the same direction, so $\theta = 0$ and $\cos \theta = 1$; in this case, $W = Fd$. For example, if you push a loaded grocery cart a distance of 50 m by exerting a horizontal force of 30 N on the cart, you do $30 \, \text{N} \times 50 \, \text{m} = 1500 \, \text{N} \cdot \text{m}$ of work on the cart.

As this example shows, in SI units work is measured in newton-meters ($\text{N} \cdot \text{m}$). A special name is given to this unit, the **joule** (J): $1 \, \text{J} = 1 \, \text{N} \cdot \text{m}$. In the cgs system, the unit of work is called the *erg* and is defined as $1 \, \text{erg} = 1 \, \text{dyne} \cdot \text{cm}$. In British units, work is measured in foot-pounds. It is easy to show that $1 \, \text{J} = 10^7 \, \text{erg} = 0.7376 \, \text{ft} \cdot \text{lb}$.

Work

Work defined (for constant force)

Units for work: the joule

FIGURE 6–1 A person pulling a crate along the floor. The work done by the force $\vec{F}$ is $W = Fd \cos \theta$, where $\vec{d}$ is the displacement.

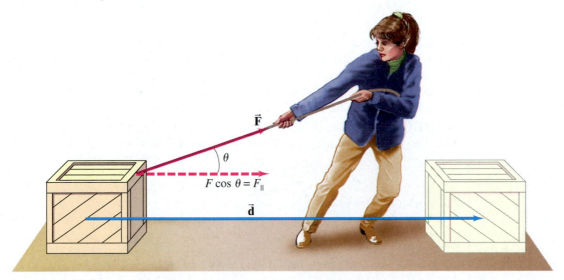

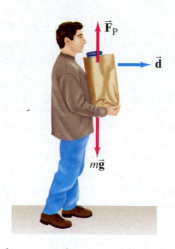

FIGURE 6–2 The person does no work on the bag of groceries since $\vec{\mathbf{F}}_\text{P}$ is perpendicular to the displacement $\vec{\mathbf{d}}$.

⚠️ **CAUTION**

Force without work

A force can be exerted on an object and yet do no work. For example, if you hold a heavy bag of groceries in your hands at rest, you do no work on it. You do exert a force on the bag, but the displacement of the bag is zero, so the work done by you on the bag is $W = 0$. You need both a force and a displacement to do work. You also do no work on the bag of groceries if you carry it as you walk horizontally across the floor at constant velocity, as shown in Fig. 6–2. No horizontal force is required to move the bag at a constant velocity. The person shown in Fig. 6–2 does exert an upward force $\vec{\mathbf{F}}_\text{P}$ on the bag equal to its weight. But this upward force is perpendicular to the horizontal displacement of the bag and thus has nothing to do with that motion. Hence, the upward force is doing no work. This conclusion comes from our definition of work, Eq. 6–1: $W = 0$, because $\theta = 90°$ and $\cos 90° = 0$. Thus, when a particular force is perpendicular to the displacement, no work is done by that force. (When you start or stop walking, there is a horizontal acceleration and you do briefly exert a horizontal force, and thus do work on the bag.)

⚠️ **CAUTION**

State that work is done on or by an object

When we deal with work, as with force, it is necessary to specify whether you are talking about work done *by* a specific object or done *on* a specific object. It is also important to specify whether the work done is due to one particular force (and which one), or the total (net) work done by the *net force* on the object.

EXAMPLE 6–1 **Work done on a crate.** A person pulls a 50-kg crate 40 m along a horizontal floor by a constant force $F_\text{P} = 100\,\text{N}$, which acts at a 37° angle as shown in Fig. 6–3. The floor is rough and exerts a friction force $F_\text{fr} = 50\,\text{N}$. Determine (a) the work done by each force acting on the crate, and (b) the net work done on the crate.

APPROACH We choose our coordinate system so that $\vec{\mathbf{x}}$ can be the vector that represents the 40-m displacement (that is, along the x axis). Four forces act on the crate, as shown in Fig. 6–3: the force exerted by the person $\vec{\mathbf{F}}_\text{P}$; the friction force $\vec{\mathbf{F}}_\text{fr}$ due to the floor; the crate's weight $m\vec{\mathbf{g}}$; and the normal force $\vec{\mathbf{F}}_\text{N}$ exerted upward by the floor. The net force on the crate is the vector sum of these four forces.

SOLUTION (a) The work done by the gravitational and normal forces is zero, since they are perpendicular to the displacement $\vec{\mathbf{x}}$ ($\theta = 90°$ in Eq. 6–1):

$$W_\text{G} = mgx \cos 90° = 0$$
$$W_\text{N} = F_\text{N} x \cos 90° = 0.$$

The work done by $\vec{\mathbf{F}}_\text{P}$ is

$$W_\text{P} = F_\text{P} x \cos \theta = (100\,\text{N})(40\,\text{m}) \cos 37° = 3200\,\text{J}.$$

The work done by the friction force is

$$W_\text{fr} = F_\text{fr} x \cos 180° = (50\,\text{N})(40\,\text{m})(-1) = -2000\,\text{J}.$$

The angle between the displacement $\vec{\mathbf{x}}$ and the force $\vec{\mathbf{F}}_\text{fr}$ is 180° because they point in opposite directions. Since the force of friction is opposing the motion (and $\cos 180° = -1$), the work done by friction on the crate is *negative*.

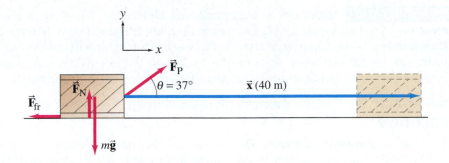

FIGURE 6–3 Example 6–1. A 50-kg crate is pulled along a floor.

(*b*) The net work can be calculated in two equivalent ways:

(1) The net work done on an object is the algebraic sum of the work done by each force, since work is a scalar:

$$W_{net} = W_G + W_N + W_P + W_{fr}$$
$$= 0 + 0 + 3200\,J - 2000\,J$$
$$= 1200\,J.$$

*W*net *is the work done by all the forces acting on the object*

(2) The net work can also be calculated by first determining the net force on the object and then taking its component along the displacement: $(F_{net})_x = F_P \cos\theta - F_{fr}$. Then the net work is

$$W_{net} = (F_{net})_x\, x = (F_P \cos\theta - F_{fr})x$$
$$= (100\,N \cos 37° - 50\,N)(40\,m)$$
$$= 1200\,J.$$

In the vertical (*y*) direction, there is no displacement and no work done.

In Example 6–1 we saw that friction did negative work. In general, the work done by a force is negative whenever the force (or the component of the force, $F_{\parallel}$) acts in the direction opposite to the direction of motion. Also, we can see that when the work done by a force on an object is negative, that force is trying to slow the object down (and would slow it down if that were the only force acting). When the work is positive, the force involved is trying to speed up the object.

⚠ **CAUTION**

Negative work

EXERCISE A A box is dragged across a floor by a force $\vec{F}_P$ which makes an angle θ with the horizontal as in Fig. 6–1 or 6–3. If the magnitude of $\vec{F}_P$ is held constant but the angle θ is increased, the work done by $\vec{F}_P$ (*a*) remains the same; (*b*) increases; (*c*) decreases; (*d*) first increases, then decreases.

PROBLEM SOLVING   Work

1. **Draw a free-body diagram** showing all the forces acting on the object you choose to study.

2. **Choose** an *xy* **coordinate system**. If the object is in motion, it may be convenient to choose one of the coordinate directions as the direction of one of the forces, or as the direction of motion. [Thus, for an object on an incline, you might choose one coordinate axis to be parallel to the incline.]

3. Apply **Newton's laws** to determine any unknown forces.

4. Find the **work done** *by a specific force* *on* the object by using $W = Fd \cos\theta$ for a constant force. Note that the work done is negative when a force tends to oppose the displacement.

5. To find the **net work** done on the object, either (*a*) find the work done by each force and add the results algebraically; or (*b*) find the net force on the object, F_{net}, and then use it to find the net work done, which for constant net force is:

$$W_{net} = F_{net}\, d \cos\theta.$$

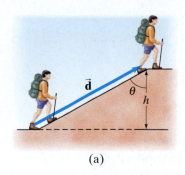

(a)

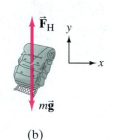

(b)

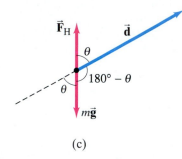

(c)

FIGURE 6–4 Example 6–2.

EXAMPLE 6–2 **Work on a backpack.** (a) Determine the work a hiker must do on a 15.0-kg backpack to carry it up a hill of height $h = 10.0$ m, as shown in Fig. 6–4a. Determine also (b) the work done by gravity on the backpack, and (c) the net work done on the backpack. For simplicity, assume the motion is smooth and at constant velocity (i.e., acceleration is negligible).

APPROACH We explicitly follow the Problem Solving Box step by step.

SOLUTION

1. **Draw a free-body diagram.** The forces on the backpack are shown in Fig. 6–4b: the force of gravity, $m\vec{g}$, acting downward; and $\vec{F}_H$, the force the hiker must exert upward to support the backpack. Since we assume there is negligible acceleration, horizontal forces on the backpack are negligible.

2. **Choose a coordinate system.** We are interested in the vertical motion of the backpack, so we choose the y coordinate as positive vertically upward.

3. **Apply Newton's laws.** Newton's second law applied in the vertical direction to the backpack gives

$$\Sigma F_y = ma_y$$
$$F_H - mg = 0.$$

Hence,

$$F_H = mg = (15.0 \text{ kg})(9.80 \text{ m/s}^2) = 147 \text{ N}.$$

4. **Find the work done by a specific force.** (a) To calculate the work done by the hiker on the backpack, we write Eq. 6–1 as

$$W_H = F_H(d \cos \theta),$$

and we note from Fig. 6–4a that $d \cos \theta = h$. So the work done by the hiker is

$$W_H = F_H(d \cos \theta) = F_H h = mgh$$
$$= (147 \text{ N})(10.0 \text{ m}) = 1470 \text{ J}.$$

Note that the work done depends only on the change in elevation and not on the angle of the hill, θ. The hiker would do the same work to lift the pack vertically the same height h.

(b) The work done by gravity on the backpack is (from Eq. 6–1 and Fig. 6–4c)

$$W_G = F_G d \cos(180° - \theta).$$

Since $\cos(180° - \theta) = -\cos \theta$, we have

$$W_G = F_G d(-\cos \theta) = mg(-d \cos \theta)$$
$$= -mgh$$
$$= -(15.0 \text{ kg})(9.80 \text{ m/s}^2)(10.0 \text{ m}) = -1470 \text{ J}.$$

NOTE The work done by gravity (which is negative here) doesn't depend on the angle of the incline, only on the vertical height h of the hill. This is because gravity acts vertically, so only the vertical component of displacement contributes to work done.

5. **Find the net work done.** (a) The *net* work done on the backpack is $W_{net} = 0$, since the net force on the backpack is zero (it is assumed not to accelerate significantly). We can also determine the net work done by adding the work done by each force:

$$W_{net} = W_G + W_H = -1470 \text{ J} + 1470 \text{ J} = 0.$$

NOTE Even though the *net* work done by all the forces on the backpack is zero, the hiker does do work on the backpack equal to 1470 J.

➡ **PROBLEM SOLVING**

Work done by gravity depends on the height of the hill and not on the angle of incline

| **CONCEPTUAL EXAMPLE 6–3** | **Does the Earth do work on the Moon?** |

The Moon revolves around the Earth in a nearly circular orbit, kept there by the gravitational force exerted by the Earth. Does gravity do (a) positive work, (b) negative work, or (c) no work on the Moon?

RESPONSE The gravitational force exerted by the Earth on the Moon (Fig. 6–5) acts toward the Earth and provides its centripetal acceleration, inward along the radius of the Moon's orbit. The Moon's displacement at any moment is tangent to the circle, in the direction of its velocity, perpendicular to the radius and perpendicular to the force of gravity. Hence the angle θ between the force $\vec{F}_G$ and the instantaneous displacement of the Moon is 90°, and the work done by the Earth's gravity on the Moon as it orbits is therefore zero ($\cos 90° = 0$). This is why the Moon, as well as artificial satellites, can stay in orbit without expenditure of fuel: no net work needs to be done against the force of gravity.

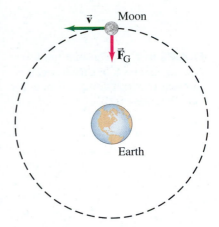

FIGURE 6–5 Example 6–3.

* 6–2 Work Done by a Varying Force

If the force acting on an object is constant, the work done by that force can be calculated using Eq. 6–1. But in many cases, the force varies in magnitude or direction during a process. For example, as a rocket moves away from Earth, work is done to overcome the force of gravity, which varies as the inverse square of the distance from the Earth's center. Other examples are the force exerted by a spring, which increases with the amount of stretch, or the work done by a varying force in pulling a box or cart up an uneven hill.

The work done by a varying force can be determined graphically. The procedure is like that for determining displacement when the velocity is known as a function of time (Section 2–8). To determine the work done by a variable force, we plot $F_{\parallel}$ ($= F \cos\theta$, the component of $\vec{F}$ parallel to the direction of motion at any point) as a function of distance d, as in Fig. 6–6a. We divide the distance into small segments Δd. For each segment, we indicate the average of $F_{\parallel}$ by a horizontal dashed line. Then the work done for each segment is $\Delta W = F_{\parallel} \Delta d$, which is the area of a rectangle Δd wide and $F_{\parallel}$ high. The total work done to move the object a total distance $d = d_B - d_A$ is the sum of the areas of the rectangles (five in the case shown in Fig. 6–6a). Usually, the average value of $F_{\parallel}$ for each segment must be estimated, and a reasonable approximation of the work done can then be made. If we subdivide the distance into many more segments, Δd can be made smaller and our estimate of the work done would be more accurate. In the limit as Δd approaches zero, the total area of the many narrow rectangles approaches the area under the curve, Fig. 6–6b. That is, *the work done by a variable force in moving an object between two points is equal to the area under the $F_{\parallel}$ vs. d curve between those two points.*

FIGURE 6–6 The work done by a force F can be calculated by taking: (a) the sum of the areas of the rectangles; (b) the area under the curve of $F_{\parallel}$ vs. d.

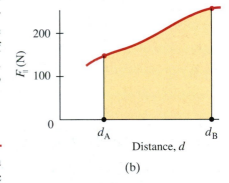

(a)

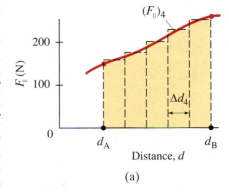

(b)

6–3 Kinetic Energy, and the Work-Energy Principle

Energy is one of the most important concepts in science. Yet we cannot give a simple general definition of energy in only a few words. Nonetheless, each specific type of energy can be defined fairly simply. In this Chapter, we define translational kinetic energy and some types of potential energy. In later Chapters, we will examine other types of energy, such as that related to heat (Chapters 14 and 15). The crucial aspect of all the types of energy is that the sum of all types, the *total energy*, is the same after any process as it was before: that is, the quantity "energy" is a conserved quantity.

For the purposes of this Chapter, we can define energy in the traditional way as "the ability to do work." This simple definition is not very precise, nor is it really valid for all types of energy.[†] It is valid, however, for mechanical energy which we discuss in this Chapter, and it serves to underscore the fundamental

---

[†]Energy associated with heat is often not available to do work, as we will discuss in Chapter 15.

SECTION 6–3   Kinetic Energy, and the Work-Energy Principle   **141**

FIGURE 6–7 A constant net force F_{net} accelerates a bus from speed v_1 to speed v_2 over a displacement d. The net work done is $W_{net} = F_{net}\,d$.

connection between work and energy. We now define and discuss one of the basic types of energy, kinetic energy.

A moving object can do work on another object it strikes. A flying cannonball does work on a brick wall it knocks down; a moving hammer does work on a nail it drives into wood. In either case, a moving object exerts a force on a second object which undergoes a displacement. An object in motion has the ability to do work and thus can be said to have energy. The energy of motion is called **kinetic energy**, from the Greek word *kinetikos*, meaning "motion."

To obtain a quantitative definition for kinetic energy, let us consider a rigid object of mass m that is moving in a straight line with an initial speed v_1. To accelerate it uniformly to a speed v_2, a constant net force F_{net} is exerted on it parallel to its motion over a displacement d, Fig. 6–7. Then the net work done on the object is $W_{net} = F_{net}\,d$. We apply Newton's second law, $F_{net} = ma$, and use Eq. 2–11c, which we now write as $v_2^2 = v_1^2 + 2ad$, with v_1 as the initial speed and v_2 the final speed. We solve for a in Eq. 2–11c,

$$a = \frac{v_2^2 - v_1^2}{2d},$$

then substitute this into $F_{net} = ma$, and determine the work done:

$$W_{net} = F_{net}\,d = mad = m\left(\frac{v_2^2 - v_1^2}{2d}\right)d = m\left(\frac{v_2^2 - v_1^2}{2}\right)$$

or

$$W_{net} = \tfrac{1}{2}mv_2^2 - \tfrac{1}{2}mv_1^2. \tag{6–2}$$

We *define* the quantity $\tfrac{1}{2}mv^2$ to be the **translational kinetic energy (KE)** of the object:

Kinetic energy defined

$$\boxed{KE = \tfrac{1}{2}mv^2.} \tag{6–3}$$

(We call this "translational" kinetic energy to distinguish it from rotational kinetic energy, which we will discuss in Chapter 8.) Equation 6–2, derived here for one-dimensional motion with a constant force, is valid in general for translational motion of an object in three dimensions and even if the force varies. We can rewrite Eq. 6–2 as:

$$W_{net} = KE_2 - KE_1$$

or

| WORK-ENERGY PRINCIPLE |

$$\boxed{W_{net} = \Delta KE.} \tag{6–4}$$

Equation 6–4 (or Eq. 6–2) is an important result known as the **work-energy principle**. It can be stated in words:

| WORK-ENERGY PRINCIPLE |

The net work done on an object is equal to the change in the object's kinetic energy.

Notice that we made use of Newton's second law, $F_{net} = ma$, where F_{net} is the *net* force—the sum of all forces acting on the object. Thus, the work-energy principle is valid only if W is the *net work* done on the object—that is, the work done by all forces acting on the object.

 **C A U T I O N**

Work-energy valid only for net work

The work-energy principle is a very useful reformulation of Newton's laws. It tells us that if (positive) net work W is done on an object, the object's kinetic energy increases by an amount W. The principle also holds true for the reverse situation: if the net work W done on an object is negative, the object's kinetic energy decreases by an amount W. That is, a net force exerted on an object opposite to the object's direction of motion decreases its speed and its kinetic energy. An example is a moving hammer (Fig. 6–8) striking a nail. The net force on the hammer ($-\vec{\mathbf{F}}$ in Fig. 6–8, where $\vec{\mathbf{F}}$ is assumed constant for simplicity) acts toward the left, whereas the displacement $\vec{\mathbf{d}}$ of the hammer is toward the right. So the net work done on the hammer, $W_h = (F)(d)(\cos 180°) = -Fd$, is negative and the hammer's kinetic energy decreases (usually to zero).

Figure 6–8 also illustrates how energy can be considered the ability to do work. The hammer, as it slows down, does positive work on the nail: if the nail exerts a force $-\vec{\mathbf{F}}$ on the hammer to slow it down, the hammer exerts a force $+\vec{\mathbf{F}}$ on the nail (Newton's third law) through the distance d. Hence the work done on the nail by the hammer is $W_n = (+F)(+d) = Fd$ and is positive. We also see that $W_n = Fd = -W_h$: the work done on the nail W_n equals the negative of the work done on the hammer. That is, the decrease in kinetic energy of the hammer is equal to the work the hammer can do on another object—which is consistent with energy being the ability to do work.

Whereas the translational kinetic energy ($= \frac{1}{2}mv^2$) is directly proportional to the mass of the object, it is proportional to the *square* of the speed. Thus, if the mass is doubled, the kinetic energy is doubled. But if the speed is doubled, the object has four times as much kinetic energy and is therefore capable of doing four times as much work.

Let us summarize the relationship between work and kinetic energy (Eq. 6–4): if the net work W done on an object is positive, then the object's kinetic energy increases. If the net work W done on an object is negative, its kinetic energy decreases. If the net work done on the object is zero, its kinetic energy remains constant (which also means its speed is constant).

Because of the direct connection between work and kinetic energy (Eq. 6–4), energy is measured in the same units as work: joules in SI units, ergs in the cgs, and foot-pounds in the British system. Like work, kinetic energy is a scalar quantity. The kinetic energy of a group of objects is the sum of the kinetic energies of the individual objects.

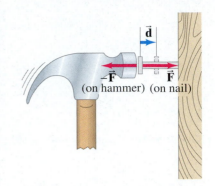

FIGURE 6–8 A moving hammer strikes a nail and comes to rest. The hammer exerts a force F on the nail; the nail exerts a force $-F$ on the hammer (Newton's third law). The work done on the nail by the hammer is positive ($W_n = Fd > 0$). The work done on the hammer by the nail is negative ($W_h = -Fd$).

If $W_{net} > 0$, KE increases
If $W_{net} < 0$, KE decreases

Energy units:
the joule

EXAMPLE 6–4 KE **and work done on a baseball.** A 145-g baseball is thrown so that it acquires a speed of 25 m/s. (*a*) What is its kinetic energy? (*b*) What was the net work done on the ball to make it reach this speed, if it started from rest?

APPROACH We use the definition of kinetic energy, Eq. 6–3, and then the work-energy principle, Eq. 6–4.

SOLUTION (*a*) The kinetic energy of the ball after the throw is

$$\text{KE} = \tfrac{1}{2}mv^2 = \tfrac{1}{2}(0.145\text{ kg})(25\text{ m/s})^2 = 45\text{ J}.$$

(*b*) Since the initial kinetic energy was zero, the net work done is just equal to the final kinetic energy, 45 J.

EXAMPLE 6–5 **Work on a car, to increase its KE.** How much net work is required to accelerate a 1000-kg car from 20 m/s to 30 m/s (Fig. 6–9)?

APPROACH To simplify a complex situation, let us treat the car as a particle or simple rigid object. We can then use the work-energy principle.

SOLUTION The net work needed is equal to the increase in kinetic energy:

$$W = \text{KE}_2 - \text{KE}_1 = \tfrac{1}{2}mv_2^2 - \tfrac{1}{2}mv_1^2$$
$$= \tfrac{1}{2}(1000\text{ kg})(30\text{ m/s})^2 - \tfrac{1}{2}(1000\text{ kg})(20\text{ m/s})^2 = 2.5 \times 10^5\text{ J}.$$

NOTE You might be tempted to work this Example by finding the force and using Eq. 6–1. That won't work, however, because we don't know how far or for how long the car was accelerated. In fact, a large force could be acting for a small distance, or a small force could be acting over a long distance; both could give the same net work.

FIGURE 6–9 Example 6–5.

$v_1 = 20$ m/s $v_2 = 30$ m/s

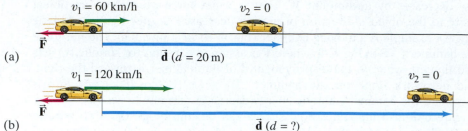

(a)

(b)

FIGURE 6–10 Example 6–6.

CONCEPTUAL EXAMPLE 6–6 **Work to stop a car.** A car traveling 60 km/h can brake to a stop within a distance d of 20 m (Fig. 6–10a). If the car is going twice as fast, 120 km/h, what is its stopping distance (Fig. 6–10b)? Assume the maximum braking force is approximately independent of speed.

RESPONSE Since the stopping force F is approximately constant, the work needed to stop the car, Fd, is proportional to the distance traveled. We apply the work-energy principle, noting that $\vec{F}$ and $\vec{d}$ are in opposite directions and that the final speed of the car is zero:

$$W_{\text{net}} = Fd \cos 180° = -Fd.$$

Then

$$-Fd = \Delta\text{KE} = \tfrac{1}{2}mv_2^2 - \tfrac{1}{2}mv_1^2$$

$$= 0 - \tfrac{1}{2}mv_1^2.$$

Thus, since the force and mass are constant, we see that the stopping distance, d, increases with the square of the speed:

$$d \propto v^2.$$

If the car's initial speed is doubled, the stopping distance is $(2)^2 = 4$ times as great, or 80 m.

PHYSICS APPLIED

Car's stopping distance $\propto$ initial speed squared

| EXERCISE B Can kinetic energy ever be negative?

6–4 Potential Energy

Potential energy

We have just discussed how an object is said to have energy by virtue of its motion, which we call kinetic energy. But it is also possible to have **potential energy**, which is the energy associated with forces that depend on the position or configuration of an object (or objects) relative to the surroundings. Various types of potential energy (PE) can be defined, and each type is associated with a particular force.

The spring of a wind-up toy is an example of an object with potential energy. The spring acquired its potential energy because work was done *on* it by the person winding the toy. As the spring unwinds, it exerts a force and does work to make the toy move.

Gravitational PE

Perhaps the most common example of potential energy is *gravitational potential energy*. A heavy brick held high in the air has potential energy because of its position relative to the Earth. The raised brick has the ability to do work, for if it is released, it will fall to the ground due to the gravitational force, and can do work on, say, a stake, driving it into the ground. Let us seek the form for the gravitational potential energy of an object near the surface of the Earth. For an object of mass m to be lifted vertically, an upward force at least equal to its weight, mg, must be exerted on it, say by a person's hand.

To lift it without acceleration a vertical displacement of height h, from position y_1 to y_2 in Fig. 6–11 (upward direction chosen positive), a person must do work equal to the product of the needed external force, $F_{ext} = mg$ upward, and the vertical displacement h. That is,

$$W_{ext} = F_{ext}\, d \cos 0° = mgh$$
$$= mg(y_2 - y_1). \tag{6–5a}$$

Gravity is also acting on the object as it moves from y_1 to y_2, and does work on it equal to

$$W_G = F_G\, d \cos \theta = mgh \cos 180°,$$

where $\theta = 180°$ because $\vec{\mathbf{F}}_G$ and $\vec{\mathbf{d}}$ point in opposite directions. So

$$W_G = -mgh$$
$$= -mg(y_2 - y_1). \tag{6–5b}$$

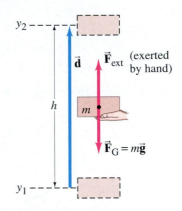

If we now allow the object to start from rest and fall freely under the action of gravity, it acquires a velocity given by $v^2 = 2gh$ (Eq. 2–11c) after falling a height h. It then has kinetic energy $\frac{1}{2}mv^2 = \frac{1}{2}m(2gh) = mgh$, and if it strikes a stake it can do work on the stake equal to mgh (work-energy principle). Thus, to raise an object of mass m to a height h *requires* an amount of work equal to mgh (Eq. 6–5a). And once at height h, the object has the *ability* to do an amount of work equal to mgh.

We therefore define the **gravitational potential energy** of an object, due to Earth's gravity, as the product of the object's weight mg and its height y above some reference level (such as the ground):

$$\text{PE}_{grav} = mgy. \tag{6–6}$$

FIGURE 6–11 A person exerts an upward force $F_{ext} = mg$ to lift a brick from y_1 to y_2.

Gravitational PE

The higher an object is above the ground, the more gravitational potential energy it has. We combine Eq. 6–5a with Eq. 6–6:

$$W_{ext} = mg(y_2 - y_1)$$
$$W_{ext} = \text{PE}_2 - \text{PE}_1 = \Delta\text{PE}. \tag{6–7a}$$

That is, the work done by an external force to move the object of mass m from point 1 to point 2 (without acceleration) is equal to the change in potential energy between positions 1 and 2.

Alternatively, we can write the change in potential energy, ΔPE, in terms of the work done by gravity itself: starting from Eq. 6–5b, we obtain

$$W_G = -mg(y_2 - y_1)$$
$$W_G = -(\text{PE}_2 - \text{PE}_1) = -\Delta\text{PE}. \tag{6–7b}$$

That is, the work done by gravity as the object of mass m moves from point 1 to point 2 is equal to the negative of the difference in potential energy between positions 1 and 2.

Potential energy belongs to a system, and not to a single object alone. Potential energy is associated with a force, and a force on one object is always exerted by some other object. Thus potential energy is a property of the system as a whole. For an object raised to a height y above the Earth's surface, the change in gravitational potential energy is mgy. The system here is the object plus the Earth, and properties of both are involved: object (m) and Earth (g).

Gravitational potential energy depends on the *vertical height* of the object *above some reference level* (Eq. 6–6). In some situations, you may wonder from what point to measure the height y. The gravitational potential energy of a book held high above a table, for example, depends on whether we measure y from the top of the table, from the floor, or from some other reference point.

⚠ CAUTION

Potential energy belongs to a system, not to a single object

What is physically important in any situation is the *change* in potential energy, ΔPE, because that is what is related to the work done, Eqs. 6–7; and it is ΔPE that can be measured. We can thus choose to measure y from any reference point that is convenient, but we must choose the reference point at the start and be consistent throughout a calculation. The *change* in potential energy between any two points does not depend on this choice.

Grav. PE depends on vertical *height*

An important result we discussed earlier (see Example 6–2 and Fig. 6–4) concerns the gravity force, which does work only in the vertical direction: the work done by gravity depends only on the vertical height h, and not on the path taken, whether it be purely vertical motion or, say, motion along an incline. Thus, from Eqs. 6–7 we see that changes in gravitational potential energy depend only on the change in vertical height and not on the path taken.

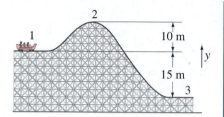

FIGURE 6–12 Example 6–7.

EXAMPLE 6–7 **Potential energy changes for a roller coaster.** A 1000-kg roller-coaster car moves from point 1, Fig. 6–12, to point 2 and then to point 3. (*a*) What is the gravitational potential energy at 2 and 3 relative to point 1? That is, take $y = 0$ at point 1. (*b*) What is the change in potential energy when the car goes from point 2 to point 3? (*c*) Repeat parts (*a*) and (*b*), but take the reference point ($y = 0$) to be at point 3.

APPROACH We are interested in the potential energy of the car–Earth system. We take upward as the positive y direction, and use the definition of gravitational potential energy to calculate PE.

SOLUTION (*a*) We measure heights from point 1, which means initially that the gravitational potential energy is zero. At point 2, where $y_2 = 10\,\text{m}$,

$$PE_2 = mgy_2 = (1000\,\text{kg})(9.8\,\text{m/s}^2)(10\,\text{m}) = 9.8 \times 10^4\,\text{J}.$$

At point 3, $y_3 = -15\,\text{m}$, since point 3 is below point 1. Therefore,

$$PE_3 = mgy_3 = (1000\,\text{kg})(9.8\,\text{m/s}^2)(-15\,\text{m}) = -1.5 \times 10^5\,\text{J}.$$

(*b*) In going from point 2 to point 3, the potential energy change $\left(PE_{\text{final}} - PE_{\text{initial}}\right)$ is

$$PE_3 - PE_2 = \left(-1.5 \times 10^5\,\text{J}\right) - \left(9.8 \times 10^4\,\text{J}\right)$$
$$= -2.5 \times 10^5\,\text{J}.$$

The gravitational potential energy decreases by $2.5 \times 10^5\,\text{J}$.

(*c*) In this instance, $y_1 = +15\,\text{m}$ at point 1, so the potential energy initially (at point 1) is

$$PE_1 = (1000\,\text{kg})(9.8\,\text{m/s}^2)(15\,\text{m}) = 1.5 \times 10^5\,\text{J}.$$

At point 2, $y_2 = 25\,\text{m}$, so the potential energy is

$$PE_2 = 2.5 \times 10^5\,\text{J}.$$

At point 3, $y_3 = 0$, so the potential energy is zero. The change in potential energy going from point 2 to point 3 is

$$PE_3 - PE_2 = 0 - 2.5 \times 10^5\,\text{J} = -2.5 \times 10^5\,\text{J},$$

which is the same as in part (*b*).

PE defined in general

There are other kinds of potential energy besides gravitational. Each form of potential energy is associated with a particular force, and can be defined analogously to gravitational potential energy. In general, the *change in potential energy associated with a particular force is equal to the negative of the work done by that force if the object is moved from one point to a second point* (as in Eq. 6–7b for gravity). Alternatively, because of Newton's third law, we can define the *change in potential energy as the work required of an external force to move the object without acceleration between the two points*, as in Eq. 6–7a.

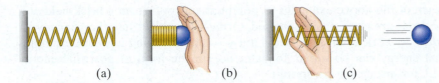

(a)   (b)   (c)

FIGURE 6–13 (a) A spring can store energy (elastic PE) when compressed as in (b) and can do work when released (c).

We now consider another type of potential energy, that associated with elastic materials. This includes a great variety of practical applications. Consider the simple coil spring shown in Fig. 6–13. The spring has potential energy when compressed (or stretched), for when it is released, it can do work on a ball as shown. To hold a spring either stretched or compressed an amount x from its natural (unstretched) length requires the hand to exert a force on the spring, F_P, that is directly proportional to x. That is,

$$F_P = kx,$$

where k is a constant, called the *spring stiffness constant*, and is a measure of the stiffness of the particular spring. The stretched or compressed spring exerts a force F_S in the opposite direction on the hand, as shown in Fig. 6–14:

$$F_S = -kx. \qquad (6-8)$$

This force is sometimes called a "restoring force" because the spring exerts its force in the direction opposite the displacement (hence the minus sign), acting to return it to its natural length. Equation 6–8 is known as the **spring equation** and also as **Hooke's law**, and is accurate for springs as long as x is not too great.

To calculate the potential energy of a stretched spring, let us calculate the work required to stretch it (Fig. 6–14b). We might expect to use Eq. 6–1 for the work done on it, $W = Fx$, where x is the amount it is stretched from its natural length. But this would be incorrect since the force $F_P (= kx)$ is not constant but varies over this distance, becoming greater the more the spring is stretched, as shown graphically in Fig. 6–15. So let us use the average force, $\bar{F}$. Since F_P varies linearly—from zero at the unstretched position to kx when stretched to x—the average force is $\bar{F} = \frac{1}{2}[0 + kx] = \frac{1}{2}kx$, where x here is the final amount stretched (shown as x_f in Fig. 6–15 for clarity). The work done is then

$$W = \bar{F}x = \left(\tfrac{1}{2}kx\right)(x) = \tfrac{1}{2}kx^2.$$

Hence the **elastic potential energy** is proportional to the square of the amount stretched:

$$\text{elastic PE} = \tfrac{1}{2}kx^2. \qquad (6-9)$$

If a spring is *compressed* a distance x from its natural length, the average force is again $\bar{F} = \frac{1}{2}kx$, and again the potential energy is given by Eq. 6–9. Thus x can be either the amount compressed or amount stretched from the spring's natural length.[†] Note that for a spring, we choose the reference point for zero PE at the spring's natural position.

[†]We can also obtain Eq. 6–9 using Section 6–2. The work done, and hence ΔPE, equals the area under the F vs. x graph of Fig. 6–15. This area is a triangle (colored in Fig. 6–15) of altitude kx and base x, and hence of area (for a triangle) equal to $\frac{1}{2}(kx)(x) = \frac{1}{2}kx^2$.

PE of elastic spring

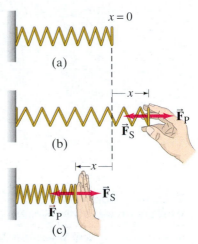

FIGURE 6–14 (a) Spring in natural (unstretched) position. (b) Spring is stretched by a person exerting a force $\vec{F}_P$ to the right (positive direction). The spring pulls back with a force $\vec{F}_S$, where $F_S = -kx$. (c) Person compresses the spring ($x < 0$) by exerting a force $\vec{F}_P$ to the left; the spring pushes back with a force $F_S = -kx$, where $F_S > 0$ because $x < 0$.

Elastic PE

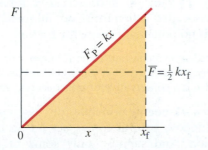

FIGURE 6–15 As a spring is stretched (or compressed), the force needed increases linearly as x increases: graph of $F = kx$ vs. x from $x = 0$ to $x = x_f$.

In each of the above examples of potential energy—from a brick held at a height y, to a stretched or compressed spring—an object has the capacity or *potential* to do work even though it is not yet actually doing it. These examples show that energy can be *stored*, for later use, in the form of potential energy (Fig. 6–13, for instance, for a spring).

Note that there is a single universal formula for the translational kinetic energy of an object, $\frac{1}{2}mv^2$, but there is no single formula for potential energy. Instead, the mathematical form of the potential energy depends on the force involved.

6–5 Conservative and Nonconservative Forces

The work done against gravity in moving an object from one point to another does not depend on the path taken. For example, it takes the same work ($= mgy$) to lift an object of mass m vertically a certain height as to carry it up an incline of the same vertical height, as in Fig. 6–4 (see Example 6–2). Forces such as gravity, for which the work done does not depend on the path taken but only on the initial and final positions, are called **conservative forces**. The elastic force of a spring (or other elastic material) in which $F = -kx$, is also a conservative force. An object that starts at a given point and returns to that same point under the action of a conservative force has no net work done on it because the potential energy is the same at the start and the finish of such a round trip.

Friction, on the other hand, is a **nonconservative force** since the work it does depends on the path. For example, when a crate is moved across a floor from one point to another, the work done depends on whether the path taken is straight, or is curved or zigzag. As shown in Fig. 6–16, if a crate is pushed from point 1 to point 2 along the longer semicircular path rather than along the straight path, more work is done against friction. That is because the distance is greater and, unlike the gravitational force, the friction force is always directed opposite to the direction of motion. (The $\cos\theta$ term in Eq. 6–1 is always $\cos 180° = -1$ at all points on the path for the friction force.) Thus the work done by friction in Fig. 6–16 does not depend *only* on points 1 and 2. Other forces that are nonconservative include the force exerted by a person and tension in a rope (see Table 6–1).

TABLE 6–1 Conservative and Nonconservative Forces

Conservative Forces	Nonconservative Forces
Gravitational	Friction
Elastic	Air resistance
Electric	Tension in cord
	Motor or rocket propulsion
	Push or pull by a person

FIGURE 6–16 A crate is pushed across the floor from position 1 to position 2 via two paths, one straight and one curved. The friction force is always in the direction exactly opposed to the direction of motion. Hence, for a constant magnitude friction force, $W_{\text{fr}} = -F_{\text{fr}}d$, so if d is greater (as for the curved path), then W is greater. The work done does not depend only on points 1 and 2.

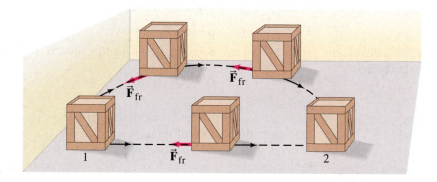

Because potential energy is energy associated with the position or configuration of objects, potential energy can only make sense if it can be stated uniquely for a given point. This cannot be done with nonconservative forces since the work done depends on the path taken (as in Fig. 6–16). Hence, *potential energy can be defined only for a conservative force.* Thus, although potential energy is always associated with a force, not all forces have a potential energy. For example, there is no potential energy for friction.

PE is defined only for a conservative force

There is no PE for friction

> **EXERCISE C** An object acted on by a constant force F moves from point 1 to point 2 and back again. The work done by the force F in this round trip is 60 J. Can you determine from this information if F is a conservative or nonconservative force?

We can now extend the **work-energy principle** (discussed in Section 6–3) to include potential energy. Suppose several forces act on an object which can undergo translational motion. And suppose only some of these forces are conservative. We write the total (net) work W_{net} as a sum of the work done by conservative forces, W_C, and the work done by nonconservative forces, W_{NC}:

$$W_{net} = W_C + W_{NC}.$$

Then, from the work-energy principle, Eq. 6–4, we have

$$W_{net} = \Delta KE$$
$$W_C + W_{NC} = \Delta KE$$

where $\Delta KE = KE_2 - KE_1$. Then

$$W_{NC} = \Delta KE - W_C.$$

Work done by a conservative force can be written in terms of potential energy, as we saw in Eq. 6–7b for gravitational potential energy:

$$W_C = -\Delta PE.$$

We combine these last two equations:

$$W_{NC} = \Delta KE + \Delta PE. \tag{6–10}$$

Thus, *the work W_{NC} done by the nonconservative forces acting on an object is equal to the total change in kinetic and potential energies.*

It must be emphasized that *all* the forces acting on an object must be included in Eq. 6–10, either in the potential energy term on the right (if it is a conservative force), or in the work term on the left (but not in both!).

6–6 Mechanical Energy and Its Conservation

If only conservative forces are acting in a system, we arrive at a particularly simple and beautiful relation involving energy.

When no nonconservative forces are present, then $W_{NC} = 0$ in Eq. 6–10, the general form of the work-energy principle. Then we have

$$\Delta KE + \Delta PE = 0 \qquad \begin{bmatrix} \text{conservative} \\ \text{forces only} \end{bmatrix} \tag{6–11a}$$

or

$$(KE_2 - KE_1) + (PE_2 - PE_1) = 0. \qquad \begin{bmatrix} \text{conservative} \\ \text{forces only} \end{bmatrix} \tag{6–11b}$$

We now define a quantity E, called the **total mechanical energy** of our system, as the sum of the kinetic and potential energies at any moment:

$$E = KE + PE.$$

Total mechanical energy defined

Now we can rewrite Eq. 6–11b as

$$KE_2 + PE_2 = KE_1 + PE_1 \qquad \begin{bmatrix} \text{conservative} \\ \text{forces only} \end{bmatrix} \tag{6–12a}$$

or

$$E_2 = E_1 = \text{constant}. \qquad \begin{bmatrix} \text{conservative} \\ \text{forces only} \end{bmatrix} \tag{6–12b}$$

Equations 6–12 express a useful and profound principle regarding the total mechanical energy of a system—namely, that it is a **conserved quantity**. The total mechanical energy E remains constant as long as no nonconservative forces act: $(KE + PE)$ at some initial time 1 is equal to the $(KE + PE)$ at any later time 2.

To say it another way, consider Eq. 6–11a which tells us $\Delta PE = -\Delta KE$; that is, if the kinetic energy KE of a system increases, then the potential energy PE must decrease by an equivalent amount to compensate. Thus, the total, $KE + PE$, remains constant:

> **If only conservative forces are acting, the total mechanical energy of a system neither increases nor decreases in any process. It stays constant—it is conserved.**

This is the **principle of conservation of mechanical energy** for conservative forces.

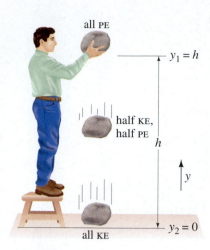

all PE

$y_1 = h$

half KE,
half PE

h

y

all KE

$y_2 = 0$

FIGURE 6–17 As it falls, the rock's potential energy changes to kinetic energy.

Conservation of mechanical energy when only gravity acts

FIGURE 6–18 Energy buckets (for Example 6–8). Kinetic energy is red and potential energy is blue. The total (KE + PE) is the same for the three points shown. The speed at $y = 0$, just before the rock hits the ground, is

$$\sqrt{2(9.8\,\text{m/s}^2)(3.0\,\text{m})} = 7.7\,\text{m/s}.$$

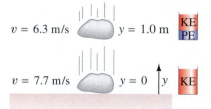

$v = 0$ $y = 3.0$ m   PE

$v = 6.3$ m/s $y = 1.0$ m   KE PE

$v = 7.7$ m/s $y = 0$ y KE

In the next Section we shall see the great usefulness of the conservation of mechanical energy principle in a variety of situations, and how it is often easier to use than the kinematic equations or Newton's laws. After that we will discuss how other forms of energy can be included in the general conservation of energy law that includes energy associated with nonconservative forces.

6–7 Problem Solving Using Conservation of Mechanical Energy

A simple example of the conservation of mechanical energy (neglecting air resistance) is a rock allowed to fall under gravity from a height h above the ground, as shown in Fig. 6–17. If the rock starts from rest, all of the initial energy is potential energy. As the rock falls, the potential energy decreases (because y decreases), but the rock's kinetic energy increases to compensate, so that the sum of the two remains constant. At any point along the path, the total mechanical energy is given by

$$E = \text{KE} + \text{PE} = \tfrac{1}{2}mv^2 + mgy$$

where y is the rock's height above the ground at a given instant and v is its speed at that point. If we let the subscript 1 represent the rock at one point along its path (for example, the initial point), and the subscript 2 represent it at some other point, then we can write

total mechanical energy at point 1 = total mechanical energy at point 2

or (see also Eq. 6–12a)

$$\tfrac{1}{2}mv_1^2 + mgy_1 = \tfrac{1}{2}mv_2^2 + mgy_2. \qquad \text{[grav. PE only]} \quad \textbf{(6–13)}$$

Just before the rock hits the ground, where we chose $y = 0$, all of the initial potential energy will have been transformed into kinetic energy.

EXAMPLE 6–8 **Falling rock.** If the original height of the rock in Fig. 6–17 is $y_1 = h = 3.0$ m, calculate the rock's speed when it has fallen to 1.0 m above the ground.

APPROACH One approach is to use the kinematic equations of Chapter 2. Let us instead apply the principle of conservation of mechanical energy, Eq. 6–13, assuming that only gravity acts on the rock. We choose the ground as our reference level ($y = 0$).

SOLUTION At the moment of release (point 1) the rock's position is $y_1 = 3.0$ m and it is at rest: $v_1 = 0$. We want to find v_2 when the rock is at position $y_2 = 1.0$ m. Equation 6–13 gives

$$\tfrac{1}{2}mv_1^2 + mgy_1 = \tfrac{1}{2}mv_2^2 + mgy_2.$$

The m's cancel out; setting $v_1 = 0$ and solving for v_2^2 we find

$$v_2^2 = 2g(y_1 - y_2)$$
$$= 2(9.8\,\text{m/s}^2)[(3.0\,\text{m}) - (1.0\,\text{m})] = 39.2\,\text{m}^2/\text{s}^2,$$

and

$$v_2 = \sqrt{39.2}\,\text{m/s} = 6.3\,\text{m/s}.$$

The rock's speed 1.0 m above the ground is 6.3 m/s downward.

NOTE The velocity at point 2 is independent of the rock's mass.

EXERCISE D Solve Example 6–8 by using the work-energy principle applied to the rock, without the concept of potential energy. Show all equations you use, starting with Eq. 6–4.

A simple way to visualize energy conservation is with an "energy bucket" as shown in Fig. 6–18. At each point in the fall of the rock, for example, the amount of kinetic energy and potential energy are shown as if they were two differently colored materials in the bucket. The total amount of material in the bucket (= total mechanical energy) remains constant.

Equation 6–13 can be applied to any object moving without friction under the action of gravity. For example, Fig. 6–19 shows a roller-coaster car starting from rest at the top of a hill, and coasting without friction to the bottom and up the hill on the other side.[†] Initially, the car has only potential energy. As it coasts down the hill, it loses potential energy and gains in kinetic energy, but the sum of the two remains constant. At the bottom of the hill it has its maximum kinetic energy; as it climbs up the other side, the kinetic energy changes back to potential energy. When the car comes to rest again, all of its energy will be potential energy. Given that the potential energy is proportional to the vertical height, energy conservation tells us that (in the absence of friction) the car comes to rest at a height equal to its original height. If the two hills are the same height, the car will just barely reach the top of the second hill when it stops. If the second hill is lower than the first, not all of the car's kinetic energy will be transformed to potential energy and the car can continue over the top and down the other side. If instead the second hill is higher, the car will only reach a height on it equal to its original height on the first hill. This is true (in the absence of friction) no matter how steep the hill is, since potential energy depends only on the vertical height (Eq. 6–6).

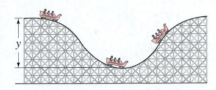

FIGURE 6–19 A roller-coaster car moving without friction illustrates the conservation of mechanical energy.

Grav. PE depends on vertical height, not path length (Eq. 6–6)

EXAMPLE 6–9 **Roller-coaster speed using energy conservation.** Assuming the height of the hill in Fig. 6–19 is 40 m, and the roller-coaster car starts from rest at the top, calculate (a) the speed of the roller-coaster car at the bottom of the hill, and (b) at what height it will have half this speed. Take $y = 0$ at the bottom of the hill.

APPROACH We choose point 1 to be where the car starts from rest $(v_1 = 0)$ at the top of the hill $(y_1 = 40\,\text{m})$. Point 2 is the bottom of the hill, which we choose as our reference level, so $y_2 = 0$. We use conservation of mechanical energy.

SOLUTION (a) We use Eq. 6–13 with $v_1 = 0$ and $y_2 = 0$. Then

$$\tfrac{1}{2}mv_1^2 + mgy_1 = \tfrac{1}{2}mv_2^2 + mgy_2$$
$$mgy_1 = \tfrac{1}{2}mv_2^2.$$

The m's cancel out and, setting $y_1 = 40\,\text{m}$, we find

$$v_2 = \sqrt{2gy_1} = \sqrt{2(9.8\,\text{m/s}^2)(40\,\text{m})} = 28\,\text{m/s}.$$

(b) We again use conservation of energy,

$$\tfrac{1}{2}mv_1^2 + mgy_1 = \tfrac{1}{2}mv_2^2 + mgy_2,$$

but now $v_2 = 14\,\text{m/s}$ (half of 28 m/s) and y_2 is unknown. We cancel the m's, set $v_1 = 0$, and solve for y_2:

$$y_2 = y_1 - \frac{v_2^2}{2g} = 30\,\text{m}.$$

That is, the car has a speed of 14 m/s when it is 30 *vertical* meters above the lowest point, both when descending the left-hand hill and when ascending the right-hand hill.

NOTE The mathematics of this Example is almost the same as that in Example 6–8. But there is an important difference between them. Example 6–8 could have been solved using force, acceleration, and the kinematic equations (Eqs. 2–11). But here, where the motion is not vertical, that approach would have been too complicated, whereas energy conservation readily gives us the answer.

[†]The forces on the car are gravity, the normal force exerted by the track, and friction (here, assumed zero). The normal force acts perpendicular to the track, and so is always perpendicular to the motion and does no work. Thus $W_{NC} = 0$ in Eq. 6–10 (so mechanical energy is conserved) and we can use Eq. 6–13 with the potential energy being only gravitational potential energy. We will see how to deal with friction, for which $W_{NC} \neq 0$, in Section 6–9.

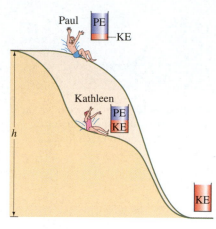

FIGURE 6–20 Example 6–10.

CONCEPTUAL EXAMPLE 6–10 **Speeds on two water slides.** Two water slides at a pool are shaped differently, but have the same length and start at the same height h (Fig. 6–20). Two riders, Paul and Kathleen, start from rest at the same time on different slides. (a) Which rider, Paul or Kathleen, is traveling faster at the bottom? (b) Which rider makes it to the bottom first? Ignore friction.

RESPONSE (a) Each rider's initial potential energy mgh gets transformed to kinetic energy, so the speed v at the bottom is obtained from $\frac{1}{2}mv^2 = mgh$. The mass cancels and so the speed will be the same, regardless of the mass of the rider. Since they descend the same vertical height, they will finish with the same speed.

(b) Note that Kathleen is consistently at a lower elevation than Paul at any instant, until the end. This means she has converted her potential energy to kinetic energy earlier. Consequently, she is traveling faster than Paul for the whole trip, except toward the end where Paul finally gets up to the same speed. Since she was going faster for the whole trip, and the distance is the same, Kathleen gets to the bottom first.

EXERCISE E Two balls are released from the same height above the floor. Ball A falls freely through the air, whereas ball B slides on a curved frictionless track to the floor. How do the speeds of the balls compare when they reach the floor?

➡ **PROBLEM SOLVING**

Whether to use energy, or Newton's laws?

🚶 **PHYSICS APPLIED**

Sports

FIGURE 6–21 Transformation of energy during a pole vault.

You may wonder sometimes whether to approach a problem using work and energy, or instead to use Newton's laws. As a rough guideline, if the force(s) involved are constant, either approach may succeed. If the forces are not constant, and/or the path is not simple, energy may be the surest approach.

There are many interesting examples of the conservation of energy in sports, such as the pole vault illustrated in Fig. 6–21. We often have to make approximations, but the sequence of events in broad outline for the pole vault is as follows. The initial kinetic energy of the running athlete is transformed into elastic potential energy of the bending pole and, as the athlete leaves the ground, into gravitational potential energy. When the vaulter reaches the top and the pole has straightened out again, the energy has all been transformed into gravitational potential energy (if we ignore the vaulter's low horizontal speed over the bar). The pole does not supply any energy, but it acts as a device to *store* energy and thus aid in the transformation of kinetic energy into gravitational potential energy, which is the net result. The energy required to pass over the bar depends on how high the center of mass (CM) of the vaulter must be raised. By bending their bodies, pole vaulters keep their CM so low that it can actually pass slightly beneath the bar (Fig. 6–22), thus enabling them to cross over a higher bar than would otherwise be possible. (Center of mass is covered in Chapter 7.)

FIGURE 6–22 By bending their bodies, pole vaulters can keep their center of mass so low that it may even pass below the bar. By changing their kinetic energy (of running) into gravitational potential energy ($= mgy$) in this way, vaulters can cross over a higher bar than if the change in potential energy were accomplished without carefully bending the body.

As another example of the conservation of mechanical energy, let us consider an object of mass m connected to a horizontal spring whose own mass can be neglected and whose spring stiffness constant is k. The mass m has speed v at any moment. The potential energy of the system (object plus spring) is given by Eq. 6–9, $\text{PE} = \frac{1}{2}kx^2$, where x is the displacement of the spring from its unstretched length. If neither friction nor any other force is acting, conservation of mechanical energy tells us that

$$\tfrac{1}{2}mv_1^2 + \tfrac{1}{2}kx_1^2 = \tfrac{1}{2}mv_2^2 + \tfrac{1}{2}kx_2^2, \qquad \text{[elastic PE only]} \quad (6\text{–}14)$$

Conservation of mechanical energy when PE is elastic

where the subscripts 1 and 2 refer to the velocity and displacement at two different moments.

EXAMPLE 6–11 **Toy dart gun.** A dart of mass 0.100 kg is pressed against the spring of a toy dart gun as shown in Fig. 6–23a. The spring (with spring stiffness constant $k = 250$ N/m) is compressed 6.0 cm and released. If the dart detaches from the spring when the spring reaches its natural length $(x = 0)$, what speed does the dart acquire?

APPROACH The dart is initially at rest (point 1), so $\text{KE}_1 = 0$. We ignore friction and use conservation of mechanical energy; the only potential energy is elastic.

SOLUTION We use Eq. 6–14 with point 1 being at the maximum compression of the spring, so $v_1 = 0$ (dart not yet released) and $x_1 = -0.060$ m. Point 2 we choose to be the instant the dart flies off the end of the spring (Fig. 6–23b), so $x_2 = 0$ and we want to find v_2. Thus Eq. 6–14 can be written

$$0 + \tfrac{1}{2}kx_1^2 = \tfrac{1}{2}mv_2^2 + 0.$$

Then

$$v_2^2 = \frac{kx_1^2}{m}$$

$$= \frac{(250\ \text{N/m})(-0.060\ \text{m})^2}{(0.100\ \text{kg})} = 9.0\ \text{m}^2/\text{s}^2$$

so $v_2 = \sqrt{v_2^2} = 3.0$ m/s.

NOTE In the horizontal direction, the only force on the dart (neglecting friction) was the force exerted by the spring. Vertically, gravity was counterbalanced by the normal force exerted on the dart by the gun barrel. After it leaves the barrel, the dart will follow a projectile's path under gravity.

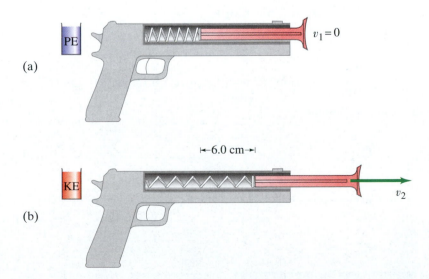

(a)

⊢—6.0 cm—⊣

(b)

FIGURE 6–23 Example 6–11. (a) A dart is pushed against a spring, compressing it 6.0 cm. The dart is then released, and in (b) it leaves the spring at velocity v_2.

Additional Example

The next Example shows how to solve a problem involving two types of potential energy.

EXAMPLE 6–12 **Two kinds of PE.** A ball of mass $m = 2.60\,\text{kg}$, starting from rest, falls a vertical distance $h = 55.0\,\text{cm}$ before striking a vertical coiled spring, which it compresses an amount $Y = 15.0\,\text{cm}$ (Fig. 6–24). Determine the spring stiffness constant of the spring. Assume the spring has negligible mass, and ignore air resistance. Measure all distances from the point where the ball first touches the uncompressed spring ($y = 0$ at this point).

APPROACH The forces acting on the ball are the gravitational pull of the Earth and the elastic force exerted by the spring. Both forces are conservative, so we can use conservation of mechanical energy, including both types of potential energy. We must be careful, however: gravity acts throughout the fall (Fig. 6–24), whereas the elastic force does not act until the ball touches the spring (Fig. 6–24b). We choose y positive upward, and $y = 0$ at the end of the spring in its natural (uncompressed) state.

SOLUTION We divide this solution into two parts. (An alternate solution follows.)
Part 1: Let us first consider the energy changes as the ball falls from a height $y_1 = h = 0.55\,\text{m}$, Fig. 6–24a, to $y_2 = 0$, just as it touches the spring, Fig. 6–24b. Our system is the ball acted on by gravity plus the spring, which up to this point doesn't do anything. Thus

$$\tfrac{1}{2}mv_1^2 + mgy_1 = \tfrac{1}{2}mv_2^2 + mgy_2$$
$$0 + mgh = \tfrac{1}{2}mv_2^2 + 0.$$

We solve for $v_2 = \sqrt{2gh} = \sqrt{2(9.80\,\text{m/s}^2)(0.550\,\text{m})} = 3.283\,\text{m/s} \approx 3.28\,\text{m/s}$. This is the speed of the ball just as it touches the top of the spring, Fig. 6–24b.
Part 2: Let's see what happens as the ball compresses the spring, Figs. 6–24b to c. Now there are two conservative forces on the ball—gravity and the spring force. So our conservation of energy equation becomes

Conservation of energy: gravity and elastic PE

$$E(\text{ball touches spring}) = E(\text{spring compresses})$$
$$\tfrac{1}{2}mv_2^2 + mgy_2 + \tfrac{1}{2}ky_2^2 = \tfrac{1}{2}mv_3^2 + mgy_3 + \tfrac{1}{2}ky_3^2.$$

We take point 2 to be the instant when the ball just touches the spring, so $y_2 = 0$ and $v_2 = 3.283\,\text{m/s}$ (keeping an extra digit for now). We take point 3 to be when the ball comes to rest (for an instant) and the spring is fully compressed, so $v_3 = 0$ and $y_3 = -Y = -0.150\,\text{m}$ (given). Substituting into the above energy equation, we get

$$\tfrac{1}{2}mv_2^2 + 0 + 0 = 0 - mgY + \tfrac{1}{2}kY^2.$$

We know m, v_2, and Y, so we can solve for k:

$$k = \frac{2}{Y^2}\left[\tfrac{1}{2}mv_2^2 + mgY\right] = \frac{m}{Y^2}\left[v_2^2 + 2gY\right]$$
$$= \frac{(2.60\,\text{kg})}{(0.150\,\text{m})^2}\left[(3.283\,\text{m/s})^2 + 2(9.80\,\text{m/s}^2)(0.150\,\text{m})\right] = 1590\,\text{N/m},$$

which is the result we sought.

FIGURE 6–24 Example 6–12.

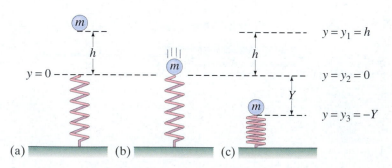

Alternate Solution Instead of dividing the solution into two parts, we can do it all at once. After all, we get to choose what two points are used on the left and right of the energy equation. Let us write the energy equation for points 1 and 3 (Fig. 6–24). Point 1 is the initial point just before the ball starts to fall (Fig. 6–24a), so $v_1 = 0$, $y_1 = h = 0.550$ m; and point 3 is when the spring is fully compressed (Fig. 6–24c), so $v_3 = 0$, $y_3 = -Y = -0.150$ m. The forces on the ball in this process are gravity and (at least part of the time) the spring. So conservation of energy tells us

$$\tfrac{1}{2}mv_1^2 + mgy_1 + \tfrac{1}{2}k(0)^2 = \tfrac{1}{2}mv_3^2 + mgy_3 + \tfrac{1}{2}ky_3^2$$
$$0 \;+\; mgh \;+\; 0 \;=\; 0 \;-\; mgY + \tfrac{1}{2}kY^2$$

where we have set $y = 0$ for the spring at point 1 because it is not acting and is not compressed or stretched at point 1. We solve for k:

$$k = \frac{2\,mg(h + Y)}{Y^2} = \frac{2(2.60\,\text{kg})(9.80\,\text{m/s}^2)(0.550\,\text{m} + 0.150\,\text{m})}{(0.150\,\text{m})^2} = 1590\,\text{N/m}$$

just as in our first method of solution.

6–8 Other Forms of Energy; Energy Transformations and the Law of Conservation of Energy

Besides the kinetic energy and potential energy of ordinary objects, other forms of energy can be defined as well. These include electric energy, nuclear energy, thermal energy, and the chemical energy stored in food and fuels. With the advent of the atomic theory, these other forms of energy have come to be considered as kinetic or potential energy at the atomic or molecular level. For example, according to the atomic theory, thermal energy is the kinetic energy of rapidly moving molecules—when an object is heated, the molecules that make up the object move faster. On the other hand, the energy stored in food and fuel such as gasoline is potential energy stored by virtue of the relative positions of the atoms within a molecule due to electric forces between the atoms (referred to as chemical bonds). For the energy in chemical bonds to be used to do work, it must be released, usually through chemical reactions. This is analogous to a compressed spring which, when released, can do work. Electric, magnetic, and nuclear energies also can be considered examples of kinetic and potential (or stored) energies. We will deal with these other forms of energy in detail in later Chapters.

Energy can be transformed from one form to another, and we have already encountered several examples of this. A rock held high in the air has potential energy; as it falls, it loses potential energy, since its height above the ground decreases. At the same time, it gains in kinetic energy, since its velocity is increasing. Potential energy is being transformed into kinetic energy.

Often the transformation of energy involves a transfer of energy from one object to another. The potential energy stored in the spring of Fig. 6–13b is transformed into the kinetic energy of the ball, Fig. 6–13c. Water at the top of a dam has potential energy, which is transformed into kinetic energy as the water falls. At the base of the dam, the kinetic energy of the water can be transferred to turbine blades and further transformed into electric energy, as we shall see in a later Chapter. The potential energy stored in a bent bow can be transformed into kinetic energy of the arrow (Fig. 6–25).

In each of these examples, the transfer of energy is accompanied by the performance of work. The spring of Fig. 6–13 does work on the ball. Water does work on turbine blades. A bow does work on an arrow. This observation gives us a further insight into the relation between work and energy: *work is done when energy is transferred from one object to another.*[†] A person throwing a ball or pushing a grocery cart provides another example. The work done is a manifestation of energy being transferred from the person (ultimately derived from the chemical energy of food) to the ball or cart.

FIGURE 6–25 Potential energy of a bent bow about to be transformed into kinetic energy of an arrow.

Work is done when energy is transferred from one object to another

[†]If the objects are at different temperatures, heat can flow between them instead, or in addition. See Chapters 14 and 15.

One of the great results of physics is that whenever energy is transferred or transformed, it is found that no energy is gained or lost in the process.

This is the **law of conservation of energy**, one of the most important principles in physics; it can be stated as:

The total energy is neither increased nor decreased in any process. Energy can be transformed from one form to another, and transferred from one object to another, but the total amount remains constant.

We have already discussed the conservation of energy for mechanical systems involving conservative forces, and we saw how it could be derived from Newton's laws and thus is equivalent to them. But in its full generality, the validity of the law of conservation of energy, encompassing all forms of energy including those associated with nonconservative forces like friction, rests on experimental observation. Even though Newton's laws are found to fail in the submicroscopic world of the atom, the law of conservation of energy has been found to hold in every experimental situation so far tested.

6–9 Energy Conservation with Dissipative Forces: Solving Problems

In our applications of energy conservation in Section 6–7, we neglected friction, a nonconservative force. But in many situations it cannot be ignored. In a real situation, the roller-coaster car in Fig. 6–19, for example, will not in fact reach the same height on the second hill as it had on the first hill because of friction. In this, and in other natural processes, the mechanical energy (sum of the kinetic and potential energies) does not remain constant but decreases. Because *Dissipative forces* frictional forces reduce the total mechanical energy (but *not* the total energy), they are called **dissipative forces**. Historically, the presence of dissipative forces hindered the formulation of a comprehensive conservation of energy law until well into the nineteenth century. It was only then that heat, which is always produced when there is friction (try rubbing your hands together), was interpreted in terms of energy. Quantitative studies by nineteenth-century scientists (discussed in Chapters 14 and 15) demonstrated that if heat is considered as a transfer of energy (thermal energy), then the total energy is conserved in any process. For example, if the roller-coaster car in Fig. 6–19 is subject to frictional forces, then the initial total energy of the car will be equal to the kinetic plus potential energy of the car at any subsequent point along its path plus the amount of thermal energy produced in the process. The thermal energy produced by a constant friction force F_{fr} is equal to the work done by friction. We now apply the general form of the work-energy principle, Eq. 6–10:

$$W_{NC} = \Delta KE + \Delta PE.$$

We can write $W_{NC} = -F_{fr} d$, where d is the distance over which the friction force acts. ($\vec{\mathbf{F}}$ and $\vec{\mathbf{d}}$ are in opposite directions, hence the minus sign.) Thus, with $KE = \frac{1}{2}mv^2$ and $PE = mgy$, we have

$$-F_{fr} d = \frac{1}{2}mv_2^2 - \frac{1}{2}mv_1^2 + mgy_2 - mgy_1$$

or

Conservation of energy with gravity and friction

$$\frac{1}{2}mv_1^2 + mgy_1 = \frac{1}{2}mv_2^2 + mgy_2 + F_{fr} d, \quad \left[\begin{array}{c}\text{gravity and} \\ \text{friction acting}\end{array}\right] \quad \textbf{(6–15)}$$

where d is the distance along the path traveled by the object in going from point 1 to point 2. Equation 6–15 can be seen to be Eq. 6–13 modified to include friction. It can be interpreted in a simple way: the initial mechanical energy of the car (point 1) equals the (reduced) final mechanical energy of the car plus the energy transformed by friction into thermal energy.

When other forms of energy are involved, such as chemical or electrical energy, the total amount of energy is always found to be conserved. Hence the law of conservation of energy is believed to be universally valid.

Work-Energy versus Energy Conservation

The work-energy principle and the law of conservation of energy are basically equivalent. The difference between them is in how you use them, and in particular on your *choice of the system* under study. If you choose as your system one or more objects on which external forces do work, then you must use the work-energy principle: the work done by the external forces on your system equals the total change in energy of your chosen system.

On the other hand, if you choose a system on which no external forces do work, then you can apply conservation of energy to that system.

Consider, for example, a spring connected to a block on a frictionless table (Fig. 6–26). If you choose the block as your system, then the work done on the block by the spring equals the change in kinetic energy of the block: the work-energy principle. (Energy conservation does not apply to this system—the block's energy changes.) If instead you choose the block plus the spring as your system, no external forces do work (since the spring is part of the chosen system). To this system you can apply conservation of energy: if you compress the spring and then release it, the spring still exerts a force on the block, but the subsequent motion can be discussed in terms of kinetic energy $(\frac{1}{2}mv^2)$ plus potential energy $(\frac{1}{2}kx^2)$, whose total remains constant.

Conservation of energy applies to any system on which no work is done by external forces.

FIGURE 6–26 A spring connected to a block on a frictionless table. If you choose your system to be the block plus spring, then
$$E = \tfrac{1}{2}mv^2 + \tfrac{1}{2}kx^2$$
is conserved.

PROBLEM SOLVING Conservation of Energy

1. **Draw a picture** of the physical situation.
2. Determine **the system** for which energy will be conserved: the object or objects and the forces acting.
3. Ask yourself what quantity you are looking for, and decide what are **the initial** (point 1) **and final** (point 2) **positions**.
4. If the object under investigation changes its height during the problem, then **choose a reference frame** with a convenient $y = 0$ level for gravitational potential energy; the lowest point in the problem is often a good choice.

   If springs are involved, choose the unstretched spring position to be x (or y) $= 0$.

5. **Apply conservation of energy.** If no friction or other nonconservative forces act, then conservation of mechanical energy holds:
   $$\text{KE}_1 + \text{PE}_1 = \text{KE}_2 + \text{PE}_2.$$
   If friction or other nonconservative forces are present, then an additional term (W_{NC}) will be needed:
   $$W_{\text{NC}} = \Delta\text{KE} + \Delta\text{PE}.$$
   To be sure which sign to give W_{NC}, you can use your intuition: is the total mechanical energy increased or decreased in the process?
6. Use the equation(s) you develop to **solve** for the unknown quantity.

EXAMPLE 6–13 **Friction on the roller coaster.** The roller-coaster car in Example 6–9 reaches a vertical height of only 25 m on the second hill before coming to a momentary stop (Fig. 6–27). It traveled a total distance of 400 m. Estimate the average friction force (assume constant) on the car, whose mass is 1000 kg.

APPROACH We explicitly follow the Problem Solving Box step by step.

SOLUTION 1. Draw a Picture. See Fig. 6–27.

2. **The system.** The system is the roller-coaster car (and the Earth since it exerts the gravitational force). The forces acting on the car are gravity and friction. (The normal force also acts on the car, but does no work, so it does not affect the energy.)

3. **Choose initial and final positions.** We take point 1 to be the instant when the car started coasting (at the top of the first hill), and point 2 to be the instant it stopped 25 m up the second hill.

4. **Choose a reference frame.** We choose the lowest point in the motion to be $y = 0$ for the gravitational potential energy.

5. **Apply conservation of energy.** There is friction acting on the car, so we use conservation of energy in the form of Eq. 6–15, with $v_1 = 0$, $y_1 = 40$ m, $v_2 = 0$, $y_2 = 25$ m, and $d = 400$ m. Thus
$$0 + (1000\,\text{kg})(9.8\,\text{m/s}^2)(40\,\text{m}) = 0 + (1000\,\text{kg})(9.8\,\text{m/s}^2)(25\,\text{m}) + F_{\text{fr}}(400\,\text{m}).$$

6. **Solve.** We can solve this equation for F_{fr}: $F_{\text{fr}} = 370$ N.

FIGURE 6–27 Example 6–13. Because of friction, a roller coaster car does not reach the original height on the second hill.

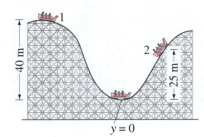

Problem solving is not a process that can be done by following a set of rules. The Problem Solving Box on page 157 is thus not a prescription, but is a *summary* of steps to help you get started in solving problems involving energy.

6–10 Power

Power defined

Power is defined as the *rate at which work is done*. Average power equals the work done divided by the time to do it. Power can also be defined as the *rate at which energy is transformed*. Thus

Average power

$$\overline{P} = \text{average power} = \frac{\text{work}}{\text{time}} = \frac{\text{energy transformed}}{\text{time}}. \quad \textbf{(6–16)}$$

The power of a horse refers to how much work it can do per unit time. The power rating of an engine refers to how much chemical or electrical energy can be transformed into mechanical energy per unit time. In SI units, power is measured in joules per second, and this unit is given a special name, the **watt** (W): $1\,\text{W} = 1\,\text{J/s}$. We are most familiar with the watt for electrical devices: the rate at which an electric lightbulb or heater changes electric energy into light or thermal energy; but the watt is used for other types of energy transformations as well. In the British system, the unit of power is the foot-pound per second ($\text{ft}\cdot\text{lb/s}$). For practical purposes, a larger unit is often used, the **horsepower**. One horsepower† (hp) is defined as $550\,\text{ft}\cdot\text{lb/s}$, which equals $746\,\text{W}$.

Power units: the watt

The horsepower

! CAUTION
Distinguish between power and energy

To see the distinction between energy and power, consider the following example. A person is limited in the work he or she can do, not only by the total energy required, but also by how fast this energy is transformed: that is, by power. For example, a person may be able to walk a long distance or climb many flights of stairs before having to stop because so much energy has been expended. On the other hand, a person who runs very quickly upstairs may fall exhausted after only a flight or two. He or she is limited in this case by power, the rate at which his or her body can transform chemical energy into mechanical energy.

FIGURE 6–28 Example 6–14.

EXAMPLE 6–14 **Stair-climbing power.** A 60-kg jogger runs up a long flight of stairs in 4.0 s (Fig. 6–28). The vertical height of the stairs is 4.5 m. (*a*) Estimate the jogger's power output in watts and horsepower. (*b*) How much energy did this require?

APPROACH The work done by the jogger is against gravity, and equals $W = mgy$. To get her power output we divide W by the time it took.

SOLUTION (*a*) The average power output was

$$\overline{P} = \frac{W}{t} = \frac{mgy}{t} = \frac{(60\,\text{kg})(9.8\,\text{m/s}^2)(4.5\,\text{m})}{4.0\,\text{s}} = 660\,\text{W}.$$

Since there are $746\,\text{W}$ in 1 hp, the jogger is doing work at a rate of just under 1 hp. A human cannot do work at this rate for very long.
(*b*) The energy required is $E = \overline{P}t$ (Eq. 6–16). Since $\overline{P} = 660\,\text{W} = 660\,\text{J/s}$, then $E = (660\,\text{J/s})(4.0\,\text{s}) = 2600\,\text{J}$. This result equals $W = mgy$.

NOTE The person had to transform more energy than this 2600 J. The total energy transformed by a person or an engine always includes some thermal energy (recall how hot you get running up stairs).

---

†The unit was chosen by James Watt (1736–1819), who needed a way to specify the power of his newly developed steam engines. He found by experiment that a good horse can work all day at an average rate of about $360\,\text{ft}\cdot\text{lb/s}$. So as not to be accused of exaggeration in the sale of his steam engines, he multiplied this by $1\frac{1}{2}$ when he defined the hp.

Automobile engines do work to overcome the force of friction (including air resistance), to climb hills, and to accelerate. A car is limited by the rate at which it can do work, which is why automobile engines are rated in horsepower. A car needs power most when climbing hills and when accelerating. In the next Example, we will calculate how much power is needed in these situations for a car of reasonable size. Even when a car travels on a level road at constant speed, it needs some power just to do work to overcome the retarding forces of internal friction and air resistance. These forces depend on the conditions and speed of the car, but are typically in the range 400 N to 1000 N.

It is often convenient to write power in terms of the net force F applied to an object and its speed v. This is readily done since $\overline{P} = W/t$ and $W = Fd$, where d is the distance traveled. Then

$$\overline{P} = \frac{W}{t} = \frac{Fd}{t} = F\overline{v}, \qquad (6\text{--}17)$$

where $\overline{v} = d/t$ is the average speed of the object.

EXAMPLE 6–15 **Power needs of a car.** Calculate the power required of a 1400-kg car under the following circumstances: (a) the car climbs a 10° hill (a fairly steep hill) at a steady 80 km/h; and (b) the car accelerates along a level road from 90 to 110 km/h in 6.0 s to pass another car. Assume the retarding force on the car is $F_R = 700\ N$ throughout. See Fig. 6–29.

APPROACH First we must be careful not to confuse $\vec{F}_R$, which is due to air resistance and friction that retards the motion, with the force $\vec{F}$ needed to accelerate the car, which is the frictional force exerted by the road on the tires—the reaction to the motor-driven tires pushing against the road. We must determine the latter force F before calculating the power.

SOLUTION (a) To move at a steady speed up the hill, the car must, by Newton's second law, exert a force F equal to the sum of the retarding force, 700 N, and the component of gravity parallel to the hill, $mg \sin 10°$. Thus

$$F = 700\ N + mg \sin 10°$$
$$= 700\ N + (1400\ kg)(9.80\ m/s^2)(0.174) = 3100\ N.$$

Since $\overline{v} = 80\ km/h = 22\ m/s$ and is parallel to $\vec{F}$, then (Eq. 6–17) the power is

$$\overline{P} = F\overline{v} = (3100\ N)(22\ m/s) = 6.80 \times 10^4\ W = 91\ hp.$$

(b) The car accelerates from 25.0 m/s to 30.6 m/s (90 to 110 km/h). Thus the car must exert a force that overcomes the 700-N retarding force plus that required to give it the acceleration

$$\overline{a}_x = \frac{(30.6\ m/s - 25.0\ m/s)}{6.0\ s} = 0.93\ m/s^2.$$

We apply Newton's second law with x being the direction of motion:

$$ma_x = \Sigma F_x = F - F_R.$$

Then the force required, F, is

$$F = ma_x + F_R$$
$$= (1400\ kg)(0.93\ m/s^2) + 700\ N$$
$$= 1300\ N + 700\ N = 2000\ N.$$

Since $\overline{P} = F\overline{v}$, the required power increases with speed and the motor must be able to provide a maximum power output of

$$\overline{P} = (2000\ N)(30.6\ m/s) = 6.12 \times 10^4\ W = 82\ hp.$$

NOTE Even taking into account the fact that only 60 to 80% of the engine's power output reaches the wheels, it is clear from these calculations that an engine of 100 to 150 hp is quite adequate from a practical point of view.

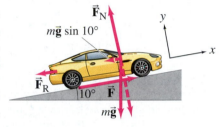

FIGURE 6–29 Example 6–15a. Calculation of power needed for a car to climb a hill.

We mentioned in Example 6–15 that only part of the energy output of a car engine reaches the wheels. Not only is some energy wasted in getting from the engine to the wheels, in the engine itself much of the input energy (from the gasoline) does not do useful work. An important characteristic of all engines is their overall *efficiency e*, defined as the ratio of the useful power output of the engine, P_{out}, to the power input, P_{in}:

Efficiency

$$e = \frac{P_{out}}{P_{in}}.$$

The efficiency is always less than 1.0 because no engine can create energy, and no engine can even transform energy from one form to another without some energy going to friction, thermal energy, and other nonuseful forms of energy. For example, an automobile engine converts chemical energy released in the burning of gasoline into mechanical energy that moves the pistons and eventually the wheels. But nearly 85% of the input energy is "wasted" as thermal energy that goes into the cooling system or out the exhaust pipe, plus friction in the moving parts. Thus car engines are roughly only about 15% efficient. We will discuss efficiency in detail in Chapter 15.

Summary

Work is done on an object by a force when the object moves through a distance d. If the direction of a constant force F makes an angle θ with the direction of motion, the work done by this force is

$$W = Fd \cos \theta. \quad \text{(6–1)}$$

Energy can be defined as the ability to do work. In SI units, work and energy are measured in **joules** ($1\,J = 1\,N \cdot m$).

Kinetic energy (KE) is energy of motion. An object of mass m and speed v has translational kinetic energy

$$\text{KE} = \tfrac{1}{2}mv^2. \quad \text{(6–3)}$$

Potential energy (PE) is energy associated with forces that depend on the position or configuration of objects. Gravitational potential energy is

$$\text{PE}_{grav} = mgy, \quad \text{(6–6)}$$

where y is the height of the object of mass m above an arbitrary reference point. Elastic potential energy is given by

$$\text{elastic PE} = \tfrac{1}{2}kx^2 \quad \text{(6–9)}$$

for a stretched or compressed spring, where x is the displacement from the unstretched position and k is the spring stiffness constant. Other potential energies include chemical,

electrical, and nuclear energy. The change in potential energy when an object changes position is equal to the external work needed to take the object from one position to the other.

The **work-energy principle** states that the *net* work done on an object (by the *net* force) equals the change in kinetic energy of that object:

$$W_{net} = \Delta\text{KE} = \tfrac{1}{2}mv_2^2 - \tfrac{1}{2}mv_1^2. \quad \text{(6–2, 6–4)}$$

The **law of conservation of energy** states that energy can be transformed from one type to another, but the total energy remains constant. It is valid even when friction is present, since the heat generated can be considered a form of energy transfer. When only *conservative forces* act, the total mechanical energy is conserved:

$$\text{KE} + \text{PE} = \text{constant}.$$

When nonconservative forces such as friction act, then

$$W_{NC} = \Delta\text{KE} + \Delta\text{PE}, \quad \text{(6–10)}$$

where W_{NC} is the work done by nonconservative forces.

Power is defined as the rate at which work is done, or the rate at which energy is transformed. The SI unit of power is the **watt** ($1\,W = 1\,J/s$).

Questions

1. In what ways is the word "work" as used in everyday language the same as that defined in physics? In what ways is it different? Give examples of both.

2. Can a centripetal force ever do work on an object? Explain.

3. Can the normal force on an object ever do work? Explain.

4. A woman swimming upstream is not moving with respect to the shore. Is she doing any work? If she stops swimming and merely floats, is work done on her?

5. Is the work done by kinetic friction forces always negative? [*Hint*: Consider what happens to the dishes when you pull a tablecloth out from under them.]

6. Why is it tiring to push hard against a solid wall even though you are doing no work?

7. You have two springs that are identical except that spring 1 is stiffer than spring 2 ($k_1 > k_2$). On which spring is more work done (*a*) if they are stretched using the same force, (*b*) if they are stretched the same distance?

8. A hand exerts a constant horizontal force on a block that is free to slide on a frictionless surface (Fig. 6–30). The block starts from rest at point A, and by the time it has traveled a distance d to point B it is traveling with speed v_B. When the block has traveled another distance d to point C, will its speed be greater than, less than, or equal to $2v_B$? Explain your reasoning.

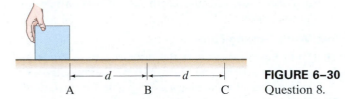

FIGURE 6–30
Question 8.

A   B   C

9. By approximately how much does your gravitational potential energy change when you jump as high as you can?

10. In Fig. 6–31, water balloons are tossed from the roof of a building, all with the same speed but with different launch angles. Which one has the highest speed on impact? Ignore air resistance.

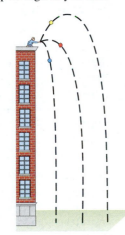

FIGURE 6–31
Question 10.

11. A pendulum is launched from a point that is a height h above its lowest point in two different ways (Fig. 6–32). During both launches, the pendulum is given an initial speed of 3.0 m/s. On the first launch, the initial velocity of the pendulum is directed upward along the trajectory, and on the second launch it is directed downward along the trajectory. Which launch will cause it to swing the largest angle from the equilibrium position? Explain.

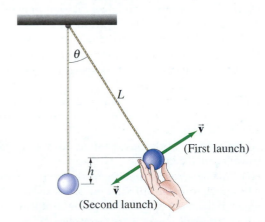

(First launch)

(Second launch)

FIGURE 6–32   Question 11.

12. A coil spring of mass m rests upright on a table. If you compress the spring by pressing down with your hand and then release it, can the spring leave the table? Explain, using the law of conservation of energy.

13. A bowling ball is hung from the ceiling by a steel wire (Fig. 6–33). The instructor pulls the ball back and stands against the wall with the ball against his nose. To avoid injury the instructor is supposed to release the ball without pushing it. Why?

FIGURE 6–33
Question 13.

14. What happens to the gravitational potential energy when water at the top of a waterfall falls to the pool below?

15. Describe the energy transformations when a child hops around on a pogo stick.

16. Describe the energy transformations that take place when a skier starts skiing down a hill, but after a time is brought to rest by striking a snowdrift.

17. A child on a sled (total mass m) starts from rest at the top of a hill of height h and slides down. Does the velocity at the bottom depend on the angle of the hill if (*a*) it is icy and there is no friction, and (*b*) there is friction (deep snow)?

18. Seasoned hikers prefer to step over a fallen log in their path rather than stepping on top and jumping down on the other side. Explain.

19. Two identical arrows, one with twice the speed of the other, are fired into a bale of hay. Assuming the hay exerts a constant frictional force on the arrows, the faster arrow will penetrate how much farther than the slower arrow? Explain.

20. Analyze the motion of a simple swinging pendulum in terms of energy, (*a*) ignoring friction, and (*b*) taking friction into account. Explain why a grandfather clock has to be wound up.

21. When a "superball" is dropped, can it rebound to a height greater than its original height? Explain.

22. Suppose you lift a suitcase from the floor to a table. The work you do on the suitcase depends on which of the following: (*a*) whether you lift it straight up or along a more complicated path, (*b*) the time it takes, (*c*) the height of the table, and (*d*) the weight of the suitcase?

23. Repeat Question 22 for the *power* needed rather than the work.

24. Why is it easier to climb a mountain via a zigzag trail than to climb straight up?

25. Recall from Chapter 4, Example 4–14, that you can use a pulley and ropes to decrease the force needed to raise a heavy load (see Fig. 6–34). But for every meter the load is raised, how much rope must be pulled up? Account for this, using energy concepts.

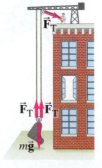

FIGURE 6–34
Question 25.

Problems

6-1 Work, Constant Force

1. (I) How much work is done by the gravitational force when a 265-kg pile driver falls 2.80 m?

2. (I) A 65.0-kg firefighter climbs a flight of stairs 20.0 m high. How much work is required?

3. (I) A 1300-N crate rests on the floor. How much work is required to move it at constant speed (a) 4.0 m along the floor against a friction force of 230 N, and (b) 4.0 m vertically?

4. (I) How much work did the movers do (horizontally) pushing a 160-kg crate 10.3 m across a rough floor without acceleration, if the effective coefficient of friction was 0.50?

5. (II) A box of mass 5.0 kg is accelerated from rest by a force across a floor at a rate of 2.0 m/s² for 7.0 s. Find the net work done on the box.

6. (II) Eight books, each 4.3 cm thick with mass 1.7 kg, lie flat on a table. How much work is required to stack them one on top of another?

7. (II) A lever such as that shown in Fig. 6–35 can be used to lift objects we might not otherwise be able to lift. Show that the ratio of output force, F_O, to input force, F_I, is related to the lengths l_I and l_O from the pivot point by $F_O/F_I = l_I/l_O$ (ignoring friction and the mass of the lever), given that the work output equals work input.

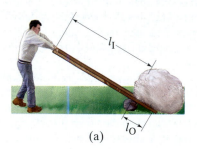
(a)

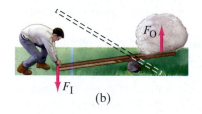
(b)

FIGURE 6–35
Problem 7.
A simple lever.

8. (II) A 330-kg piano slides 3.6 m down a 28° incline and is kept from accelerating by a man who is pushing back on it *parallel to the incline* (Fig. 6–36). The effective coefficient of kinetic friction is 0.40. Calculate: (a) the force exerted by the man, (b) the work done by the man on the piano, (c) the work done by the friction force, (d) the work done by the force of gravity, and (e) the net work done on the piano.

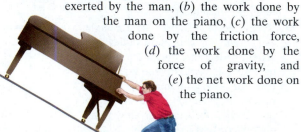

FIGURE 6–36
Problem 8.

9. (II) (a) Find the force required to give a helicopter of mass M an acceleration of $0.10\,g$ upward. (b) Find the work done by this force as the helicopter moves a distance h upward.

10. (II) What is the minimum work needed to push a 950-kg car 810 m up along a 9.0° incline? (a) Ignore friction. (b) Assume the effective coefficient of friction retarding the car is 0.25.

*6-2 Work, Varying Force

*11. (II) In Fig. 6–6a, assume the distance axis is linear and that $d_A = 10.0$ m and $d_B = 35.0$ m. Estimate the work done by force F in moving a 2.80-kg object from d_A to d_B.

*12. (II) The force on an object, acting along the x axis, varies as shown in Fig. 6–37. Determine the work done by this force to move the object (a) from $x = 0.0$ to $x = 10.0$ m, and (b) from $x = 0.0$ to $x = 15.0$ m.

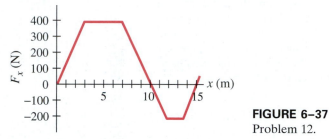

FIGURE 6–37
Problem 12.

*13. (II) A spring has $k = 88$ N/m. Use a graph to determine the work needed to stretch it from $x = 3.8$ cm to $x = 5.8$ cm, where x is the displacement from its unstretched length.

*14. (II) The net force exerted on a particle acts in the $+x$ direction. Its magnitude increases linearly from zero at $x = 0$, to 24.0 N at $x = 3.0$ m. It remains constant at 24.0 N from $x = 3.0$ m to $x = 8.0$ m, and then decreases linearly to zero at $x = 13.0$ m. Determine the work done to move the particle from $x = 0$ to $x = 13.0$ m graphically by determining the area under the F_x vs. x graph.

6-3 Kinetic Energy; Work-Energy Principle

15. (I) At room temperature, an oxygen molecule, with mass of 5.31×10^{-26} kg, typically has a KE of about 6.21×10^{-21} J. How fast is the molecule moving?

16. (I) (a) If the KE of an arrow is doubled, by what factor has its speed increased? (b) If its speed is doubled, by what factor does its KE increase?

17. (I) How much work is required to stop an electron $(m = 9.11 \times 10^{-31}$ kg) which is moving with a speed of 1.90×10^6 m/s?

18. (I) How much work must be done to stop a 1250-kg car traveling at 105 km/h?

19. (II) An 88-g arrow is fired from a bow whose string exerts an average force of 110 N on the arrow over a distance of 78 cm. What is the speed of the arrow as it leaves the bow?

20. (II) A baseball $(m = 140$ g) traveling 32 m/s moves a fielder's glove backward 25 cm when the ball is caught. What was the average force exerted by the ball on the glove?

21. (II) If the speed of a car is increased by 50%, by what factor will its minimum braking distance be increased, assuming all else is the same? Ignore the driver's reaction time.

22. (II) At an accident scene on a level road, investigators measure a car's skid mark to be 88 m long. The accident occurred on a rainy day, and the coefficient of kinetic friction was estimated to be 0.42. Use these data to determine the speed of the car when the driver slammed on (and locked) the brakes. (Why does the car's mass not matter?)

23. (II) A softball having a mass of 0.25 kg is pitched at 95 km/h. By the time it reaches the plate, it may have slowed by 10%. Neglecting gravity, estimate the average force of air resistance during a pitch, if the distance between the plate and the pitcher is about 15 m.

24. (II) How high will a 1.85-kg rock go if thrown straight up by someone who does 80.0 J of work on it? Neglect air resistance.

25. (III) A 285-kg load is lifted 22.0 m vertically with an acceleration $a = 0.160\,g$ by a single cable. Determine (a) the tension in the cable, (b) the net work done on the load, (c) the work done by the cable on the load, (d) the work done by gravity on the load, and (e) the final speed of the load assuming it started from rest.

6–4 and 6–5 Potential Energy

26. (I) A spring has a spring stiffness constant, k, of 440 N/m. How much must this spring be stretched to store 25 J of potential energy?

27. (I) A 7.0-kg monkey swings from one branch to another 1.2 m higher. What is the change in potential energy?

28. (I) By how much does the gravitational potential energy of a 64-kg pole vaulter change if his center of mass rises about 4.0 m during the jump?

29. (II) A 1200-kg car rolling on a horizontal surface has speed $v = 65$ km/h when it strikes a horizontal coiled spring and is brought to rest in a distance of 2.2 m. What is the spring stiffness constant of the spring?

30. (II) A 1.60-m tall person lifts a 2.10-kg book from the ground so it is 2.20 m above the ground. What is the potential energy of the book relative to (a) the ground, and (b) the top of the person's head? (c) How is the work done by the person related to the answers in parts (a) and (b)?

31. (II) A 55-kg hiker starts at an elevation of 1600 m and climbs to the top of a 3300-m peak. (a) What is the hiker's change in potential energy? (b) What is the minimum work required of the hiker? (c) Can the actual work done be more than this? Explain why.

32. (II) A spring with $k = 53$ N/m hangs vertically next to a ruler. The end of the spring is next to the 15-cm mark on the ruler. If a 2.5-kg mass is now attached to the end of the spring, where will the end of the spring line up with the ruler marks?

6–6 and 6–7 Conservation of Mechanical Energy

33. (I) Jane, looking for Tarzan, is running at top speed (5.3 m/s) and grabs a vine hanging vertically from a tall tree in the jungle. How high can she swing upward? Does the length of the vine affect your answer?

34. (I) A novice skier, starting from rest, slides down a frictionless 35.0° incline whose vertical height is 185 m. How fast is she going when she reaches the bottom?

35. (I) A sled is initially given a shove up a frictionless 28.0° incline. It reaches a maximum vertical height 1.35 m higher than where it started. What was its initial speed?

36. (II) In the high jump, Fran's kinetic energy is transformed into gravitational potential energy without the aid of a pole. With what minimum speed must Fran leave the ground in order to lift her center of mass 2.10 m and cross the bar with a speed of 0.70 m/s?

37. (II) A 65-kg trampoline artist jumps vertically upward from the top of a platform with a speed of 5.0 m/s. (a) How fast is he going as he lands on the trampoline, 3.0 m below (Fig. 6–38)? (b) If the trampoline behaves like a spring with spring stiffness constant 6.2×10^4 N/m, how far does he depress it?

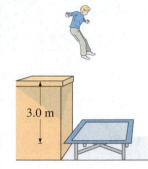

FIGURE 6–38
Problem 37.

38. (II) A projectile is fired at an upward angle of 45.0° from the top of a 265-m cliff with a speed of 185 m/s. What will be its speed when it strikes the ground below? (Use conservation of energy.)

39. (II) A vertical spring (ignore its mass), whose spring stiffness constant is 950 N/m, is attached to a table and is compressed down 0.150 m. (a) What upward speed can it give to a 0.30-kg ball when released? (b) How high above its original position (spring compressed) will the ball fly?

40. (II) A block of mass m slides without friction along the looped track shown in Fig. 6–39. If the block is to remain on the track, even at the top of the circle (whose radius is r), from what minimum height h must it be released?

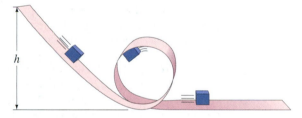

FIGURE 6–39 Problems 40 and 75.

41. (II) A block of mass m is attached to the end of a spring (spring stiffness constant k), Fig. 6–40. The block is given an initial displacement x_0, after which it oscillates back and forth. Write a formula for the total mechanical energy (ignore friction and the mass of the spring) in terms of x_0, position x, and speed v.

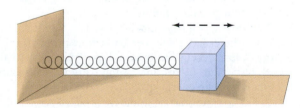

FIGURE 6–40 Problems 41, 55, and 56.

42. (II) A 62-kg bungee jumper jumps from a bridge. She is tied to a bungee cord whose unstretched length is 12 m, and falls a total of 31 m. (a) Calculate the spring stiffness constant k of the bungee cord, assuming Hooke's law applies. (b) Calculate the maximum acceleration she experiences.

43. (II) The roller-coaster car shown in Fig. 6–41 is dragged up to point 1 where it is released from rest. Assuming no friction, calculate the speed at points 2, 3, and 4.

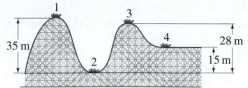

FIGURE 6–41
Problems 43 and 53.

44. (II) A 0.40-kg ball is thrown with a speed of 12 m/s at an angle of 33°. (*a*) What is its speed at its highest point, and (*b*) how high does it go? (Use conservation of energy, and ignore air resistance.)

45. (III) An engineer is designing a spring to be placed at the bottom of an elevator shaft. If the elevator cable should break when the elevator is at a height h above the top of the spring, calculate the value that the spring stiffness constant k should have so that passengers undergo an acceleration of no more than $5.0\,g$ when brought to rest. Let M be the total mass of the elevator and passengers.

46. (III) A cyclist intends to cycle up a 7.8° hill whose vertical height is 150 m. Assuming the mass of bicycle plus cyclist is 75 kg, (*a*) calculate how much work must be done against gravity. (*b*) If each complete revolution of the pedals moves the bike 5.1 m along its path, calculate the average force that must be exerted on the pedals tangent to their circular path. Neglect work done by friction and other losses. The pedals turn in a circle of diameter 36 cm.

6–8 and 6–9 Law of Conservation of Energy

47. (I) Two railroad cars, each of mass 7650 kg and traveling 95 km/h in opposite directions, collide head-on and come to rest. How much thermal energy is produced in this collision?

48. (II) A 21.7-kg child descends a slide 3.5 m high and reaches the bottom with a speed of 2.2 m/s. How much thermal energy due to friction was generated in this process?

49. (II) A ski starts from rest and slides down a 22° incline 75 m long. (*a*) If the coefficient of friction is 0.090, what is the ski's speed at the base of the incline? (*b*) If the snow is level at the foot of the incline and has the same coefficient of friction, how far will the ski travel along the level? Use energy methods.

50. (II) A 145-g baseball is dropped from a tree 13.0 m above the ground. (*a*) With what speed would it hit the ground if air resistance could be ignored? (*b*) If it actually hits the ground with a speed of 8.00 m/s, what is the average force of air resistance exerted on it?

51. (II) You drop a ball from a height of 2.0 m, and it bounces back to a height of 1.5 m. (*a*) What fraction of its initial energy is lost during the bounce? (*b*) What is the ball's speed just as it leaves the ground after the bounce? (*c*) Where did the energy go?

52. (II) A 110-kg crate, starting from rest, is pulled across a floor with a constant horizontal force of 350 N. For the first 15 m the floor is frictionless, and for the next 15 m the coefficient of friction is 0.30. What is the final speed of the crate?

53. (II) Suppose the roller coaster in Fig. 6–41 passes point 1 with a speed of 1.70 m/s. If the average force of friction is equal to one-fifth of its weight, with what speed will it reach point 2? The distance traveled is 45.0 m.

54. (II) A skier traveling 12.0 m/s reaches the foot of a steady upward 18.0° incline and glides 12.2 m up along this slope before coming to rest. What was the average coefficient of friction?

55. (III) A 0.620-kg wood block is firmly attached to a very light horizontal spring ($k = 180\,\text{N/m}$) as shown in Fig. 6–40. It is noted that the block–spring system, when compressed 5.0 cm and released, stretches out 2.3 cm beyond the equilibrium position before stopping and turning back. What is the coefficient of kinetic friction between the block and the table?

56. (III) A 280-g wood block is firmly attached to a very light horizontal spring, Fig. 6–40. The block can slide along a table where the coefficient of friction is 0.30. A force of 22 N compresses the spring 18 cm. If the spring is released from this position, how far beyond its equilibrium position will it stretch on its first cycle?

57. (III) Early test flights for the space shuttle used a "glider" (mass of 980 kg including pilot) that was launched horizontally at 500 km/h from a height of 3500 m. The glider eventually landed at a speed of 200 km/h. (*a*) What would its landing speed have been in the absence of air resistance? (*b*) What was the average force of air resistance exerted on it if it came in at a constant glide of 10° to the Earth?

6–10 Power

58. (I) How long will it take a 1750-W motor to lift a 315-kg piano to a sixth-story window 16.0 m above?

59. (I) If a car generates 18 hp when traveling at a steady 88 km/h, what must be the average force exerted on the car due to friction and air resistance?

60. (I) A 1400-kg sports car accelerates from rest to 95 km/h in 7.4 s. What is the average power delivered by the engine?

61. (I) (*a*) Show that one British horsepower (550 ft·lb/s) is equal to 746 W. (*b*) What is the horsepower rating of a 75-W lightbulb?

62. (II) Electric energy units are often expressed in the form of "kilowatt-hours." (*a*) Show that one kilowatt-hour (kWh) is equal to 3.6×10^6 J. (*b*) If a typical family of four uses electric energy at an average rate of 520 W, how many kWh would their electric bill be for one month, and (*c*) how many joules would this be? (*d*) At a cost of $0.12 per kWh, what would their monthly bill be in dollars? Does the monthly bill depend on the *rate* at which they use the electric energy?

63. (II) A driver notices that her 1150-kg car slows down from 85 km/h to 65 km/h in about 6.0 s on the level when it is in neutral. Approximately what power (watts and hp) is needed to keep the car traveling at a constant 75 km/h?

64. (II) How much work can a 3.0-hp motor do in 1.0 h?

65. (II) A shot-putter accelerates a 7.3-kg shot from rest to 14 m/s. If this motion takes 1.5 s, what average power was developed?

66. (II) A pump is to lift 18.0 kg of water per minute through a height of 3.60 m. What output rating (watts) should the pump motor have?

67. (II) During a workout, the football players at State U. ran up the stadium stairs in 66 s. The stairs are 140 m long and inclined at an angle of 32°. If a typical player has a mass of 95 kg, estimate the average power output on the way up. Ignore friction and air resistance.

68. (II) How fast must a cyclist climb a 6.0° hill to maintain a power output of 0.25 hp? Neglect work done by friction, and assume the mass of cyclist plus bicycle is 68 kg.

69. (II) A 1200-kg car has a maximum power output of 120 hp. How steep a hill can it climb at a constant speed of 75 km/h if the frictional forces add up to 650 N?

70. (II) What minimum horsepower must a motor have to be able to drag a 310-kg box along a level floor at a speed of 1.20 m/s if the coefficient of friction is 0.45?

71. (III) A bicyclist coasts down a 7.0° hill at a steady speed of 5.0 m/s. Assuming a total mass of 75 kg (bicycle plus rider), what must be the cyclist's power output to climb the same hill at the same speed?

General Problems

72. Designers of today's cars have built "5 mi/h (8 km/h) bumpers" that are designed to compress and rebound elastically without any physical damage at speeds below 8 km/h. If the material of the bumpers permanently deforms after a compression of 1.5 cm, but remains like an elastic spring up to that point, what must the effective spring stiffness constant of the bumper be, assuming the car has a mass of 1300 kg and is tested by ramming into a solid wall?

73. In a certain library the first shelf is 10.0 cm off the ground, and the remaining four shelves are each spaced 30.0 cm above the previous one. If the average book has a mass of 1.5 kg with a height of 21 cm, and an average shelf holds 25 books, how much work is required to fill all the shelves, assuming the books are all laying flat on the floor to start?

74. A film of Jesse Owens's famous long jump (Fig. 6–42) in the 1936 Olympics shows that his center of mass rose 1.1 m from launch point to the top of the arc. What minimum speed did he need at launch if he was traveling at 6.5 m/s at the top of the arc?

FIGURE 6–42
Problem 74.

75. The block of mass m sliding without friction along the looped track shown in Fig. 6–39 is to remain on the track at all times, even at the very top of the loop of radius r. (a) In terms of the given quantities, determine the minimum release height h (as in Problem 40). Next, if the actual release height is $2h$, calculate (b) the normal force exerted by the track at the bottom of the loop, (c) the normal force exerted by the track at the top of the loop, and (d) the normal force exerted by the track after the block exits the loop onto the flat section.

76. An airplane pilot fell 370 m after jumping from an aircraft without his parachute opening. He landed in a snowbank, creating a crater 1.1 m deep, but survived with only minor injuries. Assuming the pilot's mass was 78 kg and his terminal velocity was 35 m/s, estimate (a) the work done by the snow in bringing him to rest; (b) the average force exerted on him by the snow to stop him; and (c) the work done on him by air resistance as he fell.

77. A ball is attached to a horizontal cord of length L whose other end is fixed (Fig. 6–43). (a) If the ball is released, what will be its speed at the lowest point of its path? (b) A peg is located a distance h directly below the point of attachment of the cord. If $h = 0.80L$, what will be the speed of the ball when it reaches the top of its circular path about the peg?

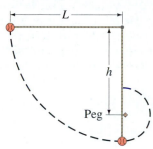

FIGURE 6–43
Problem 77.

78. A 65-kg hiker climbs to the top of a 3700-m-high mountain. The climb is made in 5.0 h starting at an elevation of 2300 m. Calculate (a) the work done by the hiker against gravity, (b) the average power output in watts and in horsepower, and (c) assuming the body is 15% efficient, what rate of energy input was required.

79. An elevator cable breaks when a 920-kg elevator is 28 m above a huge spring $(k = 2.2 \times 10^5 \text{ N/m})$ at the bottom of the shaft. Calculate (a) the work done by gravity on the elevator before it hits the spring, (b) the speed of the elevator just before striking the spring, and (c) the amount the spring compresses (note that work is done by both the spring and gravity in this part).

80. Squaw Valley ski area in California claims that its lifts can move 47,000 people per hour. If the average lift carries people about 200 m (vertically) higher, estimate the power needed.

81. Water flows $(v \approx 0)$ over a dam at the rate of 650 kg/s and falls vertically 81 m before striking the turbine blades. Calculate (a) the speed of the water just before striking the turbine blades (neglect air resistance), and (b) the rate at which mechanical energy is transferred to the turbine blades, assuming 58% efficiency.

82. Show that on a roller coaster with a circular vertical loop (Fig. 6–44), the difference in your apparent weight at the top of the circular loop and the bottom of the circular loop is 6 g's—that is, six times your weight. Ignore friction. Show also that as long as your speed is above the minimum needed, this answer doesn't depend on the size of the loop or how fast you go through it.

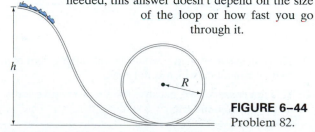

FIGURE 6–44
Problem 82.

83. (a) If the human body could convert a candy bar directly into work, how high could an 82-kg man climb a ladder if he were fueled by one bar ($=1100$ kJ)? (b) If the man then jumped off the ladder, what will be his speed when he reaches the bottom?

84. A projectile is fired at an upward angle of 45.0° from the top of a 165-m cliff with a speed of 175 m/s. What will be its speed when it strikes the ground below? (Use conservation of energy and neglect air resistance.)

85. If you stand on a bathroom scale, the spring inside the scale compresses 0.60 mm, and it tells you your weight is 710 N. Now if you jump on the scale from a height of 1.0 m, what does the scale read at its peak?

86. A 65-kg student runs at 5.0 m/s, grabs a rope, and swings out over a lake (Fig. 6–45). He releases the rope when his velocity is zero. (a) What is the angle θ when he releases the rope? (b) What is the tension in the rope just before he releases it? (c) What is the maximum tension in the rope?

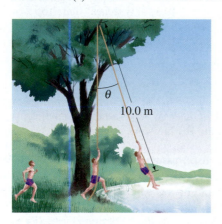

FIGURE 6–45
Problem 86.

87. In the rope climb, a 72-kg athlete climbs a vertical distance of 5.0 m in 9.0 s. What minimum power output was used to accomplish this feat?

88. Some electric-power companies use water to store energy. Water is pumped by reversible turbine pumps from a low to a high reservoir. To store the energy produced in 1.0 hour by a 120-MW $(120 \times 10^6$ W$)$ electric-power plant, how many cubic meters of water will have to be pumped from the lower to the upper reservoir? Assume the upper reservoir is 520 m above the lower and we can neglect the small change in depths within each. Water has a mass of 1000 kg for every 1.0 m³.

89. A spring with spring stiffness constant k is cut in half. What is the spring stiffness constant for each of the two resulting springs?

90. A 6.0-kg block is pushed 8.0 m up a rough 37° inclined plane by a horizontal force of 75 N. If the initial speed of the block is 2.2 m/s up the plane and a constant kinetic friction force of 25 N opposes the motion, calculate (a) the initial kinetic energy of the block; (b) the work done by the 75-N force; (c) the work done by the friction force; (d) the work done by gravity; (e) the work done by the normal force; (f) the final kinetic energy of the block.

91. If a 1500-kg car can accelerate from 35 km/h to 55 km/h in 3.2 s, how long will it take to accelerate from 55 km/h to 75 km/h? Assume the power stays the same, and neglect frictional losses.

92. In a common test for cardiac function (the "stress test"), the patient walks on an inclined treadmill (Fig. 6–46). Estimate the power required from a 75-kg patient when the treadmill is sloping at an angle of 15° and the velocity is 3.3 km/h. (How does this power compare to the power rating of a lightbulb?)

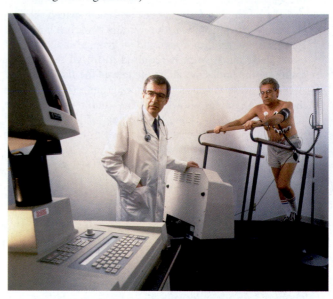

FIGURE 6–46   Problem 92.

93. (a) If a volcano spews a 500-kg rock vertically upward a distance of 500 m, what was its velocity when it left the volcano? (b) If the volcano spews the equivalent of 1000 rocks of this size every minute, what is its power output?

94. Water falls onto a water wheel from a height of 2.0 m at a rate of 95 kg/s. (a) If this water wheel is set up to provide electricity output, what is its maximum power output? (b) What is the speed of the water as it hits the wheel?

Answers to Exercises

A: (c).

B: No, because the speed v would be the square root of a negative number, which is not real.

C: It is nonconservative, because for a conservative force $W = 0$ in a round trip.

D: $W_{net} = \Delta KE$, where $W_{net} = mg(y_1 - y_2)$ and $\Delta KE = \frac{1}{2}mv_2^2 - \frac{1}{2}mv_1^2 = \frac{1}{2}mv_2^2$. Then $v_2^2 = 2g(y_1 - y_2)$.

E: Equal speeds.

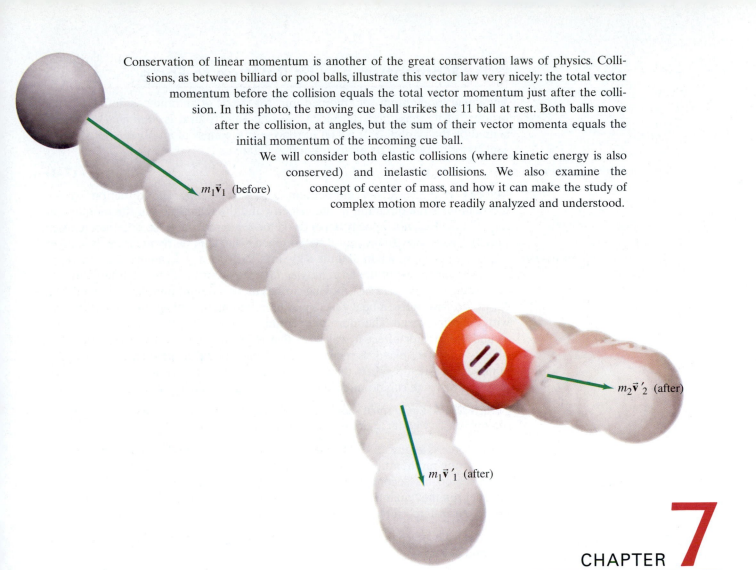

Conservation of linear momentum is another of the great conservation laws of physics. Collisions, as between billiard or pool balls, illustrate this vector law very nicely: the total vector momentum before the collision equals the total vector momentum just after the collision. In this photo, the moving cue ball strikes the 11 ball at rest. Both balls move after the collision, at angles, but the sum of their vector momenta equals the initial momentum of the incoming cue ball.

We will consider both elastic collisions (where kinetic energy is also conserved) and inelastic collisions. We also examine the concept of center of mass, and how it can make the study of complex motion more readily analyzed and understood.

$m_1\vec{v}_1$ (before)

$m_2\vec{v}'_2$ (after)

$m_1\vec{v}'_1$ (after)

CHAPTER **7**

Linear Momentum

The law of conservation of energy, which we discussed in the previous Chapter, is one of several great conservation laws in physics. Among the other quantities found to be conserved are linear momentum, angular momentum, and electric charge. We will eventually discuss all of these because the conservation laws are among the most important ideas in science. In this Chapter, we discuss linear momentum, and its conservation. The law of conservation of momentum is essentially a reworking of Newton's laws that gives us tremendous physical insight and problem-solving power.

We make use of the laws of conservation of linear momentum and of energy to analyze collisions. Indeed, the law of conservation of momentum is particularly useful when dealing with a system of two or more objects that interact with each other, such as in collisions.

Our focus up to now has been mainly on the motion of a single object, often thought of as a "particle" in the sense that we have ignored any rotation or internal motion. In this Chapter we will deal with systems of two or more objects, and toward the end of the Chapter, the concept of center of mass.

7-1 Momentum and Its Relation to Force

The **linear momentum** (or "momentum" for short) of an object is defined as the product of its mass and its velocity. Momentum (plural is *momenta*) is represented by the symbol $\vec{p}$. If we let m represent the mass of an object and $\vec{v}$ represent its velocity, then its momentum $\vec{p}$ is defined as

Linear momentum defined

$$\vec{p} = m\vec{v}. \tag{7-1}$$

Velocity is a vector, so momentum too is a vector. The direction of the momentum is the direction of the velocity, and the magnitude of the momentum is $p = mv$. Because velocity depends on the reference frame, so does momentum; thus the reference frame must be specified. The unit of momentum is that of *Units of momentum* mass × velocity, which in SI units is $\text{kg} \cdot \text{m/s}$. There is no special name for this unit.

Everyday usage of the term *momentum* is in accord with the definition above. According to Eq. 7-1, a fast-moving car has more momentum than a slow-moving car of the same mass; a heavy truck has more momentum than a small car moving with the same speed. The more momentum an object has, the harder it is to stop it, and the greater effect it will have if it is brought to rest by striking another object. A football player is more likely to be stunned if tackled by a heavy opponent running at top speed than by a lighter or slower-moving tackler. A heavy, fast-moving truck can do more damage than a slow-moving motorcycle.

EXERCISE A Can a small sports car ever have the same momentum as a large sport-utility vehicle with three times the sports car's mass? Explain.

A force is required to change the momentum of an object, whether it is to increase the momentum, to decrease it, or to change its direction. Newton originally stated his second law in terms of momentum (although he called the product mv the "quantity of motion"). Newton's statement of the **second law of motion**, translated into modern language, is as follows:

| NEWTON'S SECOND LAW |

The rate of change of momentum of an object is equal to the net force applied to it.

We can write this as an equation,

| NEWTON'S SECOND LAW |

$$\Sigma\vec{F} = \frac{\Delta\vec{p}}{\Delta t}, \tag{7-2}$$

⚠ **CAUTION**

The change in the momentum vector is in the direction of the net force

where $\Sigma\vec{F}$ is the net force applied to the object (the vector sum of all forces acting on it) and $\Delta\vec{p}$ is the resulting momentum change that occurs during the time interval† Δt.

We can readily derive the familiar form of the second law, $\Sigma\vec{F} = m\vec{a}$, from Eq. 7-2 for the case of constant mass. If $\vec{v}_1$ is the initial velocity of an object and $\vec{v}_2$ is its velocity after a time interval Δt has elapsed, then

$$\Sigma\vec{F} = \frac{\Delta\vec{p}}{\Delta t} = \frac{m\vec{v}_2 - m\vec{v}_1}{\Delta t} = \frac{m(\vec{v}_2 - \vec{v}_1)}{\Delta t}$$
$$= m\frac{\Delta\vec{v}}{\Delta t}.$$

By definition, $\vec{a} = \Delta\vec{v}/\Delta t$, so

Newton's second law for constant mass

$$\Sigma\vec{F} = m\vec{a}. \qquad \text{[constant mass]}$$

Newton's statement, Eq. 7-2, is more general than the more familiar version because it includes the situation in which the mass may change. A change in mass occurs in certain circumstances, such as for rockets which lose mass as they burn fuel, and also in the theory of relativity (Chapter 26).

---

†Normally we think of Δt as being a small time interval. If it is not small, then Eq. 7-2 is valid if $\Sigma\vec{F}$ is constant during that time interval, or if $\Sigma\vec{F}$ is the average net force during that time interval.

EXAMPLE 7–1 **ESTIMATE** **Force of a tennis serve.** For a top player, a tennis ball may leave the racket on the serve with a speed of 55 m/s (about 120 mi/h), Fig. 7–1. If the ball has a mass of 0.060 kg and is in contact with the racket for about 4 ms $(4 \times 10^{-3}\,\text{s})$, estimate the average force on the ball. Would this force be large enough to lift a 60-kg person?

APPROACH The tennis ball is hit when its initial velocity is very nearly zero at the top of the throw, so we take $v_1 = 0$. We use Newton's second law, Eq. 7–2, to calculate the force, ignoring all other forces such as gravity in comparison to that exerted by the tennis racket.

SOLUTION The force exerted on the ball by the racket is

$$F = \frac{\Delta p}{\Delta t} = \frac{mv_2 - mv_1}{\Delta t}$$

where $v_2 = 55\,\text{m/s}$, $v_1 = 0$, and $\Delta t = 0.004\,\text{s}$. Thus

$$F = \frac{\Delta p}{\Delta t} = \frac{(0.060\,\text{kg})(55\,\text{m/s}) - 0}{0.004\,\text{s}}$$

$$\approx 800\,\text{N}.$$

This is a large force, larger than the weight of a 60-kg person, which would require a force $mg = (60\,\text{kg})(9.8\,\text{m/s}^2) \approx 600\,\text{N}$ to lift.

NOTE The force of gravity acting on the tennis ball is $mg = (0.060\,\text{kg})(9.8\,\text{m/s}^2) = 0.59\,\text{N}$, which justifies our ignoring it compared to the enormous force the racket exerts.

NOTE High-speed photography and radar can give us an estimate of the contact time and the velocity of the ball leaving the racket. But a direct measurement of the force is not practical. Our calculation shows a handy technique for determining an unknown force in the real world.

FIGURE 7–1 Example 7–1.

Measuring force

EXAMPLE 7–2 **Washing a car: momentum change and force.** Water leaves a hose at a rate of 1.5 kg/s with a speed of 20 m/s and is aimed at the side of a car, which stops it, Fig. 7–2. (That is, we ignore any splashing back.) What is the force exerted by the water on the car?

APPROACH The water leaving the hose has mass and velocity, so it has a momentum p_{initial}. When the water hits the car, the water loses this momentum $(p_{\text{final}} = 0)$. We use Newton's second law in the momentum form, Eq. 7–2, to find the force that the car exerts on the water to stop it. By Newton's third law, the force exerted by the water on the car is equal and opposite. We have a continuing process: 1.5 kg of water leaves the hose in each 1.0-s time interval. So let us choose $\Delta t = 1.0\,\text{s}$, and $m = 1.5\,\text{kg}$ in Eq. 7–2.

SOLUTION We take the x direction positive to the right. In each 1.0-s time interval, water with a momentum of $p_x = mv_x = (1.5\,\text{kg})(20\,\text{m/s}) = 30\,\text{kg}\cdot\text{m/s}$ is brought to rest when it hits the car. The magnitude of the force (assumed constant) that the car must exert to change the momentum of the water by this amount is

$$F = \frac{\Delta p}{\Delta t} = \frac{p_{\text{final}} - p_{\text{initial}}}{\Delta t} = \frac{0 - 30\,\text{kg}\cdot\text{m/s}}{1.0\,\text{s}} = -30\,\text{N}.$$

The minus sign indicates that the force on the water is opposite to the water's original velocity. The car exerts a force of 30 N to the left to stop the water, so by Newton's third law, the water exerts a force of 30 N to the right on the car.

NOTE Keep track of signs, although common sense helps too. The water is moving to the right, so common sense tells us the force on the car must be to the right.

FIGURE 7–2 Example 7–2.

$v = 20\,\text{m/s}$

x

EXERCISE B If the water splashes back from the car in Example 7–2, would the force on the car be larger or smaller?

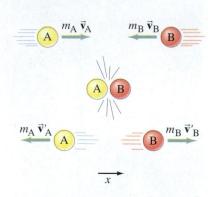

FIGURE 7–3 Momentum is conserved in a collision of two balls, labelled A and B.

CONSERVATION OF MOMENTUM
(for two objects colliding)

Momentum conservation related
to Newton's laws

FIGURE 7–4 Forces on the balls during the collision of Fig. 7–3.

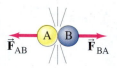

7–2 Conservation of Momentum

The concept of momentum is particularly important because, under certain circumstances, momentum is a conserved quantity. Consider, for example, the head-on collision of two billiard balls, as shown in Fig. 7–3. We assume the net external force on this system of two balls is zero—that is, the only significant forces during the collision are the forces that each ball exerts on the other. Although the momentum of each of the two balls changes as a result of the collision, the *sum* of their momenta is found to be the same before as after the collision. If $m_A \vec{v}_A$ is the momentum of ball A and $m_B \vec{v}_B$ the momentum of ball B, both measured just before the collision, then the total momentum of the two balls before the collision is the vector sum $m_A \vec{v}_A + m_B \vec{v}_B$. Immediately after the collision, the balls each have a different velocity and momentum, which we designate by a "prime" on the velocity: $m_A \vec{v}'_A$ and $m_B \vec{v}'_B$. The total momentum after the collision is the vector sum $m_A \vec{v}'_A + m_B \vec{v}'_B$. No matter what the velocities and masses are, experiments show that the total momentum before the collision is the same as afterward, whether the collision is head-on or not, as long as no net external force acts:

$$\text{momentum before} = \text{momentum after}$$

$$m_A \vec{v}_A + m_B \vec{v}_B = m_A \vec{v}'_A + m_B \vec{v}'_B. \qquad (7\text{–}3)$$

That is, the total vector momentum of the system of two colliding balls is conserved: it stays constant.

Although the law of conservation of momentum was discovered experimentally, it is closely connected to Newton's laws of motion and they can be shown to be equivalent. We will do a derivation for the head–on collision illustrated in Fig. 7–3. We assume the force F that one ball exerts on the other during the collision is constant over the brief time interval of the collision Δt. We use Newton's second law as expressed in Eq. 7–2, and rewrite it by multiplying both sides by Δt:

$$\Delta \vec{p} = \vec{F} \Delta t. \qquad (7\text{–}4)$$

We apply this to ball B alone, noting that the force $\vec{F}_{BA}$ on ball B exerted by ball A during the collision is to the right ($+x$ direction—see Fig. 7–4):

$$\Delta \vec{p}_B = \vec{F}_{BA} \Delta t$$

$$m_B \vec{v}'_B - m_B \vec{v}_B = \vec{F}_{BA} \Delta t.$$

By Newton's third law, the force $\vec{F}_{AB}$ on ball A due to ball B is $\vec{F}_{AB} = -\vec{F}_{BA}$ and acts to the left. Then applying Newton's second law in the same way to ball A yields

$$\Delta \vec{p}_A = \vec{F}_{AB} \Delta t$$

or

$$m_A \vec{v}'_A - m_A \vec{v}_A = \vec{F}_{AB} \Delta t$$

$$= -\vec{F}_{BA} \Delta t.$$

We combine these two $\Delta \vec{p}$ equations (their right sides differ only by a minus sign):

$$m_A \vec{v}'_A - m_A \vec{v}_A = -(m_B \vec{v}'_B - m_B \vec{v}_B)$$

or

$$m_A \vec{v}_A + m_B \vec{v}_B = m_A \vec{v}'_A + m_B \vec{v}'_B$$

which is Eq. 7–3, the conservation of momentum.

The above derivation can be extended to include any number of interacting objects. To show this, we let $\vec{p}$ in Eq. 7–2 represent the total momentum of a system—that is, the vector sum of the momenta of all objects in the system. (For our two-object system above, $\vec{p} = m_A \vec{v}_A + m_B \vec{v}_B$.) If the net force $\Sigma \vec{F}$ on the system is zero [as it was above for our two-object system, $\vec{F} + (-\vec{F}) = 0$,] then from Eq. 7–2, $\Delta \vec{\mathbf{p}} = \vec{\mathbf{F}} \Delta t = 0$, so the total momentum doesn't change. Thus the general statement of the **law of conservation of momentum** is

The total momentum of an isolated system of objects remains constant.

LAW OF CONSERVATION OF MOMENTUM

By a **system**, we simply mean a set of objects that we choose, and which may interact with each other. An **isolated system** is one in which the only (significant) forces are those between the objects in the system. The sum of all these "internal" forces within the system will be zero because of Newton's third law. If there are *external forces*—by which we mean forces exerted by objects outside the system—and they don't add up to zero (vectorially), then the total momentum of the system won't be conserved. However, if the system can be redefined so as to include the other objects exerting these forces, then the conservation of momentum principle can apply. For example, if we take as our system a rock falling under gravity, the momentum of this system (the rock) is not conserved: an external force, the force of gravity exerted by the Earth, is acting on it and changes its momentum. However, if we include the Earth in the system, the total momentum of rock plus Earth is conserved. (This means that the Earth comes up to meet the rock. But the Earth's mass is so great, its upward velocity is very tiny.)

Systems

Isolated system

EXAMPLE 7–3 **Railroad cars collide: momentum conserved.** A 10,000-kg railroad car, A, traveling at a speed of 24.0 m/s strikes an identical car, B, at rest. If the cars lock together as a result of the collision, what is their common speed just afterward? See Fig. 7–5.

APPROACH We choose our system to be the two railroad cars. We consider a very brief time interval, from just before the collision until just after, so that external forces such as friction can be ignored. Then we apply conservation of momentum.

SOLUTION The initial total momentum is

$$p_{\text{initial}} = m_A v_A + m_B v_B = m_A v_A$$

because car B is at rest initially $(v_B = 0)$. The direction is to the right in the $+x$ direction. After the collision, the two cars become attached, so they will have the same speed, call it v'. Then the total momentum after the collision is

$$p_{\text{final}} = (m_A + m_B)v'.$$

We have assumed there are no external forces, so momentum is conserved:

$$p_{\text{initial}} = p_{\text{final}}$$
$$m_A v_A = (m_A + m_B)v'.$$

Solving for v', we obtain

$$v' = \frac{m_A}{m_A + m_B} v_A = \left(\frac{10,000 \text{ kg}}{10,000 \text{ kg} + 10,000 \text{ kg}} \right)(24.0 \text{ m/s}) = 12.0 \text{ m/s},$$

to the right. Their mutual speed after collision is half the initial speed of car A.

NOTE We kept symbols until the very end, so we have an equation we can use in other (related) situations.

FIGURE 7–5 Example 7–3.

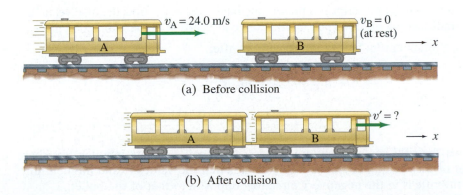

$v_A = 24.0$ m/s

$v_B = 0$
(at rest)

(a) Before collision

$v' = ?$

(b) After collision

EXERCISE C In Example 7–3, $m_A = m_B$, so in the last equation, $m_A/(m_A + m_B) = \frac{1}{2}$. Hence $v' = \frac{1}{2}v_A$. What result do you get if (a) $m_B = 3m_A$, (b) m_B is much larger than m_A ($m_B \gg m_A$), and (c) $m_B \ll m_A$?

As long as no external forces act on our chosen system, conservation of momentum is valid. In the real world, external forces do act: friction on billiard balls, gravity acting on a baseball, and so on. So it may seem that conservation of momentum cannot be applied. Or can it? In a collision, the force each object exerts on the other acts only over a very brief time interval, and is very strong. When a racket hits a tennis ball (or a bat hits a baseball), both before and after the "collision" the ball moves as a projectile under the action of gravity and air resistance. During the brief time of the collision, however, when the racket hits the ball, external forces (gravity, air resistance) are insignificant compared to the collision forces that the racket and ball exert on each other. So if we measure the momenta just before and just after the collision, we can apply momentum conservation with high accuracy.

The law of conservation of momentum is particularly useful when we are dealing with fairly simple systems such as colliding objects and certain types of "explosions". For example, *rocket propulsion*, which we saw in Chapter 4 can be understood on the basis of action and reaction, can also be explained on the basis of the conservation of momentum. We can consider the rocket and fuel as an isolated system if it is far out in space (no external forces). In the reference frame of the rocket, the total momentum of rocket plus fuel is zero. When the fuel burns, the total momentum remains unchanged: the backward momentum of the expelled gases is just balanced by the forward momentum gained by the rocket itself (see Fig. 7–6). Thus, a rocket can accelerate in empty space. There is no need for the expelled gases to push against the Earth or the air (as is sometimes erroneously thought). Similar examples of (nearly) isolated systems where momentum is conserved are the recoil of a gun when a bullet is fired, and the movement of a rowboat just after a package is thrown from it.

PHYSICS APPLIED
Rocket propulsion

⚠ CAUTION

A rocket pushes on the gases released by the fuel, not on the Earth or other objects

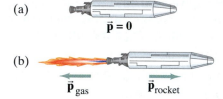

FIGURE 7–6 (a) A rocket, containing fuel, at rest in some reference frame. (b) In the same reference frame, the rocket fires and gases are expelled at high speed out the rear. The total vector momentum, $\vec{p}_{gas} + \vec{p}_{rocket}$, remains zero.

(a) $\vec{p} = 0$

(b) $\vec{p}_{gas}$ $\vec{p}_{rocket}$

EXAMPLE 7–4 **Rifle recoil.** Calculate the recoil velocity of a 5.0-kg rifle that shoots a 0.020-kg bullet at a speed of 620 m/s, Fig. 7–7.

FIGURE 7–7 Example 7–4.

(a) Before shooting (at rest)

$\vec{v}'_R$ $\vec{v}'_B$
$\vec{p}'_R$ $\vec{p}'_B$

(b) After shooting

APPROACH Our system is the rifle and the bullet, both at rest initially, just before the trigger is pulled. The trigger is pulled, an explosion occurs, and we look at the rifle and bullet just as the bullet leaves the barrel. The bullet moves to the right ($+x$), and the gun recoils to the left. During the very short time interval of the explosion, we can assume the external forces are small compared to the forces exerted by the exploding gunpowder. Thus we can apply conservation of momentum, at least approximately.

SOLUTION Let subscript B represent the bullet and R the rifle; the final velocities are indicated by primes. Then momentum conservation in the x direction gives

$$\text{momentum before} = \text{momentum after}$$
$$m_B v_B + m_R v_R = m_B v'_B + m_R v'_R$$
$$0 + 0 = m_B v'_B + m_R v'_R$$

so

$$v'_R = -\frac{m_B v'_B}{m_R} = -\frac{(0.020 \text{ kg})(620 \text{ m/s})}{(5.0 \text{ kg})} = -2.5 \text{ m/s}.$$

Since the rifle has a much larger mass, its (recoil) velocity is much less than that of the bullet. The minus sign indicates that the velocity (and momentum) of the rifle is in the negative x direction, opposite to that of the bullet.

CONCEPTUAL EXAMPLE 7–5 | **Falling on or off a sled.** (*a*) An empty sled is sliding on frictionless ice when Susan drops vertically from a tree above onto the sled. When she lands, does the sled speed up, slow down, or keep the same speed? (*b*) Later. Susan falls sideways off the sled. When she drops off, does the sled speed up, slow down, or keep the same speed?

RESPONSE (*a*) Because Susan falls vertically onto the sled, she has no initial horizontal momentum. Thus the total horizontal momentum afterward equals the momentum of the sled initially. Since the mass of the system (sled + person) has increased, the speed must decrease.
(*b*) At the instant Susan falls off, she is moving with the same horizontal speed as she was while on the sled. At the moment she leaves the sled, she has the same momentum she had an instant before. Because momentum is conserved, the sled keeps the same speed.

7–3 Collisions and Impulse

Collisions are a common occurrence in everyday life: a tennis racket or a baseball bat striking a ball, billiard balls colliding, a hammer hitting a nail. When a collision occurs, the interaction between the objects involved is usually far stronger than any interaction between our system of objects and their environment. We can then ignore the effects of any other forces during the brief time interval of the collision.

During a collision of two ordinary objects, both objects are deformed, often considerably, because of the large forces involved (Fig. 7–8). When the collision occurs, the force usually jumps from zero at the moment of contact to a very large force within a very short time, and then rapidly returns to zero again. A graph of the magnitude of the force that one object exerts on the other during a collision, as a function of time, is something like the red curve in Fig. 7–9. The time interval Δt is usually very distinct and very small.

From Newton's second law, Eq. 7–2, the *net* force on one object is equal to the rate of change of its momentum:

$$\vec{\mathbf{F}} = \frac{\Delta \vec{\mathbf{p}}}{\Delta t}.$$

(We have written $\vec{\mathbf{F}}$ instead of $\Sigma\vec{\mathbf{F}}$ for the net force, which we assume is entirely due to the brief but large average force that acts during the collision.) This equation applies to *each* of the two objects in a collision. We multiply both sides of this equation by the time interval Δt, and obtain

$$\vec{\mathbf{F}}\,\Delta t = \Delta \vec{\mathbf{p}}. \qquad (7\text{–}5)$$

The quantity on the left, the product of the force $\vec{\mathbf{F}}$ times the time Δt over which the force acts, is called the **impulse**:

$$\text{Impulse} = \vec{\mathbf{F}}\,\Delta t.$$

We see that the total change in momentum is equal to the impulse. The concept of impulse is useful mainly when dealing with forces that act during a short time interval, as when a bat hits a baseball. The force is generally not constant, and often its variation in time is like that graphed in Figs. 7–9 and 7–10. We can often approximate such a varying force as an average force $\overline{F}$ acting during a time interval Δt, as indicated by the dashed line in Fig. 7–10. $\overline{F}$ is chosen so that the area shown shaded in Fig. 7–10 (equal to $\overline{F} \times \Delta t$) is equal to the area under the actual curve of F vs. t, Fig. 7–9 (which represents the actual impulse).

EXERCISE D Suppose Fig. 7–9 illustrates the force on a golf ball vs. the time when the ball hits a wall. How would the shape of this curve change if a softer rubber ball with the same mass and speed hit the same wall?

FIGURE 7–8 Tennis racket striking a ball. Both the ball and the racket strings are deformed due to the large force each exerts on the other.

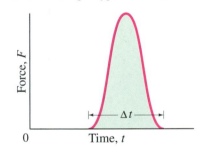

FIGURE 7–9 Force as a function of time during a typical collision.

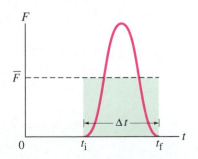

FIGURE 7–10 The average force $\overline{F}$ acting over an interval of time Δt gives the same impulse ($\overline{F}\,\Delta t$) as the actual force.

EXAMPLE 7–6 Bend your knees when landing.
(a) Calculate the impulse experienced when a 70-kg person lands on firm ground after jumping from a height of 3.0 m. (b) Estimate the average force exerted on the person's feet by the ground if the landing is stiff-legged, and again (c) with bent legs. With stiff legs, assume the body moves 1.0 cm during impact, and when the legs are bent, about 50 cm.

APPROACH We consider the short time interval that starts just before the person hits the ground and ends when he is brought to rest. During this time interval, the ground exerts a force on him and gives him an impulse which equals his change in momentum (Eq. 7–5). For part (a) we know his final speed (zero, when he comes to rest), but we need to calculate his "initial" speed just before impact with the ground. The latter is found using kinematics and his drop from a height of 3.0 m. Then Eq. 7–5 gives us $\bar{F}\Delta t$. In parts (b) and (c) we calculate how long, Δt, it takes him to slow down as he hits the ground, using kinematics, and then obtain F because we know $\bar{F}\Delta t$.

SOLUTION (a) First we need to determine the velocity of the person just before striking the ground, which we do by considering the earlier time period between the initial jump from a height of 3.0 m until just before he touches the ground. The person falls under gravity, so we can use the kinematic Eq. 2–11c, $v^2 = v_0^2 + 2a(y - y_0)$ with $a = -g$ and $v_0 = 0$, so

$$v^2 = 2g(y_0 - y)$$

or

$$v = \sqrt{2g(y_0 - y)} = \sqrt{2(9.8 \text{ m/s}^2)(3.0 \text{ m})} = 7.7 \text{ m/s}.$$

This $v = 7.7$ m/s is his speed just before hitting the ground, and so it is the initial speed for the short time interval of the impact with the ground, Δt. Now we can determine the impulse by examining this brief time interval as the person hits the ground and is brought to rest (Fig. 7–11). We don't know F and thus can't calculate the impulse $\bar{F}\Delta t$ directly; but we can use Eq. 7–5: the impulse equals the change in momentum of the object

$$\bar{F}\,\Delta t = \Delta p = m\,\Delta v$$
$$= (70 \text{ kg})(0 - 7.7 \text{ m/s}) = -540 \text{ N·s}.$$

The negative sign tells us that the force is opposed to the original (downward) momentum; that is, the force acts upward.

(b) In coming to rest, the person decelerates from 7.7 m/s to zero in a distance $d = 1.0$ cm $= 1.0 \times 10^{-2}$ m. If we assume the upward force exerted on him by the ground is constant, then the average speed during this brief period is

$$\bar{v} = \frac{(7.7 \text{ m/s} + 0 \text{ m/s})}{2} = 3.9 \text{ m/s}.$$

Thus the collision with the ground lasts for a time interval (recall the definition of speed, $\bar{v} = d/\Delta t$):

$$\Delta t = \frac{d}{\bar{v}} = \frac{(1.0 \times 10^{-2} \text{ m})}{(3.9 \text{ m/s})} = 2.6 \times 10^{-3} \text{ s}.$$

Since the magnitude of the impulse is $\bar{F}\Delta t = 540$ N·s, and $\Delta t = 2.6 \times 10^{-3}$ s, the average net force $\bar{F}$ on the person has magnitude

$$\bar{F} = \frac{540 \text{ N·s}}{2.6 \times 10^{-3} \text{ s}} = 2.1 \times 10^5 \text{ N}.$$

We are almost there. $\bar{F}$ equals the vector sum of the average force upward on the legs exerted by the ground, F_{grd}, which we take as positive, plus the downward force of gravity, $-mg$ (see Fig. 7–12):

$$\bar{F} = F_{grd} - mg.$$

Since $mg = (70 \text{ kg})(9.8 \text{ m/s}^2) = 690$ N, then

$$F_{grd} = \bar{F} + mg = (2.1 \times 10^5 \text{ N}) + (0.690 \times 10^3 \text{ N}) \approx 2.1 \times 10^5 \text{ N}.$$

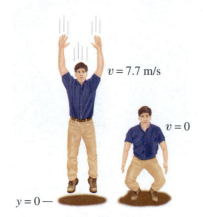

$v = 7.7$ m/s

$v = 0$

$y = 0$—

FIGURE 7–11 Example 7–6. Time interval Δt during which the impulse acts.

➡ PROBLEM SOLVING
Free-body diagrams are always useful!

(c) This is just like part (b), except $d = 0.50\,\text{m}$, so

$$\Delta t = \frac{d}{v} = \frac{0.50\,\text{m}}{3.9\,\text{m/s}} = 0.13\,\text{s}$$

and

$$\overline{F} = \frac{540\,\text{N}\cdot\text{s}}{0.13\,\text{s}} = 4.2 \times 10^3\,\text{N}.$$

The upward force exerted on the person's feet by the ground is, as in part (b):

$$F_{\text{grd}} = \overline{F} + mg = (4.2 \times 10^3\,\text{N}) + (0.69 \times 10^3\,\text{N}) = 4.9 \times 10^3\,\text{N}.$$

Clearly, the force on the feet and legs is much less now with the knees bent, and the impulse occurs over a longer time interval. In fact, the ultimate strength of the leg bone (see Chapter 9, Table 9–2) is not great enough to support the force calculated in part (b), so the leg would likely break in such a stiff landing, whereas it probably wouldn't in part (c) with bent legs.

EXERCISE E In part (b) of Example 7–6, we calculated the force exerted by the ground on the person during the collision, F_{grd}. Was F_{grd} much greater than the "external" force of gravity on the person? By what factor?

FIGURE 7–12 Example 7–6. When the person lands on the ground, the average net force during impact is $\overline{F} = F_{\text{grd}} - mg$, where F_{grd} is the force the ground exerts upward on the person.

7–4 Conservation of Energy and Momentum in Collisions

During most collisions, we usually don't know how the collision force varies over time, and so analysis using Newton's second law becomes difficult or impossible. But by making use of the conservation laws for momentum and energy, we can still determine a lot about the motion after a collision, given the motion before the collision. We saw in Section 7–2 that in the collision of two objects such as billiard balls, the total momentum is conserved. If the two objects are very hard and no heat or other form of energy is produced in the collision, then kinetic energy is conserved as well. By this we mean that the sum of the kinetic energies of the two objects is the same after the collision as before. For the brief moment during which the two objects are in contact, some (or all) of the energy is stored momentarily in the form of elastic potential energy. But if we compare the total kinetic energy just before the collision with the total kinetic energy just after the collision, they are found to be the same. Such a collision, in which the total kinetic energy is conserved, is called an **elastic collision**. If we use the subscripts A and B to represent the two objects, we can write the equation for conservation of total kinetic energy as

$$\text{total KE before} = \text{total KE after}$$
$$\tfrac{1}{2}m_A v_A^2 + \tfrac{1}{2}m_B v_B^2 = \tfrac{1}{2}m_A v_A'^2 + \tfrac{1}{2}m_B v_B'^2. \quad \text{[elastic collision]} \quad \textbf{(7–6)}$$

Here, primed quantities ($'$) mean after the collision and unprimed mean before the collision, just as in Eq. 7–3 for conservation of momentum.

At the atomic level the collisions of atoms and molecules are often elastic. But in the "macroscopic" world of ordinary objects, an elastic collision is an ideal that is never quite reached, since at least a little thermal energy (and perhaps sound and other forms of energy) is always produced during a collision. The collision of two hard elastic balls, such as billiard balls, however, is very close to being perfectly elastic, and we often treat it as such.

We do need to remember that even when the kinetic energy is not conserved, the *total* energy is always conserved.

Collisions in which kinetic energy is not conserved are said to be **inelastic collisions**. The kinetic energy that is lost is changed into other forms of energy, often thermal energy, so that the total energy (as always) is conserved. In this case,

$$\text{KE}_A + \text{KE}_B = \text{KE}_A' + \text{KE}_B' + \text{thermal and other forms of energy}.$$

See Fig. 7–13, and the details in its caption.

FIGURE 7–13 Two equal-mass objects (a) approach each other with equal speeds, (b) collide, and then (c) bounce off with equal speeds in the opposite directions if the collision is elastic, or (d) bounce back much less or not at all if the collision is inelastic.

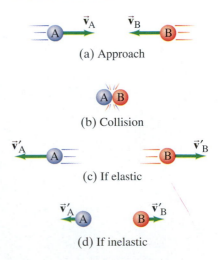

(a) Approach

(b) Collision

(c) If elastic

(d) If inelastic

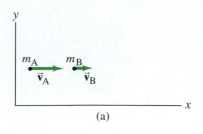

(a)

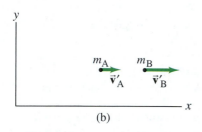

(b)

FIGURE 7–14 Two small objects of masses m_A and m_B, (a) before the collision and (b) after the collision.

Relative speeds (one dimension only)

7–5 Elastic Collisions in One Dimension

We now apply the conservation laws for momentum and kinetic energy to an elastic collision between two small objects that collide head-on, so all the motion is along a line. Let us assume that the two objects are moving with velocities v_A and v_B along the x axis before the collision, Fig. 7–14a. After the collision, their velocities are v'_A and v'_B, Fig. 7–14b. For any $v > 0$, the object is moving to the right (increasing x), whereas for $v < 0$, the object is moving to the left (toward decreasing values of x).

From conservation of momentum, we have

$$m_A v_A + m_B v_B = m_A v'_A + m_B v'_B.$$

Because the collision is assumed to be elastic, kinetic energy is also conserved:

$$\tfrac{1}{2} m_A v_A^2 + \tfrac{1}{2} m_B v_B^2 = \tfrac{1}{2} m_A v_A'^2 + \tfrac{1}{2} m_B v_B'^2.$$

We have two equations, so we can solve for two unknowns. If we know the masses and velocities before the collision, then we can solve these two equations for the velocities after the collision, v'_A and v'_B. We derive a helpful result by rewriting the momentum equation as

$$m_A(v_A - v'_A) = m_B(v'_B - v_B), \tag{i}$$

and we rewrite the kinetic energy equation as

$$m_A(v_A^2 - v_A'^2) = m_B(v_B'^2 - v_B^2).$$

Noting that algebraically $(a - b)(a + b) = a^2 - b^2$, we write this last equation as

$$m_A(v_A - v'_A)(v_A + v'_A) = m_B(v'_B - v_B)(v'_B + v_B). \tag{ii}$$

We divide Eq. (ii) by Eq. (i), and (assuming $v_A \neq v'_A$ and $v_B \neq v'_B$) obtain

$$v_A + v'_A = v'_B + v_B.$$

We can rewrite this equation as

$$v_A - v_B = v'_B - v'_A$$

or

$$v_A - v_B = -(v'_A - v'_B). \qquad \text{[head-on elastic collision]} \tag{7–7}$$

This is an interesting result: it tells us that for any elastic head-on collision, the relative speed of the two objects after the collision has the same magnitude (but opposite direction) as before the collision, no matter what the masses are.

Equation 7–7 was derived from conservation of kinetic energy for elastic collisions, and can be used in place of it. Because the v's are not squared in Eq. 7–7, it is simpler to use in calculations than the conservation of kinetic energy equation (Eq. 7–6) directly.

EXAMPLE 7–7 **Pool or billiards.** Billiard ball A of mass m moving with speed v collides head-on with ball B of equal mass at rest $(v_B = 0)$. What are the speeds of the two balls after the collision, assuming it is elastic?

APPROACH There are two unknowns, v'_A and v'_B, so we need two independent equations. We focus on the time interval from just before the collision until just after. No net external force acts on our system of two balls (mg and the normal force cancel), so momentum is conserved. Conservation of kinetic energy applies as well because the collision is elastic.

SOLUTION Given $v_A = v$ and $v_B = 0$, and $m_A = m_B = m$, then conservation of momentum gives

$$mv = mv'_A + mv'_B$$

or, since the m's cancel out,

$$v = v'_A + v'_B.$$

We have two unknowns (v'_A and v'_B) and need a second equation, which could be the conservation of kinetic energy or the simpler Eq. 7–7 we derived from it:

$$v_A - v_B = v'_B - v'_A, \qquad \text{or} \qquad v = v'_B - v'_A,$$

since $v_A = v$ and $v_B = 0$. We subtract $v = v'_B - v'_A$ from our momentum equation $(v = v'_A + v'_B)$ and obtain

$$0 = 2v'_A.$$

Hence $v'_A = 0$. We can now solve for the other unknown (v'_B) since $v = v'_B - v'_A$:

$$v'_B = v + v'_A = v + 0 = v.$$

To summarize, before the collision we have

$$v_A = v, \qquad v_B = 0$$

and after the collision

$$v'_A = 0, \qquad v'_B = v.$$

That is, ball A is brought to rest by the collision, whereas ball B acquires the original velocity of ball A. See Fig. 7–15.

NOTE Our result is often observed by billiard and pool players, and is valid only if the two balls have equal masses (and no spin is given to the balls).

FIGURE 7–15 In this multiflash photo of a head-on collision between two balls of equal mass, the white cue ball is accelerated from rest by the cue stick and then strikes the red ball, initially at rest. The white ball stops in its tracks, and the (equal mass) red ball moves off with the same speed as the white ball had before the collision. See Example 7–7.

EXAMPLE 7–8 **A nuclear collision.** A proton (p) of mass 1.01 u (unified atomic mass units) traveling with a speed of 3.60×10^4 m/s has an elastic head-on collision with a helium (He) nucleus $(m_{He} = 4.00\text{ u})$ initially at rest. What are the velocities of the proton and helium nucleus after the collision? (As mentioned in Chapter 1, $1\text{ u} = 1.66 \times 10^{-27}$ kg, but we won't need this fact.) Assume the collision takes place in nearly empty space.

APPROACH Like Example 7–7, this is an elastic head-on collision, but now the masses of our two-particle system are not equal. The only external force is Earth's gravity, but it is insignificant compared to the strong force during the collision. So again we use the conservation laws of momentum and of kinetic energy, and apply them to our system of two particles.

SOLUTION Let the proton (p) be particle A and the helium nucleus (He) be particle B. We have $v_B = v_{He} = 0$ and $v_A = v_p = 3.60 \times 10^4$ m/s. We want to find the velocities v'_p and v'_{He} after the collision. From conservation of momentum,

$$m_p v_p + 0 = m_p v'_p + m_{He} v'_{He}.$$

Because the collision is elastic, the kinetic energy of our system of two particles is conserved and we can use Eq. 7–7, which becomes

$$v_p - 0 = v'_{He} - v'_p.$$

Thus

$$v'_p = v'_{He} - v_p,$$

and substituting this into our momentum equation displayed above, we get

$$m_p v_p = m_p v'_{He} - m_p v_p + m_{He} v'_{He}.$$

Solving for v'_{He}, we obtain

$$v'_{He} = \frac{2m_p v_p}{m_p + m_{He}} = \frac{2(1.01\text{ u})(3.60 \times 10^4\text{ m/s})}{5.01\text{ u}} = 1.45 \times 10^4\text{ m/s}.$$

The other unknown is v'_p, which we can now obtain from

$$v'_p = v'_{He} - v_p = (1.45 \times 10^4\text{ m/s}) - (3.60 \times 10^4\text{ m/s}) = -2.15 \times 10^4\text{ m/s}.$$

The minus sign for v'_p tells us that the proton reverses direction upon collision, and we see that its speed is less than its initial speed (see Fig. 7–16).

NOTE This result makes sense: the lighter proton would be expected to "bounce back" from the more massive helium nucleus, but not with its full original velocity as from a rigid wall (which corresponds to extremely large, or infinite, mass).

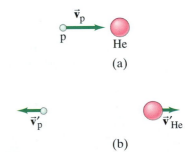

FIGURE 7–16 Example 7–8: (a) before collision, (b) after collision.

7–6 Inelastic Collisions

Collisions in which kinetic energy is not conserved are called *inelastic collisions*. Some of the initial kinetic energy is transformed into other types of energy, such as thermal or potential energy, so the total kinetic energy after the collision is less than the total kinetic energy before the collision. The inverse can also happen when potential energy (such as chemical or nuclear) is released, in which case the total kinetic energy after the interaction can be greater than the initial kinetic energy. Explosions are examples of this type.

Typical macroscopic collisions are inelastic, at least to some extent, and often to a large extent. If two objects stick together as a result of a collision, the colli-

Completely inelastic collision

sion is said to be **completely inelastic**. Two colliding balls of putty that stick together or two railroad cars that couple together when they collide are examples of completely inelastic collisions. The kinetic energy in some cases is all transformed to other forms of energy in an inelastic collision, but in other cases only part of it is. In Example 7–3, for instance, we saw that when a traveling railroad car collided with a stationary one, the coupled cars traveled off with some kinetic energy. In a completely inelastic collision, the maximum amount of kinetic energy is transformed to other forms consistent with conservation of momentum. Even though kinetic energy is not conserved in inelastic collisions, the total energy is always conserved, and the total vector momentum is also conserved.

EXAMPLE 7–9 Railroad cars again. For the completely inelastic collision of two railroad cars that we considered in Example 7–3, calculate how much of the initial kinetic energy is transformed to thermal or other forms of energy.

APPROACH The railroad cars stick together after the collision, so this is a completely inelastic collision. By subtracting the total kinetic energy after the collision from the total initial kinetic energy, we can find how much energy is transformed to other types of energy.

SOLUTION Before the collision, only car A is moving, so the total initial kinetic energy is

$$\tfrac{1}{2} m_A v_A^2 = \tfrac{1}{2}(10{,}000\,\text{kg})(24.0\,\text{m/s})^2 = 2.88 \times 10^6\,\text{J}.$$

After the collision, both cars are moving with a speed of 12.0 m/s, by conservation of momentum (Example 7–3). So the total kinetic energy afterward is

$$\tfrac{1}{2}(20{,}000\,\text{kg})(12.0\,\text{m/s})^2 = 1.44 \times 10^6\,\text{J}.$$

Hence the energy transformed to other forms is

$$\left(2.88 \times 10^6\,\text{J}\right) - \left(1.44 \times 10^6\,\text{J}\right) = 1.44 \times 10^6\,\text{J},$$

which is just half the original kinetic energy.

Ballistic pendulum

EXAMPLE 7–10 Ballistic pendulum. The *ballistic pendulum* is a device used to measure the speed of a projectile, such as a bullet. The projectile, of mass m, is fired into a large block (of wood or other material) of mass M, which is suspended like a pendulum. (Usually, M is somewhat greater than m.) As a result of the collision, the pendulum and projectile together swing up to a maximum height h, Fig. 7–17. Determine the relationship between the initial horizontal speed of the projectile, v, and the maximum height h.

APPROACH We can analyze the process by dividing it into two parts or two time intervals: (1) the time interval from just before to just after the collision itself, and (2) the subsequent time interval in which the pendulum moves from the vertical hanging position to the maximum height h.

In part (1), Fig. 7–17a, we assume the collision time is very short, so the projectile comes to rest in the block before the block has moved significantly from its position directly below its support. Thus there is effectively no net external force, and we can apply conservation of momentum to this completely inelastic collision. In part (2), Fig. 7–17b, the pendulum begins to move, subject to a net external force (gravity, tending to pull it back to the vertical position); so for part (2), we cannot use conservation of momentum. But we can use conservation of mechanical energy because gravity is a conservative force (Chapter 6). The kinetic energy immediately after the collision is changed entirely to gravitational potential energy when the pendulum reaches its maximum height, h.

SOLUTION In part (1) momentum is conserved:

$$\text{total } p \text{ before} = \text{total } p \text{ after}$$

$$mv = (m + M)v', \qquad \textbf{(i)}$$

where v' is the speed of the block and embedded projectile just after the collision, before they have moved significantly.

In part (2), mechanical energy is conserved. We choose $y = 0$ when the pendulum hangs vertically, and then $y = h$ when the pendulum–projectile system reaches its maximum height. Thus we write

$$(\text{KE} + \text{PE}) \text{ just after collision} = (\text{KE} + \text{PE}) \text{ at pendulum's maximum height}$$

or

$$\tfrac{1}{2}(m + M)v'^2 + 0 = 0 + (m + M)gh. \qquad \textbf{(ii)}$$

We solve for v':

$$v' = \sqrt{2gh}.$$

Inserting this result for v' into Eq. (i) above, and solving for v, gives

$$v = \frac{m + M}{m} v' = \frac{m + M}{m}\sqrt{2gh},$$

which is our final result.

NOTE The separation of the process into two parts was crucial. Such an analysis is a powerful problem-solving tool. But how do you decide how to make such a division? Think about the conservation laws. They are your *tools*. Start a problem by asking yourself whether the conservation laws apply in the given situation. Here, we determined that momentum is conserved only during the brief collision, which we called part (1). But in part (1), because the collision is inelastic, the conservation of mechanical energy is not valid. Then in part (2), conservation of mechanical energy is valid, but not conservation of momentum.

Note, however, that if there had been significant motion of the pendulum during the deceleration of the projectile in the block, then there *would* have been an external force (gravity) during the collision, so conservation of momentum would not have been valid in part (1).

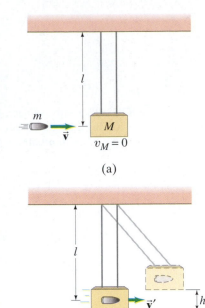

FIGURE 7–17 Ballistic pendulum. Example 7–10.

➡ **PROBLEM SOLVING**

Use the conservation laws to analyze a problem

FIGURE 7–18 A recent color-enhanced version of a cloud-chamber photograph made in the early days (1920s) of nuclear physics. Green lines are paths of helium nuclei (He) coming from the left. One He, highlighted in yellow, strikes a proton of the hydrogen gas in the chamber, and both scatter at an angle; the scattered proton's path is shown in red.

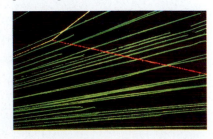

*7–7 Collisions in Two or Three Dimensions

Conservation of momentum and energy can also be applied to collisions in two or three dimensions, where the vector nature of momentum is especially important. One common type of non-head-on collision is that in which a moving object (called the "projectile") strikes a second object initially at rest (the "target"). This is the common situation in games such as billiards and pool, and for experiments in atomic and nuclear physics (the projectiles, from radioactive decay or a high-energy accelerator, strike a stationary target nucleus; Fig. 7–18).

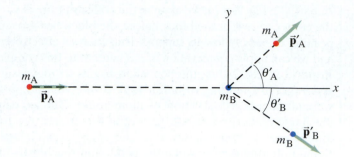

FIGURE 7–19 Object A, the projectile, collides with object B, the target. After the collision, they move off with momenta $\vec{p}'_A$ and $\vec{p}'_B$ at angles θ'_A and θ'_B.

Figure 7–19 shows the incoming projectile, m_A, heading along the x axis toward the target object, m_B, which is initially at rest. If these are billiard balls, m_A strikes m_B and they go off at the angles θ'_A and θ'_B, respectively, which are measured relative to m_A's initial direction (the x axis).[†]

Let us apply the law of conservation of momentum to a collision like that of Fig. 7–19. We choose the xy plane to be the plane in which the initial and final momenta lie. Momentum is a vector, and because the total momentum is conserved, its components in the x and y directions also are conserved. The x component of momentum conservation gives

$$p_{Ax} + p_{Bx} = p'_{Ax} + p'_{Bx}$$

or, with $p_{Bx} = m_B v_{Bx} = 0$,

p_x conserved
$$m_A v_A = m_A v'_A \cos \theta'_A + m_B v'_B \cos \theta'_B, \qquad (7\text{–}8a)$$

where the primes (') refer to quantities *after* the collision. Because there is no motion in the y direction initially, the y component of the total momentum is zero before the collision. The y component equation of momentum conservation is then

$$p_{Ay} + p_{By} = p'_{Ay} + p'_{By}$$

or

p_y conserved
$$0 = m_A v'_A \sin \theta'_A + m_B v'_B \sin \theta'_B. \qquad (7\text{–}8b)$$

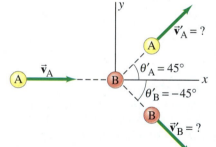

FIGURE 7–20 Example 7–11.

EXAMPLE 7–11 **Billiard ball collision in 2-D.** Billiard ball A moving with speed $v_A = 3.0$ m/s in the $+x$ direction (Fig. 7–20) strikes an equal-mass ball B initially at rest. The two balls are observed to move off at 45° to the x axis, ball A above the x axis and ball B below. That is, $\theta'_A = 45°$ and $\theta'_B = -45°$ in Fig. 7–20. What are the speeds of the two balls after the collision?

APPROACH There is no net external force on our system of two balls, assuming the table is level (the normal force balances gravity). Thus momentum conservation applies, and we apply it to both the x and y components using the xy coordinate system shown in Fig. 7–20. We get two equations, and we have two unknowns, v'_A and v'_B. From symmetry we might guess that the two balls have the same speed. But let us not assume that now. Even though we aren't told whether the collision is elastic or inelastic, we can still use conservation of momentum.

SOLUTION We apply conservation of momentum, Eqs. 7–8a and b, and we solve for v'_A and v'_B. We are given $m_A = m_B (= m)$, so

(for x) $mv_A = mv'_A \cos(45°) + mv'_B \cos(-45°)$

and

(for y) $0 = mv'_A \sin(45°) + mv'_B \sin(-45°)$.

The m's cancel out in both equations (the masses are equal).

[†]The objects may begin to deflect even before they touch if electric, magnetic, or nuclear forces act between them. You might think, for example, of two magnets oriented so that they repel each other: when one moves toward the other, the second moves away before the first one touches it.

The second equation yields [recall that $\sin(-\theta) = -\sin\theta$]:

$$v'_B = -v'_A \frac{\sin(45°)}{\sin(-45°)} = -v'_A \left(\frac{\sin 45°}{-\sin 45°} \right) = v'_A.$$

So they do have equal speeds as we guessed at first. The x component equation gives [recall that $\cos(-\theta) = \cos\theta$]:

$$v_A = v'_A \cos(45°) + v'_B \cos(45°) = 2v'_A \cos(45°),$$

so

$$v'_A = v'_B = \frac{v_A}{2\cos(45°)} = \frac{3.0\,\text{m/s}}{2(0.707)} = 2.1\,\text{m/s}.$$

NOTE When we have two independent equations, we can solve for, at most, two unknowns.

EXERCISE F Make a calculation to see if kinetic energy was conserved in the collision of Example 7–11.

If we know that a collision is elastic, we can also apply conservation of kinetic energy and obtain a third equation in addition to Eqs. 7–8a and b:

$$\text{KE}_A + \text{KE}_B = \text{KE}'_A + \text{KE}'_B$$

or, for the collision shown in Fig. 7–20,

$$\tfrac{1}{2}m_A v_A^2 = \tfrac{1}{2}m_A v'^2_A + \tfrac{1}{2}m_B v'^2_B. \qquad \text{[elastic collision]} \quad \textbf{(7–8c)} \qquad \textit{KE conserved}$$

If the collision is elastic, we have three independent equations and can solve for three unknowns. If we are given m_A, m_B, v_A (and v_B, if it is not zero), we cannot, for example, predict the final variables, v'_A, v'_B, θ'_A, and θ'_B, because there are four of them. However, if we measure one of these variables, say θ'_A, then the other three variables (v'_A, v'_B, and θ'_B) are uniquely determined, and we can determine them using Eqs. 7–8a, b, c.

A note of caution: Eq. 7–7 does *not* apply for two-dimensional collisions. It works only when a collision occurs along a line.

CAUTION

Equation 7–7 applies only in 1-D

PROBLEM SOLVING Momentum Conservation and Collisions

1. Choose your **system.** If the situation is complex, think about how you might break it up into separate parts when one or more conservation laws apply.

2. Consider whether a significant **net external force** acts on your chosen system; if it does, be sure the time interval Δt is so short that the effect on momentum is negligible. That is, the forces that act between the interacting objects must be the significant ones if momentum conservation is to be used. [Note: If this is valid for a portion of the problem, you can use momentum conservation only for that portion.]

3. Draw a **diagram** of the initial situation, just before the interaction (collision, explosion) takes place, and represent the momentum of each object with an arrow and a label. Do the same for the final situation, just after the interaction.

4. Choose a **coordinate system** and "+" and "−" directions. (For a head-on collision, you will need only an x axis.) It is often convenient to choose the $+x$ axis in the direction of one object's initial velocity.

5. Apply the **momentum conservation** equation(s):

   total initial momentum = total final momentum.

   You have one equation for each component (x, y, z): only one equation for a head-on collision. [Don't forget that it is the *total* momentum of the system that is conserved, not the momenta of individual objects.]

6. If the collision is elastic, you can also write down a **conservation of kinetic energy** equation:

   total initial KE = total final KE.

   [Alternately, you could use Eq. 7–7: $v_A - v_B = v'_B - v'_A$, if the collision is one dimensional (head-on).]

7. Solve for the **unknown(s).**

8. **Check** your work, check the units, and ask yourself whether the results are reasonable.

7–8 Center of Mass (CM)

Momentum is a powerful concept not only for analyzing collisions but also for analyzing the translational motion of real extended objects. Until now, whenever we have dealt with the motion of an extended object (that is, an object that has size), we have assumed that it could be approximated as a point particle or that it undergoes only translational motion. Real extended objects, however, can undergo rotational and other types of motion as well. For example, the diver in Fig. 7–21a undergoes only translational motion (all parts of the object follow the same path), whereas the diver in Fig. 7–21b undergoes both translational and rotational motion. We will refer to motion that is not pure translation as *general motion*.

FIGURE 7–21 The motion of the diver is pure translation in (a), but is translation plus rotation in (b). The black dot represents the diver's CM at each moment.

(a)   (b)

Observations indicate that even if an object rotates, or several parts of a system of objects move relative to one another, there is one point that moves in the same path that a particle would move if subjected to the same net force. This point is called the **center of mass** (abbreviated CM). The general motion of an extended object (or system of objects) can be considered as *the sum of the translational motion of the CM, plus rotational, vibrational, or other types of motion about the CM.*

Center of mass
General motion

As an example, consider the motion of the center of mass of the diver in Fig. 7–21; the CM follows a parabolic path even when the diver rotates, as shown in Fig. 7–21b. This is the same parabolic path that a projected particle follows when acted on only by the force of gravity (that is, projectile motion). Other points in the rotating diver's body, such as her feet or head, follow more complicated paths.

Figure 7–22 shows a wrench acted on by zero net force, translating and rotating along a horizontal surface. Note that its CM, marked by a red cross, moves in a straight line, as shown by the dashed white line.

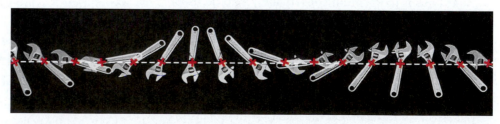

FIGURE 7–22 Translation plus rotation: a wrench moving over a horizontal surface. The CM, marked with a red cross, moves in a straight line.

We will show in Section 7–10 that the important properties of the CM follow from Newton's laws if the CM is defined in the following way. We can consider any extended object as being made up of many tiny particles. But first we consider a system made up of only two particles (or small objects), of masses m_A and m_B. We choose a coordinate system so that both particles lie on the x axis at positions x_A and x_B, Fig. 7–23. The center of mass of this system is defined to be at the position x_{CM}, given by

$$x_{CM} = \frac{m_A x_A + m_B x_B}{m_A + m_B} = \frac{m_A x_A + m_B x_B}{M},$$

x coordinate of CM (2 particles)

where $M = m_A + m_B$ is the total mass of the system. The center of mass lies on the line joining m_A and m_B. If the two masses are equal $(m_A = m_B = m)$, then x_{CM} is midway between them, since in this case

$$x_{CM} = \frac{m(x_A + x_B)}{2m} = \frac{(x_A + x_B)}{2}.$$

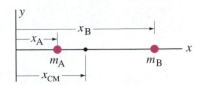

If one mass is greater than the other, say, $m_A > m_B$, then the CM is closer to the larger mass. If all the mass is concentrated at x_B, say, so $m_A = 0$, then $x_{CM} = (0 x_A + m_B x_B)/(0 + m_B) = x_B$, as we would expect.

FIGURE 7–23 The center of mass of a two-particle system lies on the line joining the two masses. Here $m_A > m_B$, so the CM is closer to m_A than to m_B.

If there are more than two particles along a line, there will be additional terms:

$$x_{CM} = \frac{m_A x_A + m_B x_B + m_C x_C + \cdots}{m_A + m_B + m_C + \cdots} = \frac{m_A x_A + m_B x_B + m_C x_C + \cdots}{M}, \quad \textbf{(7–9a)}$$

x coordinate of CM (many particles)

where M is the total mass of all the particles.

EXAMPLE 7–12 **CM of three guys on a raft.** Three people of roughly equal masses m on a lightweight (air-filled) banana boat sit along the x axis at positions $x_A = 1.0\,\text{m}$, $x_B = 5.0\,\text{m}$, and $x_C = 6.0\,\text{m}$, measured from the left-hand end as shown in Fig. 7–24. Find the position of the CM. Ignore mass of boat.

APPROACH We are given the mass and location of the three people, so we use three terms in Eq. 7–9a. We approximate each person as a point particle. Equivalently, the location of each person is the position of that person's own CM.

SOLUTION We use Eq. 7–9a with three terms:

$$
\begin{aligned}
x_{CM} &= \frac{mx_A + mx_B + mx_C}{m + m + m} = \frac{m(x_A + x_B + x_C)}{3m} \\
&= \frac{(1.0\,\text{m} + 5.0\,\text{m} + 6.0\,\text{m})}{3} = \frac{12.0\,\text{m}}{3} = 4.0\,\text{m}.
\end{aligned}
$$

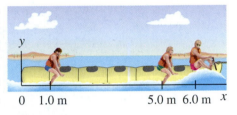

FIGURE 7–24 Example 7–12.

The CM is 4.0 m from the left-hand end of the boat.

NOTE The coordinates of the CM depend on the reference frame or coordinate system chosen. But the physical location of the CM is independent of that choice.

EXERCISE G Calculate the CM of the three people in Example 7–12 taking the origin at the driver $(x_C = 0)$ on the right. Is the physical location of the CM the same?

If the particles are spread out in two or three dimensions, then we must specify not only the x coordinate of the CM (x_{CM}), but also the y and z coordinates, which will be given by formulas like Eq. 7–9a. For example, the y coordinate of the CM will be:

$$y_{CM} = \frac{m_A y_A + m_B y_B + \cdots}{m_A + m_B + \cdots} = \frac{m_A y_A + m_B y_B + \cdots}{M}. \quad \textbf{(7–9b)}$$

y coordinate of CM

A concept similar to *center of mass* is **center of gravity** (CG). The CG of an object is that point at which the force of gravity can be considered to act. The force of gravity actually acts on *all* the different parts or particles of an object, but for purposes of determining the translational motion of an object as a whole, we can assume that the entire weight of the object (which is the sum of the weights of all its parts) acts at the CG. There is a conceptual difference between the center of gravity and the center of mass, but for nearly all practical purposes, they are at the same point.[†]

It is often easier to determine the CM or CG of an extended object experimentally rather than analytically. If an object is suspended from any point, it will swing (Fig. 7–25) unless it is placed so its CG lies on a vertical line directly below the point from which it is suspended. If the object is two dimensional, or has a plane of symmetry, it need only be hung from two different pivot points and the respective vertical (plumb) lines drawn. Then the center of gravity will be at the intersection

Center of gravity

FIGURE 7–25 The force of gravity, considered to act at the CG, causes this object to rotate about the pivot point; if the CG were on a vertical line directly below the pivot, the object would remain at rest.

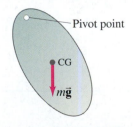

[†]There would be a difference between the CM and CG only in the unusual case of an object so large that the acceleration due to gravity, g, was different at different parts of the object.

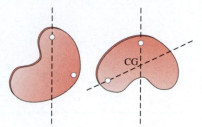

FIGURE 7–26 Finding the CG.

of the two lines, as in Fig. 7–26. If the object doesn't have a plane of symmetry, the CG with respect to the third dimension is found by suspending the object from at least three points whose plumb lines do not lie in the same plane. For symmetrically shaped objects such as uniform cylinders (wheels), spheres, and rectangular solids, the CM is located at the geometric center of the object.

For some objects, the CM may actually lie outside the object. The CM of a donut, for example, lies at the center of the hole.

*7–9 CM for the Human Body

If we have a group of extended objects, each of whose CM is known, we can find the CM of the group using Eqs. 7–9a and b. As an example, we consider the human body. Table 7–1 indicates the CM and hinge points (joints) for the different components of a "representative" person. Of course, there are wide variations among people, so these data represent only a very rough average. The numbers represent a percentage of the total height, which is regarded as 100 units; similarly, the total mass is 100 units. For example, if a person is 1.70 m tall, his or her shoulder joint would be $(1.70 \text{ m})(81.2/100) = 1.38 \text{ m}$ above the floor.

TABLE 7–1 Center of Mass of Parts of Typical Human Body
(full height and mass = 100 units)

Distance Above Floor of Hinge Points (%)	Hinge Points (•) (Joints)	Center of Mass (×) (% Height Above Floor)		Percent Mass
91.2	Base of skull	Head	93.5	6.9
81.2	Shoulder joint	Trunk and neck	71.1	46.1
		Upper arms	71.7	6.6
elbow 62.2		Lower arms	55.3	4.2
wrist 46.2		Hands	43.1	1.7
52.1	Hip joint	Upper legs (thighs)	42.5	21.5
28.5	Knee joint	Lower legs	18.2	9.6
4.0	Ankle joint	Feet	1.8	3.4
		Body CM =	58.0	100.0

FIGURE 7–27 Example 7–13: finding the CM of a leg in two different positions using percentages from Table 7–1. (⊗ represents the calculated CM).

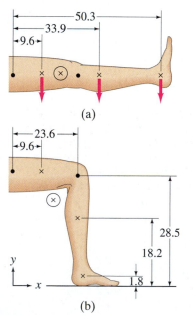

EXAMPLE 7–13 A leg's CM. Determine the position of the CM of a whole leg (a) when stretched out, and (b) when bent at 90°. See Fig. 7–27. Assume the person is 1.70 m tall.

APPROACH Our system consists of three objects: upper leg, lower leg, and foot. The location of the CM of each object, as well as the mass of each, is given in Table 7–1, where they are expressed in percentage units. To express the results in meters, these percentage values need to be multiplied by $(1.70 \text{ m}/100)$. When the leg is stretched out, the problem is one dimensional and we can solve for the x coordinate of the CM. When the leg is bent, the problem is two dimensional and we need to find both the x and y coordinates.

SOLUTION (a) We determine the distances from the hip joint using Table 7–1 and obtain the numbers (%) shown in Fig. 7–27a. Using Eq. 7–9a, we obtain

$$x_{CM} = \frac{(21.5)(9.6) + (9.6)(33.9) + (3.4)(50.3)}{21.5 + 9.6 + 3.4} = 20.4 \text{ units.}$$

Thus, the center of mass of the leg and foot is 20.4 units from the hip joint, or $52.1 - 20.4 = 31.7$ units from the base of the foot. Since the person is 1.70 m tall, this is $(1.70 \text{ m})(31.7/100) = 0.54 \text{ m}$ above the bottom of the foot. (b) We use an xy coordinate system, as shown in Fig. 7–27b. First, we calculate how far to the right of the hip joint the CM lies, accounting for all three parts:

$$x_{CM} = \frac{(21.5)(9.6) + (9.6)(23.6) + (3.4)(23.6)}{21.5 + 9.6 + 3.4} = 14.9 \text{ units.}$$

For our 1.70-m-tall person, this is $(1.70 \, \text{m})(14.9/100) = 0.25 \, \text{m}$ from the hip joint. Next, we calculate the distance, y_{CM}, of the CM above the floor:

$$y_{CM} = \frac{(3.4)(1.8) + (9.6)(18.2) + (21.5)(28.5)}{21.5 + 9.6 + 3.4} = 23.0 \, \text{units},$$

or $(1.70 \, \text{m})(23.0/100) = 0.39 \, \text{m}$. Thus, the CM is located 39 cm above the floor and 25 cm to the right of the hip joint.

NOTE The CM actually lies *outside* the body in (b).

Knowing the CM of the body when it is in various positions is of great use in studying body mechanics. One simple example from athletics is shown in Fig. 7–28. If high jumpers can get into the position shown, their CM can pass below the bar which their bodies go over, meaning that for a particular takeoff speed, they can clear a higher bar. This is indeed what they try to do.

FIGURE 7–28 A high jumper's CM may actually pass beneath the bar.

PHYSICS APPLIED

High jumping

*7–10 Center of Mass and Translational Motion

As mentioned in Section 7–8, a major reason for the importance of the concept of center of mass is that the motion of the CM for a system of particles (or an extended object) is directly related to the net force acting on the system as a whole. We now show this, taking the simple case of one-dimensional motion (x direction) and only three particles, but the extension to more objects and to three dimensions follows the same lines.

Suppose the three particles lie on the x axis and have masses m_A, m_B, m_C, and positions x_A, x_B, x_C. From Eq. 7–9a for the center of mass, we can write

$$Mx_{CM} = m_A x_A + m_B x_B + m_C x_C, \qquad \text{(i)}$$

where $M = m_A + m_B + m_C$ is the total mass of the system. If these particles are in motion (say, along the x axis with velocities v_A, v_B, and v_C, respectively), then in a short time interval Δt they each will have traveled a distance

$$\Delta x_A = x'_A - x_A = v_A \Delta t$$
$$\Delta x_B = x'_B - x_B = v_B \Delta t$$
$$\Delta x_C = x'_C - x_C = v_C \Delta t,$$

where x'_A, x'_B, and x'_C represent their new positions after time interval Δt. The position of the new CM is given by

$$Mx'_{CM} = m_A x'_A + m_B x'_B + m_C x'_C. \qquad \text{(ii)}$$

If we subtract from this equation (ii) the previous CM equation (i), we get

$$M \Delta x_{CM} = m_A \Delta x_A + m_B \Delta x_B + m_C \Delta x_C.$$

During time interval Δt, the center of mass will have moved a distance

$$\Delta x_{CM} = x'_{CM} - x_{CM} = v_{CM} \Delta t,$$

where v_{CM} is the velocity of the center of mass. We now substitute the relations for all the Δx's into the equation just before the last one:

$$Mv_{CM} \Delta t = m_A v_A \Delta t + m_B v_B \Delta t + m_C v_C \Delta t.$$

We cancel Δt and get

$$Mv_{CM} = m_A v_A + m_B v_B + m_C v_C. \qquad \text{(7–10)}$$

Since $m_A v_A + m_B v_B + m_C v_C$ is the sum of the momenta of the particles of the system, it represents the *total momentum* of the system. Thus we see from Eq. 7–10 that *the total (linear) momentum of a system of particles is equal to the product of the total mass M and the velocity of the center of mass of the system.* Or, *the linear momentum of an extended object is the product of the object's mass and the velocity of its CM.*

Total momentum, and velocity of CM

If forces are acting on the particles, then the particles may be accelerating. In a short time interval Δt, each particle's velocity will change by an amount

$$\Delta v_A = a_A \, \Delta t, \qquad \Delta v_B = a_B \, \Delta t, \qquad \Delta v_C = a_C \, \Delta t.$$

If we now use the same reasoning as we did to derive Eq. 7–10, we obtain

$$M a_{CM} = m_A a_A + m_B a_B + m_C a_C.$$

According to Newton's second law, $m_A a_A = F_A$, $m_B a_B = F_B$, and $m_C a_C = F_C$, where F_A, F_B, and F_C are the net forces on the three particles, respectively. Thus we get for the system as a whole $M a_{CM} = F_A + F_B + F_C$, or

Newton's second law for a system of particles or an extended object

$$M a_{CM} = F_{net}. \tag{7–11}$$

That is, *the sum of all the forces acting on the system is equal to the total mass of the system times the acceleration of its center of mass.* This is **Newton's second law** for a system of particles, and it also applies to an extended object (which can be thought of as a collection of particles). Thus we conclude that the *center of mass of a system of particles (or of an extended object) with total mass M moves like a single particle of mass M acted on by the same net external force.* That is, the system moves as if all its mass were concentrated at the center of mass and all the external forces acted at that point. We can thus treat the translational motion of any object or system of objects as the motion of a particle (see Figs. 7–21 and 7–22). This result simplifies our analysis of the motion of complex systems and extended objects. Although the motion of various parts of the system may be complicated, we may often be satisfied with knowing the motion of the center of mass. This result also allows us to solve certain types of problems very easily, as illustrated by the following Example.

CONCEPTUAL EXAMPLE 7–14 **A two-stage rocket.** A rocket is shot into the air as shown in Fig. 7–29. At the moment the rocket reaches its highest point, a horizontal distance d from its starting point, a prearranged explosion separates it into two parts of equal mass. Part I is stopped in midair by the explosion, and it falls vertically to Earth. Where does part II land? Assume $\vec{g}$ = constant.

RESPONSE After the rocket is fired, the path of the CM of the system continues to follow the parabolic trajectory of a projectile acted on by only a constant gravitational force. The CM will thus land at a point $2d$ from the starting point. Since the masses of I and II are equal, the CM must be midway between them at any time. Therefore, part II lands a distance $3d$ from the starting point.

NOTE If part I had been given a kick up or down, instead of merely falling, the solution would have been more complicated.

EXERCISE H A woman stands up in a rowboat and walks from one end of the boat to the other. How does the boat move, as seen from the shore?

FIGURE 7–29 Example 7–14.

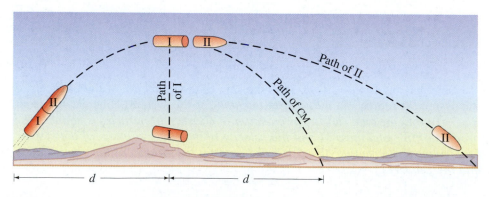

Summary

The **momentum**, $\vec{\mathbf{p}}$, of an object is defined as the product of its mass times its velocity,

$$\vec{\mathbf{p}} = m\vec{\mathbf{v}}. \qquad (7\text{–}1)$$

In terms of momentum, **Newton's second law** can be written as

$$\Sigma\vec{\mathbf{F}} = \frac{\Delta\vec{\mathbf{p}}}{\Delta t}. \qquad (7\text{–}2)$$

That is, the rate of change of momentum equals the net applied force.

The **law of conservation of momentum** states that the total momentum of an isolated system of objects remains constant. An **isolated system** is one on which the net external force is zero.

The law of conservation of momentum is very useful in dealing with **collisions**. In a collision, two (or more) objects interact with each other over a very short time interval, and the forces between them during this time interval are very large.

The **impulse** of a force on an object is defined as $\vec{\mathbf{F}}\,\Delta t$, where $\vec{\mathbf{F}}$ is the average force acting during the (usually short) time interval Δt. The impulse is equal to the change in momentum of the object:

$$\text{Impulse} = \vec{\mathbf{F}}\,\Delta t = \Delta\vec{\mathbf{p}}. \qquad (7\text{–}5)$$

Total momentum is conserved in *any* collision as long as any net external force is zero or negligible. If $m_A\vec{\mathbf{v}}_A$ and $m_B\vec{\mathbf{v}}_B$ are the momenta of two objects before the collision and $m_A\vec{\mathbf{v}}'_A$ and $m_B\vec{\mathbf{v}}'_B$ are their momenta after, then momentum conservation tell us that

$$m_A\vec{\mathbf{v}}_A + m_B\vec{\mathbf{v}}_B = m_A\vec{\mathbf{v}}'_A + m_B\vec{\mathbf{v}}'_B \qquad (7\text{–}3)$$

for this two-object system.

Total energy is also conserved, but this may not be helpful in problem solving unless the only type of energy transformation involves kinetic energy. In that case kinetic energy is conserved and the collision is called an **elastic collision**, and we can write

$$\tfrac{1}{2}m_A v_A^2 + \tfrac{1}{2}m_B v_B^2 = \tfrac{1}{2}m_A v_A'^2 + \tfrac{1}{2}m_B v_B'^2. \qquad (7\text{–}6)$$

If kinetic energy is not conserved, the collision is called **inelastic**. A **completely inelastic** collision is one in which the colliding objects stick together after the collision.

The **center of mass** (CM) of an extended object (or group of objects) is that point at which the net force can be considered to act, for purposes of determining the translational motion of the object as a whole. The x component of the CM for objects with mass $m_A, m_B, \ldots$, is given by

$$x_{CM} = \frac{m_A x_A + m_B x_B + \cdots}{m_A + m_B + \cdots}. \qquad (7\text{–}9a)$$

[*The complete motion of an object can be described as the translational motion of its center of mass plus rotation (or other internal motion) about its center of mass.]

Questions

1. We claim that momentum is conserved, yet most moving objects eventually slow down and stop. Explain.

2. When a person jumps from a tree to the ground, what happens to the momentum of the person upon striking the ground?

3. When you release an inflated but untied balloon, why does it fly across the room?

4. It is said that in ancient times a rich man with a bag of gold coins froze to death while stranded on a frozen lake. Because the ice was frictionless, he could not push himself to shore. What could he have done to save himself had he not been so miserly?

5. How can a rocket change direction when it is far out in space and is essentially in a vacuum?

6. According to Eq. 7–5, the longer the impact time of an impulse, the smaller the force can be for the same momentum change, and hence the smaller the deformation of the object on which the force acts. On this basis, explain the value of air bags, which are intended to inflate during an automobile collision and reduce the possibility of fracture or death.

7. Cars used to be built as rigid as possible to withstand collisions. Today, though, cars are designed to have "crumple zones" that collapse upon impact. What is the advantage of this new design?

8. Why can a batter hit a pitched baseball further than a ball tossed in the air by the batter?

9. Is it possible for an object to receive a larger impulse from a small force than from a large force? Explain.

10. A light object and a heavy object have the same kinetic energy. Which has the greater momentum? Explain.

11. Describe a collision in which all kinetic energy is lost.

12. At a hydroelectric power plant, water is directed at high speed against turbine blades on an axle that turns an electric generator. For maximum power generation, should the turbine blades be designed so that the water is brought to a dead stop, or so that the water rebounds?

13. A squash ball hits a wall at a 45° angle as shown in Fig. 7–30. What is the direction (*a*) of the change in momentum of the ball, (*b*) of the force on the wall?

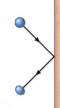

FIGURE 7–30
Question 13.

14. A Superball is dropped from a height h onto a hard steel plate (fixed to the Earth), from which it rebounds at very nearly its original speed. (*a*) Is the momentum of the ball conserved during any part of this process? (*b*) If we consider the ball and Earth as our system, during what parts of the process is momentum conserved? (*c*) Answer part (*b*) for a piece of putty that falls and sticks to the steel plate.

15. Why do you tend to lean backward when carrying a heavy load in your arms?

16. Why is the CM of a 1-m length of pipe at its mid-point, whereas this is not true for your arm or leg?

* 17. Show on a diagram how your CM shifts when you change from a lying position to a sitting position.

* 18. If only an external force can change the momentum of the center of mass of an object, how can the internal force of an engine accelerate a car?

* 19. A rocket following a parabolic path through the air suddenly explodes into many pieces. What can you say about the motion of this system of pieces?

Problems

7–1 and 7–2 Momentum and Its Conservation

1. (I) What is the magnitude of the momentum of a 28-g sparrow flying with a speed of 8.4 m/s?

2. (I) A constant friction force of 25 N acts on a 65-kg skier for 20 s. What is the skier's change in velocity?

3. (II) A 0.145-kg baseball pitched at 39.0 m/s is hit on a horizontal line drive straight back toward the pitcher at 52.0 m/s. If the contact time between bat and ball is 3.00×10^{-3} s, calculate the average force between the ball and bat during contact.

4. (II) A child in a boat throws a 6.40-kg package out horizontally with a speed of 10.0 m/s, Fig. 7–31. Calculate the velocity of the boat immediately after, assuming it was initially at rest. The mass of the child is 26.0 kg, and that of the boat is 45.0 kg. Ignore water resistance.

FIGURE 7–31 Problem 4.

5. (II) Calculate the force exerted on a rocket, given that the propelling gases are expelled at a rate of 1500 kg/s with a speed of 4.0×10^4 m/s (at the moment of takeoff).

6. (II) A 95-kg halfback moving at 4.1 m/s on an apparent breakaway for a touchdown is tackled from behind. When he was tackled by an 85-kg cornerback running at 5.5 m/s in the same direction, what was their mutual speed immediately after the tackle?

7. (II) A 12,600-kg railroad car travels alone on a level frictionless track with a constant speed of 18.0 m/s. A 5350-kg load, initially at rest, is dropped onto the car. What will be the car's new speed?

8. (II) A 9300-kg boxcar traveling at 15.0 m/s strikes a second boxcar at rest. The two stick together and move off with a speed of 6.0 m/s. What is the mass of the second car?

9. (II) During a Chicago storm, winds can whip horizontally at speeds of 100 km/h. If the air strikes a person at the rate of 40 kg/s per square meter and is brought to rest, estimate the force of the wind on a person. Assume the person is 1.50 m high and 0.50 m wide. Compare to the typical maximum force of friction ($\mu \approx 1.0$) between the person and the ground, if the person has a mass of 70 kg.

10. (II) A 3800-kg open railroad car coasts along with a constant speed of 8.60 m/s on a level track. Snow begins to fall vertically and fills the car at a rate of 3.50 kg/min. Ignoring friction with the tracks, what is the speed of the car after 90.0 min?

11. (II) An atomic nucleus initially moving at 420 m/s emits an alpha particle in the direction of its velocity, and the remaining nucleus slows to 350 m/s. If the alpha particle has a mass of 4.0 u and the original nucleus has a mass of 222 u, what speed does the alpha particle have when it is emitted?

12. (II) A 23-g bullet traveling 230 m/s penetrates a 2.0-kg block of wood and emerges cleanly at 170 m/s. If the block is stationary on a frictionless surface when hit, how fast does it move after the bullet emerges?

13. (III) A 975-kg two-stage rocket is traveling at a speed of 5.80×10^3 m/s with respect to Earth when a pre-designed explosion separates the rocket into two sections of equal mass that then move at a speed of 2.20×10^3 m/s relative to each other along the original line of motion. (a) What are the speed and direction of each section (relative to Earth) after the explosion? (b) How much energy was supplied by the explosion? [Hint: What is the change in KE as a result of the explosion?]

14. (III) A rocket of total mass 3180 kg is traveling in outer space with a velocity of 115 m/s. To alter its course by 35.0°, its rockets can be fired briefly in a direction perpendicular to its original motion. If the rocket gases are expelled at a speed of 1750 m/s, how much mass must be expelled?

7–3 Collisions and Impulse

15. (II) A golf ball of mass 0.045 kg is hit off the tee at a speed of 45 m/s. The golf club was in contact with the ball for 3.5×10^{-3} s. Find (a) the impulse imparted to the golf ball, and (b) the average force exerted on the ball by the golf club.

16. (II) A 12-kg hammer strikes a nail at a velocity of 8.5 m/s and comes to rest in a time interval of 8.0 ms. (a) What is the impulse given to the nail? (b) What is the average force acting on the nail?

17. (II) A tennis ball of mass $m = 0.060$ kg and speed $v = 25$ m/s strikes a wall at a 45° angle and rebounds with the same speed at 45° (Fig. 7–32). What is the impulse (magnitude and direction) given to the ball?

FIGURE 7–32
Problem 17.

18. (II) You are the design engineer in charge of the crashworthiness of new automobile models. Cars are tested by smashing them into fixed, massive barriers at 50 km/h (30 mph). A new model of mass 1500 kg takes 0.15 s from the time of impact until it is brought to rest. (a) Calculate the average force exerted on the car by the barrier. (b) Calculate the average deceleration of the car.

19. (II) A 95-kg fullback is running at 4.0 m/s to the east and is stopped in 0.75 s by a head-on tackle by a tackler running due west. Calculate (a) the original momentum of the fullback, (b) the impulse exerted on the fullback, (c) the impulse exerted on the tackler, and (d) the average force exerted on the tackler.

20. (II) Suppose the force acting on a tennis ball (mass 0.060 kg) points in the $+x$ direction and is given by the graph of Fig. 7–33 as a function of time. Use graphical methods to estimate (a) the total impulse given the ball, and (b) the velocity of the ball after being struck, assuming the ball is being served so it is nearly at rest initially.

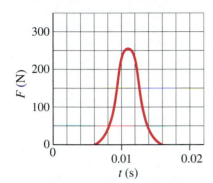

FIGURE 7–33
Problem 20.

21. (III) From what maximum height can a 75-kg person jump without breaking the lower leg bone of either leg? Ignore air resistance and assume the CM of the person moves a distance of 0.60 m from the standing to the seated position (that is, in breaking the fall). Assume the breaking strength (force per unit area) of bone is 170×10^6 N/m², and its smallest cross-sectional area is 2.5×10^{-4} m². [*Hint*: Do not try this experimentally.]

7–4 and 7–5 Elastic Collisions

22. (II) A ball of mass 0.440 kg moving east ($+x$ direction) with a speed of 3.30 m/s collides head-on with a 0.220-kg ball at rest. If the collision is perfectly elastic, what will be the speed and direction of each ball after the collision?

23. (II) A 0.450-kg ice puck, moving east with a speed of 3.00 m/s, has a head-on collision with a 0.900-kg puck initially at rest. Assuming a perfectly elastic collision, what will be the speed and direction of each object after the collision?

24. (II) Two billiard balls of equal mass undergo a perfectly elastic head-on collision. If one ball's initial speed was 2.00 m/s, and the other's was 3.00 m/s in the opposite direction, what will be their speeds after the collision?

25. (II) A 0.060-kg tennis ball, moving with a speed of 2.50 m/s, collides head-on with a 0.090-kg ball initially moving away from it at a speed of 1.15 m/s. Assuming a perfectly elastic collision, what are the speed and direction of each ball after the collision?

26. (II) A softball of mass 0.220 kg that is moving with a speed of 8.5 m/s collides head-on and elastically with another ball initially at rest. Afterward the incoming softball bounces backward with a speed of 3.7 m/s. Calculate (a) the velocity of the target ball after the collision, and (b) the mass of the target ball.

27. (II) Two bumper cars in an amusement park ride collide elastically as one approaches the other directly from the rear (Fig. 7–34). Car A has a mass of 450 kg and car B 550 kg, owing to differences in passenger mass. If car A approaches at 4.50 m/s and car B is moving at 3.70 m/s, calculate (a) their velocities after the collision, and (b) the change in momentum of each.

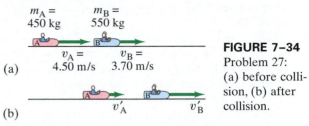

FIGURE 7–34
Problem 27:
(a) before collision, (b) after collision.

28. (II) A 0.280-kg croquet ball makes an elastic head-on collision with a second ball initially at rest. The second ball moves off with half the original speed of the first ball. (a) What is the mass of the second ball? (b) What fraction of the original kinetic energy (ΔKE/KE) gets transferred to the second ball?

29. (III) In a physics lab, a cube slides down a frictionless incline as shown in Fig. 7–35, and elastically strikes another cube at the bottom that is only one-half its mass. If the incline is 30 cm high and the table is 90 cm off the floor, where does each cube land? [*Hint*: Both leave the incline moving horizontally.]

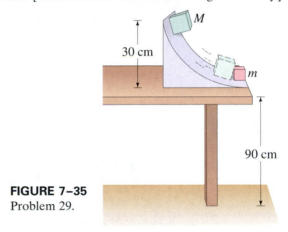

FIGURE 7–35
Problem 29.

30. (III) Take the general case of an object of mass m_A and velocity v_A elastically striking a stationary $(v_B = 0)$ object of mass m_B head-on. (a) Show that the final velocities v'_A and v'_B are given by

$$v'_A = \left(\frac{m_A - m_B}{m_A + m_B}\right) v_A,$$

$$v'_B = \left(\frac{2m_A}{m_A + m_B}\right) v_A.$$

(b) What happens in the extreme case when m_A is much smaller than m_B? Cite a common example of this. (c) What happens in the extreme case when m_A is much larger than m_B? Cite a common example of this. (d) What happens in the case when $m_A = m_B$? Cite a common example.

7–6 Inelastic Collisions

31. (I) In a ballistic pendulum experiment, projectile 1 results in a maximum height h of the pendulum equal to 2.6 cm. A second projectile causes the the pendulum to swing twice as high, $h_2 = 5.2$ cm. The second projectile was how many times faster than the first?

32. (II) A 28-g rifle bullet traveling 230 m/s buries itself in a 3.6-kg pendulum hanging on a 2.8-m-long string, which makes the pendulum swing upward in an arc. Determine the vertical and horizontal components of the pendulum's displacement.

33. (II) (a) Derive a formula for the fraction of kinetic energy lost, $\Delta\text{KE}/\text{KE}$, for the ballistic pendulum collision of Example 7–10. (b) Evaluate for $m = 14.0$ g and $M = 380$ g.

34. (II) An internal explosion breaks an object, initially at rest, into two pieces, one of which has 1.5 times the mass of the other. If 7500 J were released in the explosion, how much kinetic energy did each piece acquire?

35. (II) A 920-kg sports car collides into the rear end of a 2300-kg SUV stopped at a red light. The bumpers lock, the brakes are locked, and the two cars skid forward 2.8 m before stopping. The police officer, knowing that the coefficient of kinetic friction between tires and road is 0.80, calculates the speed of the sports car at impact. What was that speed?

36. (II) A ball is dropped from a height of 1.50 m and rebounds to a height of 1.20 m. Approximately how many rebounds will the ball make before losing 90% of its energy?

37. (II) A measure of inelasticity in a head-on collision of two objects is the *coefficient of restitution, e,* defined as

$$e = \frac{v'_A - v'_B}{v_B - v_A},$$

where $v'_A - v'_B$ is the relative velocity of the two objects after the collision and $v_B - v_A$ is their relative velocity before it. (a) Show that $e = 1$ for a perfectly elastic collision, and $e = 0$ for a completely inelastic collision. (b) A simple method for measuring the coefficient of restitution for an object colliding with a very hard surface like steel is to drop the object onto a heavy steel plate, as shown in Fig. 7–36. Determine a formula for e in terms of the original height h and the maximum height h' reached after one collision.

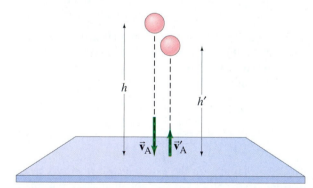

FIGURE 7–36 Problem 37. Measurement of the coefficient of restitution.

38. (II) A wooden block is cut into two pieces, one with three times the mass of the other. A depression is made in both faces of the cut, so that a firecracker can be placed in it with the block reassembled. The reassembled block is set on a rough-surfaced table, and the fuse is lit. When the firecracker explodes, the two blocks separate and slide apart. What is the ratio of distances each block travels?

39. (III) A 15.0-kg object moving in the $+x$ direction at 5.5 m/s collides head-on with a 10.0-kg object moving in the $-x$ direction at 4.0 m/s. Find the final velocity of each mass if: (a) the objects stick together; (b) the collision is elastic; (c) the 15.0-kg object is at rest after the collision; (d) the 10.0-kg object is at rest after the collision; (e) the 15.0-kg object has a velocity of 4.0 m/s in the $-x$ direction after the collision. Are the results in (c), (d), and (e) "reasonable"? Explain.

* 7–7 Collisions in Two Dimensions

* 40. (II) A radioactive nucleus at rest decays into a second nucleus, an electron, and a neutrino. The electron and neutrino are emitted at right angles and have momenta of 9.30×10^{-23} kg·m/s and 5.40×10^{-23} kg·m/s, respectively. What are the magnitude and direction of the momentum of the second (recoiling) nucleus?

* 41. (II) An eagle $(m_A = 4.3$ kg) moving with speed $v_A = 7.8$ m/s is on a collision course with a second eagle $(m_B = 5.6$ kg) moving at $v_B = 10.2$ m/s in a direction perpendicular to the first. After they collide, they hold onto one another. In what direction, and with what speed, are they moving after the collision?

* 42. (II) Billiard ball A of mass $m_A = 0.400$ kg moving with speed $v_A = 1.80$ m/s strikes ball B, initially at rest, of mass $m_B = 0.500$ kg. As a result of the collision, ball A is deflected off at an angle of $30.0°$ with a speed $v'_A = 1.10$ m/s. (a) Taking the x axis to be the original direction of motion of ball A, write down the equations expressing the conservation of momentum for the components in the x and y directions separately. (b) Solve these equations for the speed v'_B and angle θ'_B of ball B. Do not assume the collision is elastic.

* 43. (III) After a completely inelastic collision between two objects of equal mass, each having initial speed v, the two move off together with speed $v/3$. What was the angle between their initial directions?

* 44. (III) Two billiard balls of equal mass move at right angles and meet at the origin of an xy coordinate system. Ball A is moving upward along the y axis at 2.0 m/s, and ball B is moving to the right along the x axis with speed 3.7 m/s. After the collision, assumed elastic, ball B is moving along the positive y axis (Fig. 7–37). What is the final direction of ball A and what are their two speeds?

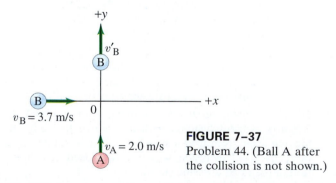

FIGURE 7–37 Problem 44. (Ball A after the collision is not shown.)

* 45. (III) A neon atom $(m = 20.0$ u) makes a perfectly elastic collision with another atom at rest. After the impact, the neon atom travels away at a $55.6°$ angle from its original direction and the unknown atom travels away at a $-50.0°$ angle. What is the mass (in u) of the unknown atom? [*Hint:* You can use the law of sines.]

7–8 Center of Mass

46. (I) Find the center of mass of the three-mass system shown in Fig. 7–38. Specify relative to the left-hand 1.00-kg mass.

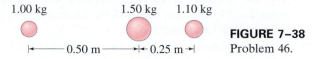

1.00 kg   1.50 kg   1.10 kg

|←——— 0.50 m ———→|← 0.25 m →|

FIGURE 7–38
Problem 46.

47. (I) The distance between a carbon atom $(m_C = 12\,u)$ and an oxygen atom $(m_O = 16\,u)$ in the CO molecule is 1.13×10^{-10} m. How far from the carbon atom is the center of mass of the molecule?

48. (I) The CM of an empty 1050-kg car is 2.50 m behind the front of the car. How far from the front of the car will the CM be when two people sit in the front seat 2.80 m from the front of the car, and three people sit in the back seat 3.90 m from the front? Assume that each person has a mass of 70.0 kg.

49. (II) A square uniform raft, 18 m by 18 m, of mass 6800 kg, is used as a ferryboat. If three cars, each of mass 1200 kg, occupy its NE, SE, and SW corners, determine the CM of the loaded ferryboat.

50. (II) Three cubes, of sides l_0, $2l_0$, and $3l_0$, are placed next to one another (in contact) with their centers along a straight line and the $l = 2l_0$ cube in the center (Fig. 7–39). What is the position, along this line, of the CM of this system? Assume the cubes are made of the same uniform material.

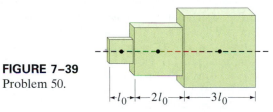

FIGURE 7–39
Problem 50.

|← l_0 →|← $2l_0$ →|← $3l_0$ →|

51. (II) A (lightweight) pallet has a load of identical cases of tomato paste (see Fig. 7–40), each of which is a cube of length l. Find the center of gravity in the horizontal plane, so that the crane operator can pick up the load without tipping it.

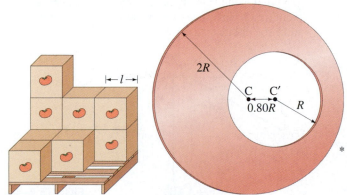

|← l →|

FIGURE 7–40 Problem 51.   **FIGURE 7–41** Problem 52.

2R
C C′
0.80R R

52. (III) A uniform circular plate of radius $2R$ has a circular hole of radius R cut out of it. The center C′ of the smaller circle is a distance $0.80R$ from the center C of the larger circle, Fig. 7–41. What is the position of the center of mass of the plate? [*Hint*: Try subtraction.]

* 7–9 CM for the Human Body

* 53. (I) Assume that your proportions are the same as those in Table 7–1, and calculate the mass of one of your legs.

* 54. (I) Determine the CM of an outstretched arm using Table 7–1.

* 55. (II) Use Table 7–1 to calculate the position of the CM of an arm bent at a right angle. Assume that the person is 155 cm tall.

* 56. (II) When a high jumper is in a position such that his arms and legs are hanging vertically, and his trunk and head are horizontal, calculate how far below the torso's median line the CM will be. Will this CM be outside the body? Use Table 7–1.

* 7–10 CM and Translational Motion

* 57. (II) The masses of the Earth and Moon are 5.98×10^{24} kg and 7.35×10^{22} kg, respectively, and their centers are separated by 3.84×10^8 m. (*a*) Where is the CM of this system located? (*b*) What can you say about the motion of the Earth–Moon system about the Sun, and of the Earth and Moon separately about the Sun?

* 58. (II) A 55-kg woman and an 80-kg man stand 10.0 m apart on frictionless ice. (*a*) How far from the woman is their CM? (*b*) If each holds one end of a rope, and the man pulls on the rope so that he moves 2.5 m, how far from the woman will he be now? (*c*) How far will the man have moved when he collides with the woman?

* 59. (II) A mallet consists of a uniform cylindrical head of mass 2.00 kg and a diameter 0.0800 m mounted on a uniform cylindrical handle of mass 0.500 kg and length 0.240 m, as shown in Fig. 7–42. If this mallet is tossed, spinning, into the air, how far above the bottom of the handle is the point that will follow a parabolic trajectory?

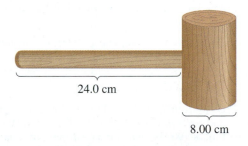

24.0 cm

8.00 cm

FIGURE 7–42 Problem 59.

* 60. (II) (*a*) Suppose that in Example 7–14 (Fig. 7–29), $m_{II} = 3m_I$. Where then would m_{II} land? (*b*) What if $m_I = 3m_{II}$?

* 61. (III) A helium balloon and its gondola, of mass M, are in the air and stationary with respect to the ground. A passenger, of mass m, then climbs out and slides down a rope with speed v, measured with respect to the balloon. With what speed and direction (relative to Earth) does the balloon then move? What happens if the passenger stops?

General Problems

62. A 0.145-kg baseball pitched horizontally at 35.0 m/s strikes a bat and is popped straight up to a height of 55.6 m. If the contact time is 1.4 ms, calculate the average force on the ball during the contact.

63. A rocket of mass m traveling with speed v_0 along the x axis suddenly shoots out fuel, equal to one-third of its mass, parallel to the y axis (perpendicular to the rocket as seen from the ground) with speed $2v_0$. Give the components of the final velocity of the rocket.

* 64. A novice pool player is faced with the corner pocket shot shown in Fig. 7–43. Relative dimensions are also shown. Should the player be worried about this being a "scratch shot," in which the cue ball will also fall into a pocket? Give details.

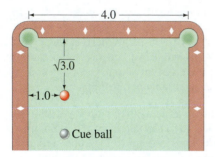

FIGURE 7–43
Problem 64.

65. A 140-kg astronaut (including space suit) acquires a speed of 2.50 m/s by pushing off with his legs from an 1800-kg space capsule. (a) What is the change in speed of the space capsule? (b) If the push lasts 0.40 s, what is the average force exerted on the astronaut by the space capsule? As the reference frame, use the position of the space capsule before the push.

66. Two astronauts, one of mass 60 kg and the other 80 kg, are initially at rest in outer space. They then push each other apart. How far apart are they when the lighter astronaut has moved 12 m?

67. A ball of mass m makes a head-on elastic collision with a second ball (at rest) and rebounds in the opposite direction with a speed equal to one-fourth its original speed. What is the mass of the second ball?

68. You have been hired as an expert witness in a court case involving an automobile accident. The accident involved car A of mass 1900 kg which crashed into stationary car B of mass 1100 kg. The driver of car A applied his brakes 15 m before he crashed into car B. After the collision, car A slid 18 m while car B slid 30 m. The coefficient of kinetic friction between the locked wheels and the road was measured to be 0.60. Show that the driver of car A was exceeding the 55-mph (90 km/h) speed limit before applying the brakes.

69. A golf ball rolls off the top of a flight of concrete steps of total vertical height 4.00 m. The ball hits four times on the way down, each time striking the horizontal part of a different step 1.0 m lower. If all collisions are perfectly elastic, what is the bounce height on the fourth bounce when the ball reaches the bottom of the stairs?

70. A bullet is fired vertically into a 1.40-kg block of wood at rest directly above it. If the bullet has a mass of 29.0 g and a speed of 510 m/s, how high will the block rise after the bullet becomes embedded in it?

71. A 25-g bullet strikes and becomes embedded in a 1.35-kg block of wood placed on a horizontal surface just in front of the gun. If the coefficient of kinetic friction between the block and the surface is 0.25, and the impact drives the block a distance of 9.5 m before it comes to rest, what was the muzzle speed of the bullet?

72. Two people, one of mass 75 kg and the other of mass 60 kg, sit in a rowboat of mass 80 kg. With the boat initially at rest, the two people, who have been sitting at opposite ends of the boat 3.2 m apart from each other, now exchange seats. How far and in what direction will the boat move?

73. A meteor whose mass was about 1.0×10^8 kg struck the Earth $(m_E = 6.0 \times 10^{24}$ kg) with a speed of about 15 km/s and came to rest in the Earth. (a) What was the Earth's recoil speed? (b) What fraction of the meteor's kinetic energy was transformed to kinetic energy of the Earth? (c) By how much did the Earth's kinetic energy change as a result of this collision?

74. An object at rest is suddenly broken apart into two fragments by an explosion. One fragment acquires twice the kinetic energy of the other. What is the ratio of their masses?

75. The force on a bullet is given by the formula $F = 580 - (1.8 \times 10^5)t$ over the time interval $t = 0$ to $t = 3.0 \times 10^{-3}$ s. In this formula, t is in seconds and F is in newtons. (a) Plot a graph of F vs. t for $t = 0$ to $t = 3.0$ ms. (b) Estimate, using graphical methods, the impulse given the bullet. (c) If the bullet achieves a speed of 220 m/s as a result of this impulse, given to it in the barrel of a gun, what must its mass be?

76. Two balls, of masses $m_A = 40$ g and $m_B = 60$ g, are suspended as shown in Fig. 7–44. The lighter ball is pulled away to a 60° angle with the vertical and released. (a) What is the velocity of the lighter ball before impact? (b) What is the velocity of each ball after the elastic collision? (c) What will be the maximum height of each ball after the elastic collision?

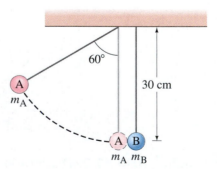

FIGURE 7–44
Problem 76.

77. An atomic nucleus at rest decays radioactively into an alpha particle and a smaller nucleus. What will be the speed of this recoiling nucleus if the speed of the alpha particle is 3.8×10^5 m/s? Assume the recoiling nucleus has a mass 57 times greater than that of the alpha particle.

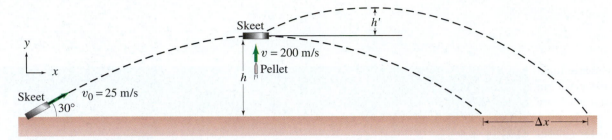

FIGURE 7–45 Problem 78.

78. A 0.25-kg skeet (clay target) is fired at an angle of 30° to the horizon with a speed of 25 m/s (Fig. 7–45). When it reaches the maximum height, it is hit from below by a 15-g pellet traveling vertically upward at a speed of 200 m/s. The pellet is embedded in the skeet. (*a*) How much higher did the skeet go up? (*b*) How much extra distance, Δx, does the skeet travel because of the collision?

79. A block of mass $m = 2.20$ kg slides down a 30.0° incline which is 3.60 m high. At the bottom, it strikes a block of mass $M = 7.00$ kg which is at rest on a horizontal surface, Fig. 7–46. (Assume a smooth transition at the bottom of the incline.) If the collision is elastic, and friction can be ignored, determine (*a*) the speeds of the two blocks after the collision, and (*b*) how far back up the incline the smaller mass will go.

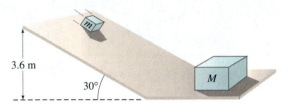

FIGURE 7–46 Problems 79 and 80.

80. In Problem 79 (Fig. 7–46), what is the upper limit on mass m if it is to rebound from M, slide up the incline, stop, slide down the incline, and collide with M again?

81. *The gravitational slingshot effect.* Figure 7–47 shows the planet Saturn moving in the negative x direction at its orbital speed (with respect to the Sun) of 9.6 km/s. The mass of Saturn is 5.69×10^{26} kg. A spacecraft with mass 825 kg approaches Saturn. When far from Saturn, it moves in the $+x$ direction at 10.4 km/s. The gravitational attraction of Saturn (a conservative force) acting on the spacecraft causes it to swing around the planet (orbit shown as dashed line) and head off in the opposite direction. Estimate the final speed of the spacecraft after it is far enough away to be considered free of Saturn's gravitational pull.

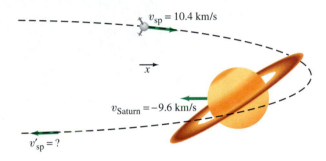

FIGURE 7–47 Problem 81.

Answers to Exercises

A: Yes, if the sports car's speed is three times greater.
B: Larger.
C: (*a*) 6.0 m/s; (*b*) almost zero; (*c*) almost 24.0 m/s.
D: The curve would be wider and less high.

E: Yes, by 300 times.
F: Yes, KE was conserved.
G: $x_{CM} = -2.0$ m; yes.
H: The boat moves in the opposite direction.

You too can experience rapid rotation—if your stomach can take the high angular velocity and centripetal acceleration of some of the faster amusement park rides. If not, try the slower merry-go-round or Ferris wheel. Rotating carnival rides have rotational KE as well as angular momentum.

8

Rotational Motion

ntil now, we have been concerned mainly with translational motion. We discussed the kinematics and dynamics of translational motion (the role of force), and the energy and momentum associated with it. In this Chapter we will deal with rotational motion. We will discuss the kinematics of rotational motion and then its dynamics (involving torque), as well as rotational kinetic energy and angular momentum (the rotational analog of linear momentum). We will find many analogies with translational motion, which will make our study easier. Our understanding of the world around us will be increased significantly—from rotating bicycle wheels and compact disks to amusement park rides, a spinning skater, the rotating Earth, and a centrifuge— and there may be a few surprises.

We will consider mainly the rotation of rigid objects. A **rigid object** is an object with a definite shape that doesn't change, so that the particles composing it stay in fixed positions relative to one another. Any real object is capable of vibrating or deforming when a force is exerted on it. But these effects are often very small, so the concept of an ideal rigid object is very useful as a good approximation.

8–1 Angular Quantities

We saw in Chapter 7 (Section 7–8) that the motion of a rigid object can be analyzed as the translational motion of the object's center of mass, plus rotational motion *about* its center of mass. We have already discussed translational motion in detail, so now we focus on purely rotational motion. By *purely rotational motion*, we mean that all points in the object move in circles, such as the point P in the rotating wheel of Fig. 8–1, and that the centers of these circles all lie on a line called the **axis of rotation**. In Fig. 8–1 the axis of rotation is perpendicular to the page and passes through point O.

Every point in an object rotating about a fixed axis moves in a circle (shown dashed in Fig. 8–1 for point P) whose center is on the axis and whose radius is r, the distance of that point from the axis of rotation. A straight line drawn from the axis to any point sweeps out the same angle θ in the same time.

To indicate the angular position of a rotating object, or how far it has rotated, we specify the angle θ of some particular line in the object (red in Fig. 8–1) with respect to a reference line, such as the x axis in Fig. 8–1. A point in the object, such as P in Fig. 8–1, moves through an angle θ when it travels the distance l measured along the circumference of its circular path. Angles are commonly measured in degrees, but the mathematics of circular motion is much simpler if we use the *radian* for angular measure. One **radian** (abbreviated rad) is defined as the angle subtended by an arc whose length is equal to the radius. For example, in Fig. 8–1b, point P is a distance r from the axis of rotation, and it has moved a distance l along the arc of a circle. The arc length l is said to "subtend" the angle θ. If $l = r$, then θ is exactly equal to 1 rad. In radians, any angle θ is given by

$$\theta = \frac{l}{r},\qquad\text{(8–1a)}$$

θ in radians

where r is the radius of the circle, and l is the arc length subtended by the angle specified in radians. If $l = r$, then $\theta = 1\,\text{rad}$.

1 rad: arc length = radius

The radian is dimensionless since it is the ratio of two lengths. Nonetheless when giving an angle in radians, we always mention rad to remind us it is not degrees. It is often useful to rewrite Eq. 8–1a in terms of arc length l:

$$l = r\theta.\qquad\text{(8–1b)}$$

Radians can be related to degrees in the following way. In a complete circle there are 360°, which must correspond to an arc length equal to the circumference of the circle, $l = 2\pi r$. Thus $\theta = l/r = 2\pi r/r = 2\pi\,\text{rad}$ in a complete circle, so

$$360° = 2\pi\,\text{rad}.$$

Conversion, degrees to rad

One radian is therefore $360°/2\pi \approx 360°/6.28 \approx 57.3°$. An object that makes one complete revolution (rev) has rotated through 360°, or 2π radians:

1 rad $\approx$ 57.3°

$$1\,\text{rev} = 360° = 2\pi\,\text{rad}.$$

EXAMPLE 8–1 Bike wheel. A bike wheel rotates 4.50 revolutions. How many radians has it rotated?

APPROACH All we need is a straightforward conversion of units using

$$1\,\text{revolution} = 360° = 2\pi\,\text{rad} = 6.28\,\text{rad}.$$

SOLUTION

$$4.50\,\text{revolutions} = (4.50\,\text{rev})\left(2\pi\,\frac{\text{rad}}{\text{rev}}\right) = 9.00\pi\,\text{rad} = 28.3\,\text{rad}.$$

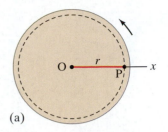

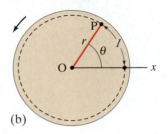

FIGURE 8–1 Looking at a wheel that is rotating counterclockwise about an axis through the wheel's center at O (axis perpendicular to the page). Each point, such as point P, moves in a circular path; l is the distance P travels as the wheel rotates through the angle θ.

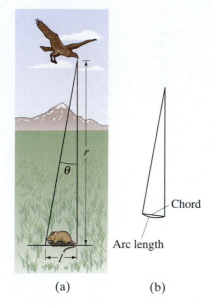

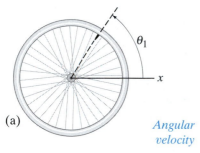

(a)   (b)

FIGURE 8–2 (a) Example 8–2.
(b) For small angles, arc length and
the chord length (straight line) are
nearly equal.

Chord

Arc length

EXAMPLE 8–2 **Birds of prey—in radians.** A particular bird's eye can just distinguish objects that subtend an angle no smaller than about 3×10^{-4} rad. (a) How many degrees is this? (b) How small an object can the bird just distinguish when flying at a height of 100 m (Fig. 8–2a)?

APPROACH For (a) we use the relation $360° = 2\pi$ rad. For (b) we use Eq. 8–1b, $l = r\theta$, to find the arc length.

SOLUTION (a) We convert 3×10^{-4} rad to degrees:

$$(3 \times 10^{-4}\,\text{rad})\left(\frac{360°}{2\pi\,\text{rad}}\right) = 0.017°.$$

(b) We use Eq. 8–1b, $l = r\theta$. For small angles, the arc length l and the chord length are approximately[†] the same (Fig. 8–2b). Since $r = 100$ m and $\theta = 3 \times 10^{-4}$ rad, we find

$$l = (100\,\text{m})(3 \times 10^{-4}\,\text{rad}) = 3 \times 10^{-2}\,\text{m} = 3\,\text{cm}.$$

A bird can distinguish a small mouse (about 3 cm long) from a height of 100 m. That is good eyesight.

NOTE Had the angle been given in degrees, we would first have had to convert it to radians to make this calculation. Equation 8–1 is valid *only* if the angle is specified in radians. Degrees (or revolutions) won't work.

To describe rotational motion, we make use of angular quantities, such as angular velocity and angular acceleration. These are defined in analogy to the corresponding quantities in linear motion, and are chosen to describe the rotating object as a whole, so they are the same for each point in the rotating object. Each point in a rotating object may also have translational velocity and acceleration, but they have different values for different points in the object.

When an object, such as the bicycle wheel in Fig. 8–3, rotates from some initial position, specified by θ_1, to some final position, θ_2, its *angular displacement* is

Angular displacement (rad)

$$\Delta\theta = \theta_2 - \theta_1.$$

FIGURE 8–3 A wheel rotates from (a) initial position θ_1 to (b) final position θ_2. The angular displacement is $\Delta\theta = \theta_2 - \theta_1$.

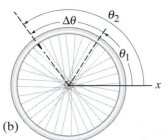

(a)

(b)

The *angular velocity* (denoted by ω, the Greek lowercase letter omega) is defined in analogy with linear (translational) velocity that was discussed in Chapter 2. Instead of linear displacement, we use the angular displacement. Thus the **average angular velocity** is defined as

Angular velocity

$$\bar{\omega} = \frac{\Delta\theta}{\Delta t}, \tag{8–2a}$$

where $\Delta\theta$ is the angle through which the object has rotated in the time interval Δt. We define the **instantaneous angular velocity** as the very small angle $\Delta\theta$, through which the object turns in the very short time interval Δt:

$$\omega = \lim_{\Delta t \to 0} \frac{\Delta\theta}{\Delta t}. \tag{8–2b}$$

Angular velocity is generally specified in radians per second (rad/s). Note that *all points in a rigid object rotate with the same angular velocity*, since every position in the object moves through the same angle in the same time interval.

An object such as the wheel in Fig. 8–3 can rotate about a fixed axis either clockwise or counterclockwise. The direction can be specified with a + or − sign, just as we did in Chapter 2 for linear motion along the +x or −x axis. The usual convention is to choose the angular displacement $\Delta\theta$ and angular velocity ω as positive when the wheel rotates counterclockwise. If the rotation is clockwise, then θ would decrease, so $\Delta\theta$ and ω would be negative.[‡]

[†] Even for an angle as large as 15°, the error in making this estimate is only 1%, but for larger angles the error increases rapidly.
[‡] The vector nature of angular velocity and other angular quantities is discussed in Section 8–9 (optional).

Angular acceleration (denoted by α, the Greek lowercase letter alpha), in analogy to linear acceleration, is defined as the change in angular velocity divided by the time required to make this change. The **average angular acceleration** is defined as

$$\bar{\alpha} = \frac{\omega_2 - \omega_1}{\Delta t} = \frac{\Delta \omega}{\Delta t},$$

(8–3a)

where ω_1 is the angular velocity initially, and ω_2 is the angular velocity after a time interval Δt. **Instantaneous angular acceleration** is defined in the usual way as the limit of this ratio as Δt approaches zero:

$$\alpha = \lim_{\Delta t \to 0} \frac{\Delta \omega}{\Delta t}.$$

(8–3b)   *Angular acceleration*

Since ω is the same for all points of a rotating object, Eq. 8–3 tells us that α also will be the same for all points. Thus, ω and α are properties of the rotating object as a whole. With ω measured in radians per second and t in seconds, α will be expressed as radians per second squared (rad/s^2).

Each point or particle of a rotating rigid object has, at any moment, a linear velocity v and a linear acceleration a. We can relate the linear quantities at each point, v and a, to the angular quantities of the rotating object, ω and α. Consider a point P located a distance r from the axis of rotation, as in Fig. 8–4. If the object rotates with angular velocity ω, any point will have a linear velocity whose direction is tangent to its circular path. The magnitude of that point's linear velocity is $v = \Delta l / \Delta t$. From Eq. 8–1b, a change in rotation angle $\Delta \theta$ (in radians) is related to the linear distance traveled by $\Delta l = r \Delta \theta$. Hence

$$v = \frac{\Delta l}{\Delta t} = r \frac{\Delta \theta}{\Delta t}$$

or

$$v = r\omega.$$

(8–4)   *Linear and angular velocity related*

Thus, although ω is the same for every point in the rotating object at any instant, the linear velocity v is greater for points farther from the axis (Fig. 8–5). Note that Eq. 8–4 is valid both instantaneously and on the average.

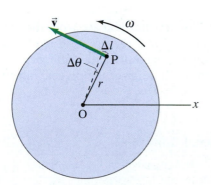

FIGURE 8–4 A point P on a rotating wheel has a linear velocity $\vec{v}$ at any moment.

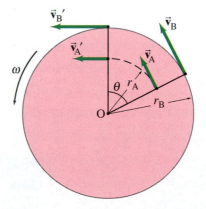

FIGURE 8–5 A wheel rotating uniformly counterclockwise. Two points on the wheel, at distances r_A and r_B from the center, have the same angular velocity ω because they travel through the same angle θ in the same time interval. But the two points have different linear velocities because they travel different distances in the same time interval. Since $r_B > r_A$, then $v_B > v_A$ $(v = r\omega)$.

CONCEPTUAL EXAMPLE 8–3 **Is the lion faster than the horse?** On a rotating carousel or merry-go-round, one child sits on a horse near the outer edge and another child sits on a lion halfway out from the center. (a) Which child has the greater linear velocity? (b) Which child has the greater angular velocity?

RESPONSE (a) The *linear* velocity is the distance traveled divided by the time interval. In one rotation the child on the outer edge travels a longer distance than the child near the center, but the time interval is the same for both. Thus the child at the outer edge, on the horse, has the greater linear velocity.
(b) The *angular* velocity is the angle of rotation divided by the time interval. In one rotation both children rotate through the same angle (360° = 2π radians). The two children have the same angular velocity.

If the angular velocity of a rotating object changes, the object as a whole—and each point in it—has an angular acceleration. Each point also has a linear acceleration whose direction is tangent to that point's circular path. We use Eq. 8–4 ($v = r\omega$) to show that the angular acceleration α is related to the tangential linear acceleration a_{tan} of a point in the rotating object by

$$a_{tan} = \frac{\Delta v}{\Delta t} = r\frac{\Delta\omega}{\Delta t}$$

or

Tangential acceleration

$$a_{tan} = r\alpha. \tag{8–5}$$

In this equation, r is the radius of the circle in which the particle is moving, and the subscript "tan" in a_{tan} stands for "tangential."

The total linear acceleration of a point is the vector sum of two components:

$$\vec{a} = \vec{a}_{tan} + \vec{a}_R,$$

where the radial[†] component, $\vec{a}_R$, is the radial or "centripetal" acceleration and its direction is toward the center of the point's circular path; see Fig. 8–6. We saw in Chapter 5 (Eq. 5–1) that $a_R = v^2/r$, and we can rewrite this in terms of ω using Eq. 8–4:

Centripetal (or radial) acceleration

$$a_R = \frac{v^2}{r} = \frac{(r\omega)^2}{r} = \omega^2 r. \tag{8–6}$$

Thus the centripetal acceleration is greater the farther you are from the axis of rotation: the children farthest out on a carousel feel the greatest acceleration. Equations 8–4, 8–5, and 8–6 relate the angular quantities describing the rotation of an object to the linear quantities for each point of the object. Table 8–1 summarizes these relationships.

FIGURE 8–6 On a rotating wheel whose angular speed is increasing, a point P has both tangential and radial (centripetal) components of linear acceleration. (See also Chapter 5.)

TABLE 8–1 Linear and Rotational Quantities

Linear	Type	Rotational	Relation
x	displacement	θ	$x = r\theta$
v	velocity	ω	$v = r\omega$
a_{tan}	acceleration	α	$a_{tan} = r\alpha$

[†]"Radial" means along the radius—that is, toward or away from the center or axis.

EXAMPLE 8–4 **Angular and linear velocities and accelerations.** A carousel is initially at rest. At $t = 0$ it is given a constant angular acceleration $\alpha = 0.060 \text{ rad/s}^2$, which increases its angular velocity for 8.0 s. At $t = 8.0$ s, determine the following quantities: (*a*) the angular velocity of the carousel; (*b*) the linear velocity of a child (Fig. 8–7a) located 2.5 m from the center, point P in Fig. 8–7b; (*c*) the tangential (linear) acceleration of that child; (*d*) the centripetal acceleration of the child; and (*e*) the total linear acceleration of the child.

APPROACH The angular acceleration α is constant, so we can use Eq. 8–3a to solve for ω after a time $t = 8.0$ s. With this ω and the given α, we determine the other quantities using the relations we just developed, Eqs. 8–4, 8–5, and 8–6.

SOLUTION (*a*) Equation 8–3a tells us

$$\bar{\alpha} = \frac{\omega_2 - \omega_1}{\Delta t}.$$

We are given $\Delta t = 8.0$ s, $\bar{\alpha} = 0.060 \text{ rad/s}^2$, and $\omega_1 = 0$. Solving for ω_2, we get

$$\omega_2 = \omega_1 + \bar{\alpha}\,\Delta t$$
$$= 0 + (0.060 \text{ rad/s}^2)(8.0 \text{ s}) = 0.48 \text{ rad/s}.$$

During the 8.0-s interval, the carousel has accelerated from $\omega_1 = 0$ (rest) to $\omega_2 = 0.48 \text{ rad/s}$.
(*b*) The linear velocity of the child with $r = 2.5$ m at time $t = 8.0$ s is found using Eq. 8–4:

$$v = r\omega = (2.5 \text{ m})(0.48 \text{ rad/s}) = 1.2 \text{ m/s}.$$

Note that the "rad" has been dropped here because it is dimensionless (and only a reminder)—it is a ratio of two distances, Eq. 8–1b.
(*c*) The child's tangential acceleration is given by Eq. 8–5:

$$a_{\text{tan}} = r\alpha = (2.5 \text{ m})(0.060 \text{ rad/s}^2) = 0.15 \text{ m/s}^2,$$

and it is the same throughout the 8.0-s acceleration interval.
(*d*) The child's centripetal acceleration at $t = 8.0$ s is given by Eq. 8–6:

$$a_{\text{R}} = \frac{v^2}{r} = \frac{(1.2 \text{ m/s})^2}{(2.5 \text{ m})} = 0.58 \text{ m/s}^2.$$

(*e*) The two components of linear acceleration calculated in parts (*c*) and (*d*) are perpendicular to each other. Thus the total linear acceleration at $t = 8.0$ s has magnitude

$$a = \sqrt{a_{\text{tan}}^2 + a_{\text{R}}^2}$$
$$= \sqrt{(0.15 \text{ m/s}^2)^2 + (0.58 \text{ m/s}^2)^2} = 0.60 \text{ m/s}^2.$$

Its direction (Fig. 8–7b) is

$$\theta = \tan^{-1}\left(\frac{a_{\text{tan}}}{a_{\text{R}}}\right) = \tan^{-1}\left(\frac{0.15 \text{ m/s}^2}{0.58 \text{ m/s}^2}\right) = 0.25 \text{ rad},$$

so $\theta \approx 15°$.

NOTE The linear acceleration is mostly centripetal, keeping the child moving in a circle with the carousel. The tangential component that speeds up the motion is smaller.

(a)

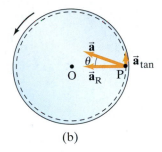

(b)

FIGURE 8–7 Example 8–4. The total acceleration vector $\vec{a} = \vec{a}_{\text{tan}} + \vec{a}_{\text{R}}$, at $t = 8.0$ s.

We can relate the angular velocity ω to the frequency of rotation, f. The **frequency** is the number of complete revolutions (rev) per second, as we saw in Chapter 5. One revolution (of a wheel, say) corresponds to an angle of 2π radians, and thus $1\,\text{rev/s} = 2\pi\,\text{rad/s}$. Hence, in general, the frequency f is related to the angular velocity ω by

Frequency

$$f = \frac{\omega}{2\pi}$$

or

$$\omega = 2\pi f. \tag{8–7}$$

The unit for frequency, revolutions per second (rev/s), is given the special name the hertz (Hz). That is

$$1\,\text{Hz} = 1\,\text{rev/s}.$$

Note that "revolution" is not really a unit, so we can also write $1\,\text{Hz} = 1\,\text{s}^{-1}$.

The time required for one complete revolution is called the **period** T, and it is related to the frequency by

Period

$$T = \frac{1}{f}. \tag{8–8}$$

If a particle rotates at a frequency of three revolutions per second, then the period of each revolution is $\frac{1}{3}\,\text{s}$.

EXERCISE A In Example 8–4, we found that the carousel, after 8.0 s, rotates at an angular velocity $\omega = 0.48\,\text{rad/s}$, and continues to do so after $t = 8.0\,\text{s}$ because the acceleration ceased. What are the frequency and period of the carousel?

Hard drive and bit speed

EXAMPLE 8–5 Hard drive. The platter of the hard drive of a computer rotates at 7200 rpm (revolutions per minute = rev/min). (*a*) What is the angular velocity of the platter? (*b*) If the reading head of the drive is located 3.00 cm from the rotation axis, what is the linear speed of the point on the platter just below it? (*c*) If a single bit requires 0.50 μm of length along the direction of motion, how many bits per second can the writing head write when it is 3.00 cm from the axis?

APPROACH We use the given frequency f to find the angular velocity ω of the platter and then the linear speed of a point on the platter ($v = r\omega$). The bit rate is found by dividing the linear speed by the length of one bit ($v = \text{distance/time}$).

SOLUTION (*a*) First we find the frequency in rev/s, given $f = 7200\,\text{rev/min}$:

$$f = \frac{(7200\,\text{rev/min})}{(60\,\text{s/min})} = 120\,\text{rev/s} = 120\,\text{Hz}.$$

Then the angular velocity is

$$\omega = 2\pi f = 754\,\text{rad/s}.$$

(*b*) The linear speed of a point 3.00 cm out from the axis is given by Eq. 8–4:

$$v = r\omega = (3.00 \times 10^{-2}\,\text{m})(754\,\text{rad/s}) = 22.6\,\text{m/s}.$$

(*c*) Each bit requires $0.50 \times 10^{-6}\,\text{m}$, so at a speed of 22.6 m/s, the number of bits passing the head per second is

$$\frac{22.6\,\text{m/s}}{0.50 \times 10^{-6}\,\text{m/bit}} = 45 \times 10^{6}\,\text{bits per second},$$

or 45 megabits/s (Mbps).

200 CHAPTER 8 Rotational Motion

8-2 Constant Angular Acceleration

In Chapter 2, we derived the useful kinematic equations (Eqs. 2–11) that relate acceleration, velocity, distance, and time for the special case of uniform linear acceleration. Those equations were derived from the definitions of linear velocity and acceleration, assuming constant acceleration. The definitions of angular velocity and angular acceleration are the same as those for their linear counterparts, except that θ has replaced the linear displacement x, ω has replaced v, and α has replaced a. Therefore, the angular equations for **constant angular acceleration** will be analogous to Eqs. 2–11 with x replaced by θ, v by ω, and a by α, and they can be derived in exactly the same way. We summarize them here, opposite their linear equivalents (we've chosen $x_0 = 0$, and $\theta_0 = 0$ at the initial time $t = 0$):

Angular	Linear			
$\omega = \omega_0 + \alpha t$	$v = v_0 + at$	[constant α, a]	**(8–9a)**	*Kinematic equations*
$\theta = \omega_0 t + \frac{1}{2}\alpha t^2$	$x = v_0 t + \frac{1}{2}at^2$	[constant α, a]	**(8–9b)**	*for constant*
$\omega^2 = \omega_0^2 + 2\alpha\theta$	$v^2 = v_0^2 + 2ax$	[constant α, a]	**(8–9c)**	*angular acceleration*
$\bar{\omega} = \dfrac{\omega + \omega_0}{2}$	$\bar{v} = \dfrac{v + v_0}{2}$	[constant α, a]	**(8–9d)**	$(x_0 = 0, \theta_0 = 0)$

Note that ω_0 represents the angular velocity at $t = 0$, whereas θ and ω represent the angular position and velocity, respectively, at time t. Since the angular acceleration is constant, $\alpha = \bar{\alpha}$.

EXAMPLE 8–6 **Centrifuge acceleration.** A centrifuge rotor is accelerated from rest to 20,000 rpm in 30 s. (a) What is its average angular acceleration? (b) Through how many revolutions has the centrifuge rotor turned during its acceleration period, assuming constant angular acceleration?

APPROACH To determine $\bar{\alpha} = \Delta\omega/\Delta t$, we need the initial and final angular velocities. For (b), we use Eqs. 8–9 (recall that one revolution corresponds to $\theta = 2\pi$ rad).

SOLUTION (a) The initial angular velocity is $\omega = 0$. The final angular velocity is

$$\omega = 2\pi f = (2\pi\,\text{rad/rev})\frac{(20{,}000\,\text{rev/min})}{(60\,\text{s/min})} = 2100\,\text{rad/s}.$$

Then, since $\bar{\alpha} = \Delta\omega/\Delta t$ and $\Delta t = 30$ s, we have

$$\bar{\alpha} = \frac{\omega - \omega_0}{\Delta t} = \frac{2100\,\text{rad/s} - 0}{30\,\text{s}} = 70\,\text{rad/s}^2.$$

That is, every second the rotor's angular velocity increases by 70 rad/s, or by $(70/2\pi) = 11$ revolutions per second.

(b) To find θ we could use either Eq. 8–9b or 8–9c, or both to check our answer. The former gives

$$\theta = 0 + \tfrac{1}{2}(70\,\text{rad/s}^2)(30\,\text{s})^2 = 3.15 \times 10^4\,\text{rad},$$

where we have kept an extra digit because this is an intermediate result. To find the total number of revolutions, we divide by 2π rad/rev and obtain

$$\frac{3.15 \times 10^4\,\text{rad}}{2\pi\,\text{rad/rev}} = 5.0 \times 10^3\,\text{rev}.$$

NOTE Let us calculate θ using Eq. 8–9c:

$$\theta = \frac{\omega^2 - \omega_0^2}{2\alpha} = \frac{(2100\,\text{rad/s})^2 - 0}{2(70\,\text{rad/s}^2)} = 3.15 \times 10^4\,\text{rad}$$

which checks our answer perfectly.

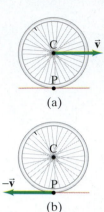

FIGURE 8–8 (a) A wheel rolling to the right. Its center C moves with velocity $\vec{v}$. Point P is at rest at this instant. (b) The same wheel as seen from a reference frame in which the axle of the wheel C is at rest—that is, we are moving to the right with velocity $\vec{v}$ relative to the ground. Point P, which was at rest in (a), here in (b) is moving to the left with velocity $-\vec{v}$ as shown. (See also Section 3–8 on relative velocity.)

8–3 Rolling Motion (Without Slipping)

The rolling motion of a ball or wheel is familiar in everyday life: a ball rolling across the floor, or the wheels and tires of a car or bicycle rolling along the pavement. Rolling *without slipping* is readily analyzed and depends on static friction between the rolling object and the ground. The friction is static because the rolling object's point of contact with the ground is at rest at each moment.

Rolling without slipping involves both rotation and translation. There is then a simple relation between the linear speed v of the axle and the angular velocity ω of the rotating wheel or sphere: namely, $v = r\omega$ (where r is the radius) as we now show. Figure 8–8a shows a wheel rolling to the right without slipping. At the moment shown, point P on the wheel is in contact with the ground and is momentarily at rest. The velocity of the axle at the wheel's center C is $\vec{v}$. In Fig. 8–8b we have put ourselves in the reference frame of the wheel—that is, we are moving to the right with velocity $\vec{v}$ relative to the ground. In this reference frame the axle C is at rest, whereas the ground and point P are moving to the left with velocity $-\vec{v}$ as shown. Here we are seeing pure rotation. So we can use Eq. 8–4 to obtain $v = r\omega$, where r is the radius of the wheel. This is the same v as in Fig. 8–8a, so we see that the linear speed v of the axle relative to the ground is related to the angular velocity ω by

$$v = r\omega. \qquad \text{[rolling without slipping]}$$

This relationship is valid only if there is no slipping.

EXAMPLE 8–7 **Bicycle.** A bicycle slows down uniformly from $v_0 = 8.40\,\text{m/s}$ to rest over a distance of 115 m, Fig. 8–9. Each wheel and tire has an overall diameter of 68.0 cm. Determine (a) the angular velocity of the wheels at the initial instant ($t = 0$); (b) the total number of revolutions each wheel rotates before coming to rest; (c) the angular acceleration of the wheel; and (d) the time it took to come to a stop.

APPROACH We assume the bicycle wheels roll without slipping and the tire is in firm contact with the ground. The speed of the bike v and the angular velocity of the wheels ω are related by $v = r\omega$. The bike slows down uniformly, so the angular acceleration is constant and we can use Eqs. 8–9.

SOLUTION (a) The initial angular velocity of the wheel, whose radius is 34.0 cm, is

$$\omega_0 = \frac{v_0}{r} = \frac{8.40\,\text{m/s}}{0.340\,\text{m}} = 24.7\,\text{rad/s}.$$

(b) In coming to a stop, the bike passes over 115 m of ground. The circumference of the wheel is $2\pi r$, so each revolution of the wheel corresponds to a distance traveled of $2\pi r = (2\pi)(0.340\,\text{m})$. Thus the number of revolutions the wheel makes in coming to a stop is

$$\frac{115\,\text{m}}{2\pi r} = \frac{115\,\text{m}}{(2\pi)(0.340\,\text{m})} = 53.8\,\text{rev}.$$

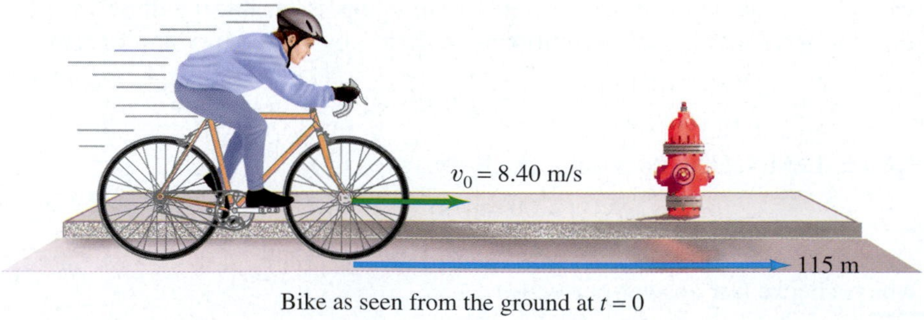

FIGURE 8–9 Example 8–7.

$v_0 = 8.40\,\text{m/s}$

115 m

Bike as seen from the ground at $t = 0$

(c) The angular acceleration of the wheel can be obtained from Eq. 8–9c, for which we set $\omega = 0$ and $\omega_0 = 24.7 \, \text{rad/s}$. Because each revolution corresponds to 2π radians of angle, then $\theta = 2\pi \, \text{rad/rev} \times 53.8 \, \text{rev} \, (= 338 \, \text{rad})$ and

$$\alpha = \frac{\omega^2 - \omega_0^2}{2\theta} = \frac{0 - (24.7 \, \text{rad/s})^2}{2(2\pi \, \text{rad/rev})(53.8 \, \text{rev})} = -0.902 \, \text{rad/s}^2.$$

(d) Equation 8–9a or b allows us to solve for the time. The first is easier:

$$t = \frac{\omega - \omega_0}{\alpha} = \frac{0 - 24.7 \, \text{rad/s}}{-0.902 \, \text{rad/s}^2} = 27.4 \, \text{s}.$$

NOTE When the bike tire completes one revolution, the bike advances linearly a distance equal to the outer circumference ($2\pi r$) of the tire, as long as there is no slipping or sliding.

8–4 Torque

We have so far discussed rotational kinematics—the description of rotational motion in terms of angle, angular velocity, and angular acceleration. Now we discuss the dynamics, or causes, of rotational motion. Just as we found analogies between linear and rotational motion for the description of motion, so rotational equivalents for dynamics exist as well.

To make an object start rotating about an axis clearly requires a force. But the direction of this force, and where it is applied, are also important. Take, for example, an ordinary situation such as the overhead view of the door in Fig. 8–10. If you apply a force $\vec{F}_A$ to the door as shown, you will find that the greater the magnitude, F_A, the more quickly the door opens. But now if you apply the same magnitude force at a point closer to the hinge—say, $\vec{F}_B$ in Fig. 8–10—the door will not open so quickly. The effect of the force is less: where the force acts, as well as its magnitude and direction, affects how quickly the door opens. Indeed, if only this one force acts, the angular acceleration of the door is proportional not only to the magnitude of the force, but is also directly proportional to *the perpendicular distance from the axis of rotation to the line along which the force acts*. This distance is called the **lever arm**, or **moment arm**, of the force, and is labeled r_A and r_B for the two forces in Fig. 8–10. Thus, if r_A in Fig. 8–10 is three times larger than r_B, then the angular acceleration of the door will be three times as great, assuming that the magnitudes of the forces are the same. To say it another way, if $r_A = 3r_B$, then F_B must be three times as large as F_A to give the same angular acceleration. (Figure 8–11 shows two examples of tools whose long lever arms are very effective.)

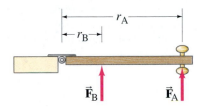

FIGURE 8–10 Applying the same force with different lever arms, r_A and r_B. If $r_A = 3r_B$, then to create the same effect (angular acceleration), F_B needs to be three times F_A, or $F_A = \frac{1}{3}F_B$.

Lever arm

FIGURE 8–11 (a) A plumber can exert greater torque using a wrench with a long lever arm. (b) A tire iron too can have a long lever arm.

(a) (b)

The angular acceleration, then, is proportional to the product of the *force times the lever arm*. This product is called the *moment of the force* about the axis, or, more commonly, it is called the **torque**, and is represented by τ (Greek lowercase letter tau). Thus, the angular acceleration α of an object is directly proportional to the net applied torque τ:

$$\alpha \propto \tau,$$

and we see that it is torque that gives rise to angular acceleration. This is the rotational analog of Newton's second law for linear motion, $a \propto F$.

We defined the lever arm as the *perpendicular* distance from the axis of rotation to the line of action of the force—that is, the distance which is perpendicular both to the axis of rotation and to an imaginary line drawn along the direction of the force. We do this to take into account the effect of forces acting at an angle. It is clear that a force applied at an angle, such as $\vec{\mathbf{F}}_C$ in Fig. 8–12, will be less effective than the same magnitude force applied perpendicular to the door, such as $\vec{\mathbf{F}}_A$ (Fig. 8–12a). And if you push on the end of the door so that the force is directed at the hinge (the axis of rotation), as indicated by $\vec{\mathbf{F}}_D$, the door will not rotate at all.

The lever arm for a force such as $\vec{\mathbf{F}}_C$ is found by drawing a line along the direction of $\vec{\mathbf{F}}_C$ (this is the "line of action" of $\vec{\mathbf{F}}_C$). Then we draw another line, perpendicular to this line of action, that goes to the axis of rotation and is perpendicular also to it. The length of this second line is the lever arm for $\vec{\mathbf{F}}_C$ and is labeled r_C in Fig. 8–12b. The lever arm is perpendicular both to the line of action of the force and, at its other end, perpendicular to the rotation axis.

The magnitude of the torque associated with $\vec{\mathbf{F}}_C$ is then $r_C F_C$. This short lever arm r_C and the corresponding smaller torque associated with $\vec{\mathbf{F}}_C$ is consistent with the observation that $\vec{\mathbf{F}}_C$ is less effective in accelerating the door than is $\vec{\mathbf{F}}_A$. When the lever arm is defined in this way, experiment shows that the relation $\alpha \propto \tau$ is valid in general. Notice in Fig. 8–12 that the line of action of the force $\vec{\mathbf{F}}_D$ passes through the hinge, and hence its lever arm is zero. Consequently, zero torque is associated with $\vec{\mathbf{F}}_D$ and it gives rise to no angular acceleration, in accord with everyday experience.

In general, then, we can write the magnitude of the torque about a given axis as

$$\tau = r_\perp F, \tag{8–10a}$$

where $r_\perp$ is the lever arm, and the perpendicular symbol ($\perp$) reminds us that we must use the distance from the axis of rotation that is perpendicular to the line of action of the force (Fig. 8–13a).

An equivalent way of determining the torque associated with a force is to resolve the force into components parallel and perpendicular to the line that connects the axis to the point of application of the force, as shown in Fig. 8–13b. The component $F_\parallel$ exerts no torque since it is directed at the rotation axis (its moment arm is zero). Hence the torque will be equal to $F_\perp$ times the distance r from the axis to the point of application of the force:

$$\tau = rF_\perp. \tag{8–10b}$$

That this gives the same result as Eq. 8–10a can be seen from the relations $F_\perp = F\sin\theta$ and $r_\perp = r\sin\theta$. [Note that θ is the angle between the directions of $\vec{\mathbf{F}}$ and r (radial line from the axis to the point where $\vec{\mathbf{F}}$ acts)]. So

$$\tau = rF\sin\theta \tag{8–10c}$$

in either case. We can use any of Eqs. 8–10 to calculate the torque, whichever is easiest.

Since torque is a distance times a force, it is measured in units of m·N in SI units,[†] cm·dyne in the cgs system, and ft·lb in the English system.

[†]Note that the units for torque are the same as those for energy. We write the unit for torque here as m·N (in SI) to distinguish it from energy (N·m) because the two quantities are very different. An obvious difference is that energy is a scalar, whereas torque has a direction and is a vector. The special name *joule* (1 J = 1 N·m) is used only for energy (and for work), *never* for torque.

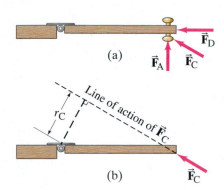

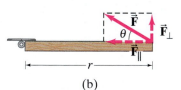

FIGURE 8–12 (a) Forces acting at different angles at the doorknob. (b) The lever arm is defined as the perpendicular distance from the axis of rotation (the hinge) to the line of action of the force.

FIGURE 8–13
Torque = $r_\perp F = rF_\perp$.

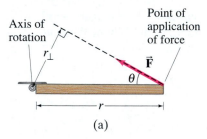

EXAMPLE 8–8 Biceps torque.
The biceps muscle exerts a vertical force on the lower arm, bent as shown in Figs. 8–14a and b. For each case, calculate the torque about the axis of rotation through the elbow joint, assuming the muscle is attached 5.0 cm from the elbow as shown.

APPROACH The force is given, and the lever arm in (a) is given. In (b) we have to take into account the angle to get the lever arm.

SOLUTION (a) $F = 700$ N and $r_\perp = 0.050$ m, so

$$\tau = r_\perp F = (0.050 \text{ m})(700 \text{ N}) = 35 \text{ m} \cdot \text{N}.$$

(b) Because the arm is at an angle below the horizontal, the lever arm is shorter (Fig. 8–14c) than in part (a): $r_\perp = (0.050 \text{ m})(\sin 60°)$, where $\theta = 60°$ is the angle between $\vec{F}$ and r. F is still 700 N, so

$$\tau = (0.050 \text{ m})(0.866)(700 \text{ N}) = 30 \text{ m} \cdot \text{N}.$$

The arm can exert less torque at this angle than when it is at 90°. Weight machines at gyms are often designed to take this variation with angle into account.

NOTE In (b), we could instead have used $\tau = r F_\perp$. As shown in Fig. 8–14d, $F_\perp = F \sin 60°$. Then $\tau = r F_\perp = r F \sin \theta = (0.050 \text{ m})(700 \text{ N})(0.866)$ gives the same result.

EXERCISE B Two forces ($F_B = 20$ N and $F_A = 30$ N) are applied to a meter stick which can rotate about its left end, Fig. 8–15. Force $\vec{F}_B$ is applied perpendicularly at the midpoint. Which force exerts the greater torque?

FIGURE 8–15 Exercise B.

When more than one torque acts on an object, the angular acceleration α is found to be proportional to the *net* torque. If all the torques acting on an object tend to rotate it in the same direction about a fixed axis of rotation, the net torque is the sum of the torques. But if, say, one torque acts to rotate an object in one direction, and a second torque acts to rotate the object in the opposite direction (as in Fig. 8–16), the net torque is the difference of the two torques. We normally assign a positive sign to torques that act to rotate the object counterclockwise, and a negative sign to torques that act to rotate the object clockwise.

EXAMPLE 8–9 Torque on a compound wheel.
Two thin disk-shaped wheels, of radii $r_A = 30$ cm and $r_B = 50$ cm, are attached to each other on an axle that passes through the center of each, as shown in Fig. 8–16. Calculate the net torque on this compound wheel due to the two forces shown, each of magnitude 50 N.

APPROACH The force $\vec{F}_A$ acts to rotate the system counterclockwise, whereas $\vec{F}_B$ acts to rotate it clockwise. So the two forces act in opposition to each other. We must choose one direction of rotation to be positive—say, counterclockwise. Then $\vec{F}_A$ exerts a positive torque, $\tau_A = r_A F_A$, since the lever arm is r_A. $\vec{F}_B$, on the other hand, produces a negative (clockwise) torque and does not act perpendicular to r_B, so we must use its perpendicular component to calculate the torque it produces: $\tau_B = -r_B F_{B\perp} = -r_B F_B \sin \theta$, where $\theta = 60°$. (Note that θ must be the angle between $\vec{F}_B$ and a radial line from the axis.)

SOLUTION The net torque is

$$\tau = r_A F_A - r_B F_B \sin 60°$$
$$= (0.30 \text{ m})(50 \text{ N}) - (0.50 \text{ m})(50 \text{ N})(0.866) = -6.7 \text{ m} \cdot \text{N}.$$

This net torque acts to accelerate the rotation of the wheel in the clockwise direction.

NOTE The two forces have the same magnitude, yet they produce a net torque because their lever arms are different.

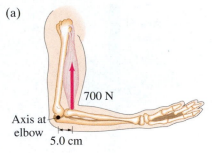
(a)

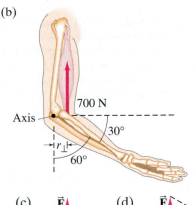
(b)

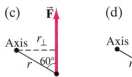
(c)   (d)

FIGURE 8–14 Example 8–8.

FIGURE 8–16 Example 8–9. The torque due to $\vec{F}_A$ tends to accelerate the wheel counterclockwise, whereas the torque due to $\vec{F}_B$ tends to accelerate the wheel clockwise.

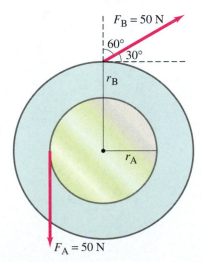

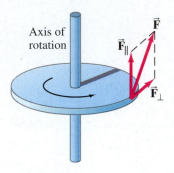

FIGURE 8–17 Only the component of $\vec{\mathbf{F}}$ that acts in the plane perpendicular to the rotation axis, $\vec{\mathbf{F}}_\perp$, acts to turn the wheel about the axis. The component parallel to the axis, $\vec{\mathbf{F}}_\parallel$, would tend to move the axis itself, which we assume is held fixed.

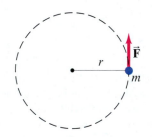

FIGURE 8–18 A mass m rotating in a circle of radius r about a fixed point.

* **Forces that Act to Tilt the Axis**

We are considering only rotation about a fixed axis, and so we consider only forces that act in a plane perpendicular to the axis of rotation. If there is a force (or component of a force) acting parallel to the axis of rotation, it will tend to tilt the axis of rotation—the component $\vec{\mathbf{F}}_\parallel$ in Fig. 8–17 is an example. Since we are assuming the axis remains fixed in direction, either there can be no such forces or else the axis must be mounted in bearings or hinges that hold the axis fixed. Thus, only a force, or component of a force ($\vec{\mathbf{F}}_\perp$ in Fig. 8–17), in a plane perpendicular to the axis will give rise to rotation about the axis, and it is only these that we consider.

8–5 Rotational Dynamics; Torque and Rotational Inertia

We have discussed that the angular acceleration α of a rotating object is proportional to the net torque τ applied to it:

$$\alpha \propto \Sigma\tau,$$

where we write $\Sigma\tau$ to remind us[†] that it is the *net* torque (sum of all torques acting on the object) that is proportional to α. This corresponds to Newton's second law for translational motion, $a \propto \Sigma F$, but here torque has taken the place of force, and, correspondingly, the angular acceleration α takes the place of the linear acceleration a. In the linear case, the acceleration is not only proportional to the net force, but it is also inversely proportional to the inertia of the object, which we call its mass, m. Thus we could write $a = \Sigma F/m$. But what plays the role of mass for the rotational case? That is what we now set out to determine. At the same time, we will see that the relation $\alpha \propto \Sigma\tau$ follows directly from Newton's second law, $\Sigma F = ma$.

We first consider a very simple case: a particle of mass m rotating in a circle of radius r at the end of a string or rod whose mass we can ignore compared to m (Fig. 8–18), and we assume a single force F acts on m as shown. The torque that gives rise to the angular acceleration is $\tau = rF$. If we use Newton's second law for linear quantities, $\Sigma F = ma$, and Eq. 8–5 relating the angular acceleration to the tangential linear acceleration, $a_{\text{tan}} = r\alpha$, then we have

$$F = ma$$
$$= mr\alpha.$$

When we multiply both sides of this equation by r, we find that the torque $\tau = rF$ is given by

$$\tau = mr^2\alpha. \qquad \text{[single particle] } \textbf{(8–11)}$$

Here at last we have a direct relation between the angular acceleration and the applied torque τ. The quantity mr^2 represents the *rotational inertia* of the particle and is called its *moment of inertia*.

Now let us consider a rotating rigid object, such as a wheel rotating about an axis through its center, which could be an axle. We can think of the wheel as consisting of many particles located at various distances from the axis of rotation. We can apply Eq. 8–11 to each particle of the object, and then sum over all the particles. The sum of the various torques is just the total torque, $\Sigma\tau$, so we obtain:

$$\Sigma\tau = (\Sigma mr^2)\alpha \qquad \textbf{(8–12)}$$

where we factored out α because it is the same for all the particles of the object. The sum Σmr^2 represents the sum of the masses of each particle in the object multiplied by the square of the distance of that particle from the axis of

[†] Recall from Chapter 4 that Σ (Greek letter sigma) means "sum of."

rotation. If we assign each particle a number $(1, 2, 3, \ldots)$, then $\Sigma mr^2 = m_1 r_1^2 + m_2 r_2^2 + m_3 r_3^2 + \cdots$. This quantity is called the **moment of inertia** (or *rotational inertia*) I of the object:

$$I = \Sigma mr^2 = m_1 r_1^2 + m_2 r_2^2 + \cdots. \qquad \text{(8-13)}$$

 *Moment of inertia*

Combining Eqs. 8–12 and 8–13, we can write

$$\Sigma \tau = I\alpha. \qquad \text{(8-14)}$$

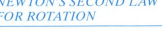
NEWTON'S SECOND LAW FOR ROTATION

This is the rotational equivalent of Newton's second law. It is valid for the rotation of a rigid object about a fixed axis.[†]

We see that the moment of inertia, I, which is a measure of the rotational inertia of an object, plays the same role for rotational motion that mass does for translational motion. As can be seen from Eq. 8–13, the rotational inertia of an object depends not only on its mass, but also on how that mass is distributed with respect to the axis. For example, a large-diameter cylinder will have greater rotational inertia than one of equal mass but smaller diameter (and therefore greater length), Fig. 8–19. The former will be harder to start rotating, and harder to stop. When the mass is concentrated farther from the axis of rotation, the rotational inertia is greater. For rotational motion, the mass of an object *cannot* be considered as concentrated at its center of mass.

⚠ **CAUTION**
Mass can not be considered concentrated at CM for rotational motion

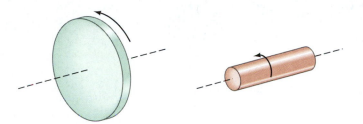

FIGURE 8–19 A large-diameter cylinder has greater rotational inertia than one of smaller diameter but equal mass.

EXAMPLE 8–10 **Two weights on a bar: different axis, different I.** Two small "weights," of mass 5.0 kg and 7.0 kg, are mounted 4.0 m apart on a light rod (whose mass can be ignored), as shown in Fig. 8–20. Calculate the moment of inertia of the system (a) when rotated about an axis halfway between the weights, Fig. 8–20a, and (b) when rotated about an axis 0.50 m to the left of the 5.0-kg mass (Fig. 8–20b).

FIGURE 8–20 Example 8–10: calculating the moment of inertia.

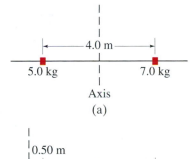

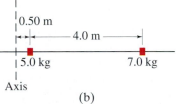

APPROACH In each case, the moment of inertia of the system is found by summing over the two parts using Eq. 8–13.

SOLUTION (a) Both weights are the same distance, 2.0 m, from the axis of rotation. Thus

$$I = \Sigma mr^2 = (5.0\,\text{kg})(2.0\,\text{m})^2 + (7.0\,\text{kg})(2.0\,\text{m})^2$$
$$= 20\,\text{kg} \cdot \text{m}^2 + 28\,\text{kg} \cdot \text{m}^2 = 48\,\text{kg} \cdot \text{m}^2.$$

(b) The 5.0-kg mass is now 0.50 m from the axis, and the 7.0-kg mass is 4.50 m from the axis. Then

$$I = \Sigma mr^2 = (5.0\,\text{kg})(0.50\,\text{m})^2 + (7.0\,\text{kg})(4.5\,\text{m})^2$$
$$= 1.3\,\text{kg} \cdot \text{m}^2 + 142\,\text{kg} \cdot \text{m}^2 = 143\,\text{kg} \cdot \text{m}^2.$$

NOTE This Example illustrates two important points. First, the moment of inertia of a given system is different for different axes of rotation. Second, we see in part (b) that mass close to the axis of rotation contributes little to the total moment of inertia; here, the 5.0-kg object contributed less than 1% to the total.

⚠ **CAUTION**
I depends on axis of rotation and on distribution of mass

[†] Equation 8–14 is also valid when the object is translating with acceleration, as long as I and α are calculated about the center of mass of the object, and the rotation axis through the CM doesn't change direction.

FIGURE 8–21 Moments of inertia for various objects of uniform composition.

Object	Location of axis		Moment of inertia
(a) **Thin hoop,** radius R	Through center	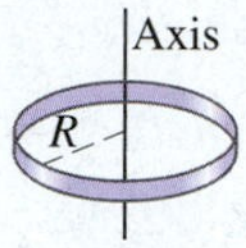	MR^2
(b) **Thin hoop,** radius R width W	Through central diameter	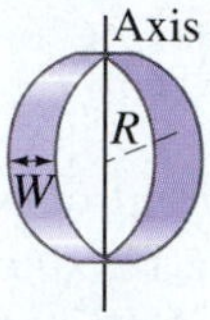	$\frac{1}{2}MR^2 + \frac{1}{12}MW^2$
(c) **Solid cylinder,** radius R	Through center	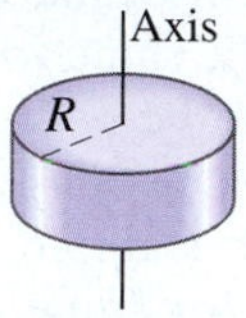	$\frac{1}{2}MR^2$
(d) **Hollow cylinder,** inner radius R_1 outer radius R_2	Through center	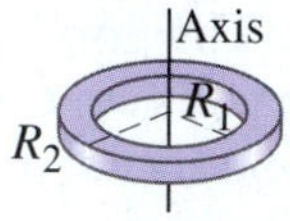	$\frac{1}{2}M(R_1^2 + R_2^2)$
(e) **Uniform sphere,** radius R	Through center	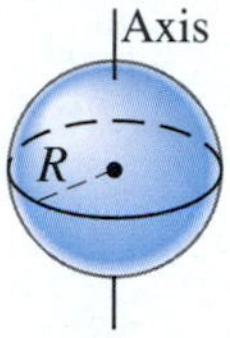	$\frac{2}{5}MR^2$
(f) **Long uniform rod,** length L	Through center	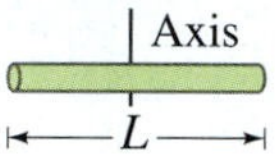	$\frac{1}{12}ML^2$
(g) **Long uniform rod,** length L	Through end	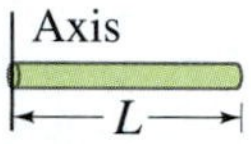	$\frac{1}{3}ML^2$
(h) **Rectangular thin plate,** length L, width W	Through center	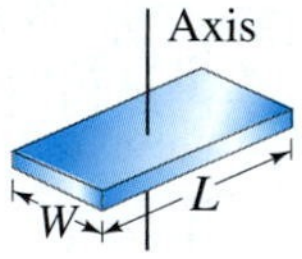	$\frac{1}{12}M(L^2 + W^2)$

For most ordinary objects, the mass is distributed continuously, and the calculation of the moment of inertia, Σmr^2, can be difficult. Expressions can, however, be worked out (using calculus) for the moments of inertia of regularly shaped objects in terms of the dimensions of the objects. Figure 8–21 gives these expressions for a number of solids rotated about the axes specified. The only one for which the result is obvious is that for the thin hoop or ring rotated about an axis passing through its center perpendicular to the plane of the hoop (Fig. 8–21a). For this object, all the mass is concentrated at the same distance from the axis, R. Thus $\Sigma mr^2 = (\Sigma m)R^2 = MR^2$, where M is the total mass of the hoop.

When calculation is difficult, I can be determined experimentally by measuring the angular acceleration α about a fixed axis due to a known net torque, $\Sigma\tau$, and applying Newton's second law, $I = \Sigma\tau/\alpha$, Eq. 8–14.

8–6 Solving Problems in Rotational Dynamics

When working with torque and angular acceleration (Eq. 8–14), it is important to use a consistent set of units, which in SI is: α in rad/s^2; τ in $\text{m}\cdot\text{N}$; and the moment of inertia, I, in $\text{kg}\cdot\text{m}^2$.

1. As always, draw a clear and complete **diagram**.

2. Choose the object or objects that will be the **system** to be studied.

3. Draw a **free-body diagram** for the object under consideration (or for each object, if more than one), showing only (and all) the forces acting on that object and exactly where they act, so you can determine the torque due to each. Gravity acts at the CG of the object (Section 7–8).

4. Identify the axis of rotation and determine the **torques** about it. Choose positive and negative

directions of rotation (counterclockwise and clockwise), and assign the correct sign to each torque.

5. Apply **Newton's second law for rotation**, $\Sigma\tau = I\alpha$. If the moment of inertia is not given, and it is not the unknown sought, you need to determine it first. Use consistent units, which in SI are: α in rad/s²; τ in m·N; and I in kg·m².

6. Also apply **Newton's second law for translation**, $\Sigma\vec{F} = m\vec{a}$, and **other** laws or principles as needed.

7. **Solve** the resulting equation(s) for the unknown(s).

8. Do a rough **estimate** to determine if your answer is reasonable.

EXAMPLE 8–11 **A heavy pulley.** A 15.0-N force (represented by $\vec{F}_T$) is applied to a cord wrapped around a pulley of mass $M = 4.00\,\text{kg}$ and radius $R = 33.0\,\text{cm}$, Fig. 8–22. The pulley accelerates uniformly from rest to an angular speed of 30.0 rad/s in 3.00 s. If there is a frictional torque $\tau_{\text{fr}} = 1.10\,\text{m·N}$ at the axle, determine the moment of inertia of the pulley. The pulley rotates about its center.

APPROACH We follow the steps of the Problem Solving Box explicitly.

SOLUTION

1. **Draw a diagram.** The pulley and the attached cord are shown in Fig. 8–22.

2. **Choose the system**: the pulley.

3. **Draw a free-body diagram.** The force that the cord exerts on the pulley is shown as $\vec{F}_T$ in Fig. 8–22. The friction force is also shown, but we are given only its torque. Two other forces could be included in the diagram: the force of gravity mg down and whatever force keeps the axle in place. They do not contribute to the torque (their lever arms are zero) and so are not shown.

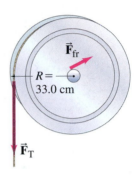

FIGURE 8–22   Example 8–11.

4. **Determine the torques.** The cord exerts a force $\vec{F}_T$ that acts at the edge of the pulley, so its lever arm is R. The torque exerted by the cord equals RF_T and is counterclockwise, which we choose to be positive. The frictional torque is given as $\tau_{\text{fr}} = 1.10\,\text{m·N}$; it opposes the motion and is negative.

5. **Apply Newton's second law for rotation.** The net torque is
$$\Sigma\tau = RF_T - \tau_{\text{fr}}$$
$$= (0.330\,\text{m})(15.0\,\text{N}) - 1.10\,\text{m·N} = 3.85\,\text{m·N}.$$

The angular acceleration α is found from the given data that it takes 3.0 s to accelerate the pulley from rest to $\omega = 30.0\,\text{rad/s}$:
$$\alpha = \frac{\Delta\omega}{\Delta t} = \frac{30.0\,\text{rad/s} - 0}{3.00\,\text{s}} = 10.0\,\text{rad/s}^2.$$

We can now solve for I in Newton's second law (see step 7).

6. **Other calculations**: None needed.

7. **Solve for unknowns.** We solve for I in Newton's second law for rotation, $\Sigma\tau = I\alpha$, and insert our values for $\Sigma\tau$ and α:
$$I = \frac{\Sigma\tau}{\alpha} = \frac{3.85\,\text{m·N}}{10.0\,\text{rad/s}^2} = 0.385\,\text{kg·m}^2.$$

8. **Do a rough estimate.** We can do a rough estimate of the moment of inertia by assuming the pulley is a uniform cylinder and using Fig. 8–21c:
$$I = \tfrac{1}{2}MR^2 = \tfrac{1}{2}(4.00\,\text{kg})(0.330\,\text{m})^2 = 0.218\,\text{kg·m}^2.$$

This is the same order of magnitude as our result, but numerically somewhat less. This makes sense, though, because a pulley is not usually a uniform cylinder but instead has more of its mass concentrated toward the outside edge. Such a pulley would be expected to have a greater moment of inertia than a solid cylinder of equal mass; a thin hoop, Fig. 8–21a, ought to have a greater I than our pulley, and indeed it does: $I = MR^2 = 0.436\,\text{kg·m}^2.$

Usefulness and power of rough estimates

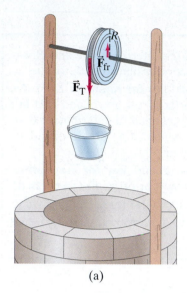

(a)

(b)

FIGURE 8–23 Example 8–12. (a) Pulley and falling bucket of mass m. (b) Free-body diagram for the bucket.

Additional Example—a bit more challenging

EXAMPLE 8–12 **Pulley and bucket.** Consider again the pulley in Fig. 8–22 and Example 8–11. But this time, instead of a constant 15.0-N force being exerted on the cord, we now have a bucket of weight $w = 15.0$ N (mass $m = w/g = 1.53$ kg) hanging from the cord. See Fig. 8–23a. We assume the cord has negligible mass and does not stretch or slip on the pulley. Calculate the angular acceleration α of the pulley and the linear acceleration a of the bucket.

APPROACH This situation looks a lot like Example 8–11, Fig. 8–22. But there is a big difference: the tension in the cord is now an unknown, and it is no longer equal to the weight of the bucket if the bucket accelerates. Our system has two parts: the bucket, which can undergo translational motion (Fig. 8–23b is its free-body diagram); and the pulley. The pulley does not translate, but it can rotate. We apply the rotational version of Newton's second law to the pulley, $\Sigma\tau = I\alpha$, and the linear version to the bucket, $\Sigma F = ma$.

SOLUTION Let F_T be the tension in the cord. Then a force F_T acts at the edge of the pulley, and we apply Newton's second law, Eq. 8–14, for the rotation of the pulley:

$$I\alpha = \Sigma\tau = RF_T - \tau_{fr}. \qquad \text{[pulley]}$$

Next we look at the (linear) motion of the bucket of mass m. Figure 8–23b, the free-body diagram for the bucket, shows that two forces act on the bucket: the force of gravity mg acts downward, and the tension of the cord F_T pulls upward. Applying Newton's second law, $\Sigma F = ma$, for the bucket, we have (taking downward as positive):

$$mg - F_T = ma. \qquad \text{[bucket]}$$

Note that the tension F_T, which is the force exerted on the edge of the pulley, is *not* equal to the weight of the bucket $(= mg = 15.0$ N). There must be a net force on the bucket if it is accelerating, so $F_T < mg$. We can also see this from the last equation above, $F_T = mg - ma$.

To obtain α, we note that the tangential acceleration of a point on the edge of the pulley is the same as the acceleration of the bucket if the cord doesn't stretch or slip. Hence we can use Eq. 8–5, $a_{tan} = a = R\alpha$. Substituting $F_T = mg - ma = mg - mR\alpha$ into the first equation above (Newton's second law for rotation of the pulley), we obtain

$$I\alpha = \Sigma\tau = RF_T - \tau_{fr} = R(mg - mR\alpha) - \tau_{fr} = mgR - mR^2\alpha - \tau_{fr}.$$

α appears in the second term on the right, so we bring that term to the left side and solve for α:

$$\alpha = \frac{mgR - \tau_{fr}}{I + mR^2}.$$

The numerator $(mgR - \tau_{fr})$ is the net torque, and the denominator $(I + mR^2)$ is the total rotational inertia of the system. Then, since $I = 0.385$ kg·m², $m = 1.53$ kg, and $\tau_{fr} = 1.10$ m·N (from Example 8–11),

$$\alpha = \frac{(15.0\text{ N})(0.330\text{ m}) - 1.10\text{ m·N}}{0.385\text{ kg·m}^2 + (1.53\text{ kg})(0.330\text{ m})^2} = 6.98\text{ rad/s}^2.$$

The angular acceleration is somewhat less in this case than the 10.0 rad/s² of Example 8–11. Why? Because F_T $(= mg - ma)$ is less than the 15.0-N weight of the bucket, mg. The linear acceleration of the bucket is

$$a = R\alpha = (0.330\text{ m})(6.98\text{ rad/s}^2) = 2.30\text{ m/s}^2.$$

NOTE The tension in the cord F_T is less than mg because the bucket accelerates.

8–7 Rotational Kinetic Energy

The quantity $\frac{1}{2}mv^2$ is the kinetic energy of an object undergoing translational motion. An object rotating about an axis is said to have **rotational kinetic energy**. By analogy with translational kinetic energy, we would expect this to be given by the expression $\frac{1}{2}I\omega^2$, where I is the moment of inertia of the object and ω is its angular velocity. We can indeed show that this is true.

Consider any rigid rotating object as made up of many tiny particles, each of mass m. If we let r represent the distance of any one particle from the axis of rotation, then its linear velocity is $v = r\omega$. The total kinetic energy of the whole object will be the sum of the kinetic energies of all its particles:

$$\text{KE} = \Sigma(\tfrac{1}{2}mv^2) = \Sigma(\tfrac{1}{2}mr^2\omega^2)$$
$$= \tfrac{1}{2}\Sigma(mr^2)\omega^2.$$

We have factored out the $\tfrac{1}{2}$ and the ω^2 since they are the same for every particle of a rigid object. Since $\Sigma mr^2 = I$, the moment of inertia, we see that the kinetic energy of a rigid rotating object is, as expected,

$$\text{rotational KE} = \tfrac{1}{2}I\omega^2. \qquad\qquad (8\text{–}15)$$

Rotational KE

The units are joules, as with all other forms of energy.

An object that rotates while its center of mass (CM) undergoes translational motion will have both translational and rotational kinetic energies. Equation 8–15 gives the rotational kinetic energy if the rotation axis is fixed. If the object is moving (such as a wheel rolling down a hill), this equation is still valid as long as the rotation axis is fixed in direction. Then the total kinetic energy is

$$\text{KE} = \tfrac{1}{2}Mv_{\text{CM}}^2 + \tfrac{1}{2}I_{\text{CM}}\omega^2, \qquad\qquad (8\text{–}16)$$

Total KE (translation + rotation)

where v_{CM} is the linear velocity of the center of mass, I_{CM} is the moment of inertia about an axis through the center of mass, ω is the angular velocity about this axis, and M is the total mass of the object.

EXAMPLE 8–13 **Sphere rolling down an incline.** What will be the speed of a solid sphere of mass M and radius R when it reaches the bottom of an incline if it starts from rest at a vertical height H and rolls without slipping? See Fig. 8–24. (Assume plenty of static friction, which does no work, so no slipping takes place.) Compare your result to that for an object *sliding* down a frictionless incline.

APPROACH We use the law of conservation of energy with gravitational potential energy, now including rotational kinetic energy as well as translational KE.

SOLUTION The total energy at any point a vertical distance y above the base of the incline is

$$\tfrac{1}{2}Mv^2 + \tfrac{1}{2}I_{\text{CM}}\omega^2 + Mgy,$$

where v is the speed of the center of mass, and Mgy is the gravitational PE. Applying conservation of energy, we equate the total energy at the top ($y = H, v = 0, \omega = 0$) to the total energy at the bottom ($y = 0$):

$$0 + 0 + MgH = \tfrac{1}{2}Mv^2 + \tfrac{1}{2}I_{\text{CM}}\omega^2 + 0.$$

The moment of inertia of a solid sphere about an axis through its center of mass is $I_{\text{CM}} = \tfrac{2}{5}MR^2$, Fig. 8–21e. Since the sphere rolls without slipping, we have $\omega = v/R$ (recall Fig. 8–8). Hence

$$MgH = \tfrac{1}{2}Mv^2 + \tfrac{1}{2}\left(\tfrac{2}{5}MR^2\right)\left(\frac{v^2}{R^2}\right).$$

Canceling the M's and R's, we obtain

$$\left(\tfrac{1}{2} + \tfrac{1}{5}\right)v^2 = gH$$

or

$$v = \sqrt{\tfrac{10}{7}gH}.$$

We can compare this result for the speed of a rolling sphere to that for an object sliding down a plane without rotating and without friction, $\tfrac{1}{2}mv^2 = mgH$ (see our energy equation above, removing the rotational term). Then $v = \sqrt{2gH}$, which is greater than our result. An object sliding without friction or rotation transforms its initial potential energy entirely into translational KE (none into rotational KE), so the speed of its center of mass is greater.

NOTE Our result for the rolling sphere shows (perhaps surprisingly) that v is independent of both the mass M and the radius R of the sphere.

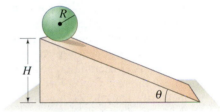

FIGURE 8–24 A sphere rolling down a hill has both translational and rotational kinetic energy. Example 8–13.

➡ **PROBLEM SOLVING**

Rotational energy adds to other forms of energy to get the total energy which is conserved

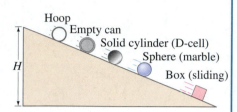

FIGURE 8–25 Example 8–14.

FIGURE 8–26 A sphere rolling to the right on a plane surface. The point in contact with the ground at any moment, point P, is momentarily at rest. Point A to the left of P is moving nearly vertically upward at the instant shown, and point B to the right is moving nearly vertically downward. An instant later, point B will touch the plane and be at rest momentarily. Thus no work is done by the force of static friction.

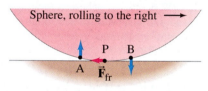

⚠ **CAUTION**

Rolling objects go slower than sliding objects because of rotational KE, *not because of friction*

CONCEPTUAL EXAMPLE 8–14 | Who's fastest? Several objects roll without slipping down an incline of vertical height H, all starting from rest at the same moment. The objects are a thin hoop (or a plain wedding band), a spherical marble, a solid cylinder (a D-cell battery), and an empty soup can. In addition, a greased box slides down without friction. In what order do they reach the bottom of the incline?

RESPONSE The sliding box wins because the potential energy loss (MgH) is transformed completely into translational KE for the box, whereas for rolling objects the initial PE is shared between translational and rotational kinetic energies, and so their linear speed is less. For each of the rolling objects we can state that the loss in potential energy equals the increase in kinetic energy:

$$MgH = \tfrac{1}{2}Mv^2 + \tfrac{1}{2}I_{CM}\,\omega^2.$$

For all our rolling objects, the moment of inertia I_{CM} is a numerical factor times the mass M and the radius R^2 (Fig. 8–21). The mass M is in each term, so the translational speed v doesn't depend on M; nor does it depend on the radius R since $\omega = v/R$, so R^2 cancels out for all the rolling objects, just as in Example 8–13. Thus the speed v at the bottom depends only on that numerical factor in I_{CM} which expresses how the mass is distributed. The hoop, with all its mass concentrated at radius R ($I_{CM} = MR^2$), has the largest moment of inertia; hence it will have the lowest speed and will arrive at the bottom behind the D-cell ($I_{CM} = \tfrac{1}{2}MR^2$), which in turn will be behind the marble ($I_{CM} = \tfrac{2}{5}MR^2$). The empty can, which is mainly a hoop plus a small disk, has most of its mass concentrated at R; so it will be a bit faster than the pure hoop but slower than the D-cell. See Fig. 8–25.

NOTE As in Example 8–13, the speed at the bottom does not depend on the object's mass M or radius R, but only on its shape (and the height of the hill H).

If there had been little or no static friction between the rolling objects and the plane in these Examples, the round objects would have slid rather than rolled, or a combination of both. Static friction must be present to make a round object roll. We did not need to take friction into account in the energy equation because it is *static* friction and does no work—the point of contact of the sphere at each instant does not slide, but moves perpendicular to the plane (first down and then up as shown in Fig. 8–26) as the sphere rolls. Thus, no work is done by the static friction force because the force and the motion (displacement) are perpendicular. The reason the rolling objects in Examples 8–13 and 8–14 move down the slope more slowly than if they were sliding is *not* because friction is doing work. Rather, it is because some of the gravitional PE is converted to rotational KE, leaving less for the translational KE.

Work Done by Torque

The work done on an object rotating about a fixed axis, such as the pulleys in Figs. 8–22 and 8–23, can be written using angular quantities. As shown in Fig. 8–27, a force F exerting a torque $\tau = rF$ on a wheel does work $W = F\Delta l$ in rotating the wheel a small distance Δl at the point of application of $\vec{F}$.

FIGURE 8–27 Torque $\tau = rF$ does work when rotating a wheel equal to $W = F\Delta l = Fr\Delta\theta = \tau\Delta\theta$.

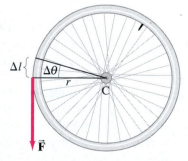

The wheel has rotated through a small angle $\Delta\theta = \Delta l / r$ (Eq. 8–1). Hence

$$W = F\Delta l = Fr\Delta\theta.$$

Since $\tau = rF$, then

$$W = \tau\Delta\theta \qquad\qquad \textbf{(8–17)} \qquad \textit{Work done by torque}$$

is the work done by the torque τ when rotating the wheel through an angle $\Delta\theta$. Finally, power P is the rate work is done: $P = W/\Delta t = \tau\Delta\theta/\Delta t = \tau\omega$.

8–8 Angular Momentum and Its Conservation

Throughout this Chapter we have seen that if we use the appropriate angular variables, the kinematic and dynamic equations for rotational motion are analogous to those for ordinary linear motion. We saw in the previous Section, for example, that rotational kinetic energy can be written as $\frac{1}{2}I\omega^2$, which is analogous to the translational kinetic energy, $\frac{1}{2}mv^2$. In like manner, the linear momentum, $p = mv$, has a rotational analog. It is called **angular momentum**, L. For an object rotating about a fixed axis, it is defined as

$$L = I\omega, \qquad\qquad \textbf{(8–18)} \qquad \textit{Angular momentum}$$

where I is the moment of inertia and ω is the angular velocity about the axis of rotation. The SI units for L are $\text{kg}\cdot\text{m}^2/\text{s}$, which has no special name.

We saw in Chapter 7 (Section 7–1) that Newton's second law can be written not only as $\Sigma F = ma$ but also more generally in terms of momentum (Eq. 7–2), $\Sigma F = \Delta p/\Delta t$. In a similar way, the rotational equivalent of Newton's second law, which we saw in Eq. 8–14 can be written as $\Sigma\tau = I\alpha$, can also be written in terms of angular momentum:

$$\Sigma\tau = \frac{\Delta L}{\Delta t}, \qquad\qquad \textbf{(8–19)}$$

where $\Sigma\tau$ is the net torque acting to rotate the object, and ΔL is the change in angular momentum in a time interval Δt. Equation 8–14, $\Sigma\tau = I\alpha$, is a special case of Eq. 8–19 when the moment of inertia is constant. This can be seen as follows. If an object has angular velocity ω_0 at time $t = 0$, and angular velocity ω after a time interval Δt, then its angular acceleration (Eq. 8–3) is

$$\alpha = \frac{\Delta\omega}{\Delta t} = \frac{\omega - \omega_0}{\Delta t}.$$

Then from Eq. 8–19, we have

$$\Sigma\tau = \frac{\Delta L}{\Delta t} = \frac{I\omega - I\omega_0}{\Delta t} = \frac{I(\omega - \omega_0)}{\Delta t} = I\frac{\Delta\omega}{\Delta t} = I\alpha,$$

which is Eq. 8–14.

Angular momentum is an important concept in physics because, under certain conditions, it is a conserved quantity. We can see from Eq. 8–19 that if the net torque $\Sigma\tau$ on an object is zero, then $\Delta L/\Delta t$ equals zero. That is, L does not change. This is the **law of conservation of angular momentum** for a rotating object:

The total angular momentum of a rotating object remains constant if the net torque acting on it is zero.

The law of conservation of angular momentum is one of the great conservation laws of physics, along with energy and linear momentum.

When there is zero net torque acting on an object, and the object is rotating about a fixed axis or about an axis through its center of mass whose direction doesn't change, we can write

$$I\omega = I_0\omega_0 = \text{constant}.$$

I_0 and ω_0 are the moment of inertia and angular velocity, respectively, about that axis at some initial time ($t = 0$), and I and ω are their values at some other time.

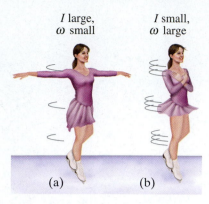

I large, ω small   I small, ω large

(a)   (b)

FIGURE 8–28 A skater doing a spin on ice, illustrating conservation of angular momentum. In (a), I is large and ω is small; in (b), I is smaller so ω is larger.

PHYSICS APPLIED
Spins in figure skating and diving

FIGURE 8–29 A diver rotates faster when arms and legs are tucked in than when they are outstretched. Angular momentum is conserved.

FIGURE 8–30 Example 8–15.

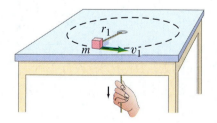

The parts of the object may alter their positions relative to one another, so I changes. But then ω changes as well to ensure that the product $I\omega$ remains constant.

Many interesting phenomena can be understood on the basis of conservation of angular momentum. Consider a skater doing a spin on the tips of her skates, Fig. 8–28. She rotates at a relatively low speed when her arms are outstretched; when she brings her arms in close to her body, she suddenly spins much faster. From the definition of moment of inertia, $I = \Sigma mr^2$, it is clear that when she pulls her arms in closer to the axis of rotation, r is reduced for the arms, so her moment of inertia is reduced. Since the angular momentum $I\omega$ remains constant (we ignore the small torque due to friction), if I decreases, then the angular velocity ω must increase. If the skater reduces her moment of inertia by a factor of 2, she will then rotate with twice the angular velocity.

EXERCISE C When a spinning figure skater pulls in her arms, her moment of inertia decreases; to conserve angular momentum, her angular velocity increases. Does her rotational kinetic energy also increase? If so, where does the energy come from?

A similar example is the diver shown in Fig. 8–29. The push as she leaves the board gives her an initial angular momentum about her center of mass. When she curls herself into the tuck position, she rotates quickly one or more times. She then stretches out again, increasing her moment of inertia which reduces the angular velocity to a small value, and then she enters the water. The change in moment of inertia from the straight position to the tuck position can be a factor of as much as $3\frac{1}{2}$.

Note that for angular momentum to be conserved, the net torque must be zero, but the net force does not necessarily have to be zero. The net force on the diver in Fig. 8–29, for example, is not zero (gravity is acting), but the net torque on her is zero because the force of gravity acts at her center of mass.

EXAMPLE 8–15 **Object rotating on a string of changing length.** A small mass m attached to the end of a string revolves in a circle on a frictionless tabletop. The other end of the string passes through a hole in the table (Fig. 8–30). Initially, the mass revolves with a speed $v_1 = 2.4\,\text{m/s}$ in a circle of radius $r_1 = 0.80\,\text{m}$. The string is then pulled slowly through the hole so that the radius is reduced to $r_2 = 0.48\,\text{m}$. What is the speed, v_2, of the mass now?

APPROACH There is no net torque on the mass m because the force exerted by the string to keep it moving in a circle is exerted toward the axis; hence the lever arm is zero. We can thus apply conservation of angular momentum.

SOLUTION Conservation of angular momentum gives

$$I_1\omega_1 = I_2\omega_2.$$

Our small mass is essentially a particle whose moment of inertia about the hole is $I = mr^2$ (Section 8–5, Eq. 8–11), so we have

$$mr_1^2\omega_1 = mr_2^2\omega_2,$$

or

$$\omega_2 = \omega_1\left(\frac{r_1^2}{r_2^2}\right).$$

Then, since $v = r\omega$, we can write

$$v_2 = r_2\omega_2 = r_2\omega_1\left(\frac{r_1^2}{r_2^2}\right) = r_2\frac{v_1}{r_1}\left(\frac{r_1^2}{r_2^2}\right) = v_1\frac{r_1}{r_2}$$

$$= (2.4\,\text{m/s})\left(\frac{0.80\,\text{m}}{0.48\,\text{m}}\right) = 4.0\,\text{m/s}.$$

The speed increases as the radius decreases.

EXERCISE D The speed of mass m in Example 8–15 increased, so its kinetic energy increased. Where did the energy come from?

EXAMPLE 8–16 **ESTIMATE** **Star collapse.** Astronomers often detect stars that are rotating extremely rapidly, known as neutron stars. These stars are believed to have formed from the inner core of a larger star that collapsed, due to its own gravitation, to a star of very small radius and very high density. Before collapse, suppose the core of such a star is the size of our Sun $(R \approx 7 \times 10^5 \text{ km})$ with mass 2.0 times as great as the Sun, and is rotating at a speed of 1.0 revolution every 10 days. If it were to undergo gravitational collapse to a neutron star of radius 10 km, what would its rotation speed be? Assume the star is a uniform sphere at all times.

APPROACH The star is isolated (no external forces), so we can use conservation of angular momentum for this process.

SOLUTION From conservation of angular momentum,

$$I_1 \omega_1 = I_2 \omega_2$$

where the subscripts 1 and 2 refer to initial (normal star) and final (neutron star), respectively. Then, assuming no mass is lost in the process,

$$\omega_2 = \left(\frac{I_1}{I_2}\right)\omega_1 = \left(\frac{\frac{2}{5} M_1 R_1^2}{\frac{2}{5} M_2 R_2^2}\right)\omega_1 = \frac{R_1^2}{R_2^2}\omega_1 .$$

The frequency $f = \omega/2\pi$, so

$$f_2 = \frac{\omega_2}{2\pi} = \frac{R_1^2}{R_2^2} f_1$$

$$= \left(\frac{7 \times 10^5 \text{ km}}{10 \text{ km}}\right)^2 \left(\frac{1.0 \text{ rev}}{10 \text{ d}(24 \text{ h/d})(3600 \text{ s/h})}\right) \approx 6 \times 10^3 \text{ rev/s}.$$

*8–9 Vector Nature of Angular Quantities

Up to now we have considered only the magnitudes of angular quantities such as ω, α, and L. But they have a vector aspect too, and now we consider the directions. In fact, we have to *define* the directions for rotational quantities, and we take first the angular velocity, $\vec{\omega}$.

Consider the rotating wheel shown in Fig. 8–31a. The linear velocities of different particles of the wheel point in all different directions. The only unique direction in space associated with the rotation is along the axis of rotation, perpendicular to the actual motion. We therefore choose the axis of rotation to be the direction of the angular velocity vector, $\vec{\omega}$. Actually, there is still an ambiguity since $\vec{\omega}$ could point in either direction along the axis of rotation (up or down in Fig. 8–31a). The convention we use, called **the right-hand rule**, is the following: When the fingers of the right hand are curled around the rotation axis and point in the direction of the rotation, then the thumb points in the direction of $\vec{\omega}$. This is shown in Fig. 8–31b. Note that $\vec{\omega}$ points in the direction a right-handed screw would move when turned in the direction of rotation. Thus, if the rotation of the wheel in Fig. 8–31b is counterclockwise, the direction of $\vec{\omega}$ is upward. If the wheel rotates clockwise, then $\vec{\omega}$ points in the opposite direction, downward. Note that no part of the rotating object moves in the direction of $\vec{\omega}$.

If the axis of rotation is fixed, then $\vec{\omega}$ can change only in magnitude. Thus $\vec{\alpha} = \Delta\vec{\omega}/\Delta t$ must also point along the axis of rotation. If the rotation is counterclockwise as in Fig. 8–31a, and if the magnitude ω is increasing, then $\vec{\alpha}$ points upward; but if ω is decreasing (the wheel is slowing down), $\vec{\alpha}$ points downward. If the rotation is clockwise, $\vec{\alpha}$ will point downward if ω is increasing, and point upward if ω is decreasing.

FIGURE 8–31 (a) Rotating wheel. (b) Right-hand rule for obtaining the direction of $\vec{\omega}$.

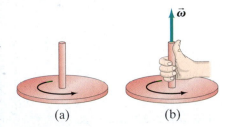

(a)   (b)

Right-hand rule

Angular momentum, like linear momentum, is a vector quantity. For a symmetrical object rotating about a symmetry axis (such as a wheel, cylinder, hoop, or sphere), we can write the vector angular momentum as

$$\vec{\mathbf{L}} = I\vec{\boldsymbol{\omega}}. \tag{8-20}$$

The angular velocity vector $\vec{\boldsymbol{\omega}}$ (and therefore also $\vec{\mathbf{L}}$) points along the axis of rotation in the direction given by the right-hand rule (Fig. 8–31b).

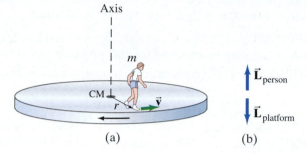

FIGURE 8–32 (a) A person standing on a circular platform, initially at rest, begins walking along the edge at speed v. The platform, assumed to be mounted on friction-free bearings, begins rotating in the opposite direction, so that the total angular momentum remains zero, as shown in (b).

The vector nature of angular momentum can be used to explain a number of interesting (and sometimes surprising) phenomena. For example, consider a person standing at rest on a circular platform capable of rotating without friction about an axis through its center (that is, a simplified merry-go-round). If the person now starts to walk along the edge of the platform, Fig. 8–32a, the platform starts rotating in the opposite direction. Why? One way to look at it is that the person's foot exerts a force on the platform. Another way to look at it (and this is the most useful analysis here) is that this is an example of the conservation of angular momentum. If the person starts walking counter-clockwise, the person's angular momentum will point upward along the axis of rotation (remember how we defined the direction of $\vec{\boldsymbol{\omega}}$ using the right-hand rule). The magnitude of the person's angular momentum will be $L = I\omega = (mr^2)(v/r)$, where v is the person's speed (relative to Earth, not to the platform), r is his distance from the rotation axis, m is his mass, and mr^2 is his moment of inertia if we consider him a particle (mass concentrated at one point). The platform rotates in the opposite direction, so its angular momentum points downward. If the initial total angular momentum of the system (person and platform) was zero (person and platform at rest), it will remain zero after the person starts walking. That is, the upward angular momentum of the person just balances the oppositely directed downward angular momentum of the platform (Fig. 8–32b), so the total vector angular momentum remains zero. Even though the person exerts a force (and torque) on the platform, the platform exerts an equal and opposite torque on the person. So the net torque on the *system* of person plus platform is zero (ignoring friction), and the total angular momentum remains constant.

FIGURE 8–33 Example 8–17.

| CONCEPTUAL EXAMPLE 8–17 | **Spinning bicycle wheel.** Your physics teacher is holding a spinning bicycle wheel while he stands on a stationary frictionless turntable (Fig. 8–33). What will happen if the teacher suddenly flips the bicycle wheel over so that it is spinning in the opposite direction?

RESPONSE We consider the system of turntable, teacher, and bicycle wheel. The total angular momentum initially is $\vec{\mathbf{L}}$ vertically upward. That is also what the system's angular momentum must be afterward, since $\vec{\mathbf{L}}$ is conserved when there is no net torque. Thus, if the wheel's angular momentum after being flipped over is $-\vec{\mathbf{L}}$ downward, then the angular momentum of teacher plus turntable will have to be $+2\vec{\mathbf{L}}$ upward. We can safely predict that the teacher will begin spinning around in the same direction the wheel was spinning originally.

Summary

When a rigid object rotates about a fixed axis, each point of the object moves in a circular path. Lines drawn perpendicularly from the rotation axis to various points in the object all sweep out the same angle θ in any given time interval.

Angles are conveniently measured in **radians**, where one radian is the angle subtended by an arc whose length is equal to the radius, or

$$2\pi \text{ rad} = 360°$$
$$1 \text{ rad} \approx 57.3°.$$

Angular velocity, ω, is defined as the rate of change of angular position:

$$\omega = \frac{\Delta\theta}{\Delta t}. \qquad (8\text{–}2)$$

All parts of a rigid object rotating about a fixed axis have the same angular velocity at any instant.

Angular acceleration, α, is defined as the rate of change of angular velocity:

$$\alpha = \frac{\Delta\omega}{\Delta t}. \qquad (8\text{–}3)$$

The linear velocity v and acceleration a of a point fixed at a distance r from the axis of rotation are related to ω and α by

$$v = r\omega, \qquad (8\text{–}4)$$
$$a_{\text{tan}} = r\alpha, \qquad (8\text{–}5)$$
$$a_R = \omega^2 r, \qquad (8\text{–}6)$$

where a_{tan} and a_R are the tangential and radial (centripetal) components of the linear acceleration, respectively.

The frequency f is related to ω by

$$\omega = 2\pi f, \qquad (8\text{–}7)$$

and to the period T by

$$T = 1/f. \qquad (8\text{–}8)$$

The equations describing uniformly accelerated rotational motion ($\alpha = $ constant) have the same form as for uniformly accelerated linear motion:

$$\omega = \omega_0 + \alpha t, \qquad \theta = \omega_0 t + \tfrac{1}{2}\alpha t^2,$$
$$\omega^2 = \omega_0^2 + 2\alpha\theta, \qquad \bar{\omega} = \frac{\omega + \omega_0}{2}. \qquad (8\text{–}9)$$

The dynamics of rotation is analogous to the dynamics of linear motion. Force is replaced by **torque** τ, which is defined as the product of force times lever arm (perpendicular distance from the line of action of the force to the axis of rotation):

$$\tau = rF\sin\theta = r_\perp F = rF_\perp. \qquad (8\text{–}10)$$

Mass is replaced by **moment of inertia** I, which depends not only on the mass of the object, but also on how the mass is distributed about the axis of rotation. Linear acceleration is replaced by angular acceleration. The rotational equivalent of Newton's second law is then

$$\Sigma\tau = I\alpha. \qquad (8\text{–}14)$$

The **rotational kinetic energy** of an object rotating about a fixed axis with angular velocity ω is

$$\text{KE} = \tfrac{1}{2}I\omega^2. \qquad (8\text{–}15)$$

For an object both translating and rotating, the total kinetic energy is the sum of the translational kinetic energy of the object's center of mass plus the rotational kinetic energy of the object about its center of mass:

$$\text{KE} = \tfrac{1}{2}Mv_{\text{CM}}^2 + \tfrac{1}{2}I_{\text{CM}}\omega^2 \qquad (8\text{–}16)$$

as long as the rotation axis is fixed in direction.

The **angular momentum** L of an object about a fixed rotation axis is given by

$$L = I\omega. \qquad (8\text{–}18)$$

Newton's second law, in terms of angular momentum, is

$$\Sigma\tau = \frac{\Delta L}{\Delta t}. \qquad (8\text{–}19)$$

If the net torque on the object is zero, $\Delta L/\Delta t = 0$, so $L = $ constant. This is the **law of conservation of angular momentum** for a rotating object.

The following Table summarizes angular (or rotational) quantities, comparing them to their translational analogs.

Translation	Rotation	Connection
x	θ	$x = r\theta$
v	ω	$v = r\omega$
a	α	$a = r\alpha$
m	I	$I = \Sigma mr^2$
F	τ	$\tau = rF\sin\theta$
$\text{KE} = \tfrac{1}{2}mv^2$	$\tfrac{1}{2}I\omega^2$	
$p = mv$	$L = I\omega$	
$W = Fd$	$W = \tau\theta$	
$\Sigma F = ma$	$\Sigma\tau = I\alpha$	
$\Sigma F = \dfrac{\Delta p}{\Delta t}$	$\Sigma\tau = \dfrac{\Delta L}{\Delta t}$	

Questions

1. A bicycle odometer (which measures distance traveled) is attached near the wheel hub and is designed for 27-inch wheels. What happens if you use it on a bicycle with 24-inch wheels?

2. Suppose a disk rotates at constant angular velocity. Does a point on the rim have radial and/or tangential acceleration? If the disk's angular velocity increases uniformly, does the point have radial and/or tangential acceleration? For which cases would the magnitude of either component of linear acceleration change?

3. Could a nonrigid body be described by a single value of the angular velocity ω? Explain.

4. Can a small force ever exert a greater torque than a larger force? Explain.

5. If a force $\vec{F}$ acts on an object such that its lever arm is zero, does it have any effect on the object's motion? Explain.

6. Why is it more difficult to do a sit-up with your hands behind your head than when your arms are stretched out in front of you? A diagram may help you to answer this.

7. A 21-speed bicycle has seven sprockets at the rear wheel and three at the pedal cranks. In which gear is it harder to pedal, a small rear sprocket or a large rear sprocket? Why? In which gear is it harder to pedal, a small front sprocket or a large front sprocket? Why?

8. Mammals that depend on being able to run fast have slender lower legs with flesh and muscle concentrated high, close to the body (Fig. 8–34). On the basis of rotational dynamics, explain why this distribution of mass is advantageous.

FIGURE 8–34 Question 8. A gazelle.

FIGURE 8–35 Question 9.

9. Why do tightrope walkers (Fig. 8–35) carry a long, narrow beam?

10. If the net force on a system is zero, is the net torque also zero? If the net torque on a system is zero, is the net force zero?

11. Two inclines have the same height but make different angles with the horizontal. The same steel ball is rolled down each incline. On which incline will the speed of the ball at the bottom be greater? Explain.

12. Two solid spheres simultaneously start rolling (from rest) down an incline. One sphere has twice the radius and twice the mass of the other. Which reaches the bottom of the incline first? Which has the greater speed there? Which has the greater total kinetic energy at the bottom?

13. A sphere and a cylinder have the same radius and the same mass. They start from rest at the top of an incline. Which reaches the bottom first? Which has the greater speed at the bottom? Which has the greater total kinetic energy at the bottom? Which has the greater rotational KE?

14. We claim that momentum and angular momentum are conserved. Yet most moving or rotating objects eventually slow down and stop. Explain.

15. If there were a great migration of people toward the Earth's equator, how would this affect the length of the day?

16. Can the diver of Fig. 8–29 do a somersault without having any initial rotation when she leaves the board?

17. The moment of inertia of a rotating solid disk about an axis through its center of mass is $\frac{1}{2}MR^2$ (Fig. 8–21c). Suppose instead that the axis of rotation passes through a point on the edge of the disk. Will the moment of inertia be the same, larger, or smaller?

18. Suppose you are sitting on a rotating stool holding a 2-kg mass in each outstretched hand. If you suddenly drop the masses, will your angular velocity increase, decrease, or stay the same? Explain.

19. Two spheres look identical and have the same mass. However, one is hollow and the other is solid. Describe an experiment to determine which is which.

* 20. In what direction is the Earth's angular velocity vector as it rotates daily about its axis?

* 21. The angular velocity of a wheel rotating on a horizontal axle points west. In what direction is the linear velocity of a point on the top of the wheel? If the angular acceleration points east, describe the tangential linear acceleration of this point at the top of the wheel. Is the angular speed increasing or decreasing?

* 22. Suppose you are standing on the edge of a large freely rotating turntable. What happens if you walk toward the center?

* 23. A shortstop may leap into the air to catch a ball and throw it quickly. As he throws the ball, the upper part of his body rotates. If you look quickly you will notice that his hips and legs rotate in the opposite direction (Fig. 8–36). Explain.

FIGURE 8–36 Question 23. A shortstop in the air, throwing the ball.

* 24. On the basis of the law of conservation of angular momentum, discuss why a helicopter must have more than one rotor (or propeller). Discuss one or more ways the second propeller can operate to keep the helicopter stable.

Problems

8–1 Angular Quantities

1. (I) Express the following angles in radians: (a) 30°, (b) 57°, (c) 90°, (d) 360°, and (e) 420°. Give as numerical values and as fractions of π.

2. (I) Eclipses happen on Earth because of an amazing coincidence. Calculate, using the information inside the Front Cover, the angular diameters (in radians) of the Sun and the Moon, as seen on Earth.

3. (I) A laser beam is directed at the Moon, 380,000 km from Earth. The beam diverges at an angle θ (Fig. 8–37) of 1.4×10^{-5} rad. What diameter spot will it make on the Moon?

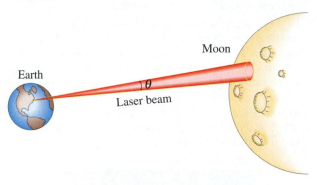

FIGURE 8–37 Problem 3.

4. (I) The blades in a blender rotate at a rate of 6500 rpm. When the motor is turned off during operation, the blades slow to rest in 3.0 s. What is the angular acceleration as the blades slow down?

5. (II) A child rolls a ball on a level floor 3.5 m to another child. If the ball makes 15.0 revolutions, what is its diameter?

6. (II) A bicycle with tires 68 cm in diameter travels 8.0 km. How many revolutions do the wheels make?

7. (II) (a) A grinding wheel 0.35 m in diameter rotates at 2500 rpm. Calculate its angular velocity in rad/s. (b) What are the linear speed and acceleration of a point on the edge of the grinding wheel?

8. (II) A rotating merry-go-round makes one complete revolution in 4.0 s (Fig. 8–38). (a) What is the linear speed of a child seated 1.2 m from the center? (b) What is her acceleration (give components)?

FIGURE 8–38 Problem 8.

9. (II) Calculate the angular velocity of the Earth (a) in its orbit around the Sun, and (b) about its axis.

10. (II) What is the linear speed of a point (a) on the equator, (b) on the Arctic Circle (latitude 66.5° N), and (c) at a latitude of 45.0° N, due to the Earth's rotation?

11. (II) How fast (in rpm) must a centrifuge rotate if a particle 7.0 cm from the axis of rotation is to experience an acceleration of 100,000 g's?

12. (II) A 70-cm-diameter wheel accelerates uniformly about its center from 130 rpm to 280 rpm in 4.0 s. Determine (a) its angular acceleration, and (b) the radial and tangential components of the linear acceleration of a point on the edge of the wheel 2.0 s after it has started accelerating.

13. (II) A turntable of radius R_1 is turned by a circular rubber roller of radius R_2 in contact with it at their outer edges. What is the ratio of their angular velocities, ω_1/ω_2?

14. (III) In traveling to the Moon, astronauts aboard the *Apollo* spacecraft put themselves into a slow rotation to distribute the Sun's energy evenly. At the start of their trip, they accelerated from no rotation to 1.0 revolution every minute during a 12-min time interval. The spacecraft can be thought of as a cylinder with a diameter of 8.5 m. Determine (a) the angular acceleration, and (b) the radial and tangential components of the linear acceleration of a point on the skin of the ship 5.0 min after it started this acceleration.

8–2 and 8–3 Constant Angular Acceleration; Rolling

15. (I) A centrifuge accelerates uniformly from rest to 15,000 rpm in 220 s. Through how many revolutions did it turn in this time?

16. (I) An automobile engine slows down from 4500 rpm to 1200 rpm in 2.5 s. Calculate (a) its angular acceleration, assumed constant, and (b) the total number of revolutions the engine makes in this time.

17. (I) Pilots can be tested for the stresses of flying high-speed jets in a whirling "human centrifuge," which takes 1.0 min to turn through 20 complete revolutions before reaching its final speed. (a) What was its angular acceleration (assumed constant), and (b) what was its final angular speed in rpm?

18. (II) A wheel 33 cm in diameter accelerates uniformly from 240 rpm to 360 rpm in 6.5 s. How far will a point on the edge of the wheel have traveled in this time?

19. (II) A cooling fan is turned off when it is running at 850 rev/min. It turns 1500 revolutions before it comes to a stop. (a) What was the fan's angular acceleration, assumed constant? (b) How long did it take the fan to come to a complete stop?

20. (II) A small rubber wheel is used to drive a large pottery wheel, and they are mounted so that their circular edges touch. The small wheel has a radius of 2.0 cm and accelerates at the rate of 7.2 rad/s², and it is in contact with the pottery wheel (radius 25.0 cm) without slipping. Calculate (a) the angular acceleration of the pottery wheel, and (b) the time it takes the pottery wheel to reach its required speed of 65 rpm.

21. (II) The tires of a car make 65 revolutions as the car reduces its speed uniformly from 95 km/h to 45 km/h. The tires have a diameter of 0.80 m. (a) What was the angular acceleration of the tires? (b) If the car continues to decelerate at this rate, how much more time is required for it to stop?

8–4 Torque

22. (I) A 55-kg person riding a bike puts all her weight on each pedal when climbing a hill. The pedals rotate in a circle of radius 17 cm. (a) What is the maximum torque she exerts? (b) How could she exert more torque?

23. (I) A person exerts a force of 55 N on the end of a door 74 cm wide. What is the magnitude of the torque if the force is exerted (a) perpendicular to the door, and (b) at a 45° angle to the face of the door?

24. (II) Calculate the net torque about the axle of the wheel shown in Fig. 8–39. Assume that a friction torque of 0.40 m·N opposes the motion.

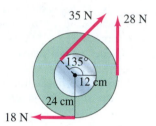

35 N   28 N
135°
12 cm
24 cm
18 N

FIGURE 8–39
Problem 24.

25. (II) Two blocks, each of mass m, are attached to the ends of a massless rod which pivots as shown in Fig. 8–40. Initially the rod is held in the horizontal position and then released. Calculate the magnitude and direction of the net torque on this system.

m L_1 L_2 m

FIGURE 8–40   Problem 25.

26. (II) The bolts on the cylinder head of an engine require tightening to a torque of 88 m·N. If a wrench is 28 cm long, what force perpendicular to the wrench must the mechanic exert at its end? If the six-sided bolt head is 15 mm in diameter, estimate the force applied near each of the six points by a socket wrench (Fig. 8–41).

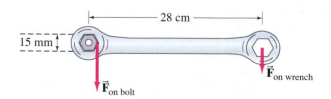

28 cm
15 mm
$\vec{F}_{\text{on wrench}}$
$\vec{F}_{\text{on bolt}}$

FIGURE 8–41   Problem 26.

8–5 and 8–6 Rotational Dynamics

27. (I) Determine the moment of inertia of a 10.8-kg sphere of radius 0.648 m when the axis of rotation is through its center.

28. (I) Calculate the moment of inertia of a bicycle wheel 66.7 cm in diameter. The rim and tire have a combined mass of 1.25 kg. The mass of the hub can be ignored (why?).

29. (II) A small 650-gram ball on the end of a thin, light rod is rotated in a horizontal circle of radius 1.2 m. Calculate (a) the moment of inertia of the ball about the center of the circle, and (b) the torque needed to keep the ball rotating at constant angular velocity if air resistance exerts a force of 0.020 N on the ball. Ignore the rod's moment of inertia and air resistance.

30. (II) A potter is shaping a bowl on a potter's wheel rotating at constant angular speed (Fig. 8–42). The friction force between her hands and the clay is 1.5 N total. (a) How large is her torque on the wheel, if the diameter of the bowl is 12 cm? (b) How long would it take for the potter's wheel to stop if the only torque acting on it is due to the potter's hand? The initial angular velocity of the wheel is 1.6 rev/s, and the moment of inertia of the wheel and the bowl is 0.11 kg·m².

FIGURE 8–42   Problem 30.

31. (II) Calculate the moment of inertia of the array of point objects shown in Fig. 8–43 about (a) the vertical axis, and (b) the horizontal axis. Assume $m = 1.8$ kg, $M = 3.1$ kg, and the objects are wired together by very light, rigid pieces of wire. The array is rectangular and is split through the middle by the horizontal axis. (c) About which axis would it be harder to accelerate this array?

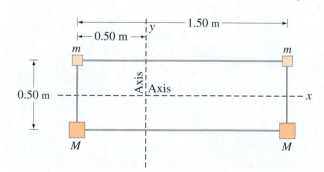

1.50 m
0.50 m
m y m
0.50 m
Axis   Axis x
M M

FIGURE 8–43   Problem 31.

32. (II) An oxygen molecule consists of two oxygen atoms whose total mass is 5.3×10^{-26} kg and whose moment of inertia about an axis perpendicular to the line joining the two atoms, midway between them, is 1.9×10^{-46} kg·m². From these data, estimate the effective distance between the atoms.

33. (II) To get a flat, uniform cylindrical satellite spinning at the correct rate, engineers fire four tangential rockets as shown in Fig. 8–44. If the satellite has a mass of 3600 kg and a radius of 4.0 m, what is the required steady force of each rocket if the satellite is to reach 32 rpm in 5.0 min?

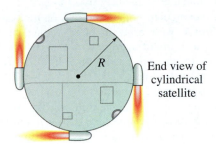

FIGURE 8–44
Problem 33.

34. (II) A grinding wheel is a uniform cylinder with a radius of 8.50 cm and a mass of 0.580 kg. Calculate (a) its moment of inertia about its center, and (b) the applied torque needed to accelerate it from rest to 1500 rpm in 5.00 s if it is known to slow down from 1500 rpm to rest in 55.0 s.

35. (II) A softball player swings a bat, accelerating it from rest to 3.0 rev/s in a time of 0.20 s. Approximate the bat as a 2.2-kg uniform rod of length 0.95 m, and compute the torque the player applies to one end of it.

36. (II) A teenager pushes tangentially on a small hand-driven merry-go-round and is able to accelerate it from rest to a frequency of 15 rpm in 10.0 s. Assume the merry-go-round is a uniform disk of radius 2.5 m and has a mass of 760 kg, and two children (each with a mass of 25 kg) sit opposite each other on the edge. Calculate the torque required to produce the acceleration, neglecting frictional torque. What force is required at the edge?

37. (II) A centrifuge rotor rotating at 10,300 rpm is shut off and is eventually brought uniformly to rest by a frictional torque of 1.20 m·N. If the mass of the rotor is 4.80 kg and it can be approximated as a solid cylinder of radius 0.0710 m, through how many revolutions will the rotor turn before coming to rest, and how long will it take?

38. (II) The forearm in Fig. 8–45 accelerates a 3.6-kg ball at 7.0 m/s² by means of the triceps muscle, as shown. Calculate (a) the torque needed, and (b) the force that must be exerted by the triceps muscle. Ignore the mass of the arm.

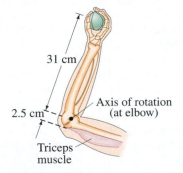

FIGURE 8–45
Problems 38 and 39.

39. (II) Assume that a 1.00-kg ball is thrown solely by the action of the forearm, which rotates about the elbow joint under the action of the triceps muscle, Fig. 8–45. The ball is accelerated uniformly from rest to 10.0 m/s in 0.350 s, at which point it is released. Calculate (a) the angular acceleration of the arm, and (b) the force required of the triceps muscle. Assume that the forearm has a mass of 3.70 kg and rotates like a uniform rod about an axis at its end.

40. (II) A helicopter rotor blade can be considered a long thin rod, as shown in Fig. 8–46. (a) If each of the three rotor helicopter blades is 3.75 m long and has a mass of 160 kg, calculate the moment of inertia of the three rotor blades about the axis of rotation. (b) How much torque must the motor apply to bring the blades up to a speed of 5.0 rev/s in 8.0 s?

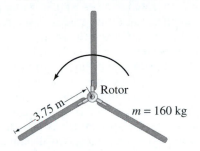

FIGURE 8–46
Problem 40.

41. (III) An *Atwood's machine* consists of two masses, m_1 and m_2, which are connected by a massless inelastic cord that passes over a pulley, Fig. 8–47. If the pulley has radius R and moment of inertia I about its axle, determine the acceleration of the masses m_1 and m_2, and compare to the situation in which the moment of inertia of the pulley is ignored. [*Hint:* The tensions F_{T1} and F_{T2} are not equal. We discussed this situation in Example 4–13, assuming $I = 0$ for the pulley.]

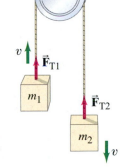

FIGURE 8–47
Problems 41 and 49.
Atwood's machine.

42. (III) A hammer thrower accelerates the hammer (mass = 7.30 kg) from rest within four full turns (revolutions) and releases it at a speed of 28.0 m/s. Assuming a uniform rate of increase in angular velocity and a horizontal circular path of radius 1.20 m, calculate (a) the angular acceleration, (b) the (linear) tangential acceleration, (c) the centripetal acceleration just before release, (d) the net force being exerted on the hammer by the athlete just before release, and (e) the angle of this force with respect to the radius of the circular motion.

8–7 Rotational Kinetic Energy

43. (I) A centrifuge rotor has a moment of inertia of 3.75×10^{-2} kg·m². How much energy is required to bring it from rest to 8250 rpm?

44. (II) An automobile engine develops a torque of 280 m·N at 3800 rpm. What is the power in watts and in horsepower?

45. (II) A bowling ball of mass 7.3 kg and radius 9.0 cm rolls without slipping down a lane at 3.3 m/s. Calculate its total kinetic energy.

46. (II) Estimate the kinetic energy of the Earth with respect to the Sun as the sum of two terms, (a) that due to its daily rotation about its axis, and (b) that due to its yearly revolution about the Sun. [Assume the Earth is a uniform sphere with mass $= 6.0 \times 10^{24}$ kg and radius $= 6.4 \times 10^6$ m, and is 1.5×10^8 km from the Sun.]

47. (II) A merry-go-round has a mass of 1640 kg and a radius of 7.50 m. How much net work is required to accelerate it from rest to a rotation rate of 1.00 revolution per 8.00 s? Assume it is a solid cylinder.

48. (II) A sphere of radius 20.0 cm and mass 1.80 kg starts from rest and rolls without slipping down a 30.0° incline that is 10.0 m long. (a) Calculate its translational and rotational speeds when it reaches the bottom. (b) What is the ratio of translational to rotational KE at the bottom? Avoid putting in numbers until the end so you can answer: (c) do your answers in (a) and (b) depend on the radius of the sphere or its mass?

49. (III) Two masses, $m_1 = 18.0$ kg and $m_2 = 26.5$ kg, are connected by a rope that hangs over a pulley (as in Fig. 8–47). The pulley is a uniform cylinder of radius 0.260 m and mass 7.50 kg. Initially, m_1 is on the ground and m_2 rests 3.00 m above the ground. If the system is now released, use conservation of energy to determine the speed of m_2 just before it strikes the ground. Assume the pulley is frictionless.

50. (III) A 2.30-m-long pole is balanced vertically on its tip. It starts to fall and its lower end does not slip. What will be the speed of the upper end of the pole just before it hits the ground? [Hint: Use conservation of energy.]

8–8 Angular Momentum

51. (I) What is the angular momentum of a 0.210-kg ball rotating on the end of a thin string in a circle of radius 1.10 m at an angular speed of 10.4 rad/s?

52. (I) (a) What is the angular momentum of a 2.8-kg uniform cylindrical grinding wheel of radius 18 cm when rotating at 1500 rpm? (b) How much torque is required to stop it in 6.0 s?

53. (II) A person stands, hands at his side, on a platform that is rotating at a rate of 1.30 rev/s. If he raises his arms to a horizontal position, Fig. 8–48, the speed of rotation decreases to 0.80 rev/s. (a) Why? (b) By what factor has his moment of inertia changed?

FIGURE 8–48
Problem 53.

54. (II) A diver (such as the one shown in Fig. 8–29) can reduce her moment of inertia by a factor of about 3.5 when changing from the straight position to the tuck position. If she makes 2.0 rotations in 1.5 s when in the tuck position, what is her angular speed (rev/s) when in the straight position?

55. (II) A figure skater can increase her spin rotation rate from an initial rate of 1.0 rev every 2.0 s to a final rate of 3.0 rev/s. If her initial moment of inertia was 4.6 kg·m², what is her final moment of inertia? How does she physically accomplish this change?

56. (II) A potter's wheel is rotating around a vertical axis through its center at a frequency of 1.5 rev/s. The wheel can be considered a uniform disk of mass 5.0 kg and diameter 0.40 m. The potter then throws a 3.1-kg chunk of clay, approximately shaped as a flat disk of radius 8.0 cm, onto the center of the rotating wheel. What is the frequency of the wheel after the clay sticks to it?

57. (II) (a) What is the angular momentum of a figure skater spinning at 3.5 rev/s with arms in close to her body, assuming her to be a uniform cylinder with a height of 1.5 m, a radius of 15 cm, and a mass of 55 kg? (b) How much torque is required to slow her to a stop in 5.0 s, assuming she does *not* move her arms?

58. (II) Determine the angular momentum of the Earth (a) about its rotation axis (assume the Earth is a uniform sphere), and (b) in its orbit around the Sun (treat the Earth as a particle orbiting the Sun). The Earth has mass $= 6.0 \times 10^{24}$ kg and radius $= 6.4 \times 10^6$ m, and is 1.5×10^8 km from the Sun.

59. (II) A nonrotating cylindrical disk of moment of inertia I is dropped onto an identical disk rotating at angular speed ω. Assuming no external torques, what is the final common angular speed of the two disks?

60. (II) A uniform disk turns at 2.4 rev/s around a frictionless spindle. A nonrotating rod, of the same mass as the disk and length equal to the disk's diameter, is dropped onto the freely spinning disk, Fig. 8–49. They then both turn around the spindle with their centers superposed. What is the angular frequency in rev/s of the combination?

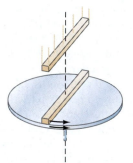

FIGURE 8–49
Problem 60.

61. (II) A person of mass 75 kg stands at the center of a rotating merry-go-round platform of radius 3.0 m and moment of inertia 920 kg·m². The platform rotates without friction with angular velocity 2.0 rad/s. The person walks radially to the edge of the platform. (a) Calculate the angular velocity when the person reaches the edge. (b) Calculate the rotational kinetic energy of the system of platform plus person before and after the person's walk.

62. (II) A 4.2-m-diameter merry-go-round is rotating freely with an angular velocity of 0.80 rad/s. Its total moment of inertia is 1760 kg·m². Four people standing on the ground, each of mass 65 kg, suddenly step onto the edge of the merry-go-round. What is the angular velocity of the merry-go-round now? What if the people were on it initially and then jumped off in a radial direction (relative to the merry-go-round)?

63. (II) Suppose our Sun eventually collapses into a white dwarf, losing about half its mass in the process, and winding up with a radius 1.0% of its existing radius. Assuming the lost mass carries away no angular momentum, what would the Sun's new rotation rate be? (Take the Sun's current period to be about 30 days.) What would be its final KE in terms of its initial KE of today?

64. (III) Hurricanes can involve winds in excess of 120 km/h at the outer edge. Make a crude estimate of (a) the energy, and (b) the angular momentum, of such a hurricane, approximating it as a rigidly rotating uniform cylinder of air (density 1.3 kg/m³) of radius 100 km and height 4.0 km.

65. (III) An asteroid of mass 1.0×10^5 kg, traveling at a speed of 30 km/s relative to the Earth, hits the Earth at the equator tangentially, and in the direction of Earth's rotation. Use angular momentum to estimate the percent change in the angular speed of the Earth as a result of the collision.

* **8–9 Angular Quantities as Vectors**

* **66.** (II) A person stands on a platform, initially at rest, that can rotate freely without friction. The moment of inertia of the person plus the platform is I_P. The person holds a spinning bicycle wheel with its axis horizontal. The wheel has moment of inertia I_W and angular velocity ω_W. What will be the angular velocity ω_P of the platform if the person moves the axis of the wheel so that it points (a) vertically upward, (b) at a 60° angle to the vertical, (c) vertically downward? (d) What will ω_P be if the person reaches up and stops the wheel in part (a)?

* **67.** (III) Suppose a 55-kg person stands at the edge of a 6.5-m diameter merry-go-round turntable that is mounted on frictionless bearings and has a moment of inertia of 1700 kg·m². The turntable is at rest initially, but when the person begins running at a speed of 3.8 m/s (with respect to the turntable) around its edge, the turntable begins to rotate in the opposite direction. Calculate the angular velocity of the turntable.

General Problems

68. A large spool of rope rolls on the ground with the end of the rope lying on the top edge of the spool. A person grabs the end of the rope and walks a distance L, holding onto it, Fig. 8–50. The spool rolls behind the person without slipping. What length of rope unwinds from the spool? How far does the spool's center of mass move?

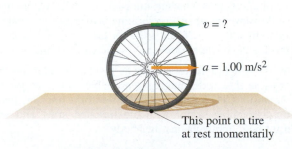

FIGURE 8–50
Problem 68.

69. The Moon orbits the Earth such that the same side always faces the Earth. Determine the ratio of the Moon's spin angular momentum (about its own axis) to its orbital angular momentum. (In the latter case, treat the Moon as a particle orbiting the Earth.)

70. A cyclist accelerates from rest at a rate of 1.00 m/s². How fast will a point on the rim of the tire (diameter = 68 cm) at the top be moving after 3.0 s? [*Hint:* At any moment, the lowest point on the tire is in contact with the ground and is at rest—see Fig. 8–51.]

$v = ?$

$a = 1.00$ m/s²

This point on tire
at rest momentarily

FIGURE 8–51 Problem 70.

71. A 1.4-kg grindstone in the shape of a uniform cylinder of radius 0.20 m acquires a rotational rate of 1800 rev/s from rest over a 6.0-s interval at constant angular acceleration. Calculate the torque delivered by the motor.

72. (a) A yo-yo is made of two solid cylindrical disks, each of mass 0.050 kg and diameter 0.075 m, joined by a (concentric) thin solid cylindrical hub of mass 0.0050 kg and diameter 0.010 m. Use conservation of energy to calculate the linear speed of the yo-yo when it reaches the end of its 1.0-m-long string, if it is released from rest. (b) What fraction of its kinetic energy is rotational?

73. (a) For a bicycle, how is the angular speed of the rear wheel (ω_R) related to that of the pedals and front sprocket (ω_F), Fig. 8–52? That is, derive a formula for ω_R/ω_F. Let N_F and N_R be the number of teeth on the front and rear sprockets, respectively. The teeth are spaced equally on all sprockets so that the chain meshes properly. (b) Evaluate the ratio ω_R/ω_F when the front and rear sprockets have 52 and 13 teeth, respectively, and (c) when they have 42 and 28 teeth.

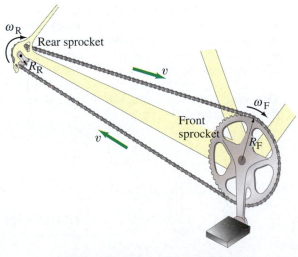

ω_R

Rear sprocket

R_R

v

v

ω_F

Front
sprocket

R_F

FIGURE 8–52 Problem 73.

74. Suppose a star the size of our Sun, but with mass 8.0 times as great, were rotating at a speed of 1.0 revolution every 12 days. If it were to undergo gravitational collapse to a neutron star of radius 11 km, losing three-quarters of its mass in the process, what would its rotation speed be? Assume that the star is a uniform sphere at all times, and that the lost mass carries off no angular momentum.

75. One possibility for a low-pollution automobile is for it to use energy stored in a heavy rotating flywheel. Suppose such a car has a total mass of 1400 kg, uses a uniform cylindrical flywheel of diameter 1.50 m and mass 240 kg, and should be able to travel 350 km without needing a flywheel "spinup." (a) Make reasonable assumptions (average frictional retarding force = 450 N, twenty acceleration periods from rest to 95 km/h, equal uphill and downhill, and that energy can be put back into the flywheel as the car goes downhill), and show that the total energy needed to be stored in the flywheel is about 1.7×10^8 J. (b) What is the angular velocity of the flywheel when it has a full "energy charge"? (c) About how long would it take a 150-hp motor to give the flywheel a full energy charge before a trip?

76. Figure 8–53 illustrates an H_2O molecule. The O–H bond length is 0.96 nm and the H–O–H bonds make an angle of 104°. Calculate the moment of inertia for the H_2O molecule about an axis passing through the center of the oxygen atom (a) perpendicular to the plane of the molecule, and (b) in the plane of the molecule, bisecting the H–O–H bonds.

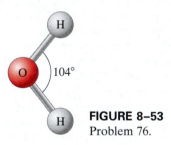

FIGURE 8–53
Problem 76.

77. A hollow cylinder (hoop) is rolling on a horizontal surface at speed $v = 3.3$ m/s when it reaches a 15° incline. (a) How far up the incline will it go? (b) How long will it be on the incline before it arrives back at the bottom?

78. A uniform rod of mass M and length L can pivot freely (i.e., we ignore friction) about a hinge attached to a wall, as in Fig. 8–54. The rod is held horizontally and then released. At the moment of release, determine (a) the angular acceleration of the rod, and (b) the linear acceleration of the tip of the rod. Assume that the force of gravity acts at the center of mass of the rod, as shown. [Hint: See Fig. 8–21g.]

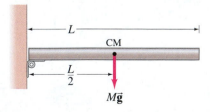

FIGURE 8–54
Problem 78.

79. A wheel of mass M has radius R. It is standing vertically on the floor, and we want to exert a horizontal force F at its axle so that it will climb a step against which it rests (Fig. 8–55). The step has height h, where $h < R$. What minimum force F is needed?

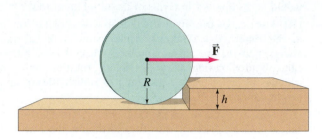

FIGURE 8–55 Problem 79.

80. A bicyclist traveling with speed $v = 4.2$ m/s on a flat road is making a turn with a radius $r = 6.4$ m. The forces acting on the cyclist and cycle are the normal force ($\vec{F}_N$) and friction force ($\vec{F}_{fr}$) exerted by the road on the tires, and $m\vec{g}$, the total weight of the cyclist and cycle (see Fig. 8–56). (a) Explain carefully why the angle θ the bicycle makes with the vertical (Fig. 8–56) must be given by $\tan \theta = F_{fr}/F_N$ if the cyclist is to maintain balance. (b) Calculate θ for the values given. (c) If the coefficient of static friction between tires and road is $\mu_s = 0.70$, what is the minimum turning radius?

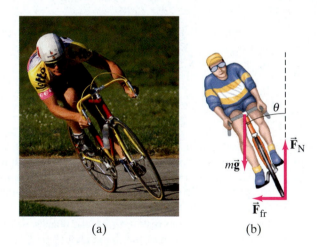

(a)   (b)

FIGURE 8–56 Problem 80.

81. Suppose David puts a 0.50-kg rock into a sling of length 1.5 m and begins whirling the rock in a nearly horizontal circle above his head, accelerating it from rest to a rate of 120 rpm after 5.0 s. What is the torque required to achieve this feat, and where does the torque come from?

82. Model a figure skater's body as a solid cylinder and her arms as thin rods, making reasonable estimates for the dimensions. Then calculate the ratio of the angular speeds for a spinning skater with outstretched arms, and with arms held tightly against her body.

83. You are designing a clutch assembly which consists of two cylindrical plates, of mass $M_A = 6.0$ kg and $M_B = 9.0$ kg, with equal radii $R = 0.60$ m. They are initially separated (Fig. 8–57). Plate M_A is accelerated from rest to an angular velocity $\omega_1 = 7.2$ rad/s in time $\Delta t = 2.0$ s. Calculate (a) the angular momentum of M_A, and (b) the torque required to have accelerated M_A from rest to ω_1. (c) Plate M_B, initially at rest but free to rotate without friction, is allowed to fall vertically (or pushed by a spring), so it is in firm contact with plate M_A (their contact surfaces are high-friction). Before contact, M_A was rotating at constant ω_1. After contact, at what constant angular velocity ω_2 do the two plates rotate?

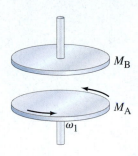

FIGURE 8–57
Problem 83.

84. A marble of mass m and radius r rolls along the looped rough track of Fig. 8–58. What is the minimum value of the vertical height h that the marble must drop if it is to reach the highest point of the loop without leaving the track? Assume $r \ll R$, and ignore frictional losses.

85. Repeat Problem 84, but do not assume $r \ll R$.

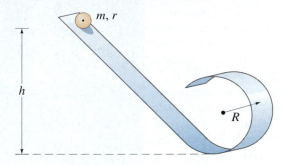

FIGURE 8–58 Problems 84 and 85.

86. The tires of a car make 85 revolutions as the car reduces its speed uniformly from 90.0 km/h to 60.0 km/h. The tires have a diameter of 0.90 m. (a) What was the angular acceleration of each tire? (b) If the car continues to decelerate at this rate, how much more time is required for it to stop?

Answers to Exercises

A: $f = 0.076$ Hz; $T = 13$ s.

B: $\vec{F}_A$.

C: Yes; she does work to pull in her arms.

D: Work was done in pulling the string and decreasing the circle's radius.

Our whole built environment, from modern bridges to skyscrapers, has required architects and engineers to determine the forces and stresses within these structures. The object is to keep these structures static—that is, not in motion, especially not falling down.

The study of statics applies equally well to the human body, including balance, the forces in muscles, joints, and bones, and ultimately the possibility of fracture.

CHAPTER **9**

Static Equilibrium; Elasticity and Fracture

This Chapter deals with forces within objects at rest

I n this Chapter, we will study a special case in mechanics—when the net force and the net torque on an object, or system of objects, are both zero. In this case both the linear acceleration and the angular acceleration of the object or system are zero. The object is either at rest, or its center of mass is moving at constant velocity. We will be concerned mainly with the first situation, in which the object or objects are all at rest. Now, you may think that the study of objects at rest is not very interesting since the objects will have neither velocity nor acceleration, and the net force and the net torque will be zero. But this does not imply that no forces at all act on the objects. In fact it is virtually impossible to find an object on which no forces act. Just how and where these forces act can be very important, both for buildings and other structures, and in the human body.

Sometimes, as we shall see in this Chapter, the forces may be so great that the object is seriously *deformed*, or it may even *fracture* (break)—and avoiding such problems gives this field of *statics* even greater importance.

Statics is concerned with the calculation of the forces acting on and within structures that are in *equilibrium*. Determination of these forces, which occupies us in the first part of this Chapter, then allows a determination of whether the structures can sustain the forces without significant deformation or fracture, subjects we discuss later in this Chapter. These techniques can be applied in a wide range of fields. Architects and engineers must be able to calculate the forces on the structural components of buildings, bridges, machines, vehicles, and other structures, since any material will buckle or break if too much force is applied (Fig. 9–1). In the human body a knowledge of the forces in muscles and joints is of great value for doctors, physical therapists, and athletes.

9–1 The Conditions for Equilibrium

Objects in daily life have at least one force acting on them (gravity). If they are at rest, then there must be other forces acting on them as well so that the net force is zero. A book at rest on a table, for example, has two forces acting on it, the downward force of gravity and the normal force the table exerts upward on it (Fig. 9–2). Since the net force on the book is zero, the upward force exerted by the table on the book must be equal in magnitude to the force of gravity acting downward on the book. Such an object is said to be in **equilibrium** (Latin for "equal forces" or "balance") under the action of these two forces.

Do not confuse the two forces in Fig. 9–2 with the equal and opposite forces of Newton's third law, which act on different objects. Here, both forces act on the same object.

EXAMPLE 9–1 **Straightening teeth.** The wire band shown in Fig. 9–3a has a tension F_T of 2.0 N along it. It therefore exerts forces of 2.0 N on the highlighted tooth (to which it is attached) in the two directions shown. Calculate the resultant force on the tooth due to the wire, F_R.

APPROACH Since the two forces F_T are equal, their sum will be directed along the line that bisects the angle between them, which we have chosen to be the y axis. The x components of the two forces add up to zero.

SOLUTION The y component of each force is $(2.0 \, \text{N})(\cos 70°) = 0.684 \, \text{N}$: adding the two together, we get a resultant force $F_R = 1.37 \, \text{N}$ as shown in Fig. 9–3b. We assume that the tooth is in equilibrium because the gums exert an equal and opposite force. Actually that is not quite so since the objective is to move the tooth ever so slowly.

NOTE If the wire is firmly attached to the tooth, the tension to the right, say, can be made larger than that to the left, and the resultant force would correspondingly be directed more toward the right.

FIGURE 9–1 Elevated walkway collapse in a Kansas City hotel in 1981. How a simple physics calculation could have prevented the tragic loss of over 100 lives is considered in Example 9–12.

PHYSICS APPLIED
Orthodontia

FIGURE 9–2 The book is in equilibrium; the net force on it is zero.

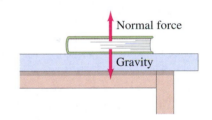

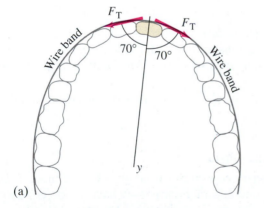

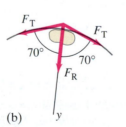

FIGURE 9–3 Forces on a tooth. Example 9–1.

(a)   (b)

The First Condition for Equilibrium

For an object to be at rest, Newton's second law tells us that the sum of the forces acting on it must add up to zero. Since force is a vector, the components of the net force must each be zero. Hence, a condition for equilibrium is that

$$\Sigma F_x = 0, \qquad \Sigma F_y = 0, \qquad \Sigma F_z = 0. \qquad (9-1)$$

We will mainly be dealing with forces that act in a plane, so we usually need only the x and y components. We must remember that if a particular force component points along the negative x or y axis, it must have a negative sign. Equations 9–1 are called the **first condition for equilibrium**.

First condition for equilibrium: the sum of all forces is zero

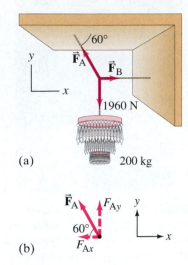

(a)   200 kg

(b)

FIGURE 9-4 Example 9-2.

EXAMPLE 9-2 **Chandelier cord tension.** Calculate the tensions $\vec{F}_A$ and $\vec{F}_B$ in the two cords that are connected to the vertical cord supporting the 200-kg chandelier in Fig. 9-4.

APPROACH We need a free-body diagram, but for which object? If we choose the chandelier, the cord supporting it must exert a force equal to the chandelier's weight $mg = (200\ \text{kg})(9.8\ \text{m/s}^2) = 1960\ \text{N}$. But the forces $\vec{F}_A$ and $\vec{F}_B$ don't get involved. Instead, let us choose as our object the point where the three cords join (it could be a knot). The free-body diagram is then as shown in Fig. 9-4a. The three forces—$\vec{F}_A$, $\vec{F}_B$, and the tension in the vertical cord equal to the weight of the 200-kg chandelier—act at this point where the three cords join. For this junction point we write $\Sigma F_x = 0$ and $\Sigma F_y = 0$, since the problem is laid out in two dimensions. The directions of $\vec{F}_A$ and $\vec{F}_B$ are known, since tension in a rope can only be along the rope—any other direction would cause the rope to bend, as already pointed out in Chapter 4. Thus, our unknowns are the magnitudes F_A and F_B.

SOLUTION We first resolve $\vec{F}_A$ into its horizontal (x) and vertical (y) components. Although we don't know the value of F_A, we can write (see Fig. 9-4b) $F_{Ax} = -F_A \cos 60°$ and $F_{Ay} = F_A \sin 60°$. $\vec{F}_B$ has only an x component. In the vertical direction, we have the downward force exerted by the vertical cord equal to the weight of the chandelier $= (200\ \text{kg})(g)$, and the vertical component of $\vec{F}_A$ upward. Since $\Sigma F_y = 0$, we have

$$\Sigma F_y = F_A \sin 60° - (200\ \text{kg})(g) = 0$$

so

$$F_A = \frac{(200\ \text{kg})g}{\sin 60°} = \frac{(200\ \text{kg})g}{0.866} = (231\ \text{kg})g = 2260\ \text{N}.$$

In the horizontal direction,

$$\Sigma F_x = F_B - F_A \cos 60° = 0.$$

Thus

$$F_B = F_A \cos 60° = (231\ \text{kg})(g)(0.500) = (115\ \text{kg})g = 1130\ \text{N}.$$

The magnitudes of $\vec{F}_A$ and $\vec{F}_B$ determine the strength of cord or wire that must be used. In this case, the wire must be able to hold more than 230 kg.

NOTE We didn't insert the value of g, the acceleration due to gravity, until the end. In this way we found the magnitude of the force in terms of g times the number of kilograms (which may be a more familiar quantity than newtons).

EXERCISE A In Example 9-2, F_A has to be greater than the chandelier's weight, mg. Why?

FIGURE 9-5 Although the net force on it is zero, the ruler will move (rotate). A pair of equal forces acting in opposite directions but at different points on an object (as shown here) is referred to as a *couple*.

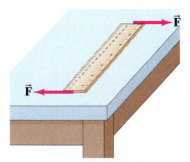

Second condition for equilibrium: the sum of all torques is zero

The Second Condition for Equilibrium

Although Eqs. 9-1 are a necessary condition for an object to be in equilibrium, they are not always a sufficient condition. Figure 9-5 shows an object on which the net force is zero. Although the two forces labeled $\vec{F}$ add up to give zero net force on the object, they do give rise to a net torque that will rotate the object. Referring to Eq. 8-14, $\Sigma\tau = I\alpha$, we see that if an object is to remain at rest, the net torque applied to it (calculated about *any* axis) must be zero. Thus we have the **second condition for equilibrium**: that the sum of the torques acting on an object, as calculated about any axis, must be zero:

$$\Sigma\tau = 0. \tag{9-2}$$

This condition will ensure that the angular acceleration, α, about any axis will be zero. If the object is not rotating initially ($\omega = 0$), it will not start rotating. Equations 9–1 and 9–2 are the only requirements for an object to be in equilibrium.

We will mainly consider cases in which the forces all act in a plane (we call it the *xy* plane). In such cases the torque is calculated about an axis that is perpendicular to the *xy* plane. *The choice of this axis is arbitrary.* If the object is at rest, then $\Sigma\tau = 0$ about any axis whatever. Therefore we can choose any axis that makes our calculation easier. Once the axis is chosen, all torques must be calculated about that axis.

CAUTION

Axis choice for $\Sigma\tau = 0$ is arbitrary. All torques must be calculated about the same axis.

CONCEPTUAL EXAMPLE 9–3 **A lever.** The bar in Fig. 9–6 is being used as a lever to pry up a large rock. The small rock acts as a fulcrum (pivot point). The force F_P required at the long end of the bar can be quite a bit smaller than the rock's weight mg, since it is the *torques* that balance in the rotation about the fulcrum. If, however, the leverage isn't sufficient, and the large rock isn't budged, what are two ways to increase the leverage?

RESPONSE One way is to increase the lever arm of the force F_P by slipping a pipe over the end of the bar and thereby pushing with a longer lever arm. A second way is to move the fulcrum closer to the large rock. This may change the long lever arm R only a little, but it changes the short lever arm r by a substantial fraction and therefore changes the ratio of R/r dramatically. In order to pry the rock, the torque due to F_P must at least balance the torque due to mg, so $mgr = F_P R$ and

$$\frac{r}{R} = \frac{F_P}{mg}.$$

With r smaller, the weight mg can be balanced with less force F_P. The ratio R/r is the **mechanical advantage** of the system. A lever is a "simple machine." We discussed another simple machine, the pulley, in Chapter 4, Example 4–14.

EXERCISE B For simplicity, we wrote the equation in Example 9–3 as if the lever were perpendicular to the forces. Would the equation be valid even for a lever at an angle as shown in Fig. 9–6?

PHYSICS APPLIED
The lever

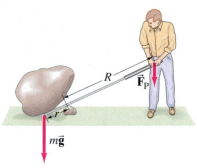

FIGURE 9–6 Example 9–3. A lever can "multiply" your force.

9–2 Solving Statics Problems

This subject of statics is important because it allows us to calculate certain forces on (or within) a structure when some of the forces on it are already known. We will mainly consider situations in which all the forces act in a plane, so we can have two force equations (*x* and *y* components) and one torque equation, for a total of three equations. Of course, you do not have to use all three equations if they are not needed. When using a torque equation, a torque that tends to rotate the object counterclockwise is usually considered positive, whereas a torque that tends to rotate it clockwise is considered negative. (But the opposite convention would not be wrong.)

One of the forces that acts on objects is the force of gravity. Our analysis in this Chapter is greatly simplified if we use the concept of center of gravity (CG) or center of mass (CM), which for practical purposes are the same point. As we discussed in Section 7–8, we can consider the force of gravity on the object as acting at its CG. For uniform symmetrically shaped objects, the CG is at the geometric center. For more complicated objects, the CG can be determined as discussed in Section 7–8.

There is no single technique for attacking statics problems, but the following procedure may be helpful.

PROBLEM SOLVING

$\tau > 0$ *counterclockwise*
$\tau < 0$ *clockwise*

1. Choose one object at a time for consideration. Make a careful **free-body diagram** by showing all the forces acting on that object and the points at which these forces act. If you aren't sure of the direction of a force, choose a direction; if the actual direction is opposite, your eventual calculation will give a result with a minus sign.

2. Choose a convenient **coordinate system**, and resolve the forces into their components.

3. Using letters to represent unknowns, write down the **equilibrium equations** for the **forces**:

$$\Sigma F_x = 0 \quad \text{and} \quad \Sigma F_y = 0,$$

assuming all the forces act in a plane.

4. For the **torque equation**,

$$\Sigma \tau = 0,$$

choose any axis perpendicular to the xy plane that

might make the calculation easier. (For example, you can reduce the number of unknowns in the resulting equation by choosing the axis so that one of the unknown forces acts through that axis; then this force will have zero lever arm and produce zero torque, and so won't appear in the equation.) Pay careful attention to determining the lever arm for each force correctly. Give each torque a + or − sign to indicate torque direction. For example, if torques tending to rotate the object counterclockwise are positive, then those tending to rotate it clockwise are negative.

5. **Solve** these equations for the unknowns. Three equations allow a maximum of three unknowns to be solved for. They can be forces, distances, or even angles.

EXAMPLE 9–4 **Balancing a seesaw.** A board of mass $M = 2.0\,\text{kg}$ serves as a seesaw for two children, as shown in Fig. 9–7a. Child A has a mass of 30 kg and sits 2.5 m from the pivot point, P (his center of gravity is 2.5 m from the pivot). At what distance x from the pivot must child B, of mass 25 kg, place herself to balance the seesaw? Assume the board is uniform and centered over the pivot.

APPROACH We follow the steps of the Problem Solving Box explicitly.

SOLUTION

1. **Free-body diagram.** We choose the board as our object, and assume it is horizontal. Its free-body diagram is shown in Fig. 9–7b. The forces acting on the board are the forces exerted downward on it by each child, $\vec{\mathbf{F}}_A$ and $\vec{\mathbf{F}}_B$, the upward force exerted by the pivot $\vec{\mathbf{F}}_N$, and the force of gravity on the board ($= Mg$) which acts at the center of the uniform board.

2. **Coordinate system.** We choose y to be vertical, with positive upward, and x horizontal to the right, with origin at the pivot.

3. **Force equation.** All the forces are in the y (vertical) direction, so

$$\Sigma F_y = 0$$
$$F_N - m_A g - m_B g - Mg = 0,$$

where $F_A = m_A g$ and $F_B = m_B g$ because each child is in equilibrium when the seesaw is balanced.

FIGURE 9–7 (a) Two children on a seesaw, Example 9–4. (b) Free-body diagram of the board.

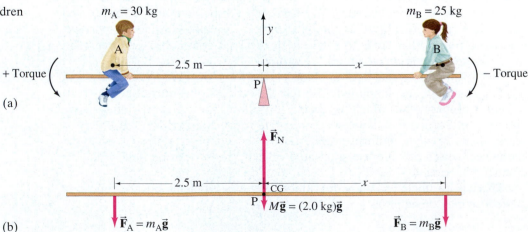

4. **Torque equation**. Let us calculate the torque about an axis through the board at the pivot point, P. Then the lever arms for F_N and for the weight of the board are zero, and they will contribute zero torque (about point P) in our torque equation. Thus the torque equation will involve only the forces $\vec{F}_A$ and $\vec{F}_B$, which are equal to the weights of the children. The torque exerted by each child will be mg times the appropriate lever arm, which here is the distance of each child from the pivot point. Hence the torque equation is

$$\Sigma\tau = 0$$

$$m_A g(2.5\ \text{m}) - m_B gx + Mg(0\ \text{m}) + F_N(0\ \text{m}) = 0$$

or

$$m_A g(2.5\ \text{m}) - m_B gx = 0,$$

where two terms were dropped because their lever arms were zero.

5. **Solve**. We solve the torque equation for x and find

$$x = \frac{m_A}{m_B}(2.5\ \text{m}) = \frac{30\ \text{kg}}{25\ \text{kg}}(2.5\ \text{m}) = 3.0\ \text{m}.$$

To balance the seesaw, child B must sit so that her CM is 3.0 m from the pivot point. This makes sense: since she is lighter, she must sit farther from the pivot than the heavier child.

EXERCISE C We did not need to use the force equation to solve Example 9–4 because of our choice of the axis. Use the force equation to find the force exerted by the pivot.

EXAMPLE 9–5 **Forces on a beam and supports.** A uniform 1500-kg beam, 20.0 m long, supports a 15,000-kg printing press 5.0 m from the right support column (Fig. 9–8). Calculate the force on each of the vertical support columns.

APPROACH We analyze the forces on the beam (the force the beam exerts on each column is equal and opposite to the force exerted by the column on the beam). We label these forces $\vec{F}_A$ and $\vec{F}_B$ in Fig. 9–8. The weight of the beam itself acts at its center of gravity, 10.0 m from either end. We choose a convenient axis for writing the torque equation: the point of application of $\vec{F}_A$ (labeled P), so $\vec{F}_A$ will not enter the equation (its lever arm will be zero) and we will have an equation in only one unknown, F_B.

SOLUTION The torque equation, $\Sigma\tau = 0$, with the counterclockwise direction as positive gives

$$\Sigma\tau = -(10.0\ \text{m})(1500\ \text{kg})g - (15.0\ \text{m})(15{,}000\ \text{kg})g + (20.0\ \text{m})F_B = 0.$$

Solving for F_B, we find $F_B = (12{,}000\ \text{kg})g = 118{,}000\ \text{N}$. To find F_A, we use $\Sigma F_y = 0$, with $+y$ upward:

$$\Sigma F_y = F_A - (1500\ \text{kg})g - (15{,}000\ \text{kg})g + F_B = 0.$$

Putting in $F_B = (12{,}000\ \text{kg})g$, we find that $F_A = (4500\ \text{kg})g = 44{,}100\ \text{N}$.

Figure 9–9 shows a uniform beam that extends beyond its support like a diving board. Such a beam is called a **cantilever**. The forces acting on the beam in Fig. 9–9 are those due to the supports, $\vec{F}_A$ and $\vec{F}_B$, and the force of gravity which acts at the CG, 5.0 m to the right of the right-hand support. If you follow the procedure of the last Example and calculate F_A and F_B, assuming they point upward as shown in Fig. 9–9, you will find that F_A comes out negative. If the beam has a mass of 1200 kg and a weight $mg = 12{,}000\ \text{N}$, then $F_B = 15{,}000\ \text{N}$ and $F_A = -3000\ \text{N}$ (see Problem 10). Whenever an unknown force comes out negative, it merely means that the force actually points in the opposite direction from what you assumed. Thus in Fig. 9–9, $\vec{F}_A$ actually points downward. With a little reflection it should become clear that the left-hand support must indeed pull downward on the beam (by means of bolts, screws, fasteners and/or glue) if the beam is to be in equilibrium; otherwise the sum of the torques about the CG (or about the point where $\vec{F}_B$ acts) could not be zero.

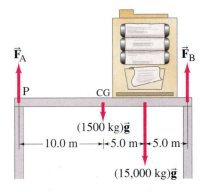

FIGURE 9–8 A 1500-kg beam supports a 15,000-kg machine. Example 9–5.

PHYSICS APPLIED
Cantilever

PROBLEM SOLVING
If a force comes out negative

FIGURE 9–9 A cantilever.

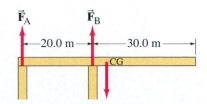

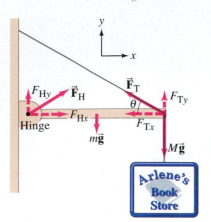

FIGURE 9–10 Example 9–6.

Our next Example involves a beam that is attached to a wall by a hinge and is supported by a cable or cord (Fig. 9–10). It is important to remember that a flexible cable can support a force only along its length. (If there were a component of force perpendicular to the cable, it would bend because it is flexible.) But for a rigid device, such as the hinge in Fig. 9–10, the force can be in any direction and we can know the direction only after solving the problem. The hinge is assumed small and smooth, so it can exert no internal torque (about its center) on the beam.

EXAMPLE 9–6 **Hinged beam and cable.** A uniform beam, 2.20 m long with mass $m = 25.0$ kg, is mounted by a hinge on a wall as shown in Fig. 9–10. The beam is held in a horizontal position by a cable that makes an angle $\theta = 30.0°$ as shown. The beam supports a sign of mass $M = 28.0$ kg suspended from its end. Determine the components of the force $\vec{F}_H$ that the hinge exerts on the beam, and the tension F_T in the supporting cable.

APPROACH Figure 9–10 is the free-body diagram for the beam, showing all the forces acting on the beam. It also shows the components of $\vec{F}_H$ and $\vec{F}_T$. We have three unknowns, F_{Hx}, F_{Hy}, and F_T (we are given θ), so we will need all three equations, $\Sigma F_x = 0$, $\Sigma F_y = 0$, $\Sigma\tau = 0$.

SOLUTION The sum of the forces in the vertical (y) direction is

$$\Sigma F_y = 0$$
$$F_{Hy} + F_{Ty} - mg - Mg = 0. \tag{i}$$

In the horizontal (x) direction, the sum of the forces is

$$\Sigma F_x = 0$$
$$F_{Hx} - F_{Tx} = 0. \tag{ii}$$

For the torque equation, we choose the axis at the point where $\vec{F}_T$ and $M\vec{g}$ act (so our equation then contains only one unknown, F_{Hy}). We choose torques that tend to rotate the beam counterclockwise as positive. The weight mg of the (uniform) beam acts at its center, so we have

$$\Sigma\tau = 0$$
$$-(F_{Hy})(2.20\text{ m}) + mg(1.10\text{ m}) = 0.$$

We solve for F_{Hy}:

$$F_{Hy} = \left(\frac{1.10\text{ m}}{2.20\text{ m}}\right)mg = (0.500)(25.0\text{ kg})(9.80\text{ m/s}^2) = 123\text{ N}. \tag{iii}$$

Next, since the tension $\vec{F}_T$ in the cable acts along the cable ($\theta = 30.0°$), we see from Fig. 9–10 that $\tan\theta = F_{Ty}/F_{Tx}$, or

$$F_{Ty} = F_{Tx}\tan\theta = F_{Tx}(\tan 30.0°) = 0.577\,F_{Tx}. \tag{iv}$$

Equation (i) above gives

$$F_{Ty} = (m + M)g - F_{Hy} = (53.0\text{ kg})(9.80\text{ m/s}^2) - 123\text{ N} = 396\text{ N};$$

Equations (iv) and (ii) give

$$F_{Tx} = F_{Ty}/0.577 = 687\text{ N};$$
$$F_{Hx} = F_{Tx} = 687\text{ N}.$$

The components of $\vec{F}_H$ are $F_{Hy} = 123$ N and $F_{Hx} = 687$ N. The tension in the wire is $F_T = \sqrt{F_{Tx}^2 + F_{Ty}^2} = 793$ N.

Alternate Solution Let us see the effect of choosing a different axis for calculating torques, such as an axis through the hinge. Then the lever arm for F_H is zero, and the torque equation $(\Sigma\tau = 0)$ becomes

$$-mg(1.10\,\text{m}) - Mg(2.20\,\text{m}) + F_{Ty}(2.20\,\text{m}) = 0.$$

We solve this for F_{Ty} and find

$$F_{Ty} = \frac{m}{2}g + Mg = (12.5\,\text{kg} + 28.0\,\text{kg})(9.80\,\text{m/s}^2) = 397\,\text{N}.$$

We get the same result, within the precision of our significant figures.

NOTE It doesn't matter which axis we choose for $\Sigma\tau = 0$. Using a second axis can serve as a check.

Additional Example—The Ladder

EXAMPLE 9–7 **Ladder.** A 5.0-m-long ladder leans against a wall at a point 4.0 m above a cement floor as shown in Fig. 9–11. The ladder is uniform and has mass $m = 12.0\,\text{kg}$. Assuming the wall is frictionless (but the floor is not), determine the forces exerted on the ladder by the floor and by the wall.

APPROACH Figure 9–11 is the free-body diagram for the ladder, showing all the forces acting on the ladder. The wall, since it is frictionless, can exert a force only perpendicular to the wall, and we label that force $\vec{F}_W$. The cement floor exerts a force $\vec{F}_C$ which has both horizontal and vertical force components: F_{Cx} is frictional and F_{Cy} is the normal force. Finally, gravity exerts a force $mg = (12.0\,\text{kg})(9.80\,\text{m/s}^2) = 118\,\text{N}$ on the ladder at its midpoint, since the ladder is uniform.

SOLUTION Again we use the equilibrium conditions, $\Sigma F_x = 0$, $\Sigma F_y = 0$, $\Sigma\tau = 0$. We will need all three since there are three unknowns: F_W, F_{Cx}, and F_{Cy}. The y component of the force equation is

$$\Sigma F_y = F_{Cy} - mg = 0,$$

so immediately we have

$$F_{Cy} = mg = 118\,\text{N}.$$

The x component of the force equation is

$$\Sigma F_x = F_{Cx} - F_W = 0.$$

To determine both F_{Cx} and F_W, we need a torque equation. If we choose to calculate torques about an axis through the point where the ladder touches the cement floor, then $\vec{F}_C$, which acts at this point, will have a lever arm of zero and so won't enter the equation. The ladder touches the floor a distance $x_0 = \sqrt{(5.0\,\text{m})^2 - (4.0\,\text{m})^2} = 3.0\,\text{m}$ from the wall. The lever arm for mg is half this, or 1.5 m, and the lever arm for F_W is 4.0 m, Fig. 9–11. We get

$$\Sigma\tau = (4.0\,\text{m})F_W - (1.5\,\text{m})mg = 0.$$

Thus

$$F_W = \frac{(1.5\,\text{m})(12.0\,\text{kg})(9.8\,\text{m/s}^2)}{4.0\,\text{m}} = 44\,\text{N}.$$

Then, from the x component of the force equation,

$$F_{Cx} = F_W = 44\,\text{N}.$$

Since the components of $\vec{F}_C$ are $F_{Cx} = 44\,\text{N}$ and $F_{Cy} = 118\,\text{N}$, then

$$F_C = \sqrt{(44\,\text{N})^2 + (118\,\text{N})^2} = 126\,\text{N} \approx 130\,\text{N}$$

(rounded off to two significant figures), and it acts at an angle to the floor of

$$\theta = \tan^{-1}(118\,\text{N}/44\,\text{N}) = 70°.$$

NOTE The force $\vec{F}_C$ does *not* have to act along the ladder's direction because the ladder is rigid and not flexible like a cord or cable.

EXERCISE D Why is it reasonable to ignore friction along the wall, but not reasonable to ignore it along the floor?

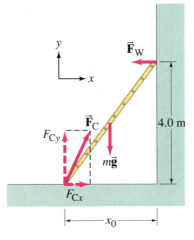

FIGURE 9–11 A ladder leaning against a wall. Example 9–7.

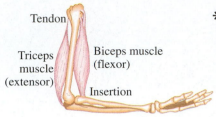

Tendon

Triceps muscle (extensor)

Biceps muscle (flexor)

Insertion

FIGURE 9–12 The biceps (flexor) and triceps (extensor) muscles in the human arm.

PHYSICS APPLIED
Forces in muscles and joints

FIGURE 9–13 Example 9–8.

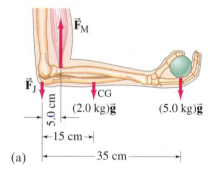

$\vec{F}_M$

$\vec{F}_J$

5.0 cm

CG

(2.0 kg)$\vec{g}$

(5.0 kg)$\vec{g}$

|←15 cm→|

(a) |←————35 cm————→|

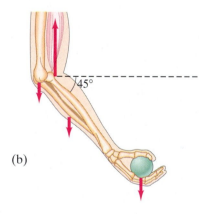

45°

(b)

PHYSICS APPLIED
Muscle insertion and lever arm

PHYSICS APPLIED
Forces on the spine, and back pain

*9–3 Applications to Muscles and Joints

The techniques we have been discussing for calculating forces on objects in equilibrium can readily be applied to the human (or animal) body, and can be of great use in studying the forces on muscles, bones, and joints for organisms in motion or at rest. Generally a muscle is attached, via tendons, to two different bones, as in Fig. 9–12. The points of attachment are called *insertions*. Two bones are flexibly connected at a *joint*, such as those at the elbow, knee, and hip. A muscle exerts a pull when its fibers contract under stimulation by a nerve, but a muscle cannot exert a push. Muscles that tend to bring two limbs closer together, such as the biceps muscle in the upper arm (Fig. 9–12) are called *flexors*; those that act to extend a limb outward, such as the triceps muscle in Fig. 9–12, are called *extensors*. You use the flexor muscle in the upper arm when lifting an object in your hand; you use the extensor muscle when throwing a ball.

EXAMPLE 9–8 **Force exerted by biceps muscle.** How much force must the biceps muscle exert when a 5.0-kg mass is held in the hand (*a*) with the arm horizontal as in Fig. 9–13a, and (*b*) when the arm is at a 45° angle as in Fig. 9–13b? Assume that the mass of forearm and hand together is 2.0 kg and their CG is as shown.

APPROACH The forces acting on the forearm are shown in Fig. 9–13 and include the weights of the arm and ball, the upward force $\vec{F}_M$ exerted by the muscle, and a force $\vec{F}_J$ exerted at the joint by the bone in the upper arm (all assumed to act vertically). We wish to find the magnitude of $\vec{F}_M$, which is done most easily by using the torque equation and by choosing our axis through the joint so that $\vec{F}_J$ contributes zero torque.

SOLUTION (*a*) We calculate torques about the point where $\vec{F}_J$ acts in Fig. 9–13a. The $\Sigma\tau = 0$ equation gives

$$(0.050\,\text{m})F_M - (0.15\,\text{m})(2.0\,\text{kg})g - (0.35\,\text{m})(5.0\,\text{kg})g = 0.$$

We solve for F_M:

$$F_M = \frac{(0.15\,\text{m})(2.0\,\text{kg})g + (0.35\,\text{m})(5.0\,\text{kg})g}{0.050\,\text{m}} = (41\,\text{kg})g = 400\,\text{N}.$$

(*b*) The lever arm, as calculated about the joint, is reduced by the factor $\sin 45°$ for all three forces. Our torque equation will look like the one just above, except that each term will have its lever arm reduced by the same factor, which will cancel out. The same result is obtained, $F_M = 400\,\text{N}$.

NOTE The force required of the muscle (400 N) is quite large compared to the weight of the object lifted (49 N). Indeed, the muscles and joints of the body are generally subjected to quite large forces.

The point of insertion of a muscle varies from person to person. A slight increase in the distance of the joint to the point of insertion of the biceps muscle from 5.0 cm to 5.5 cm can be a considerable advantage for lifting and throwing. Champion athletes are often found to have muscle insertions farther from the joint than the average person, and if this applies to one muscle, it usually applies to all.

As another example of the large forces acting within the human body, we consider the muscles used to support the trunk when a person bends forward (Fig. 9–14a). The lowest vertebra on the spinal column (fifth lumbar vertebra) acts as a fulcrum for this bending position. The "erector spinae" muscles in the back that support the trunk act at an effective angle of about 12° to the axis of the spine. Figure 9–14b is a simplified schematic drawing showing the forces on the upper body. We assume the trunk makes an angle of 30° with the horizontal.

The force exerted by the back muscles is represented by $\vec{F}_M$, the force exerted on the base of the spine at the lowest vertebra is $\vec{F}_V$, and $\vec{w}_H$, $\vec{w}_A$, and $\vec{w}_T$ represent the weights of the **h**ead, freely hanging **a**rms, and **t**runk, respectively. The values shown are approximations taken from Table 7–1. The distances (in cm) refer to a person 180 cm tall, but are approximately in the same ratio of 1:2:3 for an average person of any height, and the result in the following Example is then independent of the height of the person.

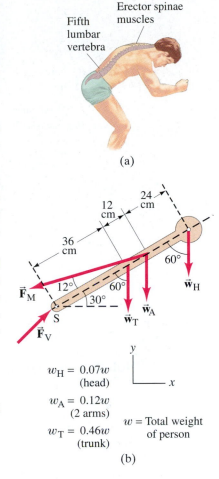
(a)

EXAMPLE 9–9 **Forces on your back.** Calculate the magnitude and direction of the force $\vec{F}_V$ acting on the fifth lumbar vertebra for the example shown in Fig. 9–14b.

APPROACH We use the model of the upper body described above and shown in Fig. 9–14b. We can calculate F_M using the torque equation if we take the axis at the base of the spine (point S); with this choice, the other unknown, F_V, doesn't appear in the equation because its lever arm is zero. To figure the lever arms, we need to use trigonometric functions.

SOLUTION For $\vec{F}_M$, the lever arm (perpendicular distance from axis to line of action of the force) will be the real distance to where the force acts (48 cm) multiplied by $\sin 12°$, as shown in Fig. 9–14c. The lever arms for $\vec{w}_H$, $\vec{w}_A$, and $\vec{w}_T$ can be seen from Fig. 9–14b to be their respective distances from S times $\sin 60°$. F_M tends to rotate the trunk counterclockwise, which we take to be positive. Then $\vec{w}_H$, $\vec{w}_A$, $\vec{w}_T$ will contribute negative torques. Thus $\Sigma\tau = 0$ gives

$$(0.48\,\text{m})(\sin 12°)(F_M) - (0.72\,\text{m})(\sin 60°)(w_H)$$
$$- (0.48\,\text{m})(\sin 60°)(w_A) - (0.36\,\text{m})(\sin 60°)(w_T) = 0.$$

Solving for F_M and putting in the values for w_H, w_A, w_T given in Fig. 9–14b, we find

$$F_M = \frac{(0.72\,\text{m})(0.07w) + (0.48\,\text{m})(0.12w) + (0.36\,\text{m})(0.46w)}{(0.48\,\text{m})(\sin 12°)}(\sin 60°)$$

$$= 2.37w \approx 2.4w,$$

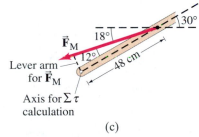

$w_H = 0.07w$
(head)

$w_A = 0.12w$
(2 arms)

$w_T = 0.46w$
(trunk)

w = Total weight of person

(b)

where w is the total weight of the body. To get the components of $\vec{F}_V$ we use the x and y components of the force equation (noting that $30° - 12° = 18°$):

$$\Sigma F_y = F_{Vy} - F_M \sin 18° - w_H - w_A - w_T = 0$$

so

$$F_{Vy} = 1.38w \approx 1.4w,$$

and

$$\Sigma F_x = F_{Vx} - F_M \cos 18° = 0$$

so

$$F_{Vx} = 2.25w \approx 2.3w,$$

where we keep 3 significant figures for calculating, but round off to 2 for giving the answer. Then

$$F_V = \sqrt{F_{Vx}^2 + F_{Vy}^2} = 2.6w.$$

The angle θ that F_V makes with the horizontal is given by $\tan\theta = F_{Vy}/F_{Vx}$ = 0.61, so $\theta = 32°$.

Lever arm for $\vec{F}_M$

Axis for $\Sigma\tau$ calculation

(c)

FIGURE 9–14 (a) A person bending over. (b) Forces on the back exerted by the back muscles ($\vec{F}_M$) and by the vertebrae ($\vec{F}_V$) when a person bends over. (c) Finding the lever arm for $\vec{F}_M$.

NOTE The force on the lowest vertebra is over $2\frac{1}{2}$ times the total body weight! This force is exerted by the "sacral" bone at the base of the spine, through the fluid-filled and somewhat flexible *intervertebral disc*. The discs at the base of the spine are clearly being compressed under very large forces. [If the body was less bent over (say, the 30° angle in Fig. 9–14b becomes 40° or 50°), then the stress on the lower back will be less (see Problem 35).]

If the person in Fig. 9–14 has a mass of 90 kg and is holding 20 kg in his hands (this increases w_A to $0.34w$), then F_V is increased to almost four times the person's weight ($3.7w$). For this 200-lb person, the force on the disc would be over 700 lb! With such strong forces acting, it is little wonder that so many people suffer from low back pain at one time or another.

9–4 Stability and Balance

An object in static equilibrium, if left undisturbed, will undergo no translational or rotational acceleration since the sum of all the forces and the sum of all the torques acting on it are zero. However, if the object is displaced slightly, three outcomes are possible: (1) the object returns to its original position, in which case it is said to be in **stable equilibrium**; (2) the object moves even farther from its original position, and it is said to be in **unstable equilibrium**; or (3) the object remains in its new position, and it is said to be in **neutral equilibrium**.

Stable and unstable equilibria

Consider the following examples. A ball suspended freely from a string is in stable equilibrium, for if it is displaced to one side, it will return to its original position (Fig. 9–15a) due to the net force and torque exerted on it. On the other hand, a pencil standing on its point is in unstable equilibrium. If its center of gravity is directly over its tip (Fig. 9–15b), the net force and net torque on it will be zero. But if it is displaced ever so slightly as shown—say, by a slight vibration or tiny air current—there will be a torque on it, and this torque acts to make the pencil continue to fall in the direction of the original displacement. Finally, an example of an object in neutral equilibrium is a sphere resting on a horizontal tabletop. If it is placed slightly to one side, it will remain in its new position—no net torque acts on it.

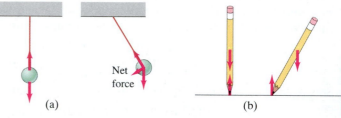

FIGURE 9–15 (a) Stable equilibrium, and (b) unstable equilibrium.

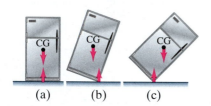

(a)   (b)   (c)

FIGURE 9–16 Equilibrium of a refrigerator resting on a flat floor.

FIGURE 9–17 Humans adjust their posture to achieve stability when carrying loads.

Total CG

Humans and balance

In most situations, such as in the design of structures and in working with the human body, we are interested in maintaining stable equilibrium, or *balance*, as we sometimes say. In general, an object whose center of gravity (CG) is below its point of support, such as a ball on a string, will be in stable equilibrium. If the CG is above the base of support, we have a more complicated situation. Consider a standing refrigerator (Fig. 9–16a). If it is tipped slightly, it will return to its original position due to the torque on it as shown in Fig. 9–16b. But if it is tipped too far, Fig. 9–16c, it will fall over. The critical point is reached when the CG shifts from one side of the pivot point to the other. When the CG is on one side, the torque pulls the object back onto its original base of support, Fig. 9–16b. If the object is tipped further, the CG goes past the pivot point and the torque causes the object to topple, Fig. 9–16c. In general, *an object whose center of gravity is above its base of support will be stable if a vertical line projected downward from the CG falls within the base of support.* This is because the normal force upward on the object (which balances out gravity) can be exerted only within the area of contact, so if the force of gravity acts beyond this area, a net torque will act to topple the object.

Stability, then, can be relative. A brick lying on its widest face is more stable than a brick standing on its end, for it will take more of an effort to tip it over. In the extreme case of the pencil in Fig. 9–15b, the base is practically a point and the slightest disturbance will topple it. In general, the larger the base and the lower the CG, the more stable the object.

In this sense, humans are much less stable than four-legged mammals, which not only have a larger base of support because of their four legs, but also have a lower center of gravity. When walking and performing other kinds of movement, a person continually shifts the body so that its CG is over the feet, although in the normal adult this requires no conscious thought. Even as simple a movement as bending over requires moving the hips backward so that the CG remains over the feet, and you do this repositioning without thinking about it. To see this, position yourself with your heels and back to a wall and try to touch your toes. You won't be able to do it without falling. Persons carrying heavy loads automatically adjust their posture so that the CG of the total mass is over their feet, Fig. 9–17.

In the first part of this Chapter we studied how to calculate the forces on objects in equilibrium. In this Section we study the effects of these forces: any object changes shape under the action of applied forces. If the forces are great enough, the object will break, or *fracture*, as we will discuss in Section 9–6.

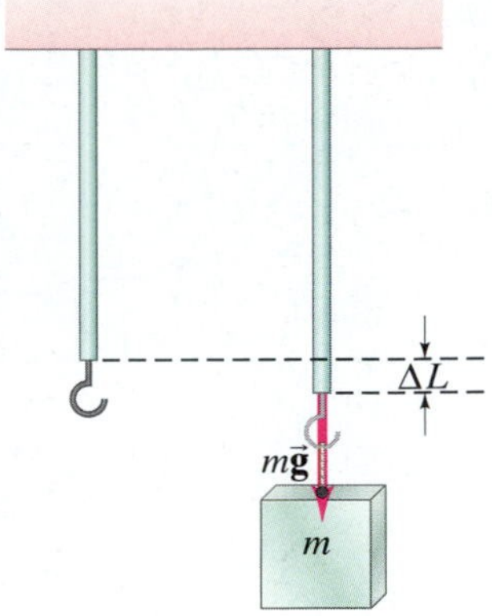

* Elasticity and Hooke's Law

If a force is exerted on an object, such as the vertically suspended metal rod shown in Fig. 9–18, the length of the object changes. If the amount of elongation, ΔL, is small compared to the length of the object, experiment shows that ΔL is proportional to the force exerted on the object. This proportionality, as we saw in Section 6–4, can be written as an equation:

$$F = k\,\Delta L. \qquad (9\text{–}3)$$

Hooke's law (again)

Here F represents the force pulling on the object, ΔL is the change in length, and k is a proportionality constant. Equation 9–3, which is sometimes called **Hooke's law**[†] after Robert Hooke (1635–1703), who first noted it, is found to be valid for almost any solid material from iron to bone—but it is valid only up to a point. For if the force is too great, the object stretches excessively and eventually breaks.

Figure 9–19 shows a typical graph of applied force versus elongation. Up to a point called the **proportional limit**, Eq. 9–3 is a good approximation for many common materials, and the curve is a straight line. Beyond this point, the graph deviates from a straight line, and no simple relationship exists between F and ΔL. Nonetheless, up to a point farther along the curve called the **elastic limit**, the object will return to its original length if the applied force is removed. The region from the origin to the elastic limit is called the *elastic region*. If the object is stretched beyond the elastic limit, it enters the *plastic region*: it does not return to the original length upon removal of the external force, but remains permanently deformed (such as a bent paper clip). The maximum elongation is reached at the *breaking point*. The maximum force that can be applied without breaking is called the **ultimate strength** of the material (actually, force per unit area as we discuss in Section 9–6).

FIGURE 9–18 Hooke's law: $\Delta L \propto$ applied force.

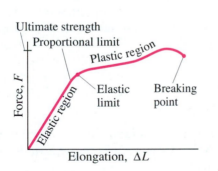

FIGURE 9–19 Applied force vs. elongation for a typical metal under tension.

* Young's Modulus

The amount of elongation of an object, such as the rod shown in Fig. 9–18, depends not only on the force applied to it, but also on the material of which it is made and on its dimensions. That is, the constant k in Eq. 9–3 can be written in terms of these factors.

[†]The term "law" applied to this relation is not really appropriate, since first of all, it is only an approximation, and secondly, it refers only to a limited set of phenomena. Most physicists prefer to reserve the word "law" for those relations that are deeper and more encompassing and precise, such as Newton's laws of motion or the law of conservation of energy.

TABLE 9–1 Elastic Moduli

Material	Young's Modulus, E (N/m²)	Shear Modulus, G (N/m²)	Bulk Modulus, B (N/m²)
Solids			
Iron, cast	100×10^9	40×10^9	90×10^9
Steel	200×10^9	80×10^9	140×10^9
Brass	100×10^9	35×10^9	80×10^9
Aluminum	70×10^9	25×10^9	70×10^9
Concrete	20×10^9		
Brick	14×10^9		
Marble	50×10^9		70×10^9
Granite	45×10^9		45×10^9
Wood (pine) (parallel to grain)	10×10^9		
(perpendicular to grain)	1×10^9		
Nylon	5×10^9		
Bone (limb)	15×10^9	80×10^9	
Liquids			
Water			2.0×10^9
Alcohol (ethyl)			1.0×10^9
Mercury			2.5×10^9
Gases[†]			
Air, H_2, He, CO_2			1.01×10^5

[†] At normal atmospheric pressure; no variation in temperature during process.

If we compare rods made of the same material but of different lengths and cross-sectional areas, it is found that for the same applied force, the amount of stretch (again assumed small compared to the total length) is proportional to the original length and inversely proportional to the cross-sectional area. That is, the longer the object, the more it elongates for a given force; and the thicker it is, the less it elongates. These findings can be combined with Eq. 9–3 to yield

$$\Delta L = \frac{1}{E}\frac{F}{A}L_0, \qquad (9\text{–}4)$$

Young's modulus

where L_0 is the original length of the object, A is the cross-sectional area, and ΔL is the change in length due to the applied force F. E is a constant of proportionality[‡] known as the **elastic modulus**, or **Young's modulus**; its value depends only on the material. The value of Young's modulus for various materials is given in Table 9–1 (the shear modulus and bulk modulus in this Table are discussed later in this Section). Because E is a property only of the material and is independent of the object's size or shape, Eq. 9–4 is far more useful for practical calculation than Eq. 9–3.

EXAMPLE 9–10 **Tension in piano wire.** A 1.60-m-long steel piano wire has a diameter of 0.20 cm. How great is the tension in the wire if it stretches 0.25 cm when tightened?

APPROACH We assume Hooke's law holds, and use it in the form of Eq. 9–4, finding E for steel in Table 9–1.

SOLUTION We solve for F in Eq. 9–4 and note that the area of the wire is $A = \pi r^2 = (3.14)(0.0010\ \text{m})^2 = 3.14 \times 10^{-6}\ \text{m}^2$. Then

$$F = E\frac{\Delta L}{L_0}A = (2.0 \times 10^{11}\ \text{N/m}^2)\left(\frac{0.0025\ \text{m}}{1.60\ \text{m}}\right)(3.14 \times 10^{-6}\ \text{m}^2) = 980\ \text{N}.$$

The large tension in all the wires in a piano must be supported by a strong frame.

[‡] The fact that E is in the denominator, so $1/E$ is the actual proportionality constant, is merely a convention. When we rewrite Eq. 9–4 to get Eq. 9–5, E is found in the numerator.

* Stress and Strain

Let us return to solid objects. From Eq. 9–4, we see that the change in length of an object is directly proportional to the product of the object's length L_0 and the force per unit area F/A applied to it. It is general practice to define the force per unit area as the **stress**:

$$\text{stress} = \frac{\text{force}}{\text{area}} = \frac{F}{A},$$

Stress defined

which has SI units of N/m². Also, the **strain** is defined to be the ratio of the change in length to the original length:

$$\text{strain} = \frac{\text{change in length}}{\text{original length}} = \frac{\Delta L}{L_0},$$

Strain defined

and is dimensionless (no units). Strain is thus the fractional change in length of the object, and is a measure of how much the rod has been deformed. Stress is applied to the material by external agents, whereas strain is the material's response to the stress. Equation 9–4 can be rewritten as

$$\frac{F}{A} = E \frac{\Delta L}{L_0} \tag{9-5}$$

or

$$E = \frac{F/A}{\Delta L/L_0} = \frac{\text{stress}}{\text{strain}}.$$

Young's modulus

Thus we see that the strain is directly proportional to the stress, in the linear (elastic) region of Fig. 9–19.

* Tension, Compression, and Shear Stress

The rod shown in Fig. 9–20a is said to be under *tension* or **tensile stress**. Not only is there a force pulling down on the rod at its lower end, but since the rod is in equilibrium we know that the support at the top is exerting an equal[†] upward force on the rod at its upper end, Fig. 9–20a. In fact, this tensile stress exists throughout the material. Consider, for example, the lower half of a suspended rod as shown in Fig. 9–20b. This lower half is in equilibrium, so there must be an upward force on it to balance the downward force at its lower end. What exerts this upward force? It must be the upper part of the rod. Thus we see that external forces applied to an object give rise to internal forces, or stress, within the material itself. (Recall also the discussion of tension in a cord, page 86.)

Tension and compression

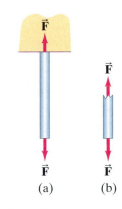

Strain or deformation due to tensile stress is but one type of stress to which materials can be subjected. There are two other common types of stress: compressive and shear. **Compressive stress** is the exact opposite of tensile stress. Instead of being stretched, the material is compressed: the forces act inwardly on the object. Columns that support a weight, such as the columns of a Greek temple (Fig. 9–21), are subjected to compressive stress. Equations 9–4 and 9–5 apply equally well to compression and tension, and the values for the modulus E are usually the same.

FIGURE 9–20 Stress exists *within* the material.

[†]If the weight of the rod can be ignored compared to F.

FIGURE 9–21 This Greek temple, in Agrigento, Sicily, built 2500 years ago, shows the post-and-beam construction. The columns are under compression.

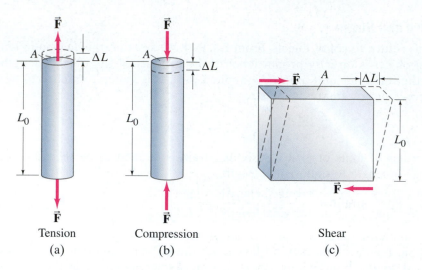

FIGURE 9–22 The three types of stress for rigid objects.

| Tension | Compression | Shear |
| (a) | (b) | (c) |

Figure 9–22 compares tensile and compressive stresses as well as the third type, shear stress. An object under **shear stress** has equal and opposite forces applied *across* its opposite faces. A simple example is a book or brick firmly attached to a tabletop, on which a force is exerted parallel to the top surface. The table exerts an equal and opposite force along the bottom surface. Although the dimensions of the object do not change significantly, the shape of the object does change, Fig. 9–22c. An equation similar to Eq. 9–4 can be applied to calculate shear strain:

Shear

$$\Delta L = \frac{1}{G}\frac{F}{A}L_0,$$ (9–6)

but ΔL, L_0, and A must be reinterpreted as indicated in Fig. 9–22c. Note that A is the area of the surface *parallel* to the applied force (and not perpendicular as for tension and compression), and ΔL is *perpendicular* to L_0. The constant of proportionality G is called the **shear modulus** and is generally one-half to one-third the value of Young's modulus E (see Table 9–1). Figure 9–23 illustrates why $\Delta L \propto L_0$: the fatter book shifts more for the same shearing force.

Shear modulus

FIGURE 9–23
The fatter book (a) shifts more than the thinner book (b) with the same applied shear force.

(a)   (b)

* Volume Change—Bulk Modulus

If an object is subjected to inward forces from all sides, its volume will decrease. A common situation is an object submerged in a fluid; in this case, the fluid exerts a pressure on the object in all directions, as we shall see in Chapter 10. *Pressure* is defined as force per unit area, and thus is the equivalent of stress. For this situation the change in volume, ΔV, is proportional to the original volume, V_0, and to the change in the pressure, ΔP. We thus obtain a relation of the same form as Eq. 9–4 but with a proportionality constant called the **bulk modulus** B:

$$\frac{\Delta V}{V_0} = -\frac{1}{B}\Delta P$$ (9–7)

or

Bulk modulus defined

$$B = -\frac{\Delta P}{\Delta V/V_0}.$$

The minus sign means the volume *decreases* with an increase in pressure.

Values for the bulk modulus are given in Table 9–1. Since liquids and gases do not have a fixed shape, only the bulk modulus (not the Young's or shear moduli) applies to them.

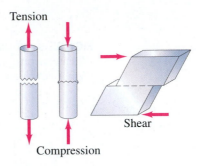
Tension
Shear
Compression

FIGURE 9–24 Fracture as a result of the three types of stress.

*9–6 Fracture

If the stress on a solid object is too great, the object fractures, or breaks (Fig. 9–24). Table 9–2 lists the ultimate strengths for tension, compression, and shear for a variety of materials. These values give the maximum force per unit area, or stress, that an object can withstand under each of these three types of stress for various types of material. They are, however, representative values only, and the actual value for a given specimen can differ considerably. It is therefore necessary to maintain a *safety factor* of from 3 to perhaps 10 or more—that is, the actual stresses on a structure should not exceed one-tenth to one-third of the values given in the Table. You may encounter tables of "allowable stresses" in which appropriate safety factors have already been included.

TABLE 9–2 Ultimate Strengths of Materials (force/area)

Material	Tensile Strength (N/m²)	Compressive Strength (N/m²)	Shear Strength (N/m²)
Iron, cast	170×10^6	550×10^6	170×10^6
Steel	500×10^6	500×10^6	250×10^6
Brass	250×10^6	250×10^6	200×10^6
Aluminum	200×10^6	200×10^6	200×10^6
Concrete	2×10^6	20×10^6	2×10^6
Brick		35×10^6	
Marble		80×10^6	
Granite		170×10^6	
Wood (pine) (parallel to grain)	40×10^6	35×10^6	5×10^6
(perpendicular to grain)		10×10^6	
Nylon	500×10^6		
Bone (limb)	130×10^6	170×10^6	

EXAMPLE 9–11 Breaking the piano wire. The steel piano wire we discussed in Example 9–10 was 1.60 m long with a diameter of 0.20 cm. Approximately what tension force would break it?

APPROACH We set the tensile stress F/A equal to the tensile strength of steel given in Table 9–2.

SOLUTION The area of the wire is $A = \pi r^2$, where $r = 0.10$ cm $= 1.0 \times 10^{-3}$ m. Then

$$\frac{F}{A} = 500 \times 10^6 \, \text{N/m}^2$$

so the wire would likely break if the force exceeded

$$F = (500 \times 10^6 \, \text{N/m}^2)(\pi)(1.0 \times 10^{-3} \, \text{m})^2 = 1600 \, \text{N}.$$

As can be seen in Table 9–2, concrete (like stone and brick) is reasonably strong under compression but extremely weak under tension. Thus concrete can be used as vertical columns placed under compression, but is of little value as a beam because it cannot withstand the tensile forces that result from the inevitable sagging of the lower edge of a beam (see Fig. 9–25).

FIGURE 9–25 A beam sags, at least a little (but is exaggerated here), even under its own weight. The beam thus changes shape: the upper edge is compressed, and the lower edge is under tension (elongated). Shearing stress also occurs within the beam.

Compression
Tension

FIGURE 9–26 Steel rods around which concrete will be poured to form a new highway.

PHYSICS APPLIED
Reinforced concrete and prestressed concrete

Reinforced concrete, in which iron rods are embedded in the concrete (Fig. 9–26), is much stronger. But the concrete on the lower edge of a loaded beam still tends to crack because it is weak under tension. This problem is solved with *prestressed concrete*, which also contains iron rods or a wire mesh, but during the pouring of the concrete, the rods or wire are held under tension. After the concrete dries, the tension on the iron is released, putting the concrete under compression. The amount of compressive stress is carefully predetermined so that when loads are applied to the beam, they reduce the compression on the lower edge, but never put the concrete into tension.

PHYSICS APPLIED
A tragic collapse

CONCEPTUAL EXAMPLE 9–12 | **A tragic substitution.** Two walkways, one above the other, are suspended from vertical rods attached to the ceiling of a high hotel lobby, Fig. 9–27a. The original design called for single rods 14 m long, but when such long rods proved to be unwieldy to install, it was decided to replace each long rod with two shorter ones as shown schematically in Fig. 9–27b. Determine the net force exerted by the rods on the supporting pin A (assumed to be the same size) for each design. Assume each vertical rod supports a mass m of each bridge.

FIGURE 9–27 Example 9–12.

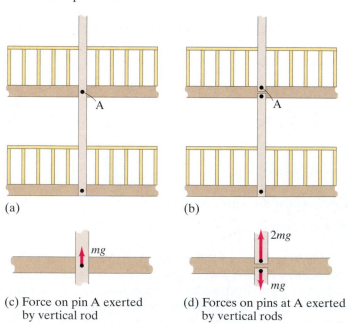

(a)

(b)

(c) Force on pin A exerted by vertical rod

mg

(d) Forces on pins at A exerted by vertical rods

$2mg$

mg

RESPONSE The single long vertical rod in Fig. 9–27a exerts an upward force equal to mg on pin A to support the mass m of the upper bridge. Why? Because the pin is in equilibrium, and the other force that balances this is the downward force mg exerted on it by the upper bridge (Fig. 9–27c). There is thus a shear stress on the pin because the rod pulls up on one end of the pin, and the bridge pulls down on the other end. The situation when two shorter rods support the bridges (Fig. 9–27b) is shown in Fig. 9–27d, in which only the connections at the upper bridge are shown. The lower rod exerts a force of mg downward on the lower of the two pins because it supports the lower bridge. The upper rod exerts a force of $2mg$ on the upper pin (pin A) because the upper rod supports both bridges. Thus we see that when the builders substituted two shorter rods for each single long one, the stress in the supporting pin A was *doubled*. What perhaps seemed like a simple substitution did, in fact, lead to a tragic collapse in 1981 with a loss of life of over 100 people (see Fig. 9–1). Having a feel for physics, and being able to make simple calculations based on physics, can have a great effect, literally, on people's lives.

*9–7 Spanning a Space: Arches and Domes

There are a great many areas where the arts and humanities overlap the sciences, and this is especially clear in architecture, where the forces in the materials that make up a structure need to be understood to avoid excessive deformation and collapse. Many of the features we admire in the architecture of the past were introduced not simply for their decorative effect, but for technical reasons. One example is the development of methods to span a space, from the simple beam to arches and domes.

The first important architectural invention was the post-and-beam (or post-and-lintel) construction, in which two upright posts support a horizontal beam. Before steel was introduced in the nineteenth century, the length of a beam was quite limited because the strongest building materials were then stone and brick. Hence the width of a span was limited by the size of available stones. Equally important, stone and brick, though strong under compression—are very weak under tension and shear; all three types of stress occur in a beam (see Fig. 9–25). The minimal space that could be spanned using stone is shown by the closely spaced columns of the great Greek temples (Fig. 9–21).

The introduction of the semicircular **arch** by the Romans (Fig. 9–28), aside from its aesthetic appeal, was a tremendous technological innovation.

🚶 PHYSICS APPLIED
Architecture: beams, arches and domes

(a)

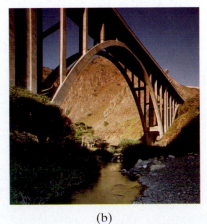
(b)

FIGURE 9–28 (a) 2000-year-old round arches in Rome. The one in the background is the Arch of Titus. (b) A modern arch used to span a chasm on the California coast.

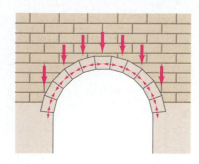

FIGURE 9–29 Stones in a round (or "true") arch are mainly under compression.

FIGURE 9–30 Flying buttresses (on the cathedral of Notre Dame, in Paris).

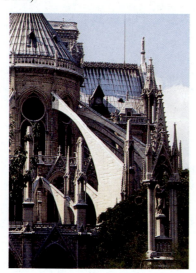

The advantage of the "true" or round (semicircular) arch is that, if well designed, its wedge-shaped stones experience stress which is mainly compressive (Fig. 9–29) even when supporting a large load such as the wall and roof of a cathedral. A round arch consisting of many well-shaped stones could span a very wide space. However, considerable buttressing on the sides was needed to support the horizontal components of the forces, which we discuss shortly.

The pointed arch came into use about A.D. 1100 and became the hallmark of the great Gothic cathedrals. It too was an important technical innovation, and was first used to support heavy loads such as the tower of a cathedral, and as the central arch. Because of the steepness of the pointed arch, the forces due to the weight above could be brought down more nearly vertically, so less horizontal buttressing would be needed. The pointed arch reduced the load on the walls, so there could be more openness and light. The smaller buttressing needed was provided on the outside by graceful flying buttresses (Fig. 9–30).

The technical innovation of the pointed arch was achieved not through calculation but through experience and intuition; it was not until much later that detailed calculations, such as those presented earlier in this Chapter, came into use. To make an accurate analysis of a stone arch is quite difficult in practice. But if we make some simplifying assumptions, we can show why the horizontal component of the force at the base is less for a pointed arch than for a round one. Figure 9–31 shows a round arch and a pointed arch, each with an 8.0-m span. The height of the round arch is thus 4.0 m, whereas that of the pointed arch is larger and has been chosen to be 8.0 m. Each arch supports a weight of 12.0×10^4 N ($\approx 12{,}000$ kg $\times$ g), which, for simplicity, we have divided into two parts (each 6.0×10^4 N) acting on the two halves of each arch as shown. For the arch to be in equilibrium, each of the supports must exert an upward force of 6.0×10^4 N. Each support also exerts a horizontal force, F_H, at the base of the arch, and it is this we want to calculate. We focus only on the right half of each arch. We set equal to zero the total torque calculated about the apex of the arch due to the forces exerted on that half arch, as if there were a hinge at the apex. For the round arch, the torque equation ($\Sigma \tau = 0$) is

$$(4.0 \text{ m})(6.0 \times 10^4 \text{ N}) - (2.0 \text{ m})(6.0 \times 10^4 \text{ N}) - (4.0 \text{ m})(F_H) = 0.$$

Thus $F_H = 3.0 \times 10^4$ N for the round arch. For the pointed arch, the torque equation is

$$(4.0 \text{ m})(6.0 \times 10^4 \text{ N}) - (2.0 \text{ m})(6.0 \times 10^4 \text{ N}) - (8.0 \text{ m})(F_H) = 0.$$

Solving, we find that $F_H = 1.5 \times 10^4$ N—only half as much as for the round arch! From this calculation we can see that the horizontal buttressing force required for a pointed arch is less because the arch is higher, and there is therefore a longer lever arm for this force. Indeed, the steeper the arch, the less the horizontal component of the force needs to be, and hence the more nearly vertical is the force exerted at the base of the arch.

FIGURE 9–31 Forces in (a) a round arch, compared with those in (b) a pointed arch.

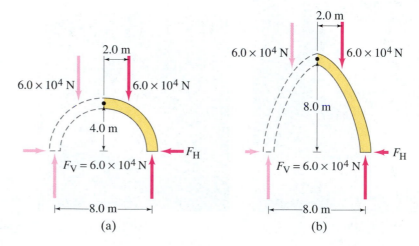

Whereas an arch spans a two-dimensional space, a **dome**—which is basically an arch rotated about a vertical axis—spans a three-dimensional space. The Romans built the first large domes. Their shape was hemispherical and some still stand, such as that of the Pantheon in Rome (Fig. 9–32), built 2000 years ago.

Fourteen centuries later, a new cathedral was being built in Florence. It was to have a dome 43 m in diameter to rival that of the Pantheon, whose construction has remained a mystery. The new dome was to rest on a "drum" with no external abutments. Filippo Brunelleschi (1377–1446) designed a pointed dome (Fig. 9–33), since a pointed dome, like a pointed arch, exerts a smaller side thrust against its base. A dome, like an arch, is not stable until all the stones are in place. To support smaller domes during construction, wooden frameworks were used. But no trees big enough or strong enough could be found to span the 43-m space required. Brunelleschi decided to try to build the dome in horizontal layers, each bonded to the previous one, holding it in place until the last stone of the circle was placed. Each closed ring was then strong enough to support the next layer. It was an amazing feat. Only in the twentieth century were larger domes built, the largest being that of the Superdome in New Orleans, completed in 1975.

FIGURE 9–32 Interior of the Pantheon in Rome, built in the first century. This view, showing the great dome and its central opening for light, was painted about 1740 by Panini. Photographs do not capture its grandeur as well as this painting does.

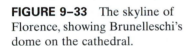
FIGURE 9–33 The skyline of Florence, showing Brunelleschi's dome on the cathedral.

EXAMPLE 9–13 **A modern dome.** The 1.2×10^6 kg dome of the Small Sports Palace in Rome (Fig. 9–34a) is supported by 36 buttresses positioned at a 38° angle so that they connect smoothly with the dome. Calculate the components of the force, F_H and F_V, that each buttress exerts on the dome so that the force acts purely in compression—that is, at a 38° angle (Fig. 9–34b).

APPROACH We can find the vertical component F_V exerted upward by each buttress because each supports $\frac{1}{36}$ of the dome's weight. We find F_H knowing that the buttress needs to be under compression so $\vec{F} = \vec{F}_V + \vec{F}_H$ acts at a 38° angle.

SOLUTION The vertical load on *each* buttress is $\frac{1}{36}$ of the total weight. Thus

$$F_V = \frac{mg}{36} = \frac{(1.2 \times 10^6 \text{ kg})(9.8 \text{ m/s}^2)}{36} = 330{,}000 \text{ N}.$$

The force must act at a 38° angle at the base of the dome in order to be purely compressive. Thus

$$\tan 38° = \frac{F_V}{F_H};$$

$$F_H = \frac{F_V}{\tan 38°} = \frac{330{,}000 \text{ N}}{\tan 38°} = 420{,}000 \text{ N}.$$

NOTE For each buttress to exert this 420,000-N horizontal force, a prestressed-concrete tension ring surrounds the base of the buttresses beneath the ground (see Problem 56 and Fig. 9–70).

FIGURE 9–34 Example 9–13. (a) The dome of the Small Sports Palace in Rome, built for the 1960 Olympics. (b) The force components each buttress exerts on the dome.

(a)

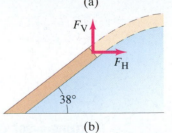
(b)

Summary

An object at rest is said to be in **equilibrium**. The subject concerned with the determination of the forces within a structure at rest is called **statics**.

The two necessary conditions for an object to be in equilibrium are that (1) the vector sum of all the forces on it must be zero, and (2) the sum of all the torques (calculated about any arbitrary axis) must also be zero:

$$\Sigma F_x = 0, \qquad \Sigma F_y = 0, \qquad \Sigma \tau = 0. \qquad \text{(9–1, 9–2)}$$

It is important when doing statics problems to apply the equilibrium conditions to only one object at a time.

[*An object in static equilibrium is said to be in (a) **stable**, (b) **unstable**, or (c) **neutral equilibrium**, depending on whether a slight displacement leads to (a) a return to the original position, (b) further movement away from the original position, or (c) rest in the new position. An object in stable equilibrium is also said to be in **balance**.]

[*Hooke's law** applies to many elastic solids, and states that the change in length of an object is proportional to the applied force:

$$F = k \, \Delta L. \qquad \text{(9–3)}$$

If the force is too great, the object will exceed its **elastic limit**, which means it will no longer return to its original shape when the distorting force is removed. If the force is even greater, the **ultimate strength** of the material can be exceeded, and the object will **fracture**. The force per unit area acting on an object is called the **stress**, and the resulting fractional change in length is called the **strain**. The stress on an object is present within the object and can be of three types: **compression**, **tension**, or **shear**. The ratio of stress to strain is called the **elastic modulus** of the material. **Young's modulus** applies for compression and tension, and the **shear modulus** for shear; **bulk modulus** applies to an object whose volume changes as a result of pressure on all sides. All three moduli are constants for a given material when distorted within the elastic region.]

Questions

1. Describe several situations in which an object is not in equilibrium, even though the net force on it is zero.

2. A bungee jumper momentarily comes to rest at the bottom of the dive before he springs back upward. At that moment, is the bungee jumper in equilibrium? Explain.

3. You can find the center of gravity of a meter stick by resting it horizontally on your two index fingers, and then slowly drawing your fingers together. First the meter stick will slip on one finger, and then on the other, but eventually the fingers meet at the CG. Why does this work?

4. Your doctor's scale has arms on which weights slide to counter your weight, Fig. 9–35. These weights are much lighter than you are. How does this work?

FIGURE 9–35
Question 4.

5. A ground retaining wall is shown in Fig. 9–36a. The ground, particularly when wet, can exert a significant force F on the wall. (a) What force produces the torque to keep the wall upright? (b) Explain why the retaining wall in Fig. 9–36b would be much less likely to overturn than that in Fig. 9–36a.

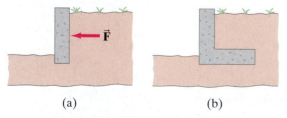

(a)   (b)

FIGURE 9–36 Question 5.

6. Explain why touching your toes while you are seated on the floor with outstretched legs produces less stress on the lower spinal column than when touching your toes from a standing position. Use a diagram.

7. A ladder, leaning against a wall, makes a 60° angle with the ground. When is it more likely to slip: when a person stands on the ladder near the top or near the bottom? Explain.

8. A uniform meter stick supported at the 25-cm mark is in equilibrium when a 1-kg rock is suspended at the 0-cm end (as shown in Fig. 9–37). Is the mass of the meter stick greater than, equal to, or less than the mass of the rock? Explain your reasoning.

FIGURE 9–37   Question 8.

9. Can the sum of the torques on an object be zero while the net force on the object is nonzero? Explain.

10. Figure 9–38 shows a cone. Explain how to lay it on a flat table so that it is in (a) stable equilibrium, (b) unstable equilibrium, (c) neutral equilibrium.

FIGURE 9–38   Question 10.

11. Which of the configurations of brick, (*a*) or (*b*) of Fig. 9–39, is the more likely to be stable? Why?

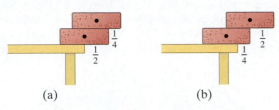

(a)   (b)

FIGURE 9–39 Question 11. The dots indicate the CG of each brick. The fractions $\frac{1}{4}$ and $\frac{1}{2}$ indicate what portion of each brick is hanging beyond its support.

12. Why do you tend to lean backward when carrying a heavy load in your arms?

13. Place yourself facing the edge of an open door. Position your feet astride the door with your nose and abdomen touching the door's edge. Try to rise on your tiptoes. Why can't this be done?

14. Why is it not possible to sit upright in a chair and rise to your feet without first leaning forward?

15. Why is it more difficult to do sit-ups when your knees are bent than when your legs are stretched out?

16. Name the type of equilibrium for each position of the ball in Fig. 9–40.

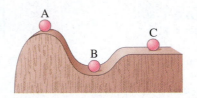

FIGURE 9–40 Question 16.

* **17.** Is the Young's modulus for a bungee cord smaller or larger than that for an ordinary rope?

* **18.** Examine how a pair of scissors or shears cuts through a piece of cardboard. Is the name "shears" justified? Explain.

* **19.** Materials such as ordinary concrete and stone are very weak under tension or shear. Would it be wise to use such a material for either of the supports of the cantilever shown in Fig. 9–9? If so, which one(s)? Explain.

Problems

9–1 and 9–2 Equilibrium

1. (I) Three forces are applied to a tree sapling, as shown in Fig. 9–41, to stabilize it. If $\vec{F}_A = 310\,N$ and $\vec{F}_B = 425\,N$, find $\vec{F}_C$ in magnitude and direction.

FIGURE 9–41 Problem 1.

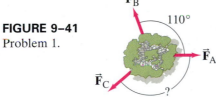

2. (I) Calculate the torque about the front support post (B) of a diving board, Fig. 9–42, exerted by a 58-kg person 3.0 m from that post.

FIGURE 9–42 Problems 2, 4, and 6.

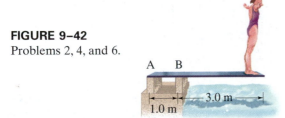

3. (I) Calculate the mass m needed in order to suspend the leg shown in Fig. 9–43. Assume the leg (with cast) has a mass of 15.0 kg, and its CG is 35.0 cm from the hip joint; the sling is 80.5 cm from the hip joint.

FIGURE 9–43 Problem 3.

4. (I) How far out on the diving board (Fig. 9–42) would a 58-kg diver have to be to exert a torque of 1100 m·N on the board, relative to the left (A) support post?

5. (II) Two cords support a chandelier in the manner shown in Fig. 9–4 except that the upper wire makes an angle of 45° with the ceiling. If the cords can sustain a force of 1550 N without breaking, what is the maximum chandelier weight that can be supported?

6. (II) Calculate the forces F_A and F_B that the supports exert on the diving board of Fig. 9–42 when a 58-kg person stands at its tip. (*a*) Ignore the weight of the board. (*b*) Take into account the board's mass of 35 kg. Assume the board's CG is at its center.

7. (II) A uniform steel beam has a mass of 940 kg. On it is resting half of an identical beam, as shown in Fig. 9–44. What is the vertical support force at each end?

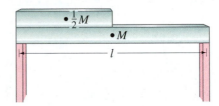

FIGURE 9–44 Problem 7.

8. (II) A 140-kg horizontal beam is supported at each end. A 320-kg piano rests a quarter of the way from one end. What is the vertical force on each of the supports?

9. (II) A 75-kg adult sits at one end of a 9.0-m-long board. His 25-kg child sits on the other end. (*a*) Where should the pivot be placed so that the board is balanced, ignoring the board's mass? (*b*) Find the pivot point if the board is uniform and has a mass of 15 kg.

10. (II) Calculate F_A and F_B for the uniform cantilever shown in Fig. 9–9 whose mass is 1200 kg.

11. (II) Find the tension in the two cords shown in Fig. 9–45. Neglect the mass of the cords, and assume that the angle θ is 33° and the mass m is 170 kg.

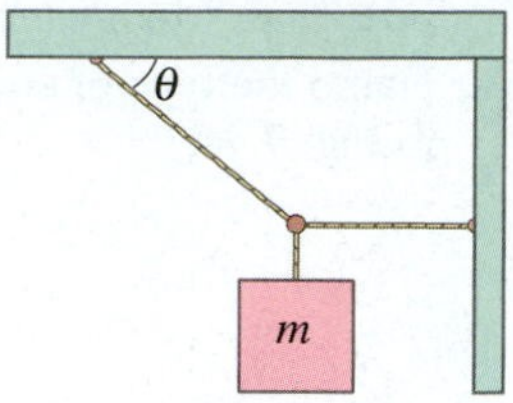

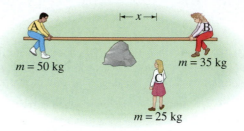

FIGURE 9–45
Problem 11.

12. (II) Find the tension in the two wires supporting the traffic light shown in Fig. 9–46.

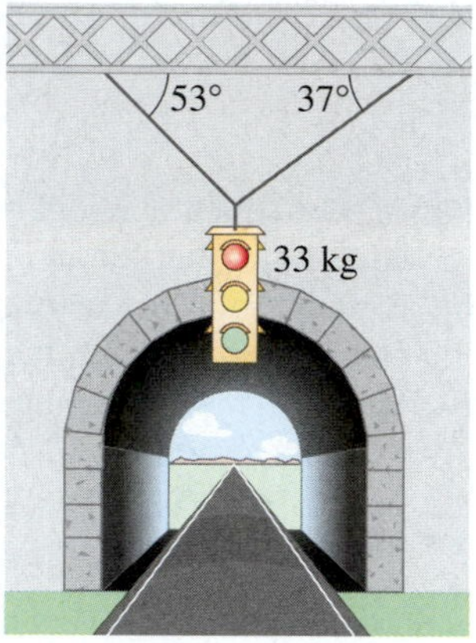

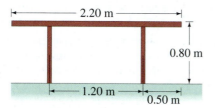

FIGURE 9–46
Problem 12.

13. (II) How close to the edge of the 20.0-kg table shown in Fig. 9–47 can a 66.0-kg person sit without tipping it over?

FIGURE 9–47 Problem 13.

14. (II) A 0.60-kg sheet hangs from a massless clothesline as shown in Fig. 9–48. The clothesline on either side of the sheet makes an angle of 3.5° with the horizontal. Calculate the tension in the clothesline on either side of the sheet. Why is the tension so much greater than the weight of the sheet?

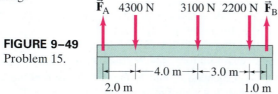

FIGURE 9–48 Problem 14.

15. (II) Calculate F_A and F_B for the beam shown in Fig. 9–49. The downward forces represent the weights of machinery on the beam. Assume the beam is uniform and has a mass of 250 kg.

FIGURE 9–49
Problem 15.

$\vec{F}_A$ 4300 N   3100 N 2200 N $\vec{F}_B$

├── 4.0 m ──┼── 3.0 m ──┤
2.0 m   1.0 m

16. (II) Three children are trying to balance on a seesaw, which consists of a fulcrum rock, acting as a pivot at the center, and a very light board 3.6 m long (Fig. 9–50). Two playmates are already on either end. Boy A has a mass of 50 kg, and girl B a mass of 35 kg. Where should girl C, whose mass is 25 kg, place herself so as to balance the seesaw?

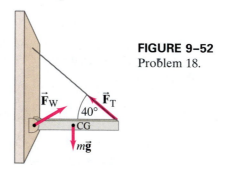

FIGURE 9–50 Problem 16.

17. (II) Figure 9–51 shows a pair of forceps used to hold a thin plastic rod firmly. If each finger squeezes with a force $F_T = F_B = 11.0$ N, what force do the forceps jaws exert on the plastic rod?

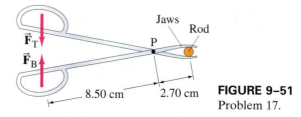

FIGURE 9–51
Problem 17.

18. (II) Calculate (a) the tension F_T in the wire that supports the 27-kg beam shown in Fig. 9–52, and (b) the force $\vec{F}_W$ exerted by the wall on the beam (give magnitude and direction).

FIGURE 9–52
Problem 18.

19. (II) A 172-cm-tall person lies on a light (massless) board which is supported by two scales, one under the top of her head and one beneath the bottom of her feet (Fig. 9–53). The two scales read, respectively, 35.1 and 31.6 kg. What distance is the center of gravity of this person from the bottom of her feet?

FIGURE 9–53 Problem 19.

20. (II) A shop sign weighing 245 N is supported by a uniform 155-N beam as shown in Fig. 9–54. Find the tension in the guy wire and the horizontal and vertical forces exerted by the hinge on the beam.

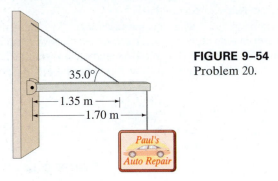

FIGURE 9–54
Problem 20.

21. (II) A traffic light hangs from a pole as shown in Fig. 9–55. The uniform aluminum pole AB is 7.50 m long and has a mass of 12.0 kg. The mass of the traffic light is 21.5 kg. Determine (a) the tension in the horizontal mass-less cable CD, and (b) the vertical and horizontal components of the force exerted by the pivot A on the aluminum pole.

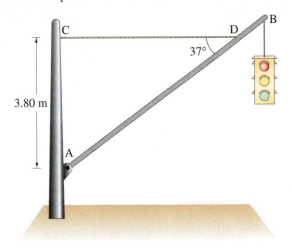

FIGURE 9–55   Problem 21.

22. (II) The 72-kg-man's hands in Fig. 9–56 are 36 cm apart. His CG is located 75% of the distance from his right hand toward his left. Find the force on each hand due to the ground.

FIGURE 9–56   Problem 22.

23. (II) A uniform meter stick with a mass of 180 g is supported horizontally by two vertical strings, one at the 0-cm mark and the other at the 90-cm mark (Fig. 9–57). What is the tension in the string (a) at 0 cm? (b) at 90 cm?

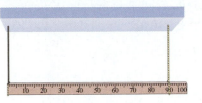

FIGURE 9–57
Problem 23.

24. (II) The two trees in Fig. 9–58 are 7.6 m apart. A backpacker is trying to lift his pack out of the reach of bears. Calculate the magnitude of the force $\vec{F}$ that he must exert downward to hold a 19-kg backpack so that the rope sags at its midpoint by (a) 1.5 m, (b) 0.15 m.

FIGURE 9–58   Problem 24.

25. (III) A door 2.30 m high and 1.30 m wide has a mass of 13.0 kg. A hinge 0.40 m from the top and another hinge 0.40 m from the bottom each support half the door's weight (Fig. 9–59). Assume that the center of gravity is at the geometrical center of the door, and determine the horizontal and vertical force components exerted by each hinge on the door.

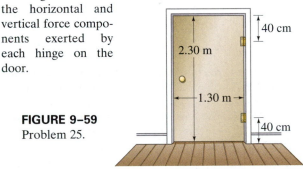

FIGURE 9–59
Problem 25.

26. (III) A uniform ladder of mass m and length l leans at an angle θ against a frictionless wall, Fig. 9–60. If the coefficient of static friction between the ladder and the ground is μ, determine a formula for the minimum angle at which the ladder will not slip.

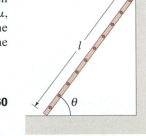

FIGURE 9–60
Problem 26.

27. (III) Consider a ladder with a painter climbing up it (Fig. 9–61). If the mass of the ladder is 12.0 kg, the mass of the painter is 55.0 kg, and the ladder begins to slip at its base when her feet are 70% of the way up the length of the ladder, what is the coefficient of static friction between the ladder and the floor? Assume the wall is frictionless.

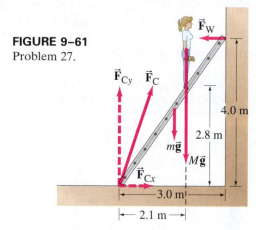

FIGURE 9–61
Problem 27.

28. (III) A person wants to push a lamp (mass 7.2 kg) across the floor, for which the coefficient of friction is 0.20. Calculate the maximum height x above the floor at which the person can push the lamp so that it slides rather than tips (Fig. 9–62).

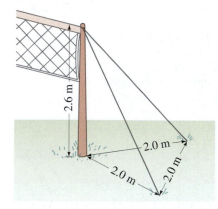

FIGURE 9–62
Problem 28.

29. (III) Two wires run from the top of a pole 2.6 m tall that supports a volleyball net. The two wires are anchored to the ground 2.0 m apart, and each is 2.0 m from the pole (Fig. 9–63). The tension in each wire is 95 N. What is the tension in the net, assumed horizontal and attached at the top of the pole?

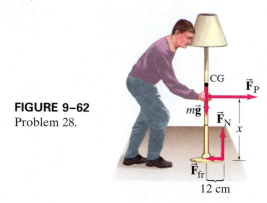

FIGURE 9–63 Problem 29.

* **9–3 Muscles and Joints**

* 30. (I) Suppose the point of insertion of the biceps muscle into the lower arm shown in Fig. 9–13a (Example 9–8) is 6.0 cm instead of 5.0 cm; how much mass could the person hold with a muscle exertion of 450 N?

* 31. (I) Approximately what magnitude force, F_M, must the extensor muscle in the upper arm exert on the lower arm to hold a 7.3-kg shot put (Fig. 9–64)? Assume the lower arm has a mass of 2.8 kg and its CG is 12 cm from the elbow-joint pivot.

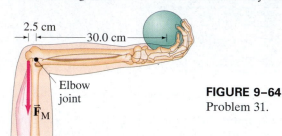

FIGURE 9–64
Problem 31.

* 32. (II) (a) Calculate the force, F_M, required of the "deltoid" muscle to hold up the outstretched arm shown in Fig. 9–65. The total mass of the arm is 3.3 kg. (b) Calculate the magnitude of the force F_J exerted by the shoulder joint on the upper arm.

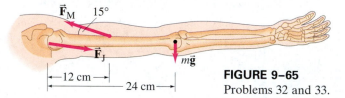

FIGURE 9–65
Problems 32 and 33.

* 33. (II) Suppose the hand in Problem 32 holds a 15-kg mass. What force, F_M, is required of the deltoid muscle, assuming the mass is 52 cm from the shoulder joint?

* 34. (II) The Achilles tendon is attached to the rear of the foot as shown in Fig. 9–66. When a person elevates himself just barely off the floor on the "ball of one foot," estimate the tension F_T in the Achilles tendon (pulling upward), and the (downward) force F_B exerted by the lower leg bone on the foot. Assume the person has a mass of 72 kg and D is twice as long as d.

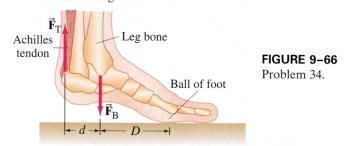

FIGURE 9–66
Problem 34.

* 35. (II) Redo Example 9–9, assuming now that the person is less bent over so that the 30° in Fig. 9–14b is instead 45°. What will be the magnitude of F_V on the vertebra?

9–4 Stability and Balance

36. (II) The Leaning Tower of Pisa is 55 m tall and about 7.0 m in diameter. The top is 4.5 m off center. Is the tower in stable equilibrium? If so, how much farther can it lean before it becomes unstable? Assume the tower is of uniform composition.

37. (III) Four bricks are to be stacked at the edge of a table, each brick overhanging the one below it, so that the top brick extends as far as possible beyond the edge of the table. (a) To achieve this, show that successive bricks must extend no more than (starting at the top) $\frac{1}{2}, \frac{1}{4}, \frac{1}{6}$, and $\frac{1}{8}$ of their length beyond the one below (Fig. 9–67a). (b) Is the top brick completely beyond the base? (c) Determine a general formula for the maximum total distance spanned by n bricks if they are to remain stable. (d) A builder wants to construct a corbeled arch (Fig. 9–67b) based on the principle of stability discussed in (a) and (c) above. What minimum number of bricks, each 0.30 m long, is needed if the arch is to span 1.0 m?

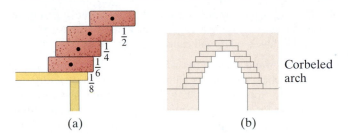

Corbeled arch

(a)   (b)

FIGURE 9–67 Problem 37.

* 9–5 Elasticity; Stress and Strain

* **38.** (I) A nylon string on a tennis racket is under a tension of 275 N. If its diameter is 1.00 mm, by how much is it lengthened from its untensioned length of 30.0 cm?

* **39.** (I) A marble column of cross-sectional area 1.2 m² supports a mass of 25,000 kg. (a) What is the stress within the column? (b) What is the strain?

* **40.** (I) By how much is the column in Problem 39 shortened if it is 9.6 m high?

* **41.** (I) A sign (mass 2100 kg) hangs from the end of a vertical steel girder with a cross-sectional area of 0.15 m². (a) What is the stress within the girder? (b) What is the strain on the girder? (c) If the girder is 9.50 m long, how much is it lengthened? (Ignore the mass of the girder itself.)

* **42.** (II) One liter of alcohol $\left(1000 \text{ cm}^3\right)$ in a flexible container is carried to the bottom of the sea, where the pressure is $2.6 \times 10^6 \text{ N/m}^2$. What will be its volume there?

* **43.** (II) A 15-cm-long tendon was found to stretch 3.7 mm by a force of 13.4 N. The tendon was approximately round with an average diameter of 8.5 mm. Calculate the Young's modulus of this tendon.

* **44.** (II) How much pressure is needed to compress the volume of an iron block by 0.10%? Express your answer in N/m², and compare it to atmospheric pressure $\left(1.0 \times 10^5 \text{ N/m}^2\right)$.

* **45.** (II) At depths of 2000 m in the sea, the pressure is about 200 times atmospheric pressure $\left(1 \text{ atm} = 1.0 \times 10^5 \text{ N/m}^2\right)$. By what percentage does the interior space of an iron bathysphere's volume change at this depth?

* **46.** (III) A scallop forces open its shell with an elastic material called abductin, whose Young's modulus is about $2.0 \times 10^6 \text{ N/m}^2$. If this piece of abductin is 3.0 mm thick and has a cross-sectional area of 0.50 cm², how much potential energy does it store when compressed 1.0 mm?

* **47.** (III) A pole projects horizontally from the front wall of a shop. A 5.1-kg sign hangs from the pole at a point 2.2 m from the wall (Fig. 9–68). (a) What is the torque due to this sign calculated about the point where the pole meets the wall? (b) If the pole is not to fall off, there must be another torque exerted to balance it. What exerts this torque? Use a diagram to show how this torque must act. (c) Discuss whether compression, tension, and/or shear play a role in part (b).

FIGURE 9–68
Problem 47.

The *Bear*

* 9–6 Fracture

* **48.** (I) The femur bone in the human leg has a minimum effective cross section of about 3.0 cm² $\left(=3.0 \times 10^{-4} \text{ m}^2\right)$. How much compressive force can it withstand before breaking?

* **49.** (II) (a) What is the maximum tension possible in a 1.00-mm-diameter nylon tennis racket string? (b) If you want tighter strings, what do you do to prevent breakage: use thinner or thicker strings? Why? What causes strings to break when they are hit by the ball?

* **50.** (II) If a compressive force of 3.6×10^4 N is exerted on the end of a 22-cm-long bone of cross-sectional area 3.6 cm², (a) will the bone break, and (b) if not, by how much does it shorten?

* **51.** (II) (a) What is the minimum cross-sectional area required of a vertical steel cable from which is suspended a 320-kg chandelier? Assume a safety factor of 7.0 (b) If the cable is 7.5 m long, how much does it elongate?

* **52.** (II) Assume the supports of the uniform cantilever shown in Fig. 9–69 (mass = 2600 kg) are made of wood. Calculate the minimum cross-sectional area required of each, assuming a safety factor of 8.5.

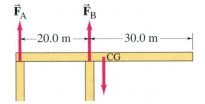

FIGURE 9–69
Problem 52.

* **53.** (II) An iron bolt is used to connect two iron plates together. The bolt must withstand shear forces up to about 3200 N. Calculate the minimum diameter for the bolt, based on a safety factor of 6.0.

* **54.** (III) A steel cable is to support an elevator whose total (loaded) mass is not to exceed 3100 kg. If the maximum acceleration of the elevator is 1.2 m/s², calculate the diameter of cable required. Assume a safety factor of 7.0.

* 9–7 Arches and Domes

* **55.** (II) How high must a pointed arch be if it is to span a space 8.0 m wide and exert one-third the horizontal force at its base that a round arch would?

* **56.** (II) The subterranean tension ring that exerts the balancing horizontal force on the abutments for the dome in Fig. 9–34 is 36-sided, so each segment makes a 10° angle with the adjacent one (Fig. 9–70). Calculate the tension F that must exist in each segment so that the required force of 4.2×10^5 N can be exerted at each corner (Example 9–13).

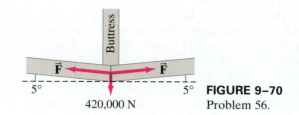

FIGURE 9–70
Problem 56.

General Problems

57. The mobile in Fig. 9–71 is in equilibrium. Object B has mass of 0.885 kg. Determine the masses of objects A, C, and D. (Neglect the weights of the crossbars.)

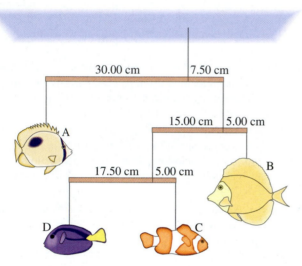

FIGURE 9–71 Problem 57.

58. A tightly stretched "high wire" is 46 m long. It sags 2.2 m when a 60.0-kg tightrope walker stands at its center. What is the tension in the wire? Is it possible to increase the tension in the wire so that there is no sag?

59. What minimum horizontal force F is needed to pull a wheel of radius R and mass M over a step of height h as shown in Fig. 9–72 $(R > h)$? (a) Assume the force is applied at the top edge as shown. (b) Assume the force is applied instead at the wheel's center.

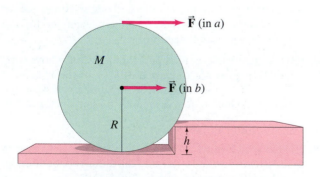

FIGURE 9–72 Problem 59.

60. A 25-kg round table is supported by three legs equal distances apart on the edge. What minimum mass, placed on the table's edge, will cause the table to overturn?

61. When a wood shelf of mass 5.0 kg is fastened inside a slot in a vertical support as shown in Fig. 9–73, the support exerts a torque on the shelf. (a) Draw a free-body diagram for the shelf, assuming three vertical forces (two exerted by the support slot—explain why). Then calculate (b) the magnitudes of the three forces and (c) the torque exerted by the support (about the left end of the shelf).

FIGURE 9–73 Problem 61.

62. A 50-story building is being planned. It is to be 200.0 m high with a base 40.0 m by 70.0 m. Its total mass will be about 1.8×10^7 kg, and its weight therefore about 1.8×10^8 N. Suppose a 200-km/h wind exerts a force of 950 N/m² over the 70.0-m-wide face (Fig. 9–74). Calculate the torque about the potential pivot point, the rear edge of the building (where $\vec{F}_E$ acts in Fig. 9–74), and determine whether the building will topple. Assume the total force of the wind acts at the midpoint of the building's face, and that the building is not anchored in bedrock. [*Hint:* $\vec{F}_E$ in Fig. 9–74 represents the force that the Earth would exert on the building in the case where the building would just begin to tip.]

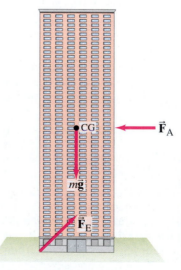

FIGURE 9–74 Forces on a building subjected to wind ($\vec{F}_A$), gravity ($m\vec{g}$), and the force $\vec{F}_E$ on the building due to the Earth if the building were just about to tip. Problem 62.

63. The center of gravity of a loaded truck depends on how the truck is packed. If it is 4.0 m high and 2.4 m wide, and its CG is 2.2 m above the ground, how steep a slope can the truck be parked on without tipping over (Fig. 9–75)?

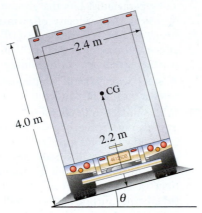

FIGURE 9–75
Problem 63.

64. In Fig. 9–76, consider the right-hand (northernmost) section of the Golden Gate Bridge, which has a length $d_1 = 343$ m. Assume the CG of this span is halfway between the tower and anchor. Determine F_{T1} and F_{T2} (which act on the northernmost cable) in terms of mg, the weight of the northernmost span, and calculate the tower height h needed for equilibrium. Assume the roadway is supported only by the suspension cables, and neglect the mass of the cables and vertical wires. [*Hint*: F_{T3} does not act on this section.]

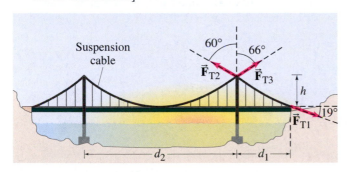

FIGURE 9–76   Problem 64.

65. When a mass of 25 kg is hung from the middle of a fixed straight aluminum wire, the wire sags to make an angle of 12° with the horizontal as shown in Fig. 9–77. Determine the radius of the wire.

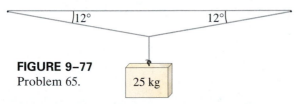

FIGURE 9–77
Problem 65.

66. The forces acting on a 67,000-kg aircraft flying at constant velocity are shown in Fig. 9–78. The engine thrust, $F_T = 5.0 \times 10^5$ N, acts on a line 1.6 m below the CM. Determine the drag force F_D and the distance above the CM that it acts. Assume $\vec{F}_D$ and $\vec{F}_T$ are horizontal.

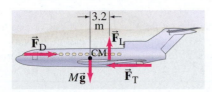

FIGURE 9–78
Problem 66.

67. A uniform flexible steel cable of weight mg is suspended between two points at the same elevation as shown in Fig. 9–79, where $\theta = 60°$. Determine the tension in the cable (*a*) at its lowest point, and (*b*) at the points of attachment. (*c*) What is the direction of the tension force in each case?

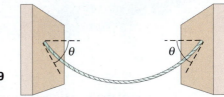

FIGURE 9–79
Problem 67.

68. A 20.0-m-long uniform beam weighing 550 N rests on walls A and B, as shown in Fig. 9–80. (*a*) Find the maximum weight of a person who can walk to the extreme end D without tipping the beam. Find the forces that the walls A and B exert on the beam when the person is standing: (*b*) at D; (*c*) at a point 2.0 m to the right of B; (*d*) 2.0 m to the right of A.

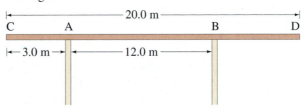

FIGURE 9–80   Problem 68.

69. A cube of side l rests on a rough floor. It is subjected to a steady horizontal pull F, exerted a distance h above the floor as shown in Fig. 9–81. As F is increased, the block will either begin to slide, or begin to tip over. Determine the coefficient of static friction μ_s so that (*a*) the block begins to slide rather than tip; (*b*) the block begins to tip. [*Hint*: Where will the normal force on the block act if it tips?]

FIGURE 9–81
Problem 69.

70. A 60.0-kg painter is on a uniform 25-kg scaffold supported from above by ropes (Fig. 9–82). There is a 4.0-kg pail of paint to one side, as shown. Can the painter walk safely to both ends of the scaffold? If not, which end(s) is dangerous, and how close to the end can he approach safely?

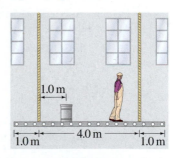

FIGURE 9–82
Problem 70.

71. A woman holds a 2.0-m-long uniform 10.0-kg pole as shown in Fig. 9–83. (*a*) Determine the forces she must exert with each hand (magnitude and direction). To what position should she move her left hand so that neither hand has to exert a force greater than (*b*) 150 N? (*c*) 85 N?

FIGURE 9–83
Problem 71.

72. A man doing push-ups pauses in the position shown in Fig. 9–84. His mass $m = 75$ kg. Determine the normal force exerted by the floor (a) on each hand; (b) on each foot.

$m\vec{g}$

30 cm

40 cm — 95 cm

FIGURE 9–84 Problem 72.

73. A 20-kg sphere rests between two smooth planes as shown in Fig. 9–85. Determine the magnitude of the force acting on the sphere exerted by each plane.

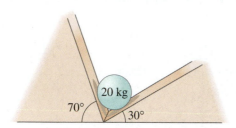

70° 20 kg 30°

FIGURE 9–85 Problem 73.

74. A 2200-kg trailer is attached to a stationary truck at point B, Fig. 9–86. Determine the normal force exerted by the road on the rear tires at A, and the vertical force exerted on the trailer by the support B.

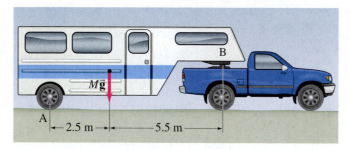

B

$M\vec{g}$

A — 2.5 m — 5.5 m

FIGURE 9–86 Problem 74.

* **75.** Parachutists whose chutes have failed to open have been known to survive if they land in deep snow. Assume that a 75-kg parachutist hits the ground with an area of impact of 0.30 m² at a velocity of 60 m/s, and that the ultimate strength of body tissue is 5×10^5 N/m². Assume that the person is brought to rest in 1.0 m of snow. Show that the person may escape serious injury.

* **76.** A steel wire 2.0 mm in diameter stretches by 0.030% when a mass is suspended from it. How large is the mass?

* **77.** In Example 7–6 in Chapter 7, we calculated the impulse and average force on the leg of a person who jumps 3.0 m down to the ground. If the legs are not bent upon landing, so that the body moves a distance d of only 1.0 cm during collision, determine (a) the stress in the tibia (a lower leg bone of area = 3.0×10^{-4} m²), and (b) whether or not the bone will break. (c) Repeat for a bent-knees landing ($d = 50.0$ cm).

* **78.** The roof over a 7.0-m × 10.0-m room in a school has a total mass of 12,600 kg. The roof is to be supported by vertical "2 × 4s" (actually about 4.0 cm × 9.0 cm) along the 10.0-m sides. How many supports are required on each side, and how far apart must they be? Consider only compression, and assume a safety factor of 12.

* **79.** A 25-kg object is being lifted by pulling on the ends of a 1.00-mm-diameter nylon string that goes over two 3.00-m-high poles that are 4.0 m apart, as shown in Fig. 9–87. How high above the floor will the object be when the string breaks?

25 kg

FIGURE 9–87 Problem 79.

* **80.** There is a maximum height of a uniform vertical column made of any material that can support itself without buckling, and it is independent of the cross-sectional area (why?). Calculate this height for (a) steel (density = mass/volume = 7.8×10^3 kg/m³), and (b) granite (density = 2.7×10^3 kg/m³).

Answers to Exercises

A: F_A also has a component to balance the sideways force F_B.

B: Yes: sin θ appears on both sides and cancels out.

C: $F_N = m_A g + m_B g + Mg = 560$ N.

D: Static friction at the cement floor $(= F_{Cx})$ is crucial, or else the ladder would slip. At the top, the ladder can move and adjust, so we wouldn't expect a strong static friction force there.

Scuba divers and sea creatures under water experience a buoyant force $(\vec{\mathbf{F}}_B)$ that almost exactly balances their weight $m\vec{\mathbf{g}}$. The buoyant force is equal to the weight of the volume of fluid displaced (Archimedes' principle) and arises because the pressure increases with depth in the fluid. Sea creatures have a density very close to that of water, so their weight very nearly equals the buoyant force. Humans have a density slightly less than water, so they can float.

When fluids flow, interesting effects occur because the pressure in the fluid is lower where the fluid velocity is higher (Bernoulli's principle).

CHAPTER 10

Fluids

I n previous Chapters we considered objects that were solid and assumed to maintain their shape except for a small amount of elastic deformation. We sometimes treated objects as point particles. Now we are going to shift our attention to materials that are very deformable and can flow. Such "fluids" include liquids and gases. We will examine fluids both at rest (fluid statics) and in motion (fluid dynamics).

10–1 Phases of Matter

The three common **phases**, or **states**, of matter are solid, liquid, and gas. We can distinguish these three phases as follows. A **solid** maintains a fixed shape and a fixed size; even if a large force is applied to a solid, it does not readily change in shape or volume. A **liquid** does not maintain a fixed shape—it takes on the shape of its container—but like a solid it is not readily compressible, and its volume can be changed significantly only by a very large force. A **gas** has neither a fixed shape nor a fixed volume—it will expand to fill its container. For example, when air is pumped into an automobile tire, the air does not all run to the bottom of the tire as a liquid would; it spreads out to fill the whole volume of the tire. Since liquids and gases do not maintain a fixed shape, they both have the ability to flow; they are thus often referred to collectively as **fluids**.

Phases of matter

The division of matter into three phases is not always simple. How, for example, should butter be classified? Furthermore, a fourth phase of matter can be distinguished, the **plasma** phase, which occurs only at very high temperatures and consists of ionized atoms (electrons separated from the nuclei). Some scientists believe that so-called colloids (suspensions of tiny particles in a liquid) should also be considered a separate phase of matter. **Liquid crystals**, which are used in laptop computer screens, calculators, digital watches, and so on can be considered a phase of matter intermediate between solids and liquids. However, for our present purposes we will mainly be interested in the three ordinary phases of matter.

10–2 Density and Specific Gravity

It is sometimes said that iron is "heavier" than wood. This cannot really be true since a large log clearly weighs more than an iron nail. What we should say is that iron is more *dense* than wood.

The **density**, ρ, of a substance (ρ is the lowercase Greek letter rho) is defined as its mass per unit volume:

Density defined

$$\rho = \frac{m}{V},$$ (10–1)

where m is the mass of a sample of the substance and V its volume. Density is a characteristic property of any pure substance. Objects made of a particular pure substance, such as pure gold, can have any size or mass, but the density will be the same for each. (We will sometimes use the concept of density, Eq. 10–1, to write the mass of an object as $m = \rho V$, and the weight of an object, mg, as $\rho V g$.)

The SI unit for density is kg/m^3. Sometimes densities are given in g/cm^3. Note that since $1 \, kg/m^3 = 1000 \, g/(100 \, cm)^3 = 10^3 \, g/10^6 \, cm^3 = 10^{-3} \, g/cm^3$, then a density given in g/cm^3 must be multiplied by 1000 to give the result in kg/m^3. Thus the density of aluminum is $\rho = 2.70 \, g/cm^3$, which is equal to $2700 \, kg/m^3$. The densities of a variety of substances are given in Table 10–1. The Table specifies temperature and atmospheric pressure because they affect the density of substances (although the effect is slight for liquids and solids).

TABLE 10–1
Densities of Substances†

Substance	Density, $\rho \, (kg/m^3)$
Solids	
Aluminum	2.70×10^3
Iron and steel	$7.8 \ \times 10^3$
Copper	$8.9 \ \times 10^3$
Lead	11.3×10^3
Gold	19.3×10^3
Concrete	$2.3 \ \times 10^3$
Granite	$2.7 \ \times 10^3$
Wood (typical)	$0.3\text{–}0.9 \times 10^3$
Glass, common	$2.4\text{–}2.8 \times 10^3$
Ice (H_2O)	0.917×10^3
Bone	$1.7\text{–}2.0 \times 10^3$
Liquids	
Water (4°C)	$1.00 \ \times 10^3$
Blood, plasma	$1.03 \ \times 10^3$
Blood, whole	$1.05 \ \times 10^3$
Sea water	1.025×10^3
Mercury	$13.6 \ \times 10^3$
Alcohol, ethyl	$0.79 \ \times 10^3$
Gasoline	$0.68 \ \times 10^3$
Gases	
Air	1.29
Helium	0.179
Carbon dioxide	1.98
Water (steam) (100°C)	0.598

†Densities are given at 0°C and 1 atm pressure unless otherwise specified.

EXAMPLE 10–1 **Mass, given volume and density.** What is the mass of a solid iron wrecking ball of radius 18 cm?

APPROACH First we use the standard formula $V = \frac{4}{3}\pi r^3$ (see inside rear cover) to obtain the volume of the sphere. Then Eq. 10–1 and Table 10–1 give us the mass m.

SOLUTION The volume of the sphere is

$$V = \tfrac{4}{3}\pi r^3 = \tfrac{4}{3}(3.14)(0.18 \, m)^3 = 0.024 \, m^3.$$

From Table 10–1, the density of iron is $\rho = 7800 \, kg/m^3$, so Eq. 10–1 gives

$$m = \rho V = (7800 \, kg/m^3)(0.024 \, m^3) = 190 \, kg.$$

The **specific gravity** of a substance is defined as the ratio of the density of that substance to the density of water at 4.0°C. Because specific gravity (abbreviated SG) is a ratio, it is a simple number without dimensions or units. The density of water is $1.00 \, g/cm^3 = 1.00 \times 10^3 \, kg/m^3$, so the specific gravity of any substance will be equal numerically to its density specified in g/cm^3, or 10^{-3} times its density specified in kg/m^3. For example (see Table 10–1), the specific gravity of lead is 11.3, and that of alcohol is 0.79.

The concepts of density and specific gravity are especially helpful in the study of fluids because we are not always dealing with a fixed volume or mass.

10–3 Pressure in Fluids

Pressure is defined as force per unit area, where the force F is understood to be the magnitude of the force acting perpendicular to the surface area A:

$$\text{pressure} = P = \frac{F}{A}. \qquad (10\text{–}2)$$

Pressure defined

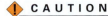

⚠ CAUTION

Pressure is a scalar, not a vector

The pascal (unit)

Although force is a vector, pressure is a scalar. Pressure has magnitude only. The SI unit of pressure is N/m^2. This unit has the official name **pascal** (Pa), in honor of Blaise Pascal (see Section 10–5); that is, $1\,Pa = 1\,N/m^2$. However, for simplicity, we will often use N/m^2. Other units sometimes used are $dynes/cm^2$, and $lb/in.^2$ (abbreviated "psi"). Several other units for pressure are discussed, along with conversions between them, in Section 10–6 (see also the Table inside the front cover).

EXAMPLE 10–2 Calculating pressure. The two feet of a 60-kg person cover an area of $500\,cm^2$. (*a*) Determine the pressure exerted by the two feet on the ground. (*b*) If the person stands on one foot, what will the pressure be under that foot?

APPROACH Assume the person is at rest. Then the ground pushes up on her with a force equal to her weight mg, and she exerts a force mg on the ground where her feet (or foot) contact it. Because $1\,cm^2 = (10^{-2}\,m)^2 = 10^{-4}\,m^2$, then $500\,cm^2 = 0.050\,m^2$.

SOLUTION (*a*) The pressure on the ground exerted by the two feet is

$$P = \frac{F}{A} = \frac{mg}{A} = \frac{(60\,kg)(9.8\,m/s^2)}{(0.050\,m^2)} = 12 \times 10^3\,N/m^2.$$

(*b*) If the person stands on one foot, the force is still equal to the person's weight, but the area will be half as much, so the pressure will be twice as much: $24 \times 10^3\,N/m^2$.

Pressure is particularly useful for dealing with fluids. It is an experimental observation that *a fluid can exert a pressure in any direction*. This is well known to swimmers and divers who feel the water pressure on all parts of their bodies. At any point in a fluid at rest, the pressure is the same in all directions at a given depth. This is illustrated in Fig. 10–1. Consider a tiny cube of the fluid which is so small that we can ignore the force of gravity on it. The pressure on one side of it must equal the pressure on the opposite side. If this weren't true, there would be a net force on the cube and it would start moving. If the fluid is not flowing, then the pressures must be equal.

Fluids exert pressure in all directions

Another important property of a fluid at rest is that the force due to fluid pressure always acts *perpendicular* to any solid surface it is in contact with. If there were a component of the force parallel to the surface, as shown in Fig. 10–2, then according to Newton's third law the solid surface would exert a force back on the fluid that also would have a component parallel to the surface. Such a component would cause the fluid to flow, in contradiction to our assumption that the fluid is at rest. Thus the force due to the pressure in a fluid at rest is always perpendicular to the surface.

FIGURE 10–1 Pressure is the same in every direction in a fluid at a given depth; if it weren't, the fluid would be in motion.

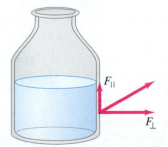

FIGURE 10–2 If there were a component of force parallel to the solid surface of the container, the liquid would move in response to it. For a liquid at rest, $F_{\parallel} = 0$.

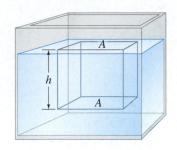

FIGURE 10–3 Calculating the pressure at a depth h in a liquid.

Pressure variation with depth

Let us now calculate quantitatively how the pressure in a liquid of uniform density varies with depth. Consider a point at a depth h below the surface of the liquid (that is, the surface is a height h above this point), as shown in Fig. 10–3. The pressure due to the liquid at this depth h is due to the weight of the column of liquid above it. Thus the force due to the weight of liquid acting on the area A is $F = mg = (\rho V)g = \rho Ahg$, where Ah is the volume of the column of liquid, ρ is the density of the liquid (assumed to be constant), and g is the acceleration of gravity. The pressure P due to the weight of liquid is then

$$P = \frac{F}{A} = \frac{\rho Ahg}{A}$$

$$P = \rho gh. \qquad \text{[liquid]} \quad \textbf{(10–3a)}$$

Note that the area A doesn't affect the pressure at a given depth. The fluid pressure is directly proportional to the density of the liquid and to the depth within the liquid. In general, the pressure at equal depths within a uniform liquid is the same.

Equation 10–3a is extremely useful. It is valid for fluids whose density is constant and does not change with depth—that is, if the fluid is *incompressible*. This is usually a good approximation for liquids (although at great depths in the ocean, the density of water is increased substantially by compression due to the great weight of water above).

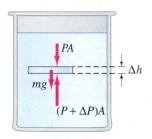

FIGURE 10–4 Forces on a thin slab of fluid (shown as a liquid, but it could instead be a gas).

Gases, on the other hand, are very compressible, and density can vary significantly with depth. For this more general case, in which ρ may vary, Eq. 10–3a may not be useful. So let us consider a thin slab of liquid of volume $V = A \Delta h$ as shown in Fig. 10–4. We choose Δh thin enough so that ρ doesn't vary significantly over the small thickness Δh. Let P be the pressure exerted downward on the top surface, and let $P + \Delta P$ be the pressure upward on the bottom surface. The forces acting on our thin slab of fluid, as shown in Fig. 10–4, are $(P + \Delta P)A$ upward and PA downward, and the downward weight of the slab, $mg = (\rho V)g = \rho A \Delta h\, g$. We assume the fluid is at rest, so the net force on the slab is zero. Then

$$(P + \Delta P)A - PA - \rho A \Delta h\, g = 0.$$

The area A cancels from each term, and when we solve for ΔP we obtain

Change in pressure with change in depth in a fluid

$$\Delta P = \rho g \Delta h. \qquad [\rho \approx \text{constant over } \Delta h] \quad \textbf{(10–3b)}$$

Equation 10–3b tells us how the pressure changes over a small change in depth (Δh) within a fluid, even if compressible.

FIGURE 10–5 Example 10–3.

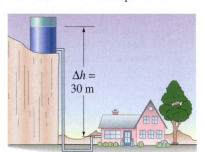

EXAMPLE 10–3 **Pressure at a faucet.** The surface of the water in a storage tank is 30 m above a water faucet in the kitchen of a house, Fig. 10–5. Calculate the difference in water pressure between the faucet and the surface of the water in the tank.

APPROACH Water is practically incompressible, so ρ is constant even for a $\Delta h = 30\,\text{m}$ when used in Eq. 10–3b. Only Δh matters; we can ignore the "route" of the pipe and its bends.

SOLUTION The same atmospheric pressure acts both at the surface of the water in the storage tank and on the water leaving the faucet. So, the water pressure difference between the faucet and the surface of the water in the tank is

$$\Delta P = \rho g \Delta h = (1.0 \times 10^3 \,\text{kg/m}^3)(9.8 \,\text{m/s}^2)(30 \,\text{m})$$
$$= 2.9 \times 10^5 \,\text{N/m}^2.$$

NOTE The height h is sometimes called the **pressure head**. In this Example, the head of water is 30 m at the faucet. The very different diameters of the tank and faucet don't affect the result—only pressure does.

EXERCISE A A dam holds back a lake that is 85 m deep at the dam. If the lake is 20 km long, how much thicker should the dam be than if the lake were smaller, only 1.0 km long?

10-4 Atmospheric Pressure and Gauge Pressure

Atmospheric Pressure

The pressure of the Earth's atmosphere, as in any fluid, changes with depth. But the Earth's atmosphere is somewhat complicated: not only does the density of air vary greatly with altitude but there is no distinct top surface to the atmosphere from which h (in Eq. 10–3a) could be measured. We can, however, calculate the approximate difference in pressure between two altitudes using Eq. 10–3b.

The pressure of the air at a given place varies slightly according to the weather. At sea level, the pressure of the atmosphere on average is $1.013 \times 10^5 \, \text{N/m}^2$ (or 14.7 lb/in.²). This value lets us define a commonly used unit of pressure, the **atmosphere** (abbreviated atm):

$$1 \, \text{atm} = 1.013 \times 10^5 \, \text{N/m}^2 = 101.3 \, \text{kPa}.$$

One atmosphere (unit of pressure)

Another unit of pressure sometimes used (in meteorology and on weather maps) is the **bar**, which is defined as

$$1 \, \text{bar} = 1.00 \times 10^5 \, \text{N/m}^2.$$

The bar (unit of pressure)

Thus standard atmospheric pressure is slightly more than 1 bar.

The pressure due to the weight of the atmosphere is exerted on all objects immersed in this great sea of air, including our bodies. How does a human body withstand the enormous pressure on its surface? The answer is that living cells maintain an internal pressure that closely equals the external pressure, just as the pressure inside a balloon closely matches the outside pressure of the atmosphere. An automobile tire, because of its rigidity, can maintain internal pressures much greater than the external pressure.

PHYSICS APPLIED
Pressure on living cells

CONCEPTUAL EXAMPLE 10–4 **Finger holds water in a straw.** You insert a straw of length L into a tall glass of water. You place your finger over the top of the straw, capturing some air above the water but preventing any additional air from getting in or out, and then you lift the straw from the water. You find that the straw retains most of the water. (See Fig. 10–6a.) Does the air in the space between your finger and the top of the water have a pressure P that is greater than, equal to, or less than, the atmospheric pressure P_A outside the straw?

RESPONSE Consider the forces on the column of water (Fig. 10–6b). Atmospheric pressure outside the straw pushes upward on the water at the bottom of the straw, gravity pulls the water downward, and the air pressure inside the top of the straw pushes downward on the water. Since the water is in equilibrium, the upward force due to atmospheric pressure must balance the two downward forces. The only way this is possible is for the air pressure inside the straw to be less than the atmosphere pressure outside the straw. (When you initially remove the straw, a little water may leave the bottom of the straw, thus increasing the volume of trapped air and reducing its density and pressure.)

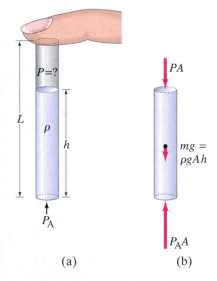

FIGURE 10–6 Example 10–4.

Gauge Pressure

It is important to note that tire gauges, and most other pressure gauges, register the pressure above and beyond atmospheric pressure. This is called **gauge pressure**. Thus, to get the **absolute pressure**, P, we must add the atmospheric pressure, P_A, to the gauge pressure, P_G:

Gauge pressure

$$P = P_A + P_G.$$

Absolute pressure = atmospheric pressure + gauge pressure

If a tire gauge registers 220 kPa, the absolute pressure within the tire is $220 \, \text{kPa} + 101 \, \text{kPa} = 321 \, \text{kPa}$, equivalent to about 3.2 atm (2.2 atm gauge pressure).

10-5 Pascal's Principle

The Earth's atmosphere exerts a pressure on all objects with which it is in contact, including other fluids. External pressure acting on a fluid is transmitted throughout that fluid. For instance, according to Eq. 10–3a, the pressure due to the water at a depth of 100 m below the surface of a lake is $P = \rho g \, \Delta h = (1000 \, \text{kg/m}^3)(9.8 \, \text{m/s}^2)(100 \, \text{m}) = 9.8 \times 10^5 \, \text{N/m}^2$, or 9.7 atm. However, the total pressure at this point is due to the pressure of water plus the pressure of the air above it. Hence the total pressure (if the lake is near sea level) is 9.7 atm + 1.0 atm = 10.7 atm. This is just one example of a general principle attributed to the French philosopher and scientist Blaise Pascal (1623–1662).

Pascal's principle states that *if an external pressure is applied to a confined fluid, the pressure at every point within the fluid increases by that amount.*

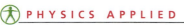
PHYSICS APPLIED

Hydraulic lift

A number of practical devices make use of Pascal's principle. One example is the hydraulic lift, illustrated in Fig. 10–7a, in which a small input force is used to exert a large output force by making the area of the output piston larger than the area of the input piston. To see how this works, we assume the input and output pistons are at the same height (at least approximately). Then the external input force F_{in}, by Pascal's principle, increases the pressure equally throughout. Therefore, at the same level (see Fig. 10–7a),

$$P_{\text{out}} = P_{\text{in}}$$

where the input quantities are represented by the subscript "in" and the output by "out." Since $P = F/A$, we write the above equality as

$$\frac{F_{\text{out}}}{A_{\text{out}}} = \frac{F_{\text{in}}}{A_{\text{in}}},$$

or

$$\frac{F_{\text{out}}}{F_{\text{in}}} = \frac{A_{\text{out}}}{A_{\text{in}}}.$$

Mechanical advantage

The quantity $F_{\text{out}}/F_{\text{in}}$ is called the *mechanical advantage* of the hydraulic lift, and it is equal to the ratio of the areas. For example, if the area of the output piston is 20 times that of the input cylinder, the force is multiplied by a factor of 20: thus a force of 200 lb could lift a 4000-lb car.

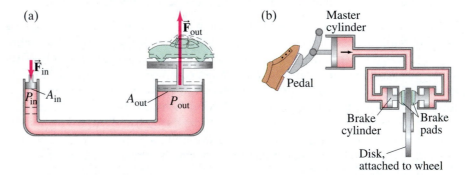

FIGURE 10–7 Applications of Pascal's principle: (a) hydraulic lift; (b) hydraulic brakes in a car.

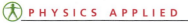
PHYSICS APPLIED

Brakes

Figure 10–7b illustrates the brake system of a car. When the driver presses the brake pedal, the pressure in the master cylinder increases. This pressure increase occurs throughout the brake fluid, thus pushing the brake pads against the disk attached to the car's wheel.

10-6 Measurement of Pressure; Gauges and the Barometer

Manometer

Many devices have been invented to measure pressure, some of which are shown in Fig. 10–8. The simplest is the open-tube *manometer* (Fig. 10–8a), which is a U-shaped tube partially filled with a liquid, usually mercury or water. The pressure P being measured is related (by Eq. 10–3b) to the difference in

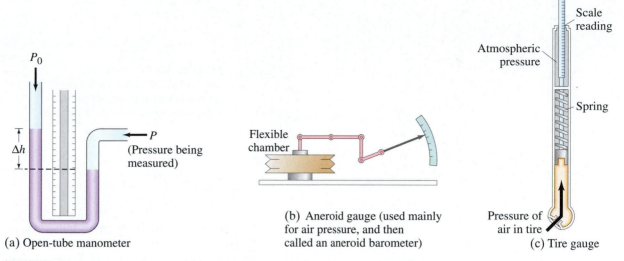

(a) Open-tube manometer

(b) Aneroid gauge (used mainly for air pressure, and then called an aneroid barometer)

(c) Tire gauge

FIGURE 10–8 Pressure gauges: (a) open-tube manometer, (b) aneroid gauge, and (c) common tire pressure gauge.

height Δh of the two levels of the liquid by the relation

$$P = P_0 + \rho g \, \Delta h, \tag{10–3c}$$

Pressure beneath surface of liquid open to the atmosphere

where P_0 is atmospheric pressure (acting on the top of the liquid in the left-hand tube), and ρ is the density of the liquid. Note that the quantity $\rho g \, \Delta h$ is the gauge pressure—the amount by which P exceeds atmospheric pressure P_0. If the liquid in the left-hand column were lower than that in the right-hand column, P would have to be less than atmospheric pressure (and Δh would be negative).

Instead of calculating the product $\rho g \, \Delta h$, sometimes only the change in height Δh is specified. In fact, pressures are sometimes specified as so many "millimeters of mercury" (mm-Hg) or "mm of water" (mm-H$_2$O). The unit mm-Hg is equivalent to a pressure of 133 N/m^2, since $\rho g \, \Delta h$ for $1 \, \text{mm} = 1.0 \times 10^{-3} \, \text{m}$ of mercury gives

$$\rho g \, \Delta h = (13.6 \times 10^3 \, \text{kg/m}^3)(9.80 \, \text{m/s}^2)(1.00 \times 10^{-3} \, \text{m}) = 1.33 \times 10^2 \, \text{N/m}^2.$$

The unit mm-Hg is also called the **torr** in honor of Evangelista Torricelli (1608–1647), a student of Galileo's who invented the barometer (see below). Conversion factors among the various units of pressure (an incredible nuisance!) are given in Table 10–2. It is important that only N/m^2 = Pa, the proper SI unit, be used in calculations involving other quantities specified in SI units.

Another type of pressure gauge is the aneroid gauge (Fig. 10–8b) in which the pointer is linked to the flexible ends of an evacuated thin metal chamber. In an electronic gauge, the pressure may be applied to a thin metal diaphragm whose resulting distortion is translated into an electrical signal by a transducer. How a common tire gauge is constructed is shown in Fig. 10–8c.

The torr (unit of pressure)

→ PROBLEM SOLVING

Use SI unit in calculations:
$1 \, \text{Pa} = 1 \, \text{N/m}^2$

TABLE 10–2 Conversion Factors Between Different Units of Pressure

In Terms of $1 \, \text{Pa} = 1 \, \text{N/m}^2$	1 atm in Different Units
$1 \, \text{atm} = 1.013 \times 10^5 \, \text{N/m}^2$	$1 \, \text{atm} = 1.013 \times 10^5 \, \text{N/m}^2$
$= 1.013 \times 10^5 \, \text{Pa} = 101.3 \, \text{kPa}$	
$1 \, \text{bar} = 1.000 \times 10^5 \, \text{N/m}^2$	$1 \, \text{atm} = 1.013 \, \text{bar}$
$1 \, \text{dyne/cm}^2 = 0.1 \, \text{N/m}^2$	$1 \, \text{atm} = 1.013 \times 10^6 \, \text{dyne/cm}^2$
$1 \, \text{lb/in.}^2 = 6.90 \times 10^3 \, \text{N/m}^2$	$1 \, \text{atm} = 14.7 \, \text{lb/in.}^2$
$1 \, \text{lb/ft}^2 = 47.9 \, \text{N/m}^2$	$1 \, \text{atm} = 2.12 \times 10^3 \, \text{lb/ft}^2$
$1 \, \text{cm-Hg} = 1.33 \times 10^3 \, \text{N/m}^2$	$1 \, \text{atm} = 76 \, \text{cm-Hg}$
$1 \, \text{mm-Hg} = 133 \, \text{N/m}^2$	$1 \, \text{atm} = 760 \, \text{mm-Hg}$
$1 \, \text{torr} = 133 \, \text{N/m}^2$	$1 \, \text{atm} = 760 \, \text{torr}$
$1 \, \text{mm-H}_2\text{O} \, (4°\text{C}) = 9.81 \, \text{N/m}^2$	$1 \, \text{atm} = 1.03 \times 10^4 \, \text{mm-H}_2\text{O} \, (4°\text{C})$

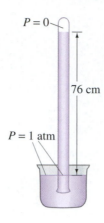

$P = 0$

76 cm

$P = 1$ atm

FIGURE 10–9 A mercury barometer, invented by Torricelli, is shown here when the air pressure is standard atmospheric, 76 cm-Hg.

Barometer

Atmospheric pressure can be measured by a modified kind of mercury manometer with one end closed, called a mercury **barometer** (Fig. 10–9). The glass tube is completely filled with mercury and then inverted into the bowl of mercury. If the tube is long enough, the level of the mercury will drop, leaving a vacuum at the top of the tube, since atmospheric pressure can support a column of mercury only about 76 cm high (exactly 76.0 cm at standard atmospheric pressure). That is, a column of mercury 76 cm high exerts the same pressure as the atmosphere[†]:

$$P = \rho g \, \Delta h$$
$$= (13.6 \times 10^3 \, \text{kg/m}^3)(9.80 \, \text{m/s}^2)(0.760 \, \text{m}) = 1.013 \times 10^5 \, \text{N/m}^2 = 1.00 \, \text{atm}.$$

Household barometers are usually of the aneroid type, either mechanical (Fig. 10–8b) or electronic.

A calculation similar to that above will show that atmospheric pressure can maintain a column of water 10.3 m high in a tube whose top is under vacuum (Fig. 10–10). No matter how good a vacuum pump is, it cannot lift water more than about 10 m. To pump water out of deep mine shafts with a vacuum pump requires multiple stages for depths greater than 10 m. Galileo studied this problem, and his student Torricelli was the first to explain it. The point is that a pump does not really suck water up a tube—it merely reduces the pressure at the top of the tube. Atmospheric air pressure *pushes* the water up the tube if the top end is at low pressure (under a vacuum), just as it is air pressure that pushes (or maintains) the mercury 76 cm high in a barometer.

FIGURE 10–10 A water barometer: a full tube of water is inserted into a tub of water, keeping the spigot at the top closed. When the bottom end of the tube is uncovered, some water flows out of the tube into the tub, leaving a vacuum between the water's upper surface and the spigot. Why? Because air pressure can not support a column of water more than 10 m high.

CONCEPTUAL EXAMPLE 10–5 **Suction.** You sit in a meeting where a novice NASA engineer proposes suction cup shoes for Space Shuttle astronauts working on the exterior of the spacecraft. Having just studied this Chapter, you gently remind him of the fallacy of this plan. What is it?

RESPONSE Suction cups work by pushing out the air underneath the cup. What holds the cup in place is the air pressure outside the cup. (This can be a substantial force when on Earth. For example, a 10-cm-diameter cup has an area of $7.9 \times 10^{-3} \, \text{m}^2$. The force of the atmosphere on it is $(7.9 \times 10^{-3} \, \text{m}^2)(1.0 \times 10^5 \, \text{N/m}^2) \approx 800 \, \text{N}$, about 180 lbs!). But in outer space, there is no air pressure to hold the suction cup onto the spacecraft.

We sometimes mistakenly think of suction as something we actively do. For example, we intuitively think that we pull the soda up through a straw. Instead, what we do is lower the pressure at the top of the straw, and the atmosphere *pushes* the soda up the straw.

[†]This calculation confirms the entry in Table 10–2, 1 atm = 76 cm-Hg.

10-7 Buoyancy and Archimedes' Principle

Objects submerged in a fluid appear to weigh less than they do when outside the fluid. For example, a large rock that you would have difficulty lifting off the ground can often be easily lifted from the bottom of a stream. When the rock breaks through the surface of the water, it suddenly seems to be much heavier. Many objects, such as wood, float on the surface of water. These are two examples of *buoyancy*. In each example, the force of gravity is acting downward. But in addition, an upward *buoyant force* is exerted by the liquid. The buoyant force on fish and underwater divers (as in the chapter-opening photo) almost exactly balances the force of gravity downward, and allows them to "hover" in equilibrium.

Rocks seem to weigh less under water

Wood floats

The buoyant force occurs because the pressure in a fluid increases with depth. Thus the upward pressure on the bottom surface of a submerged object is greater than the downward pressure on its top surface. To see this effect, consider a cylinder of height Δh whose top and bottom ends have an area A and which is completely submerged in a fluid of density ρ_F, as shown in Fig. 10–11. The fluid exerts a pressure $P_1 = \rho_F g h_1$ at the top surface of the

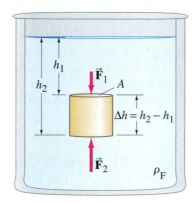

FIGURE 10–11 Determination of the buoyant force.

cylinder (Eq. 10–3a). The force due to this pressure on top of the cylinder is $F_1 = P_1 A = \rho_F g h_1 A$, and it is directed downward. Similarly, the fluid exerts an upward force on the bottom of the cylinder equal to $F_2 = P_2 A = \rho_F g h_2 A$. The net force on the cylinder exerted by the fluid pressure, which is the **buoyant force**, $\vec{F}_B$, acts upward and has the magnitude

$$
\begin{aligned}
F_B = F_2 - F_1 &= \rho_F g A (h_2 - h_1) \\
&= \rho_F g A \, \Delta h \\
&= \rho_F V g \\
&= m_F g,
\end{aligned}
$$

where $V = A \, \Delta h$ is the volume of the cylinder, the product $\rho_F V$ is its mass, and $\rho_F V g = m_F g$ is the weight of fluid which takes up a volume equal to the volume of the cylinder. Thus the buoyant force on the cylinder is equal to the weight of fluid displaced by the cylinder. This result is valid no matter what the shape of the object. Its discovery is credited to Archimedes (287?–212 B.C.), and it is called **Archimedes' principle**: *the buoyant force on an object immersed in a fluid is equal to the weight of the fluid displaced by that object.*

Archimedes' principle

By "fluid displaced," we mean a volume of fluid equal to the volume of the submerged object, or that part of the object submerged if it floats or is only partly submerged (the fluid that used to be where the object is). If the object is placed in a glass or tub initially filled to the brim with water, the water that flows over the top represents the water displaced by the object.

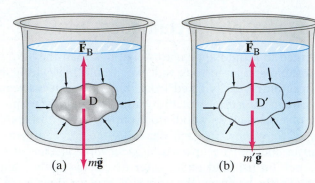

(a) $m\vec{g}$ (b) $m'\vec{g}$

FIGURE 10–12 Archimedes' principle.

We can derive Archimedes' principle in general by the following simple but elegant argument. The irregularly shaped object D shown in Fig. 10–12a is acted on by the force of gravity (its weight, $m\vec{g}$, downward) and the buoyant force, $\vec{F}_B$, upward. We wish to determine F_B. To do so, we next consider a body (D′ in Fig. 10–12b), this time made of the fluid itself, with the same shape and size as the original object, and located at the same depth. You might think of this body of fluid as being separated from the rest of the fluid by an imaginary membrane. The buoyant force F_B on this body of fluid will be exactly the same as that on the original object since the surrounding fluid, which exerts F_B, is in exactly the same configuration. This body of fluid D′ is in equilibrium (the fluid as a whole is at rest). Therefore, $F_B = m'g$, where $m'g$ is the weight of the body of fluid. Hence the buoyant force F_B is equal to the weight of the body of fluid whose volume equals the volume of the original submerged object, which is Archimedes' principle.

Archimedes' discovery was made by experiment. What we have done in the last two paragraphs is show that Archimedes' principle can be derived from Newtons' laws.

CONCEPTUAL EXAMPLE 10–6 **Two pails of water.** Consider two identical pails of water filled to the brim. One pail contains only water, the other has a piece of wood floating in it. Which pail has the greater weight?

RESPONSE Both pails weigh the same. Recall Archimedes' principle: the wood displaces a volume of water with weight equal to the weight of the wood. Some water will overflow the pail, but Archimedes' principle tells us the spilled water has weight equal to that of the wood; so the pails have the same weight.

EXAMPLE 10–7 **Recovering a submerged statue.** A 70-kg ancient statue lies at the bottom of the sea. Its volume is $3.0 \times 10^4 \text{ cm}^3$. How much force is needed to lift it?

APPROACH The force F needed to lift the statue is equal to the statue's weight mg minus the buoyant force F_B. Figure 10–13 is the free-body diagram.

SOLUTION The buoyant force on the statue due to the water is equal to the weight of $3.0 \times 10^4 \text{ cm}^3 = 3.0 \times 10^{-2} \text{ m}^3$ of water (for seawater, $\rho = 1.025 \times 10^3 \text{ kg/m}^3$):

$$F_B = m_{H_2O}\, g = \rho_{H_2O} V g$$
$$= (1.025 \times 10^3 \text{ kg/m}^3)(3.0 \times 10^{-2} \text{ m}^3)(9.8 \text{ m/s}^2)$$
$$= 3.0 \times 10^2 \text{ N}.$$

The weight of the statue is $mg = (70 \text{ kg})(9.8 \text{ m/s}^2) = 6.9 \times 10^2 \text{ N}$. Hence the force F needed to lift it is $690 \text{ N} - 300 \text{ N} = 390 \text{ N}$. It is as if the statue had a mass of only $(390 \text{ N})/(9.8 \text{ m/s}^2) = 40 \text{ kg}$.

NOTE Here $F = 390 \text{ N}$ is the force needed to lift the statue without acceleration when it is under water. As the statue comes *out* of the water, the force F increases, reaching 690 N when the statue is fully out of the water.

FIGURE 10–13 Example 10–7. The force needed to lift the statue is $\vec{F}$.

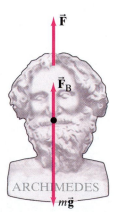

Archimedes is said to have discovered his principle in his bath while thinking how he might determine whether the king's new crown was pure gold or a fake. Gold has a specific gravity of 19.3, somewhat higher than that of most metals, but a determination of specific gravity or density is not readily done directly because, even if the mass is known, the volume of an irregularly shaped object is not easily calculated. However, if the object is weighed in air ($=w$) and also "weighed" while it is under water ($=w'$), the density can be determined using Archimedes' principle, as the following Example shows. The quantity w' is called the *apparent weight* in water, and is what a scale reads when the object is submerged in water (see Fig. 10–14); w' equals the true weight ($w = mg$) minus the buoyant force.

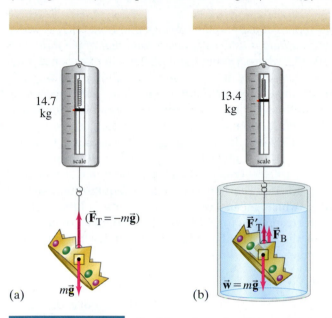

(a)   (b)

FIGURE 10–14 (a) A scale reads the mass of an object in air—in this case the crown of Example 10–8. All objects are at rest, so the tension F_T in the connecting cord equals the weight w of the object: $F_T = mg$. We show the free-body diagram of the crown, and F_T is what causes the scale reading (it's equal to the net downward force on the scale, by Newton's third law).
(b) Submerged, the object has an additional force on it, the buoyant force F_B. The net force is zero, so $F'_T + F_B = mg$ ($=w$). The scale now reads $m' = 13.4$ kg, where m' is related to the effective weight by $w' = m'g$. Thus $F'_T = w' = w - F_B$.

EXAMPLE 10–8 **Archimedes: Is the crown gold?** When a crown of mass 14.7 kg is submerged in water, an accurate scale reads only 13.4 kg. Is the crown made of gold?

APPROACH If the crown is gold, its density and specific gravity must be very high, SG = 19.3 (see Section 10–2 and Table 10–1). We determine the specific gravity using Archimedes' principle and the two free-body diagrams shown in Fig. 10–14.

SOLUTION The apparent weight of the submerged object (the crown) is w', and it equals F'_T in Fig. 10–14b. The sum of the forces on the object is zero, so w' equals the actual weight w ($=mg$) minus the buoyant force F_B:

$$w' = F'_T = w - F_B$$

so

$$w - w' = F_B.$$

Let V be the volume of the completely submerged object and ρ_O its density (so $\rho_O V$ is its mass), and let ρ_F be the density of the fluid (water). Then $(\rho_F V)g$ is the weight of fluid displaced ($= F_B$). Now we can write   *Recall $m = \rho V$ (Eq. 10–1)*

$$w = mg = \rho_O V g$$
$$w - w' = F_B = \rho_F V g.$$

We divide these two equations and obtain

$$\frac{w}{w - w'} = \frac{\rho_O V g}{\rho_F V g} = \frac{\rho_O}{\rho_F}.$$

We see that $w/(w - w')$ is equal to the specific gravity of the object if the fluid in which it is submerged is water ($\rho_F = 1.00 \times 10^3$ kg/m³). Thus

$$\frac{\rho_O}{\rho_{H_2O}} = \frac{w}{w - w'} = \frac{(14.7 \text{ kg})g}{(14.7 \text{ kg} - 13.4 \text{ kg})g} = \frac{14.7 \text{ kg}}{1.3 \text{ kg}} = 11.3.$$

This corresponds to a density of 11,300 kg/m³. The crown seems to be made of lead (see Table 10–1)!

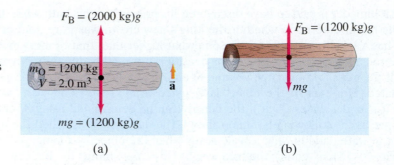

FIGURE 10–15 (a) The fully submerged log accelerates upward because $F_B > mg$. It comes to equilibrium (b) when $\Sigma F = 0$, so $F_B = mg = (1200\text{ kg})g$. Thus 1200 kg, or 1.2 m³, of water is displaced.

Archimedes' principle applies equally well to objects that float, such as wood. *Floating* In general, *an object floats on a fluid if its density is less than that of the fluid.* This is readily seen from Fig. 10–15a, where a submerged object will experience a net upward force and float to the surface if $F_B > mg$; that is, if $\rho_F V g > \rho_O V g$ or $\rho_F > \rho_O$. At equilibrium—that is, when floating—the buoyant force on an object has magnitude equal to the weight of the object. For example, a log whose specific gravity is 0.60 and whose volume is 2.0 m³ has a mass $m = \rho_O V = (0.60 \times 10^3\text{ kg/m}^3)(2.0\text{ m}^3) = 1200$ kg. If the log is fully submerged, it will displace a mass of water $m_F = \rho_F V = (1000\text{ kg/m}^3)(2.0\text{ m}^3) = 2000$ kg. Hence the buoyant force on the log will be greater than its weight, and it will float upward to the surface (Fig. 10–15). The log will come to equilibrium when it displaces 1200 kg of water, which means that 1.2 m³ of its volume will be submerged. This 1.2 m³ corresponds to 60% of the volume of the log $(1.2/2.0 = 0.60)$, so 60% of the log is submerged. In general when an object floats, we have $F_B = mg$, which we can write as (see Fig. 10–16)

$$\rho_F V_{\text{displ}}\, g = \rho_O V_O g,$$

where V_O is the full volume of the object and V_{displ} is the volume of fluid it displaces (= volume submerged). Thus

$$\frac{V_{\text{displ}}}{V_O} = \frac{\rho_O}{\rho_F}.$$

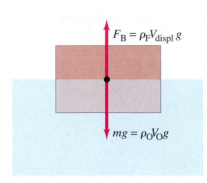

FIGURE 10–16 An object floating in equilibrium: $F_B = mg$.

Fraction of floating object submerged in water = its SG

That is, the fraction of the object submerged is given by the ratio of the object's density to that of the fluid. If the fluid is water, this fraction equals the specific gravity of the object.

FIGURE 10–17 A hydrometer. Example 10–9.

EXAMPLE 10–9 **Hydrometer calibration.** A **hydrometer** is a simple instrument used to measure the specific gravity of a liquid by indicating how deeply the instrument sinks in the liquid. A particular hydrometer (Fig. 10–17) consists of a glass tube, weighted at the bottom, which is 25.0 cm long and 2.00 cm² in cross-sectional area, and has a mass of 45.0 g. How far from the end should the 1.000 mark be placed?

APPROACH The hydrometer will float in water if its density ρ is less than $\rho_w = 1.000\text{ g/cm}^3$, the density of water. The fraction of the hydrometer submerged $(V_{\text{displaced}}/V_{\text{total}})$ is equal to the density ratio ρ/ρ_w.

SOLUTION The hydrometer has an overall density

$$\rho = \frac{m}{V} = \frac{45.0\text{ g}}{(2.00\text{ cm}^2)(25.0\text{ cm})} = 0.900\text{ g/cm}^3.$$

Thus, when placed in water, it will come to equilibrium when 0.900 of its volume is submerged. Since it is of uniform cross section, $(0.900)(25.0\text{ cm}) = 22.5$ cm of its length will be submerged. The specific gravity of water is defined to be 1.000, so the mark should be placed 22.5 cm from the end.

EXERCISE B On the hydrometer of Example 10–9, will the marks above the 1.000 mark represent higher or lower values of density of the liquid in which it is submerged?

Archimedes principle is also useful in geology. According to the theories of plate tectonics and continental drift, the continents float on a fluid "sea" of slightly deformable rock (mantle rock). Some interesting calculations can be done using very simple models, which we consider in the Problems at the end of the Chapter.

Air is a fluid, and it too exerts a buoyant force. Ordinary objects weigh less in air than they do if weighed in a vacuum. Because the density of air is so small, the effect for ordinary solids is slight. There are objects, however, that *float* in air—helium-filled balloons, for example, because the density of helium is less than the density of air.

FIGURE 10–18 Example 10–10.

EXAMPLE 10–10 **Helium balloon.** What volume V of helium is needed if a balloon is to lift a load of 180 kg (including the weight of the empty balloon)?

APPROACH The buoyant force on the helium balloon, F_B, which is equal to the weight of displaced air, must be at least equal to the weight of the helium plus the weight of the balloon and load (Fig. 10–18). Table 10–1 gives the density of helium as $0.179 \, \text{kg/m}^3$.

SOLUTION The buoyant force must have a minimum value of

$$F_B = (m_{He} + 180 \, \text{kg})g.$$

This equation can be written in terms of density using Archimedes' principle:

$$\rho_{air} V g = (\rho_{He} V + 180 \, \text{kg})g.$$

Solving now for V, we find

$$V = \frac{180 \, \text{kg}}{\rho_{air} - \rho_{He}}$$

$$= \frac{180 \, \text{kg}}{(1.29 \, \text{kg/m}^3 - 0.179 \, \text{kg/m}^3)} = 160 \, \text{m}^3.$$

NOTE This is the minimum volume needed near the Earth's surface, where $\rho_{air} = 1.29 \, \text{kg/m}^3$. To reach a high altitude, a greater volume would be needed since the density of air decreases with altitude.

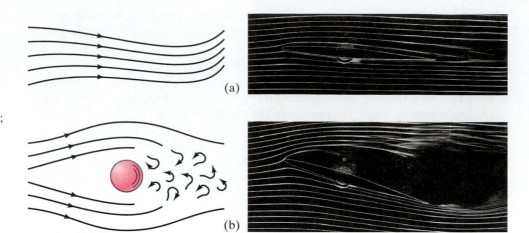

(a)

(b)

FIGURE 10–19
(a) Streamline, or laminar, flow;
(b) turbulent flow.

10–8 Fluids in Motion; Flow Rate and the Equation of Continuity

We now turn to the subject of fluids in motion, which is called **fluid dynamics**, or (especially if the fluid is water) **hydrodynamics**. Many aspects of fluid motion are still being studied (for example, turbulence as a manifestation of chaos is a "hot" topic today). Nonetheless, with certain simplifying assumptions, we can understand a lot about this subject.

We can distinguish two main types of fluid flow. If the flow is smooth, such that neighboring layers of the fluid slide by each other smoothly, the flow is said to be **streamline** or **laminar flow**.[†] In streamline flow, each particle of the fluid follows a smooth path, called a **streamline**, and these paths do not cross one another (Fig. 10–19a). Above a certain speed, the flow becomes turbulent. **Turbulent flow** is characterized by erratic, small, whirlpool-like circles called *eddy currents* or *eddies* (Fig. 10–19b). Eddies absorb a great deal of energy, and although a certain amount of internal friction called **viscosity** is present even during streamline flow, it is much greater when the flow is turbulent. A few tiny drops of ink or food coloring dropped into a moving liquid can quickly reveal whether the flow is streamline or turbulent.

Let us consider the steady laminar flow of a fluid through an enclosed tube or pipe as shown in Fig. 10–20. First we determine how the speed of the fluid changes when the size of the tube changes. The mass **flow rate** is defined as the mass Δm of fluid that passes a given point per unit time Δt:

$$\text{mass flow rate} = \frac{\Delta m}{\Delta t}.$$

FIGURE 10–20 Fluid flow through a pipe of varying diameter.

In Fig. 10–20, the volume of fluid passing point 1 (that is, through area A_1) in a time Δt is $A_1 \Delta l_1$, where Δl_1 is the distance the fluid moves in time Δt. Since the velocity[‡] of fluid passing point 1 is $v_1 = \Delta l_1/\Delta t$, the mass flow rate $\Delta m_1/\Delta t$ through area A_1 is

$$\frac{\Delta m_1}{\Delta t} = \frac{\rho_1 \Delta V_1}{\Delta t} = \frac{\rho_1 A_1 \Delta l_1}{\Delta t} = \rho_1 A_1 v_1,$$

where $\Delta V_1 = A_1 \Delta l_1$ is the volume of mass Δm_1, and ρ_1 is the fluid density. Similarly, at point 2 (through area A_2), the flow rate is $\rho_2 A_2 v_2$. Since no fluid flows in or out the sides, the flow rates through A_1 and A_2 must be equal. Thus, since

$$\frac{\Delta m_1}{\Delta t} = \frac{\Delta m_2}{\Delta t},$$

Equation of continuity (general)

then

$$\rho_1 A_1 v_1 = \rho_2 A_2 v_2. \tag{10–4a}$$

This is called the **equation of continuity**.

[†]The word laminar means "in layers."

[‡]If there were no viscosity, the velocity would be the same across a cross section of the tube. Real fluids have viscosity, and this internal friction causes different layers of the fluid to flow at different speeds. In this case v_1 and v_2 represent the average speeds at each cross section.

If the fluid is incompressible (ρ doesn't change with pressure), which is an excellent approximation for liquids under most circumstances (and sometimes for gases as well), then $\rho_1 = \rho_2$, and the equation of continuity becomes

$$A_1 v_1 = A_2 v_2. \qquad [\rho = \text{constant}] \quad \textbf{(10–4b)}$$

Equation of continuity
(ρ = constant)

The product Av represents the *volume rate of flow* (volume of fluid passing a given point per second), since $\Delta V/\Delta t = A\,\Delta l/\Delta t = Av$, which in SI units is m^3/s. Equation 10–4b tells us that where the cross-sectional area is large, the velocity is small, and where the area is small, the velocity is large. That this is reasonable can be seen by looking at a river. A river flows slowly through a meadow where it is broad, but speeds up to torrential speed when passing through a narrow gorge.

EXAMPLE 10–11 **ESTIMATE** **Blood flow.** In humans, blood flows from the heart into the aorta, from which it passes into the major arteries. These branch into the small arteries (arterioles), which in turn branch into myriads of tiny capillaries, Fig. 10–21. The blood returns to the heart via the veins. The radius of the aorta is about 1.2 cm, and the blood passing through it has a speed of about 40 cm/s. A typical capillary has a radius of about 4×10^{-4} cm, and blood flows through it at a speed of about 5×10^{-4} m/s. Estimate the number of capillaries that are in the body.

APPROACH We assume the density of blood doesn't vary significantly from the aorta to the capillaries. By the equation of continuity, the volume flow rate in the aorta must equal the volume flow rate through *all* the capillaries. The total area of all the capillaries is given by the area of one capillary multiplied by the total number N of capillaries.

SOLUTION Let A_1 be the area of the aorta and A_2 be the area of *all* the capillaries through which blood flows. Then $A_2 = N\pi r_{\text{cap}}^2$, where $r_{\text{cap}} \approx 4 \times 10^{-4}$ cm is the estimated average radius of one capillary. From the equation of continuity (Eq. 10–4), we have

$$v_2 A_2 = v_1 A_1$$
$$v_2 N\pi r_{\text{cap}}^2 = v_1 \pi r_{\text{aorta}}^2$$

so

$$N = \frac{v_1}{v_2}\frac{r_{\text{aorta}}^2}{r_{\text{cap}}^2} = \left(\frac{0.40\ \text{m/s}}{5 \times 10^{-4}\ \text{m/s}}\right)\left(\frac{1.2 \times 10^{-2}\ \text{m}}{4 \times 10^{-6}\ \text{m}}\right)^2 \approx 7 \times 10^9,$$

or on the order of 10 billion capillaries.

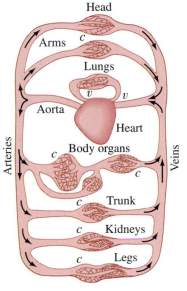

v = valves
c = capillaries

FIGURE 10–21
Human circulatory system.

EXAMPLE 10–12 **Heating duct to a room.** What area must a heating duct have if air moving 3.0 m/s along it can replenish the air every 15 minutes in a room of volume 300 m^3? Assume the air's density remains constant.

APPROACH We apply the equation of continuity at constant density, Eq. 10–4, to the air that flows through the duct (point 1 in Fig. 10–22) and then into the room (point 2). The volume flow rate in the room equals the volume of the room divided by the 15-minute replenishing time.

SOLUTION Consider the room as a large section of the duct, Fig. 10–22, and think of air equal to the volume of the room as passing by point 2 in $t = 15$ minutes $= 900$ s. Reasoning in the same way we did to obtain Eq. 10–4a (changing Δt to t), we write $v_2 = l_2/t$ so $A_2 v_2 = A_2 l_2/t = V_2/t$, where V_2 is the volume of the room. Then the equation of continuity becomes $A_1 v_1 = A_2 v_2 = V_2/t$ and

$$A_1 = \frac{V_2}{v_1 t} = \frac{300\ \text{m}^3}{(3.0\ \text{m/s})(900\ \text{s})} = 0.11\ \text{m}^2.$$

If the duct is square, then each side has length $l = \sqrt{A} = 0.33$ m, or 33 cm. A rectangular duct 20 cm $\times$ 55 cm will also do.

PHYSICS APPLIED
Heating duct

FIGURE 10–22 Example 10–12.

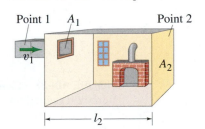

10–9 Bernoulli's Equation

Have you ever wondered why an airplane can fly, or how a sailboat can move against the wind? These are examples of a principle worked out by Daniel Bernoulli (1700–1782) concerning fluids in motion. In essence, **Bernoulli's principle** states that *where the velocity of a fluid is high, the pressure is low, and where the velocity is low, the pressure is high.* For example, if the pressures at points 1 and 2 in Fig. 10–20 are measured, it will be found that the pressure is lower at point 2, where the velocity is greater, than it is at point 1, where the velocity is smaller. At first glance, this might seem strange; you might expect that the greater speed at point 2 would imply a higher pressure. But this cannot be the case. For if the pressure at point 2 were higher than at 1, this higher pressure would slow the fluid down, whereas in fact it has sped up in going from point 1 to point 2. Thus the pressure at point 2 must be less than at point 1, to be consistent with the fact that the fluid accelerates. [To help clarify any misconceptions, a faster fluid *would* exert a greater force on an obstacle placed in its path. But that is not what we mean by pressure of a fluid, and besides we are not considering obstacles that interrupt the flow. We are examining smooth streamline flow. The fluid pressure is exerted on the walls of a pipe or surface of any material the fluid passes over.]

Bernoulli's principle

Bernoulli developed an equation that expresses this principle quantitatively. To derive Bernoulli's equation, we assume the flow is steady and laminar, the fluid is incompressible, and the viscosity is small enough to be ignored. To be general, we assume the fluid is flowing in a tube of nonuniform cross section that varies in height above some reference level, Fig. 10–23. We will consider the volume of fluid shown in color and calculate the work done to move it from the position shown in Fig. 10–23a to that shown in Fig. 10–23b. In this process, fluid at point 1 flows a distance Δl_1 and forces the fluid at point 2 to move a distance Δl_2. The fluid to the left of point 1 exerts a pressure P_1 on our section of fluid and does an amount of work

$$W_1 = F_1 \Delta l_1 = P_1 A_1 \Delta l_1.$$

At point 2, the work done on our cross section of fluid is

$$W_2 = -P_2 A_2 \Delta l_2.$$

The negative sign is present because the force exerted on the fluid is opposite to the motion (thus the fluid shown in color does work on the fluid to the right of point 2). Work is also done on the fluid by the force of gravity. The net effect of the process shown in Fig. 10–23 is to move a mass m of volume $A_1 \Delta l_1$ ($= A_2 \Delta l_2$, since the fluid is incompressible) from point 1 to point 2, so the work done by gravity is

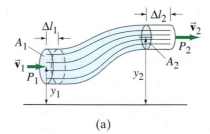

(a)

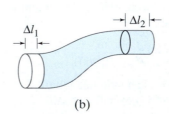

(b)

FIGURE 10–23 Fluid flow: for derivation of Bernoulli's equation.

$$W_3 = -mg(y_2 - y_1),$$

where y_1 and y_2 are heights of the center of the tube above some (arbitrary) reference level. In the case shown in Fig. 10–23, this term is negative since the motion is uphill against the force of gravity. The net work W done on the fluid is thus

$$W = W_1 + W_2 + W_3$$
$$W = P_1 A_1 \Delta l_1 - P_2 A_2 \Delta l_2 - mgy_2 + mgy_1.$$

According to the work-energy principle (Section 6–3), the net work done on a system is equal to its change in kinetic energy. Hence

$$\tfrac{1}{2}mv_2^2 - \tfrac{1}{2}mv_1^2 = P_1 A_1 \Delta l_1 - P_2 A_2 \Delta l_2 - mgy_2 + mgy_1.$$

The mass m has volume $A_1 \Delta l_1 = A_2 \Delta l_2$. Thus we can substitute $m = \rho A_1 \Delta l_1 = \rho A_2 \Delta l_2$, and then divide through by $A_1 \Delta l_1 = A_2 \Delta l_2$, to obtain

$$\tfrac{1}{2}\rho v_2^2 - \tfrac{1}{2}\rho v_1^2 = P_1 - P_2 - \rho g y_2 + \rho g y_1,$$

which we rearrange to get

$$P_1 + \tfrac{1}{2}\rho v_1^2 + \rho g y_1 = P_2 + \tfrac{1}{2}\rho v_2^2 + \rho g y_2. \qquad \textbf{(10–5)} \qquad \textit{Bernoulli's equation}$$

This is **Bernoulli's equation**. Since points 1 and 2 can be any two points along a tube of flow, Bernoulli's equation can be written as

$$P + \tfrac{1}{2}\rho v^2 + \rho g y = \text{constant}$$

at every point in the fluid, where y is the height of the center of the tube above a fixed reference level. [Note that if there is no flow $(v_1 = v_2 = 0)$, then Eq. 10–5 reduces to the hydrostatic equation, Eq. 10–3b or c.]

Bernoulli's equation is an expression of the law of energy conservation, since we derived it from the work-energy principle.

EXERCISE C As water in a level pipe passes from a narrow cross section of pipe to a wider cross section, how does the pressure change?

EXAMPLE 10–13 **Flow and pressure in a hot-water heating system.**
Water circulates throughout a house in a hot-water heating system. If the water is pumped at a speed of 0.50 m/s through a 4.0-cm-diameter pipe in the basement under a pressure of 3.0 atm, what will be the flow speed and pressure in a 2.6-cm-diameter pipe on the second floor 5.0 m above? Assume the pipes do not divide into branches.

PHYSICS APPLIED
Hot-water heating system

APPROACH We use the equation of continuity at constant density to determine the flow speed on the second floor, and then Bernoulli's equation to find the pressure.

SOLUTION We take v_2 in the equation of continuity, Eq. 10–4, as the flow speed on the second floor, and v_1 as the flow speed in the basement. Noting that the areas are proportional to the radii squared $(A = \pi r^2)$, we obtain

$$v_2 = \frac{v_1 A_1}{A_2} = \frac{v_1 \pi r_1^2}{\pi r_2^2} = (0.50 \text{ m/s}) \frac{(0.020 \text{ m})^2}{(0.013 \text{ m})^2} = 1.2 \text{ m/s}.$$

To find the pressure on the second floor, we use Bernoulli's equation:

$$P_2 = P_1 + \rho g(y_1 - y_2) + \tfrac{1}{2}\rho(v_1^2 - v_2^2)$$
$$= (3.0 \times 10^5 \text{ N/m}^2) + (1.0 \times 10^3 \text{ kg/m}^3)(9.8 \text{ m/s}^2)(-5.0 \text{ m})$$
$$+ \tfrac{1}{2}(1.0 \times 10^3 \text{ kg/m}^3)[(0.50 \text{ m/s})^2 - (1.2 \text{ m/s})^2]$$
$$= (3.0 \times 10^5 \text{ N/m}^2) - (4.9 \times 10^4 \text{ N/m}^2) - (6.0 \times 10^2 \text{ N/m}^2)$$
$$= 2.5 \times 10^5 \text{ N/m}^2 = 2.5 \text{ atm}.$$

NOTE The velocity term contributes very little in this case.

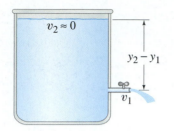

FIGURE 10–24 Torricelli's theorem: $v_1 = \sqrt{2g(y_2 - y_1)}$.

Torricelli's theorem

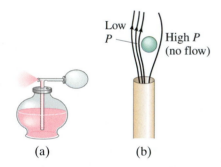

(a)   (b)

FIGURE 10–25 Examples of Bernoulli's principle: (a) atomizer, (b) Ping-Pong ball in jet of air.

FIGURE 10–26 Lift on an airplane wing. We are in the reference frame of the wing, seeing the air flow by.

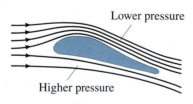

PHYSICS APPLIED
Airplanes and dynamic lift

10–10 Applications of Bernoulli's Principle: from Torricelli to Airplanes, Baseballs, and TIA

Bernoulli's equation can be applied to many situations. One example is to calculate the velocity, v_1, of a liquid flowing out of a spigot at the bottom of a reservoir, Fig. 10–24. We choose point 2 in Eq. 10–5 to be the top surface of the liquid. Assuming the diameter of the reservoir is large compared to that of the spigot, v_2 will be almost zero. Points 1 (the spigot) and 2 (top surface) are open to the atmosphere, so the pressure at both points is equal to atmospheric pressure: $P_1 = P_2$. Then Bernoulli's equation becomes

$$\tfrac{1}{2}\rho v_1^2 + \rho g y_1 = \rho g y_2$$

or

$$v_1 = \sqrt{2g(y_2 - y_1)}. \tag{10–6}$$

This result is called **Torricelli's theorem**. Although it is seen to be a special case of Bernoulli's equation, it was discovered a century earlier by Evangelista Torricelli. Equation 10–6 tells us that the liquid leaves the spigot with the same speed that a freely falling object would attain if falling from the same height. This should not be too surprising since the derivation of Bernoulli's equation relies on the conservation of energy.

Another special case of Bernoulli's equation arises when a fluid is flowing horizontally with no appreciable change in height; that is, $y_1 = y_2$. Then Eq. 10–5 becomes

$$P_1 + \tfrac{1}{2}\rho v_1^2 = P_2 + \tfrac{1}{2}\rho v_2^2, \tag{10–7}$$

which tells us quantitatively that the speed is high where the pressure is low, and vice versa. It explains many common phenomena, some of which are illustrated in Figs. 10–25 to 10–31. The pressure in the air blown at high speed across the top of the vertical tube of a perfume atomizer (Fig. 10–25a) is less than the normal air pressure acting on the surface of the liquid in the bowl. Thus atmospheric pressure in the bowl pushes the perfume up the tube because of the lower pressure at the top. A Ping-Pong ball can be made to float above a blowing jet of air (some vacuum cleaners can blow air), Fig. 10–25b; if the ball begins to leave the jet of air, the higher pressure in the still air outside the jet pushes the ball back in.

Airplane Wings and Dynamic Lift

Airplanes experience a "lift" force on their wings, keeping them up in the air, if they are moving at a sufficiently high speed relative to the air and the wing is tilted upward at a small angle (the "attack angle"), as in Fig. 10–26, where streamlines of air are shown rushing by the wing. (We are in the reference frame of the wing, as if sitting on the wing.) The upward tilt, as well as the rounded upper surface of the wing, causes the streamlines to be forced upward and to be crowded together above the wing. The area for air flow between any two streamlines is reduced as the streamlines get closer together, so from the equation of continuity $(A_1 v_1 = A_2 v_2)$, the air speed increases above the wing where the streamlines are squished together. (Recall also how the crowded streamlines in a pipe constriction, Fig. 10–20, indicate the velocity is higher in the constriction.) Because the air speed is greater above the wing than below it, the pressure above the wing is less than the pressure below the wing (Bernoulli's principle). Hence there is a net upward force on the wing called **dynamic lift**. Experiments show that the speed of air above the wing can even be double the speed of the air below it. (Friction between the air and wing exerts a *drag force*, toward the rear, which must be overcome by the plane's engines.)

A flat wing, or one with symmetric cross section, will experience lift as long as the front of the wing is tilted upward (attack angle). The wing shown in Fig. 10–26 can experience lift even if the attack angle is zero, because the rounded upper surface deflects air up, squeezing the streamlines together. Airplanes can fly upside down, experiencing lift, if the attack angle is sufficient to deflect streamlines up and closer together.

Our picture considers streamlines; but if the attack angle is larger than about 15°, turbulence sets in (Fig. 10–19b) leading to greater drag and less lift, causing the wing to "stall" and the plane to drop.

From another point of view, the upward tilt of a wing means the air moving horizontally in front of the wing is deflected downward; the change in momentum of the rebounding air molecules results in an upward force on the wing (Newtons' third law).

Sailboats

A sailboat can move *against* the wind, with the aid of the Bernoulli effect, by setting the sails at an angle, as shown in Fig. 10–27. The air travels rapidly over the bulging front surface of the sail, and the relatively still air behind the sail exerts a greater pressure, resulting in a net force on the sail, $\vec{\mathbf{F}}_{wind}$. This force would tend to make the boat move sideways if it weren't for the keel that extends vertically downward beneath the water: the water exerts a force ($\vec{\mathbf{F}}_{water}$) on the keel nearly perpendicular to the keel. The resultant of these two forces ($\vec{\mathbf{F}}_R$) is almost directly forward as shown.

Baseball Curve

Why a spinning pitched baseball (or tennis ball) curves can also be explained using Bernoulli's principle. It is simplest if we put ourselves in the reference frame of the ball, with the air rushing by, just as we did for the airplane wing. Suppose the ball is rotating counterclockwise as seen from above, Fig. 10–28. A thin layer of air ("boundary layer") is being dragged around by the ball. We are looking down on the ball, and at point A in Fig. 10–28, this boundary layer tends to slow down the oncoming air. At point B, the air rotating with the ball adds its speed to that of the oncoming air, so the air speed is higher at B than at A. The higher speed at B means the pressure is lower at B than at A, resulting in a net force toward B. The ball's path curves toward the left (as seen by the pitcher).

Lack of Blood to the Brain—TIA

In medicine, one of many applications of Bernoulli's principle is to explain a TIA, a transient ischemic attack (meaning a temporary lack of blood supply to the brain). A person suffering a TIA may experience symptoms such as dizziness, double vision, headache, and weakness of the limbs. A TIA can occur as follows. Blood normally flows up to the brain at the back of the head via the two vertebral arteries—one going up each side of the neck—which meet to form the basilar artery just below the brain, as shown in Fig. 10–29. The vertebral arteries issue from the subclavian arteries, as shown, before the latter pass to the arms. When an arm is exercised vigorously, blood flow increases to meet the needs of the arm's muscles. If the subclavian artery on one side of the body is partially blocked, however, as in arteriosclerosis (hardening of the arteries), the blood velocity will have to be higher on that side to supply the needed blood. (Recall the equation of continuity: smaller area means larger velocity for the same flow rate, Eq. 10–4.) The increased blood velocity past the opening to the vertebral artery results in lower pressure (Bernoulli's principle). Thus, blood rising in the vertebral artery on the "good" side at normal pressure can be *diverted down* into the other vertebral artery because of the low pressure on that side, instead of passing upward to the brain. Hence the blood supply to the brain is reduced.

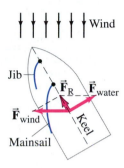

FIGURE 10–27 Sailboat sailing against the wind.

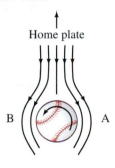

FIGURE 10–28 Looking down on a pitched baseball heading toward home plate. We are in the reference frame of the baseball, with the air flowing by.

FIGURE 10–29 Rear of the head and shoulders showing arteries leading to the brain and to the arms. High blood velocity past the constriction in the left subclavian artery causes low pressure in the left vertebral artery, in which a reverse (downward) blood flow can then occur, resulting in a TIA.

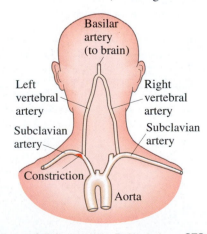

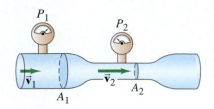

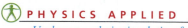

FIGURE 10–30 Venturi meter.

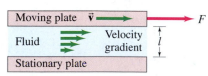

PHYSICS APPLIED

Smoke up a chimney

PHYSICS APPLIED

Underground air circulation for burrowing animals

Other Applications

A **venturi tube** is essentially a pipe with a narrow constriction (the throat). The flowing air speeds up as it passes through this constriction, so the pressure is lower in the throat. A *venturi meter*, Fig. 10–30, is used to measure the flow speed of gases and liquids, including blood velocity in arteries.

Why does smoke go up a chimney? It's partly because hot air rises (it's less dense and therefore buoyant). But Bernoulli's principle also plays a role. When wind blows across the top of a chimney, the pressure is less there than inside the house. Hence, air and smoke are pushed up the chimney by the higher indoor pressure. Even on an apparently still night there is usually enough ambient air flow at the top of a chimney to assist upward flow of smoke.

If gophers, prairie dogs, rabbits, and other animals that live underground are to avoid suffocation, the air must circulate in their burrows. The burrows always have at least two entrances (Fig. 10–31). The speed of air flow across different holes will usually be slightly different. This results in a slight pressure difference, which forces a flow of air through the burrow a la Bernoulli's principle. The flow of air is enhanced if one hole is higher than the other (animals often build mounds) since wind speed tends to increase with height.

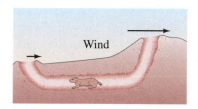

FIGURE 10–31 Bernoulli's principle explains air flow in underground burrows.

Wind

Limitations on Bernoulli's equation

Bernoulli's equation ignores the effects of friction (viscosity) and the compressibility of the fluid. The energy that is transformed to internal (or potential) energy due to compression and to thermal energy by friction can be taken into account by adding terms to Eq. 10–5. These terms are difficult to calculate theoretically and are normally determined empirically. They do not significantly alter the explanations for the phenomena described above.

* 10–11 Viscosity

Real fluids have a certain amount of internal friction called **viscosity**, as mentioned in Section 10–8. Viscosity exists in both liquids and gases, and is essentially a frictional force between adjacent layers of fluid as the layers move past one another. In liquids, viscosity is due to the electrical cohesive forces between the molecules. In gases, it arises from collisions between the molecules.

The viscosity of different fluids can be expressed quantitatively by a *coefficient of viscosity*, η (the Greek lowercase letter eta), which is defined in the following way. A thin layer of fluid is placed between two flat plates. One plate is stationary and the other is made to move, Fig. 10–32. The fluid directly in contact with each plate is held to the surface by the adhesive force between the molecules of the liquid and those of the plate. Thus the upper surface of the fluid moves with the same speed v as the upper plate, whereas the fluid in contact with the stationary plate remains stationary. The stationary layer of fluid retards the flow of the layer just above it, which in turn retards the flow of the next layer, and so on. Thus the velocity varies continuously from 0 to v, as shown. The increase in velocity divided by the distance over which this change is made—equal to v/l—is called the *velocity gradient*. To move the upper plate requires a force, which you can verify by moving a flat plate across a puddle of syrup on a table. For a given fluid, it is found that the force required, F, is proportional to the area of fluid in contact with each plate, A, and to the speed, v, and is inversely proportional to the separation, l, of the plates: $F \propto vA/l$. For different fluids, the more viscous the fluid, the greater is the required force.

FIGURE 10–32
Determination of viscosity.

Moving plate $\vec{v}$ —— F

Fluid | Velocity gradient | l

Stationary plate

Hence the proportionality constant for this equation is defined as the coefficient of viscosity, η:

$$F = \eta A \frac{v}{l}. \qquad (10\text{–}8)$$

Solving for η, we find $\eta = Fl/vA$. The SI unit for η is $\text{N·s/m}^2 = \text{Pa·s}$ (pascal·second). In the cgs system, the unit is dyne·s/cm^2, which is called a *poise* (P). Viscosities are often given in centipoise $(1\text{ cP} = 10^{-2}\text{ P})$. Table 10–3 lists the coefficient of viscosity for various fluids. The temperature is also specified, since it has a strong effect; the viscosity of liquids such as motor oil, for example, decreases rapidly as temperature increases.[‡]

TABLE 10–3
Coefficients of Viscosity

Fluid (temperature in C°)	Coefficient of Viscosity, η (Pa · s)[†]
Water (0°)	1.8×10^{-3}
(20°)	1.0×10^{-3}
(100°)	0.3×10^{-3}
Whole blood (37°)	$\approx 4 \times 10^{-3}$
Blood plasma (37°)	$\approx 1.5 \times 10^{-3}$
Ethyl alcohol (20°)	1.2×10^{-3}
Engine oil (30°) (SAE 10)	200×10^{-3}
Glycerine (20°)	1500×10^{-3}
Air (20°)	0.018×10^{-3}
Hydrogen (0°)	0.009×10^{-3}
Water vapor (100°)	0.013×10^{-3}

[†]$1\text{ Pa·s} = 10\text{ P} = 1000\text{ cP}.$

*10–12 Flow in Tubes: Poiseuille's Equation, Blood Flow

If a fluid had no viscosity, it could flow through a level tube or pipe without a force being applied. Viscosity acts like a sort of friction, so a pressure difference between the ends of a level tube is necessary for the steady flow of any real fluid, be it water or oil in a pipe, or blood in the circulatory system of a human.

The rate of flow of a fluid in a round tube depends on the viscosity of the fluid, the pressure difference, and the dimensions of the tube. The French scientist J. L. Poiseuille (1799–1869), who was interested in the physics of blood circulation (and after whom the "poise" is named), determined how the variables affect the flow rate of an incompressible fluid undergoing laminar flow in a cylindrical tube. His result, known as *Poiseuille's equation*, is:

$$Q = \frac{\pi R^4 (P_1 - P_2)}{8\eta L}, \qquad (10\text{–}9)$$

Poiseuille's equation
for flow rate in a tube

where R is the inside radius of the tube, L is its length, $P_1 - P_2$ is the pressure difference between the ends, η is the coefficient of viscosity, and Q is the volume rate of flow (volume of fluid flowing past a given point per unit time which in SI has units of m^3/s). Equation 10–9 applies only to laminar flow.

Poiseuille's equation tells us that the flow rate Q is directly proportional to the "pressure gradient," $(P_1 - P_2)/L$, and it is inversely proportional to the viscosity of the fluid. This is just what we might expect. It may be surprising, however, that Q also depends on the *fourth* power of the tube's radius. This means that for the same pressure gradient, if the tube radius is halved, the flow rate is decreased by a factor of 16! Thus the rate of flow, or alternately the pressure required to maintain a given flow rate, is greatly affected by only a small change in tube radius.

An interesting example of this R^4 dependence is *blood flow* in the human body. Poiseuille's equation is valid only for the streamline flow of an incompressible fluid with constant viscosity η. So it cannot be precisely accurate for blood whose flow is not without turbulence and that contains blood cells (whose diameter is almost equal to that of a capillary). Hence η depends to a certain extent on the blood flow speed v. Nonetheless, Poiseuille's equation does give a reasonable first approximation. The body controls the flow of blood by means of tiny bands of muscle surrounding the arteries. Contraction of these muscles reduces the diameter of an artery and, because of the R^4 term in Eq. 10–9, the flow rate is greatly reduced for only a small change in radius. Very small actions by these muscles can thus control precisely the flow of blood to

PHYSICS APPLIED
Medicine—
blood flow
and
heart disease

[‡]The Society of Automotive Engineers assigns numbers to represent the viscosity of oils: 30 weight (SAE 30) is more viscous than 10 weight. Multigrade oils, such as 20–50, are designed to maintain viscosity as temperature increases; 20–50 means the oil is 20 wt when cool but is like a 50-wt pure oil when it is hot (engine running temperature).

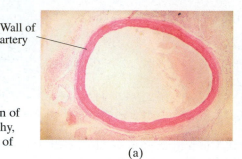
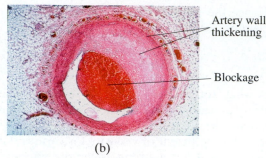

Wall of artery

Artery wall thickening

Blockage

(a)   (b)

FIGURE 10–33 A cross section of a human artery that (a) is healthy, (b) is partly blocked as a result of arteriosclerosis.

different parts of the body. Another aspect is that the radius of arteries is reduced as a result of arteriosclerosis (thickening and hardening of artery walls, Fig. 10–33) and by cholesterol buildup. When this happens, the pressure gradient must be increased to maintain the same flow rate. If the radius is reduced by half, the heart would have to increase the pressure by a factor of about $2^4 = 16$ in order to maintain the same blood-flow rate. The heart must work much harder under these conditions, but usually cannot maintain the original flow rate. Thus, high blood pressure is an indication both that the heart is working harder and that the blood-flow rate is reduced.

**10–13* Surface Tension and Capillarity

FIGURE 10–34 Spherical water droplets, dew on a blade of grass.

Surface tension

FIGURE 10–35 U-shaped wire apparatus holding a film of liquid to measure surface tension ($\gamma = F/2L$).

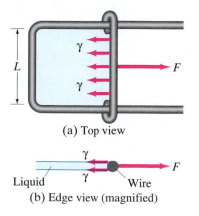

(a) Top view

Liquid   Wire
(b) Edge view (magnified)

The *surface* of a liquid at rest behaves in an interesting way, almost as if it were a stretched membrane under tension. For example, a drop of water on the end of a dripping faucet, or hanging from a thin branch in the early morning dew (Fig. 10–34), forms into a nearly spherical shape as if it were a tiny balloon filled with water. A steel needle can be made to float on the surface of water even though it is denser than the water. The surface of a liquid acts like it is under tension, and this tension, acting along the surface, arises from the attractive forces between the molecules. This effect is called **surface tension**. More specifically, a quantity called the *surface tension*, γ (the Greek letter gamma), is defined as the force F per unit length L that acts perpendicular to any line or cut in a liquid surface, tending to pull the surface closed:

$$\gamma = \frac{F}{L}. \qquad (10\text{–}10)$$

To understand this, consider the U-shaped apparatus shown in Fig. 10–35 which encloses a thin film of liquid. Because of surface tension, a force F is required to pull the movable wire and thus increase the surface area of the liquid. The liquid contained by the wire apparatus is a thin film having both a top and a bottom surface. Hence the length of the surface being increased is $2L$, and the surface tension is $\gamma = F/2L$. A delicate apparatus of this type can be used to measure the surface tension of various liquids. The surface tension of water is 0.072 N/m at 20°C. Table 10–4 gives the values for several substances. Note that temperature has a considerable effect on the surface tension.

Because of surface tension, some insects (Fig. 10–36) can walk on water, and objects more dense than water, such as a steel needle, can float on the

FIGURE 10–36 A water strider.

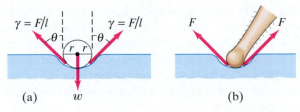

FIGURE 10–37 Surface tension acting on (a) a sphere, and (b) an insect leg. Example 10–14.

TABLE 10–4
Surface Tension of Some Substances

Substance	Surface Tension (N/m)
Mercury (20°C)	0.44
Blood, whole (37°C)	0.058
Blood, plasma (37°C)	0.073
Alcohol, ethyl (20°C)	0.023
Water (0°C)	0.076
(20°C)	0.072
(100°C)	0.059
Benzene (20°C)	0.029
Soap solution (20°C)	≈0.025
Oxygen (−193°C)	0.016

surface. Figure 10–37a shows how the surface tension can support the weight w of an object. Actually, the object sinks slightly into the fluid, so w is the "effective weight" of that object—its true weight less the buoyant force.

EXAMPLE 10–14 ESTIMATE Insect walks on water. The base of an insect's leg is approximately spherical in shape, with a radius of about 2.0×10^{-5} m. The 0.0030-g mass of the insect is supported equally by its six legs. Estimate the angle θ (see Fig. 10–37) for an insect on the surface of water. Assume the water temperature is 20°C.

APPROACH Since the insect is in equilibrium, the upward surface tension force is equal to the effective pull of gravity downward on each leg.

SOLUTION For each leg, we assume the surface tension force acts all around a circle of radius r, at an angle θ, as shown in Fig. 10–37. Only the vertical component, $\gamma \cos \theta$, acts to balance the weight mg. So we set the length L in Eq. 10–10 equal to the circumference of the circle, $L \approx 2\pi r$. Then the net upward force due to surface tension is $F_y \approx (\gamma \cos \theta)L \approx 2\pi r \gamma \cos \theta$. We set this surface tension force equal to one-sixth the weight of the insect since it has six legs:

$$2\pi r \gamma \cos \theta \approx \tfrac{1}{6} mg$$

$$(6.28)(2.0 \times 10^{-5}\,\text{m})(0.072\,\text{N/m}) \cos \theta \approx \tfrac{1}{6}(3.0 \times 10^{-6}\,\text{kg})(9.8\,\text{m/s}^2)$$

$$\cos \theta \approx \frac{0.49}{0.90} = 0.54.$$

So $\theta \approx 57°$. If $\cos \theta$ had come out greater than 1, the surface tension would not be great enough to support the insect's weight.

NOTE Our estimate ignored the buoyant force and ignored any difference between the radius of the insect's "foot" and the radius of the surface depression.

Soaps and detergents lower the surface tension of water. This is desirable for washing and cleaning since the high surface tension of pure water prevents it from penetrating easily between the fibers of material and into tiny crevices. Substances that reduce the surface tension of a liquid are called *surfactants*.

Surface tension plays a role in another interesting phenomenon, capillarity. It is a common observation that water in a glass container rises up slightly where it touches the glass, Fig. 10–38a. The water is said to "wet" the glass. Mercury, on the other hand, is depressed when it touches the glass, Fig. 10–38b; the mercury does not wet the glass. Whether a liquid wets a solid surface is determined by the relative strength of the cohesive forces between the molecules of the liquid compared to the adhesive forces between the molecules of the liquid and those of the container. *Cohesion* refers to the force between molecules of the same type, whereas *adhesion* refers to the force between molecules of different types. Water wets glass because the water molecules are more strongly attracted to the glass molecules than they are to other water molecules. The opposite is true for mercury: the cohesive forces are stronger than the adhesive forces.

PHYSICS APPLIED
Soaps and detergents

Capillarity

FIGURE 10–38 Water (a) "wets" the surface of glass, whereas (b) mercury does not "wet" the glass.

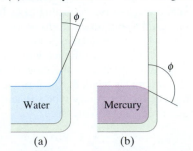

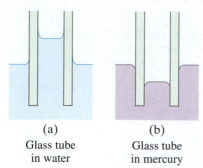

(a)
Glass tube
in water

(b)
Glass tube
in mercury

FIGURE 10–39 Capillarity.

In tubes having very small diameters, liquids are observed to rise or fall relative to the level of the surrounding liquid. This phenomenon is called **capillarity**, and such thin tubes are called **capillaries**. Whether the liquid rises or falls (Fig. 10–39) depends on the relative strengths of the adhesive and cohesive forces. Thus water rises in a glass tube, whereas mercury falls. The actual amount of rise (or fall) depends on the surface tension—which is what keeps the liquid surface from breaking apart.

* 10–14 Pumps, and the Heart

We conclude this Chapter with a brief discussion of pumps of various types, including the heart. Pumps can be classified into categories according to their function. A *vacuum pump* is designed to reduce the pressure (usually of air) in a given vessel. A *force pump*, on the other hand, is a pump that is intended to increase the pressure—for example, to lift a liquid (such as water from a well) or to push a fluid through a pipe. Figure 10–40 illustrates the principle behind a simple reciprocating pump. It could be a vacuum pump, in which case the intake is connected to the vessel to be evacuated. A similar mechanism is used in some force pumps, and in this case the fluid is forced under increased pressure through the outlet.

FIGURE 10–40 One kind of pump: the intake valve opens and air (or fluid that is being pumped) fills the empty space when the piston moves to the left. When the piston moves to the right (not shown), the outlet valve opens and fluid is forced out.

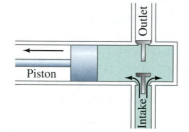

Other kinds of pumps are illustrated in Fig. 10–41. The centrifugal pump, or any force pump, can be used as a *circulating pump*—that is, to circulate a fluid around a closed path, such as the cooling water or lubricating oil in an automobile.

FIGURE 10–41 (a) Centrifugal pump: the rotating blades force fluid through the outlet pipe; this kind of pump is used in vacuum cleaners and as a water pump in automobiles. (b) Rotary oil-seal pump, used to obtain vacuums as low as 10^{-4} mm-Hg: gas (usually air) from the vessel to be evacuated diffuses into the space G via the intake pipe I; the rotating off-center cylinder C traps the gas in G and carries it around to push it out the exhaust valve E, in the meantime allowing more gas to diffuse into G for the next cycle. The sliding valve V is kept in contact with C by a spring S, and this prevents the exhaust gas from returning to G. (c) Diffusion pump, used to obtain vacuums as low as 10^{-8} mm-Hg: air molecules from the vessel to be evacuated diffuse into the jet, where a rapidly moving jet of oil sweeps the molecules away. A "forepump" is needed, which is a mechanical pump, such as the rotary type (b), and acts as a first stage in reducing the pressure.

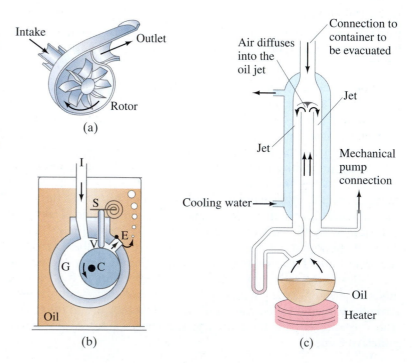

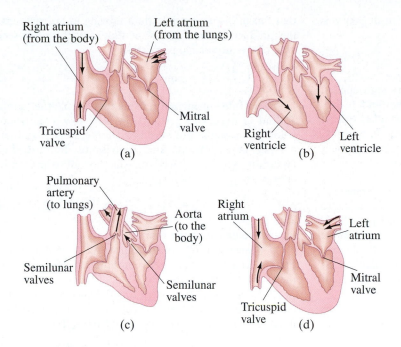

Right atrium
(from the body)

Left atrium
(from the lungs)

Tricuspid
valve

Mitral
valve

(a)

Right
ventricle

Left
ventricle

(b)

Pulmonary
artery
(to lungs)

Aorta
(to the
body)

Semilunar
valves

Semilunar
valves

(c)

Right
atrium

Left
atrium

Mitral
valve

Tricuspid
valve

(d)

FIGURE 10–42 (a) In the diastole phase, the heart relaxes between beats. Blood moves into the heart; both atria fill rapidly. (b) When the atria contract, the systole or pumping phase begins. The contraction pushes the blood through the mitral and tricuspid valves into the ventricles. (c) The contraction of the ventricles forces the blood through the semilunar valves into the pulmonary artery, which leads to the lungs, and to the aorta (the body's largest artery), which leads to the arteries serving all the body. (d) When the heart relaxes, the semilunar valves close; blood fills the atria, beginning the cycle again.

The heart of a human (and of other animals as well) is essentially a circulating pump. The action of a human heart is shown in Fig. 10–42. There are actually two separate paths for blood flow. The longer path takes blood to the parts of the body, via the arteries, bringing oxygen to body tissues and picking up carbon dioxide, which it carries back to the heart via veins. This blood is then pumped to the lungs (the second path), where the carbon dioxide is released and oxygen is taken up. The oxygen-laden blood is returned to the heart, where it is again pumped to the tissues of the body.

Blood pressure is measured using either a mercury-filled manometer or one of the other types of gauge mentioned earlier (Section 10–6), and it is usually calibrated in mm-Hg. The gauge is attached to a closed, air-filled jacket that is wrapped around the upper arm at the level of the heart, Fig. 10–43. Two values of blood pressure are measured: the maximum pressure when the heart is pumping, called *systolic pressure*; and the pressure when the heart is in the resting part of the cycle, called *diastolic pressure*. Initially, the air pressure in the jacket is increased high above the systolic pressure by means of a hand pump, and this compresses the main (brachial) artery in the arm and briefly cuts off the flow of blood. The air pressure is then reduced slowly until blood again begins to flow into the arm; it is detected by listening with a stethoscope to the characteristic tapping sound[†] of the blood returning to the forearm. At this point, systolic pressure is just equal to the air pressure in the jacket which can be read off the gauge. The air pressure is subsequently reduced further, and the tapping sound disappears when blood at low pressure can enter the artery. At this point, the gauge indicates the diastolic pressure. Normal systolic pressure is around 120 mm-Hg, whereas normal diastolic pressure is around 80 mm-Hg.

[†]When the blood starts flowing through the constriction caused by the tight jacket, its velocity is high and the flow is turbulent. It is the turbulence that causes the tapping sound.

PHYSICS APPLIED
Heart as a pump

PHYSICS APPLIED
Blood pressure

FIGURE 10–43 Device for measuring blood pressure.

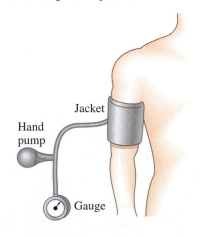

Jacket

Hand
pump

Gauge

Summary

The three common phases of matter are **solid**, **liquid**, and **gas**. Liquids and gases are collectively called **fluids**, meaning they have the ability to flow. The **density** of a material is defined as its mass per unit volume:

$$\rho = \frac{m}{V}. \tag{10–1}$$

Specific gravity is the ratio of the density of the material to the density of water (at 4°C).

Pressure is defined as force per unit area:

$$P = \frac{F}{A}. \tag{10–2}$$

The pressure at a depth h in a liquid of constant density ρ is given by

$$P = \rho g h, \qquad (10\text{-}3a)$$

where g is the acceleration due to gravity.

Pascal's principle says that an external pressure applied to a confined fluid is transmitted throughout the fluid.

Pressure is measured using a manometer or other type of gauge. A **barometer** is used to measure atmospheric pressure. Standard **atmospheric pressure** (average at sea level) is $1.013 \times 10^5 \, \text{N/m}^2$. **Gauge pressure** is the total (absolute) pressure less atmospheric pressure.

Archimedes' principle states that an object submerged wholly or partially in a fluid is buoyed up by a force equal to the weight of fluid it displaces ($F_B = m_F g = \rho_F V_{displ} g$).

Fluid flow can be characterized either as **streamline** (sometimes called **laminar**), in which the layers of fluid move smoothly and regularly along paths called **streamlines**, or as **turbulent**, in which case the flow is not smooth and regular but is characterized by irregularly shaped whirlpools.

Fluid flow rate is the mass or volume of fluid that passes a given point per unit time. The **equation of continuity** states that for an incompressible fluid flowing in an enclosed tube, the product of the velocity of flow and the cross-sectional area of the tube remains constant:

$$Av = \text{constant}. \qquad (10\text{-}4)$$

Bernoulli's principle tells us that where the velocity of a fluid is high, the pressure in it is low, and where the velocity is low, the pressure is high. For steady laminar flow of an incompressible and nonviscous fluid, **Bernoulli's equation**, which is based on the law of conservation of energy, is

$$P_1 + \tfrac{1}{2}\rho v_1^2 + \rho g y_1 = P_2 + \tfrac{1}{2}\rho v_2^2 + \rho g y_2, \qquad (10\text{-}5)$$

for two points along the flow.

[***Viscosity** refers to friction within a fluid and is essentially a frictional force between adjacent layers of fluid as they move past one another.]

[*Liquid surfaces hold together as if under tension (**surface tension**), allowing drops to form and objects like needles and insects to stay on the surface.]

Questions

1. If one material has a higher density than another, must the molecules of the first be heavier than those of the second? Explain.

2. Airplane travelers sometimes note that their cosmetics bottles and other containers have leaked during a flight. What might cause this?

3. The three containers in Fig. 10–44 are filled with water to the same height and have the same surface area at the base; hence the water pressure, and the total force on the base of each, is the same. Yet the total weight of water is different for each. Explain this "hydrostatic paradox."

FIGURE 10–44
Question 3.

4. Consider what happens when you push both a pin and the blunt end of a pen against your skin with the same force. Decide what determines whether your skin is cut—the net force applied to it or the pressure.

5. A small amount of water is boiled in a 1-gallon metal can. The can is removed from the heat and the lid put on. Shortly thereafter the can collapses. Explain.

6. When blood pressure is measured, why must the jacket be held at the level of the heart?

7. An ice cube floats in a glass of water filled to the brim. What can you say about the density of ice? As the ice melts, will the water overflow? Explain.

8. Will an ice cube float in a glass of alcohol? Why or why not?

9. A submerged can of Coke® will sink, but a can of Diet Coke® will float. (Try it!) Explain.

10. Why don't ships made of iron sink?

11. Explain how the tube in Fig. 10–45, known as a **siphon**, can transfer liquid from one container to a lower one even though the liquid must flow uphill for part of its journey. (Note that the tube must be filled with liquid to start with.)

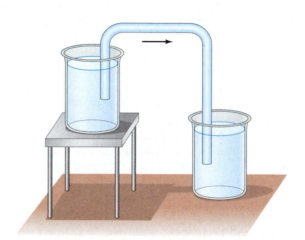

FIGURE 10–45 Question 11. A siphon.

12. A barge filled high with sand approaches a low bridge over the river and cannot quite pass under it. Should sand be added to, or removed from, the barge? [*Hint*: Consider Archimedes' principle.]

13. Will an empty balloon have precisely the same apparent weight on a scale as a balloon filled with air? Explain.

14. Explain why helium weather balloons, which are used to measure atmospheric conditions at high altitudes, are normally released while filled to only 10%–20% of their maximum volume.

15. A small wooden boat floats in a swimming pool, and the level of the water at the edge of the pool is marked. Consider the following situations and explain whether the level of the water will rise, fall, or stay the same. (*a*) The boat is removed from the water. (*b*) The boat in the water holds an iron anchor which is removed from the boat and placed on the shore. (*c*) The iron anchor is removed from the boat and dropped in the pool.

16. Why do you float higher in salt water than in fresh?

17. If you dangle two pieces of paper vertically, a few inches apart (Fig. 10–46), and blow between them, how do you think the papers will move? Try it and see. Explain.

FIGURE 10–46
Question 17.

18. Why does the canvas top of a convertible bulge out when the car is traveling at high speed? [*Hint*: the windshield deflects air upward, pushing streamlines closer together.]

19. Roofs of houses are sometimes "blown" off (or are they pushed off?) during a tornado or hurricane. Explain, using Bernoulli's principle.

20. Children are told to avoid standing too close to a rapidly moving train because they might get sucked under it. Is this possible? Explain.

21. A tall Styrofoam cup is filled with water. Two holes are punched in the cup near the bottom, and water begins rushing out. If the cup is dropped so it falls freely, will the water continue to flow from the holes? Explain.

22. Why do airplanes normally take off into the wind?

23. Why does the stream of water from a faucet become narrower as it falls (Fig. 10–47)?

FIGURE 10–47
Question 23 and Problem 82.
Water coming from a faucet.

24. Two ships moving in parallel paths close to one another risk colliding. Why?

Problems

10–2 Density and Specific Gravity

1. (I) The approximate volume of the granite monolith known as El Capitan in Yosemite National Park (Fig. 10–48) is about $10^8 \, \text{m}^3$. What is its approximate mass?

FIGURE 10–48 Problem 1.

2. (I) What is the approximate mass of air in a living room $4.8 \, \text{m} \times 3.8 \, \text{m} \times 2.8 \, \text{m}$?

3. (I) If you tried to smuggle gold bricks by filling your backpack, whose dimensions are $60 \, \text{cm} \times 28 \, \text{cm} \times 18 \, \text{cm}$, what would its mass be?

4. (I) State your mass and then estimate your volume. [*Hint*: Because you can swim on or just under the surface of the water in a swimming pool, you have a pretty good idea of your density.]

5. (II) A bottle has a mass of 35.00 g when empty and 98.44 g when filled with water. When filled with another fluid, the mass is 88.78 g. What is the specific gravity of this other fluid?

6. (II) If 5.0 L of antifreeze solution (specific gravity = 0.80) is added to 4.0 L of water to make a 9.0-L mixture, what is the specific gravity of the mixture?

10–3 to 10–6 Pressure; Pascal's Principle

7. (I) Estimate the pressure exerted on a floor by (*a*) one pointed chair leg (60 kg on all four legs) of area = $0.020 \, \text{cm}^2$, and (*b*) a 1500-kg elephant standing on one foot (area = $800 \, \text{cm}^2$).

8. (I) What is the difference in blood pressure (mm-Hg) between the top of the head and bottom of the feet of a 1.60-m-tall person standing vertically?

9. (I) (*a*) Calculate the total force of the atmosphere acting on the top of a table that measures $1.6 \, \text{m} \times 2.9 \, \text{m}$. (*b*) What is the total force acting upward on the underside of the table?

10. (II) In a movie, Tarzan evades his captors by hiding underwater for many minutes while breathing through a long, thin reed. Assuming the maximum pressure difference his lungs can manage and still breathe is $-85 \, \text{mm-Hg}$, calculate the deepest he could have been.

11. (II) The gauge pressure in each of the four tires of an automobile is 240 kPa. If each tire has a "footprint" of $220 \, \text{cm}^2$, estimate the mass of the car.

12. (II) The maximum gauge pressure in a hydraulic lift is 17.0 atm. What is the largest size vehicle (kg) it can lift if the diameter of the output line is 28.0 cm?

13. (II) How high would the level be in an alcohol barometer at normal atmospheric pressure?

14. (II) (*a*) What are the total force and the absolute pressure on the bottom of a swimming pool 22.0 m by 8.5 m whose uniform depth is 2.0 m? (*b*) What will be the pressure against the *side* of the pool near the bottom?

15. (II) How high would the atmosphere extend if it were of uniform density throughout, equal to half the present density at sea level?

16. (II) Water and then oil (which don't mix) are poured into a U-shaped tube, open at both ends. They come to equilibrium as shown in Fig. 10–49. What is the density of the oil? [*Hint:* Pressures at points a and b are equal. Why?]

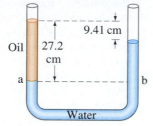

FIGURE 10–49
Problem 16.

17. (II) A house at the bottom of a hill is fed by a full tank of water 5.0 m deep and connected to the house by a pipe that is 110 m long at an angle of 58° from the horizontal (Fig. 10–50). (*a*) Determine the water gauge pressure at the house. (*b*) How high could the water shoot if it came vertically out of a broken pipe in front of the house?

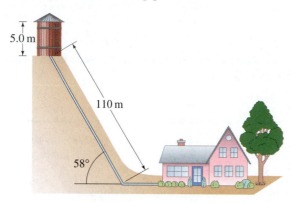

FIGURE 10–50 Problem 17.

18. (II) Determine the minimum gauge pressure needed in the water pipe leading into a building if water is to come out of a faucet on the twelfth floor, 38 m above that pipe.

19. (II) An open-tube mercury manometer is used to measure the pressure in an oxygen tank. When the atmospheric pressure is 1040 mbar, what is the absolute pressure (in Pa) in the tank if the height of the mercury in the open tube is (*a*) 28.0 cm higher, (*b*) 4.2 cm lower, than the mercury in the tube connected to the tank?

20. (II) In working out his principle, Pascal showed dramatically how force can be multiplied with fluid pressure. He placed a long, thin tube of radius $r = 0.30$ cm vertically into a wine barrel of radius $R = 21$ cm, Fig. 10–51. He found that when the barrel was filled with water and the tube filled to a height of 12 m, the barrel burst. Calculate (*a*) the mass of water in the tube, and (*b*) the net force exerted by the water in the barrel on the lid just before rupture.

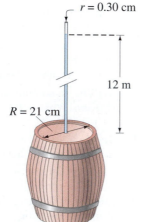

FIGURE 10–51
Problem 20
(not to scale).

282 CHAPTER 10 Fluids

*****21.** (III) Estimate the density of the water 6.0 km deep in the sea. (See Table 9–1 and Section 9–5 regarding bulk modulus.) By what fraction does it differ from the density at the surface?

10–7 Buoyancy and Archimedes' Principle

22. (I) A geologist finds that a Moon rock whose mass is 9.28 kg has an apparent mass of 6.18 kg when submerged in water. What is the density of the rock?

23. (I) What fraction of a piece of aluminum will be submerged when it floats in mercury?

24. (II) A crane lifts the 18,000-kg steel hull of a ship out of the water. Determine (*a*) the tension in the crane's cable when the hull is submerged in the water, and (*b*) the tension when the hull is completely out of the water.

25. (II) A spherical balloon has a radius of 7.35 m and is filled with helium. How large a cargo can it lift, assuming that the skin and structure of the balloon have a mass of 930 kg? Neglect the buoyant force on the cargo volume itself.

26. (II) A 78-kg person has an apparent mass of 54 kg (because of buoyancy) when standing in water that comes up to the hips. Estimate the mass of each leg. Assume the body has SG = 1.00.

27. (II) What is the likely identity of a metal (see Table 10–1) if a sample has a mass of 63.5 g when measured in air and an apparent mass of 55.4 g when submerged in water?

28. (II) Calculate the true mass (in vacuum) of a piece of aluminum whose apparent mass is 2.0000 kg when weighed in air.

29. (II) An undersea research chamber is spherical with an external diameter of 5.20 m. The mass of the chamber, when occupied, is 74,400 kg. It is anchored to the sea bottom by a cable. What is (*a*) the buoyant force on the chamber, and (*b*) the tension in the cable?

30. (II) A scuba diver and her gear displace a volume of 65.0 L and have a total mass of 68.0 kg. (*a*) What is the buoyant force on the diver in sea water? (*b*) Will the diver sink or float?

31. (II) Archimedes' principle can be used not only to determine the specific gravity of a solid using a known liquid (Example 10–8); the reverse can be done as well. (*a*) As an example, a 3.40-kg aluminum ball has an apparent mass of 2.10 kg when submerged in a particular liquid: calculate the density of the liquid. (*b*) Derive a formula for determining the density of a liquid using this procedure.

32. (II) A 0.48-kg piece of wood floats in water but is found to sink in alcohol (SG = 0.79), in which it has an apparent mass of 0.047 kg. What is the SG of the wood?

33. (II) The specific gravity of ice is 0.917, whereas that of seawater is 1.025. What fraction of an iceberg is above the surface of the water?

34. (III) A 5.25-kg piece of wood (SG = 0.50) floats on water. What minimum mass of lead, hung from the wood by a string, will cause it to sink?

10–8 to 10–10 Fluid Flow; Bernoulli's Equation

35. (I) Using the data of Example 10–11, calculate the average speed of blood flow in the major arteries of the body, which have a total cross-sectional area of about 2.0 cm².

36. (I) A 15-cm-radius air duct is used to replenish the air of a room 9.2 m × 5.0 m × 4.5 m every 16 min. How fast does air flow in the duct?

37. (I) Show that Bernoulli's equation reduces to the hydro-static variation of pressure with depth (Eq. 10–3b) when there is no flow $(v_1 = v_2 = 0)$.

38. (I) How fast does water flow from a hole at the bottom of a very wide, 4.6-m-deep storage tank filled with water? Ignore viscosity.

39. (II) A $\frac{5}{8}$-inch (inside) diameter garden hose is used to fill a round swimming pool 6.1 m in diameter. How long will it take to fill the pool to a depth of 1.2 m if water issues from the hose at a speed of 0.40 m/s?

40. (II) What gauge pressure in the water mains is necessary if a firehose is to spray water to a height of 15 m?

41. (II) A 6.0-cm-diameter horizontal pipe gradually narrows to 4.0 cm. When water flows through this pipe at a certain rate, the gauge pressure in these two sections is 32.0 kPa and 24.0 kPa, respectively. What is the volume rate of flow?

42. (II) What is the volume rate of flow of water from a 1.85-cm-diameter faucet if the pressure head is 15.0 m?

43. (II) If wind blows at 35 m/s over a house, what is the net force on the roof if its area is 240 m^2 and is flat?

44. (II) What is the lift (in newtons) due to Bernoulli's principle on a wing of area 78 m^2 if the air passes over the top and bottom surfaces at speeds of 260 m/s and 150 m/s, respectively?

45. (II) Estimate the air pressure inside a category 5 hurricane, where the wind speed is 300 km/h (Fig. 10–52).

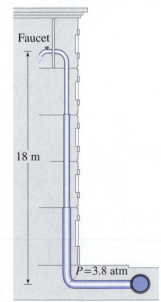

FIGURE 10–52 Problem 45.

46. (II) Water at a gauge pressure of 3.8 atm at street level flows into an office building at a speed of 0.60 m/s through a pipe 5.0 cm in diameter. The pipe tapers down to 2.6 cm in diameter by the top floor, 18 m above (Fig. 10–53), where the faucet has been left open. Calculate the flow velocity and the gauge pressure in such a pipe on the top floor. Assume no branch pipes and ignore viscosity.

FIGURE 10–53 Problem 46.

47. (III) (a) Show that the flow velocity measured by a venturi meter (see Fig. 10–30) is given by the relation

$$v_1 = A_2\sqrt{\frac{2(P_1 - P_2)}{\rho(A_1^2 - A_2^2)}}.$$

(b) A venturi tube is measuring the flow of water; it has a main diameter of 3.0 cm tapering down to a throat diameter of 1.0 cm. If the pressure difference is measured to be 18 mm-Hg, what is the velocity of the water?

48. (III) In Fig. 10–54, take into account the speed of the top surface of the tank and show that the speed of fluid leaving the opening at the bottom is

$$v_1 = \sqrt{\frac{2gh}{(1 - A_1^2/A_2^2)}},$$

where $h = y_2 - y_1$, and A_1 and A_2 are the areas of the opening and of the top surface, respectively. Assume $A_1 \ll A_2$ so that the flow remains nearly steady and laminar.

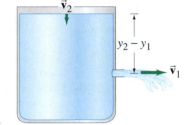

FIGURE 10–54 Problems 48 and 49.

49. (III) Suppose the opening in the tank of Fig. 10–54 is a height h_1 above the base and the liquid surface is a height h_2 above the base. The tank rests on level ground. (a) At what horizontal distance from the base of the tank will the fluid strike the ground? (b) At what other height, h_1', can a hole be placed so that the emerging liquid will have the same "range"? Assume $v_2 \approx 0$.

* 10–11 Viscosity

* **50. (II)** A *viscometer* consists of two concentric cylinders, 10.20 cm and 10.60 cm in diameter. A particular liquid fills the space between them to a depth of 12.0 cm. The outer cylinder is fixed, and a torque of 0.024 m·N keeps the inner cylinder turning at a steady rotational speed of 62 rev/min. What is the viscosity of the liquid?

* 10–12 Flow in Tubes; Poiseuille's Equation

* **51. (I)** A gardener feels it is taking him too long to water a garden with a $\frac{3}{8}$-in.-diameter hose. By what factor will his time be cut if he uses a $\frac{5}{8}$-in.-diameter hose? Assume nothing else is changed.

* **52. (II)** Engine oil (assume SAE 10, Table 10–3) passes through a 1.80-mm-diameter tube in a prototype engine. The tube is 5.5 cm long. What pressure difference is needed to maintain a flow rate of 5.6 mL/min?

* **53. (II)** What must be the pressure difference between the two ends of a 1.9-km section of pipe, 29 cm in diameter, if it is to transport oil $(\rho = 950 \text{ kg/m}^3, \eta = 0.20 \text{ Pa·s})$ at a rate of 450 cm^3/s?

* **54. (II)** What diameter must a 21.0-m-long air duct have if the ventilation and heating system is to replenish the air in a room 9.0 m × 12.0 m × 4.0 m every 10 min? Assume the pump can exert a gauge pressure of 0.71 × 10^{-3} atm.

* **55. (II)** Calculate the pressure drop per cm along the aorta using the data of Example 10–11 and Table 10–3.

*56. (II) Assuming a constant pressure gradient, if blood flow is reduced by 75%, by what factor is a blood vessel decreased in radius?

*57. (II) Poiseuille's equation does not hold if the flow velocity is high enough that turbulence sets in. The onset of turbulence occurs when the **Reynolds number**, Re, exceeds approximately 2000. Re is defined as

$$Re = \frac{2\bar{v}r\rho}{\eta},$$

where $\bar{v}$ is the average speed of the fluid, ρ is its density, η is its viscosity, and r is the radius of the tube in which the fluid is flowing. (a) Determine if blood flow through the aorta is laminar or turbulent when the average speed of blood in the aorta ($r = 1.2$ cm) during the resting part of the heart's cycle is about 40 cm/s. (b) During exercise, the blood-flow speed approximately doubles. Calculate the Reynolds number in this case, and determine if the flow is laminar or turbulent.

*58. (III) A patient is to be given a blood transfusion. The blood is to flow through a tube from a raised bottle to a needle inserted in the vein (Fig. 10–55). The inside diameter of the 4.0-cm-long needle is 0.40 mm, and the required flow rate is 4.0 cm³ of blood per minute. How high h should the bottle be placed above the needle? Obtain ρ and η from the Tables. Assume the blood pressure is 18 torr above atmospheric pressure.

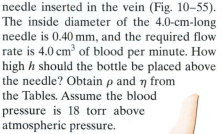

FIGURE 10–55
Problems 58 and 63.

* 10–13 Surface Tension and Capillarity

*59. (I) If the force F needed to move the wire in Fig. 10–35 is 5.1×10^{-3} N, calculate the surface tension γ of the enclosed fluid. Assume $L = 0.070$ m.

*60. (I) Calculate the force needed to move the wire in Fig. 10–35 if it is immersed in a soapy solution and the wire is 18.2 cm long.

*61. (II) If the base of an insect's leg has a radius of about 3.0×10^{-5} m and the insect's mass is 0.016 g, would you expect the six-legged insect to remain on top of the water? Why or why not?

*62. (II) The surface tension of a liquid can be determined by measuring the force F needed to just lift a circular platinum ring of radius r from the surface of the liquid. (a) Find a formula for γ in terms of F and r. (b) At 30°C, if $F = 8.40 \times 10^{-3}$ N and $r = 2.8$ cm, calculate γ for the tested liquid.

General Problems

63. Intravenous infusions are often made under gravity, as shown in Fig. 10–55. Assuming the fluid has a density of 1.00 g/cm³, at what height h should the bottle be placed so the liquid pressure is (a) 55 mm-Hg, and (b) 650 mm-H₂O? (c) If the blood pressure is 18 mm-Hg above atmospheric pressure, how high should the bottle be placed so that the fluid just barely enters the vein?

64. A 2.4-N force is applied to the plunger of a hypodermic needle. If the diameter of the plunger is 1.3 cm and that of the needle 0.20 mm, (a) with what force does the fluid leave the needle? (b) What force on the plunger would be needed to push fluid into a vein where the gauge pressure is 18 mm-Hg? Answer for the instant just before the fluid starts to move.

65. A bicycle pump is used to inflate a tire. The initial tire (gauge) pressure is 210 kPa (30 psi). At the end of the pumping process, the final pressure is 310 kPa (45 psi). If the diameter of the plunger in the cylinder of the pump is 3.0 cm, what is the range of the force that needs to be applied to the pump handle from beginning to end?

66. Estimate the pressure on the mountains underneath the Antarctic ice sheet, which is typically 3 km thick.

67. What is the approximate difference in air pressure between the top and the bottom of the Empire State building in New York City? It is 380 m tall and is located at sea level. Express as a fraction of atmospheric pressure at sea level.

68. A hydraulic lift is used to jack a 970-kg car 12 cm off the floor. The diameter of the output piston is 18 cm, and the input force is 250 N. (a) What is the area of the input piston? (b) What is the work done in lifting the car 12 cm? (c) If the input piston moves 13 cm in each stroke, how high does the car move up for each stroke? (d) How many strokes are required to jack the car up 12 cm? (e) Show that energy is conserved.

69. Giraffes are a wonder of cardiovascular engineering. Calculate the difference in pressure (in atmospheres) that the blood vessels in a giraffe's head have to accommodate as the head is lowered from a full upright position to ground level for a drink. The height of an average giraffe is about 6 m.

70. When you ascend or descend a great deal when driving in a car, your ears "pop," which means that the pressure behind the eardrum is being equalized to that outside. If this did not happen, what would be the approximate force on an eardrum of area 0.50 cm² if a change in altitude of 950 m takes place?

71. One arm of a U-shaped tube (open at both ends) contains water, and the other alcohol. If the two fluids meet at exactly the bottom of the U, and the alcohol is at a height of 18.0 cm, at what height will the water be?

72. A simple model (Fig. 10–56) considers a continent as a block (density ≈ 2800 kg/m³) floating in the mantle rock around it (density ≈ 3300 kg/m³). Assuming the continent is 35 km thick (the average thickness of the Earth's continental crust), estimate the height of the continent above the surrounding rock.

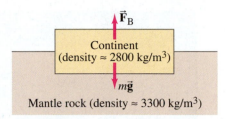

FIGURE 10–56 Problem 72.

73. The contraction of the left ventricle (chamber) of the heart pumps blood to the body. Assuming that the inner surface of the left ventricle has an area of 82 cm² and the maximum pressure in the blood is 120 mm-Hg, estimate the force exerted by that ventricle at maximum pressure.

74. Estimate the total mass of the Earth's atmosphere, using the known value of atmospheric pressure at sea level.

75. Suppose a person can reduce the pressure in his lungs to −80 mm-Hg gauge pressure. How high can water then be sucked up a straw?

76. A ship, carrying fresh water to a desert island in the Caribbean, has a horizontal cross-sectional area of 2650 m² at the waterline. When unloaded, the ship rises 8.50 m higher in the sea. How much water was delivered?

77. A copper (Cu) weight is placed on top of a 0.50-kg block of wood (density = 0.60 × 10³ kg/m³) floating in water, as shown in Fig. 10–57. What is the mass of the copper if the top of the wood block is exactly at the water's surface?

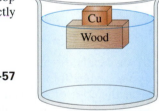

FIGURE 10–57
Problem 77.

78. A raft is made of 10 logs lashed together. Each is 56 cm in diameter and has a length of 6.1 m. How many people can the raft hold before they start getting their feet wet, assuming the average person has a mass of 68 kg? Do *not* neglect the weight of the logs. Assume the specific gravity of wood is 0.60.

79. During each heartbeat, approximately 70 cm³ of blood is pushed from the heart at an average pressure of 105 mm-Hg. Calculate the power output of the heart, in watts, assuming 70 beats per minute.

80. A bucket of water is accelerated upward at 2.4g. What is the buoyant force on a 3.0-kg granite rock (SG = 2.7) submerged in the water? Will the rock float? Why or why not?

81. How high should the pressure head be if water is to come from a faucet at a speed of 9.5 m/s? Ignore viscosity.

82. The stream of water from a faucet decreases in diameter as it falls (Fig. 10–47). Derive an equation for the diameter of the stream as a function of the distance y below the faucet, given that the water has speed v_0 when it leaves the faucet, whose diameter is d.

83. Four lawn sprinkler heads are fed by a 1.9-cm-diameter pipe. The water comes out of the heads at an angle of 35° to the horizontal and covers a radius of 8.0 m. (a) What is the velocity of the water coming out of each sprinkler head? (Assume zero air resistance.) (b) If the output diameter of each head is 3.0 mm, how many liters of water do the four heads deliver per second? (c) How fast is the water flowing inside the 1.9-cm-diameter pipe?

84. You need to siphon water from a clogged sink. The sink has an area of 0.48 m² and is filled to a height of 4.0 cm. Your siphon tube rises 50 cm above the bottom of the sink and then descends 100 cm to a pail as shown in Fig. 10–58. The siphon tube has a diameter of 2.0 cm. (a) Assuming that the water enters the siphon tube with almost zero velocity, calculate its velocity when it enters the pail. (b) Estimate how long it will take to empty the sink.

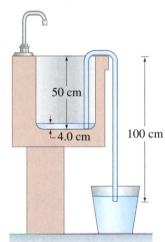

FIGURE 10–58
Problems 84 and 85.

85. Consider a siphon which transfers water from one vessel to a second (lower) one, as in Fig. 10–58. Determine the rate of flow if the tube has a diameter of 1.2 cm and the difference in water levels of the two containers is 64 cm.

86. An airplane has a mass of 2.0 × 10⁶ kg, and the air flows past the lower surface of the wings at 95 m/s. If the wings have a surface area of 1200 m², how fast must the air flow over the upper surface of the wing if the plane is to stay in the air? Consider only the Bernoulli effect.

* 87. Blood from an animal is placed in a bottle 1.70 m above a 3.8-cm-long needle, of inside diameter 0.40 mm, from which it flows at a rate of 4.1 cm³/min. What is the viscosity of this blood?

* 88. If cholesterol build-up reduces the diameter of an artery by 15%, what will be the effect on blood flow?

Answers to Exercises

A: The same. Pressure depends on depth, not on length.
B: Lower.

C: Increases.

The pendulum of a clock is an example of oscillatory motion. Many kinds of oscillatory motion are sinusoidal in time, or nearly so, and are referred to as being simple harmonic motion. Real systems generally have at least some friction, causing the motion to be "damped." When an external sinusoidal force is exerted on a system able to oscillate, resonance occurs if the driving force is at or near the natural frequency of vibration.

Vibrations can give rise to waves—such as water waves or waves traveling along a cord—which travel outward from their source.

CHAPTER 11

Vibrations and Waves

Many objects vibrate or oscillate—an object on the end of a spring, a tuning fork, the balance wheel of an old watch, a pendulum, a plastic ruler held firmly over the edge of a table and gently struck, the strings of a guitar or piano. Spiders detect prey by the vibrations of their webs; cars oscillate up and down when they hit a bump; buildings and bridges vibrate when heavy trucks pass or the wind is fierce. Indeed, because most solids are elastic (see Section 9–5), they vibrate (at least briefly) when given an impulse. Electrical oscillations occur in radio and television sets. At the atomic level, atoms vibrate within a molecule, and the atoms of a solid vibrate about their relatively fixed positions. Because it is so common in everyday life and occurs in so many areas of physics, oscillatory (or vibrational) motion is of great importance. Mechanical vibrations are fully described on the basis of Newtonian mechanics.

Vibrations and wave motion are intimately related subjects. Waves—whether ocean waves, waves on a string, earthquake waves, or sound waves in air—have as their source a vibration. In the case of sound, not only is the source a vibrating object, but so is the detector—the eardrum or the membrane of a microphone. Indeed, when a wave travels through a medium, the medium vibrates (such as air for sound waves). In the second half of this Chapter, after we discuss vibrations, we will discuss simple waves such as those on water or on a string. In Chapter 12 we will study sound waves, and in later Chapters we will encounter other forms of wave motion, including electromagnetic waves and light.

11-1 Simple Harmonic Motion

When an object **vibrates** or **oscillates** back and forth, over the same path, each vibration taking the same amount of time, the motion is **periodic**. The simplest form of periodic motion is represented by an object oscillating on the end of a uniform coil spring. Because many other types of vibrational motion closely resemble this system, we will look at it in detail. We assume that the mass of the spring can be ignored, and that the spring is mounted horizontally, as shown in Fig. 11–1a, so that the object of mass m slides without friction on the horizontal surface. Any spring has a natural length at which it exerts no force on the mass m. The position of the mass at this point is called the **equilibrium position**. If the mass is moved either to the left, which compresses the spring, or to the right, which stretches it, the spring exerts a force on the mass that acts in the direction of returning the mass to the equilibrium position; hence it is called a *restoring force*. We consider the common situation where we can assume the magnitude of the restoring force F is directly proportional to the displacement x the spring has been stretched (Fig. 11–1b) or compressed (Fig. 11–1c) from the equilibrium position:

$$F = -kx. \qquad \text{[force exerted by spring]} \quad \textbf{(11–1)}$$

Equilibrium position

Note that the equilibrium position has been chosen at $x = 0$. Equation 11–1, which is often referred to as Hooke's law (see Sections 6–4 and 9–5), is accurate as long as the spring is not compressed to the point where the coils are close to touching, or stretched beyond the elastic region (see Fig. 9–19).

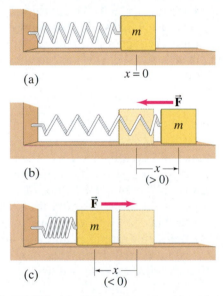

(a) $x = 0$

(b) x (>0) $\vec{F}$

(c) x (<0) $\vec{F}$

FIGURE 11–1 A mass vibrating at the end of a uniform spring.

The minus sign in Eq. 11–1 indicates that the restoring force is always in the direction opposite to the displacement x. For example, if we choose the positive direction to the right in Fig. 11–1, x is positive when the spring is stretched, but the direction of the restoring force is to the left (negative direction). If the spring is compressed, x is negative (to the left) but the force F acts toward the right (Fig. 11–1c).

The proportionality constant k in Eq. 11–1 is called the *spring constant* or *spring stiffness constant*. To stretch the spring a distance x, one has to exert an (external) force on the free end of the spring at least equal to

$$F = +kx. \qquad \text{[external force on spring]}$$

The greater the value of k, the greater the force needed to stretch a spring a given distance. That is, the stiffer the spring, the greater the spring constant k.

Note that the force F in Eq. 11–1 is *not* a constant, but varies with position. Therefore the acceleration of the mass m is not constant, so we *cannot* use the equations for constant acceleration developed in Chapter 2.

 **CAUTION**

Force and acceleration are not *constant; Eqs. 2–11 are not useful here*

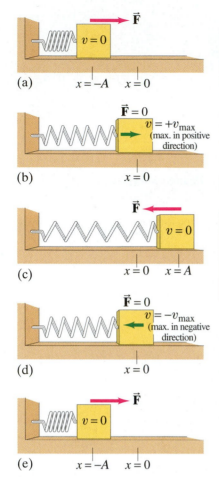

(a) $x = -A$ $x = 0$

(b) $x = 0$

(c) $x = 0$ $x = A$

(d) $x = 0$

(e) $x = -A$ $x = 0$

FIGURE 11–2 Force on, and velocity of, a mass at different positions of its oscillation cycle on a frictionless surface.

Let us examine what happens when our uniform spring is initially compressed a distance $x = -A$, as shown in Fig. 11–2a, and then released. The spring exerts a force on the mass that pushes it toward the equilibrium position. But because the mass has been accelerated by the force, it passes the equilibrium position with considerable speed. Indeed, as the mass reaches the equilibrium position, the force on it decreases to zero, but its speed at this point is a maximum, v_{max}, Fig. 11–2b. As the mass moves farther to the right, the force on it acts to slow it down, and it stops momentarily at $x = A$, Fig. 11–2c. It then begins moving back in the opposite direction, accelerating until it passes the equilibrium point, Fig. 11–2d, and then slows down until it reaches zero speed at the original starting point, $x = -A$, Fig. 11–2e. It then repeats the motion, moving back and forth symmetrically between $x = A$ and $x = -A$.

EXERCISE A An object is oscillating back and forth. Which of the following statements are true at some time during the course of the motion? (*a*) The object can have zero velocity and, simultaneously, nonzero acceleration. (*b*) The object can have zero velocity and, simultaneously, zero acceleration. (*c*) The object can have zero acceleration and, simultaneously, nonzero velocity. (*d*) The object can have nonzero velocity and nonzero acceleration simultaneously.

To discuss vibrational motion, we need to define a few terms. The distance x of the mass from the equilibrium point at any moment is called the **displacement**. The maximum displacement—the greatest distance from the equilibrium point—is called the **amplitude**, A. One **cycle** refers to the complete to-and-fro motion from some initial point back to that same point—say, from $x = -A$ to $x = A$ and back to $x = -A$. The **period**, T, is defined as the time required to complete one cycle. Finally, the **frequency**, f, is the number of complete cycles per second. Frequency is generally specified in hertz (Hz), where $1\,Hz = 1$ cycle per second (s^{-1}). It is easy to see, from their definitions, that frequency and period are inversely related, as we saw earlier (Eqs. 5–2 and 8–8):

$$f = \frac{1}{T} \quad \text{and} \quad T = \frac{1}{f}; \tag{11–2}$$

for example, if the frequency is 5 cycles per second, then each cycle takes $\frac{1}{5}$ s.

The oscillation of a spring hung vertically is essentially the same as that of a horizontal spring. Because of gravity, the length of a vertical spring with a mass m on the end will be longer at equilibrium than when that same spring is horizontal, as shown in Fig. 11–3. The spring is in equilibrium when $\Sigma F = 0 = mg - kx_0$, so the spring stretches an extra amount $x_0 = mg/k$ to be in equilibrium. If x is measured from this new equilibrium position, Eq. 11–1 can be used directly with the same value of k.

FIGURE 11–3
(a) Free spring, hung vertically.
(b) Mass m attached to spring in new equilibrium position, which occurs when $\Sigma F = 0 = mg - kx_0$.

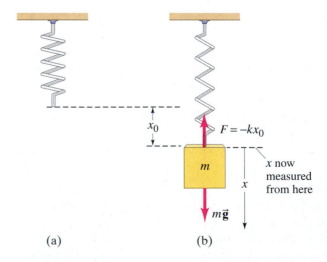

(a)   (b)

EXAMPLE 11–1 **Car springs.** When a family of four with a total mass of 200 kg step into their 1200-kg car, the car's springs compress 3.0 cm. (a) What is the spring constant of the car's springs (Fig. 11–4), assuming they act as a single spring? (b) How far will the car lower if loaded with 300 kg rather than 200 kg?

APPROACH We use Hooke's law. The extra force equal to the weight of the people, mg, causes a 3.0-cm displacement.

SOLUTION (a) The added force of $(200 \text{ kg})(9.8 \text{ m/s}^2) = 1960 \text{ N}$ causes the springs to compress 3.0×10^{-2} m. Therefore (Eq. 11–1), the spring constant is

$$k = \frac{F}{x} = \frac{1960 \text{ N}}{3.0 \times 10^{-2} \text{ m}} = 6.5 \times 10^4 \text{ N/m}.$$

(b) If the car is loaded with 300 kg, Hooke's law gives

$$x = \frac{F}{k} = \frac{(300 \text{ kg})(9.8 \text{ m/s}^2)}{(6.5 \times 10^4 \text{ N/m})} = 4.5 \times 10^{-2} \text{ m},$$

or 4.5 cm.

NOTE We could have obtained x without solving for k: since x is proportional to F, if 200 kg compresses the spring 3.0 cm, then 1.5 times the force will compress the spring 1.5 times as much, or 4.5 cm.

FIGURE 11–4 Photo of a car's spring. (Also visible is the shock absorber, in red—see Section 11–5.)

Any vibrating system for which the restoring force is directly proportional to the negative of the displacement (as in Eq. 11–1, $F = -kx$) is said to exhibit **simple harmonic motion** (SHM).[†] Such a system is often called a **simple harmonic oscillator** (SHO). We saw in Section 9–5 that most solid materials stretch or compress according to Eq. 11–1 as long as the displacement is not too great. Because of this, many natural vibrations are simple harmonic, or sufficiently close to it that they can be treated using this SHM model.

SHM
SHO

CONCEPTUAL EXAMPLE 11–2 **Is the motion simple harmonic?** Which of the following represent a simple harmonic oscillator: (a) $F = -0.5x^2$, (b) $F = -2.3y$, (c) $F = 8.6x$, (d) $F = -4\theta$?

RESPONSE Both (b) and (d) represent simple harmonic oscillators because they give the force as minus a constant times a displacement. The displacement need not be x, but the minus sign is required to restore the system to equilibrium, which is why (c) is not a SHO.

11–2 Energy in the Simple Harmonic Oscillator

With forces that are not constant, such as here with simple harmonic motion, it is often convenient and useful to use the energy approach, as we saw in Chapter 6.

To stretch or compress a spring, work has to be done. Hence potential energy is stored in a stretched or compressed spring. Indeed, we have already seen in Section 6–4 that elastic potential energy is given by

$$\text{PE} = \tfrac{1}{2}kx^2.$$

The total mechanical energy E of a mass–spring system is the sum of the kinetic and potential energies,

$$E = \tfrac{1}{2}mv^2 + \tfrac{1}{2}kx^2, \tag{11–3}$$

Total energy of SHO

where v is the velocity of the mass m when it is a distance x from the equilibrium position. As long as there is no friction, the total mechanical energy E

[†]The word "harmonic" refers to the motion being *sinusoidal*, which we discuss in Section 11–3. It is "simple" when there is sinusoidal motion of a single frequency.

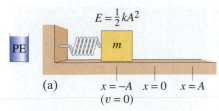

$E = \frac{1}{2}kA^2$

(a) $x = -A$ $x = 0$ $x = A$
        $(v = 0)$

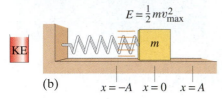

$E = \frac{1}{2}mv_{max}^2$

(b) $x = -A$ $x = 0$ $x = A$

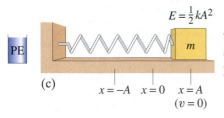

$E = \frac{1}{2}kA^2$

(c) $x = -A$ $x = 0$ $x = A$
        $(v = 0)$

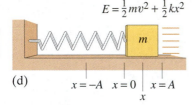

$E = \frac{1}{2}mv^2 + \frac{1}{2}kx^2$

(d) $x = -A$ $x = 0$ $x = A$
                    x

FIGURE 11–5 Energy changes from potential energy to kinetic energy and back again as the spring oscillates. Energy "buckets" (on the left) are described in Section 6–7.

remains constant. As the mass oscillates back and forth, the energy continuously changes from potential energy to kinetic energy, and back again (Fig. 11–5). At the extreme points, $x = -A$ and $x = A$ (Fig. 11–5a, c), all the energy is stored in the spring as potential energy (and is the same whether the spring is compressed or stretched to the full amplitude). At these extreme points, the mass stops momentarily as it changes direction, so $v = 0$ and

$$E = \tfrac{1}{2}m(0)^2 + \tfrac{1}{2}kA^2 = \tfrac{1}{2}kA^2. \tag{11–4a}$$

Thus, the **total mechanical energy of a simple harmonic oscillator is proportional to the square of the amplitude**. At the equilibrium point, $x = 0$ (Fig. 11–5b), all the energy is kinetic:

$$E = \tfrac{1}{2}mv_{max}^2 + \tfrac{1}{2}k(0)^2 = \tfrac{1}{2}mv_{max}^2, \tag{11–4b}$$

where v_{max} represents the *maximum* velocity during the motion (which occurs at $x = 0$). At intermediate points (Fig. 11–5d), the energy is part kinetic and part potential; because energy is conserved (we use Eqs. 11–3 and 11–4a),

$$\tfrac{1}{2}mv^2 + \tfrac{1}{2}kx^2 = \tfrac{1}{2}kA^2. \tag{11–4c}$$

From this conservation of energy equation, we can obtain the velocity as a function of position. Solving for v^2, we have

$$v^2 = \frac{k}{m}(A^2 - x^2) = \frac{k}{m}A^2\left(1 - \frac{x^2}{A^2}\right).$$

From Eqs. 11–4a and 11–4b, we have $\tfrac{1}{2}mv_{max}^2 = \tfrac{1}{2}kA^2$, so $v_{max}^2 = (k/m)A^2$. Inserting this into the equation above and taking the square root, we have

$$v = \pm v_{max}\sqrt{1 - \frac{x^2}{A^2}}. \tag{11–5}$$

This gives the velocity of the object at any position x. The object moves back and forth, so its velocity can be either in the $+$ or $-$ direction, but its magnitude depends only on the magnitude of x.

CONCEPTUAL EXAMPLE 11–3 | **Doubling the amplitude.** Suppose the spring in Fig. 11–5 is stretched twice as far (to $x = 2A$). What happens to (a) the energy of the system, (b) the maximum velocity of the oscillating mass, (c) the maximum acceleration of the mass?

RESPONSE (a) From Eq. 11–4a, the total energy is proportional to the square of the amplitude A, so stretching it twice as far quadruples the energy ($2^2 = 4$). You may protest, "I did work stretching the spring from $x = 0$ to $x = A$. Don't I do the same work stretching it from A to $2A$?" No. The force you exert is proportional to the displacement x, so for the second displacement, from $x = A$ to $2A$, you do more work than for the first displacement ($x = 0$ to A). (b) From Eq. 11–4b, we can see that since the energy is quadrupled, the maximum velocity must be doubled. $\left[v_{max} \propto \sqrt{E} \propto A.\right]$
(c) Since the force is twice as great when we stretch the spring twice as far, the acceleration is also twice as great: $a \propto F \propto x$.

EXERCISE B Suppose the spring in Fig. 11–5 is compressed to $x = -A$, but is given a push to the right so that the initial speed of the mass m is v_0. What effect does this push have on (a) the energy of the system, (b) the maximum velocity, (c) the maximum acceleration?

EXAMPLE 11–4 **Spring calculations.** A spring stretches 0.150 m when a 0.300-kg mass is gently lowered on it as in Fig. 11–3b. The spring is then set up horizontally with the 0.300-kg mass resting on a frictionless table as in Fig. 11–5. The mass is pulled so that the spring is stretched 0.100 m from the equilibrium point, and released from rest. Determine (a) the spring stiffness constant k, (b) the amplitude of the horizontal oscillation A, (c) the magnitude of the maximum velocity v_{max}, (d) the magnitude of the velocity v when the mass is 0.050 m from equilibrium, and (e) the magnitude of the maximum acceleration a_{max} of the mass.

APPROACH When the 0.300-kg mass hangs at rest from the spring as in Fig. 11–3b, we apply Newton's second law for the vertical forces: $\Sigma F = 0 = mg - kx_0$, so $k = mg/x_0$. For the horizontal oscillations, the amplitude is given, the velocities are found using conservation of energy, and the acceleration from $F = ma$.

SOLUTION (a) The spring stretches 0.150 m due to the 0.300-kg load, so

$$k = \frac{F}{x_0} = \frac{mg}{x_0} = \frac{(0.300 \text{ kg})(9.80 \text{ m/s}^2)}{0.150 \text{ m}} = 19.6 \text{ N/m}.$$

(b) The spring is now horizontal (on a table). It is stretched 0.100 m from equilibrium and is given no initial speed, so $A = 0.100$ m.

(c) The maximum velocity v_{max} is attained as the mass passes through the equilibrium point where all the energy is kinetic. By comparing the total energy (see Eq. 11–3) at equilibrium with that at full extension, conservation of energy tells us that

$$\tfrac{1}{2}mv_{max}^2 + 0 = 0 + \tfrac{1}{2}kA^2,$$

where $A = 0.100$ m. (Or, compare Eqs. 11–4a and b.) Solving for v_{max}, we have

$$v_{max} = A\sqrt{\frac{k}{m}} = (0.100 \text{ m})\sqrt{\frac{19.6 \text{ N/m}}{0.300 \text{ kg}}} = 0.808 \text{ m/s}.$$

(d) We use conservation of energy, or Eq. 11–5 derived from it, and find that

$$v = v_{max}\sqrt{1 - \frac{x^2}{A^2}} = (0.808 \text{ m/s})\sqrt{1 - \frac{(0.050 \text{ m})^2}{(0.100 \text{ m})^2}} = 0.70 \text{ m/s}.$$

(e) By Newton's second law, $F = ma$. So the maximum acceleration occurs where the force is greatest—that is, when $x = A = 0.100$ m. Thus

$$a_{max} = \frac{F_{max}}{m} = \frac{kA}{m} = \frac{(19.6 \text{ N/m})(0.100 \text{ m})}{0.300 \text{ kg}} = 6.53 \text{ m/s}^2.$$

NOTE We cannot use the kinematic equations, Eqs. 2–11, because the acceleration is not constant in SHM.

EXAMPLE 11–5 **More spring calculations—energy.** For the simple harmonic oscillator of Example 11–4, determine (a) the total energy, and (b) the kinetic and potential energies at half amplitude ($x = \pm A/2$).

APPROACH We use conservation of energy for a mass–spring system, Eqs. 11–3 and 11–4.

SOLUTION (a) With $k = 19.6$ N/m and $A = 0.100$ m, the total energy E from Eq. 11–4a is

$$E = \tfrac{1}{2}kA^2 = \tfrac{1}{2}(19.6 \text{ N/m})(0.100 \text{ m})^2 = 9.80 \times 10^{-2} \text{ J}.$$

(b) At $x = A/2 = 0.050$ m, we have

$$\text{PE} = \tfrac{1}{2}kx^2 = \tfrac{1}{2}(19.6 \text{ N/m})(0.050 \text{ m})^2 = 2.5 \times 10^{-2} \text{ J}.$$

By conservation of energy, the kinetic energy must be

$$\text{KE} = E - \text{PE} = 7.3 \times 10^{-2} \text{ J}.$$

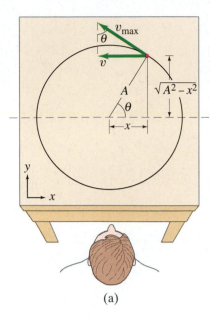

(a)

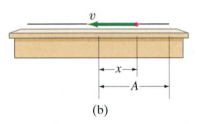

(b)

FIGURE 11–6 (a) Circular motion of a small (red) object. (b) Side view of circular motion (x component) is simple harmonic motion.

11–3 The Period and Sinusoidal Nature of SHM

The period of a simple harmonic oscillator is found to depend on the stiffness of the spring and also on the mass m that is oscillating. But—strange as it may seem—the *period does not depend on the amplitude*. You can find this out for yourself by using a watch and timing 10 or 20 cycles of an oscillating spring for a small amplitude and then for a large amplitude.

We can derive a formula for the period of simple harmonic motion (SHM) by comparing SHM to an object rotating in a circle. From this same "reference circle" we can obtain a second useful result—a formula for the position of an oscillating mass as a function of time. There is nothing actually rotating in a circle when a spring oscillates linearly, but it is the mathematical similarity that we find useful.

Period and Frequency

Consider a small object of mass m revolving counterclockwise in a circle of radius A, with constant speed v_{max}, on top of a table as shown in Fig. 11–6. As viewed from above, the motion is a circle in the xy plane. But a person who looks at the motion from the edge of the table sees an oscillatory motion back and forth, and this one-dimensional motion corresponds precisely to simple harmonic motion, as we shall now see.

What the person sees, and what we are interested in, is the projection of the circular motion onto the x axis (Fig. 11–6b). To see that this x-motion is analogous to SHM, let us calculate the magnitude of the x component of the velocity v_{max}, which is labeled v in Fig. 11–6. The two triangles involving θ in Fig. 11–6a are similar, so

$$\frac{v}{v_{max}} = \frac{\sqrt{A^2 - x^2}}{A}$$

or

$$v = v_{max}\sqrt{1 - \frac{x^2}{A^2}}.$$

This is exactly the equation for the speed of a mass oscillating with SHM, as we saw in Eq. 11–5. Thus the projection on the x axis of an object revolving in a circle has the same motion as a mass at the end of a spring.

We can now determine the period of SHM because it is equal to that of the revolving object making one complete revolution. First we note that the velocity v_{max} is equal to the circumference of the circle (distance) divided by the period T:

$$v_{max} = \frac{2\pi A}{T} = 2\pi A f. \tag{11–6}$$

We solve for the period T:

$$T = \frac{2\pi A}{v_{max}}.$$

From energy conservation, Eqs. 11–4a and b, we have $\frac{1}{2}kA^2 = \frac{1}{2}mv_{max}^2$, so $A/v_{max} = \sqrt{m/k}$. Thus

Period T of SHM

$$T = 2\pi\sqrt{\frac{m}{k}}. \tag{11–7a}$$

Period and frequency of SHM don't depend on amplitude

This is the formula we were looking for. The period depends on the mass m and the spring stiffness constant k, but not on the amplitude A. We see from Eq. 11–7a that the larger the mass, the longer the period; and the stiffer the spring (larger k), the shorter the period. This makes sense since a larger mass means more inertia and therefore slower response (smaller acceleration). And larger k means greater force and therefore quicker response (larger acceleration). Notice that Eq. 11–7a is not a direct proportion: the period varies as the *square root* of m/k. For example, the mass must be quadrupled to double the period.

Equation 11–7a is fully in accord with experiment and is valid not only for a spring, but for all kinds of simple harmonic motion—that is, for motion subject to a restoring force proportional to displacement, Eq. 11–1.

We can write the frequency using $f = 1/T$ (Eq. 11–2):

$$f = \frac{1}{T} = \frac{1}{2\pi}\sqrt{\frac{k}{m}}.$$ (11–7b) *Frequency f of SHM*

EXERCISE C Does a car bounce faster on its springs when empty or fully loaded?

EXAMPLE 11–6 | ESTIMATE | Spider web. A spider of mass 0.30 g waits in its web of negligible mass (Fig. 11–7). A slight movement causes the web to vibrate with a frequency of about 15 Hz. (*a*) Estimate the value of the spring stiffness constant k for the web. (*b*) At what frequency would you expect the web to vibrate if an insect of mass 0.10 g were trapped in addition to the spider?

APPROACH We can only make a rough estimate because a spider's web is fairly complicated and may vibrate with a mixture of frequencies. We use SHM as an approximate model.

SOLUTION (*a*) The frequency of SHM is given by Eq. 11–7b,

$$f = \frac{1}{2\pi}\sqrt{\frac{k}{m}}.$$

We solve for k:

$$k = (2\pi f)^2 m$$
$$= (6.28 \times 15\,\text{s}^{-1})^2 (3.0 \times 10^{-4}\,\text{kg}) = 2.7\,\text{N/m}.$$

(*b*) The total mass is now $0.10\,\text{g} + 0.30\,\text{g} = 4.0 \times 10^{-4}\,\text{kg}$. We could substitute $m = 4.0 \times 10^{-4}\,\text{kg}$ into Eq. 11–7b. Instead, we notice that the frequency decreases with the square root of the mass. Since the new mass is 4/3 times the first mass, the frequency changes by a factor of $1/\sqrt{4/3} = \sqrt{3/4}$. Thus $f = (15\,\text{Hz})(\sqrt{3/4}) = 13\,\text{Hz}$.

NOTE Check this result by direct substitution of k, found in part (*a*), and the new mass m into Eq. 11–7b.

FIGURE 11–7 A spider waits for its prey (Example 11–6).

Position as a Function of Time

We now use the reference circle to find the position of a mass undergoing simple harmonic motion as a function of time. From Fig. 11–6, we see that $\cos\theta = x/A$, so the projection of the object's position on the x axis is

$$x = A\cos\theta.$$

Because the mass is rotating with angular velocity ω, we can write $\theta = \omega t$, where θ is in radians (Section 8–1). Thus

$$x = A\cos\omega t.$$ (11–8a) *Position*

Furthermore, since the angular velocity ω (specified in radians per second) can be written as $\omega = 2\pi f$, where f is the frequency (Eq. 8–7), we then write   *as a*

$$x = A\cos(2\pi f t),$$ (11–8b) *function*

or in terms of the period T,   *of time (SHM)*

$$x = A\cos(2\pi t/T).$$ (11–8c)

Notice in Eq. 11–8c that when $t = T$ (that is, after a time equal to one period), we have the cosine of 2π, which is the same as the cosine of zero. This makes sense since the motion repeats itself after a time $t = T$.

⚠ **C A U T I O N**

t is a variable (time);
T is a constant for a given situation

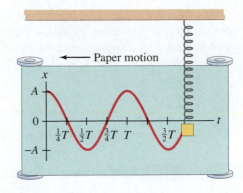

FIGURE 11–8 Position as a function of time $x = A\cos(2\pi t/T)$.

As we have seen, the x component of a uniformly rotating object's motion corresponds precisely to the motion of a simple harmonic oscillator. Thus Eqs. 11–8 give the position of an object undergoing simple harmonic motion. Since the cosine function varies between 1 and -1, x varies between A and $-A$, as it must. If a pen is attached to a vibrating mass as a sheet of paper is moved at a steady rate beneath it (Fig. 11–8), a curve will be drawn that accurately follows Eqs. 11–8.

EXAMPLE 11–7 **Starting with** $x = A\cos\omega t.$ The displacement of an object is described by the following equation, where x is in meters and t is in seconds:

$$x = (0.30\,\text{m})\cos(8.0\,t).$$

Determine the oscillating object's (a) amplitude, (b) frequency, (c) period, (d) maximum speed, and (e) maximum acceleration.

APPROACH We start by comparing the given equation for x with Eq. 11–8b, $x = A\cos(2\pi ft)$.

SOLUTION From $x = A\cos(2\pi ft)$, we see by inspection that (a) the amplitude $A = 0.30\,\text{m}$, and (b) $2\pi f = 8.0\,\text{s}^{-1}$; so $f = (8.0\,\text{s}^{-1}/2\pi) = 1.27\,\text{Hz}$. (c) Then $T = 1/f = 0.79\,\text{s}$. (d) The maximum speed (see Eq. 11–6) is

$$v_{\text{max}} = 2\pi Af = (2\pi)(0.30\,\text{m})(1.27\,\text{s}^{-1}) = 2.4\,\text{m/s}.$$

(e) The maximum acceleration, by Newton's second law, is $a_{\text{max}} = F_{\text{max}}/m = kA/m$, because $F\,(= kx)$ is greatest when x is greatest. From Eq. 11–7b we see that $k/m = (2\pi f)^2$. Hence

$$a_{\text{max}} = \frac{k}{m}A = (2\pi f)^2 A = (2\pi)^2(1.27\,\text{s}^{-1})^2(0.30\,\text{m}) = 19\,\text{m/s}^2.$$

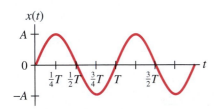

FIGURE 11–9 Sinusoidal nature of SHM as a function of time; in this case, $x = A\sin(2\pi t/T)$ because at $t = 0$ the mass is at the equilibrium position $x = 0$, but it also has (or is given) an initial speed at $t = 0$ that carries it to $x = A$ at $t = \frac{1}{4}T$.

SHM is sinusoidal

Sinusoidal Motion

Equation 11–8 assumes that the oscillating object starts from rest ($v = 0$) at its maximum displacement ($x = A$) at $t = 0$. Other equations for simple harmonic motion are also possible, depending on the initial conditions (when you choose t to be zero). For example, if at $t = 0$ the object is at the equilibrium position and the oscillations are begun by giving the object a push to the right ($+x$), the equation would be

$$x = A\sin\omega t = A\sin(2\pi t/T).$$

This curve (Fig. 11–9) has the same shape as the cosine curve shown in Fig. 11–8, except it is shifted to the right by a quarter cycle. Hence at $t = 0$ it starts out at $x = 0$ instead of at $x = A$.

Both sine and cosine curves are referred to as being **sinusoidal** (having the shape of a sine function). Thus simple harmonic motion[†] is said to be sinusoidal because the position varies as a sinusoidal function of time.

[†]Simple harmonic motion can be *defined* as motion that is sinusoidal. This definition is fully consistent with our earlier definition in Section 11–1.

* Velocity and Acceleration as Functions of Time

Figure 11–10a, like Fig. 11–8, shows a graph of displacement x vs. time t, as given by Eqs. 11–8. We can also find the velocity v as a function of time from Fig. 11–6a. For the position shown (red dot in Fig. 11–6a), we see that the magnitude of v is $v_{max} \sin\theta$, but $\vec{v}$ points to the left, so $v = -v_{max} \sin\theta$. Again setting $\theta = \omega t = 2\pi f t = 2\pi t/T$, we have

$$v = -v_{max} \sin\omega t = -v_{max}\sin(2\pi f t) = -v_{max}\sin(2\pi t/T). \quad \textbf{(11–9)}$$

Just after $t = 0$, the velocity is negative (points to the left) and remains so until $t = \frac{1}{2}T$ (corresponding to $\theta = 180° = \pi$ radians). After $t = \frac{1}{2}T$ until $t = T$ the velocity is positive. The velocity as a function of time (Eq. 11–9) is plotted in Fig. 11–10b. From Eqs. 11–6 and 11–7b,

$$v_{max} = 2\pi A f = A\sqrt{\frac{k}{m}}.$$

For a given spring–mass system, the maximum speed v_{max} is higher if the amplitude is larger, and always occurs as the mass passes the equilibrium point.

The acceleration as a function of time is found from Newton's second law:

$$a = \frac{F}{m} = \frac{-kx}{m} = -\left(\frac{kA}{m}\right)\cos\omega t = -a_{max}\cos(2\pi t/T) \quad \textbf{(11–10)}$$

where the maximum acceleration is

$$a_{max} = kA/m.$$

Equation 11–10 is plotted in Fig. 11–10c. Because the acceleration of a SHO is *not* constant, the equations for uniformly accelerated motion do *not* apply to SHM.

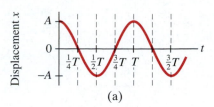

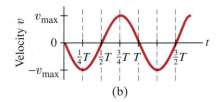

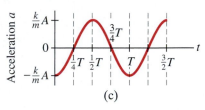

FIGURE 11–10 Graphs showing (a) displacement x as a function of time t: $x = A\cos(2\pi t/T)$; (b) velocity as a function of time: $v = -v_{max}\sin(2\pi t/T)$; (c) acceleration as a function of time: $a = -(kA/m)\cos(2\pi t/T)$.

EXAMPLE 11–8 **Loudspeaker.** The cone of a loudspeaker vibrates in SHM at a frequency of 262 Hz ("middle C"). The amplitude at the center of the cone is $A = 1.5 \times 10^{-4}$ m, and at $t = 0$, $x = A$. (a) What equation describes the motion of the center of the cone? (b) What are the velocity and acceleration as a funcion of time? (c) What is the position of the cone at $t = 1.00$ ms $(= 1.00 \times 10^{-3}$ s)?

APPROACH The motion begins $(t = 0)$ with the cone at its maximum displacement $(x = A$ at $t = 0)$. So we use the cosine function, $x = A\cos\omega t$, to describe this SHM.

SOLUTION (a) Here

$$\omega = 2\pi f = (6.28\,\text{rad})(262\,\text{s}^{-1}) = 1650\,\text{rad/s}.$$

The motion is described as

$$x = A\cos(2\pi f t) = (1.5 \times 10^{-4}\,\text{m})\cos(1650 t).$$

(b) The maximum velocity, from Eq. 11–6, is $v_{max} = 2\pi A f = 2\pi(1.5 \times 10^{-4}\,\text{m})(262\,\text{s}^{-1}) = 0.25$ m/s. Then by Eq. 11–9,

$$v = -(0.25\,\text{m/s})\sin(1650 t).$$

From Eqs. 11–10 and 11–7b, the maximum acceleration is $a_{max} = (k/m)A = (2\pi f)^2 A = 4\pi^2(262\,\text{s}^{-1})^2(1.5 \times 10^{-4}\,\text{m}) = 410$ m/s^2, which is more than 40 g's. So

$$a = -(410\,\text{m/s}^2)\cos(1650 t).$$

(c) At $t = 1.00 \times 10^{-3}$ s, Eq. 11–8a gives us

$$x = A\cos\omega t = (1.5 \times 10^{-4}\,\text{m})\cos\big[(1650\,\text{rad/s})(1.00 \times 10^{-3}\,\text{s})\big]$$

$$= (1.5 \times 10^{-4}\,\text{m})\cos(1.65\,\text{rad}) = -1.2 \times 10^{-5}\,\text{m}.$$

NOTE Be sure your calculator is set in RAD mode, not DEG mode, for these $\cos\omega t$ calculations.

CAUTION

Always be sure your calculator is in the correct mode for angles

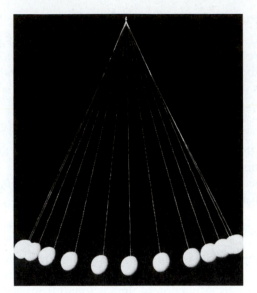

FIGURE 11–11 Strobe-light photo of an oscillating simple pendulum.

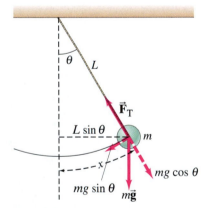

FIGURE 11–12 Simple pendulum, and a free-body diagram.

TABLE 11–1
Sin θ at Small Angles

θ (degrees)	θ (radians)	$\sin\theta$	% Difference
0	0	0	0
1°	0.01745	0.01745	0.005%
5°	0.08727	0.08716	0.1%
10°	0.17453	0.17365	0.5%
15°	0.26180	0.25882	1.1%
20°	0.34907	0.34202	2.0%
30°	0.52360	0.50000	4.7%

11–4 The Simple Pendulum

A **simple pendulum** consists of a small object (the pendulum bob) suspended from the end of a lightweight cord, Fig. 11–11. We assume that the cord doesn't stretch and that its mass can be ignored relative to that of the bob. The motion of a simple pendulum swinging back and forth with negligible friction resembles simple harmonic motion: the pendulum bob oscillates along the arc of a circle with equal amplitude on either side of its equilibrium point, and as it passes through the equilibrium point (where it would hang vertically) it has its maximum speed. But is it really undergoing SHM? That is, is the restoring force proportional to its displacement? Let us find out.

The displacement of the pendulum along the arc is given by $x = L\theta$, where θ is the angle the cord makes with the vertical and L is the length of the cord (Fig. 11–12). If the restoring force is proportional to x or to θ, the motion will be simple harmonic. The restoring force is the net force on the bob, equal to the component of the weight, mg, tangent to the arc:

$$F = -mg\sin\theta,$$

where g is the acceleration of gravity. The minus sign here, as in Eq. 11–1, means the force is in the direction opposite to the angular displacement θ. Since F is proportional to the sine of θ and not to θ itself, the motion is *not* SHM. However, if θ is small, then $\sin\theta$ is very nearly equal to θ when the latter is specified in radians. This can be seen by noting in Fig. 11–12 that the arc length $x \, (= L\theta)$ is nearly the same length as the chord $(= L\sin\theta)$ indicated by the horizontal straight dashed line, *if θ is small*. For angles less than 15°, the difference between θ (in radians) and $\sin\theta$ is less than 1%—see Table 11–1. Thus, to a very good approximation for small angles,

$$F = -mg\sin\theta \approx -mg\theta.$$

Substituting $x = L\theta$, or $\theta = x/L$, we have

$$F \approx -\frac{mg}{L}x.$$

Thus, for small displacements, the motion is essentially simple harmonic, since this equation fits Hooke's law, $F = -kx$. The effective force constant is $k = mg/L$.

If we substitute $k = mg/L$ into Eq. 11–7a, we obtain the period of a simple pendulum:

$$T = 2\pi \sqrt{\frac{m}{k}} = 2\pi \sqrt{\frac{m}{mg/L}}$$

or

$$T = 2\pi \sqrt{\frac{L}{g}}. \qquad \text{[θ small]} \quad \textbf{(11–11a)} \qquad \textit{Period, simple pendulum}$$

The frequency is $f = 1/T$, so

$$f = \frac{1}{2\pi} \sqrt{\frac{g}{L}}. \qquad \text{[θ small]} \quad \textbf{(11–11b)} \qquad \textit{Frequency, simple pendulum}$$

The mass m of the pendulum bob does not appear in these formulas for T and f. Thus we have the surprising result that the period and frequency of a simple pendulum do not depend on the mass of the pendulum bob. You may have noticed this if you pushed a small child and a large one on the same swing.

We also see from Eq. 11–11a that the period of a pendulum does not depend on the amplitude (like any SHM, Section 11–3), as long as the amplitude θ is small. Galileo is said to have first noted this fact while watching a swinging lamp in the cathedral at Pisa (Fig. 11–13). This discovery led to the invention of the pendulum clock, the first really precise timepiece, which became the standard for centuries.

Because a pendulum does not undergo *precisely* SHM, the period does depend slightly on the amplitude—the more so for large amplitudes. The accuracy of a pendulum clock would be affected, after many swings, by the decrease in amplitude due to friction. But the mainspring in a pendulum clock (or the falling weight in a grandfather clock) supplies energy to compensate for the friction and to maintain the amplitude constant, so that the timing remains precise.

EXAMPLE 11–9 **Measuring g.** A geologist uses a simple pendulum that has a length of 37.10 cm and a frequency of 0.8190 Hz at a particular location on the Earth. What is the acceleration of gravity at this location?

APPROACH We can use the length L and frequency f of the pendulum in Eq. 11–11b, which contains our unknown, g.

SOLUTION We solve Eq. 11–11b for g and obtain

$$g = (2\pi f)^2 L = (6.283 \times 0.8190 \text{ s}^{-1})^2 (0.3710 \text{ m}) = 9.824 \text{ m/s}^2.$$

EXERCISE D (*a*) Estimate the length of the pendulum in a grandfather clock that ticks once per second. (*b*) What would be the period of a clock with a 1.0-m-long pendulum?

Equations 11–11 apply to a simple pendulum—a concentrated mass at the end of a string of negligible mass—but not to the oscillation of, say, a baseball bat suspended from one end.

FIGURE 11–13 The swinging motion of this lamp, hanging by a very long cord from the ceiling of the cathedral at Pisa, is said to have been observed by Galileo and to have inspired him to the conclusion that the period of a pendulum does not depend on amplitude.

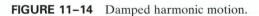

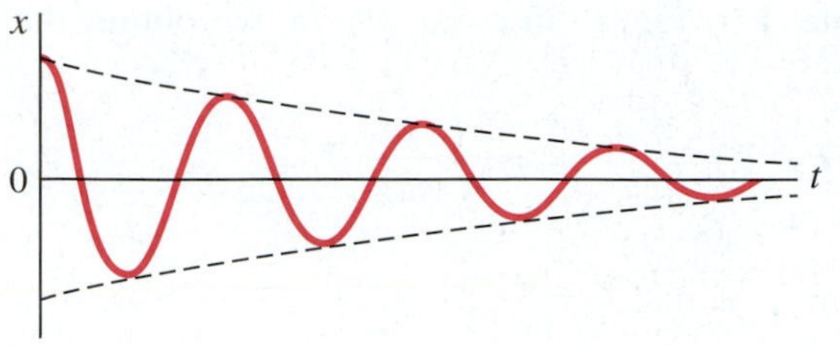

FIGURE 11–14 Damped harmonic motion.

11–5 Damped Harmonic Motion

The amplitude of any real oscillating spring or swinging pendulum will slowly decrease in time until the oscillations stop altogether. Figure 11–14 shows a typical graph of the displacement as a function of time. This is called **damped harmonic motion**. The damping[†] is generally due to the resistance of air and to internal friction within the oscillating system. The energy that is dissipated to thermal energy results in a decreased amplitude of oscillation.

Since natural oscillating systems are damped in general, why do we even talk about (undamped) simple harmonic motion? The answer is that SHM is much easier to deal with mathematically. And if the damping is not large, the oscillations can be thought of as simple harmonic motion on which the damping is superposed. The decrease in amplitude shown by the dashed curves in Fig. 11–14 represents the damping. Although frictional damping does alter the frequency of vibration, the effect is usually small unless the damping is large; thus Eqs. 11–7 can still be used in most cases.

Sometimes the damping is so large, however, that the motion no longer resembles simple harmonic motion. Three common cases of heavily damped systems are shown in Fig. 11–15. Curve A represents an **underdamped** situation, in which the system makes several swings before coming to rest, and corresponds to a more heavily damped version of Fig. 11–14. Curve C represents the **overdamped** situation, for which the damping is so large that it takes a long time to reach equilibrium. Curve B represents **critical damping**: in this case equilibrium is reached in the shortest time. These terms all derive from the use of practical damped systems such as door-closing mechanisms and shock absorbers in a car (Fig. 11–16). Such devices are usually designed to give critical damping. But as they wear out, underdamping occurs: a door slams or a car bounces up and down several times each time it hits a bump.

In many systems, the oscillatory motion is what counts, as in clocks and watches, and damping needs to be minimized. In other systems, oscillations are the problem, such as a car's springs, so a proper amount of damping (i.e., critical) is desired. Well-designed damping is needed for all kinds of applications. Large buildings, especially in California, are now built (or retrofitted) with huge dampers to reduce earthquake damage (Fig. 11–17).

[†]To "damp" means to diminish, restrain, or extinguish, as to "dampen one's spirits."

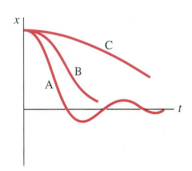

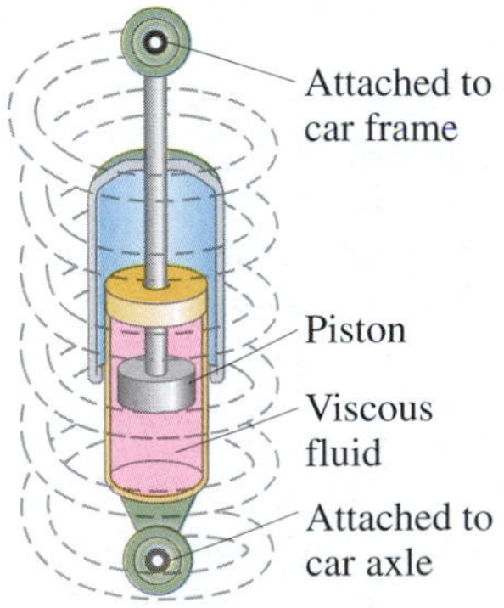

FIGURE 11–15 Graphs that represent (A) underdamped, (B) critically damped, and (C) overdamped oscillatory motion.

PHYSICS APPLIED

Shock absorbers and building dampers

FIGURE 11–16 Automobile spring and shock absorber to provide damping so that a car won't bounce up and down so much.

Attached to car frame

Piston

Viscous fluid

Attached to car axle

FIGURE 11–17 These huge dampers placed in a building look a lot like huge automobile shock absorbers, and they serve a similar purpose—to reduce the amplitude and the acceleration of movement when the shock of an earthquake hits.

11–6 Forced Vibrations; Resonance

When a vibrating system is set into motion, it vibrates at its natural frequency (Eqs. 11–7b and 11–11b). However, a system may have an external force applied to it that has its own particular frequency. Then we have a **forced vibration**. For example, we might pull the mass on the spring of Fig. 11–1 back and forth at an externally applied frequency f. The mass then vibrates at the external frequency f of the external force, even if this frequency is different from the **natural frequency** of the spring, which we will now denote by f_0, where (see Eq. 11–7b)

$$f_0 = \frac{1}{2\pi}\sqrt{\frac{k}{m}}.$$

For a forced vibration, the amplitude of vibration is found to depend on the difference between f and f_0, and is a maximum when the frequency of the external force equals the natural frequency of the system—that is, when $f = f_0$. The amplitude is plotted in Fig. 11–18 as a function of the external frequency f. Curve A represents light damping and curve B heavy damping. The amplitude can become large when the external driving frequency f is near the natural frequency, $f \approx f_0$, as long as the damping is not too large. When the damping is small, the increase in amplitude near $f = f_0$ is very large (and often dramatic). This effect is known as **resonance**. The natural vibrating frequency f_0 of a system is also called its **resonant frequency**.

A simple illustration of resonance is pushing a child on a swing. A swing, like any pendulum, has a natural frequency of oscillation. If you push on the swing at a random frequency, the swing bounces around and reaches no great amplitude. But if you push with a frequency equal to the natural frequency of the swing, the amplitude increases greatly. At resonance, relatively little effort is required to obtain a large amplitude.

The great tenor Enrico Caruso was said to be able to shatter a crystal goblet by singing a note of just the right frequency at full voice. This is an example of resonance, for the sound waves emitted by the voice act as a forced vibration on the glass. At resonance, the resulting vibration of the goblet may be large enough in amplitude that the glass exceeds its elastic limit and breaks.

Since material objects are, in general, elastic, resonance is an important phenomenon in a variety of situations. It is particularly important in building, although the effects are not always foreseen. For example, it has been reported that a railway bridge collapsed because a nick in one of the wheels of a crossing train set up a resonant vibration in the bridge. Marching soldiers break step when crossing a bridge to avoid the possibility that their rhythmic march might match a resonant frequency of the bridge. The collapse of the Tacoma Narrows Bridge (Fig. 11–19a) in 1940 occurred as a result of gusting winds whose approximate frequency matched that of a natural frequency of the bridge, thus driving the span into large-amplitude oscillatory motion. Bridges and tall buildings are now designed with more inherent damping. The Oakland freeway collapse in the 1989 California earthquake (Fig. 11–19b) involved resonant oscillation of a section built on mudfill.

Resonance can be very useful, too, and we will meet important examples later, such as in musical instruments and tuning a radio. We will also see that vibrating objects often have not one, but many resonant frequencies.

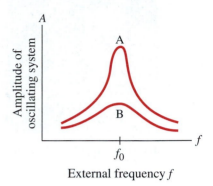

FIGURE 11–18 Resonance for lightly damped (A) and heavily damped (B) systems.

PHYSICS APPLIED
Swinging

PHYSICS APPLIED
Shattering glass via resonance

PHYSICS APPLIED
Resonant collapse

FIGURE 11–19 (a) Large-amplitude oscillations of the Tacoma Narrows Bridge, due to gusty winds, led to its collapse (November 7, 1940). (b) Collapse of a freeway in California, due to the 1989 earthquake, in which resonance played a part.

(a)　　　(b)

FIGURE 11–20 Water waves spreading outward from a source.

When you throw a stone into a lake or pool of water, circular waves form and move outward, Fig. 11–20. Waves will also travel along a cord that is stretched out straight on a table if you vibrate one end back and forth as shown in Fig. 11–21. Water waves and waves on a cord are two common examples of wave motion. We will discuss other kinds of waves later, but for now we will concentrate on these **mechanical waves**.

If you have ever watched ocean waves moving toward shore before they break[†], you may have wondered if the waves were carrying water from far out at sea into the beach. They don't. Water waves move with a recognizable velocity. But each particle (or molecule) of the water itself merely oscillates about an equilibrium point. This is clearly demonstrated by observing leaves on a pond as waves move by. The leaves (or a cork) are not carried forward by the waves, but simply oscillate about an equilibrium point because this is the motion of the water itself.

FIGURE 11–21 Wave traveling on a cord. The wave travels to the right along the cord. Particles of the cord oscillate back and forth on the tabletop.

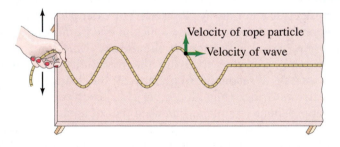

Velocity of rope particle
Velocity of wave

CONCEPTUAL EXAMPLE 11–10 **Wave vs. particle velocity.** Is the velocity of a wave moving along a cord the same as the velocity of a particle of the cord? See Fig. 11–21.

RESPONSE No. The two velocities are different, both in magnitude and direction. The wave on the rope of Fig. 11–21 moves to the right along the tabletop, but each piece of the rope only vibrates to and fro. (The rope clearly does not travel in the direction that the wave on it does.)

Waves can move over large distances, but the medium (the water or the rope) itself has only a limited movement, oscillating about an equilibrium point

[†]Do not be confused by the "breaking" of ocean waves, which occurs when a wave interacts with the ground in shallow water and hence is no longer a simple wave.

as in simple harmonic motion. Thus, although a wave is not matter, the wave pattern can travel in matter. A wave consists of oscillations that move without carrying matter with them.

Waves are moving oscillations, not carrying matter along

Waves carry energy from one place to another. Energy is given to a water wave, for example, by a rock thrown into the water, or by wind far out at sea. The energy is transported by waves to the shore. The oscillating hand in Fig. 11–21 transfers energy to the rope, and that energy is transported down the rope and can be transferred to an object at the other end. All forms of traveling waves transport energy.

Let us look a little more closely at how a wave is formed and how it comes to "travel." We first look at a single wave bump, or **pulse**. A single pulse can be formed on a rope by a quick up-and-down motion of the hand, Fig. 11–22. The hand pulls up on one end of the rope. Because the end section is attached to adjacent sections, these also feel an upward force and they too begin to move upward. As each succeeding section of rope moves upward, the wave crest moves outward along the rope. Meanwhile, the end section of rope has been returned to its original position by the hand. As each succeeding section of rope reaches its peak position, it too is pulled back down again by the adjacent section of rope. Thus the source of a traveling wave pulse is a disturbance, and cohesive forces between adjacent sections of rope cause the pulse to travel outward. Waves in other media are created and propagate outward in a similar fashion.

Wave pulse

A **continuous** or **periodic wave**, such as that shown in Fig. 11–21, has as its source a disturbance that is continuous and oscillating; that is, the source is a *vibration* or *oscillation*. In Fig. 11–21, a hand oscillates one end of the rope. Water waves may be produced by any vibrating object at the surface, such as your hand; or the water itself is made to vibrate when wind blows across it or a rock is thrown into it. A vibrating tuning fork or drum membrane gives rise to sound waves in air. And we will see later that oscillating electric charges give rise to light waves. Indeed, almost any vibrating object sends out waves.

Periodic wave

The source of any wave, then, is a vibration. And it is a *vibration* that propagates outward and thus constitutes the wave. If the source vibrates sinusoidally in SHM, then the wave itself—if the medium is perfectly elastic—will have a sinusoidal shape both in space and in time. (1) In space: if you take a picture of the wave in space at a given instant of time, the wave will have the shape of a sine or cosine as a function of position. (2) In time: if you look at the motion of the medium at one place over a long period of time—for example, if you look between two closely spaced posts of a pier or out of a ship's porthole as water waves pass by—the up-and-down motion of that small segment of water will be simple harmonic motion. The water moves up and down sinusoidally in time.

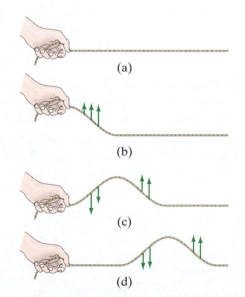

(a)

(b)

(c)

(d)

FIGURE 11–22 Motion of a wave pulse to the right. Arrows indicate velocity of cord particles.

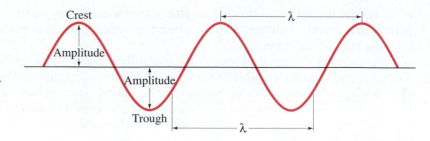

FIGURE 11–23 Characteristics of a single-frequency continuous wave.

Some of the important quantities used to describe a periodic sinusoidal wave are shown in Fig. 11–23. The high points on a wave are called *crests*; the low points, *troughs*. The **amplitude**, *A*, is the maximum height of a crest, or depth of a trough, relative to the normal (or equilibrium) level. The total swing from a crest to a trough is twice the amplitude. The distance between two successive crests is called the **wavelength**, λ (the Greek letter lambda). The wavelength is also equal to the distance between *any* two successive identical points on the wave. The **frequency**, *f*, is the number of crests—or complete cycles—that pass a given point per unit time. The **period**, *T*, equals $1/f$ and is the time elapsed between two successive crests passing by the same point in space.

The **wave velocity**, *v*, is the velocity at which wave crests (or any other part of the waveform) move. The wave velocity must be distinguished from the velocity of a particle of the medium itself as we saw in Example 11–10.

A wave crest travels a distance of one wavelength, λ, in a time equal to one period, *T*. Thus the wave velocity is $v = \lambda/T$. Then, since $1/T = f$,

Amplitude, A

Wavelength, λ

Frequency, f
Period, T

Wave velocity

v = λf (sinusoidal waves)

$$v = \lambda f. \tag{11–12}$$

For example, suppose a wave has a wavelength of 5 m and a frequency of 3 Hz. Since three crests pass a given point per second, and the crests are 5 m apart, the first crest (or any other part of the wave) must travel a distance of 15 m during the 1 s. So its speed is 15 m/s.

The magnitude of the velocity of a wave, or its speed, depends on the properties of the medium in which it travels. The speed of a wave on a stretched string or cord, for example, depends on the tension in the cord, F_T, and on the cord's mass per unit length, m/L. For waves of small amplitude, the relationship is

Speed of wave on a cord

$$v = \sqrt{\frac{F_T}{m/L}}. \tag{11–13}$$

This formula makes sense qualitatively on the basis of Newtonian mechanics. That is, we expect the tension to be in the numerator and the mass per unit length in the denominator. Why? Because when the tension is greater, we expect the velocity to be greater since each segment of cord is in tighter contact with its neighbor; and the greater the mass per unit length, the more inertia the cord has and the more slowly the wave would be expected to propagate.

EXAMPLE 11–11 **Wave on a wire.** A wave whose wavelength is 0.30 m is traveling down a 300-m-long wire whose total mass is 15 kg. If the wire is under a tension of 1000 N, what are the speed and frequency of this wave?

APPROACH We assume the velocity of this wave on a wire is given by Eq. 11–13. We get the frequency from Eq. 11–12, $f = v/\lambda$.

SOLUTION From Eq. 11–13, the velocity is

$$v = \sqrt{\frac{1000\ \text{N}}{(15\ \text{kg})/(300\ \text{m})}} = \sqrt{\frac{1000\ \text{N}}{(0.050\ \text{kg/m})}} = 140\ \text{m/s}.$$

The frequency is

$$f = \frac{v}{\lambda} = \frac{140\ \text{m/s}}{0.30\ \text{m}} = 470\ \text{Hz}.$$

NOTE A higher tension would increase both *v* and *f*, whereas a thicker, denser wire would reduce *v* and *f*.

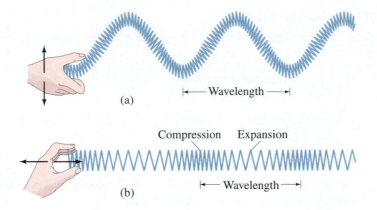

(a)

←— Wavelength —→

Compression   Expansion

(b)

←— Wavelength —→

FIGURE 11–24
(a) Transverse wave;
(b) longitudinal wave.

11–8 Types of Waves: Transverse and Longitudinal

When a wave travels down a rope—say, from left to right as in Fig. 11–21—the particles of the rope vibrate up and down in a direction transverse (that is, perpendicular) to the motion of the wave itself. Such a wave is called a **transverse wave** (Fig. 11–24a). There exists another type of wave known as a **longitudinal wave**. In a longitudinal wave, the vibration of the particles of the medium is *along* the direction of the wave's motion. Longitudinal waves are readily formed on a stretched spring or Slinky by alternately compressing and expanding one end. This is shown in Fig. 11–24b, and can be compared to the transverse wave in Fig. 11–24a. A series of compressions and expansions propagate along the spring. The *compressions* are those areas where the coils are momentarily close together. *Expansions* (sometimes called *rarefactions*) are regions where the coils are momentarily far apart. Compressions and expansions correspond to the crests and troughs of a transverse wave.

An important example of a longitudinal wave is a sound wave in air. A vibrating drumhead, for instance, alternately compresses and rarefies the air in contact with it, producing a longitudinal wave that travels outward in the air, as shown in Fig. 11–25.

As in the case of transverse waves, each section of the medium in which a longitudinal wave passes oscillates over a very small distance, whereas the wave itself can travel large distances. Wavelength, frequency, and wave velocity all have meaning for a longitudinal wave. The wavelength is the distance between successive compressions (or between successive expansions), and frequency is the number of compressions that pass a given point per second. The wave velocity is the velocity with which each compression appears to move; it is equal to the product of wavelength and frequency, $v = \lambda f$ (Eq. 11–12).

A longitudinal wave can be represented graphically by plotting the density of air molecules (or coils of a Slinky) versus position at a given instant, as shown in Fig. 11–26. Such a graphical representation makes it easy to illustrate what is happening. Note that the graph looks much like a transverse wave.

Transverse and longitudinal waves

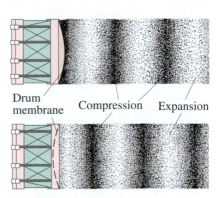

Drum   Compression   Expansion
membrane

FIGURE 11–25   Production of a sound wave, which is longitudinal, shown at two moments in time about a half period $\left(\frac{1}{2}T\right)$ apart.

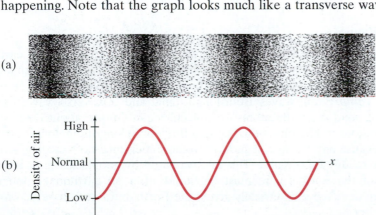

(a)

(b)   Density of air   High · Normal · Low x

FIGURE 11–26
(a) A longitudinal wave with
(b) its graphical representation at a particular instant in time.

* Speed of Longitudinal Waves

The speed of a longitudinal wave has a form similar to that for a transverse wave on a cord (Eq. 11–13):

$$v = \sqrt{\frac{\text{elastic force factor}}{\text{inertia factor}}}.$$

In particular, for a longitudinal wave traveling down a long solid rod,

Longitudinal wave speed in a long solid rod

$$v = \sqrt{\frac{E}{\rho}}, \tag{11–14a}$$

where E is the elastic modulus (Section 9–5) of the material and ρ is its density. For a longitudinal wave traveling in a liquid or gas,

Longitudinal wave speed in a fluid

$$v = \sqrt{\frac{B}{\rho}}, \tag{11–14b}$$

where B is the bulk modulus (Section 9–5) and ρ is the density.

PHYSICS APPLIED

Space perception by animals using sound waves

FIGURE 11–27 A toothed whale (Example 11–12).

EXAMPLE 11–12 Echolocation. Echolocation is a form of sensory perception used by animals such as bats, toothed whales, and porpoises. The animal emits a pulse of sound (a longitudinal wave) which, after reflection from objects, is detected by the animal. Echolocation waves emitted by whales (Fig. 11–27) have frequencies of about 200,000 Hz. (*a*) What is the wavelength of the whale's echolocation wave? (*b*) If an obstacle is 100 m from the whale, how long after the whale emits a wave is its reflection detected?

APPROACH We first compute the speed of longitudinal (sound) waves in sea water, using Eq. 11–14b and Tables 9–1 and 10–1. The wavelength is $\lambda = v/f$.

SOLUTION (*a*) The speed of longitudinal waves in sea water, which is slightly more dense than pure water, is

$$v = \sqrt{\frac{B}{\rho}} = \sqrt{\frac{2.0 \times 10^9\,\text{N/m}^2}{1.025 \times 10^3\,\text{kg/m}^3}} = 1.40 \times 10^3\,\text{m/s}.$$

Then, using Eq. 11–12, we find

$$\lambda = \frac{v}{f} = \frac{(1.40 \times 10^3\,\text{m/s})}{(2.0 \times 10^5\,\text{Hz})} = 7.0\,\text{mm}.$$

(*b*) The time required for the round-trip between the whale and the object is

$$t = \frac{\text{distance}}{\text{speed}} = \frac{2(100\,\text{m})}{1.40 \times 10^3\,\text{m/s}} = 0.14\,\text{s}.$$

NOTE We shall see later that waves can "resolve" (or detect) objects only if the wavelength is comparable to or smaller than the object. Thus, a whale can resolve objects on the order of a centimeter or larger in size.

Other Waves

PHYSICS APPLIED

Earthquake waves

Both transverse and longitudinal waves are produced when an **earthquake** occurs. The transverse waves that travel through the body of the Earth are called S waves (S for shear), and the longitudinal waves are called P waves (P for pressure) or *compression* waves. Both longitudinal and transverse waves can travel through a solid since the atoms or molecules can vibrate about their relatively fixed positions in any direction. But in a fluid, only longitudinal waves can propagate, because any transverse motion would experience no restoring force since a fluid is readily deformable. This fact was used by geophysicists to infer that a portion of the Earth's core must be liquid: after an earthquake, longitudinal waves are detected diametrically across the Earth, but not transverse waves.

Besides these two types of waves, *surface waves* can travel along the boundary between two materials. A wave on water is actually a surface wave that moves on the boundary between water and air. The motion of each particle of water at the surface is circular or elliptical (Fig. 11–28), so it is a combination of transverse and longitudinal motions. Below the surface, there is also transverse plus longitudinal wave motion, as shown. At the bottom, the motion is only longitudinal. When a wave approaches shore, the water drags at the bottom and is slowed down, while the crests move ahead at higher speed (Fig. 11–29) and "spill" over the top.

Surface waves are also set up on the Earth when an earthquake occurs. The waves that travel along the surface are mainly responsible for the damage caused by earthquakes.

Waves traveling along a line, as on a stretched string, are *one-dimensional waves*. Surface waves, such as the water waves of Fig. 11–20, are *two-dimensional waves*. Waves that move out from a source in all directions in a medium, such as sound from a speaker or earthquake waves through the Earth, are *three-dimensional waves*.

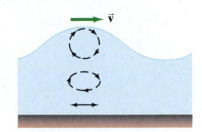

FIGURE 11–28 A water wave is an example of a *surface wave*, which is a combination of transverse and longitudinal wave motions.

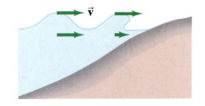

FIGURE 11–29 How a wave breaks. The green arrows represent the local velocity of water molecules.

11–9 Energy Transported by Waves

Waves transport energy from one place to another. As waves travel through a medium, the energy is transferred as vibrational energy from particle to particle of the medium. For a sinusoidal wave of frequency f, the particles move in SHM as a wave passes, so each particle has an energy $E = \frac{1}{2}kA^2$, where A is the amplitude of its motion, either transversely or longitudinally. (See Eq. 11–4a.)

Thus, we have the important result that the **energy transported by a wave is proportional to the square of the amplitude**. The **intensity** I of a wave is defined as the power (energy per unit time) transported across unit area perpendicular to the direction of energy flow:

$$I = \frac{\text{energy/time}}{\text{area}} = \frac{\text{power}}{\text{area}}.$$

The SI unit of intensity is watts per square meter $\left(\text{W/m}^2\right)$. Since the energy is proportional to the wave amplitude squared, so too is the intensity:

$$I \propto A^2. \tag{11–15}$$

If a wave flows out from the source in all directions, it is a three-dimensional wave. Examples are sound traveling in open air, earthquake waves, and light waves. If the medium is isotropic (same in all directions), the wave is a *spherical wave* (Fig. 11–30). As the wave moves outward, the energy it carries is spread over a larger and larger area since the surface area of a sphere of radius r is $4\pi r^2$. Thus the intensity of a spherical wave is

$$I = \frac{\text{power}}{\text{area}} = \frac{P}{4\pi r^2}. \qquad \text{[spherical wave]} \quad \textbf{(11–16a)}$$

If the power output P of the source is constant, then the intensity decreases as the inverse square of the distance from the source:

$$I \propto \frac{1}{r^2}. \tag{11–16b}$$

If we consider two points at distances r_1 and r_2 from the source, as in Fig. 11–30, then $I_1 = P/4\pi r_1^2$ and $I_2 = P/4\pi r_2^2$, so

$$\frac{I_2}{I_1} = \frac{r_1^2}{r_2^2}. \tag{11–16c}$$

Thus, for example, when the distance doubles $(r_2/r_1 = 2)$, the intensity is reduced to $\frac{1}{4}$ its earlier value: $I_2/I_1 = \left(\frac{1}{2}\right)^2 = \frac{1}{4}$.

Wave energy $\propto$ (amplitude)^2

Intensity (defined)

Intensity $\propto$ (amplitude)^2

FIGURE 11–30 A wave traveling outward in three dimensions from a source is spherical. Two crests (or compressions) are shown, of radii r_1 and r_2.

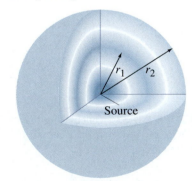

$I \propto \dfrac{1}{r^2}$

Sounds are quieter farther from the source

The amplitude of a wave also decreases with distance. Since the intensity is proportional to the square of the amplitude (Eq. 11–15), the amplitude A must decrease as $1/r$ so that $I \propto A^2$ will be proportional to $1/r^2$ (as in Eq. 11–16b). Hence

$$A \propto \frac{1}{r}.$$

If we consider again two distances from the source, r_1 and r_2, then

$$\frac{A_2}{A_1} = \frac{r_1}{r_2}.$$

When the wave is twice as far from the source, the amplitude is half as large, and so on (ignoring damping due to friction).

EXAMPLE 11–13 **Earthquake intensity.** The intensity of an earthquake P wave traveling through the Earth and detected 100 km from the source is 1.0×10^6 W/m². What is the intensity of that wave if detected 400 km from the source?

APPROACH We assume the wave is spherical, so the intensity decreases as the square of the distance from the source.

SOLUTION At 400 km the distance is 4 times greater than at 100 km, so the intensity will be $\left(\frac{1}{4}\right)^2 = \frac{1}{16}$ of its value at 100 km, or $(1.0 \times 10^6$ W/m²$)/16 = 6.3 \times 10^4$ W/m².

NOTE Using Eq. 11–16c directly gives:

$$I_2 = I_1 r_1^2/r_2^2 = \left(1.0 \times 10^6 \text{ W/m}^2\right)\left(100 \text{ km}\right)^2/\left(400 \text{ km}\right)^2 = 6.3 \times 10^4 \text{ W/m}^2.$$

The situation is different for a one-dimensional wave, such as a transverse wave on a string or a longitudinal wave pulse traveling down a thin uniform metal rod. The area remains constant, so the amplitude A also remains constant (ignoring friction). Thus the amplitude and the intensity do not decrease with distance.

In practice, frictional damping is generally present, and some of the energy is transformed into thermal energy. Thus the amplitude and intensity of a one-dimensional wave will decrease with distance from the source. For a three-dimensional wave, the decrease will be greater than that discussed above, although the effect may often be small.

*11–10 Intensity Related to Amplitude and Frequency

We can obtain an explicit relation between the energy carried by a wave, or the wave's intensity I, and the amplitude and frequency of the wave. For a sinusoidal wave of frequency f, the particles move in SHM as a wave passes, so each particle has an energy $E = \frac{1}{2}kA^2$, where A is the amplitude of its motion, either transversely or longitudinally. Using Eq. 11–7b, we can write k in terms of the frequency: $k = 4\pi^2mf^2$, where m is the mass of a particle (or small volume) of the medium. Then

$$E = \tfrac{1}{2}kA^2 = 2\pi^2mf^2A^2.$$

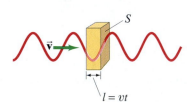

FIGURE 11–31 Calculating the energy carried by a wave moving with velocity v.

The mass $m = \rho V$, where ρ is the density of the medium and V the volume of a small slice of the medium as shown in Fig. 11–31. The volume $V = Sl$, where S is the cross-sectional surface area through which the wave travels. (We use S instead of A for area because we are using A for amplitude.) We can write l as the distance the wave travels in a time t as $l = vt$, where v is the speed of the wave. Thus $m = \rho V = \rho Sl = \rho Svt$, and

$$E = 2\pi^2 \rho Svt f^2 A^2. \tag{11–17a}$$

From this equation, we see again the important result that the energy transported by a wave is proportional to the square of the amplitude. The power transported, $P = E/t$, is

$$P = \frac{E}{t} = 2\pi^2 \rho S v f^2 A^2. \qquad \textbf{(11–17b)}$$

Finally, the **intensity** I of a wave is the power transported across unit area perpendicular to the direction of energy flow:

$$I = \frac{P}{S} = 2\pi^2 v \rho f^2 A^2. \qquad \textbf{(11–18)}$$

This relation shows explicitly that the intensity of a wave is proportional both to the square of the wave amplitude A at any point and to the square of the frequency f.

11–11 Reflection and Transmission of Waves

When a wave strikes an obstacle, or comes to the end of the medium it is traveling in, at least a part of the wave is reflected. You have probably seen water waves reflect off a rock or the side of a swimming pool. And you may have heard a shout reflected from a distant cliff—which we call an "echo."

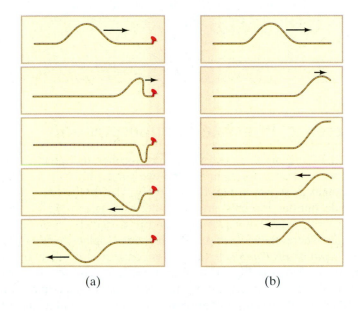

(a) (b)

FIGURE 11–32 Reflection of a wave pulse on a rope lying on a table top. (a) The end of the rope is fixed to a peg. (b) The end of the rope is free to move.

A wave pulse traveling down a rope is reflected as shown in Fig. 11–32. The reflected pulse returns inverted as in Fig. 11–32a if the end of the rope is fixed; it returns right side up if the end is free as in Fig. 11–32b. When the end is fixed to a support, as in Fig. 11–32a, the pulse reaching that fixed end exerts a force (upward) on the support. The support exerts an equal but opposite force downward on the rope (Newton's third law). This downward force on the rope is what "generates" the inverted reflected pulse.

Consider next a pulse that travels down a rope which consists of a light section and a heavy section, as shown in Fig. 11–33. When the wave pulse reaches the boundary between the two sections, part of the pulse is reflected and part is transmitted, as shown. The heavier the second section of rope, the less the energy that is transmitted. (When the second section is a wall or rigid support, very little is transmitted and most is reflected, as in Fig. 11–32a.) For a periodic wave, the frequency of the transmitted wave does not change across the boundary because the boundary point oscillates at that frequency. Thus if the transmitted wave has a lower speed, its wavelength is also shorter ($\lambda = v/f$).

FIGURE 11–33 When a wave pulse traveling to the right along a thin cord (a) reaches a discontinuity where the rope becomes thicker and heavier, then part is reflected and part is transmitted (b).

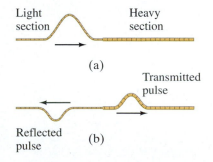

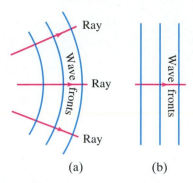

(a)   (b)

FIGURE 11–34 Rays, signifying the direction of wave motion, are always perpendicular to the wave fronts (wave crests). (a) Circular or spherical waves near the source. (b) Far from the source, the wave fronts are nearly straight or flat, and are called plane waves.

For a two- or three-dimensional wave, such as a water wave, we are concerned with **wave fronts**, by which we mean all the points along the wave forming the wave crest (what we usually refer to simply as a "wave" at the seashore). A line drawn in the direction of wave motion, perpendicular to the wave front, is called a **ray**, as shown in Fig. 11–34. Wave fronts far from the source have lost almost all their curvature (Fig. 11–34b) and are nearly straight, as ocean waves often are; they are then called **plane waves**.

For reflection of a two- or three-dimensional plane wave, as shown in Fig. 11–35, the angle that the incoming or *incident wave* makes with the reflecting surface is equal to the angle made by the reflected wave. This is the **law of reflection: the angle of reflection equals the angle of incidence**. The "angle of incidence" is defined as the angle the incident ray makes with the perpendicular to the reflecting surface (or the wave front makes with a tangent to the surface). The "angle of reflection" is the corresponding angle for the reflected wave.

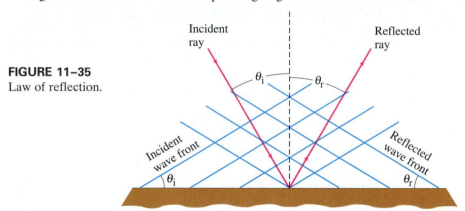

FIGURE 11–35
Law of reflection.

11–12 Interference; Principle of Superposition

Interference refers to what happens when two waves pass through the same region of space at the same time. Consider, for example, the two wave pulses on a string traveling toward each other as shown in Fig. 11–36. In Fig. 11–36a the two pulses have the same amplitude, but one is a crest and the other a trough; in Fig. 11–36b they are both crests. In both cases, the waves meet and pass right by each other. However, in the region where they overlap, the resultant displacement is the *algebraic sum of their separate displacements* (a crest is considered positive and a trough negative). This is called the **principle of superposition**. In Fig. 11–36a, the two waves have opposite displacements at the instant they pass one another, and they add to zero. The result is called **destructive interference**. In Fig. 11–36b, at the instant the two pulses overlap, they produce a resultant displacement that is greater than the displacement of either separate pulse, and the result is **constructive interference**.

Superposition principle

Destructive interference

Constructive interference

FIGURE 11–36 Two wave pulses pass each other. Where they overlap, interference occurs: (a) destructive, and (b) constructive.

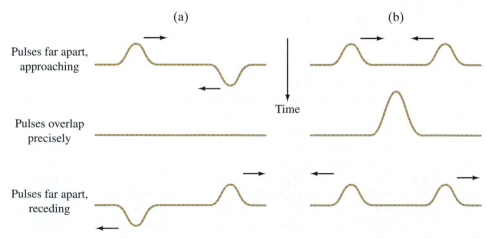

(a)

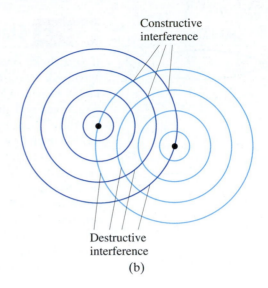

Constructive
interference

Destructive
interference

(b)

FIGURE 11–37 Interference of water waves.

When two rocks are thrown into a pond simultaneously, the two sets of circular waves interfere with one another as shown in Fig. 11–37a. In some areas of overlap, crests of one wave repeatedly meet crests of the other (and troughs meet troughs); see Fig. 11–37b. Constructive interference is occurring at these points, and the water continuously oscillates up and down with greater amplitude than either wave separately. In other areas, destructive interference occurs where the water does not move up and down at all over time. This is where crests of one wave meet troughs of the other, and vice versa. Figure 11–38a shows the displacement of two waves graphically as a function of time, as well as their sum, for the case of constructive interference. For any two such waves, we use the term **phase** to describe the relative positions of *Phase* their crests. When the crests and troughs are aligned as in Fig. 11–38a, for constructive interference, the two waves are **in phase**. At points where destructive interference occurs—see Fig. 11–38b—crests of one wave repeatedly meet troughs of the other wave and the two waves are said to be completely **out of phase** or, more precisely, out of phase by one-half wavelength. That is, the crests of one wave occur a half wavelength behind the crests of the other wave. The relative phase of the two water waves in Fig. 11–37 in most areas is intermediate between these two extremes, resulting in *partially* destructive interference, as illustrated in Fig. 11–38c. If the amplitudes of two interfering waves are not equal, fully destructive interference (as in Fig. 11–38b) does not occur.

FIGURE 11–38 Graphs showing two waves, and their sum, as a function of time at three locations. In (a) the two waves interfere constructively, in (b) destructively, and in (c) partially destructively.

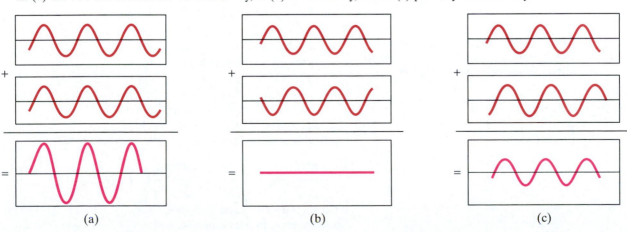

(a) (b) (c)

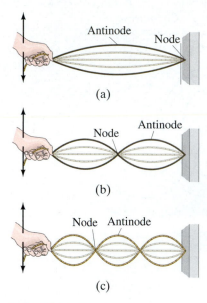

FIGURE 11–39 Standing waves corresponding to three resonant frequencies.

(a)

Antinode — Node

(b)

Node — Antinode

(c)

Node Antinode

Resonant frequencies

11–13 Standing Waves; Resonance

If you shake one end of a cord and the other end is kept fixed, a continuous wave will travel down to the fixed end and be reflected back, inverted, as we saw in Fig. 11–32a. As you continue to vibrate the cord, waves will travel in both directions, and the wave traveling along the cord, away from your hand, will interfere with the reflected wave coming back. Usually there will be quite a jumble. But if you vibrate the cord at just the right frequency, the two traveling waves will interfere in such a way that a large-amplitude **standing wave** will be produced, Fig. 11–39. It is called a "standing wave" because it doesn't appear to be traveling. The cord simply appears to have segments that oscillate up and down in a fixed pattern. The points of destructive interference, where the cord remains still at all times, are called **nodes**. Points of constructive interference, where the cord oscillates with maximum amplitude, are called **antinodes**. The nodes and antinodes remain in fixed positions for a particular frequency.

Standing waves can occur at more than one frequency. The lowest frequency of vibration that produces a standing wave gives rise to the pattern shown in Fig. 11–39a. The standing waves shown in Figs. 11–39b and 11–39c are produced at precisely twice and three times the lowest frequency, respectively, assuming the tension in the cord is the same. The cord can also vibrate with four loops (four antinodes) at four times the lowest frequency, and so on.

The frequencies at which standing waves are produced are the **natural frequencies** or **resonant frequencies** of the cord, and the different standing wave patterns shown in Fig. 11–39 are different "resonant modes of vibration." A standing wave on a cord is the result of the interference of two waves traveling in opposite directions. A standing wave is also a vibrating object at resonance. Standing waves represent the same phenomenon as the resonance of a vibrating spring or pendulum, which we discussed in Section 11–6. The only difference is that a spring or pendulum has only one resonant frequency, whereas the cord has an infinite number of resonant frequencies, each of which is a whole-number multiple of the lowest resonant frequency.

Consider a string stretched between two supports that is plucked like a guitar or violin string, Fig. 11–40a. Waves of a great variety of frequencies will

FIGURE 11–40 (a) A string is plucked. (b) Only standing waves corresponding to resonant frequencies persist for long.

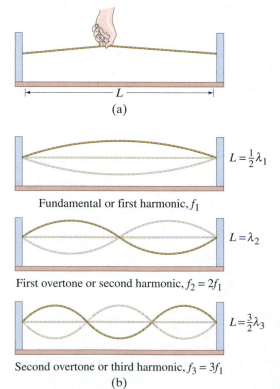

(a)

$L = \frac{1}{2}\lambda_1$

Fundamental or first harmonic, f_1

$L = \lambda_2$

First overtone or second harmonic, $f_2 = 2f_1$

$L = \frac{3}{2}\lambda_3$

Second overtone or third harmonic, $f_3 = 3f_1$

(b)

travel in both directions along the string, will be reflected at the ends, and will travel back in the opposite direction. Most of these waves interfere with each other and quickly die out. However, those waves that correspond to the resonant frequencies of the string will persist. The ends of the string, since they are fixed, will be nodes. There may be other nodes as well. Some of the possible resonant modes of vibration (standing waves) are shown in Fig. 11–40b. Generally, the motion will be a combination of these different resonant modes, but only those frequencies that correspond to a resonant frequency will be present.

To determine the resonant frequencies, we first note that the wavelengths of the standing waves bear a simple relationship to the length L of the string. The lowest frequency, called the **fundamental frequency**, corresponds to one antinode (or loop). And as can be seen in Fig. 11–40b, the whole length corresponds to one-half wavelength. Thus $L = \frac{1}{2}\lambda_1$, where λ_1 stands for the wavelength of the fundamental frequency. The other natural frequencies are called **overtones**; for a vibrating string they are whole-number (integral) multiples of the fundamental, and then are also called **harmonics**, with the fundamental being referred to as the **first harmonic**.[†] The next mode of vibration after the fundamental has two loops and is called the **second harmonic** (or first overtone), Fig. 11–40b. The length of the string L at the second harmonic corresponds to one complete wavelength: $L = \lambda_2$. For the third and fourth harmonics, $L = \frac{3}{2}\lambda_3$, and $L = 2\lambda_4$, respectively, and so on. In general, we can write

Fundamental frequency

Overtones and harmonics

$$ L = \frac{n\lambda_n}{2}, \qquad \text{where } n = 1, 2, 3, \cdots. $$

The integer n labels the number of the harmonic: $n = 1$ for the fundamental, $n = 2$ for the second harmonic, and so on. We solve for λ_n and find

$$ \lambda_n = \frac{2L}{n}, \qquad n = 1, 2, 3, \cdots. \tag{11–19a} $$

To find the frequency f of each vibration we use Eq. 11–12, $f = v/\lambda$, and we see that

$$ f_n = \frac{v}{\lambda_n} = n\frac{v}{2L} = nf_1, \qquad n = 1, 2, 3, \cdots, \tag{11–19b} $$

where $f_1 = v/\lambda_1 = v/2L$ is the fundamental frequency. We see that each resonant frequency is an integer multiple of the fundamental frequency.

Because a standing wave is equivalent to two traveling waves moving in opposite directions, the concept of wave velocity still makes sense and is given by Eq. 11–13 in terms of the tension F_T in the string and its mass per unit length (m/L). That is, $v = \sqrt{F_T/(m/L)}$ for waves traveling in both directions.

EXAMPLE 11–14 **Piano string.** A piano string is 1.10 m long and has a mass of 9.00 g. (*a*) How much tension must the string be under if it is to vibrate at a fundamental frequency of 131 Hz? (*b*) What are the frequencies of the first four harmonics?

APPROACH To determine the tension, we need to find the wave speed using Eq. 11–12 ($v = \lambda f$), and then use Eq. 11–13, solving it for F_T.

SOLUTION (*a*) The wavelength of the fundamental is $\lambda = 2L = 2.20$ m (Eq. 11–19a with $n = 1$). The speed of the wave on the string is $v = \lambda f = (2.20 \text{ m})(131 \text{ s}^{-1}) = 288$ m/s. Then we have (Eq. 11–13)

$$ F_T = \frac{m}{L}v^2 = \left(\frac{9.00 \times 10^{-3} \text{ kg}}{1.10 \text{ m}} \right)(288 \text{ m/s})^2 = 679 \text{ N}. $$

(*b*) The frequencies of the second, third, and fourth harmonics are two, three, and four times the fundamental frequency: 262, 393, and 524 Hz.

NOTE The speed of the wave on the string is *not* the same as the speed of the sound that is produced in the air (as we shall see in Chapter 12).

[†]The term "harmonic" comes from music, because such integral multiples of frequencies "harmonize."

Standing waves are produced not only on strings, but on any object that is struck, such as a drum membrane or an object made of metal or wood. The resonant frequencies depend on the dimensions of the object, just as for a string they depend on its length. Large objects have lower resonant frequencies than small objects. All musical instruments, from stringed instruments to wind instruments (in which a column of air vibrates as a standing wave) to drums and other percussion instruments, depend on standing waves to produce their musical sounds, as we shall see in Chapter 12.

*11–14 Refraction[†]

When any wave strikes a boundary, some of the energy is reflected and some is transmitted or absorbed. When a two- or three-dimensional wave traveling in one medium crosses a boundary into a medium where its speed is different, the transmitted wave may move in a different direction than the incident wave, as shown in Fig. 11–41. This phenomenon is known as **refraction**. One example is a water wave; the velocity decreases in shallow water and the waves refract, as shown in Fig. 11–42 below. [When the wave velocity changes gradually, as in Fig. 11–42, without a sharp boundary, the waves change direction (refract) gradually.]

In Fig. 11–41, the velocity of the wave in medium 2 is less than in medium 1. In this case, the wave front bends so it travels more nearly parallel to the boundary. That is, the *angle of refraction*, θ_r, is less than the *angle of incidence*, θ_i. To see why this is so, and to help us get a quantitative relation between θ_r and θ_i, let us think of each wave front as a row of soldiers. The soldiers are marching from firm ground (medium 1) into mud (medium 2) and hence are slowed down after the boundary. The soldiers that reach the mud first are slowed down first, and the row bends as shown in Fig. 11–43a. Let us consider the wave front (or row of soldiers) labeled A in Fig. 11–43b. In the same time t that A_1 moves a distance $l_1 = v_1 t$, we see that A_2 moves a distance $l_2 = v_2 t$. The two right triangles in Fig. 11–43b, shaded yellow and green, have the side labeled a in common. Thus

$$\sin \theta_1 = \frac{l_1}{a} = \frac{v_1 t}{a}$$

since a is the hypotenuse, and

$$\sin \theta_2 = \frac{l_2}{a} = \frac{v_2 t}{a}.$$

Dividing these two equations, we obtain the *law of refraction*:

Law of refraction

$$\frac{\sin \theta_2}{\sin \theta_1} = \frac{v_2}{v_1}. \tag{11–20}$$

Since θ_1 is the angle of incidence (θ_i), and θ_2 is the angle of refraction (θ_r), Eq. 11–20 gives the quantitative relation between the two. If the wave were

[†]This Section and the next are covered in more detail in Chapters 23 to 25, on optics.

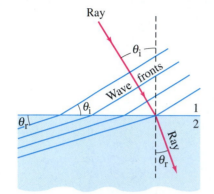

FIGURE 11–41 Refraction of waves passing a boundary.

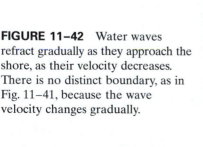

FIGURE 11–42 Water waves refract gradually as they approach the shore, as their velocity decreases. There is no distinct boundary, as in Fig. 11–41, because the wave velocity changes gradually.

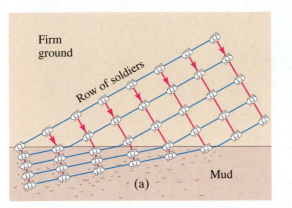

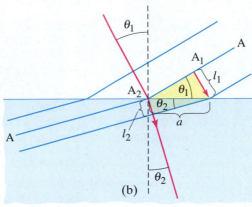

FIGURE 11–43 (a) Soldier analogy to derive (b) law of refraction for waves.

going in the opposite direction, the geometry would not change; only θ_1 and θ_2 would change roles: θ_1 would be the angle of refraction and θ_2 the angle of incidence. Clearly then, if the wave travels into a medium where it can move faster, it will bend the opposite way, $\theta_r > \theta_i$. We see from Eq. 11–20 that if the velocity increases, the angle increases, and vice versa.

Earthquake waves refract within the Earth as they travel through rock layers of different densities (and therefore the velocity is different) just as water waves do. Light waves refract as well, and when we discuss light, we shall find Eq. 11–20 very useful.

PHYSICS APPLIED
Earthquake wave refraction

EXAMPLE 11–15 **Refraction of an earthquake wave.** An earthquake P wave passes across a boundary in rock where its velocity increases from 6.5 km/s to 8.0 km/s. If it strikes this boundary at 30°, what is the angle of refraction?

APPROACH We apply the law of refraction, Eq. 11–20.

SOLUTION Since $\sin 30° = 0.50$, Eq. 11–20 yields

$$\sin \theta_2 = \frac{(8.0 \text{ m/s})}{(6.5 \text{ m/s})}(0.50) = 0.62.$$

So $\theta_2 = \sin^{-1}(0.62) = 38°$.

NOTE Be careful with angles of incidence and refraction. As we discussed in Section 11–11 (Fig. 11–35), these angles are between the wave front and the boundary line, or—equivalently—between the ray (direction of wave motion) and the line perpendicular to the boundary. Inspect Fig. 11–43b carefully.

*11–15 Diffraction

Waves spread as they travel. When they encounter an obstacle, they bend around it somewhat and pass into the region behind as shown in Fig. 11–44 for water waves. This phenomenon is called **diffraction**.

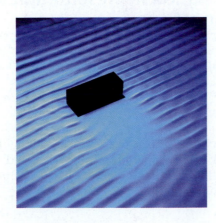

FIGURE 11–44 Wave diffraction. The waves are coming from the upper left. Note how the waves, as they pass the obstacle, bend around it, into the "shadow region" behind it.

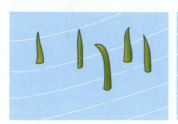

(a) Water waves passing blades of grass

(b) Stick in water

(c) Short-wavelength waves passing log

(d) Long-wavelength waves passing log

FIGURE 11–45 Water waves passing objects of various sizes. Note that the longer the wavelength compared to the size of the object, the more diffraction there is into the "shadow region."

The amount of diffraction depends on the wavelength of the wave and on the size of the obstacle, as shown in Fig. 11–45. If the wavelength is much larger than the object, as with the grass blades of Fig. 11–45a, the wave bends around them almost as if they are not there. For larger objects, parts (b) and (c), there is more of a "shadow" region behind the obstacle where we might not expect the waves to penetrate—but they do, at least a little. Then notice in part (d), where the obstacle is the same as in part (c) but the wavelength is longer, that there is more diffraction into the shadow region. As a rule of thumb, *only if the wavelength is smaller than the size of the object will there be a significant shadow region*. This rule applies to *reflection* from an obstacle as well. Very little of a wave is reflected unless the wavelength is smaller than the size of the obstacle.

A rough guide to the amount of diffraction is

$$\theta(\text{radians}) \approx \frac{\lambda}{L},$$

where θ is roughly the angular spread of waves after they have passed through an opening of width L or around an obstacle of width L.

That waves can bend around obstacles, and thus can carry energy to areas behind obstacles, is very different from energy carried by material particles. A clear example is the following: if you are standing around a corner on one side of a building, you can't be hit by a baseball thrown from the other side, but you can hear a shout or other sound because the sound waves diffract around the edges.

CONCEPTUAL EXAMPLE 11–16 **Cell phones.** Cellular phones operate by radio waves with frequencies of about 1 or 2 GHz $\left(1 \text{ gigahertz} = 10^9 \text{ Hz}\right)$. These waves cannot penetrate objects that conduct electricity, such as a tree trunk or a sheet of metal. The connection is best if the transmitting antenna is within clear view of the handset. Yet it is possible to carry on a phone conversation even if the tower is blocked by trees, or if the handset is inside a car. Why?

RESPONSE If the radio waves have a frequency of about 2 GHz, and the speed of propagation is equal to the speed of light, 3×10^8 m/s (Section 1–5), then the wavelength is $\lambda = v/f = (3 \times 10^8 \text{ m/s})/(2 \times 10^9 \text{ Hz}) = 0.15 \text{ m}$. The waves can diffract readily around objects 15 cm in diameter or smaller.

*11–16 Mathematical Representation of a Traveling Wave

FIGURE 11–46 The characteristics of a single-frequency wave at $t = 0$ (just as in Fig. 11–23).

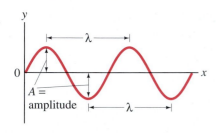

A simple wave with a single frequency, as in Fig. 11–46, is sinusoidal. To express such a wave mathematically, we assume it has a particular wavelength λ and frequency f. At $t = 0$, the wave shape shown is

$$y = A \sin \frac{2\pi}{\lambda} x, \tag{11–21}$$

where y is the **displacement** of the wave (be it a longitudinal or transverse wave) at position x; A is the **amplitude** of the wave, and λ is the wavelength. [Equation 11–21 works because it repeats itself every wavelength: when $x = \lambda$, $y = \sin 2\pi = \sin 0$.]

Suppose the wave is moving to the right with velocity v. After a time t, each part of the wave (indeed, the whole wave "shape") has moved to the right a distance vt. Figure 11–47 shows the wave at $t = 0$ as a solid curve, and at a later time t as a dashed curve. Consider any point on the wave at $t = 0$: say, a crest at some position x. After a time t, that crest will have traveled a distance vt, so its new position is a distance vt greater than its old position. To describe this same point on the wave shape, the argument of the sine function must have the same numerical value, so we replace x in Eq. 11–21 by $(x - vt)$:

$$y = A \sin\left[\frac{2\pi}{\lambda}(x - vt)\right]. \tag{11–22}$$

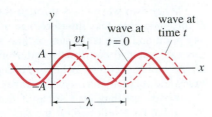

FIGURE 11–47 A traveling wave. In time t, the wave moves a distance vt.

1-D wave, moving
in positive x direction

Said another way, if you are on a crest, as t increases, x must increase at the same rate so that $(x - vt)$ remains constant.

For a wave traveling along the x axis to the left, toward decreasing values of x, v becomes $-v$, so

$$y = A \sin\left[\frac{2\pi}{\lambda}(x + vt)\right].$$

1-D wave, traveling in
negative x direction (to the left)

Summary

A vibrating object undergoes **simple harmonic motion** (SHM) if the restoring force is proportional to the displacement,

$$F = -kx. \tag{11–1}$$

The maximum displacement is called the **amplitude**.

The **period**, T, is the time required for one complete cycle (back and forth), and the **frequency**, f, is the number of cycles per second; they are related by

$$f = \frac{1}{T}. \tag{11–2}$$

The period of vibration for a mass m on the end of a spring is given by

$$T = 2\pi\sqrt{\frac{m}{k}}. \tag{11–7a}$$

SHM is **sinusoidal**, which means that the displacement as a function of time follows a sine or cosine curve.

During SHM, the total energy

$$E = \tfrac{1}{2}mv^2 + \tfrac{1}{2}kx^2 \tag{11–3}$$

is continually changing from potential to kinetic and back again.

A **simple pendulum** of length L approximates SHM if its amplitude is small and friction can be ignored. For small amplitudes, its period is then given by

$$T = 2\pi\sqrt{\frac{L}{g}}, \tag{11–11a}$$

where g is the acceleration of gravity.

When friction is present (for all real springs and pendulums), the motion is said to be **damped**. The maximum displacement decreases in time, and the energy is eventually all transformed to thermal energy.

If an oscillating force is applied to a system capable of vibrating, the system's amplitude of vibration can be very large if the frequency of the applied force matches the **natural** (or **resonant**) **frequency** of the oscillator. This effect is called **resonance**.

Vibrating objects act as sources of **waves** that travel outward from the source. Waves on water and on a string are examples. The wave may be a **pulse** (a single crest), or it may be continuous (many crests and troughs).

The **wavelength** of a continuous sinusoidal wave is the distance between two successive crests.

The **frequency** is the number of wavelengths (or crests) that pass a given point per unit time.

The **amplitude** of a wave is the maximum height of a crest, or depth of a trough, relative to the normal (or equilibrium) level.

The **wave velocity** (how fast a crest moves) is equal to the product of wavelength and frequency,

$$v = \lambda f. \tag{11–12}$$

In a **transverse wave**, the oscillations are perpendicular to the direction in which the wave travels. An example is a wave on a string.

In a **longitudinal wave**, the oscillations are along (parallel to) the line of travel; sound is an example.

The **intensity** of a wave is the energy per unit time carried across unit area (in watts/m^2). For three-dimensional waves traveling in open space, the intensity decreases inversely as the distance from the source squared:

$$I \propto \frac{1}{r^2}. \tag{11–16b}$$

[*Wave intensity is proportional to the amplitude squared and to the frequency squared.]

Waves reflect off objects in their path. When the **wave front** (of a two- or three-dimensional wave) strikes an object, the *angle of reflection* is equal to the *angle of incidence*. When a wave strikes a boundary between two materials in which it can travel, part of the wave is reflected and part is transmitted.

When two waves pass through the same region of space at the same time, they **interfere**. The resultant displacement at any point and time is the sum of their separate displacements; this can result in **constructive interference, destructive interference**, or something in between, depending on the amplitudes and relative phases of the waves.

Waves traveling on a string of fixed length interfere with waves that have reflected off the end and are traveling back in the opposite direction. At certain frequencies, **standing waves** can be produced in which the waves seem to be standing still rather than traveling. The string (or other medium) is vibrating as a whole. This is a resonance phenomenon, and the frequencies at which standing waves occur are called **resonant frequencies**.

Points of destructive interference (no vibration) are called **nodes**. Points of constructive interference (maximum amplitude of vibration) are called **antinodes**.

[*Waves change direction, or **refract**, when traveling from one medium into a second medium where their speed is different. Waves spread, or **diffract**, as they travel and encounter obstacles. A rough guide to the amount of diffraction is $\theta \approx \lambda/L$, where λ is the wavelength and L the width of an opening or obstacle. There is a significant "shadow region" only if the wavelength λ is smaller than the size of the obstacle.]

[*A traveling wave can be represented mathematically as $y = A \sin\{(2\pi/\lambda)(x - vt)\}$.]

Questions

1. Give some examples of everyday vibrating objects. Which exhibit SHM, at least approximately?

2. Is the acceleration of a simple harmonic oscillator ever zero? If so, where?

3. Explain why the motion of a piston in an automobile engine is approximately simple harmonic.

4. Real springs have mass. Will the true period and frequency be larger or smaller than given by the equations for a mass oscillating on the end of an idealized massless spring? Explain.

5. How could you double the maximum speed of a simple harmonic oscillator (SHO)?

6. A 5.0-kg trout is attached to the hook of a vertical spring scale, and then is released. Describe the scale reading as a function of time.

7. If a pendulum clock is accurate at sea level, will it gain or lose time when taken to high altitude? Why?

8. A tire swing hanging from a branch reaches nearly to the ground (Fig. 11–48). How could you estimate the height of the branch using only a stopwatch?

9. Why can you make water slosh back and forth in a pan only if you shake the pan at a certain frequency?

10. Give several everyday examples of resonance.

11. Is a rattle in a car ever a resonance phenomenon? Explain.

12. Is the frequency of a simple periodic wave equal to the frequency of its source? Why or why not?

13. Explain the difference between the speed of a transverse wave traveling down a cord and the speed of a tiny piece of the cord.

14. Why do the strings used for the lowest-frequency notes on a piano normally have wire wrapped around them?

15. What kind of waves do you think will travel down a horizontal metal rod if you strike its end (a) vertically from above and (b) horizontally parallel to its length?

* 16. Since the density of air decreases with an increase in temperature, but the bulk modulus B is nearly independent of temperature, how would you expect the speed of sound waves in air to vary with temperature?

17. Give two reasons why circular water waves decrease in amplitude as they travel away from the source.

* 18. Two linear waves have the same amplitude and speed, and otherwise are identical, except one has half the wavelength of the other. Which transmits more energy? By what factor?

19. When a sinusoidal wave crosses the boundary between two sections of cord as in Fig. 11–33, the frequency does not change (although the wavelength and velocity do change). Explain why.

20. If a string is vibrating in three segments, are there any places you could touch it with a knife blade without disturbing the motion?

21. When a standing wave exists on a string, the vibrations of incident and reflected waves cancel at the nodes. Does this mean that energy was destroyed? Explain.

* 22. If we knew that energy was being transmitted from one place to another, how might we determine whether the energy was being carried by particles (material bodies) or by waves?

FIGURE 11–48 Question 8.

Problems

11–1 to 11–3 Simple Harmonic Motion

1. (I) If a particle undergoes SHM with amplitude 0.18 m, what is the total distance it travels in one period?

2. (I) An elastic cord is 65 cm long when a weight of 75 N hangs from it but is 85 cm long when a weight of 180 N hangs from it. What is the "spring" constant k of this elastic cord?

3. (I) The springs of a 1500-kg car compress 5.0 mm when its 68-kg driver gets into the driver's seat. If the car goes over a bump, what will be the frequency of vibrations?

4. (II) A fisherman's scale stretches 3.6 cm when a 2.7-kg fish hangs from it. (a) What is the spring stiffness constant and (b) what will be the amplitude and frequency of vibration if the fish is pulled down 2.5 cm more and released so that it vibrates up and down?

5. (II) An elastic cord vibrates with a frequency of 3.0 Hz when a mass of 0.60 kg is hung from it. What is its frequency if only 0.38 kg hangs from it?

6. (II) Construct a Table indicating the position x of the mass in Fig. 11–2 at times $t = 0, \frac{1}{4}T, \frac{1}{2}T, \frac{3}{4}T, T,$ and $\frac{5}{4}T,$ where T is the period of oscillation. On a graph of x vs. $t,$ plot these six points. Now connect these points with a smooth curve. Based on these simple considerations, does your curve resemble that of a cosine or sine wave (Fig. 11–8a or 11–9)?

7. (II) A small fly of mass 0.25 g is caught in a spider's web. The web vibrates predominately with a frequency of 4.0 Hz. (a) What is the value of the effective spring stiffness constant k for the web? (b) At what frequency would you expect the web to vibrate if an insect of mass 0.50 g were trapped?

8. (II) A mass m at the end of a spring vibrates with a frequency of 0.88 Hz. When an additional 680-g mass is added to $m,$ the frequency is 0.60 Hz. What is the value of m?

9. (II) A 0.60-kg mass at the end of a spring vibrates 3.0 times per second with an amplitude of 0.13 m. Determine (a) the velocity when it passes the equilibrium point, (b) the velocity when it is 0.10 m from equilibrium, (c) the total energy of the system, and (d) the equation describing the motion of the mass, assuming that x was a maximum at $t = 0.$

10. (II) At what displacement from equilibrium is the speed of a SHO half the maximum value?

11. (II) A mass attached to the end of a spring is stretched a distance x_0 from equilibrium and released. At what distance from equilibrium will it have acceleration equal to half its maximum acceleration?

12. (II) A mass of 2.62 kg stretches a vertical spring 0.315 m. If the spring is stretched an additional 0.130 m and released, how long does it take to reach the (new) equilibrium position again?

13. (II) An object with mass 3.0 kg is attached to a spring with spring stiffness constant $k = 280$ N/m and is executing simple harmonic motion. When the object is 0.020 m from its equilibrium position, it is moving with a speed of 0.55 m/s. (a) Calculate the amplitude of the motion. (b) Calculate the maximum velocity attained by the object. [*Hint*: Use conservation of energy.]

14. (II) It takes a force of 80.0 N to compress the spring of a toy popgun 0.200 m to "load" a 0.180-kg ball. With what speed will the ball leave the gun?

15. (II) A mass sitting on a horizontal, frictionless surface is attached to one end of a spring; the other end is fixed to a wall. 3.0 J of work is required to compress the spring by 0.12 m. If the mass is released from rest with the spring compressed, the mass experiences a maximum acceleration of 15 m/s². Find the value of (a) the spring stiffness constant and (b) the mass.

16. (II) A 0.60-kg mass vibrates according to the equation $x = 0.45 \cos 6.40t,$ where x is in meters and t is in seconds. Determine (a) the amplitude, (b) the frequency, (c) the total energy, and (d) the kinetic energy and potential energies when $x = 0.30$ m.

17. (II) At what displacement from equilibrium is the energy of a SHO half KE and half PE?

18. (II) If one vibration has 7.0 times the energy of a second, but their frequencies and masses are the same, what is the ratio of their amplitudes?

19. (II) A 2.00-kg pumpkin oscillates from a vertically hanging light spring once every 0.65 s. (a) Write down the equation giving the pumpkin's position y (+ upward) as a function of time $t,$ assuming it started by being compressed 18 cm from the equilibrium position (where $y = 0$), and released. (b) How long will it take to get to the equilibrium position for the first time? (c) What will be the pumpkin's maximum speed? (d) What will be its maximum acceleration, and where will that first be attained?

20. (II) A block of mass m is supported by two identical parallel vertical springs, each with spring stiffness constant k (Fig. 11–49). What will be the frequency of vibration?

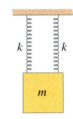

FIGURE 11–49
Problem 20.

21. (II) A 300-g mass vibrates according to the equation $x = 0.38 \sin 6.50t,$ where x is in meters and t is in seconds. Determine (a) the amplitude, (b) the frequency, (c) the period, (d) the total energy, and (e) the KE and PE when x is 9.0 cm. (f) Draw a careful graph of x vs. t showing the correct amplitude and period.

22. (II) Figure 11–50 shows two examples of SHM, labeled A and B. For each, what is (a) the amplitude, (b) the frequency, and (c) the period? (d) Write the equations for both A and B in the form of a sine or cosine.

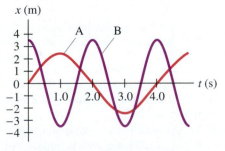

FIGURE 11–50 Problem 22.

23. (II) At $t = 0$, a 755-g mass at rest on the end of a horizontal spring ($k = 124\,\text{N/m}$) is struck by a hammer, which gives the mass an initial speed of 2.96 m/s. Determine (a) the period and frequency of the motion, (b) the amplitude, (c) the maximum acceleration, (d) the position as a function of time, and (e) the total energy.

24. (II) A vertical spring with spring stiffness constant 305 N/m vibrates with an amplitude of 28.0 cm when 0.260 kg hangs from it. The mass passes through the equilibrium point ($y = 0$) with positive velocity at $t = 0$. (a) What equation describes this motion as a function of time? (b) At what times will the spring have its maximum and minimum extensions?

25. (II) A mass m is connected to two springs, with spring stiffness constants k_1 and k_2, as shown in Fig. 11–51. Ignore friction. Show that the period is given by

$$T = 2\pi \sqrt{\frac{m}{k_1 + k_2}}.$$

FIGURE 11–51 Problem 25.

26. (III) A 25.0-g bullet strikes a 0.600-kg block attached to a fixed horizontal spring whose spring stiffness constant is $7.70 \times 10^3\,\text{N/m}$. The block is set into vibration with an amplitude of 21.5 cm. What was the speed of the bullet before impact if the bullet and block move together after impact?

27. (III) A bungee jumper with mass 65.0 kg jumps from a high bridge. After reaching his lowest point, he oscillates up and down, hitting a low point eight more times in 38.0 s. He finally comes to rest 25.0 m below the level of the bridge. Calculate the spring stiffness constant and the unstretched length of the bungee cord.

11–4 Simple Pendulum

28. (I) A pendulum makes 36 vibrations in exactly 60 s. What is its (a) period, and (b) frequency?

29. (I) How long must a simple pendulum be if it is to make exactly one swing per second? (That is, one complete vibration takes exactly 2.0 s.)

30. (I) A pendulum has a period of 0.80 s on Earth. What is its period on Mars, where the acceleration of gravity is about 0.37 that on Earth?

31. (II) What is the period of a simple pendulum 80 cm long (a) on the Earth, and (b) when it is in a freely falling elevator?

32. (II) The length of a simple pendulum is 0.760 m, the pendulum bob has a mass of 365 grams, and it is released at an angle of 12.0° to the vertical. (a) With what frequency does it vibrate? Assume SHM. (b) What is the pendulum bob's speed when it passes through the lowest point of the swing? (c) What is the total energy stored in this oscillation, assuming no losses?

33. (II) Your grandfather clock's pendulum has a length of 0.9930 m. If the clock loses half a minute per day, how should you adjust the length of the pendulum?

34. (II) Derive a formula for the maximum speed v_{max} of a simple pendulum bob in terms of g, the length L, and the angle of swing θ_0.

35. (III) A clock pendulum oscillates at a frequency of 2.5 Hz. At $t = 0$, it is released from rest starting at an angle of 15° to the vertical. Ignoring friction, what will be the position (angle) of the pendulum at (a) $t = 0.25\,\text{s}$, (b) $t = 1.60\,\text{s}$, and (c) $t = 500\,\text{s}$? [*Hint*: Do not confuse the angle of swing θ of the pendulum with the angle that appears as the argument of the cosine.]

11–7 and 11–8 Waves

36. (I) A fisherman notices that wave crests pass the bow of his anchored boat every 3.0 s. He measures the distance between two crests to be 6.5 m. How fast are the waves traveling?

37. (I) A sound wave in air has a frequency of 262 Hz and travels with a speed of 343 m/s. How far apart are the wave crests (compressions)?

38. (I) (a) AM radio signals have frequencies between 550 kHz and 1600 kHz (kilohertz) and travel with a speed of 3.00×10^8 m/s. What are the wavelengths of these signals? (b) On FM, the frequencies range from 88.0 MHz to 108 MHz (megahertz) and travel at the same speed; what are their wavelengths?

* 39. (I) Calculate the speed of longitudinal waves in (a) water, (b) granite, and (c) steel.

* 40. (II) Two solid rods have the same elastic modulus, but one is twice as dense as the other. In which rod will the speed of longitudinal waves be greater, and by what factor?

41. (II) A cord of mass 0.65 kg is stretched between two supports 28 m apart. If the tension in the cord is 150 N, how long will it take a pulse to travel from one support to the other?

42. (II) A ski gondola is connected to the top of a hill by a steel cable of length 620 m and diameter 1.5 cm. As the gondola comes to the end of its run, it bumps into the terminal and sends a wave pulse along the cable. It is observed that it took 16 s for the pulse to return. (a) What is the speed of the pulse? (b) What is the tension in the cable?

* 43. (II) A sailor strikes the side of his ship just below the surface of the sea. He hears the echo of the wave reflected from the ocean floor directly below 3.0 s later. How deep is the ocean at this point?

44. (II) P and S waves from an earthquake travel at different speeds, and this difference helps in locating the earthquake "epicenter" (where the disturbance took place). (a) Assuming typical speeds of 8.5 km/s and 5.5 km/s for P and S waves, respectively, how far away did the earthquake occur if a particular seismic station detects the arrival of these two types of waves 2.0 min apart? (b) Is one seismic station sufficient to determine the position of the epicenter? Explain.

45. (III) An earthquake-produced surface wave can be approximated by a sinusoidal transverse wave. Assuming a frequency of 0.50 Hz (typical of earthquakes, which actually include a mixture of frequencies), what amplitude is needed so that objects begin to leave contact with the ground? [*Hint*: Set the acceleration $a > g$.]

11–9 Wave Energy

46. (II) What is the ratio of (a) the intensities, and (b) the amplitudes, of an earthquake P wave passing through the Earth and detected at two points 10 km and 20 km from the source.

47. (II) The intensity of an earthquake wave passing through the Earth is measured to be 2.0×10^6 J/m²·s at a distance of 48 km from the source. (a) What was its intensity when it passed a point only 1.0 km from the source? (b) At what rate did energy pass through an area of 5.0 m² at 1.0 km?

* 11–10 Intensity Related to A and f

* 48. (I) Two earthquake waves of the same frequency travel through the same portion of the Earth, but one is carrying twice the energy. What is the ratio of the amplitudes of the two waves?

* 49. (I) Two waves traveling along a stretched string have the same frequency, but one transports three times the power of the other. What is the ratio of the amplitudes of the two waves?

* 50. (II) A bug on the surface of a pond is observed to move up and down a total vertical distance of 6.0 cm, from the lowest to the highest point, as a wave passes. If the ripples decrease to 4.5 cm, by what factor does the bug's maximum KE change?

11–12 Interference

51. (I) The two pulses shown in Fig. 11–52 are moving toward each other. (a) Sketch the shape of the string at the moment they directly overlap. (b) Sketch the shape of the string a few moments later. (c) In Fig. 11–36a, at the moment the pulses pass each other, the string is straight. What has happened to the energy at this moment?

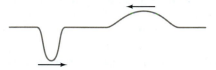

FIGURE 11–52
Problem 51.

11–13 Standing Waves; Resonance

52. (I) If a violin string vibrates at 440 Hz as its fundamental frequency, what are the frequencies of the first four harmonics?

53. (I) A violin string vibrates at 294 Hz when unfingered. At what frequency will it vibrate if it is fingered one-third of the way down from the end? (That is, only two-thirds of the string vibrates as a standing wave.)

54. (I) A particular string resonates in four loops at a frequency of 280 Hz. Name at least three other frequencies at which it will resonate.

55. (II) The velocity of waves on a string is 92 m/s. If the frequency of standing waves is 475 Hz, how far apart are two adjacent nodes?

56. (II) If two successive overtones of a vibrating string are 280 Hz and 350 Hz, what is the frequency of the fundamental?

57. (II) A guitar string is 90 cm long and has a mass of 3.6 g. The distance from the bridge to the support post is $L = 62$ cm, and the string is under a tension of 520 N. What are the frequencies of the fundamental and first two overtones?

58. (II) A particular guitar string is supposed to vibrate at 200 Hz, but it is measured to vibrate at 205 Hz. By what percent should the tension in the string be changed to correct the frequency?

59. (II) One end of a horizontal string is attached to a small-amplitude mechanical 60-Hz vibrator. The string's mass per unit length is 3.9×10^{-4} kg/m. The string passes over a pulley, a distance $L = 1.50$ m away, and weights are hung from this end, Fig. 11–53. What mass m must be hung from this end of the string to produce (a) one loop, (b) two loops, and (c) five loops of a standing wave? Assume the string at the vibrator is a node, which is nearly true.

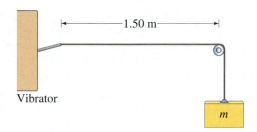

FIGURE 11–53 Problems 59 and 60.

60. (II) In Problem 59, the length of the string may be adjusted by moving the pulley. If the hanging mass m is fixed at 0.080 kg, how many different standing wave patterns may be achieved by varying L between 10 cm and 1.5 m?

61. (II) When you slosh the water back and forth in a tub at just the right frequency, the water alternately rises and falls at each end, remaining relatively calm at the center. Suppose the frequency to produce such a standing wave in a 65-cm-wide tub is 0.85 Hz. What is the speed of the water wave?

* 11–14 Refraction

* 62. (I) An earthquake P wave traveling at 8.0 km/s strikes a boundary within the Earth between two kinds of material. If it approaches the boundary at an incident angle of 47° and the angle of refraction is 35°, what is the speed in the second medium?

* 63. (I) Water waves approach an underwater "shelf" where the velocity changes from 2.8 m/s to 2.1 m/s. If the incident wave crests make a 34° angle with the shelf, what will be the angle of refraction?

* 64. (II) A sound wave is traveling in warm air when it hits a layer of cold, dense air. If the sound wave hits the cold air interface at an angle of 25°, what is the angle of refraction? Assume that the cold air temperature is −10°C and the warm air temperature is +10°C. The speed of sound as a function of temperature can be approximated by $v = (331 + 0.60\,T)$ m/s, where T is in °C.

* 65. (III) A longitudinal earthquake wave strikes a boundary between two types of rock at a 38° angle. As the wave crosses the boundary, the specific gravity of the rock changes from 3.6 to 2.8. Assuming that the elastic modulus is the same for both types of rock, determine the angle of refraction.

* 11–15 Diffraction

* 66. (II) A satellite dish is about 0.5 m in diameter. According to the user's manual, the dish has to be pointed in the direction of the satellite, but an error of about 2° is allowed without loss of reception. Estimate the wavelength of the electromagnetic waves received by the dish.

General Problems

67. A tsunami of wavelength 250 km and velocity 750 km/h travels across the Pacific Ocean. As it approaches Hawaii, people observe an unusual decrease of sea level in the harbors. Approximately how much time do they have to run to safety? (In the absence of knowledge and warning, people have died during tsunamis, some of them attracted to the shore to see stranded fishes and boats.)

68. An energy-absorbing car bumper has a spring stiffness constant of 550 kN/m. Find the maximum compression of the bumper if the car, with mass 1500 kg, collides with a wall at a speed of 2.2 m/s (approximately 5 mi/h). [*Hint*: Use conservation of energy.]

69. A 65-kg person jumps from a window to a fire net 18 m below, which stretches the net 1.1 m. Assume that the net behaves like a simple spring, and (*a*) calculate how much it would stretch if the same person were lying in it. (*b*) How much would it stretch if the person jumped from 35 m?

70. A mass *m* is gently placed on the end of a freely hanging spring. The mass then falls 33 cm before it stops and begins to rise. What is the frequency of the oscillation?

71. A 950-kg car strikes a huge spring at a speed of 22 m/s (Fig. 11–54), compressing the spring 5.0 m. (*a*) What is the spring stiffness constant of the spring? (*b*) How long is the car in contact with the spring before it bounces off in the opposite direction?

900 kg

FIGURE 11–54
Problem 71.

72. When you walk with a cup of coffee (diameter 8 cm) at just the right pace of about 1 step per second, the coffee sloshes more and more until eventually it starts to spill over the top (Fig. 11–55). Estimate the speed of waves in the coffee.

FIGURE 11–55 Problem 72.

73. The ripples in a certain groove 10.8 cm from the center of a 33-rpm phonograph record have a wavelength of 1.70 mm. What will be the frequency of the sound emitted?

74. A 2.00-kg mass vibrates according to the equation $x = 0.650 \cos 7.40t$, where *x* is in meters and *t* in seconds. Determine (*a*) the amplitude, (*b*) the frequency, (*c*) the total energy, and (*d*) the kinetic energy and potential energy when $x = 0.260$ m.

75. A simple pendulum oscillates with frequency *f*. What is its frequency if it accelerates at $0.50g$ (*a*) upward, and (*b*) downward?

76. A 220-kg wooden raft floats on a lake. When a 75-kg man stands on the raft, it sinks 4.0 cm deeper into the water. When he steps off, the raft vibrates for a while. (*a*) What is the frequency of vibration? (*b*) What is the total energy of vibration (ignoring damping)?

77. Two strings on a musical instrument are tuned to play at 392 Hz (G) and 440 Hz (A). (*a*) What are the frequencies of the first two overtones for each string? (*b*) If the two strings have the same length and are under the same tension, what is the ratio of their masses (m_G/m_A)? (*c*) If the strings instead have the same mass per unit length and are under the same tension, what is the ratio of their lengths (L_G/L_A)? (*d*) If their masses and lengths are the same, what must be the ratio of the tensions in the two strings?

78. Consider a sine wave traveling down the stretched two-part cord of Fig. 11–33. Determine a formula (*a*) for the ratio of the speeds of the wave in the heavy section versus that in the lighter section, v_H/v_L, and (*b*) for the ratio of the wavelengths in the two sections. (The frequency is the same in both sections. Why?) (*c*) Is the wavelength greater in the heavier section of cord or the lighter?

79. A tuning fork vibrates at a frequency of 264 Hz, and the tip of each prong moves 1.8 mm to either side of center. Calculate (*a*) the maximum speed and (*b*) the maximum acceleration of the tip of a prong.

80. A diving board oscillates with simple harmonic motion of frequency 1.5 cycles per second. What is the maximum amplitude with which the end of the board can vibrate in order that a pebble placed there (Fig. 11–56) will not lose contact with the board during the oscillation?

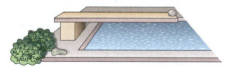

FIGURE 11–56
Problem 80.

81. A string can have a "free" end if that end is attached to a ring that can slide without friction on a vertical pole (Fig. 11–57). Determine the wavelengths of the resonant vibrations of such a string with one end fixed and the other free.

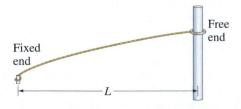
Fixed
end
Free
end
L

FIGURE 11–57 Problem 81.

82. A "seconds" pendulum has a period of exactly 2.000 s— each one-way swing takes 1.000 s. (a) What is the length of a seconds pendulum in Austin, Texas, where $g = 9.793 \text{ m/s}^2$? (b) If the pendulum is moved to Paris, where $g = 9.809 \text{ m/s}^2$, by how many millimeters must we lengthen the pendulum? (c) What would be the length of a seconds pendulum on the Moon, where $g = 1.62 \text{ m/s}^2$?

83. A mass hanging from a spring can oscillate in the vertical direction or can swing as a pendulum of small amplitude, but not both at the same time. Which one is longer, the period of the vertical oscillations or the period of the horizontal swings, and by what amount? [*Hint:* Let l_0 be the length of the unstretched spring, and L be its length with the mass attached at rest.]

84. A block with mass $M = 5.0 \text{ kg}$ rests on a frictionless table and is attached by a horizontal spring ($k = 130 \text{ N/m}$) to a wall. A second block, of mass $m = 1.25 \text{ kg}$, rests on top of M. The coefficient of static friction between the two blocks is 0.30. What is the maximum possible amplitude of oscillation such that m will not slip off M?

85. A 10.0-m-long wire of mass 123 g is stretched under a tension of 255 N. A pulse is generated at one end, and 20.0 ms later a second pulse is generated at the opposite end. Where will the two pulses first meet?

86. A block of mass M is suspended from a ceiling by a spring with spring stiffness constant k. A penny of mass m is placed on top of the block. What is the maximum amplitude of oscillations that will allow the penny to just stay on top of the block? (Assume $m \ll M$.)

* 87. A crane has hoisted a 1200-kg car at the junkyard. The steel crane cable is 22 m long and has a diameter of 6.4 mm. A breeze starts the car bouncing at the end of the cable. What is the period of the bouncing? [*Hint:* Refer to Table 9–1.]

* 88. A block of jello rests on a plate as shown in Fig. 11–58 (which also gives the dimensions of the block). You push it sideways as shown, and then you let go. The jello springs back and begins to vibrate. In analogy to a mass vibrating on a spring, estimate the frequency of this vibration, given that the shear modulus (Section 9–5) of jello is 520 N/m^2 and its density is 1300 kg/m^3.

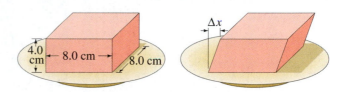

FIGURE 11–58 Problem 88.

If music be the food of physics, play on. [See Shakespeare, *Twelfth Night*, line 1.]

An orchestra contains stringed instruments, whose sound depends on transverse standing waves on strings, and wind instruments whose sound originates in longitudinal standing waves of an air column. Percussion instruments create more complicated standing waves.

Besides examining sources of sound, we also study the decibel scale of sound level, the ear's response, sound wave interference and beats, the Doppler effect, shock waves and sonic booms, and ultrasound imaging.

CHAPTER **12**

Sound

Sound is associated with our sense of hearing and, therefore, with the physiology of our ears and the psychology of our brain, which interprets the sensations that reach our ears. The term *sound* also refers to the physical sensation that stimulates our ears: namely, longitudinal waves.

We can distinguish three aspects of any sound. First, there must be a *source* for a sound; as with any wave, the source of a sound wave is a vibrating object. Second, the energy is transferred from the source in the form of longitudinal sound *waves*. And third, the sound is *detected* by an ear or by a microphone. We start this Chapter by looking at some aspects of sound waves themselves.

12–1 Characteristics of Sound

We saw in Chapter 11, Fig. 11–25, how a vibrating drumhead produces a sound wave in air. Indeed, we usually think of sound waves traveling in the air, for normally it is the vibrations of the air that force our eardrums to vibrate. But sound waves can also travel in other materials.

Two stones struck together under water can be heard by a swimmer beneath the surface, for the vibrations are carried to the ear by the water. When you put your ear flat against the ground, you can hear an approaching train or truck. In this case the ground does not actually touch your eardrum, but the longitudinal wave transmitted by the ground is called a sound wave just the same, for its vibrations cause the outer ear and the air within it to vibrate. Clearly, sound cannot travel in the absence of matter. For example, a bell ringing inside an evacuated jar cannot be heard, and sound cannot travel through the empty reaches of outer space.

The **speed of sound** is different in different materials. In air at 0°C and 1 atm, sound travels at a speed of 331 m/s. The speed of sound in various materials is given in Table 12–1. The values depend somewhat on temperature, especially for gases. For example, in air near room temperature, the speed increases approximately 0.60 m/s for each Celsius degree increase in temperature:

$$v \approx (331 + 0.60T)\ \text{m/s},$$

where T is the temperature in °C. Unless stated otherwise, we will assume in this Chapter that $T = 20°C$, so $v = \left[331 + (0.60)(20)\right]\ \text{m/s} = 343\ \text{m/s}$.

TABLE 12–1 Speed of Sound in Various Materials (20°C and 1 atm)	
Material	**Speed (m/s)**
Air	343
Air (0°C)	331
Helium	1005
Hydrogen	1300
Water	1440
Sea water	1560
Iron and steel	≈ 5000
Glass	≈ 4500
Aluminum	≈ 5100
Hardwood	≈ 4000
Concrete	≈ 3000

Speed of sound in air

| CONCEPTUAL EXAMPLE 12–1 | **Distance from a lightning strike.** A rule of thumb that tells how close lightning has hit is, "one mile for every five seconds before the thunder is heard." Justify, noting that the speed of light is so high (3×10^8 m/s, almost a million times faster than sound) that the time for light to travel is negligible compared to the time for sound. |

RESPONSE The speed of sound in air is about 340 m/s, so to travel 1 km = 1000 m takes about 3 seconds. One mile is about 1.6 kilometers, so the time for the thunder to travel a mile is about $(1.6)(3) \approx 5$ seconds.

 **PHYSICS APPLIED**
How far away is the lightning?

| EXERCISE A What would be the rule of thumb of Example 12–1 in terms of kilometers?

Two aspects of any sound are immediately evident to a human listener: "loudness" and "pitch." Each refers to a sensation in the consciousness of the listener. But to each of these subjective sensations there corresponds a physically measurable quantity. **Loudness** is related to the intensity (energy per unit time crossing unit area) in the sound wave, and we shall discuss it in the next Section.

Loudness

The **pitch** of a sound refers to whether it is high, like the sound of a piccolo or violin, or low, like the sound of a bass drum or string bass. The physical quantity that determines pitch is the frequency, as was first noted by Galileo. The lower the frequency, the lower the pitch; the higher the frequency, the higher the pitch.[†] The best human ears can respond to frequencies from about 20 Hz to almost 20,000 Hz. (Recall that 1 Hz is 1 cycle per second.) This frequency range is called the **audible range**. These limits vary somewhat from one individual to another. One general trend is that as people age, they are less able to hear high frequencies, so the high-frequency limit may be 10,000 Hz or less.

Pitch

Audible frequency range

Sound waves whose frequencies are outside the audible range may reach the ear, but we are not generally aware of them. Frequencies above 20,000 Hz are called **ultrasonic** (do not confuse with *supersonic*, which is used for an object moving with a speed faster than the speed of sound). Many animals can hear ultrasonic frequencies; dogs, for example, can hear sounds as high as 50,000 Hz, and bats can detect frequencies as high as 100,000 Hz. Ultrasonic waves have a number of applications in medicine and other fields, which we will discuss later in this Chapter.

 **CAUTION**
Do not confuse ultrasonic (high frequency) with supersonic (high speed)

[†]Although pitch is determined mainly by frequency, it also depends to a slight extent on loudness. For example, a very loud sound may seem slightly lower in pitch than a quiet sound of the same frequency.

FIGURE 12–1 Example 12–2. Autofocusing camera emits an ultrasonic pulse. Solid lines represent the wave front of the outgoing wave pulse moving to the right; dashed lines represent the wave front of the pulse reflected off the person's face, returning to the camera. The time information allows the camera mechanism to adjust the lens to focus at the proper distance.

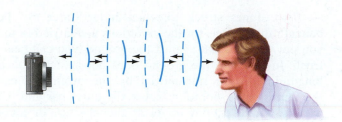

PHYSICS APPLIED

Autofocusing camera

EXAMPLE 12–2 Autofocusing with sound waves. Autofocusing cameras emit a pulse of very high frequency (ultrasonic) sound that travels to the object being photographed, and include a sensor that detects the returning reflected sound, as shown in Fig. 12–1. To get an idea of the time sensitivity of the detector, calculate the travel time of the pulse for an object (*a*) 1.0 m away, and (*b*) 20 m away.

APPROACH If we assume the temperature is about 20°C, then the speed of sound is 343 m/s. Using this speed v and the total distance d back and forth in each case, we can obtain the time $(v = d/t)$.

SOLUTION (*a*) The pulse travels 1.0 m to the object and 1.0 m back, for a total of 2.0 m. We solve for t in $v = d/t$:

$$t = \frac{d}{v} = \frac{2.0\text{ m}}{343\text{ m/s}} = 0.0058\text{ s} = 5.8\text{ ms.}$$

(*b*) The total distance now is $2 \times 20\text{ m} = 40\text{ m}$, so

$$t = \frac{40\text{ m}}{343\text{ m/s}} = 0.12\text{ s} = 120\text{ ms.}$$

NOTE These times are very short, so the wait for the camera to focus is not noticeable.

Sound waves whose frequencies are below the audible range (that is, less than 20 Hz) are called **infrasonic**. Sources of infrasonic waves include earthquakes, thunder, volcanoes, and waves produced by vibrating heavy machinery. This last source can be particularly troublesome to workers, for infrasonic waves—even though inaudible—can cause damage to the human body. These low-frequency waves act in a resonant fashion, causing motion and irritation of the body's organs.

FIGURE 12–2 The membrane of a drum, as it vibrates, alternately compresses the air and, as it recedes (moves to the left), leaves a rarefaction or expansion of air. See also Fig. 11–25.

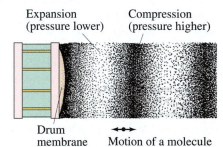

Expansion (pressure lower) Compression (pressure higher)

Drum membrane   Motion of a molecule

FIGURE 12–3 Representation of a sound wave in space at a given instant in terms of (a) displacement, and (b) pressure.

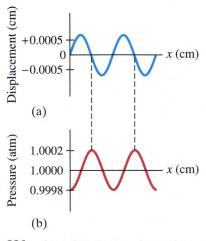

(a)

(b)

We often describe a sound wave in terms of the vibration of the molecules of the medium in which it travels—that is, in terms of the motion or displacement of the molecules. But sound waves can also be analyzed from the point of view of pressure. Indeed, longitudinal waves are often called **pressure waves**. The pressure variation is usually easier to measure than the displacement. As Figure 12–2 shows, in a wave "compression" (where molecules are closest together), the pressure is higher than normal, whereas in an expansion (or *rarefaction*) the pressure is less than normal. Figure 12–3 shows a graphical representation of a sound wave in air in terms of (a) displacement and (b) pressure. Note that the displacement wave is a quarter wavelength out of phase with the pressure wave: where the pressure is a maximum or minimum, the displacement from equilibrium is zero; and where the pressure variation is zero, the displacement is a maximum or minimum.

12–2 Intensity of Sound: Decibels

Like pitch, **loudness** is a sensation in the consciousness of a human being. It too is related to a physically measurable quantity, the **intensity** of the wave. Intensity is defined as the energy transported by a wave per unit time across a unit area perpendicular to the energy flow. As we saw in Chapter 11, intensity is proportional to the square of the wave amplitude. Intensity has units of power per unit area, or watts/meter^2 (W/m^2).

The human ear can detect sounds with an intensity as low as $10^{-12}\,\text{W/m}^2$ and as high as $1\,\text{W/m}^2$ (and even higher, although above this it is painful). This is an incredibly wide range of intensity, spanning a factor of 10^{12} from lowest to highest. Presumably because of this wide range, what we perceive as loudness is not directly proportional to the intensity. To produce a sound that sounds about twice as loud requires a sound wave that has about 10 times the intensity. This is roughly valid at any sound level for frequencies near the middle of the audible range. For example, a sound wave of intensity $10^{-2}\,\text{W/m}^2$ sounds to an average human being like it is about twice as loud as one whose intensity is $10^{-3}\,\text{W/m}^2$, and four times as loud as $10^{-4}\,\text{W/m}^2$.

Sound Level

Because of this relationship between the subjective sensation of loudness and the physically measurable quantity "intensity," sound intensity levels are usually specified on a logarithmic scale. The unit on this scale is a **bel**, after the inventor Alexander Graham Bell, or much more commonly, the **decibel** (dB), which is $\frac{1}{10}$ bel (10 dB = 1 bel). The **sound level**, β, of any sound is defined in terms of its intensity, I, as

$$\beta\ (\text{in dB}) = 10 \log \frac{I}{I_0}, \qquad (12\text{–}1)$$

where I_0 is the intensity of a chosen reference level, and the logarithm is to the base 10. I_0 is usually taken as the minimum intensity audible to a good ear—the "threshold of hearing," which is $I_0 = 1.0 \times 10^{-12}\,\text{W/m}^2$. Thus, for example, the sound level of a sound whose intensity $I = 1.0 \times 10^{-10}\,\text{W/m}^2$ will be

$$\beta = 10 \log \left(\frac{1.0 \times 10^{-10}\,\text{W/m}^2}{1.0 \times 10^{-12}\,\text{W/m}^2} \right) = 10 \log 100 = 20\,\text{dB},$$

since log 100 is equal to 2.0. Notice that the sound level at the threshold of hearing is 0 dB. That is, $\beta = 10 \log 10^{-12}/10^{-12} = 10 \log 1 = 0$ since $\log 1 = 0$. Notice too that an increase in intensity by a factor of 10 corresponds to a sound level increase of 10 dB. An increase in intensity by a factor of 100 corresponds to a sound level increase of 20 dB. Thus a 50-dB sound is 100 times more intense than a 30-dB sound, and so on.

Intensities and sound levels for a number of common sounds are listed in Table 12–2.

EXAMPLE 12–3 **Sound intensity on the street.** At a busy street corner, the sound level is 70 dB. What is the intensity of sound there?

APPROACH We have to solve Eq. 12–1 for intensity I, remembering that $I_0 = 1.0 \times 10^{-12}\,\text{W/m}^2$.

SOLUTION From Eq. 12–1

$$\log \frac{I}{I_0} = \frac{\beta}{10},$$

so

$$\frac{I}{I_0} = 10^{\beta/10}.$$

With $\beta = 70$, then

$$I = I_0 10^{\beta/10} = (1.0 \times 10^{-12}\,\text{W/m}^2)(10^7) = 1.0 \times 10^{-5}\,\text{W/m}^2.$$

NOTE Recall (Appendix A) that $x = \log y$ is the same as $y = 10^x$.

PHYSICS APPLIED
Wide range of human hearing

The dB unit

Sound level (decibels)

CAUTION
0 dB does not mean zero intensity

Each 10 dB corresponds to a 10-fold change in intensity

TABLE 12–2 Intensity of Various Sounds

Source of the Sound	Sound Level (dB)	Intensity (W/m²)
Jet plane at 30 m	140	100
Threshold of pain	120	1
Loud rock concert	120	1
Siren at 30 m	100	1×10^{-2}
Auto interior, at 90 km/h	75	3×10^{-5}
Busy street traffic	70	1×10^{-5}
Talk, at 50 cm	65	3×10^{-6}
Quiet radio	40	1×10^{-8}
Whisper	20	1×10^{-10}
Rustle of leaves	10	1×10^{-11}
Threshold of hearing	0	1×10^{-12}

EXAMPLE 12–4 **Loudspeaker response.** A high-quality loudspeaker is advertised to reproduce, at full volume, frequencies from 30 Hz to 18,000 Hz with uniform sound level ± 3 dB. That is, over this frequency range, the sound level output does not vary by more than 3 dB for a given input level. By what factor does the intensity change for the maximum output sound level change of 3 dB?

APPROACH Let us call the average intensity I_1 and the average sound level β_1. Then the maximum intensity, I_2, corresponds to a level $\beta_2 = \beta_1 + 3$ dB. We then use the relation between intensity and sound level, Eq. 12–1.

SOLUTION Equation 12–1 gives

$$\beta_2 - \beta_1 = 10 \log \frac{I_2}{I_0} - 10 \log \frac{I_1}{I_0}$$

$$3 \text{ dB} = 10 \left(\log \frac{I_2}{I_0} - \log \frac{I_1}{I_0} \right)$$

$$= 10 \log \frac{I_2}{I_1}$$

because $(\log a - \log b) = \log a/b$ (see Appendix A). This last equation gives

$$\log \frac{I_2}{I_1} = 0.30,$$

or

$$\frac{I_2}{I_1} = 10^{0.30} = 2.0.$$

x = log y means y = 10^x

So ± 3 dB corresponds to a doubling or halving of the intensity.

EXERCISE B If an increase of 3 dB means "twice as intense," what does an increase of 6 dB mean?

It is worth noting that a sound-level difference of 3 dB (which corresponds to a doubled intensity, as we just saw) corresponds to only a very small change in the subjective sensation of apparent loudness. Indeed, the average human can distinguish a difference in sound level of only about 1 or 2 dB.

Normally, the loudness or intensity of a sound decreases as you get farther from the source of the sound. In interior rooms, this effect is reduced because of reflections from the walls. However, if a source is in the open so that sound can radiate out freely in all directions, the intensity decreases as the inverse square of the distance,

$$I \propto \frac{1}{r^2},$$

as we saw in Section 11–9. Over large distances, the intensity decreases faster than $1/r^2$ because some of the energy is transferred into irregular motion of air molecules. This loss happens more for higher frequencies, so any sound of mixed frequencies will be less "bright" at a distance.

EXAMPLE 12–5 **Airplane roar.** The sound level measured 30 m from a jet plane is 140 dB. What is the sound level at 300 m? (Ignore reflections from the ground.)

APPROACH Given the sound level, we can determine the intensity at 30 m using Eq. 12–1. Because intensity decreases as the square of the distance, ignoring reflections, we can find I at 300 m and again apply Eq. 12–1 to obtain the sound level.

SOLUTION The intensity I at 30 m is

$$140 \text{ dB} = 10 \log\left(\frac{I}{10^{-12} \text{ W/m}^2}\right).$$

Reversing the log equation to solve for I, we have

$$10^{14} = \frac{I}{10^{-12} \text{ W/m}^2},$$

so $I = (10^{14})(10^{-12} \text{ W/m}^2) = 10^2 \text{ W/m}^2$. At 300 m, 10 times as far, the intensity will be $\left(\frac{1}{10}\right)^2 = 1/100$ as much, or 1 W/m². Hence, the sound level is

$$\beta = 10 \log\left(\frac{1 \text{ W/m}^2}{10^{-12} \text{ W/m}^2}\right) = 120 \text{ dB}.$$

Even at 300 m, the sound is at the threshold of pain. This is why workers at airports wear ear covers to protect their ears from damage (Fig. 12–4).

NOTE Here is a simpler approach that avoids Eq. 12–1: because the intensity decreases as the square of the distance, at 10 times the distance the intensity decreases by $\left(\frac{1}{10}\right)^2 = \frac{1}{100}$. We can use the result that 10 dB corresponds to an intensity change by a factor of 10 (see just before Example 12–3). Then an intensity change by a factor of 100 corresponds to a sound-level change of $(2)(10 \text{ dB}) = 20 \text{ dB}$. This confirms our result above: $140 \text{ dB} - 20 \text{ dB} = 120 \text{ dB}$.

FIGURE 12–4 Example 12–5. Airport worker with sound-intensity-reducing ear covers (headphones).

EXERCISE C If you double your distance from a source of sound that is radiating freely in all directions, how does the intensity that you hear change? By how many dB does the sound level change?

* Intensity Related to Amplitude

The intensity I of a wave is proportional to the square of the wave amplitude, A, as discussed in Sections 11–9 and 11–10. We can then relate the amplitude quantitatively to the intensity I or level β, as the following Example shows.

EXAMPLE 12–6 How tiny the displacement is. Calculate the displacement of air molecules for a sound having a frequency of 1000 Hz at the threshold of hearing.

APPROACH In Section 11–10 we found a relation between intensity I and displacement amplitude A of a wave, Eq. 11–18. The amplitude of oscillation of air molecules is what we want to solve for, given the intensity.

SOLUTION At the threshold of hearing, $I = 1 \times 10^{-12} \text{ W/m}^2$ (Table 12–2). We solve for the amplitude A in Eq. 11–18:

$$A = \frac{1}{\pi f}\sqrt{\frac{I}{2\rho v}}$$

$$= \frac{1}{(3.14)(1.0 \times 10^3 \text{ s}^{-1})}\sqrt{\frac{1.0 \times 10^{-12} \text{ W/m}^2}{(2)(1.29 \text{ kg/m}^3)(343 \text{ m/s})}}$$

$$= 1.1 \times 10^{-11} \text{ m},$$

PHYSICS APPLIED
Incredible sensitivity of the ear

where we have taken the density of air to be 1.29 kg/m³ and the speed of sound in air (assumed 20°C) as 343 m/s.

NOTE We see how incredibly sensitive the human ear is: it can detect displacements of air molecules which are actually less than the diameter of atoms (about 10^{-10} m).

Ear detects displacements smaller than size of atoms

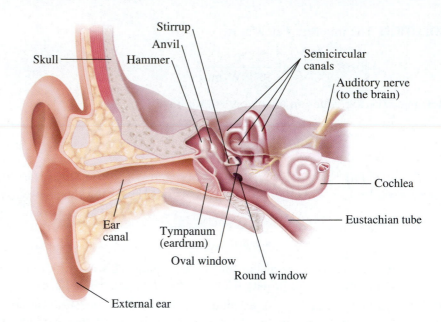

FIGURE 12–5 Diagram of the human ear.

The human ear is a remarkably sensitive detector of sound. Mechanical detectors of sound (microphones) can barely match the ear in detecting low-intensity sounds.

The function of the ear is to transform the vibrational energy of waves into electrical signals which are carried to the brain by way of nerves. A microphone performs a similar task. Sound waves striking the diaphragm of a microphone set it into vibration, and these vibrations are transformed into an electrical signal with the same frequencies, which can then be amplified and sent to a loudspeaker or tape recorder. We shall discuss the operation of microphones when we study electricity and magnetism in later Chapters. Here we shall discuss the structure and response of the ear.

PHYSICS APPLIED

Human ear

Figure 12–5 is a diagram of the human ear. The ear consists of three main divisions: the outer ear, middle ear, and inner ear. In the outer ear, sound waves from the outside travel down the ear canal to the eardrum (the tympanum), which vibrates in response to the impinging waves. The middle ear consists of three small bones known as the hammer, anvil, and stirrup, which transfer the vibrations of the eardrum to the inner ear at the oval window. This delicate system of levers, coupled with the relatively large area of the eardrum compared to the area of the oval window, results in the pressure being amplified by a factor of about 40. The inner ear consists of the semicircular canals, which are important for controlling balance, and the liquid-filled cochlea where the vibrational energy of sound waves is transformed into electrical energy and sent to the brain.

* The Ear's Response

Sensitivity of the ear

Loudness (in "phons")

The ear is not equally sensitive to all frequencies. To hear the same loudness for sounds of different frequencies requires different intensities. Studies averaged over large numbers of people have produced the curves shown in Fig. 12–6. On this graph, each curve represents sounds that seemed to be equally loud. The number labeling each curve represents the **loudness level** (the units are called *phons*), which is numerically equal to the sound level in dB at 1000 Hz. For example, the curve labeled 40 represents sounds that are heard by an average person to have the same loudness as a 1000-Hz sound with a sound level of 40 dB. From this 40-phon curve, we see that a 100-Hz tone must be at a level of about 62 dB to be perceived as loud as a 1000-Hz tone of only 40 dB.

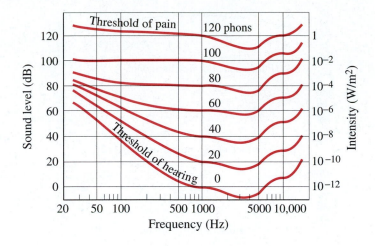

FIGURE 12–6
Sensitivity of the human ear
as a function of frequency (see text).
Note that the frequency scale
is "logarithmic" in order to
cover a wide range of frequencies.

The lowest curve in Fig. 12–6 (labeled 0) represents the sound level, as a function of frequency, for the *threshold of hearing*, the softest sound that is just audible by a very good ear. Note that the ear is most sensitive to sounds of frequency between 2000 and 4000 Hz, which are common in speech and music. Note too that whereas a 1000-Hz sound is audible at a level of 0 dB, a 100-Hz sound must be nearly 40 dB to be heard. The top curve in Fig. 12–6, labeled 120 phons, represents the *threshold of pain*. Sounds above this level can actually be felt and cause pain.

Figure 12–6 shows that at lower sound levels, our ears are less sensitive to the high and low frequencies relative to middle frequencies. The "loudness" control on stereo systems is intended to compensate for this low-volume insensitivity. As the volume is turned down, the loudness control boosts the high and low frequencies relative to the middle frequencies so that the sound will have a more "normal-sounding" frequency balance. Many listeners, however, find the sound more pleasing or natural without the loudness control.

12–4 Sources of Sound: Vibrating Strings and Air Columns

The source of any sound is a vibrating object. Almost any object can vibrate and hence be a source of sound. We now discuss some simple sources of sound, particularly musical instruments. In musical instruments, the source is set into vibration by striking, plucking, bowing, or blowing. Standing waves are produced and the source vibrates at its natural resonant frequencies. The vibrating source is in contact with the air (or other medium) and pushes on it to produce sound waves that travel outward. The frequencies of the waves are the same as those of the source, but the speed and wavelengths can be different. A drum has a stretched membrane that vibrates. Xylophones and marimbas have metal or wood bars that can be set into vibration. Bells, cymbals, and gongs also make use of a vibrating metal. The most widely used instruments make use of vibrating strings, such as the violin, guitar, and piano, or make use of vibrating columns of air, such as the flute, trumpet, and pipe organ. We have already seen that the pitch of a pure sound is determined by the frequency. Typical frequencies for musical notes on the "equally tempered chromatic scale" are given in Table 12–3 for the octave beginning with middle C. Note that one octave corresponds to a doubling of frequency. For example, middle C has frequency of 262 Hz whereas C′ (C above middle C) has twice that frequency, 524 Hz. [Middle C is the C or "do" note at the middle of a piano keyboard.]

TABLE 12–3 Equally Tempered Chromatic Scale†

Note	Frequency (Hz)
C	262
C♯ or D♭	277
D	294
D♯ or E♭	311
E	330
F	349
F♯ or G♭	370
G	392
G♯ or A♭	415
A	440
A♯ or B♭	466
B	494
C′	524

†Only one octave is included.

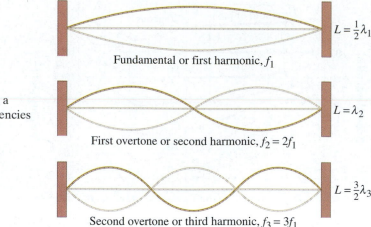

$L = \frac{1}{2}\lambda_1$

Fundamental or first harmonic, f_1

$L = \lambda_2$

First overtone or second harmonic, $f_2 = 2f_1$

$L = \frac{3}{2}\lambda_3$

Second overtone or third harmonic, $f_3 = 3f_1$

FIGURE 12–7 Standing waves on a string—only the lowest three frequencies are shown.

Stringed Instruments

We saw in Chapter 11, Fig. 11–40, how standing waves are established on a string, and we show this again here in Fig. 12–7. Such standing waves are the basis for all stringed instruments. The pitch is normally determined by the lowest resonant frequency, the **fundamental**, which corresponds to nodes occurring only at the ends. The string vibrating up and down as a whole corresponds to a half wavelength as shown at the top of Fig. 12–7; so the wavelength of the fundamental on the string is equal to twice the length of the string. Therefore, the fundamental frequency is $f_1 = v/\lambda = v/2L$, where v is the velocity of the wave on the string. The possible frequencies for standing waves on a stretched string are whole-number multiples of the fundamental frequency:

$$f_n = nf_1 = n\frac{v}{2L}, \qquad n = 1, 2, 3, \cdots$$

(just as in Eq. 11–19b), where $n = 1$ refers to the fundamental and $n = 2, 3, \cdots$ are the overtones. All of the standing waves, $n = 1, 2, 3, \cdots$, are called harmonics[†], as we saw in Section 11–13.

When a finger is placed on the string of a guitar or violin, the effective length of the string is shortened. So its fundamental frequency, and pitch, is higher since the wavelength of the fundamental is shorter (Fig. 12–8). The strings on a guitar or violin are all the same length. They sound at a different pitch because the strings have different mass per unit length, m/L, which affects the velocity as seen in Eq. 11–13,

$$v = \sqrt{F_T/(m/L)}. \qquad \text{[stretched string]}$$

Thus the velocity on a heavier string is less and the frequency will be less for the same wavelength. The tension F_T may also be different. Adjusting the tension is the means for tuning the pitch of each string. In pianos and harps, on the other hand, the strings are of different lengths. For the lower notes the strings are not only longer, but heavier as well, and the reason is illustrated in the following Example.

FIGURE 12–8 The wavelength of (a) an unfingered string is longer than that of (b) a fingered string. Hence, the frequency of the fingered string is higher. Only one string is shown on this guitar, and only the simplest standing wave, the fundamental, is shown.

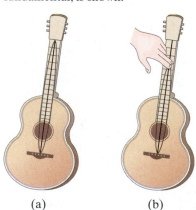

(a)   (b)

EXAMPLE 12–7 Piano strings. The highest key on a piano corresponds to a frequency about 150 times that of the lowest key. If the string for the highest note is 5.0 cm long, how long would the string for the lowest note have to be if it had the same mass per unit length and was under the same tension?

APPROACH Since $v = \sqrt{F_T/(m/L)}$, the velocity would be the same on each string. So the frequency is inversely proportional to the length L of the string ($f = v/\lambda = v/2L$).

[†]When the resonant frequencies above the fundamental (that is, the overtones) are integral multiples of the fundamental, as here, they are called harmonics. But if the overtones are not integral multiples of the fundamental, as is the case for a vibrating drumhead, for example, they are not harmonics.

SOLUTION We can write, for the fundamental frequencies of each string, the ratio

$$\frac{L_L}{L_H} = \frac{f_H}{f_L},$$

where the subscripts L and H refer to the lowest and highest notes, respectively. Thus $L_L = L_H(f_H/f_L) = (5.0\text{ cm})(150) = 750\text{ cm}$, or 7.5 m. This would be ridiculously long (≈ 25 ft) for a piano.

NOTE The longer strings of lower frequency are made heavier, so even on grand pianos the strings are less than 3 m long.

(a)

EXAMPLE 12–8 **Frequencies and wavelengths in the violin.** A 0.32-m-long violin string is tuned to play A above middle C at 440 Hz. (*a*) What is the wavelength of the fundamental string vibration, and (*b*) what are the frequency and wavelength of the sound wave produced? (*c*) Why is there a difference?

APPROACH The wavelength of the fundamental string vibration equals twice the length of the string (Fig. 12–7). As the string vibrates, it pushes on the air, which is thus forced to oscillate at the same frequency as the string.

SOLUTION (*a*) From Fig. 12–7 the wavelength of the fundamental is

$$\lambda = 2L = 2(0.32\text{ m}) = 0.64\text{ m} = 64\text{ cm}.$$

This is the wavelength of the standing wave on the string.
(*b*) The sound wave that travels outward in the air (to reach our ears) has the same frequency, 440 Hz. Its wavelength is

$$\lambda = \frac{v}{f} = \frac{343\text{ m/s}}{440\text{ Hz}} = 0.78\text{ m} = 78\text{ cm},$$

where v is the speed of sound in air (assumed at 20°C), Section 12–1.
(*c*) The wavelength of the sound wave is different from that of the standing wave on the string because the speed of sound in air (343 m/s at 20°C) is different from the speed of the wave on the string ($= f\lambda = 440\text{ Hz} \times 0.64\text{ m} = 280\text{ m/s}$) which depends on the tension in the string and its mass per unit length.

NOTE The frequencies on the string and in the air are the same: the string and air are in contact, and the string "forces" the air to vibrate at the same frequency. But the wavelengths are different because the wave speed on the string is different than that in air.

(b)

FIGURE 12–9 (a) Piano, showing sounding board to which the strings are attached; (b) sounding box (guitar).

 **CAUTION**

Speed of standing wave on string ≠ speed of sound wave in air

Stringed instruments would not be very loud if they relied on their vibrating strings to produce the sound waves since the strings are too thin to compress and expand much air. Stringed instruments therefore make use of a kind of mechanical amplifier known as a *sounding board* (piano) or *sounding box* (guitar, violin), which acts to amplify the sound by putting a greater surface area in contact with the air (Fig. 12–9). When the strings are set into vibration, the sounding board or box is set into vibration as well. Since it has much greater area in contact with the air, it can produce a more intense sound wave. On an electric guitar, the sounding box is not so important since the vibrations of the strings are amplified electronically.

FIGURE 12–10 Wind instruments: clarinet (left) and flute.

Wind Instruments

Instruments such as woodwinds, the brasses, and the pipe organ produce sound from the vibrations of standing waves in a column of air within a tube or pipe (Fig. 12–10). Standing waves can occur in the air of any cavity, but the frequencies present are complicated for any but very simple shapes such as the uniform, narrow tube of a flute or an organ pipe. In some instruments, a vibrating reed or the vibrating lip of the player helps to set up vibrations of the air column. In others, a stream of air is directed against one edge of the opening or mouthpiece, leading to turbulence which sets up the vibrations. Because of the disturbance, whatever its source, the air within the tube vibrates with a variety of frequencies, but only frequencies that correspond to standing waves will persist.

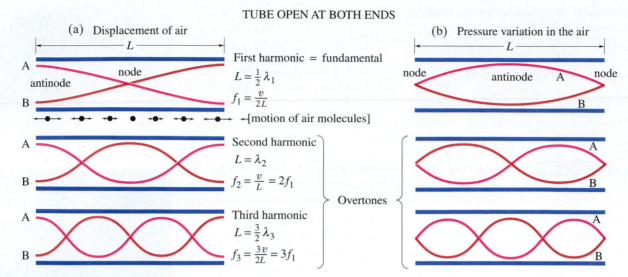

(a) Displacement of air

First harmonic = fundamental
$L = \frac{1}{2}\lambda_1$
$f_1 = \frac{v}{2L}$

—[motion of air molecules]

Second harmonic
$L = \lambda_2$
$f_2 = \frac{v}{L} = 2f_1$

Overtones

Third harmonic
$L = \frac{3}{2}\lambda_3$
$f_3 = \frac{3v}{2L} = 3f_1$

(b) Pressure variation in the air

FIGURE 12–11 Graphs of the three simplest modes of vibration (standing waves) for a uniform tube open at both ends ("open tube"). These simplest modes of vibration are shown in (a), on the left, in terms of the motion of the air (displacement), and in (b), on the right, in terms of air pressure. Each graph shows the wave format at two times, A and B, a half period apart. The actual motion of molecules for one case, the fundamental, is shown just below the tube at top left.

PHYSICS APPLIED

Wind instruments

Open tube

Open tubes produce all harmonics

For a string fixed at both ends, Fig. 12–7, the standing waves have nodes (no movement) at the two ends, and one or more antinodes (large amplitude of vibration) in between. A node separates successive antinodes. The lowest-frequency standing wave, the *fundamental*, corresponds to a single antinode. The higher-frequency standing waves are called **overtones** or **harmonics**, as we saw in Section 11–13. Specifically, the first harmonic is the fundamental, the second harmonic (= first overtone) has twice the frequency of the fundamental, and so on.

The situation is similar for a column of air in a tube of uniform diameter, but we must remember that it is now air itself that is vibrating. We can describe the waves either in terms of the flow of the air—that is, in terms of the *displacement* of air—or in terms of the *pressure* in the air (see Figs. 12–2 and 12–3). In terms of displacement, the air at the closed end of a tube is a displacement node since the air is not free to move there, whereas near the open end of a tube there will be an antinode since the air can move freely in and out. The air within the tube vibrates in the form of longitudinal standing waves. The possible modes of vibration for a tube open at both ends (called an **open tube**) are shown graphically in Fig. 12–11. They are shown for a tube that is open at one end but closed at the other (called a **closed tube**) in Fig. 12–12. [A tube closed at *both* ends, having no connection to the outside air, would be useless as an instrument.] The graphs in part (a) of each Figure (left-hand sides) represent the displacement amplitude of the vibrating air in the tube. Note that these are graphs, and that the air molecules themselves oscillate *horizontally*, parallel to the tube length, as shown by the small arrows in the top diagram of Fig. 12–11a (on the left). The exact position of the antinode near the open end of a tube depends on the diameter of the tube, but if the diameter is small compared to the length, which is the usual case, the antinode occurs very close to the end as shown. We assume this is the case in what follows. (The position of the antinode may also depend slightly on the wavelength and other factors.)

Let us look in detail at the open tube, in Fig. 12–11a, which might be an organ pipe or a flute. An open tube has displacement antinodes at both ends since the air is free to move at open ends. There must be at least one node within an open tube if there is to be a standing wave at all. A single node corresponds to the *fundamental frequency* of the tube. Since the distance between two successive nodes, or between two successive antinodes, is $\frac{1}{2}\lambda$, there is one-half of a wavelength within the length of the tube for the simplest case of the fundamental (top diagram in Fig. 12–11a): $L = \frac{1}{2}\lambda$, or $\lambda = 2L$. So the fundamental frequency is $f_1 = v/\lambda = v/2L$, where v is the velocity of sound in air (the air in the tube). The standing wave with two nodes is the *first overtone* or *second harmonic* and has half the wavelength ($L = \lambda$) and twice the frequency of the fundamental. Indeed, in a uniform tube open at both ends, the frequency of each overtone is an integral multiple of the fundamental frequency, as shown in Fig. 12–11a. This is just what is found for a string.

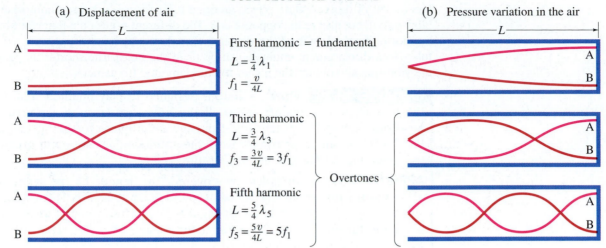

(a) Displacement of air

First harmonic = fundamental
$$L = \tfrac{1}{4}\lambda_1$$
$$f_1 = \frac{v}{4L}$$

Third harmonic
$$L = \tfrac{3}{4}\lambda_3$$
$$f_3 = \frac{3v}{4L} = 3f_1$$

Overtones

Fifth harmonic
$$L = \tfrac{5}{4}\lambda_5$$
$$f_5 = \frac{5v}{4L} = 5f_1$$

(b) Pressure variation in the air

FIGURE 12–12 Modes of vibration (standing waves) for a tube closed at one end ("closed tube"). See caption for Fig. 12–11.

Closed tube

For a closed tube, shown in Fig. 12–12a, which could be an organ pipe, there is always a displacement node at the closed end (because the air is not free to move) and an antinode at the open end (where the air can move freely). Since the distance between a node and the nearest antinode is $\tfrac{1}{4}\lambda$, we see that the fundamental in a closed tube corresponds to only one-fourth of a wavelength within the length of the tube: $L = \lambda/4$, and $\lambda = 4L$. The fundamental frequency is thus $f_1 = v/4L$, or half that for an open pipe of the same length. There is another difference, for as we can see from Fig. 12–12a, only the odd harmonics are present in a closed tube: the overtones have frequencies equal to $3, 5, 7, \cdots$ times the fundamental frequency. There is no way for waves with $2, 4, 6, \cdots$ times the fundamental frequency to have a node at one end and an antinode at the other, and thus they cannot exist as standing waves in a closed tube.

Closed tubes produce only odd harmonics

Another way to analyze the vibrations in a uniform tube is to consider a description in terms of the *pressure* in the air, shown in part (b) of Figs. 12–11 and 12–12 (right-hand sides). Where the air in a wave is compressed, the pressure is higher, whereas in a wave expansion (or rarefaction), the pressure is less than normal. The open end of a tube is open to the atmosphere. Hence the pressure variation at an open end must be a *node*: the pressure doesn't alternate, but remains at the outside atmospheric pressure. If a tube has a closed end, the pressure at that closed end can readily alternate to be above or below atmospheric pressure. Hence there is a pressure *antinode* at a closed end of a tube. There can be pressure nodes and antinodes within the tube. Some of the possible vibrational modes in terms of pressure for an open tube are shown in Fig. 12–11b, and for a closed tube are shown in Fig. 12–12b.

EXAMPLE 12–9 **Organ pipes.** What will be the fundamental frequency and first three overtones for a 26-cm-long organ pipe at 20°C if it is (a) open and (b) closed?

APPROACH All our calculations can be based on Figs. 12–11a and 12–12a.

SOLUTION (a) For the open pipe, Fig. 12–11a, the fundamental frequency is

$$f_1 = \frac{v}{2L} = \frac{343 \text{ m/s}}{2(0.26 \text{ m})} = 660 \text{ Hz}.$$

The speed v is the speed of sound in air (the air vibrating in the pipe). The overtones include all harmonics: 1320 Hz, 1980 Hz, 2640 Hz, and so on. (b) For a closed pipe, Fig. 12–12a, the fundamental frequency is

$$f_1 = \frac{v}{4L} = \frac{343 \text{ m/s}}{4(0.26 \text{ m})} = 330 \text{ Hz}.$$

Only odd harmonics are present: the first three overtones are 990 Hz, 1650 Hz, and 2310 Hz.

NOTE The closed pipe plays 330 Hz, which, from Table 12–3, is E above middle C, whereas the open pipe of the same length plays 660 Hz, an octave higher.

Pipe organs use both open and closed pipes, with lengths from a few centimeters to 5 m or more. A flute acts as an open tube, for it is open not only where you blow into it, but also at the opposite end. The different notes on a flute are obtained by shortening the length of the vibrating air column, by uncovering holes along the tube (so a displacement antinode can occur at the hole). The shorter the length of the vibrating air column, the higher the fundamental frequency.

EXAMPLE 12–10 **Flute.** A flute is designed to play middle C (262 Hz) as the fundamental frequency when all the holes are covered. Approximately how long should the distance be from the mouthpiece to the far end of the flute? (This is only approximate since the antinode does not occur precisely at the mouthpiece.) Assume the temperature is 20°C.

APPROACH When all holes are covered, the length of the vibrating air column is the full length. The speed of sound in air at 20°C is 343 m/s. Because a flute is open at both ends, we use Fig. 12–11: the fundamental frequency f_1 is related to the length L of the vibrating air column by $f = v/2L$.

SOLUTION Solving for L, we find

$$L = \frac{v}{2f} = \frac{343 \text{ m/s}}{2(262 \text{ s}^{-1})} = 0.655 \text{ m}.$$

EXERCISE D To see why players of wind instruments "warm up" their instruments (so they will be in tune), determine the fundamental frequency of the flute of Example 12–10 when all holes are covered and the temperature is 10°C instead of 20°C.

EXAMPLE 12–11 **ESTIMATE** **Wind noise frequencies.** Wind can be noisy—it can "howl" in trees; it can "moan" in chimneys. What is causing the noise, and about what range of frequencies would you expect to hear?

APPROACH Gusts of air in the wind cause vibrations or oscillations of the tree limb (or air column in the chimney), which produce sound waves of the same frequency. The end of a tree limb fixed to the tree trunk is a node, whereas the other end is free to move and therefore is an antinode; the tree limb is thus about $\frac{1}{4}\lambda$ (Fig. 12–13).

SOLUTION We estimate $v \approx 4000$ m/s for the speed of sound in wood (Table 12–1). Suppose that a tree limb has length $L \approx 2$ m; then $\lambda = 4L = 8$ m and $f = v/\lambda = (4000 \text{ m/s})/(8 \text{ m}) \approx 500$ Hz.

NOTE Wind can excite air oscillations in a chimney, much like in an organ pipe or flute. A chimney is a fairly long tube, perhaps 3 m in length, acting like a tube open at either one end or even both ends. If open at both ends ($\lambda = 2L$), with $v \approx 340$ m/s, we find $f_1 \approx v/2L \approx 56$ Hz, which is a fairly low note—no wonder chimneys "moan"!

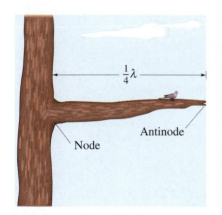

FIGURE 12–13 Example 12–11.

FIGURE 12–14 The amplitudes of the fundamental and first two overtones are added at each point to get the "sum," or composite waveform.

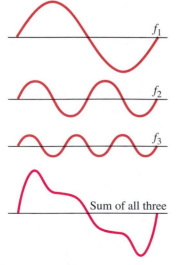

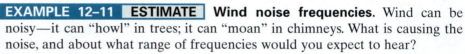

* 12–5 Quality of Sound, and Noise; Superposition

Whenever we hear a sound, particularly a musical sound, we are aware of its loudness, its pitch, and also of a third aspect called "quality." For example, when a piano and then a flute play a note of the same loudness and pitch (say, middle C), there is a clear difference in the overall sound. We would never mistake a piano for a flute. This is what is meant by the **quality** of a sound. For musical instruments, the terms *timbre* and *tone color* are also used.

Just as loudness and pitch can be related to physically measurable quantities, so too can quality. The quality of a sound depends on the presence of overtones—their number and their relative amplitudes. Generally, when a note is played on a musical instrument, the fundamental as well as overtones are present simultaneously. Figure 12–14 illustrates how the *principle of superposition* (Section 11–12) applies to three wave forms, in this case the fundamental and first two overtones (with particular amplitudes): they add together at each point to give a composite *waveform*. Of course, more than two overtones are

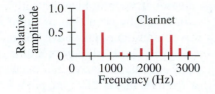

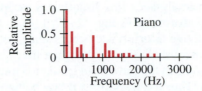

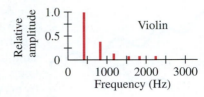

FIGURE 12–15 Sound spectra for different instruments. The spectra change when the instruments play different notes. The clarinet is a bit complicated: it acts like a closed tube at lower frequencies, having only odd harmonics, but at higher frequencies all harmonics occur as for an open tube.

usually present. [Any complex wave can be analyzed into a superposition of sinusoidal waves of appropriate amplitudes, wavelengths, and frequencies. Such an analysis is called a *Fourier analysis*.]

The relative amplitudes of the overtones for a given note are different for different musical instruments, which is what gives each instrument its characteristic quality or timbre. A bar graph showing the relative amplitudes of the harmonics for a given note produced by an instrument is called a *sound spectrum*. Several typical examples for different musical instruments are shown in Fig. 12–15. The fundamental usually has the greatest amplitude, and its frequency is what is heard as the pitch.

The manner in which an instrument is played strongly influences the sound quality. Plucking a violin string, for example, makes a very different sound than pulling a bow across it. The sound spectrum at the very start (or end) of a note (as when a hammer strikes a piano string) can be very different from the subsequent sustained tone. This too affects the subjective tone quality of an instrument.

An ordinary sound, like that made by striking two stones together, is a noise that has a certain quality, but a clear pitch is not discernible. Such a noise is a mixture of many frequencies which bear little relation to one another. A sound spectrum made of that noise would not show discrete lines like those of Fig. 12–15. Instead it would show a continuous, or nearly continuous, spectrum of frequencies. Such a sound we call "noise" in comparison with the more harmonious sounds which contain frequencies that are simple multiples of the fundamental.

FIGURE 12–16 Sound waves from two loudspeakers interfere.

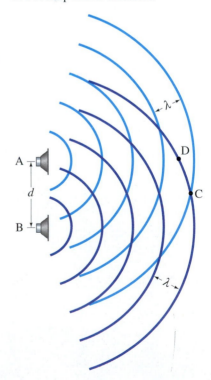

12–6 Interference of Sound Waves; Beats

Interference in Space

We saw in Section 11–12 that when two waves simultaneously pass through the same region of space, they interfere with one another. Interference also occurs with sound waves.

Consider two large loudspeakers, A and B, a distance d apart on the stage of an auditorium as shown in Fig. 12–16. Let us assume the two speakers are emitting sound waves of the same single frequency and that they are in phase: that is, when one speaker is forming a compression, so is the other. (We ignore reflections from walls, floor, etc.) The curved lines in the diagram represent the crests of sound waves from each speaker at one instant in time. We must remember that for a sound wave, a crest is a compression in the air whereas a trough—which falls between two crests—is a rarefaction. A person or detector at a point such as C, which is the same distance from each speaker, will experience a loud sound because the interference will be constructive—two crests reach it at one moment, two troughs reach it a moment later. On the other hand, at a point such as D in the diagram, little if any sound will be heard because destructive interference occurs—compressions of one wave meet rarefactions of the other and vice versa (see Fig. 11–37 and the related discussion on water waves in Section 11–12).

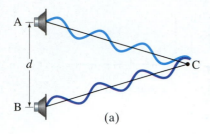

(a)

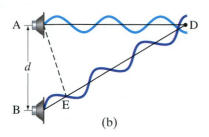

(b)

FIGURE 12–17 Sound waves of a single frequency from loudspeakers A and B (see Fig. 12–16) constructively interfere at C and destructively interfere at D. [Shown here are graphical representations, not the actual longitudinal sound waves.]

An analysis of this situation is perhaps clearer if we graphically represent the waveforms as in Fig. 12–17. In Fig. 12–17a it can be seen that at point C, constructive interference occurs since both waves simultaneously have crests or simultaneously have troughs when they arrive at C. In Fig. 12–17b we see that, to reach point D, the wave from speaker B must travel a greater distance than the wave from A. Thus the wave from B lags behind that from A. In this diagram, point E is chosen so that the distance ED is equal to AD. Thus we see that if the distance BE is equal to precisely one-half the wavelength of the sound, the two waves will be exactly out of phase when they reach D, and destructive interference occurs. This then is the criterion for determining at what points destructive interference occurs: destructive interference occurs at any point whose distance from one speaker is greater than its distance from the other speaker by one-half wavelength. Notice that if this extra distance (BE in Fig. 12–17b) is equal to a whole wavelength (or 2, 3, ··· wavelengths), then the two waves will be in phase and *constructive interference* occurs. If the distance BE equals $\frac{1}{2}, 1\frac{1}{2}, 2\frac{1}{2}, \cdots$ wavelengths, *destructive interference* occurs.

It is important to realize that a person sitting at point D in Fig. 12–16 or 12–17 hears nothing at all (or nearly so), yet sound is coming from both speakers. Indeed, if one of the speakers is turned off, the sound from the other speaker will be clearly heard.

If a loudspeaker emits a whole range of frequencies, only specific wavelengths will destructively interfere completely at a given point.

EXAMPLE 12–12 Loudspeakers' interference. Two loudspeakers are 1.00 m apart. A person stands 4.00 m from one speaker. How far must this person be from the second speaker to detect destructive interference when the speakers emit an 1150-Hz sound? Assume the temperature is 20°C.

APPROACH To sense destructive interference, the person must be one-half wavelength closer to or farther from one speaker than from the other—that is, at a distance = 4.00 m ± $\lambda/2$. We can determine λ since we know f and v.

SOLUTION The speed of sound at 20°C is 343 m/s, so the wavelength of this sound is (Eq. 11–12)

$$\lambda = \frac{v}{f} = \frac{343 \text{ m/s}}{1150 \text{ Hz}} = 0.30 \text{ m}.$$

For destructive interference to occur, the person must be one-half wavelength farther from one loudspeaker than from the other, or 0.15 m. Thus the person must be 3.85 m or 4.15 m from the second speaker.

NOTE If the speakers are less than 0.15 m apart, there will be no point that is 0.15 m farther from one speaker than the other, and there will be no point where destructive interference could occur.

Beats—Interference in Time

We have been discussing interference of sound waves that takes place in space. An interesting and important example of interference that occurs in time is the *Beats* phenomenon known as **beats**: If two sources of sound—say, two tuning forks—are close in frequency but not exactly the same, sound waves from the two sources interfere with each other. The sound level at a given position alternately rises and falls in time, because the two waves are sometimes in phase and sometimes out of phase due to their different wavelengths. The regularly spaced intensity changes are called beats.

To see how beats arise, consider two equal-amplitude sound waves of frequency $f_A = 50 \text{ Hz}$ and $f_B = 60 \text{ Hz}$, respectively. In 1.00 s, the first source makes 50 vibrations whereas the second makes 60. We now examine the waves at one point in space equidistant from the two sources. The waveforms for each wave as a function of time, at a fixed position, are shown on the top graph of Fig. 12–18; the magenta line represents the 50-Hz wave, and the blue line represents the 60-Hz wave.

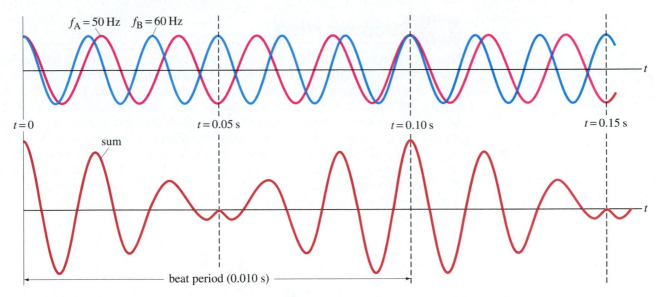

FIGURE 12–18 Beats occur as a result of the superposition of two sound waves of slightly different frequency.

The lower graph in Fig. 12–18 shows the sum of the two waves as a function of time. At time $t = 0$ the two waves are shown to be in phase and interfere constructively. Because the two waves vibrate at different rates, at time $t = 0.05$ s they are completely out of phase and interfere destructively. At $t = 0.10$ s, they are again in phase and the resultant amplitude again is large. Thus the resultant amplitude is large every 0.10 s and drops drastically in between. This rising and falling of the intensity is what is heard as beats.[†] In this case the beats are 0.10 s apart. That is, the **beat frequency** is ten per second, or 10 Hz. This result, that the beat frequency equals the difference in frequency of the two waves is valid in general.

Beat frequency = difference in the two wave frequencies

The phenomenon of beats can occur with any kind of wave and is a very sensitive method for comparing frequencies. For example, to tune a piano, a piano tuner listens for beats produced between his standard tuning fork and that of a particular string on the piano, and knows it is in tune when the beats disappear. The members of an orchestra tune up by listening for beats between their instruments and that of a standard tone (usually A above middle C at 440 Hz) produced by a piano or an oboe.

PHYSICS APPLIED
Tuning a piano

EXAMPLE 12–13 **Beats.** A tuning fork produces a steady 400-Hz tone. When this tuning fork is struck and held near a vibrating guitar string, twenty beats are counted in five seconds. What are the possible frequencies produced by the guitar string?

APPROACH For beats to occur, the string must vibrate at a frequency different from 400 Hz by whatever the beat frequency is.

SOLUTION The beat frequency is

$$f_{beat} = 20 \text{ vibrations}/5 \text{ s} = 4 \text{ Hz.}$$

This is the difference of the frequencies of the two waves. Because one wave is known to be 400 Hz, the other must be either 404 Hz or 396 Hz.

EXERCISE E What is the beat frequency for the tuning fork and guitar of Example 12–13 when 500-Hz and 506-Hz sounds are heard together?

[†]Beats will be heard even if the amplitudes are not equal, as long as the difference in amplitude is not great.

12-7 Doppler Effect

You may have noticed that you hear the pitch of the siren on a speeding firetruck drop abruptly as it passes you. Or you may have noticed the change in pitch of a blaring horn on a fast-moving car as it passes by you. The pitch of the engine noise of a race car changes as the car passes an observer. When a source of sound is moving toward an observer, the pitch the observer hears is higher than when the source is at rest; and when the source is traveling away from the observer, the pitch is lower. This phenomenon is known as the **Doppler effect**[†] and occurs for all types of waves. Let us now see why it occurs, and calculate the difference between the perceived and source frequencies when there is relative motion between source and observer.

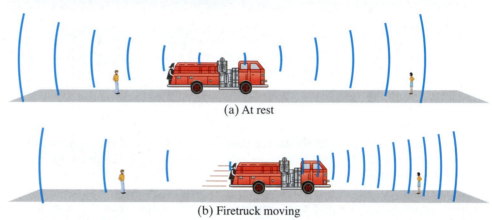

(a) At rest

(b) Firetruck moving

FIGURE 12–19   (a) Both observers on the sidewalk hear the same frequency from the firetruck at rest. (b) Doppler effect: observer toward whom the firetruck moves hears a higher-frequency sound, and observer behind the firetruck hears a lower-frequency sound.

FIGURE 12–20   Determination of the frequency shift in the Doppler effect (see text). The red dot is the source.

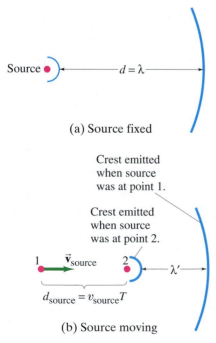

(a) Source fixed

Crest emitted when source was at point 1.

Crest emitted when source was at point 2.

(b) Source moving

Consider the siren of a firetruck at rest, which is emitting sound of a particular frequency in all directions as shown in Fig. 12–19a. The sound waves are moving at the speed of sound in air, v_{snd}, which is independent of the velocity of the source or observer. If our source, the firetruck, is moving, the siren emits sound at the same frequency as it does at rest. But the sound wavefronts it emits forward, in front of it, are closer together than when the firetruck is at rest, as shown in Fig. 12–19b. This is because the firetruck, as it moves, is "chasing" the previously emitted wavefronts, and emits each crest closer to the previous one. Thus an observer on the sidewalk in front of the truck will detect more wave crests passing per second, so the frequency heard is higher. The wavefronts emitted behind the truck, on the other hand, are farther apart than when the truck is at rest because the truck is speeding away from them. Hence, fewer wave crests per second pass by an observer behind the moving truck (Fig. 12–19b) and the perceived pitch is lower.

We can calculate the frequency shift perceived by making use of Fig. 12–20, and we assume the air (or other medium) is at rest in our reference frame. (The stationary observer is off to the right.) In Fig. 12–20a, the source of the sound is shown as a red dot, and is at rest. Two successive wave crests are shown, the second of which has just been emitted and so is still near the source. The distance between these crests is λ, the wavelength. If the frequency of the source is f, then the time between emissions of wave crests is

$$T = \frac{1}{f} = \frac{\lambda}{v_{snd}}.$$

In Fig. 12–20b, the source is moving with a velocity v_{source} toward the observer.

[†]After J. C. Doppler (1803–1853).

In a time T (as just defined), the first wave crest has moved a distance $d = v_{snd} T = \lambda$, where v_{snd} is the velocity of the sound wave in air (which is the same whether the source is moving or not). In this same time, the source has moved a distance $d_{source} = v_{source} T$. Then the distance between successive wave crests, which is the wavelength λ' the observer will perceive, is

Frequency change, moving source, fixed observer

$$\lambda' = d - d_{source}$$
$$= \lambda - v_{source} T$$
$$= \lambda - v_{source} \frac{\lambda}{v_{snd}}$$
$$= \lambda \left(1 - \frac{v_{source}}{v_{snd}} \right).$$

We subtract λ from both sides of this equation and find that the shift in wavelength, $\Delta\lambda$, is

$$\Delta\lambda = \lambda' - \lambda = -\lambda \frac{v_{source}}{v_{snd}}.$$

So the shift in wavelength is directly proportional to the source speed v_{source}. The frequency f' that will be perceived by our stationary observer on the ground is given by (Eq. 11–12)

$$f' = \frac{v_{snd}}{\lambda'} = \frac{v_{snd}}{\lambda \left(1 - \dfrac{v_{source}}{v_{snd}} \right)}.$$

Since $v_{snd}/\lambda = f$, then

$$f' = \frac{f}{\left(1 - \dfrac{v_{source}}{v_{snd}} \right)}. \qquad \begin{bmatrix} \text{source moving toward} \\ \text{stationary observer} \end{bmatrix} \quad \textbf{(12–2a)}$$

Because the denominator is less than 1, the observed frequency f' is greater than the source frequency f. That is, $f' > f$. For example, if a source emits a sound of frequency 400 Hz when at rest, then when the source moves toward a fixed observer with a speed of 30 m/s, the observer hears a frequency (at 20°C) of

$$f' = \frac{400\ \text{Hz}}{1 - \dfrac{30\ \text{m/s}}{343\ \text{m/s}}} = 438\ \text{Hz}.$$

Now consider a source moving *away* from the stationary observer at a speed v_{source}. Using the same arguments as above, the wavelength λ' perceived by our observer will have the minus sign on d_{source} (at top of this page) changed to plus:

$$\lambda' = d + d_{source}$$
$$= \lambda \left(1 + \frac{v_{source}}{v_{snd}} \right).$$

The difference between the observed and emitted wavelengths will be $\Delta\lambda = \lambda' - \lambda = +\lambda(v_{source}/v_{snd})$. The observed frequency of the wave, $f' = v_{snd}/\lambda'$, will be

$$f' = \frac{f}{\left(1 + \dfrac{v_{source}}{v_{snd}} \right)}. \qquad \begin{bmatrix} \text{source moving away from} \\ \text{stationary observer} \end{bmatrix} \quad \textbf{(12–2b)}$$

If a source emitting at 400 Hz is moving away from a fixed observer at 30 m/s, the observer hears a frequency $f' = (400\ \text{Hz})/\left[1 + (30\ \text{m/s})/(343\ \text{m/s})\right] = 368\ \text{Hz}$.

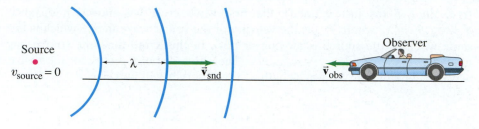

FIGURE 12–21 Observer moving with speed v_{obs} toward a stationary source detects wave crests passing at speed $v' = v_{snd} + v_{obs}$ where v_{snd} is the speed of the sound waves in air.

Frequency change, fixed source, moving observer

The Doppler effect also occurs when the source is at rest and the observer is in motion. If the observer is traveling *toward* the source, the pitch heard is higher than that of the emitted source frequency. If the observer is traveling *away* from the source, the pitch heard is lower. Quantitatively the change in frequency is different than for the case of a moving source. With a fixed source and a moving observer, the distance between wave crests, the wavelength λ, is not changed. But the velocity of the crests with respect to the observer *is* changed. If the observer is moving toward the source, Fig. 12–21, the speed v' of the waves relative to the observer is a simple addition of velocities: $v' = v_{snd} + v_{obs}$, where v_{snd} is the velocity of sound in air (we assume the air is still) and v_{obs} is the velocity of the observer. Hence, the frequency heard is

$$f' = \frac{v'}{\lambda} = \frac{v_{snd} + v_{obs}}{\lambda}.$$

Because $\lambda = v_{snd}/f$, then

$$f' = \frac{(v_{snd} + v_{obs})f}{v_{snd}},$$

or

$$f' = \left(1 + \frac{v_{obs}}{v_{snd}}\right)f. \qquad \begin{bmatrix} \text{observer moving toward} \\ \text{stationary source} \end{bmatrix} \quad \textbf{(12–3a)}$$

If the observer is moving away from the source, the relative velocity is $v' = v_{snd} - v_{obs}$, so

$$f' = \left(1 - \frac{v_{obs}}{v_{snd}}\right)f. \qquad \begin{bmatrix} \text{observer moving away} \\ \text{from stationary source} \end{bmatrix} \quad \textbf{(12–3b)}$$

EXAMPLE 12–14 **A moving siren.** The siren of a police car at rest emits at a predominant frequency of 1600 Hz. What frequency will you hear if you are at rest and the police car moves at 25.0 m/s (*a*) toward you, and (*b*) away from you?

APPROACH The observer is fixed, and the source moves, so we use Eqs. 12–2. The frequency you (the observer) hear is the emitted frequency f divided by the factor $(1 \pm v_{source}/v_{snd})$ where v_{source} is the speed of the police car. Use the minus sign when the car moves toward you (giving a higher frequency); use the plus sign when the car moves away from you (lower frequency).

SOLUTION (*a*) The car is moving toward you, so (Eq. 12–2a)

$$f' = \frac{f}{\left(1 - \dfrac{v_{source}}{v_{snd}}\right)} = \frac{1600 \text{ Hz}}{\left(1 - \dfrac{25.0 \text{ m/s}}{343 \text{ m/s}}\right)} = 1726 \text{ Hz}.$$

(*b*) The car is moving away from you, so

$$f' = \frac{f}{\left(1 + \dfrac{v_{source}}{v_{snd}}\right)} = \frac{1600 \text{ Hz}}{\left(1 + \dfrac{25.0 \text{ m/s}}{343 \text{ m/s}}\right)} = 1491 \text{ Hz}.$$

EXERCISE F Suppose the police car of Example 12–14 is at rest and emits still at 1600 Hz. What frequency would you hear if you were moving at 25.0 m/s (*a*) toward it, and (*b*) away from it?

When a sound wave is reflected from a moving obstacle, the frequency of the reflected wave will, because of the Doppler effect, be different from that of the incident wave. This is illustrated in the following Example.

EXAMPLE 12–15 Two Doppler shifts. A 5000-Hz sound wave is emitted by a stationary source. This sound wave reflects from an object moving 3.50 m/s toward the source (Fig. 12–22). What is the frequency of the wave reflected by the moving object as detected by a detector at rest near the source?

APPROACH There are actually two Doppler shifts in this situation. First, the moving object acts like an observer moving toward the source with speed $v_{obs} = 3.50$ m/s (Fig. 12–22a) and so "detects" a sound wave of frequency (Eq. 12–3a) $f' = f[1 + (v_{obs}/v_{snd})]$. Second, reflection of the wave from the moving object is equivalent to the object reemitting the wave, acting effectively as a moving source with speed $v_{source} = 3.50$ m/s (Fig. 12–22b). The final frequency detected, f'', is given by $f'' = f'/[1 - v_{source}/v_{snd}]$, Eq.12–2a.

SOLUTION The frequency f' that is "detected" by the moving object is (Eq. 12–3a):

$$f' = \left(1 + \frac{v_{obs}}{v_{snd}}\right)f = \left(1 + \frac{3.50 \text{ m/s}}{343 \text{ m/s}}\right)(5000 \text{ Hz}) = 5051 \text{ Hz}.$$

The moving object now "emits" (reflects) a sound of frequency (Eq. 12–2a)

$$f'' = \frac{f'}{\left(1 - \dfrac{v_{source}}{v_{snd}}\right)} = \frac{5051 \text{ Hz}}{\left(1 - \dfrac{3.50 \text{ m/s}}{343 \text{ m/s}}\right)} = 5103 \text{ Hz}.$$

Thus the frequency shifts by 103 Hz.

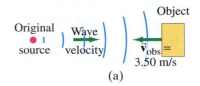

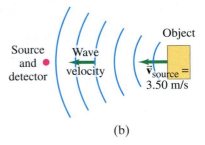

(b)

FIGURE 12–22 Example 12–15.

The incident wave and the reflected wave in Example 12–15, when mixed together (say, electronically), interfere with one another and beats are produced. The beat frequency is equal to the difference in the two frequencies, 103 Hz. This Doppler technique is used in a variety of medical applications, usually with ultrasonic waves in the megahertz frequency range. For example, ultrasonic waves reflected from red blood cells can be used to determine the velocity of blood flow. Similarly, the technique can be used to detect the movement of the chest of a young fetus and to monitor its heartbeat.

PHYSICS APPLIED
Doppler blood-flow meter and other medical uses

For convenience, we can write Eqs. 12–2 and 12–3 as a single equation that covers all cases of both source and observer in motion:

$$f' = f\left(\frac{v_{snd} \pm v_{obs}}{v_{snd} \mp v_{source}}\right). \tag{12–4}$$

Source and observer moving

PROBLEM SOLVING
Getting the signs right

To get the signs right, recall from your own experience that the frequency is higher when observer and source approach each other, and lower when they move apart. Thus the upper signs in numerator and denominator apply if source and/or observer move toward each other; the lower signs apply if they are moving apart.

* Doppler Effect for Light

The Doppler effect occurs for other types of waves as well. Light and other types of electromagnetic waves (such as radar) exhibit the Doppler effect: although the formulas for the frequency shift are not identical to Eqs. 12–2 and 12–3, as we shall see in Chapter 33, the effect is similar. One important application is for weather forecasting using radar. The time delay between the emission of radar pulses and their reception after being reflected off raindrops gives the position of precipitation. Measuring the Doppler shift in frequency (as in Example 12–15) tells how fast the storm is moving and in which direction.

PHYSICS APPLIED
Doppler effect for EM waves and weather forecasting

Another important application is to astronomy, where the velocities of distant galaxies can be determined from the Doppler shift. Light from distant galaxies is shifted toward lower frequencies, indicating that the galaxies are moving away from us. This is called the **redshift** since red has the lowest frequency of visible light. The greater the frequency shift, the greater the velocity of recession. It is found that the farther the galaxies are from us, the faster they move away. This observation is the basis for the idea that the universe is expanding, and is one basis for the idea that the universe began as a great explosion, affectionately called the "Big Bang" (see Chapter 33).

*12–8 Shock Waves and the Sonic Boom

An object such as an airplane traveling faster than the speed of sound is said to have a **supersonic speed**. Such a speed is often given as a **Mach**[†] **number**, which is defined as the ratio of the speed of the object to the speed of sound in the surrounding medium. For example, a plane traveling 600 m/s high in the atmosphere, where the speed of sound is only 300 m/s, has a speed of Mach 2.

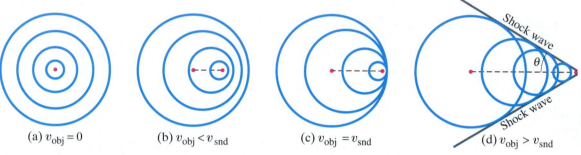

(a) $v_{obj} = 0$ (b) $v_{obj} < v_{snd}$ (c) $v_{obj} = v_{snd}$ (d) $v_{obj} > v_{snd}$

FIGURE 12–23 Sound waves emitted by an object (a) at rest or (b, c, and d) moving. (b) If the object's velocity is less than the velocity of sound, the Doppler effect occurs; (d) if its velocity is greater than the velocity of sound, a shock wave is produced.

Shock wave

When a source of sound moves at subsonic speeds (less than the speed of sound), the pitch of the sound is altered as we have seen (the Doppler effect); see also Fig. 12–23a and b. But if a source of sound moves faster than the speed of sound, a more dramatic effect known as a **shock wave** occurs. In this case the source is actually "outrunning" the waves it produces. As shown in Fig. 12–23c, when the source is traveling *at* the speed of sound, the wave fronts it emits in the forward direction "pile up" directly in front of it. When the object moves faster, at a supersonic speed, the wave fronts pile up on one another along the sides, as shown in Fig. 12–23d. The different wave crests overlap one another and form a single very large crest which is the shock wave. Behind this very large crest there is usually a very large trough. A shock wave is essentially the result of constructive interference of a large number of wave fronts. A shock wave in air is analogous to the bow wave of a boat traveling faster than the speed of the water waves it produces, Fig. 12–24.

When an airplane travels at supersonic speeds, the noise it makes and its disturbance of the air form into a shock wave containing a tremendous amount of sound energy. When the shock wave passes a listener, it is heard as a loud *sonic boom*. A sonic boom lasts only a fraction of a second, but the energy it contains is often sufficient to break windows and cause other damage. Actually, a sonic boom is made up of two or more booms since major shock waves can form at the front and the rear of the aircraft, as well as at the wings, etc. (Fig. 12–25). Bow waves of a boat are also multiple, as can be seen in Fig. 12–24.

FIGURE 12–24 Bow waves produced by a boat.

[†] After the Austrian physicist Ernst Mach (1838–1916).

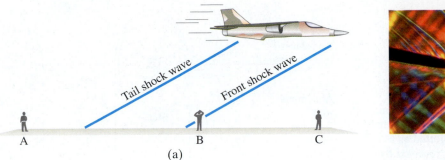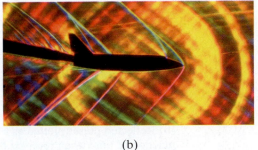

(a)   (b)

FIGURE 12–25 (a) The (double) sonic boom has already been heard by person A on the left. It is just being heard by person B in the center. And it will shortly be heard by person C on the right. (b) Special photo of supersonic aircraft showing shock waves produced in the air. (Several closely spaced shock waves are produced by different parts of the aircraft.)

When an aircraft approaches the speed of sound, it encounters a barrier of sound waves in front of it (see Fig. 12–23c). To exceed the speed of sound, the aircraft needs extra thrust to pass through this "sound barrier." This is called "breaking the sound barrier." Once a supersonic speed is attained, this barrier no longer impedes the motion. It is sometimes erroneously thought that a sonic boom is produced only at the moment an aircraft is breaking through the sound barrier. Actually, a shock wave follows the aircraft at all times it is traveling at supersonic speeds. A series of observers on the ground will each hear a loud "boom" as the shock wave passes, Fig. 12–25. The shock wave consists of a cone whose apex is at the aircraft. The angle of this cone, θ (see Fig. 12–23d), is given by

$$\sin \theta = \frac{v_{\text{snd}}}{v_{\text{obj}}}, \tag{12-5}$$

where v_{obj} is the velocity of the object (the aircraft) and v_{snd} is the velocity of sound in the medium. (The proof is left as Problem 63.)

* 12–9 Applications: Sonar, Ultrasound, and Medical Imaging

* Sonar

The reflection of sound is used in many applications to determine distance. The **sonar**[†] or pulse-echo technique is used to locate underwater objects. A transmitter sends out a sound pulse through the water, and a detector receives its reflection, or echo, a short time later. This time interval is carefully measured, and from it the distance to the reflecting object can be determined since the speed of sound in water is known. The depth of the sea and the location of reefs, sunken ships, submarines, or schools of fish can be determined in this way. The interior structure of the Earth is studied in a similar way by detecting reflections of waves traveling through the Earth whose source was a deliberate explosion (called "soundings"). An analysis of waves reflected from various structures and boundaries within the Earth reveals characteristic patterns that are also useful in the exploration for oil and minerals.

PHYSICS APPLIED
Sonar: depth finding, Earth soundings

Sonar generally makes use of **ultrasonic** frequencies: that is, waves whose frequencies are above 20 kHz, beyond the range of human detection. For sonar, the frequencies are typically in the range 20 kHz to 100 kHz. One reason for using ultrasound waves, other than the fact that they are inaudible, is that for shorter wavelengths there is less diffraction (Section 11–15) so the beam spreads less and smaller objects can be detected.

[†]Sonar stands for "*so*und *na*vigation *ra*nging."

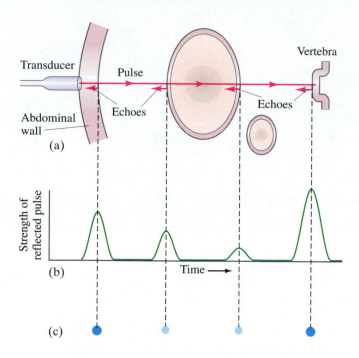

FIGURE 12–26 (a) Ultrasound pulse passes through the abdomen, reflecting from surfaces in its path. (b) Reflected pulses plotted as a function of time when received by transducer. The vertical dashed lines point out which reflected pulse goes with which surface. (c) Dot display for the same echoes: brightness of each dot is related to signal strength.

* Ultrasound Medical Imaging

The diagnostic use of ultrasound in medicine, in the form of images (sometimes called *sonograms*) is an important and interesting application of physical principles. A **pulse-echo technique** is used, much like sonar, except that the frequencies used are in the range of 1 to 10 MHz ($1\,\text{MHz} = 10^6\,\text{Hz}$). A high-frequency sound pulse is directed into the body, and its reflections from boundaries or interfaces between organs and other structures and lesions in the body are then detected. Tumors and other abnormal growths, or pockets of fluid, can be distinguished; the action of heart valves and the development of a fetus can be examined; and information about various organs of the body, such as the brain, heart, liver, and kidneys, can be obtained. Although ultrasound does not replace X-rays, for certain kinds of diagnosis it is more helpful. Some kinds of tissue or fluid are not detected in X-ray photographs, but ultrasound waves are reflected from their boundaries. "Real-time" ultrasound images are like a movie of a section of the interior of the body.

The pulse-echo technique for medical imaging works as follows. A brief pulse of ultrasound is emitted by a transducer that transforms an electrical pulse into a sound-wave pulse. Part of the pulse is reflected as echoes at each interface in the body, and most of the pulse (usually) continues on, Fig. 12–26a. The detection of reflected pulses by the same transducer can then be displayed on the screen of a display terminal or monitor. The time elapsed from when the pulse is emitted to when each reflection (echo) is received is proportional to the distance to the reflecting surface. For example, if the distance from transducer to the vertebra is 25 cm, the pulse travels a round-trip distance of $2 \times 25\,\text{cm} = 0.50\,\text{m}$. The speed of sound in human tissue is about 1540 m/s (close to that of sea water), so the time taken is

$$t = \frac{d}{v} = \frac{(0.50\,\text{m})}{(1540\,\text{m/s})} = 320\,\mu\text{s}.$$

The *strength* of a reflected pulse depends mainly on the difference in density of the two materials on either side of the interface and can be displayed as a pulse or as a dot (Figs. 12–26b and c). Each echo dot (Fig. 12–26c) can be represented as a point whose position is given by the time delay and whose brightness depends on the strength of the echo. A two-dimensional image can then be formed out of these dots from a series of scans. The transducer is

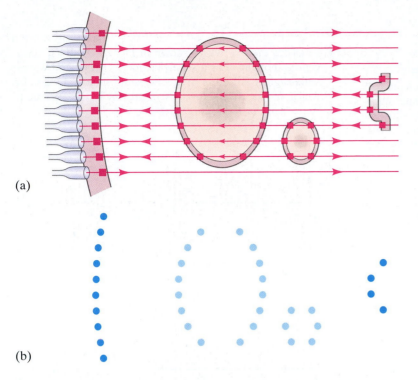

(a)

(b)

FIGURE 12–27 (a) Ten traces are made across the abdomen by moving the transducer, or by using an array of transducers. (b) The echoes are plotted as dots to produce the image. More closely spaced traces would give a more detailed image.

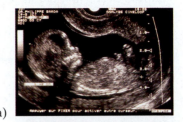

(a)

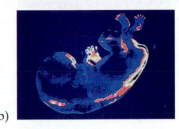

(b)

FIGURE 12–28 (a) Ultrasound image of a human fetus (with head at the left) within the uterus. (b) False-color high-resolution ultrasound image of a fetus. (Different colors represent different intensities of reflected pulses.)

moved, or an array of transducers is used, each of which sends out a pulse at each position and receives echoes as shown in Fig. 12–27. Each trace can be plotted, spaced appropriately one below the other, to form an image on a display terminal as shown in Fig. 12–27b. Only 10 lines are shown in Fig. 12–27, so the image is crude. More lines give a more precise image.[†] Photographs of ultrasound images are shown in Fig. 12–28.

[†] *Radar* used for aircraft involves a similar pulse-echo technique except that it uses electromagnetic (EM) waves, which, like light, travel with a speed of 3×10^8 m/s.

Summary

Sound travels as a longitudinal wave in air and other materials. In air, the speed of sound increases with temperature; at 20°C, it is about 343 m/s.

The **pitch** of a sound is determined by the frequency; the higher the frequency, the higher the pitch.

The **audible range** of frequencies for humans is roughly 20 Hz to 20,000 Hz (1 Hz = 1 cycle per second).

The **loudness** or **intensity** of a sound is related to the amplitude squared of the wave. Because the human ear can detect sound intensities from 10^{-12} W/m² to over 1 W/m², sound levels are specified on a logarithmic scale. The **sound level** β, specified in decibels, is defined in terms of intensity I as

$$\beta = 10 \log\left(\frac{I}{I_0}\right), \tag{12–1}$$

where the reference intensity I_0 is usually taken to be 10^{-12} W/m².

Musical instruments are simple sources of sound in which *standing waves* are produced.

The strings of a stringed instrument may vibrate as a whole with nodes only at the ends; the frequency at which this standing wave occurs is called the **fundamental**. The fundamental frequency corresponds to a wavelength equal to twice the length of the string, $\lambda_1 = 2L$. The string can also vibrate at higher frequencies, called **overtones** or **harmonics**, in which there are one or more additional nodes. The frequency of each harmonic is a whole-number multiple of the fundamental.

In wind instruments, standing waves are set up in the column of air within the tube.

The vibrating air in an **open tube** (open at both ends) has displacement antinodes at both ends. The fundamental frequency corresponds to a wavelength equal to twice the tube length: $\lambda_1 = 2L$. The harmonics have frequencies that are 1, 2, 3, 4, ⋯ times the fundamental frequency, just as for strings.

For a **closed tube** (closed at one end), the fundamental corresponds to a wavelength four times the length of the tube: $\lambda_1 = 4L$. Only the odd harmonics are present, equal to 1, 3, 5, 7, ⋯ times the fundamental frequency.

Sound waves from different sources can interfere with each other. If two sounds are at slightly different frequencies, **beats** can be heard at a frequency equal to the difference in frequency of the two sources.

The **Doppler effect** refers to the change in pitch of a sound due to the motion either of the source or of the listener. If source and listener are approaching each other, the perceived pitch is higher; if they are moving apart, the perceived pitch is lower.

[*Shock waves and a sonic boom occur when an object moves at a supersonic speed—faster than the speed of sound. Ultrasonic-frequency (higher than 20 kHz) sound waves are used in many applications, including sonar and medical imaging.]

Questions

1. What is the evidence that sound travels as a wave?

2. What is the evidence that sound is a form of energy?

3. Children sometimes play with a homemade "telephone" by attaching a string to the bottoms of two paper cups. When the string is stretched and a child speaks into one cup, the sound can be heard at the other cup (Fig. 12–29). Explain clearly how the sound wave travels from one cup to the other.

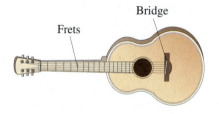

FIGURE 12–29 Question 3.

4. When a sound wave passes from air into water, do you expect the frequency or wavelength to change?

5. What evidence can you give that the speed of sound in air does not depend significantly on frequency?

6. The voice of a person who has inhaled helium sounds very high-pitched. Why?

7. How will the air temperature in a room affect the pitch of organ pipes?

8. Explain how a tube might be used as a filter to reduce the amplitude of sounds in various frequency ranges. (An example is a car muffler.)

9. Why are the frets on a guitar (Fig. 12–30) spaced closer together as you move up the fingerboard toward the bridge?

FIGURE 12–30
Question 9.

10. A noisy truck approaches you from behind a building. Initially you hear it but cannot see it. When it emerges and you do see it, its sound is suddenly "brighter"—you hear more of the high-frequency noise. Explain. [*Hint*: See Section 11–15 on diffraction.]

11. Standing waves can be said to be due to "interference in space," whereas beats can be said to be due to "interference in time." Explain.

12. In Fig. 12–16, if the frequency of the speakers were lowered, would the points D and C (where destructive and constructive interference occur) move farther apart or closer together?

13. Traditional methods of protecting the hearing of people who work in areas with very high noise levels have consisted mainly of efforts to block or reduce noise levels. With a relatively new technology, headphones are worn that do not block the ambient noise. Instead, a device is used which detects the noise, inverts it electronically, then feeds it to the headphones *in addition to* the ambient noise. How could adding *more* noise reduce the sound levels reaching the ears?

14. Consider the two waves shown in Fig. 12–31. Each wave can be thought of as a superposition of two sound waves with slightly different frequencies, as in Fig. 12–18. In which of the waves, (a) or (b), are the two component frequencies farther apart? Explain.

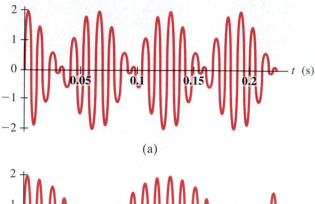

(a)

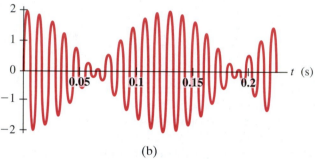

(b)

FIGURE 12–31 Question 14.

15. Is there a Doppler shift if the source and observer move in the same direction, with the same velocity? Explain.

16. If a wind is blowing, will this alter the frequency of the sound heard by a person at rest with respect to the source? Is the wavelength or velocity changed?

17. Figure 12–32 shows various positions of a child in motion on a swing. A monitor is blowing a whistle in front of the child on the ground. At which position, A through E, will the child hear the highest frequency for the sound of the whistle? Explain your reasoning.

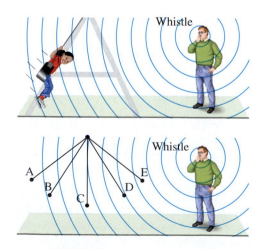

FIGURE 12–32 Question 17.

Problems

[Unless stated otherwise, assume $T = 20°C$ and $v_{sound} = 343 \, m/s$ in air.]

12–1 Characteristics of Sound

1. (I) A hiker determines the length of a lake by listening for the echo of her shout reflected by a cliff at the far end of the lake. She hears the echo 2.0 s after shouting. Estimate the length of the lake.

2. (I) A sailor strikes the side of his ship just below the waterline. He hears the echo of the sound reflected from the ocean floor directly below 2.5 s later. How deep is the ocean at this point? Assume the speed of sound in seawater is 1560 m/s (Table 12–1) and does not vary significantly with depth.

3. (I) (a) Calculate the wavelengths in air at 20°C for sounds in the maximum range of human hearing, 20 Hz to 20,000 Hz. (b) What is the wavelength of a 10-MHz ultrasonic wave?

4. (II) An ocean fishing boat is drifting just above a school of tuna on a foggy day. Without warning, an engine backfire occurs on another boat 1.0 km away (Fig. 12–33). How much time elapses before the backfire is heard (a) by the fish, and (b) by the fishermen?

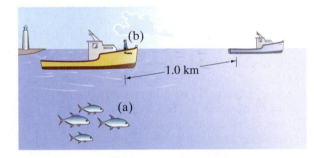

FIGURE 12–33 Problem 4.

5. (II) A stone is dropped from the top of a cliff. The splash it makes when striking the water below is heard 3.5 s later. How high is the cliff?

6. (II) A person, with his ear to the ground, sees a huge stone strike the concrete pavement. A moment later two sounds are heard from the impact: one travels in the air and the other in the concrete, and they are 1.1 s apart. How far away did the impact occur? See Table 12–1.

7. (II) Calculate the percent error made over one mile of distance by the "5-second rule" for estimating the distance from a lightning strike if the temperature is (a) 30°C, and (b) 10°C.

12–2 Intensity of Sound; Decibels

8. (I) What is the intensity of a sound at the pain level of 120 dB? Compare it to that of a whisper at 20 dB.

9. (I) What is the sound level of a sound whose intensity is $2.0 \times 10^{-6} \, W/m^2$?

10. (II) If two firecrackers produce a sound level of 95 dB when fired simultaneously at a certain place, what will be the sound level if only one is exploded? [*Hint*: Add intensities, not dB's.]

11. (II) A person standing a certain distance from an airplane with four equally noisy jet engines is experiencing a sound level bordering on pain, 120 dB. What sound level would this person experience if the captain shut down all but one engine? [*Hint*: Add intensities, not dB's.]

12. (II) A cassette player is said to have a signal-to-noise ratio of 58 dB, whereas for a CD player it is 95 dB. What is the ratio of intensities of the signal and the background noise for each device?

13. (II) (a) Estimate the power output of sound from a person speaking in normal conversation. Use Table 12–2. Assume the sound spreads roughly uniformly over a sphere centered on the mouth. (b) How many people would it take to produce a total sound output of 100 W of ordinary conversation? [*Hint*: Add intensities, not dB's.]

14. (II) A 50-dB sound wave strikes an eardrum whose area is $5.0 \times 10^{-5} \, m^2$. (a) How much energy is absorbed by the eardrum per second? (b) At this rate, how long would it take your eardrum to receive a total energy of 1.0 J?

15. (II) Expensive amplifier A is rated at 250 W, while the more modest amplifier B is rated at 40 W. (a) Estimate the sound level in decibels you would expect at a point 3.5 m from a loudspeaker connected in turn to each amp. (b) Will the expensive amp sound twice as loud as the cheaper one?

16. (II) At a rock concert, a dB meter registered 130 dB when placed 2.8 m in front of a loudspeaker on the stage. (a) What was the power output of the speaker, assuming uniform spherical spreading of the sound and neglecting absorption in the air? (b) How far away would the sound level be a somewhat reasonable 90 dB?

17. (II) Human beings can typically detect a difference in sound level of 2.0 dB. What is the ratio of the amplitudes of two sounds whose levels differ by this amount? [*Hint*: See Section 11–9.]

18. (II) If the amplitude of a sound wave is tripled, (a) by what factor will the intensity increase? (b) By how many dB will the sound level increase?

* 19. (II) Two sound waves have equal displacement amplitudes, but one has twice the frequency of the other. What is the ratio of their intensities?

* 20. (II) What would be the sound level (in dB) of a sound wave in air that corresponds to a displacement amplitude of vibrating air molecules of 0.13 mm at 300 Hz?

* 12–3 Loudness

* 21. (I) A 6000-Hz tone must have what sound level to seem as loud as a 100-Hz tone that has a 50-dB sound level? (See Fig. 12–6.)

* 22. (I) What are the lowest and highest frequencies that an ear can detect when the sound level is 30 dB? (See Fig. 12–6.)

* 23. (II) Your auditory system can accommodate a huge range of sound levels. What is the ratio of highest to lowest intensity at (a) 100 Hz, (b) 5000 Hz? (See Fig. 12–6.)

12–4 Sources of Sound: Strings and Air Columns

24. (I) The A string on a violin has a fundamental frequency of 440 Hz. The length of the vibrating portion is 32 cm, and it has a mass of 0.35 g. Under what tension must the string be placed?

25. (I) An organ pipe is 112 cm long. What are the fundamental and first three audible overtones if the pipe is (a) closed at one end, and (b) open at both ends?

26. (I) (a) What resonant frequency would you expect from blowing across the top of an empty soda bottle that is 18 cm deep, if you assumed it was a closed tube? (b) How would that change if it was one-third full of soda?

27. (I) If you were to build a pipe organ with open-tube pipes spanning the range of human hearing (20 Hz to 20 kHz), what would be the range of the lengths of pipes required?

28. (II) A tight guitar string has a frequency of 540 Hz as its third harmonic. What will be its fundamental frequency if it is fingered at a length of only 60% of its original length?

29. (II) An unfingered guitar string is 0.73 m long and is tuned to play E above middle C (330 Hz). (a) How far from the end of this string must a fret (and your finger) be placed to play A above middle C (440 Hz)? (b) What is the wavelength on the string of this 440-Hz wave? (c) What are the frequency and wavelength of the sound wave produced in air at 20°C by this fingered string?

30. (II) (a) Determine the length of an open organ pipe that emits middle C (262 Hz) when the temperature is 21°C. (b) What are the wavelength and frequency of the fundamental standing wave in the tube? (c) What are λ and f in the traveling sound wave produced in the outside air?

31. (II) An organ is in tune at 20°C. By what percent will the frequency be off at 5.0°C?

32. (II) How far from the mouthpiece of the flute in Example 12–10 should the hole be that must be uncovered to play D above middle C at 294 Hz?

33. (II) (a) At $T = 20°C$, how long must an open organ pipe be to have a fundamental frequency of 294 Hz? (b) If this pipe is filled with helium, what is its fundamental frequency?

34. (II) A particular organ pipe can resonate at 264 Hz, 440 Hz, and 616 Hz, but not at any other frequencies in between. (a) Show why this is an open or a closed pipe. (b) What is the fundamental frequency of this pipe?

35. (II) A uniform narrow tube 1.80 m long is open at both ends. It resonates at two successive harmonics of frequencies 275 Hz and 330 Hz. What is (a) the fundamental frequency, and (b) the speed of sound in the gas in the tube?

36. (II) A pipe in air at 20°C is to be designed to produce two successive harmonics at 240 Hz and 280 Hz. How long must the pipe be, and is it open or closed?

37. (II) How many overtones are present within the audible range for a 2.14-m-long organ pipe at 20°C (a) if it is open, and (b) if it is closed?

38. (III) The human ear canal is approximately 2.5 cm long. It is open to the outside and is closed at the other end by the eardrum. Estimate the frequencies (in the audible range) of the standing waves in the ear canal. What is the relationship of your answer to the information in the graph of Fig. 12–6?

12–6 Interference; Beats

39. (I) A piano tuner hears one beat every 2.0 s when trying to adjust two strings, one of which is sounding 440 Hz. How far off in frequency is the other string?

40. (I) What is the beat frequency if middle C (262 Hz) and C# (277 Hz) are played together? What if each is played two octaves lower (each frequency reduced by a factor of 4)?

41. (I) A certain dog whistle operates at 23.5 kHz, while another (brand X) operates at an unknown frequency. If neither whistle can be heard by humans when played separately, but a shrill whine of frequency 5000 Hz occurs when they are played simultaneously, estimate the operating frequency of brand X.

42. (II) A guitar string produces 4 beats/s when sounded with a 350-Hz tuning fork and 9 beats/s when sounded with a 355-Hz tuning fork. What is the vibrational frequency of the string? Explain your reasoning.

43. (II) Two violin strings are tuned to the same frequency, 294 Hz. The tension in one string is then decreased by 2.0%. What will be the beat frequency heard when the two strings are played together? [Hint: Recall Eq. 11–13.]

44. (II) How many beats will be heard if two identical flutes each try to play middle C (262 Hz), but one is at 5.0°C and the other at 25.0°C?

45. (II) You have three tuning forks, A, B, and C. Fork B has a frequency of 441 Hz; when A and B are sounded together, a beat frequency of 3 Hz is heard. When B and C are sounded together, the beat frequency is 4 Hz. What are the possible frequencies of A and C? What beat frequencies are possible when A and C are sounded together?

46. (II) Two loudspeakers are 1.80 m apart. A person stands 3.00 m from one speaker and 3.50 m from the other. (a) What is the lowest frequency at which destructive interference will occur at this point? (b) Calculate two other frequencies that also result in destructive interference at this point (give the next two highest). Let $T = 20°C$.

47. (III) Two piano strings are supposed to be vibrating at 132 Hz, but a piano tuner hears three beats every 2.0 s when they are played together. (a) If one is vibrating at 132 Hz, what must be the frequency of the other (is there only one answer)? (b) By how much (in percent) must the tension be increased or decreased to bring them in tune?

48. (III) A source emits sound of wavelengths 2.64 m and 2.76 m in air. How many beats per second will be heard? (Assume $T = 20°C$.)

12–7 Doppler Effect

49. (I) The predominant frequency of a certain fire engine's siren is 1550 Hz when at rest. What frequency do you detect if you move with a speed of 30.0 m/s (a) toward the fire engine, and (b) away from it?

50. (I) You are standing still. What frequency do you detect if a fire engine whose siren emits at 1550 Hz moves at a speed of 32 m/s (a) toward you, or (b) away from you?

51. (II) (a) Compare the shift in frequency if a 2000-Hz source is moving toward you at 15 m/s, versus you moving toward it at 15 m/s. Are the two frequencies exactly the same? Are they close? (b) Repeat the calculation for 150 m/s and then again (c) for 300 m/s. What can you conclude about the asymmetry of the Doppler formulas?

52. (II) Two automobiles are equipped with the same single-frequency horn. When one is at rest and the other is moving toward the first at 15 m/s, the driver at rest hears a beat frequency of 5.5 Hz. What is the frequency the horns emit? Assume $T = 20°C$.

53. (II) A bat at rest sends out ultrasonic sound waves at 50.0 kHz and receives them returned from an object moving directly away from it at 25.0 m/s. What is the received sound frequency?

54. (II) A bat flies toward a wall at a speed of 5.0 m/s. As it flies, the bat emits an ultrasonic sound wave with frequency 30.0 kHz. What frequency does the bat hear in the reflected wave?

55. (II) In one of the original Doppler experiments, a tuba was played on a moving flat train car at a frequency of 75 Hz, and a second identical tuba played the same tone while at rest in the railway station. What beat frequency was heard if the train car approached the station at a speed of 10.0 m/s?

56. (II) A *Doppler flow meter* uses ultrasound waves to measure blood-flow speeds. Suppose the device emits sound at 3.5 MHz, and the speed of sound in human tissue is taken to be 1540 m/s. What is the expected beat frequency if blood is flowing in large leg arteries at 2.0 cm/s directly away from the sound source?

57. (III) The Doppler effect using ultrasonic waves of frequency 2.25×10^6 Hz is used to monitor the heartbeat of a fetus. A (maximum) beat frequency of 500 Hz is observed. Assuming that the speed of sound in tissue is 1.54×10^3 m/s, calculate the maximum velocity of the surface of the beating heart.

58. (III) A factory whistle emits sound of frequency 570 Hz. When the wind velocity is 12.0 m/s from the north, what frequency will observers hear who are located, at rest, (a) due north, (b) due south, (c) due east, and (d) due west, of the whistle? What frequency is heard by a cyclist heading (e) north or (f) west, toward the whistle at 15.0 m/s? Assume $T = 20°C$.

* 12–8 Shock Waves; Sonic Boom

* 59. (I) (a) How fast is an object moving on land if its speed at 20°C is Mach 0.33? (b) A high-flying jet cruising at 3000 km/h displays a Mach number of 3.2 on a screen. What is the speed of sound at that altitude?

* 60. (II) An airplane travels at Mach 2.3 where the speed of sound is 310 m/s. (a) What is the angle the shock wave makes with the direction of the airplane's motion? (b) If the plane is flying at a height of 7100 m, how long after it is directly overhead will a person on the ground hear the shock wave?

* 61. (II) A space probe enters the thin atmosphere of a planet where the speed of sound is only about 35 m/s. (a) What is the probe's Mach number if its initial speed is 15,000 km/h? (b) What is the angle of the shock wave relative to the direction of motion?

* 62. (II) A meteorite traveling 8500 m/s strikes the ocean. Determine the shock wave angle it produces (a) in the air just before entering the ocean, and (b) in the water just after entering. Assume $T = 20°C$.

* 63. (II) Show that the angle θ a sonic boom makes with the path of a supersonic object is given by Eq. 12–5.

* 64. (II) You look directly overhead and see a plane exactly 1.5 km above the ground flying faster than the speed of sound. By the time you hear the sonic boom, the plane has traveled a horizontal distance of 2.0 km. See Fig. 12–34. Determine (a) the angle of the shock cone, θ, and (b) the speed of the plane (the Mach number). Assume the speed of sound is 330 m/s.

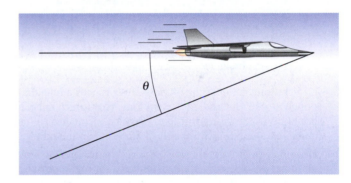

FIGURE 12–34 Problem 64.

General Problems

65. A fish finder uses a sonar device that sends 20,000-Hz sound pulses downward from the bottom of the boat, and then detects echoes. If the maximum depth for which it is designed to work is 200 m, what is the minimum time between pulses (in fresh water)?

66. Approximately how many octaves are there in the human audible range?

67. A science museum has a display called a sewer pipe symphony. It consists of many plastic pipes of various lengths, which are open on both ends. (a) If the pipes have lengths of 3.0 m, 2.5 m, 2.0 m, 1.5 m and 1.0 m, what frequencies will be heard by a visitor's ear placed near the ends of the pipes? (b) Why does this display work better on a noisy day than on a quiet day?

68. A single mosquito 5.0 m from a person makes a sound close to the threshold of human hearing (0 dB). What will be the sound level of 1000 such mosquitoes?

69. What is the resultant sound level when an 82-dB sound and an 87-dB sound are heard simultaneously?

70. The sound level 12.0 m from a loudspeaker, placed in the open, is 105 dB. What is the acoustic power output (W) of the speaker, assuming it radiates equally in all directions?

71. A stereo amplifier is rated at 150 W output at 1000 Hz. The power output drops by 10 dB at 15 kHz. What is the power output in watts at 15 kHz?

72. Workers around jet aircraft typically wear protective devices over their ears. Assume that the sound level of a jet airplane engine, at a distance of 30 m, is 140 dB, and that the average human ear has an effective radius of 2.0 cm. What would be the power intercepted by an unprotected ear at a distance of 30 m from a jet airplane engine?

73. In audio and communications systems, the *gain*, β, in decibels is defined as

$$\beta = 10 \log\left(\frac{P_{\text{out}}}{P_{\text{in}}}\right),$$

where P_{in} is the power input to the system and P_{out} is the power output. A particular stereo amplifier puts out 100 W of power for an input of 1 mW. What is its gain in dB?

74. Each string on a violin is tuned to a frequency $1\frac{1}{2}$ times that of its neighbor. The four equal-length strings are to be placed under the same tension; what must be the mass per unit length of each string relative to that of the lowest string?

75. The A string of a violin is 32 cm long between fixed points with a fundamental frequency of 440 Hz and a mass per unit length of 6.1×10^{-4} kg/m. (*a*) What are the wave speed and tension in the string? (*b*) What is the length of the tube of a simple wind instrument (say, an organ pipe) closed at one end whose fundamental is also 440 Hz if the speed of sound is 343 m/s in air? (*c*) What is the frequency of the first overtone of each instrument?

76. A tuning fork is set into vibration above a vertical open tube filled with water (Fig. 12–35). The water level is allowed to drop slowly. As it does so, the air in the tube above the water level is heard to resonate with the tuning fork when the distance from the tube opening to the water level is 0.125 m and again at 0.395 m. What is the frequency of the tuning fork?

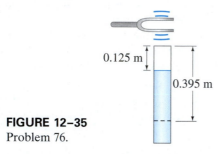

FIGURE 12–35
Problem 76.

77. A 75-cm-long guitar string of mass 2.10 g is near a tube that is open at one end and also 75 cm long. How much tension should be in the string if it is to produce resonance (in its fundamental mode) with the third harmonic in the tube?

78. (II) A highway overpass was observed to resonate as one full loop $\left(\frac{1}{2}\lambda\right)$ when a small earthquake shook the ground vertically at 4.0 Hz. The highway department put a support at the center of the overpass, anchoring it to the ground as shown in Fig. 12–36. What resonant frequency would you now expect for the overpass? Earthquakes rarely do significant shaking above 5 or 6 Hz. Did the modifications do any good?

Before modification

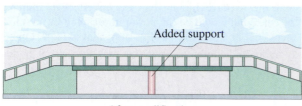

Added support

After modification

FIGURE 12–36   Problem 78.

79. A person hears a pure tone in the 500–1000-Hz range coming from two sources. The sound is loudest at points equidistant from the two sources. To determine exactly what the frequency is, the person moves about and finds that the sound level is minimal at a point 0.34 m farther from one source than the other. What is the frequency of the sound?

80. Two trains emit 424-Hz whistles. One train is stationary. The conductor on the stationary train hears a 3.0-Hz beat frequency when the other train approaches. What is the speed of the moving train?

81. The frequency of a steam train whistle as it approaches you is 538 Hz. After it passes you, its frequency is measured as 486 Hz. How fast was the train moving (assume constant velocity)?

82. At a race track, you can estimate the speed of cars just by listening to the difference in pitch of the engine noise between approaching and receding cars. Suppose the sound of a certain car drops by a full octave (frequency halved) as it goes by on the straightaway. How fast is it going?

83. Two open organ pipes, sounding together, produce a beat frequency of 11 Hz. The shorter one is 2.40 m long. How long is the other?

84. Two loudspeakers are at opposite ends of a railroad car as it moves past a stationary observer at 10.0 m/s, as shown in Fig. 12–37. If the speakers have identical sound frequencies of 212 Hz, what is the beat frequency heard by the observer when (*a*) he listens from the position A, in front of the car, (*b*) he is between the speakers, at B, and (*c*) he hears the speakers after they have passed him, at C?

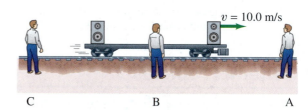

FIGURE 12–37   Problem 84.

85. If the velocity of blood flow in the aorta is normally about 0.32 m/s, what beat frequency would you expect if 5.50-MHz ultrasound waves were directed along the flow and reflected from the red blood cells? Assume that the waves travel with a speed of 1.54×10^3 m/s.

86. A bat flies toward a moth at speed 6.5 m/s while the moth is flying toward the bat at speed 5.0 m/s. The bat emits a sound wave of 51.35 kHz. What is the frequency of the wave detected by the bat after that wave reflects off the moth?

87. A bat emits a series of high frequency sound pulses as it approaches a moth. The pulses are approximately 70.0 ms apart, and each is about 3.0 ms long. How far away can the moth be detected by the bat so that the echo from one chirp returns before the next chirp is emitted?

88. The "alpenhorn" (Fig. 12–38) was once used to send signals from one Alpine village to another. Since lower frequency sounds are less susceptible to intensity loss, long horns were used to create deep sounds. When played as a musical instrument, the alpenhorn must be blown in such a way that only one of the overtones is resonating. The most popular alpenhorn is about 3.4 m long, and it is called the F sharp (or G flat) horn. What is the fundamental frequency of this horn, and which overtone is close to F sharp? (See Table 12–3.) Model as an open tube.

FIGURE 12–38 Problem 88.

89. Room acoustics for stereo listening can be compromised by the presence of standing waves, which can cause acoustic "dead spots" at the locations of the pressure nodes. Consider a living room 5.0 m long, 4.0 m wide, and 2.8 m high. Calculate the fundamental frequencies for the standing waves in this room.

90. A dramatic demonstration, called "singing rods," involves a long, slender aluminum rod held in the hand near the rod's midpoint. The rod is stroked with the other hand. With a little practice, the rod can be made to "sing," or emit a clear, loud, ringing sound. For a 90-cm-long rod, (a) what is the fundamental frequency of the sound? (b) What is its wavelength in the rod, and (c) what is the traveling wavelength in air at 20°C?

* 91. The intensity at the threshold of hearing for the human ear at a frequency of about 1000 Hz is $I_0 = 1.0 \times 10^{-12} \text{ W/m}^2$, for which β, the sound level, is 0 dB. The threshold of pain at the same frequency is about 120 dB, or $I = 1.0 \text{ W/m}^2$, corresponding to an increase of intensity by a factor of 10^{12}. By what factor does the displacement amplitude, A, vary?

* 92. A plane is traveling at Mach 2.0. An observer on the ground hears the sonic boom 1.5 min after the plane passes directly overhead. What is the plane's altitude?

* 93. The wake of a speedboat is 15° in a lake where the speed of the water wave is 2.2 km/h. What is the speed of the boat?

Answers to Exercises

A: 1 km for every 3 s before the thunder is heard.
B: 4 times as intense.
C: One-quarter its original value; 6 dB.

D: 257 Hz.
E: 6 Hz.
F: (a) 1717 Hz, (b) 1483 Hz.

Heating the air inside a "hot-air" balloon raises the air's temperature, causing it to expand and forcing air out the opening at the bottom. The reduced amount of gas inside means its density is lower, so there is a net buoyant force upward on the balloon. In this Chapter we study temperature and its effects on matter: thermal expansion and the gas laws. Most important is the ideal gas law and its expression in terms of molecules.

Temperature and Kinetic Theory

This Chapter is the first of three (Chapters 13, 14, and 15) that are devoted to the subjects of temperature, heat, and thermodynamics. Much of this Chapter will be devoted to an investigation of the theory that matter is made up of atoms and that these atoms are in continuous random motion. This theory is called the *kinetic theory*. ("Kinetic," you may recall, is Greek for "moving.")

We also discuss the concept of temperature and how it is measured, as well as the experimentally measured properties of gases which serve as a foundation for the kinetic theory.

13–1 Atomic Theory of Matter

The idea that matter is made up of atoms dates back to the ancient Greeks. According to the Greek philosopher Democritus, if a pure substance—say, a piece of iron—were cut into smaller and smaller bits, eventually a smallest piece of that substance would be obtained which could not be divided further. This smallest piece was called an **atom**, which in Greek means "indivisible."[†]

Atomic theory—the evidence

Today the atomic theory is generally accepted. The experimental evidence in its favor, however, came mainly in the eighteenth, nineteenth, and twentieth centuries, and much of it was obtained from the analysis of chemical reactions.

[†]Today we don't consider the atom as indivisible, but rather as consisting of a nucleus (containing protons and neutrons) and electrons.

We will often speak of the relative masses of atoms and molecules—what we call the **atomic mass** or **molecular mass**, respectively.[†] These are based on arbitrarily assigning the abundant carbon atom, ^{12}C, the value of exactly 12.0000 **unified atomic mass units** (u). In terms of kilograms,

$$1\,u = 1.6605 \times 10^{-27}\,kg.$$

The atomic mass of hydrogen is then 1.0078 u, and the values for other atoms are as listed in the periodic table inside the back cover of this book, and also in Appendix B. The molecular mass of a compound is the sum of atomic masses of the atoms making up the molecules of that compound.[‡]

An important piece of evidence for the atomic theory is called **Brownian motion**, named after the biologist Robert Brown, who is credited with its discovery in 1827. While he was observing tiny pollen grains suspended in water under his microscope, Brown noticed that the tiny grains moved about in tortuous paths (Fig. 13–1), even though the water appeared to be perfectly still. The atomic theory easily explains Brownian motion if the further reasonable assumption is made that the atoms of any substance are continually in motion. Then Brown's tiny pollen grains are jostled about by the vigorous barrage of rapidly moving molecules of water.

In 1905, Albert Einstein examined Brownian motion from a theoretical point of view and was able to calculate from the experimental data the approximate size and mass of atoms and molecules. His calculations showed that the diameter of a typical atom is about 10^{-10} m.

At the start of Chapter 10, we distinguished the three common states, or phases, of matter—solid, liquid, gas—based on **macroscopic**, or "large-scale," properties. Now let us see how these three phases of matter differ, from the atomic or **microscopic** point of view. Clearly, atoms and molecules must exert attractive forces on each other. For how else could a brick or a piece of aluminum hold together in one piece? The attractive forces between molecules are of an electrical nature (more on this in later Chapters). When molecules come too close together, the force between them must become repulsive (electric repulsion between their outer electrons), for how else could matter take up space? Thus molecules maintain a minimum distance from each other. In a solid material, the attractive forces are strong enough that the atoms or molecules move only slightly (oscillate) about relatively fixed positions, often in an array known as a crystal lattice, as shown in Fig. 13–2a. In a liquid, the atoms or molecules are moving more rapidly, or the forces between them are weaker, so that they are sufficiently free to pass over one another, as in Fig. 13–2b. In a gas, the forces are so weak, or the speeds so high, that the molecules do not even stay close together. They move rapidly every which way, Fig. 13–2c, filling any container and occasionally

FIGURE 13–1 Path of a tiny particle (pollen grain, for example) suspended in water. The straight lines connect observed positions of the particle at equal time intervals.

Atomic and molecular masses

Phases of matter

Macroscopic vs. microscopic properties

[†]The terms *atomic weight* and *molecular weight* are sometimes used for these quantities, but properly speaking we are comparing masses.

[‡]An *element* is a substance, such as gold, iron, or copper, that cannot be broken down into simpler substances by chemical means. *Compounds* are substances made up of elements, and can be broken down into them; examples are carbon dioxide and water. The smallest piece of an element is an atom; the smallest piece of a compound is a molecule. Molecules are made up of atoms; a molecule of water, for example, is made up of two atoms of hydrogen and one of oxygen; its chemical formula is H_2O.

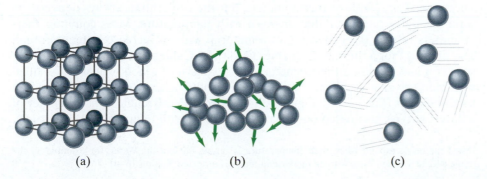

(a)   (b)   (c)

FIGURE 13–2 Atomic arrangements in (a) a crystalline solid, (b) a liquid, and (c) a gas.

colliding with one another. On the average, the speeds are sufficiently high in a gas that when two molecules collide, the force of attraction is not strong enough to keep them close together and they fly off in new directions.

EXAMPLE 13–1 ESTIMATE **Distance between atoms.** The density of copper is $8.9 \times 10^3 \, \text{kg/m}^3$, and each copper atom has a mass of 63 u. Estimate the average distance between neighboring copper atoms.

APPROACH We consider a cube of copper 1 m on a side. From the given density we can calculate the mass of a 1-m^3 cube. We divide this by the mass of one atom (63 u) to obtain the number of atoms in 1 m^3. Let N be the number of atoms in a 1-m length; then $(N)(N)(N) = N^3$ equals this total number of atoms in 1 m^3.

SOLUTION The mass of 1 copper atom is $63 \, \text{u} = 63 \times 1.66 \times 10^{-27} \, \text{kg} = 1.05 \times 10^{-25} \, \text{kg}$. This means that in a cube of copper 1 m on a side (volume = 1 m^3), there are

$$\frac{8.9 \times 10^3 \, \text{kg/m}^3}{1.05 \times 10^{-25} \, \text{kg/atom}} = 8.5 \times 10^{28} \, \text{atoms/m}^3.$$

The volume of a cube of side l is $V = l^3$, so on one edge of the 1-m-long cube there are $(8.5 \times 10^{28})^{\frac{1}{3}}$ atoms $= 4.4 \times 10^9$ atoms. Hence the distance between neighboring atoms is

$$\frac{1 \, \text{m}}{4.4 \times 10^9 \, \text{atoms}} = 2.3 \times 10^{-10} \, \text{m}.$$

NOTE Watch out for units. Even though "atoms" is not a unit, it is helpful to include it to make sure you calculate correctly.

13–2 Temperature and Thermometers

In everyday life, **temperature** is a measure of how hot or cold something is. A hot oven is said to have a high temperature, whereas the ice of a frozen lake is said to have a low temperature.

Many properties of matter change with temperature. For example, most materials expand when heated.[†] An iron beam is longer when hot than when cold. Concrete roads and sidewalks expand and contract slightly according to temperature, which is why compressible spacers or expansion joints (Fig. 13–3) are placed at regular intervals. The electrical resistance of matter changes with temperature (see Chapter 18). So too does the color radiated by objects, at least at high temperatures: you may have noticed that the heating element of an electric stove glows with a red color when hot. At higher temperatures, solids such as iron glow orange or even white. The white light from an ordinary incandescent lightbulb comes from an extremely hot tungsten wire. The surface temperatures of the Sun and other stars can be measured by the predominant color (more precisely, wavelengths) of light they emit.

Instruments designed to measure temperature are called **thermometers**. There are many kinds of thermometers, but their operation always depends on some property of matter that changes with temperature. Most common thermometers rely on the expansion of a material with an increase in temperature. The first idea for a thermometer (Fig. 13–4a), by Galileo, made use of the expansion of a gas. Common thermometers today consist of a hollow glass tube filled with mercury or with alcohol colored with a red dye, as were the earliest usable thermometers (Fig. 13–4b). Figure 13–4c shows an early clinical thermometer of a different type, also based on a change in density with temperature.

FIGURE 13–3 Expansion joint on a bridge.

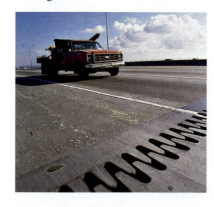

[†] Most materials expand when their temperature is raised, but not all. Water, for example, in the range 0°C to 4°C contracts with an increase in temperature (see Section 13–4).

(a)

(b)

(c)

FIGURE 13–4 (a) Model of Galileo's original idea for a thermometer. (b) Actual thermometers built by the Accademia del Cimento (1657–1667) in Florence are among the earliest known. These sensitive and exquisite instruments contained alcohol, sometimes colored, like many thermometers today. (c) Clinical thermometers in the shape of a frog, also built by the Accademia del Cimento, could be tied to a patient's wrist. The small spheres suspended in the liquid each have a slightly different density. The number of spheres that would sink was a measure of the patient's fever.

In the common liquid-in-glass thermometer, the liquid expands more than the glass when the temperature is increased, so the liquid level rises in the tube (Fig. 13–5a). Although metals also expand with temperature, the change in length of a metal rod, say, is generally too small to measure accurately for ordinary changes in temperature. However, a useful thermometer can be made by bonding together two dissimilar metals whose rates of expansion are different (Fig. 13–5b). When the temperature is increased, the different amounts of expansion cause the bimetallic strip to bend. Often the bimetallic strip is in the form of a coil, one end of which is fixed while the other is attached to a pointer, Fig. 13–6. This kind of thermometer is used as ordinary air thermometers, oven thermometers, automatic off switches in electric coffeepots, and in room thermostats for determining when the heater or air conditioner should go on or off. Very precise thermometers make use of electrical properties (Chapter 18), such as resistance thermometers, thermo-couples, and thermistors, often with a digital readout.

Temperature Scales

In order to measure temperature quantitatively, some sort of numerical scale must be defined. The most common scale today is the **Celsius** scale, sometimes called the **centigrade** scale. In the United States, the **Fahrenheit** scale is also common. The most important scale in scientific work is the absolute, or Kelvin, scale, and it will be discussed later in this Chapter.

One way to define a temperature scale is to assign arbitrary values to two readily reproducible temperatures. For both the Celsius and Fahrenheit scales these two fixed points are chosen to be the freezing point and the boiling point[†] of water, both taken at atmospheric pressure. On the Celsius scale, the freezing

---

[†]The freezing point of a substance is defined as that temperature at which the solid and liquid phases coexist in equilibrium—that is, without any net liquid changing into the solid or vice versa. Experimentally, this is found to occur at only one definite temperature, for a given pressure. Similarly, the boiling point is defined as that temperature at which the liquid and gas coexist in equilibrium. Since these points vary with pressure, the pressure must be specified (usually it is 1 atm).

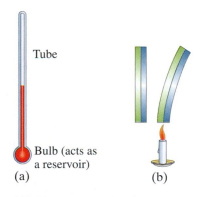
Tube

Bulb (acts as a reservoir)

(a)   (b)

FIGURE 13–5 (a) Mercury- or alcohol-in-glass thermometer; (b) bimetallic strip.

FIGURE 13–6 Photograph of a thermometer using a coiled bimetallic strip.

point of water is chosen to be 0°C ("zero degrees Celsius") and the boiling point 100°C. On the Fahrenheit scale, the freezing point is defined as 32°F and the boiling point 212°F. A practical thermometer is calibrated by placing it in carefully prepared environments at each of the two temperatures and marking the position of the liquid or pointer. For a Celsius scale, the distance between the two marks is divided into one hundred equal intervals representing each degree between 0°C and 100°C (hence the name "centigrade scale" meaning "hundred steps"). For a Fahrenheit scale, the two points are labeled 32°F and 212°F and the distance between them is divided into 180 equal intervals. For temperatures below the freezing point of water and above the boiling point of water, the scales may be extended using the same equally spaced intervals. However, thermometers can be used only over a limited temperature range because of their own limitations—for example, the liquid mercury in a mercury-in-glass thermometer solidifies at some point, below which the thermometer will be useless. It is also rendered useless above temperatures where the fluid vaporizes. For very low or very high temperatures, specialized thermometers are required, some of which we will mention later.

Every temperature on the Celsius scale corresponds to a particular temperature on the Fahrenheit scale, Fig. 13–7. It is easy to convert from one to the other if you remember that 0°C corresponds to 32°F and that a range of 100° on the Celsius scale corresponds to a range of 180° on the Fahrenheit scale. Thus, one Fahrenheit degree (1 F°) corresponds to $100/180 = \frac{5}{9}$ of a Celsius degree (1 C°). That is, $1\,\text{F}° = \frac{5}{9}\,\text{C}°$. (Notice that when we refer to a specific temperature, we say "degrees Celsius," as in 20°C; but when we refer to a *change* in temperature or a temperature interval, we say "Celsius degrees," as in "2 C°.") The conversion between the two temperature scales can be written

$$T(°C) = \tfrac{5}{9}\big[T(°F) - 32\big] \quad \text{or} \quad T(°F) = \tfrac{9}{5}T(°C) + 32.$$

Rather than memorizing these relations (it would be easy to confuse them), it is usually easier simply to remember that $0°C = 32°F$ and that a change of $5\,\text{C}° = $ a change of $9\,\text{F}°$.

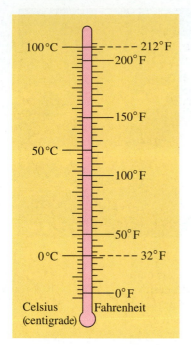

FIGURE 13–7 Celsius and Fahrenheit scales compared.

EXAMPLE 13–2 **Taking your temperature.** Normal body temperature is 98.6°F. What is this on the Celsius scale?

APPROACH We recall that $0°C = 32°F$ and $5\,\text{C}° = 9\,\text{F}°$.

SOLUTION First we relate the given temperature to the freezing point of water (0°C). That is, 98.6°F is $98.6 - 32.0 = 66.6\,\text{F}°$ above the freezing point of water. Since each F° is equal to $\frac{5}{9}\text{C}°$, this corresponds to $66.6 \times \frac{5}{9} = 37.0$ Celsius degrees above the freezing point. The freezing point is 0°C, so the temperature is 37.0°C.

EXERCISE A Determine the temperature at which both scales agree $(T_C = T_F)$.

Different materials do not expand in quite the same way over a wide temperature range. Consequently, if we calibrate different kinds of thermometers exactly as described above, they will not usually agree precisely. Because of how we calibrated them, they will agree at 0°C and at 100°C. But because of different expansion properties, they may not agree precisely at intermediate temperatures (remember we arbitrarily divided the thermometer scale into 100 equal divisions between 0°C and 100°C). Thus a carefully calibrated mercury-in-glass thermometer might register 52.0°C, whereas a carefully calibrated thermometer of another type might read 52.6°C.

Because of this discrepancy, some standard kind of thermometer must be chosen so that these intermediate temperatures can be precisely defined. The chosen standard for this purpose is the **constant-volume gas thermometer**. As shown in the simplified diagram of Fig. 13–8, this thermometer consists of a

FIGURE 13–8 Constant-volume gas thermometer.

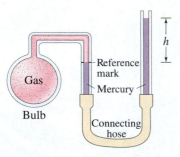

bulb filled with a dilute gas connected by a thin tube to a mercury manometer. The volume of the gas is kept constant by raising or lowering the right-hand tube of the manometer so that the mercury in the left-hand tube coincides with the reference mark. An increase in temperature causes a proportional increase in pressure in the bulb. Thus the tube must be lifted higher to keep the gas volume constant. The height of the mercury in the right-hand column is then a measure of the temperature. This thermometer gives the same results for all gases in the limit of reducing the gas pressure in the bulb toward zero. The resulting scale serves as a basis for the standard temperature scale.

* 13–3 Thermal Equilibrium and the Zeroth Law of Thermodynamics

We are all familiar with the fact that if two objects at different temperatures are placed in thermal contact (meaning thermal energy can transfer from one to the other), the two objects will eventually reach the same temperature. They are then said to be in **thermal equilibrium**. For example, you leave a fever thermometer in your mouth until it comes into thermal equilibrium with that environment, and then you read it. Two objects are defined to be in thermal equilibrium if, when placed in thermal contact, no energy flows from one to the other, and their temperatures don't change. Experiments indicate that *if two systems are in thermal equilibrium with a third system, then they are in thermal equilibrium with each other*. This postulate is called the **zeroth law of thermodynamics**. It has this unusual name since it was not until after the great first and second laws of thermodynamics (Chapter 15) were worked out that scientists realized that this apparently obvious postulate needed to be stated first.

Temperature is a property of a system that determines whether the system will be in thermal equilibrium with other systems. When two systems are in thermal equilibrium, their temperatures are, by definition, equal, and no net thermal energy will be exchanged between them. This is consistent with our everyday notion of temperature, since when a hot object and a cold one are put into contact, they eventually come to the same temperature. Thus the importance of the zeroth law is that it allows a useful definition of temperature.

13–4 Thermal Expansion

Most substances expand when heated and contract when cooled. However, the amount of expansion or contraction varies, depending on the material.

Linear Expansion

Experiments indicate that the change in length ΔL of almost all solids is, to a good approximation, directly proportional to the change in temperature ΔT, as long as ΔT is not too large. As might be expected, the change in length is also proportional to the original length of the object, L_0, Fig. 13–9. That is, for the same temperature change, a 4-m-long iron rod will increase in length twice as much as a 2-m-long iron rod. We can write this proportionality as an equation:

$$\Delta L = \alpha L_0 \, \Delta T, \tag{13–1a}$$

where α, the proportionality constant, is called the *coefficient of linear expansion* for the particular material and has units of $(C°)^{-1}$. We set $L = L_0 + \Delta L$, and rewrite this equation as

$$L = L_0(1 + \alpha \, \Delta T), \tag{13–1b}$$

where L_0 is the length initially, at temperature T_0, and L is the length after heating or cooling to a temperature T. If the temperature change $\Delta T = T - T_0$ is negative, then $\Delta L = L - L_0$ is also negative; thus the length shortens as the temperature decreases.

FIGURE 13–9 A thin rod of length L_0 at temperature T_0 is heated to a new uniform temperature T and acquires length L, where $L = L_0 + \Delta L$.

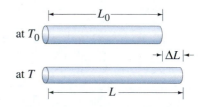

Linear expansion

TABLE 13–1 Coefficients of Expansion, near 20°C

Material	Coefficient of Linear Expansion, α $(C°)^{-1}$	Coefficient of Volume Expansion, β $(C°)^{-1}$
Solids		
Aluminum	25×10^{-6}	75×10^{-6}
Brass	19×10^{-6}	56×10^{-6}
Copper	17×10^{-6}	50×10^{-6}
Gold	14×10^{-6}	42×10^{-6}
Iron or steel	12×10^{-6}	35×10^{-6}
Lead	29×10^{-6}	87×10^{-6}
Glass (Pyrex®)	3×10^{-6}	9×10^{-6}
Glass (ordinary)	9×10^{-6}	27×10^{-6}
Quartz	0.4×10^{-6}	1×10^{-6}
Concrete and brick	$\approx 12 \times 10^{-6}$	$\approx 36 \times 10^{-6}$
Marble	$1.4 - 3.5 \times 10^{-6}$	$4 - 10 \times 10^{-6}$
Liquids		
Gasoline		950×10^{-6}
Mercury		180×10^{-6}
Ethyl alcohol		1100×10^{-6}
Glycerin		500×10^{-6}
Water		210×10^{-6}
Gases		
Air (and most other gases at atmospheric pressure)		3400×10^{-6}

The values of α for various materials at 20°C are listed in Table 13–1. Actually, α does vary slightly with temperature (which is why thermometers made of different materials do not agree precisely). However, if the temperature range is not too great, the variation can usually be ignored.

PHYSICS APPLIED

Expansion in structures

EXAMPLE 13–3 **Bridge expansion.** The steel bed of a suspension bridge is 200 m long at 20°C. If the extremes of temperature to which it might be exposed are $-30°C$ to $+40°C$, how much will it contract and expand?

APPROACH We assume the bridge bed will expand and contract linearly with temperature, as given by Eq. 13–1a.

SOLUTION From Table 13–1, we find that $\alpha = 12 \times 10^{-6}(C°)^{-1}$ for steel. The increase in length when it is at 40°C will be

$$\Delta L = \alpha L_0 \Delta T = (12 \times 10^{-6}/C°)(200 \text{ m})(40°C - 20°C) = 4.8 \times 10^{-2} \text{ m},$$

or 4.8 cm. When the temperature decreases to $-30°C$, $\Delta T = -50 \, C°$. Then

$$\Delta L = (12 \times 10^{-6}/C°)(200 \text{ m})(-50 \, C°) = -12.0 \times 10^{-2} \text{ m},$$

or a decrease in length of 12 cm. The total range the expansion joints must accommodate is 12 cm + 4.8 cm $\approx$ 17 cm.

CONCEPTUAL EXAMPLE 13–4 **Do holes expand or contract?** If you heat a thin, circular ring (Fig. 13–10a) in the oven, does the ring's hole get larger or smaller?

RESPONSE You might guess that the metal expands into the hole, making the hole smaller. But it is not so. Imagine the ring is solid, like a coin (Fig. 13–10b). Draw a circle on it with a pen as shown. When the metal expands, the material inside the circle will expand along with the rest of the metal; so the circle expands. Cutting the metal where the circle is makes clear to us that the hole increases in diameter.

FIGURE 13–10 Example 13–4.

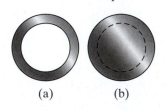

(a)   (b)

EXAMPLE 13–5 **Ring on a rod.** An iron ring is to fit snugly on a cylindrical iron rod. At 20°C, the diameter of the rod is 6.445 cm and the inside diameter of the ring is 6.420 cm. To slip over the rod, the ring must be slightly larger than the rod diameter by about 0.008 cm. To what temperature must the ring be brought if its hole is to be large enough so it will slip over the rod?

APPROACH The hole in the ring must be increased from a diameter of 6.420 cm to 6.445 cm + 0.008 cm = 6.453 cm. The ring must be heated since the hole diameter will increase linearly with temperature (as in Example 13–4).

SOLUTION We solve for ΔT in Eq. 13–1a and find

$$\Delta T = \frac{\Delta L}{\alpha L_0} = \frac{6.453 \text{ cm} - 6.420 \text{ cm}}{(12 \times 10^{-6}/\text{C}°)(6.420 \text{ cm})} = 430 \text{ C}°.$$

So it must be raised at least to $T = (20°\text{C} + 430 \text{ C}°) = 450°\text{C}$.

NOTE In doing Problems, don't forget the last step, adding in the initial temperature (20°C here).

CONCEPTUAL EXAMPLE 13–6 **Opening a tight jar lid.** When the lid of a glass jar is tight, holding the lid under hot water for a short time will often make it easier to open. Why?

PHYSICS APPLIED
Opening a tight lid

RESPONSE The lid may be struck by the hot water more directly than the glass and so expand sooner. But even if not, metals generally expand more than glass for the same temperature change (α is greater—see Table 13–1).

Volume Expansion

The change in *volume* of a material which undergoes a temperature change is given by a relation similar to Eq. 13–1a, namely,

$$\Delta V = \beta V_0 \Delta T, \tag{13–2}$$

Volume expansion

where ΔT is the change in temperature, V_0 is the original volume, ΔV is the change in volume, and β is the *coefficient of volume expansion*. The units of β are $(\text{C}°)^{-1}$.

Values of β for various materials are given in Table 13–1. Notice that for solids, β is normally equal to approximately 3α (work Problem 19 to see why). For solids that are not isotropic (that is, not having the same properties in all directions), the relation $\beta \approx 3\alpha$ is not valid. (Note that linear expansion has no meaning for liquids and gases since they do not have fixed shapes.)

$\beta \approx 3\alpha$

EXAMPLE 13–7 **Gas tank in the sun.** The 70-L steel gas tank of a car is filled to the top with gasoline at 20°C. The car sits in the sun and the tank reaches a temperature of 40°C (104°F). How much gasoline do you expect to overflow from the tank?

PHYSICS APPLIED
Gas tank overflow

APPROACH Both the gasoline and the tank expand as the temperature increases, and we assume they do so linearly as described by Eq. 13–2. The volume of overflowing gasoline equals the volume increase of the gasoline minus the increase in volume of the tank.

SOLUTION The gasoline expands by

$$\Delta V = \beta V_0 \Delta T = (950 \times 10^{-6} \text{ C}°^{-1})(70 \text{ L})(40°\text{C} - 20°\text{C}) = 1.3 \text{ L}.$$

The tank also expands. We can think of it as a steel shell that undergoes volume expansion $(\beta \approx 3\alpha = 36 \times 10^{-6} \text{ C}°^{-1})$. If the tank were solid, the surface layer (the shell) would expand just the same. Thus the tank increases in volume by

$$\Delta V = (36 \times 10^{-6} \text{ C}°^{-1})(70 \text{ L})(40°\text{C} - 20°\text{C}) = 0.050 \text{ L},$$

so the tank expansion has little effect. More than a liter of gas could spill out.

NOTE Want to save a few pennies? Fill your gas tank when it is cool and the gas is denser—more molecules for the same price. But don't fill the tank all the way.

Equations 13–1 and 13–2 are accurate only if ΔL (or ΔV) is small compared to L_0 (or V_0). This is of particular concern for liquids and even more so for gases because of the large values of β. Furthermore, β itself varies substantially with temperature for gases. Therefore, a better description of volume changes for gases is needed, as we will discuss starting in Section 13–6.

Anomalous Behavior of Water Below 4°C

Most substances expand more or less uniformly with an increase in temperature, as long as no phase change occurs. Water, however, does not follow the usual pattern. If water at 0°C is heated, it actually *decreases* in volume until it reaches 4°C. Above 4°C water behaves normally and expands in volume as the temperature is increased, Fig. 13–11. Water thus has its greatest density at 4°C. This anomalous behavior of water is of great importance for the survival of aquatic life during cold winters. When the water in a lake or river is above 4°C and begins to cool by contact with cold air, the water at the surface sinks because of its greater density. It is replaced by warmer water from below. This mixing continues until the temperature reaches 4°C. As the surface water cools further, it remains on the surface because it is less dense than the 4°C water below. Water then freezes first at the surface, and the ice remains on the surface since ice (specific gravity = 0.917) is less dense than water. The water at the bottom remains liquid unless it is so cold that the whole body of water freezes. If water were like most substances, becoming more dense as it cools, the water at the bottom of a lake would be frozen first. Lakes would freeze solid more easily since circulation would bring the warmer water to the surface to be efficiently cooled. The complete freezing of a lake would cause severe damage to its plant and animal life. Because of the unusual behavior of water below 4°C, it is rare for any large body of water to freeze completely, and this is helped by the layer of ice on the surface which acts as an insulator to reduce the flow of heat out of the water into the cold air above. Without this peculiar but wonderful property of water, life on this planet as we know it might not have been possible.

PHYSICS APPLIED

Life under ice

Not only does water expand as it cools from 4°C to 0°C, it expands even more as it freezes to ice. This is why ice cubes float in water and pipes break when water inside them freezes.

FIGURE 13–11 Behavior of water as a function of temperature near 4°C. (a) Volume of 1.00000 gram of water as a function of temperature. (b) Density vs. temperature. [Note the breaks in each axis.]

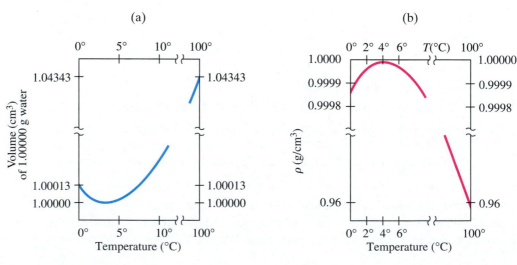

* 13–5 Thermal Stresses

In many situations, such as in buildings and roads, the ends of a beam or slab of material are rigidly fixed, which greatly limits expansion or contraction. If the temperature should change, large compressive or tensile stresses, called *thermal stresses*, will occur. The magnitude of such stresses can be calculated using the concept of elastic modulus developed in Chapter 9. To calculate the internal stress, we can think of this process as occurring in two steps. The beam tries to expand (or contract) by an amount ΔL given by Eq. 13–1; secondly, the solid in contact with the beam exerts a force to compress (or expand) it, keeping it at its original length. The force F required is given by Eq. 9–4:

$$\Delta L = \frac{1}{E}\frac{F}{A}L_0,$$

where E is Young's modulus for the material. To calculate the internal stress, F/A, we then set ΔL in Eq. 13–1a equal to ΔL in the equation above and find

$$\alpha L_0 \, \Delta T = \frac{1}{E}\frac{F}{A}L_0.$$

Hence, the stress

$$\frac{F}{A} = \alpha E \, \Delta T.$$

EXAMPLE 13–8 **Stress in concrete on a hot day.** A highway is to be made of blocks of concrete 10 m long placed end to end with no space between them to allow for expansion. If the blocks were placed at a temperature of 10°C, what compressive stress would occur if the temperature reached 40°C? The contact area between each block is 0.20 m². Will fracture occur?

APPROACH We use the expression for the stress F/A we just derived, and find the value of E from Table 9–1. To see if fracture occurs, we compare this stress to the ultimate strength of concrete in Table 9–2.

SOLUTION

$$\frac{F}{A} = \alpha \, E \, \Delta T = \left(12 \times 10^{-6}/\text{C}°\right)\left(20 \times 10^9 \, \text{N/m}^2\right)\left(30 \, \text{C}°\right) = 7.2 \times 10^6 \, \text{N/m}^2.$$

This stress is not far from the ultimate strength of concrete under compression (Table 9–2) and exceeds it for tension and shear. If the concrete is not perfectly aligned, part of the force will act in shear, and fracture is likely. This is why soft spacers or expansion joints (Fig. 13–3) are used in concrete sidewalks, highways, and bridges.

PHYSICS APPLIED
Highway buckling

13–6 The Gas Laws and Absolute Temperature

Equation 13–2 is not very useful for describing the expansion of a gas, partly because the expansion can be so great, and partly because gases generally expand to fill whatever container they are in. Indeed, Eq. 13–2 is meaningful only if the pressure is kept constant. The volume of a gas depends very much on the pressure as well as on the temperature. It is therefore valuable to determine a relation between the volume, the pressure, the temperature, and the mass of a gas. Such a relation is called an **equation of state**. (By the word *state*, we mean the physical condition of the system.)

If the state of a system is changed, we will always wait until the pressure and temperature have reached the same values throughout. We thus consider only **equilibrium states** of a system—when the variables that describe it (such as temperature and pressure) are the same throughout the system and are not changing in time. We also note that the results of this Section are accurate only for gases that are not too dense (the pressure is not too high, on the order of an atmosphere or so) and not close to the liquefaction (boiling) point.

For a given quantity of gas it is found experimentally that, to a good approximation, *the volume of a gas is inversely proportional to the absolute pressure applied to it when the temperature is kept constant.* That is,

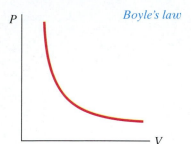

Boyle's law

$$V \propto \frac{1}{P}, \qquad \text{[constant } T\text{]}$$

where P is the absolute pressure (*not* "gauge pressure"—see Section 10–4). For example, if the pressure on a gas is doubled, the volume is reduced to half its original volume. This relation is known as **Boyle's law**, after Robert Boyle (1627–1691), who first stated it on the basis of his own experiments. A graph of P vs. V for a fixed temperature is shown in Fig. 13–12. Boyle's law can also be written

$$PV = \text{constant}. \qquad \text{[constant } T\text{]}$$

FIGURE 13–12 Pressure vs. volume of a fixed amount of gas at a constant temperature, showing the inverse relationship as given by Boyle's law: as the pressure decreases, the volume increases.

That is, at constant temperature, if either the pressure or volume of the gas is allowed to vary, the other variable also changes so that the product PV remains constant.

Temperature also affects the volume of a gas, but a quantitative relationship between V and T was not found until more than a century after Boyle's work. The Frenchman Jacques Charles (1746–1823) found that when the pressure is not too high and is kept constant, the volume of a gas increases with temperature at a nearly constant rate, as in Fig. 13–13a. However, all gases liquefy at low temperatures (for example, oxygen liquefies at $-183°C$), so the graph cannot be extended below the liquefaction point. Nonetheless, the graph is essentially a straight line and if projected to lower temperatures, as shown by the dashed line, it crosses the axis at about $-273°C$.

Such a graph can be drawn for any gas, and the straight line always projects back to $-273°C$ at zero volume. This seems to imply that if a gas could be cooled to $-273°C$, it would have zero volume, and at lower temperatures a negative volume, which makes no sense. It could be argued that $-273°C$ is the lowest temperature possible; indeed, many other more recent experiments indicate that this is so. This temperature is called the **absolute zero** of temperature. Its value has been determined to be $-273.15°C$.

Absolute zero

FIGURE 13–13 Volume of a fixed amount of gas as a function of (a) Celsius temperature, and (b) Kelvin temperature, when the pressure is kept constant.

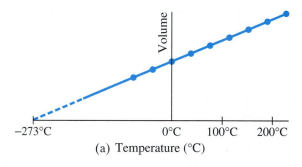

(a) Temperature (°C)

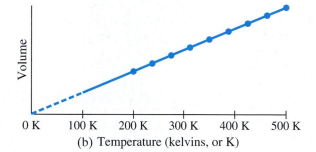

(b) Temperature (kelvins, or K)

Kelvin scale

Absolute zero forms the basis of a temperature scale known as the **absolute scale** or **Kelvin scale**, and it is used extensively in scientific work. On this scale the temperature is specified as degrees Kelvin or, preferably, simply as *kelvins* (K) without the degree sign. The intervals are the same as for the Celsius scale, but the zero on this scale (0 K) is chosen as absolute zero. Thus the freezing point of water (0°C) is 273.15 K, and the boiling point of water is 373.15 K. Indeed, any temperature on the Celsius scale can be changed to kelvins by adding 273.15 to it:

Conversion between Kelvin (absolute) and Celsius scales

$$T(\text{K}) = T(°C) + 273.15.$$

Now let us look at Fig. 13–13b, where the graph of the volume of a gas versus absolute temperature is a straight line that passes through the origin. Thus, to a good approximation, *the volume of a given amount of gas is directly proportional to the absolute temperature when the pressure is kept constant.* This is known as **Charles's law**, and is written

$$V \propto T. \qquad \text{[constant } P\text{]}$$

Charles's law

A third gas law, known as **Gay-Lussac's law**, after Joseph Gay-Lussac (1778–1850), states that *at constant volume, the absolute pressure of a gas is directly proportional to the absolute temperature*:

$$P \propto T. \qquad \text{[constant } V\text{]}$$

Gay-Lussac's law

A familiar example is that a closed jar or an aerosol can thrown into a hot fire will explode due to the increase in gas pressure inside that results from the temperature increase.

The laws of Boyle, Charles, and Gay-Lussac are not really laws in the sense that we use this term today (precise, deep, wide-ranging validity). They are really only approximations that are accurate for real gases only as long as the pressure and density of the gas are not too high, and the gas is not too close to liquefaction (condensation). The term *law* applied to these three relationships has become traditional, however, so we have stuck with that usage.

CONCEPTUAL EXAMPLE 13–9 | **Don't throw a closed glass jar into a campfire.** What can happen if you did throw an empty glass jar, with lid on tight, into a fire, and why?

RESPONSE The inside of the jar is not empty. It is filled with air. As the fire heats the air inside, its temperature rises. The volume of the glass jar changes only slightly due to the heating. According to Gay-Lussac's law the pressure P of the air inside the jar can increase dramatically, enough to cause the jar to explode, throwing glass pieces outward.

13–7 The Ideal Gas Law

The gas laws of Boyle, Charles, and Gay-Lussac were obtained by means of a technique that is very useful in science: namely, holding one or more variables constant to see clearly the effects on one variable due to changing one other variable. These laws can now be combined into a single more general relation between the absolute pressure, volume, and absolute temperature of a fixed quantity of gas:

$$PV \propto T.$$

This relation indicates how any of the quantities P, V, or T will vary when the other two quantities change. This relation reduces to Boyle's, Charles's, or Gay-Lussac's law when either the temperature, the pressure, or the volume, respectively, is held constant.

Finally, we must incorporate the effect of the amount of gas present. Anyone who has blown up a balloon knows that the more air forced into the balloon, the bigger it gets (Figure 13–14). Indeed, careful experiments show that at constant temperature and pressure, the volume V of an enclosed gas increases in direct proportion to the mass m of gas present. Hence we write

$$PV \propto mT.$$

This proportion can be made into an equation by inserting a constant of proportionality. Experiment shows that this constant has a different value for different gases. However, the constant of proportionality turns out to be the same for all gases if, instead of the mass m, we use the number of *moles*.

FIGURE 13–14 Blowing up a balloon means putting more air (more air molecules) into the balloon, which increases its volume. The pressure is nearly constant (atmospheric) except for the small effect of the balloon's elasticity.

One **mole** (abbreviated mol) is defined as the amount of substance that contains as many atoms or molecules as there are in precisely 12 grams of carbon 12 (whose atomic mass is exactly 12 u). A simpler but equivalent definition is this: 1 mol is that number of grams of a substance numerically equal to the molecular mass (Section 13–1) of the substance. For example, the molecular mass of hydrogen gas (H_2) is 2.0 u (since each molecule contains two atoms of hydrogen and each atom has an atomic mass of 1.0 u). Thus 1 mol of H_2 has a mass of 2.0 g. Similarly, 1 mol of neon gas has a mass of 20 g, and 1 mol of CO_2 has a mass of $[12 + (2 \times 16)] = 44$ g since oxygen has atomic mass of 16 (see periodic Table inside the rear cover). The mole is the official unit of amount of substance in the SI system. In general, the number of moles, n, in a given sample of a pure substance is equal to the mass of the sample in grams divided by the molecular mass specified as grams per mole:

$$n \,(\text{mol}) = \frac{\text{mass (grams)}}{\text{molecular mass (g/mol)}}.$$

For example, the number of moles in 132 g of CO_2 (molecular mass 44 u) is

$$n = \frac{132 \text{ g}}{44 \text{ g/mol}} = 3.0 \text{ mol}.$$

We can now write the proportion discussed above as an equation:

$$PV = nRT, \tag{13–3}$$

where n represents the number of moles and R is the constant of proportionality. R is called the **universal gas constant** because its value is found experimentally to be the same for all gases. The value of R, in several sets of units (only the first is the proper SI unit), is

$$
\begin{aligned}
R &= 8.314 \text{ J/(mol·K)} && \text{[SI units]} \\
&= 0.0821 \text{ (L·atm)/(mol·K)} \\
&= 1.99 \text{ calories/(mol·K).}^{\dagger}
\end{aligned}
$$

Equation 13–3 is called the **ideal gas law**, or the **equation of state for an ideal gas**. We use the term "ideal" because real gases do not follow Eq. 13–3 precisely, particularly at high pressure (and density) or when the gas is near the liquefaction point (= boiling point). However, at pressures less than an atmosphere or so, and when T is not close to the liquefaction point of the gas, Eq. 13–3 is quite accurate and useful for real gases.

Always remember, when using the ideal gas law, that temperatures must be given in kelvins (K) and that the pressure P must always be *absolute* pressure, not gauge pressure (Section 10–4).

13–8 Problem Solving with the Ideal Gas Law

The ideal gas law is an extremely useful tool, and we now consider some Examples. We will often refer to "standard conditions" or "standard temperature and pressure" (STP), which means:

$$T = 273 \text{ K} \;(0°\text{C}) \quad \text{and} \quad P = 1.00 \text{ atm} = 1.013 \times 10^5 \text{ N/m}^2 = 101.3 \text{ kPa.}$$

EXAMPLE 13–10 **Volume of one mol at STP.** Determine the volume of 1.00 mol of any gas, assuming it behaves like an ideal gas, at STP.

APPROACH We use the ideal gas law, solving for V.

SOLUTION We solve for V in Eq. 13–3:

$$V = \frac{nRT}{P} = \frac{(1.00 \text{ mol})(8.314 \text{ J/mol·K})(273 \text{ K})}{(1.013 \times 10^5 \text{ N/m}^2)} = 22.4 \times 10^{-3} \text{ m}^3.$$

Since 1 liter is $1000 \text{ cm}^3 = 1 \times 10^{-3} \text{ m}^3$, 1 mol of any gas has $V = 22.4$ L at STP.

---

†Calories will be defined in Section 14–1; sometimes it is useful to use R as given in terms of calories.

The value of 22.4 L for the volume of 1 mol of an ideal gas at STP is worth remembering, for it sometimes makes calculation simpler.

EXERCISE B What is the volume of 1.00 mol of ideal gas at 20°C?

EXAMPLE 13–11 **Helium balloon.** A helium party balloon, assumed to be a perfect sphere, has a radius of 18.0 cm. At room temperature (20°C), its internal pressure is 1.05 atm. Find the number of moles of helium in the balloon and the mass of helium needed to inflate the balloon to these values.

APPROACH We can use the ideal gas law to find n, since we are given P and T, and can find V from the given radius.

SOLUTION We get the volume V from the formula for a sphere:

$$V = \tfrac{4}{3}\pi r^3$$
$$= \tfrac{4}{3}\pi (0.180 \text{ m})^3 = 0.0244 \text{ m}^3.$$

The pressure is given as $1.05 \text{ atm} = 1.064 \times 10^5 \text{ N/m}^2$. The temperature must be expressed in kelvins, so we change 20°C to $(20 + 273)\text{K} = 293 \text{ K}$. Finally, we choose the value of R to be $R = 8.314 \text{ J/(mol·K)}$ because we are using SI units. Thus

$$n = \frac{PV}{RT} = \frac{(1.064 \times 10^5 \text{ N/m}^2)(0.0244 \text{ m}^3)}{(8.314 \text{ J/mol·K})(293 \text{ K})} = 1.066 \text{ mol}.$$

The mass of helium (atomic mass = 4.00 g/mol as given in Appendix B or the periodic Table) can be obtained from

$$\text{mass} = n \times \text{molecular mass} = (1.066 \text{ mol})(4.00 \text{ g/mol}) = 4.26 \text{ g}.$$

EXAMPLE 13–12 **ESTIMATE** **Mass of air in a room.** Estimate the mass of air in a room whose dimensions are 5 m × 3 m × 2.5 m high, at STP.

PHYSICS APPLIED
Mass (and weight) of the air in a room

APPROACH First we determine the number of moles n using the given volume. Then we can multiply by the mass of one mole to get the total mass.

SOLUTION Example 13–10 told us that 1 mol at 0°C has a volume of 22.4 L. The room's volume is 5 m × 3 m × 2.5 m, so

$$n = \frac{(5 \text{ m})(3 \text{ m})(2.5 \text{ m})}{22.4 \times 10^{-3} \text{ m}^3} \approx 1700 \text{ mol}.$$

Air is a mixture of about 20% oxygen (O_2) and 80% nitrogen (N_2). The molecular masses are $2 \times 16 \text{ u} = 32 \text{ u}$ and $2 \times 14 \text{ u} = 28 \text{ u}$, respectively, for an average of about 29 u. Thus, 1 mol of air has a mass of about $29 \text{ g} = 0.029 \text{ kg}$, so our room has a mass of air

$$m \approx (1700 \text{ mol})(0.029 \text{ kg/mol}) \approx 50 \text{ kg}.$$

NOTE That is roughly 100 lbs of air!

EXERCISE C At 20°C, would there be more or less air mass in a room than at 0°C?

Frequently, volume is specified in liters and pressure in atmospheres. Rather than convert these to SI units, we can instead use the value of R given in Section 13–7 as 0.0821 L·atm/mol·K.

In many situations it is not necessary to use the value of R at all. For example, many problems involve a change in the pressure, temperature, and volume of a fixed amount of gas. In this case, $PV/T = nR =$ constant, since n and R remain constant. If we now let P_1, V_1, and T_1 represent the appropriate variables initially, and P_2, V_2, T_2 represent the variables after the change is made, then we can write

$$\frac{P_1 V_1}{T_1} = \frac{P_2 V_2}{T_2}.$$

If we know any five of the quantities in this equation, we can solve for the sixth. Or, if one of the three variables is constant ($V_1 = V_2$, or $P_1 = P_2$, or $T_1 = T_2$) then we can use this equation to solve for one unknown when given the other three quantities.

FIGURE 13–15 Example 13–13.

EXAMPLE 13–13 **Check tires cold.** An automobile tire is filled (Fig. 13–15) to a gauge pressure of 200 kPa at 10°C. After a drive of 100 km, the temperature within the tire rises to 40°C. What is the pressure within the tire now?

APPROACH We don't know the number of moles of gas, or the volume of the tire, but we assume they are constant. We use the ratio form of the ideal gas law.

SOLUTION Since $V_1 = V_2$, then

$$\frac{P_1}{T_1} = \frac{P_2}{T_2}.$$

This is, incidentally, a statement of Gay-Lussac's law. Since the pressure given is the gauge pressure (Section 10–4), we must add atmospheric pressure (= 101 kPa) to get the absolute pressure $P_1 = (200\ \text{kPa} + 101\ \text{kPa}) = 301\ \text{kPa}$. We convert temperatures to kelvins by adding 273 and solve for P_2:

$$P_2 = P_1\left(\frac{T_2}{T_1}\right) = (3.01 \times 10^5\ \text{Pa})\left(\frac{313\ \text{K}}{283\ \text{K}}\right) = 333\ \text{kPa}.$$

Subtracting atmospheric pressure, we find the resulting gauge pressure to be 232 kPa, which is a 16% increase. This Example shows why car manuals suggest checking tire pressure when the tires are cold.

NOTE When using the ideal gas law, temperatures must be given in kelvins (K) and the pressure P must always be *absolute* pressure, not gauge pressure.

13–9 Ideal Gas Law in Terms of Molecules: Avogadro's Number

The fact that the gas constant, R, has the same value for all gases is a remarkable reflection of simplicity in nature. It was first recognized, although in a slightly different form, by the Italian scientist Amedeo Avogadro (1776–1856). Avogadro stated that *equal volumes of gas at the same pressure and temperature* *Avogadro's hypothesis* *contain equal numbers of molecules*. This is sometimes called **Avogadro's hypothesis**. That this is consistent with R being the same for all gases can be seen as follows. First of all, from Eq. 13–3 we see that for the same number of moles, n, and the same pressure and temperature, the volume will be the same for all gases as long as R is the same. Second, the number of molecules in 1 mole is the same for all gases.[†] Thus Avogadro's hypothesis is equivalent to R being the same for all gases.

The number of molecules in one mole of any pure substance is known as **Avogadro's number**, N_A. Although Avogadro conceived the notion, he was not able to actually determine the value of N_A. Indeed, precise measurements were not done until the twentieth century.

[†]For example, the molecular mass of H_2 gas is 2.0 atomic mass units (u), whereas that of O_2 gas is 32.0 u. Thus 1 mol of H_2 has a mass of 0.0020 kg and 1 mol of O_2 gas, 0.0320 kg. The number of molecules in a mole is equal to the total mass M of a mole divided by the mass m of one molecule; since this ratio (M/m) is the same for all gases by definition of the mole, a mole of any gas must contain the same number of molecules.

A number of methods have been devised to measure N_A, and the accepted value today is

$$N_A = 6.02 \times 10^{23}. \qquad \text{[molecules/mole]}$$

Avogadro's number

Since the total number of molecules, N, in a gas is equal to the number per mole times the number of moles $(N = nN_A)$, the ideal gas law, Eq. 13–3, can be written in terms of the number of molecules present:

$$PV = nRT = \frac{N}{N_A} RT,$$

or

$$PV = NkT, \qquad\qquad (13\text{–}4)$$

IDEAL GAS LAW
(in terms of molecules)

where $k = R/N_A$ is called **Boltzmann's constant** and has the value

$$k = \frac{R}{N_A} = \frac{8.314 \text{ J/mol} \cdot \text{K}}{6.02 \times 10^{23}/\text{mol}} = 1.38 \times 10^{-23} \text{ J/K}.$$

Boltzmann's constant

EXAMPLE 13–14 **Hydrogen atom mass.** Use Avogadro's number to determine the mass of a hydrogen atom.

APPROACH The mass of one atom equals the mass of 1 mol divided by the number of atoms in 1 mol, N_A.

SOLUTION One mole of hydrogen atoms (atomic mass $= 1.008$ u, Section 13–1 or Appendix B) has a mass of 1.008×10^{-3} kg and contains 6.02×10^{23} atoms. Thus one atom has a mass

$$m = \frac{1.008 \times 10^{-3} \text{ kg}}{6.02 \times 10^{23}} = 1.67 \times 10^{-27} \text{ kg.}$$

NOTE Historically, the reverse process was one method used to obtain N_A: that is, a precise value of N_A can be obtained from a precise measurement of the mass of the hydrogen atom.

EXAMPLE 13–15 **ESTIMATE** **How many molecules in one breath?** Estimate how many molecules you breathe in with a 1.0-L breath of air.

PHYSICS APPLIED
Molecules in a breath

APPROACH We determine what fraction of a mole 1.0 L is using the result of Example 13–10 that 1 mole has a volume of 22.4 L at STP, and then multiply that by N_A to get the number of molecules in this number of moles.

SOLUTION One mole corresponds to 22.4 L at STP, so 1.0 L of air is $(1.0 \text{ L})/(22.4 \text{ L/mol}) = 0.045$ mol. Then 1.0 L of air contains

$$(0.045 \text{ mol})(6.02 \times 10^{23} \text{ molecules/mole}) \approx 3 \times 10^{22} \text{ molecules.}$$

13–10 Kinetic Theory and the Molecular Interpretation of Temperature

The analysis of matter in terms of atoms in continuous random motion is called the **kinetic theory**. We now investigate the properties of a gas from the point of view of kinetic theory, which is based on the laws of classical mechanics. But to apply Newton's laws to each of the vast number of molecules in a gas $(>10^{25}/\text{m}^3$ at STP) is far beyond the capability of any present computer. Instead we take a statistical approach and determine averages of certain quantities, and these averages correspond to macroscopic variables. We will, of course, demand that our microscopic description correspond to the macroscopic properties of gases; otherwise our theory would be of little value. Most importantly, we will arrive at an important relation between the average kinetic energy of molecules in a gas and the absolute temperature.

We make the following assumptions about the molecules in a gas. These assumptions reflect a simple view of a gas, but nonetheless the results they predict correspond well to the essential features of real gases that are at low pressures and far from the liquefaction point. Under these conditions real gases follow the ideal gas law quite closely, and indeed the gas we now describe is referred to as an **ideal gas**. The assumptions, which represent the basic postulates of the kinetic theory, are:

Postulates of kinetic theory

1. There are a large number of molecules, N, each of mass m, moving in random directions with a variety of speeds. This assumption is in accord with our observation that a gas fills its container and, in the case of air on Earth, is kept from escaping only by the force of gravity.

2. The molecules are, on the average, far apart from one another. That is, their average separation is much greater than the diameter of each molecule.

3. The molecules are assumed to obey the laws of classical mechanics, and are assumed to interact with one another only when they collide. Although molecules exert weak attractive forces on each other between collisions, the potential energy associated with these forces is small compared to the kinetic energy, and we ignore it for now.

4. Collisions with another molecule or the wall of the vessel are assumed to be perfectly elastic, like the collisions of perfectly elastic billiard balls (Chapter 7). We assume the collisions are of very short duration compared to the time between collisions. Then we can ignore the potential energy associated with collisions in comparison to the kinetic energy between collisions.

Boyle's law explained

We can see immediately how this kinetic view of a gas can explain Boyle's law (Section 13–6). The pressure exerted on a wall of a container of gas is due to the constant bombardment of molecules. If the volume is reduced by (say) half, the molecules are closer together and twice as many will be striking a given area of the wall per second. Hence we expect the pressure to be twice as great, in agreement with Boyle's law.

Now let us calculate quantitatively the pressure a gas exerts on its container as based on kinetic theory. We imagine that the molecules are inside a rectangular container (at rest) whose ends have area A and whose length is l, as shown in Fig. 13–16a. The pressure exerted by the gas on the walls of its container is, according to our model, due to the collisions of the molecules with the walls. Let us focus our attention on the wall, of area A, at the left end of the container and examine what happens when one molecule strikes this wall, as shown in Fig. 13–16b. This molecule exerts a force on the wall, and according to Newton's third law the wall exerts an equal and opposite force back on the molecule. The magnitude of this force on the molecule, according to Newton's second law, is equal to the molecule's rate of change of momentum, $F = \Delta(mv)/\Delta t$ (Eq. 7–2). Assuming the collision is elastic, only the x component of the molecule's momentum changes, and it changes from $-mv_x$ (it is moving in the negative x direction) to $+mv_x$. Thus the change in the molecule's momentum, $\Delta(mv)$, which is the final momentum minus the initial momentum, is

$$\Delta(mv) = mv_x - (-mv_x) = 2mv_x$$

for one collision. This molecule will make many collisions with the wall, each separated by a time Δt, which is the time it takes the molecule to travel across the container and back again, a distance (x component) equal to $2l$. Thus $2l = v_x \Delta t$, or

$$\Delta t = \frac{2l}{v_x}.$$

The time Δt between collisions is very small, so the number of collisions per second is very large. Thus the average force—averaged over many collisions—

FIGURE 13–16 (a) Molecules of a gas moving about in a rectangular container. (b) Arrows indicate the momentum of one molecule as it rebounds from the end wall.

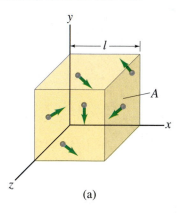

(a)

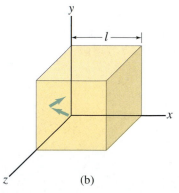

(b)

will be equal to the momentum change during one collision divided by the time between collisions (Newton's second law):

$$F = \frac{\Delta(mv)}{\Delta t} = \frac{2mv_x}{2l/v_x} = \frac{mv_x^2}{l}. \qquad \text{[due to one molecule]}$$

During its passage back and forth across the container, the molecule may collide with the tops and sides of the container, but this does not alter its x component of momentum and thus does not alter our result. It may also collide with other molecules, which may change its v_x. However, any loss (or gain) of momentum is acquired by other molecules, and because we will eventually sum over all the molecules, this effect will be included. So our result above is not altered.

The actual force due to one molecule is intermittent, but because a huge number of molecules are striking the wall per second, the force is, on average, nearly constant. To calculate the force due to *all* the molecules in the container, we have to add the contributions of each. Thus the net force on the wall is

$$F = \frac{m}{l}\left(v_{x1}^2 + v_{x2}^2 + \cdots + v_{xN}^2\right),$$

where v_{x1} means v_x for molecule number 1 (we arbitrarily assign each molecule a number) and the sum extends over the total number of molecules N in the container. The average value of the square of the x component of velocity is

$$\overline{v_x^2} = \frac{v_{x1}^2 + v_{x2}^2 + \cdots + v_{xN}^2}{N}. \qquad (13\text{--}5)$$

Thus we can write the force as

$$F = \frac{m}{l}\,N\overline{v_x^2}.$$

We know that the square of any vector is equal to the sum of the squares of its components (theorem of Pythagoras). Thus $v^2 = v_x^2 + v_y^2 + v_z^2$ for any velocity v. Taking averages, we obtain

$$\overline{v^2} = \overline{v_x^2} + \overline{v_y^2} + \overline{v_z^2}.$$

Since the velocities of the molecules in our gas are assumed to be random, there is no preference to one direction or another. Hence

$$\overline{v_x^2} = \overline{v_y^2} = \overline{v_z^2}.$$

Combining this relation with the one just above, we get

$$\overline{v^2} = 3\overline{v_x^2}.$$

We substitute this into the equation for net force F:

$$F = \frac{m}{l}\,N\,\frac{\overline{v^2}}{3}.$$

The pressure on the wall is then

$$P = \frac{F}{A} = \frac{1}{3}\frac{Nm\overline{v^2}}{Al}$$

or

$$P = \frac{1}{3}\frac{Nm\overline{v^2}}{V}, \qquad (13\text{--}6) \qquad \textit{Pressure in a gas}$$

where $V = lA$ is the volume of the container. This is the result we were seeking, the pressure exerted by a gas on its container expressed in terms of molecular properties.

Equation 13–6, $P = \frac{1}{3}Nm\overline{v^2}/V$, can be rewritten in a clearer form by multiplying both sides by V and rearranging the right-hand side:

$$PV = \frac{2}{3}N\left(\frac{1}{2}m\overline{v^2}\right). \qquad (13\text{–}7)$$

The quantity $\frac{1}{2}m\overline{v^2}$ is the average kinetic energy $(\overline{\text{KE}})$ of the molecules in the gas. If we compare Eq. 13–7 with Eq. 13–4, the ideal gas law $PV = NkT$, we see that the two agree if

$$\frac{2}{3}\left(\frac{1}{2}m\overline{v^2}\right) = kT,$$

or

$$\boxed{\overline{\text{KE}} = \frac{1}{2}m\overline{v^2} = \frac{3}{2}kT.} \qquad \text{[ideal gas]} \quad (13\text{–}8)$$

TEMPERATURE RELATED TO AVERAGE KINETIC ENERGY OF MOLECULES

This equation tells us that

the average translational kinetic energy of molecules in random motion in an ideal gas is directly proportional to the absolute temperature of the gas.

The higher the temperature, according to kinetic theory, the faster the molecules are moving on the average. This relation is one of the triumphs of the kinetic theory.

EXAMPLE 13–16 **Molecular kinetic energy.** What is the average translational kinetic energy of molecules in an ideal gas at 37°C?

APPROACH We use the absolute temperature in Eq. 13–8.

SOLUTION We change 37°C to 310 K and insert into Eq. 13–8:

$$\overline{\text{KE}} = \frac{3}{2}kT = \frac{3}{2}\left(1.38 \times 10^{-23}\,\text{J/K}\right)(310\,\text{K}) = 6.42 \times 10^{-21}\,\text{J}.$$

NOTE A mole of molecules would have a total translational kinetic energy equal to $\left(6.42 \times 10^{-21}\,\text{J}\right)\left(6.02 \times 10^{23}\right) = 3900\,\text{J}$, which equals the kinetic energy of a 1-kg stone traveling faster than 85 m/s.

Equation 13–8 holds not only for gases, but also applies reasonably accurately to liquids and solids. Thus the result of Example 13–16 would apply to molecules within living cells at body temperature (37°C).

We can use Eq. 13–8 to calculate how fast molecules are moving on the average. Notice that the average in Eqs. 13–5 through 13–8 is over the *square* of the speed. The square root of $\overline{v^2}$ is called the **root-mean-square** speed, v_{rms} (since we are taking the square *root* of the *mean* of the *square* of the speed):

Root-mean-square (rms) speed

rms speed of molecules

$$v_{\text{rms}} = \sqrt{\overline{v^2}} = \sqrt{\frac{3kT}{m}}. \qquad (13\text{–}9)$$

EXAMPLE 13–17 **Speeds of air molecules.** What is the rms speed of air molecules (O_2 and N_2) at room temperature (20°C)?

APPROACH To obtain v_{rms}, we need the masses of O_2 and N_2 molecules and then apply Eq. 13–9 to oxygen and nitrogen separately, since they have different masses.

SOLUTION The masses of one molecule of O_2 (molecular mass = 32 u) and N_2 (molecular mass = 28 u) are (where $1\,\text{u} = 1.66 \times 10^{-27}\,\text{kg}$)

$$m(O_2) = (32)\left(1.66 \times 10^{-27}\,\text{kg}\right) = 5.3 \times 10^{-26}\,\text{kg},$$
$$m(N_2) = (28)\left(1.66 \times 10^{-27}\,\text{kg}\right) = 4.6 \times 10^{-26}\,\text{kg}.$$

Thus, for oxygen

$$v_{\text{rms}} = \sqrt{\frac{3kT}{m}} = \sqrt{\frac{(3)\left(1.38 \times 10^{-23}\,\text{J/K}\right)(293\,\text{K})}{\left(5.3 \times 10^{-26}\,\text{kg}\right)}} = 480\,\text{m/s},$$

and for nitrogen the result is $v_{\text{rms}} = 510\,\text{m/s}$. These speeds[†] are more than 1700 km/h or 1000 mi/h.

[†]The speed v_{rms} is a magnitude only. The *velocity* of molecules averages to zero: the velocity has direction, and as many molecules move to the right as to the left.

370 CHAPTER 13 Temperature and Kinetic Theory

EXERCISE D What speed would a 1-gram paper clip have if it had the same KE as a molecule of Example 13–17?

Equation 13–8, $\overline{KE} = \frac{3}{2}kT$, implies that as the temperature approaches absolute zero, the kinetic energy of molecules approaches zero. Modern quantum theory, however, tells us this is not quite so. Instead, as absolute zero is approached, the kinetic energy approaches a very small nonzero minimum value. Even though all real gases become liquid or solid near 0 K, molecular motion does not cease, even at absolute zero.

* 13–11 Distribution of Molecular Speeds

The molecules in a gas are assumed to be in random motion, which means that many molecules have speeds less than the rms speed and others have greater speeds. In 1859, James Clerk Maxwell (1831–1879) derived, on the basis of kinetic theory, that the speeds of molecules in a gas are distributed according to the graph shown in Fig. 13–17. This is known as the **Maxwell distribution of speeds**.[†] The speeds vary from zero to many times the rms speed, but as the graph shows, most molecules have speeds that are not far from the average. Less than 1% of the molecules exceed four times v_{rms}.

Maxwell distribution of speeds of molecules in a gas

Experiments to determine the distribution in real gases, starting in the 1920s, confirmed with considerable accuracy the Maxwell distribution and the direct proportion between average kinetic energy and absolute temperature, Eq. 13–8.

Figure 13–18 shows the Maxwell distribution for two different temperatures; just as v_{rms} increases with temperature, so the whole distribution curve shifts to the right at higher temperatures. This Figure illustrates how kinetic theory can explain why many chemical reactions, including those in biological cells, take place more rapidly as the temperature increases. Two molecules may chemically react only if their kinetic energy is above some minimum value (called the *activation energy*), E_A, so that when they collide, they penetrate into each other somewhat. Figure 13–18 shows that at a higher temperature, many more molecules have a speed and kinetic energy KE above the needed threshold E_A.

PHYSICS APPLIED

How chemical reactions depend on temperature

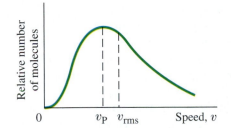

FIGURE 13–17 Distribution of speeds of molecules in an ideal gas. Note that v_{rms} is not at the peak of the curve (that speed is called the "most probable speed," v_P). This is because the curve is skewed to the right: it is not symmetrical.

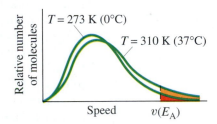

FIGURE 13–18 Distribution of molecular speeds for two different temperatures.

* 13–12 Real Gases and Changes of Phase

The ideal gas law is an accurate description of the behavior of a real gas as long as the pressure is not too high and as long as the temperature is far from the liquefaction point. But what happens when these two criteria are not satisfied? First we discuss real gas behavior and then we examine how kinetic theory can help us understand this behavior.

[†]Mathematically, the distribution is given by $\Delta N = Cv^2 \exp\left(-\frac{1}{2}mv^2/kT\right)\Delta v$, where ΔN is the number of molecules with speed between v and $v + \Delta v$, C is a constant, and exp means the expression in parentheses is an exponent on the natural number $e = 2.718\ldots$.

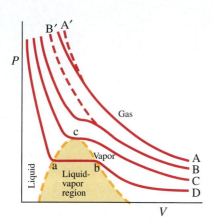

PV diagram

FIGURE 13–19 *PV* diagram for a real substance. Curves A, B, C, and D represent the same gas at different fixed temperatures $(T_A > T_B > T_C > T_D)$.

Let us look at a graph of pressure plotted against volume for a given amount of gas. On such a "*PV* diagram," Fig. 13–19, each point represents an equilibrium state of the given substance. The various curves (labeled A, B, C, and D) show how the pressure varies as the volume is changed at constant temperature for several different values of the temperature. The dashed curve A′ represents the behavior of a gas as predicted by the ideal gas law; that is, $PV = $ constant. The solid curve A represents the behavior of a real gas at the same temperature. Notice that at high pressure, the volume of a real gas is less than that predicted by the ideal gas law. The curves B and C in Fig. 13–19 represent the gas at successively lower temperatures, and we see that the behavior deviates even more from the curves predicted by the ideal gas law (for example, B′), and the deviation is greater the closer the gas is to liquefying.

To explain this, we note that at higher pressure we expect the molecules to be closer together. And, particularly at lower temperatures, the potential energy associated with the attractive forces between the molecules (which we ignored before) is no longer negligible compared to the now reduced kinetic energy of the molecules. These attractive forces tend to pull the molecules closer together so at a given pressure, the volume is less than expected from the ideal gas law. At still lower temperatures, these forces cause liquefaction, and the molecules become very close together.

Curve D represents the situation when liquefaction occurs. At low pressure on curve D (on the right in Fig. 13–19), the substance is a gas and occupies a large volume. As the pressure is increased, the volume decreases until point b is reached. Beyond b, the volume decreases with no change in pressure; the substance is gradually changing from the gas to the liquid phase. At point a, all of the substance has changed to liquid. Further increase in pressure reduces the volume only slightly—liquids are nearly incompressible—so on the left the curve is very steep as shown. The shaded area under the dashed line represents the region where the gas and liquid phases exist together in equilibrium.

Curve C in Fig. 13–19 represents the behavior of the substance at its **critical temperature**; the point c (the one point where this curve is horizontal) is called *Critical point* the **critical point**. At temperatures less than the critical temperature (and this is the definition of the term), a gas will change to the liquid phase if sufficient pressure is applied. Above the critical temperature, no amount of pressure can cause a gas to change phase and become a liquid: no liquid surface forms. The critical temperatures for various gases are given in Table 13–2. Scientists tried for many years to liquefy oxygen without success. Only after the discovery of the critical point was it realized that oxygen can be liquefied only if first cooled below its critical temperature of $-118°C$.

Often a distinction is made between the terms "gas" and "vapor": a *Vapor vs. gas* substance below its critical temperature in the gaseous state is called a **vapor**; above the critical temperature, it is called a **gas**.

The behavior of a substance can be diagrammed not only on a *PV* diagram *Phase diagram (PT)* but also on a *PT* diagram. A *PT* diagram, often called a **phase diagram**, is

TABLE 13–2 Critical Temperatures and Pressures

Substance	Critical Temperature		Critical Pressure (atm)
	°C	K	
Water	374	647	218
CO_2	31	304	72.8
Oxygen	−118	155	50
Nitrogen	−147	126	33.5
Hydrogen	−239.9	33.3	12.8
Helium	−267.9	5.3	2.3

particularly convenient for comparing the different phases of a substance. Figure 13–20 is the phase diagram for water. The curve labeled *l-v* represents those points where the liquid and vapor phases are in equilibrium—it is thus a graph of the boiling point versus pressure. Note that the curve correctly shows that at a pressure of 1 atm the boiling point is 100°C and that the boiling point is lowered for a decreased pressure. The curve *s-l* represents points where solid and liquid exist in equilibrium and thus is a graph of the freezing point versus pressure. At 1 atm, the freezing point of water is 0°C, as shown. Notice also in Fig. 13–20 that at a pressure of 1 atm, the substance is in the liquid phase if the temperature is between 0°C and 100°C, but is in the solid or vapor phase if the temperature is below 0°C or above 100°C. The curve labeled *s-v* is the *sublimation point* versus pressure curve. **Sublimation** refers to the process whereby at low pressures a solid changes directly into the vapor phase without passing through the liquid phase. For water, sublimation occurs if the pressure of the water vapor is less than 0.0060 atm. Carbon dioxide, which in the solid phase is called dry ice, sublimates even at atmospheric pressure.

The intersection of the three curves (in Fig. 13–20) is the **triple point**. For water this occurs at $T = 273.16$ K and $P = 6.03 \times 10^{-3}$ atm. It is only at the triple point that the three phases can exist together in equilibrium. Because the triple point corresponds to a unique value of temperature and pressure, it is precisely reproducible and is often used as a point of reference. For example, the standard of temperature is usually specified as exactly 273.16 K at the triple point of water, rather than 273.15 K at the freezing point of water at 1 atm.

Notice that the *s-l* curve for water slopes upward to the left. This is true only of substances that *expand* upon freezing: at a higher pressure, a lower temperature is needed to cause the liquid to freeze. More commonly, substances contract upon freezing and the *s-l* curve slopes upward to the right, as shown for carbon dioxide (CO_2) in Fig. 13–21.

The phase transitions we have been discussing are the common ones. Some substances, however, can exist in several forms in the solid phase. A transition from one phase to another occurs at a particular temperature and pressure, just like ordinary phase changes. For example, ice has been observed in at least eight forms at very high pressure. Ordinary helium has two distinct liquid phases, called helium I and II. They exist only at temperatures within a few degrees of absolute zero. Helium II exhibits very unusual properties referred to as **superfluidity**. It has essentially zero viscosity and exhibits strange properties such as climbing up the sides of an open container.

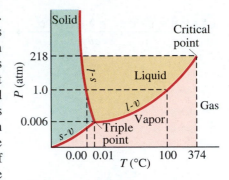

FIGURE 13–20 Phase diagram for water (note that the scales are not linear).

Triple point

FIGURE 13–21 Phase diagram for carbon dioxide.

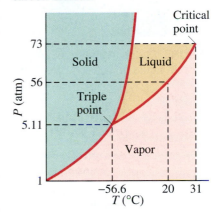

Superfluidity

* 13–13 Vapor Pressure and Humidity

Evaporation

If a glass of water is left out overnight, the water level will have dropped by morning. We say the water has evaporated, meaning that some of the water has changed to the vapor or gas phase.

This process of **evaporation** can be explained on the basis of kinetic theory. The molecules in a liquid move past one another with a variety of speeds that follow, approximately, the Maxwell distribution. There are strong attractive forces between these molecules, which is what keeps them close together in the liquid phase. A molecule near the surface of the liquid may, because of its speed, leave the liquid momentarily. But just as a rock thrown into the air returns to the Earth, so the attractive forces of the other molecules can pull the vagabond molecule back to the liquid surface—that is, if its velocity is not too large. A molecule with a high enough velocity, however, will escape the liquid entirely, like a rocket escaping the Earth, and become part of the gas phase. Only those molecules that have kinetic energy above a particular value can escape to the gas phase. We have already seen that kinetic theory predicts that the relative number of molecules with kinetic energy above a particular value (such as E_A in Fig. 13–18) increases with temperature. This is in accord with the well-known observation that the evaporation rate is greater at higher temperatures.

Evaporation

Because it is the fastest molecules that escape from the surface, the average speed of those remaining is less. When the average speed is less, the absolute temperature is less. Thus kinetic theory predicts that *evaporation is a cooling process*. You have no doubt noticed this effect when you stepped out of a warm shower and felt cold as the water on your body began to evaporate; and after working up a sweat on a hot day, even a slight breeze makes you feel cool through evaporation.

FIGURE 13–22 Vapor appears above a liquid in a closed container.

Vapor Pressure

Air normally contains water vapor (water in the gas phase), and it comes mainly from evaporation. To look at this process in a little more detail, consider a closed container that is partially filled with water (or another liquid) and from which the air has been removed (Fig. 13–22). The fastest moving molecules quickly evaporate into the empty space above the liquid's surface. As they move about, some of these molecules strike the liquid surface and again become part of the liquid phase: this is called **condensation**. The number of molecules in the vapor increases until a point is reached when the number of molecules returning to the liquid equals the number leaving in the same time interval. Equilibrium then exists, and the space above the liquid surface is said to be *saturated*. The pressure of the vapor when it is saturated is called the **saturated vapor pressure** (or sometimes simply the vapor pressure).

The saturated vapor pressure does not depend on the volume of the container. If the volume above the liquid were reduced suddenly, the density of molecules in the vapor phase would be increased temporarily. More molecules would then be striking the liquid surface per second. There would be a net flow of molecules back to the liquid phase until equilibrium was again reached, and this would occur at the same value of the saturated vapor pressure, as long as the temperature had not changed.

The saturated vapor pressure of any substance depends on the temperature. At higher temperatures, more molecules have sufficient kinetic energy to break from the liquid surface into the vapor phase. Hence equilibrium will be reached at a higher pressure. The saturated vapor pressure of water at various temperatures is given in Table 13–3. Notice that even solids—for example, ice—have a measurable saturated vapor pressure.

In everyday situations, evaporation from a liquid takes place into the air above it rather than into a vacuum. This does not materially alter the discussion above relating to Fig. 13–22. Equilibrium will still be reached when there are sufficient molecules in the gas phase that the number reentering the liquid equals the number leaving. The concentration of particular molecules (such as water) in the gas phase is not affected by the presence of air, although collisions with air molecules may lengthen the time needed to reach equilibrium. Thus equilibrium occurs at the same value of the saturated vapor pressure as if air weren't there.

If the container is large or is not closed, all the liquid may evaporate before saturation is reached. And if the container is not sealed—as, for example, a room in your house—it is not likely that the air will become saturated with water vapor (unless it is raining outside).

TABLE 13–3 Saturated Vapor Pressure of Water

Temp-erature (°C)	Saturated Vapor Pressure	
	torr (= mm-Hg)	Pa (= N/m²)
−50	0.030	4.0
−10	1.95	2.60×10^2
0	4.58	6.11×10^2
5	6.54	8.72×10^2
10	9.21	1.23×10^3
15	12.8	1.71×10^3
20	17.5	2.33×10^3
25	23.8	3.17×10^3
30	31.8	4.24×10^3
40	55.3	7.37×10^3
50	92.5	1.23×10^4
60	149	1.99×10^4
70[†]	234	3.12×10^4
80	355	4.73×10^4
90	526	7.01×10^4
100[‡]	760	1.01×10^5
120	1489	1.99×10^5
150	3570	4.76×10^5

[†] Boiling point on summit of Mt. Everest.
[‡] Boiling point at sea level.

Boiling

The saturated vapor pressure of a liquid increases with temperature. When the temperature is raised to the point where the saturated vapor pressure at that temperature equals the external pressure, **boiling** occurs (Fig. 13–23). As the boiling point is approached, tiny bubbles tend to form in the liquid, which indicate a change from the liquid to the gas phase. However, if the vapor pressure inside the bubbles is less than the external pressure, the bubbles immediately are crushed. As the temperature is increased, the saturated vapor pressure inside a bubble eventually becomes equal to or exceeds the external air pressure. The bubble will then not collapse but can rise to the surface. Boiling has then begun. *A liquid boils when its saturated vapor pressure equals the external pressure.* This occurs for water at a pressure of 1 atm (760 torr) at 100°C, as can be seen from Table 13–3.

At boiling, saturated vapor pressure equals external pressure

The boiling point of a liquid clearly depends on the external pressure. At high elevations, the boiling point of water is somewhat less than at sea level since the air pressure is less up there. For example, on the summit of Mt. Everest (8850 m) the air pressure is about one-third of what it is at sea level, and from Table 13–3 we can see that water will boil at about 70°C. Cooking food by boiling takes longer at high elevations, since the temperature is less. Pressure cookers, however, reduce cooking time, because they build up a pressure as high as 2 atm, allowing higher boiling temperatures to be attained.

Partial Pressure and Humidity

When we refer to the weather as being dry or humid, we are referring to the water vapor content of the air. In a gas such as air, which is a mixture of several types of gases, the total pressure is the sum of the *partial pressures* of each gas present.[†] By **partial pressure**, we mean the pressure each gas would exert if it alone were present. The partial pressure of water in the air can be as low as zero and can vary up to a maximum equal to the saturated vapor pressure of water at the given temperature. Thus, at 20°C, the partial pressure of water cannot exceed 17.5 torr (see Table 13–3). The **relative humidity** is defined as the ratio of the partial pressure of water vapor to the saturated vapor pressure at a given temperature. It is usually expressed as a percentage:

$$\text{Relative humidity} = \frac{\text{partial pressure of } H_2O}{\text{saturated vapor pressure of } H_2O} \times 100\%.$$

FIGURE 13–23 Boiling: bubbles of water vapor float upward from the bottom (where the temperature is highest).

Relative humidity

Thus, when the humidity is close to 100%, the air holds nearly all the water vapor it can.

EXAMPLE 13–18 **Relative humidity.** On a particular hot day, the temperature is 30°C and the partial pressure of water vapor in the air is 21.0 torr. What is the relative humidity?

APPROACH From Table 13–3, we see that the saturated vapor pressure of water at 30°C is 31.8 torr.

SOLUTION The relative humidity is thus

$$\frac{21.0 \text{ torr}}{31.8 \text{ torr}} \times 100\% = 66\%.$$

Humans are sensitive to humidity. A relative humidity of 40–50% is generally optimum for both health and comfort. High humidity, particularly on a hot day, reduces the evaporation of moisture from the skin, which is one of the body's vital mechanisms for regulating body temperature. Very low humidity, on the other hand, can dry the skin and mucous membranes.

PHYSICS APPLIED
Humidity and comfort

[†]For example, 78% (by volume) of air molecules are nitrogen and 21% oxygen, with much smaller amounts of water vapor, argon, and other gases. At an air pressure of 1 atm, oxygen exerts a partial pressure of 0.21 atm and nitrogen 0.78 atm.

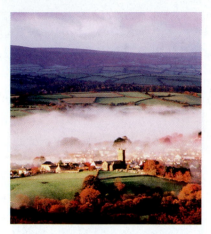

FIGURE 13–24 Fog or mist settling over a low-lying village where the temperature has dropped below the dew point.

Air is saturated with water vapor when the partial pressure of water in the air is equal to the saturated vapor pressure at that temperature. If the partial pressure of water exceeds the saturated vapor pressure, the air is said to be **supersaturated**. This situation can occur when a temperature decrease occurs. For example, suppose the temperature is 30°C and the partial pressure of water is 21 torr, which represents a humidity of 66% as we saw in Example 13–18. Suppose now that the temperature falls to, say, 20°C, as might happen at nightfall. From Table 13–3 we see that the saturated vapor pressure of water at 20°C is 17.5 torr. Hence the relative humidity would be greater than 100%, and the supersaturated air cannot hold this much water. The excess water may condense and appear as dew, or as fog or rain (Fig. 13–24).

When air containing a given amount of water is cooled, a temperature is reached where the partial pressure of water equals the saturated vapor pressure. This is called the **dew point**. Measurement of the dew point is the most accurate means of determining the relative humidity. One method uses a polished metal surface in contact with air, which is gradually cooled down. The temperature at which moisture begins to appear on the surface is the dew point, and the partial pressure of water can then be obtained from saturated vapor pressure tables. If, for example, on a given day the temperature is 20°C and the dew point is 5°C, then the partial pressure of water (Table 13–3) in the 20°C air was 6.54 torr, whereas its saturated vapor pressure was 17.5 torr; hence the relative humidity was 6.54/17.5 = 37%.

* 13–14 Diffusion

If you carefully place a few drops of food coloring in a container of water as in Fig. 13–25, you will find that the color spreads throughout the water. The process may take several hours (assuming you don't shake the glass), but eventually the color will become uniform. This mixing, known as **diffusion**, is further evidence for the random movement of molecules. Diffusion occurs in gases too. Common examples include perfume or smoke diffusing in air, including the odor of something cooking, although convection (moving air currents) often plays a greater role in spreading odors than does diffusion. Diffusion depends on *concentration*, by which we mean the number of molecules or moles per unit volume. In general, *the diffusing substance moves from a region where its concentration is high to one where its concentration is low.*

Diffusion occurs from high to low concentration

Diffusion can be readily understood on the basis of kinetic theory and the random motion of molecules. Consider a tube of cross-sectional area A containing molecules in a higher concentration on the left than on the right,

FIGURE 13–25 A few drops of food coloring spreads slowly throughout the water, eventually becoming uniform.

(a)

(b)

(c)

Fig. 13–26. We assume the molecules are in random motion. Yet there will be a net flow of molecules to the right. To see why this is true, let us consider the small section of tube of length Δx as shown. Molecules from both regions 1 and 2 cross into this central section as a result of their random motion. The more molecules there are in a region, the more will strike a given area or cross a boundary. Since there is a greater concentration of molecules in region 1 than in region 2, more molecules cross into the central section from region 1 than from region 2. There is, then, a net flow of molecules from left to right, from high concentration toward low concentration. The net flow becomes zero only when the concentrations become equal.

You might expect that the greater the difference in concentration, the greater the flow rate. Indeed, the rate of diffusion, J (number of molecules or moles or kg per second), is directly proportional to the change in concentration per unit distance, $(C_1 - C_2)/\Delta x$ (which is called the **concentration gradient**), and to the cross-sectional area A (see Fig. 13–26):

$$J = DA\frac{C_1 - C_2}{\Delta x}.$$ (13–10) *Diffusion equation*

D is a constant of proportionality called the **diffusion constant**. Equation 13–10 is known as the **diffusion equation**, or **Fick's law**. If the concentrations are given in mol/m^3, then J is the number of moles passing a given point per second. If the concentrations are given in kg/m^3, then J is the mass movement per second (kg/s). The length Δx is given in meters. The values of D for a variety of substances are given in Table 13–4.

EXAMPLE 13–19 **ESTIMATE** **Diffusion of ammonia in air.** To get an idea of the time required for diffusion, estimate how long it might take for ammonia (NH_3) to be detected 10 cm from a bottle after it is opened, assuming only diffusion is occurring.

APPROACH This will be an order-of-magnitude calculation. The rate of diffusion J can be set equal to the number of molecules N diffusing across area A in a time t: $J = N/t$. Then the time $t = N/J$, where J is given by Eq. 13–10. We will have to make some assumptions and rough approximations about concentrations to use Eq. 13–10.

SOLUTION Using Eq. 13–10, we get

$$t = \frac{N}{J} = \frac{N}{DA}\frac{\Delta x}{\Delta C}.$$

The average concentration (midway between bottle and nose) can be approximated by $\overline{C} \approx N/V$, where V is the volume over which the molecules move and is roughly of the order of $V \approx A\,\Delta x$, where Δx is 10 cm = 0.10 m. We substitute $N = \overline{C}V = \overline{C}A\,\Delta x$ into the above equation:

$$t \approx \frac{(\overline{C}A\,\Delta x)\Delta x}{DA\,\Delta C} = \frac{\overline{C}}{\Delta C}\frac{(\Delta x)^2}{D}.$$

The concentration of ammonia is high near the bottle and low near the detecting nose, so $\overline{C} \approx \Delta C/2$, or $(\overline{C}/\Delta C) \approx \frac{1}{2}$. Since NH_3 molecules have a size somewhere between H_2 and O_2, from Table 13–4 we can estimate $D \approx 4 \times 10^{-5}\,m^2/s$. Then

$$t \approx \frac{1}{2}\frac{(0.10\,m)^2}{(4 \times 10^{-5}\,m^2/s)} \approx 100\,s,$$

or about a minute or two.

NOTE This result seems rather long from experience, suggesting that air currents (convection) are more important than diffusion for transmitting odors.

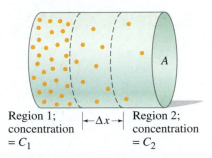

FIGURE 13–26 Diffusion occurs from a region of high concentration to one of lower concentration. (Only one type of molecule is shown.)

Region 1; concentration $= C_1$ $\vert\!\leftarrow\!\Delta x\!\rightarrow\!\vert$ Region 2; concentration $= C_2$

PHYSICS APPLIED
Diffusion time

TABLE 13–4 Diffusion Constants, D (20°C, 1 atm)

Diffusing Molecules	Medium	D (m^2/s)
H_2	Air	6.3×10^{-5}
O_2	Air	1.8×10^{-5}
O_2	Water	100×10^{-11}
Blood hemoglobin	Water	6.9×10^{-11}
Glycine (an amino acid)	Water	95×10^{-11}
DNA (mass $6 \times 10^6\,u$)	Water	0.13×10^{-11}

Diffusion is extremely important for living organisms. For example, molecules produced in certain chemical reactions within cells diffuse to other areas where they take part in other reactions.

Gas diffusion is important too. Plants require carbon dioxide for photosynthesis. The CO_2 diffuses into leaves from the outside air through tiny openings (stomata). As CO_2 is utilized by the cells, its concentration drops below that in the air outside, and more diffuses inward. Water vapor and oxygen produced by the cells diffuse outward into the air.

Animals also exchange oxygen and CO_2 with the environment. Oxygen is required for energy-producing reactions and must diffuse into cells. CO_2 is produced as an end product of many metabolic reactions and must diffuse out of cells. But diffusion is slow over longer distances, so only the smallest organisms in the animal world could survive without having developed complex respiratory and circulatory systems. In humans, oxygen is taken into the lungs, where it diffuses short distances across lung tissue and into the blood. Then the blood circulates it to cells throughout the body. The blood also carries CO_2 produced by the cells back to the lungs, where it diffuses outward.

Summary

The atomic theory of matter postulates that all matter is made up of tiny entities called **atoms**, which are typically 10^{-10} m in diameter.

Atomic and **molecular masses** are specified on a scale where ordinary carbon (^{12}C) is arbitrarily given the value 12.0000 u (atomic mass units).

The distinction between solids, liquids, and gases can be attributed to the strength of the attractive forces between the atoms or molecules and to their average speed.

Temperature is a measure of how hot or cold something is. **Thermometers** are used to measure temperature on the **Celsius** (°C), **Fahrenheit** (°F), and **Kelvin** (K) scales. Two standard points on each scale are the freezing point of water (0°C, 32°F, 273.15 K) and the boiling point of water (100°C, 212°F, 373.15 K). A one-kelvin change in temperature equals a change of one Celsius degree or $\frac{9}{5}$ Fahrenheit degrees. Kelvins are related to °C by

$$T(\text{K}) = T(°\text{C}) + 273.15.$$

The change in length, ΔL, of a solid, when its temperature changes by an amount ΔT, is directly proportional to the temperature change and to its original length L_0. That is,

$$\Delta L = \alpha L_0 \, \Delta T, \tag{13–1a}$$

where α is the *coefficient of linear expansion*.

The change in volume of most solids, liquids, and gases is proportional to the temperature change and to the original volume V_0:

$$\Delta V = \beta V_0 \, \Delta T. \tag{13–2}$$

The *coefficient of volume expansion*, β, is approximately equal to 3α for uniform solids.

Water is unusual because, unlike most materials whose volume increases with temperature, its volume actually decreases as the temperature increases in the range from 0°C to 4°C.

The **ideal gas law**, or **equation of state for an ideal gas**, relates the pressure P, volume V, and temperature T (in kelvins) of n moles of gas by

$$PV = nRT, \tag{13–3}$$

where $R = 8.314 \, \text{J/mol·K}$ for all gases. Real gases obey the ideal gas law quite accurately if they are not at too high a pressure or near their liquefaction point.

One **mole** of a substance is defined as the number of grams which is numerically equal to the atomic or molecular mass.

Avogadro's number, $N_A = 6.02 \times 10^{23}$, is the number of atoms or molecules in 1 mol of any pure substance.

The ideal gas law can be written in terms of the number of molecules N in the gas as

$$PV = NkT, \tag{13–4}$$

where $k = R/N_A = 1.38 \times 10^{-23} \, \text{J/K}$ is Boltzmann's constant.

According to the **kinetic theory** of gases, which is based on the idea that a gas is made up of molecules that are moving rapidly and randomly, the average kinetic energy of molecules is proportional to the Kelvin temperature T:

$$\overline{\text{KE}} = \tfrac{1}{2} m\overline{v^2} = \tfrac{3}{2} kT, \tag{13–8}$$

where k is Boltzmann's constant. At any moment, there exists a wide distribution of molecular speeds within a gas.

[*The behavior of real gases at high pressure, and/or near their liquefaction point, deviates from the ideal gas law, due to molecular size and the attractive forces between molecules. Below the **critical temperature**, a gas can change to a liquid if sufficient pressure is applied; but if the temperature is higher than the critical temperature, no amount of pressure will cause a liquid surface to form. The **triple point** of a substance is that unique temperature and pressure at which all three phases—solid, liquid, and gas—can coexist in equilibrium.]

[***Evaporation** of a liquid is the result of the fastest moving molecules escaping from the surface. **Saturated vapor pressure** refers to the pressure of the vapor above a liquid when the two phases are in equilibrium. The vapor pressure of a substance at its boiling point is equal to atmospheric pressure. **Relative humidity** of air at a given place is the ratio of the partial pressure of water vapor in the air to the saturated vapor pressure at that temperature; it is usually expressed as a percentage.]

[***Diffusion** is the process whereby molecules of a substance move (on average) from one area to another because of a difference in that substance's concentration.]

Questions

1. Which has more atoms: 1 kg of iron or 1 kg of aluminum? See the periodic table or Appendix B.

2. Name several properties of materials that could be exploited to make a thermometer.

3. Which is larger, $1\,C°$ or $1\,F°$?

*4. If system A is in thermal equilibrium with system B, but B is not in thermal equilibrium with system C, what can you say about the temperatures of A, B, and C?

5. A flat bimetallic strip consists of aluminum riveted to a strip of iron. When heated, the strip will bend. Which metal will be on the outside of the curve? [*Hint*: See Table 13–1.] Why?

6. In the relation $\Delta L = \alpha L_0 \Delta T$, should L_0 be the initial length, the final length, or does it matter? Explain.

7. The units for the coefficient of linear expansion α are $(C°)^{-1}$, and there is no mention of a length unit such as meters. Would the expansion coefficient change if we used feet or millimeters instead of meters? Explain.

8. Figure 13–27 shows a diagram of a simple *thermostat* used to control a furnace (or other heating or cooling system). The bimetallic strip consists of two strips of different metals bonded together. The electric switch is a glass vessel containing liquid mercury that conducts electricity when it can flow to touch both contact wires. Explain how this device controls the furnace and how it can be set at different temperatures.

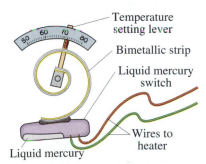

Temperature setting lever
Bimetallic strip
Liquid mercury switch
Wires to heater
Liquid mercury

FIGURE 13–27
A thermostat
(Question 8).

9. Long steam pipes that are fixed at the ends often have a section in the shape of a ∪. Why?

10. A flat, uniform cylinder of lead floats in mercury at 0°C. Will the lead float higher or lower when the temperature is raised? Explain.

11. When a cold mercury-in-glass thermometer is first placed in a hot tub of water, the mercury initially descends a bit and then rises. Explain.

12. A glass container may break if one part of it is heated or cooled more rapidly than adjacent parts. Explain.

13. The principal virtue of Pyrex glass is that its coefficient of linear expansion is much smaller than that for ordinary glass (Table 13–1). Explain why this gives rise to the increased heat resistance of Pyrex.

14. Will a grandfather clock, accurate at 20°C, run fast or slow on a hot day (30°C)? Explain. The clock uses a pendulum supported on a long, thin brass rod.

15. Freezing a can of soda will cause its bottom and top to bulge so badly the can will not stand up. What has happened?

16. When a gas is rapidly compressed (say, by pushing down a piston), its temperature increases. When a gas expands against a piston, it cools. Explain these changes in temperature using the kinetic theory, in particular noting what happens to the momentum of molecules when they strike the moving piston.

17. Will the buoyant force on an aluminum sphere submerged in water increase or decrease if the temperature is increased from 20°C to 40°C? Explain.

18. Explain in words how Charles's law follows from kinetic theory and the relation between average kinetic energy and the absolute temperature.

19. Explain in words how Gay-Lussac's law follows from kinetic theory.

20. As you go higher in the Earth's atmosphere, the ratio of N_2 molecules to O_2 molecules increases. Why?

* 21. Escape velocity for the Earth refers to the minimum speed an object must have to leave the Earth and never return. The escape velocity for the Moon is about one-fifth what it is for the Earth due to the Moon's smaller mass. Explain why the Moon has practically no atmosphere.

* 22. Alcohol evaporates more quickly than water at room temperature. What can you infer about the molecular properties of one relative to the other?

* 23. Explain why a hot humid day is far more uncomfortable than a hot dry day at the same temperature.

* 24. Is it possible to boil water at room temperature (20°C) without heating it? Explain.

* 25. Consider two days when the air temperature is the same but the humidity is different. Which is more dense, the dry air or the humid air at the same T? Explain.

* 26. Explain why it is dangerous to open the radiator cap of an overheated automobile engine.

* 27. Why does exhaled air appear as a little white cloud in the winter (Fig. 13–28)?

FIGURE 13–28 Question 27.

Problems

13–1 Atomic Theory

1. (I) How many atoms are there in a 3.4-gram copper penny?
2. (I) How does the number of atoms in a 26.5-gram gold ring compare to the number in a silver ring of the same mass?

13–2 Temperature and Thermometers

3. (I) (a) "Room temperature" is often taken to be 68°F. What is this on the Celsius scale? (b) The temperature of the filament in a lightbulb is about 1800°C. What is this on the Fahrenheit scale?
4. (I) Among the highest and lowest temperatures recorded are 136°F in the Libyan desert and −129°F in Antarctica. What are these temperatures on the Celsius scale?
5. (I) (a) 15° below zero on the Celsius scale is what Fahrenheit temperature? (b) 15° below zero on the Fahrenheit scale is what Celsius temperature?
6. (II) In an alcohol-in-glass thermometer, the alcohol column has length 11.82 cm at 0.0°C and length 22.85 cm at 100.0°C. What is the temperature if the column has length (a) 16.70 cm, and (b) 20.50 cm?

13–4 Thermal Expansion

7. (I) A concrete highway is built of slabs 12 m long (20°C). How wide should the expansion cracks between the slabs be (at 20°C) to prevent buckling if the range of temperature is −30°C to +50°C?
8. (I) Super Invar™, an alloy of iron and nickel, is a strong material with a very low coefficient of linear expansion $[0.2 \times 10^{-6}\,(\text{C}°)^{-1}]$. A 2.0-m-long tabletop made of this alloy is used for sensitive laser measurements where extremely high tolerances are required. How much will this table expand along its length if the temperature increases 5.0 C°? Compare to tabletops made of steel.
9. (I) The Eiffel Tower (Fig. 13–29) is built of wrought iron approximately 300 m tall. Estimate how much its height changes between July (average temperature of 25°C) and January (average temperature of 2°C). Ignore the angles of the iron beams, and treat the tower as a vertical beam.

FIGURE 13–29
Problem 9.
The Eiffel Tower in Paris.

10. (II) To make a secure fit, rivets that are larger than the rivet hole are often used and the rivet is cooled (usually in dry ice) before it is placed in the hole. A steel rivet 1.871 cm in diameter is to be placed in a hole 1.869 cm in diameter at 20°C. To what temperature must the rivet be cooled if it is to fit in the hole?
11. (II) The density of water at 4°C is $1.00 \times 10^3\,\text{kg/m}^3$. What is water's density at 94°C?
12. (II) A quartz sphere is 8.75 cm in diameter. What will be its change in volume if it is heated from 30°C to 200°C?

13. (II) An ordinary glass is filled to the brim with 350.0 mL of water at 100.0°C. If the temperature decreased to 20.0°C, how much water could be added to the glass?
14. (II) It is observed that 55.50 mL of water at 20°C completely fills a container to the brim. When the container and the water are heated to 60°C, 0.35 g of water is lost. (a) What is the coefficient of volume expansion of the container? (b) What is the most likely material of the container? The density of water at 60°C is 0.98324 g/mL.
15. (II) (a) A brass plug is to be placed in a ring made of iron. At 20°C, the diameter of the plug is 8.753 cm and that of the inside of the ring is 8.743 cm. They must both be brought to what common temperature in order to fit? (b) What if the plug were iron and the ring brass?
16. (II) If a fluid is contained in a long, narrow vessel so it can expand in essentially one direction only, show that the effective coefficient of linear expansion α is approximately equal to the coefficient of volume expansion β.
17. (II) (a) Show that the change in the density ρ of a substance, when the temperature changes by ΔT, is given by $\Delta\rho = -\beta\rho\,\Delta T$. (b) What is the fractional change in density of a lead sphere whose temperature decreases from 25°C to −40°C?
18. (II) A uniform rectangular plate of length l and width w has coefficient of linear expansion α. Show that, if we neglect very small quantities, the change in area of the plate due to a temperature change ΔT is $\Delta A = 2\alpha l w\,\Delta T$. See Fig. 13–30.

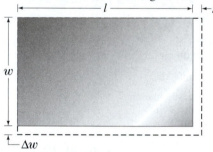

FIGURE 13–30
Problem 18.
A rectangular plate is heated.

19. (III) Show that for an isotropic solid, $\beta = 3\alpha$, if the amount of expansion is small. β and α are the coefficients of volume and linear expansion, respectively. [*Hint:* Consider a cubical solid, and neglect very small quantities. See also Problem 18 and Fig. 13–30.]
20. (III) The pendulum in a grandfather clock is made of brass and keeps perfect time at 17°C. How much time is gained or lost in a year if the clock is kept at 25°C? (Assume the frequency dependence on length for a simple pendulum applies.)
21. (III) (a) The tube of a mercury thermometer has an inside diameter of 0.140 mm. The bulb has a volume of 0.255 cm³. How far will the thread of mercury move when the temperature changes from 11.5°C to 33.0°C? Take into account expansion of the Pyrex glass. (b) Determine a formula for the change in length of the mercury column in terms of relevant variables. Ignore tube volume compared to bulb volume.
22. (III) A 23.4-kg solid aluminum cylindrical wheel of radius 0.41 m is rotating about its axle on frictionless bearings with angular velocity $\omega = 32.8$ rad/s. If its temperature is now raised from 20.0°C to 75.0°C, what is the fractional change in ω?

* 23. (II) An aluminum bar has the desired length when at 15°C. How much stress is required to keep it at this length if the temperature increases to 35°C?

* 24. (II) (a) A horizontal steel I-beam of cross-sectional area 0.041 m² is rigidly connected to two vertical steel girders. If the beam was installed when the temperature was 30°C, what stress is developed in the beam when the temperature drops to −30°C? (b) Is the ultimate strength of the steel exceeded? (c) What stress is developed if the beam is concrete and has a cross-sectional area of 0.13 m²? Will it fracture?

* 25. (III) A barrel of diameter 134.122 cm at 20°C is to be enclosed by an iron band. The circular band has an inside diameter of 134.110 cm at 20°C. It is 7.4 cm wide and 0.65 cm thick. (a) To what temperature must the band be heated so that it will fit over the barrel? (b) What will be the tension in the band when it cools to 20°C?

13–6 Gas Laws; Absolute Temperature

26. (I) What are the following temperatures on the Kelvin scale: (a) 86°C, (b) 78°F, (c) −100°C, (d) 5500°C, (e) −459°F?

27. (I) Absolute zero is what temperature on the Fahrenheit scale?

28. (II) Typical temperatures in the interior of the Earth and Sun are about 4000°C and 15×10^6 °C, respectively. (a) What are these temperatures in kelvins? (b) What percent error is made in each case if a person forgets to change °C to K?

13–7 and 13–8 Ideal Gas Law

29. (I) If 3.00 m³ of a gas initially at STP is placed under a pressure of 3.20 atm, the temperature of the gas rises to 38.0°C. What is the volume?

30. (I) In an internal combustion engine, air at atmospheric pressure and a temperature of about 20°C is compressed in the cylinder by a piston to $\frac{1}{9}$ of its original volume (compression ratio = 9.0). Estimate the temperature of the compressed air, assuming the pressure reaches 40 atm.

31. (II) Calculate the density of oxygen at STP using the ideal gas law.

32. (II) A storage tank contains 21.6 kg of nitrogen (N_2) at an absolute pressure of 3.65 atm. What will the pressure be if the nitrogen is replaced by an equal mass of CO_2?

33. (II) A storage tank at STP contains 18.5 kg of nitrogen (N_2). (a) What is the volume of the tank? (b) What is the pressure if an additional 15.0 kg of nitrogen is added without changing the temperature?

34. (II) If 18.75 mol of helium gas is at 10.0°C and a gauge pressure of 0.350 atm, (a) calculate the volume of the helium gas under these conditions. (b) Calculate the temperature if the gas is compressed to precisely half the volume at a gauge pressure of 1.00 atm.

35. (II) What is the pressure inside a 35.0-L container holding 105.0 kg of argon gas at 385 K?

36. (II) A tank contains 26.0 kg of O_2 gas at a gauge pressure of 8.70 atm. If the oxygen is replaced by helium, how many kilograms of the latter will be needed to produce a gauge pressure of 7.00 atm?

37. (II) A hot-air balloon achieves its buoyant lift by heating the air inside the balloon, which makes it less dense than the air outside. Suppose the volume of a balloon is 1800 m³ and the required lift is 2700 N (rough estimate of the weight of the equipment and passenger). Calculate the temperature of the air inside the balloon which will produce the required lift. Assume that the outside air temperature is 0°C and that air is an ideal gas under these conditions. What factors limit the maximum altitude attainable by this method for a given load? (Neglect variables like wind.)

38. (II) A tire is filled with air at 15°C to a gauge pressure of 220 kPa. If the tire reaches a temperature of 38°C, what fraction of the original air must be removed if the original pressure of 220 kPa is to be maintained?

39. (II) If 61.5 L of oxygen at 18.0°C and an absolute pressure of 2.45 atm are compressed to 48.8 L and at the same time the temperature is raised to 50.0°C, what will the new pressure be?

40. (III) A helium-filled balloon escapes a child's hand at sea level and 20.0°C. When it reaches an altitude of 3000 m, where the temperature is 5.0°C and the pressure is only 0.70 atm, how will its volume compare to that at sea level?

13–9 Ideal Gas Law in Terms of Molecules; Avogadro's Number

41. (I) Calculate the number of molecules/m³ in an ideal gas at STP.

42. (I) How many moles of water are there in 1.000 L? How many molecules?

43. (II) Estimate the number of (a) moles, and (b) molecules of water in all the Earth's oceans. Assume water covers 75% of the Earth to an average depth of 3 km.

44. (II) A cubic box of volume 5.1×10^{-2} m³ is filled with air at atmospheric pressure at 20°C. The box is closed and heated to 180°C. What is the net force on each side of the box?

45. (III) Estimate how many molecules of air are in each 2.0-L breath you inhale that were also in the last breath Galileo took. [Hint: Assume the atmosphere is about 10 km high and of constant density.]

13–10 Molecular Interpretation of Temperature

46. (I) (a) What is the average translational kinetic energy of an oxygen molecule at STP? (b) What is the total translational kinetic energy of 2.0 mol of O_2 molecules at 20°C?

47. (I) Calculate the rms speed of helium atoms near the surface of the Sun at a temperature of about 6000 K.

48. (I) By what factor will the rms speed of gas molecules increase if the temperature is increased from 0°C to 100°C?

49. (I) A gas is at 20°C. To what temperature must it be raised to double the rms speed of its molecules?

50. (I) Twelve molecules have the following speeds, given in units of km/s: 6, 2, 4, 6, 0, 4, 1, 8, 5, 3, 7, and 8. Calculate the rms speed.

51. (II) The rms speed of molecules in a gas at 20.0°C is to be increased by 1.0%. To what temperature must it be raised?

52. (II) If the pressure of a gas is doubled while its volume is held constant, by what factor does v_{rms} change?

53. (II) Show that the rms speed of molecules in a gas is given by $v_{rms} = \sqrt{3P/\rho}$, where P is the pressure in the gas, and ρ is the gas density.

54. (II) Show that for a mixture of two gases at the same temperature, the ratio of their rms speeds is equal to the inverse ratio of the square roots of their molecular masses.

55. (II) What is the rms speed of nitrogen molecules contained in an 8.5-m^3 volume at 2.1 atm if the total amount of nitrogen is 1300 mol?

56. (II) Calculate (a) the rms speed of an oxygen molecule at 0°C and (b) determine how many times per second it would move back and forth across a 7.0-m-long room on the average, assuming it made very few collisions with other molecules.

57. (II) What is the average distance between nitrogen molecules at STP?

58. (II) (a) Estimate the rms speed of an amino acid whose molecular mass is 89 u in a living cell at 37°C. (b) What would be the rms speed of a protein of molecular mass 50,000 u at 37°C?

59. (II) Show that the pressure P of a gas can be written $P = \frac{1}{3}\rho v^2$, where ρ is the density of the gas and v is the rms speed of the molecules.

60. (III) The two isotopes of uranium, ^{235}U and ^{238}U (the superscripts refer to their atomic mass), can be separated by a gas-diffusion process by combining them with fluorine to make the gaseous compound UF_6. Calculate the ratio of the rms speeds of these molecules for the two isotopes, at constant T.

* 13–12 Real Gases; Phase Changes

* 61. (I) (a) At atmospheric pressure, in what phases can CO_2 exist? (b) For what range of pressures and temperatures can CO_2 be a liquid? Refer to Fig. 13–21.

* 62. (I) Water is in which phase when the pressure is 0.01 atm and the temperature is (a) 90°C, (b) −20°C?

* 13–13 Vapor Pressure; Humidity

* 63. (I) What is the dew point (approximately) if the humidity is 50% on a day when the temperature is 25°C?

* 64. (I) What is the air pressure at a place where water boils at 90°C?

* 65. (I) If the air pressure at a particular place in the mountains is 0.72 atm, estimate the temperature at which water boils.

* 66. (I) What is the temperature on a day when the partial pressure of water is 530 Pa and the relative humidity is 40%?

* 67. (I) What is the partial pressure of water on a day when the temperature is 25°C and the relative humidity is 35%?

* 68. (I) What is the approximate pressure inside a pressure cooker if the water is boiling at a temperature of 120°C? Assume no air escaped during the heating process, which started at 20°C.

* 69. (II) If the humidity in a room of volume 680 m^3 at 25°C is 80%, what mass of water can still evaporate from an open pan?

* 70. (III) Air that is at its dew point of 5°C is drawn into a building where it is heated to 25°C. What will be the relative humidity at this temperature? Assume constant pressure of 1.0 atm. Take into account the expansion of the air.

* 13–14 Diffusion

* 71. (II) Estimate the time needed for a glycine molecule (see Table 13–4) to diffuse a distance of 15 μm in water at 20°C if its concentration varies over that distance from 1.00 mol/m^3 to 0.40 mol/m^3. Compare this "speed" to its rms (thermal) speed. The molecular mass of glycine is about 75 u.

* 72. (II) Oxygen diffuses from the surface of insects to the interior through tiny tubes called tracheae. An average trachea is about 2 mm long and has cross-sectional area of $2 \times 10^{-9} m^2$. Assuming the concentration of oxygen inside is half what it is outside in the atmosphere, (a) show that the concentration of oxygen in the air (assume 21% is oxygen) at 20°C is about 8.7 mol/m^3, then (b) calculate the diffusion rate J, and (c) estimate the average time for a molecule to diffuse in. Assume the diffusion constant is $1 \times 10^{-5} m^2/s$.

General Problems

73. A precise steel tape measure has been calibrated at 20°C. At 34°C, (a) will it read high or low, and (b) what will be the percentage error?

74. A Pyrex measuring cup was calibrated at normal room temperature. How much error will be made in a recipe calling for 300 mL of cool water, if the water and the cup are hot, at 80°C, instead of at 20°C? Neglect the glass expansion.

75. The gauge pressure in a helium gas cylinder is initially 28 atm. After many balloons have been blown up, the gauge pressure has decreased to 5 atm. What fraction of the original gas remains in the cylinder?

76. Estimate the number of air molecules in a room of length 6.5 m, width 3.1 m, and height 2.5 m. Assume the temperature is 22°C. How many moles does that correspond to?

77. In outer space the density of matter is about one atom per cm^3, mainly hydrogen atoms, and the temperature is about 2.7 K. Calculate the rms speed of these hydrogen atoms, and the pressure (in atmospheres).

78. The lowest pressure attainable using the best available vacuum techniques is about $10^{-12} N/m^2$. At such a pressure, how many molecules are there per cm^3 at 0°C?

79. If a scuba diver fills his lungs to full capacity of 5.5 L when 10 m below the surface, to what volume would his lungs expand if he quickly rose to the surface? Is this advisable?

80. A space vehicle returning from the Moon enters Earth's atmosphere at a speed of about 40,000 km/h. Molecules (assume nitrogen) striking the nose of the vehicle with this speed correspond to what temperature? (Because of this high temperature, the nose of a space vehicle must be made of special materials; indeed, part of it does vaporize, and this is seen as a bright blaze upon reentry.)

81. The temperature of an ideal gas is increased from 110°C to 360°C while the volume and the number of moles stay constant. By what factor does the pressure change? By what factor does v_{rms} change?

82. A house has a volume of 770 m³. (a) What is the total mass of air inside the house at 20°C? (b) If the temperature drops to −10°C, what mass of air enters or leaves the house?

83. From the known value of atmospheric pressure at the surface of the Earth, estimate the total number of air molecules in the Earth's atmosphere.

84. What is the rms speed of nitrogen molecules contained in a 7.6-m³ volume at 4.2 atm if the total amount of nitrogen is 1800 mol?

85. A standard cylinder of oxygen used in a hospital has gauge pressure = 2000 psi (13,800 kPa) and volume = 16 L (0.016 m³) at T = 295 K. How long will the cylinder last if the flow rate, measured at atmospheric pressure, is constant at 2.4 L/min?

86. An iron cube floats in a bowl of liquid mercury at 0°C. (a) If the temperature is raised to 25°C, will the cube float higher or lower in the mercury? (b) By what percent will the fraction of volume submerged change?

87. The density of gasoline at 0°C is 0.68 × 10³ kg/m³. What is the density on a hot day, when the temperature is 38°C? What is the percentage change?

88. If a steel band were to fit snugly around the Earth's equator at 25°C, but then was heated to 45°C, how high above the Earth would the band be (assume equal everywhere)?

89. A brass lid screws tightly onto a glass jar at 20°C. To help open the jar, it can be placed into a bath of hot water. After this treatment, the temperatures of the lid and the jar are both 60°C. The inside diameter of the lid is 8.0 cm at 20°C. Find the size of the gap (difference in radius) that develops by this procedure.

90. The first length standard, adopted in the 18th century, was a platinum bar with two very fine marks separated by what was defined to be exactly 1 m. If this standard bar was to be accurate to within ± 1.0 μm, how carefully would the trustees have needed to control the temperature? The coefficient of linear expansion for platinum is $9 \times 10^{-6} \, C^{\circ -1}$.

91. A scuba tank, when fully charged, has a pressure of 195 atm at 20°C. The volume of the tank is 11.3 L. (a) What would the volume of the air be at 1.00 atm and at the same temperature? (b) Before entering the water, a person consumes 2.0 L of air in each breath, and breathes 12 times a minute. At this rate, how long would the tank last? (c) At a depth of 20.0 m of sea water and temperature of 10°C, how long would the same tank last assuming the breathing rate does not change?

92. The escape speed from the Earth is 1.12×10^4 m/s, so a gas molecule travelling away from Earth near the outer boundary of the Earth's atmosphere would, at this speed, be able to escape from the Earth's gravitational field. At what temperature is the average speed of (a) oxygen molecules, and (b) helium atoms equal to 1.12×10^4 m/s? (c) Can you see why our atmosphere contains oxygen but not helium?

93. A 1.0-kg trash-can lid is suspended against gravity by tennis balls thrown vertically upward at it. How many tennis balls per second must rebound from the lid elastically, assuming they have a mass of 0.060 kg and are thrown at 12 m/s?

94. A scuba diver releases a 3.00-cm-diameter (spherical) bubble of air from a depth of 14.0 m in a lake. Assume the temperature is constant at 298 K, and the air behaves as a perfect gas. How large is the bubble when it reaches the surface?

* 95. Calculate the total water vapor pressure in the air on the following two days: (a) a hot summer day, with the temperature 30°C and the relative humidity at 40%; (b) a cold winter day, with the temperature 5°C and the relative humidity at 80%.

* 96. A sauna has 7.0 m³ of air volume, and the temperature is 90°C. The air is perfectly dry. How much water (in kg) should be evaporated if we want to increase the relative humidity from 0% to 10%? (See Table 13–3.)

* 97. Estimate the percent difference in the density of iron at STP, and when it is a solid deep in the Earth where the temperature is 2000°C and under 5000 atm of pressure. Assume the bulk modulus (90 × 10⁹ N/m²) and the coefficient of volume expansion do not vary with temperature and are the same as at STP.

* 98. (a) Use the ideal gas law to show that, for an ideal gas at constant pressure, the coefficient of volume expansion is equal to $\beta = 1/T$, where T is the temperature in kelvins. Compare to Table 13–1 for gases at T = 293 K. (b) Show that the bulk modulus (Section 9–5) for an ideal gas held at constant temperature is $B = P$, where P is the pressure.

* 99. In humid climates, people constantly dehumidify their cellars to prevent rot and mildew. If the cellar in a house (kept at 20°C) has 95 m² of floor space and a ceiling height of 2.8 m, what is the mass of water that must be removed from it to drop the humidity from 95% to a more reasonable 30%?

Answers to Exercises

A: −40°C = −40°F.
B: 24.0 L.

C: Less.
D: 3.5×10^{-9} m/s.

When it is cold, warm clothes act as insulators to reduce heat loss from the body to the outside by conduction and convection. Heat radiation from a campfire can warm you and your clothes. The fire can also transfer energy directly by heat conduction to what you are cooking. Heat, like work, represents a transfer of energy. Heat is defined as a transfer of energy due to a difference of temperature. Another useful concept is the internal energy U, which is the sum total of all the energies of the molecules of the system.

CHAPTER **14**

Heat

W hen a pot of cold water is placed on a hot burner of a stove, the temperature of the water increases. We say that heat "flows" from the hot burner to the cold water. When two objects at different temperatures are put in contact, heat spontaneously flows from the hotter one to the colder one. The spontaneous flow of heat is in the direction tending to equalize the temperature. If the two objects are kept in contact long enough for their temperatures to become equal, the objects are said to be in thermal equilibrium, and there is no further heat flow between them. For example, when a fever thermometer is first placed in your mouth, heat flows from your mouth to the thermometer. When the thermometer reaches the same temperature as the inside of your mouth, the thermometer and your mouth are then in equilibrium, and no more heat flows.

Heat and temperature are often confused. They are very different concepts, and in this Chapter we will make a clear distinction between them. We begin by defining and using the concept of heat. We also discuss how heat is used in calorimetry, how it is involved in changes of state of matter, and the processes of heat transfer—conduction, convection, and radiation.

14–1 Heat as Energy Transfer

We use the term "heat" in everyday life as if we knew what we meant. But the term is often used inconsistently, so it is important for us to define heat clearly, and to clarify the phenomena and concepts related to heat.

We commonly speak of the flow of heat—heat flows from a stove burner to a pot of soup, from the Sun to the Earth, from a person's mouth into a fever thermometer. Heat flows spontaneously from an object at higher temperature to one at lower temperature. Indeed, an eighteenth-century model of heat pictured heat flow as movement of a fluid substance called *caloric*. However, the caloric fluid was never able to be detected. In the nineteenth century, it was found that the various phenomena associated with heat could be described consistently using a new model that views heat as being akin to work, as we will discuss in a moment. First we note that a common unit for heat, still in use today, is named after caloric. It is called the **calorie** (cal) and is defined as *the amount of heat necessary to raise the temperature of 1 gram of water by 1 Celsius degree.* [To be precise, the particular temperature range from 14.5°C to 15.5°C is specified because the heat required is very slightly different at different temperatures. The difference is less than 1% over the range 0 to 100°C, and we will ignore it for most purposes.] More often used than the calorie is the **kilocalorie** (kcal), which is 1000 calories. Thus *1 kcal is the heat needed to raise 1 kg of water by 1 C°.* Often a kilocalorie is called a **Calorie** (with a capital C), and it is by this unit that the energy value of food is specified. In the British system of units, heat is measured in British thermal units (Btu). One Btu is defined as the heat needed to raise the temperature of 1 lb of water by 1 F°. It can be shown (Problem 4) that 1 Btu = 0.252 kcal = 1055 J.

The idea that heat is related to energy was pursued by a number of scientists in the 1800s, particularly by an English brewer, James Prescott Joule (1818–1889). Joule and others performed a number of experiments that were crucial for establishing our present-day view that heat, like work, represents a transfer of energy. One of Joule's experiments is shown (simplified) in Fig. 14–1. The falling weight causes the paddle wheel to turn. The friction between the water and the paddle wheel causes the temperature of the water to rise slightly (barely measurable, in fact, by Joule). The same temperature rise could also be obtained by heating the water on a hot stove. In this and many other experiments (some involving electrical energy), Joule determined that a given amount of work done was always equivalent to a particular amount of heat input. Quantitatively, 4.186 joules (J) of work was found to be equivalent to 1 calorie (cal) of heat. This is known as the **mechanical equivalent of heat**:

$$4.186 \text{ J} = 1 \text{ cal};$$
$$4.186 \text{ kJ} = 1 \text{ kcal}.$$

As a result of these and other experiments, scientists came to interpret heat not as a substance, and not exactly as a form of energy. Rather, heat refers to a *transfer of energy*: when heat flows from a hot object to a cooler one, it is energy that is being transferred from the hot to the cold object. Thus, **heat** is *energy transferred from one object to another because of a difference in temperature.* In SI units, the unit for heat, as for any form of energy, is the joule. Nonetheless, calories and kcal are still sometimes used. Today the calorie is *defined* in terms of the joule (via the mechanical equivalent of heat, above), rather than in terms of the properties of water, as given previously. The latter is still handy to remember: 1 cal raises 1 g of water by 1 C°, or 1 kcal raises 1 kg of water by 1 C°.

Whenever we use the word "heat," we mean an energy transfer from one place or object to another at a lower temperature.

CAUTION
Heat is not a fluid

The calorie (unit)

Kilocalorie (= dietary Calorie)

BTU

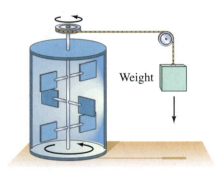

FIGURE 14–1 Joule's experiment on the mechanical equivalent of heat.

Mechanical equivalent of heat

Heat defined: energy transferred because of a ΔT

CAUTION
Heat refers to a transfer of energy—not to energy itself

Joule's result was crucial because it extended the work-energy principle to include processes involving heat. It also led to the establishment of the law of conservation of energy, which we shall discuss more fully in the next Chapter.

PHYSICS APPLIED
Working off Calories

EXAMPLE 14–1 **ESTIMATE** **Working off the extra Calories.** Suppose you throw caution to the wind and eat too much ice cream and cake on the order of 500 Calories. To compensate, you want to do an equivalent amount of work climbing stairs or a mountain. How much total height must you climb? For this calculation, take your mass to be about 60 kg.

APPROACH The work W you need to do in climbing stairs equals the change in gravitational potential energy: $W = \Delta PE = mgh$, where h is the vertical height climbed.

SOLUTION 500 Calories is 500 kcal, which in joules is

$$(500 \text{ kcal})(4.186 \times 10^3 \text{ J/kcal}) = 2.1 \times 10^6 \text{ J}.$$

The work done to climb a vertical height h is $W = mgh$. We solve for h:

$$h = \frac{W}{mg} = \frac{2.1 \times 10^6 \text{ J}}{(60 \text{ kg})(9.80 \text{ m/s}^2)} = 3600 \text{ m}.$$

This is a huge elevation change (over 11,000 ft).

NOTE The human body does not transform energy with 100% efficiency— it is more like 20% efficient. As we'll discuss in the next Chapter, some energy is always "wasted," so you would actually have to climb only about $(0.2)(3600 \text{ m}) \approx 700 \text{ m}$, which is still a lot (about 2300 ft of elevation gain).

14–2 Internal Energy

Internal energy

The sum total of all the energy of all the molecules in an object is called its **internal energy**. (Sometimes **thermal energy** is used to mean the same thing.) We introduce the concept of internal energy now since it will help clarify ideas about heat.

Distinguishing Temperature, Heat, and Internal Energy

CAUTION
Distinguish heat from internal energy and from temperature

Using the kinetic theory, we can make a clear distinction between temperature, heat, and internal energy. Temperature (in kelvins) is a measure of the *average* kinetic energy of individual molecules. Internal energy refers to the *total* energy of all the molecules in the object. (Thus two equal-mass hot ingots of iron may have the same temperature, but two of them have twice as much thermal energy as one does.) Heat, finally, refers to a *transfer* of energy from one object to another because of a difference in temperature.

CAUTION
Direction of heat flow depends on temperature (not on amount of internal energy)

Notice that the direction of heat flow between two objects depends on their temperatures, not on how much internal energy each has. Thus, if 50 g of water at 30°C is placed in contact (or mixed) with 200 g of water at 25°C, heat flows *from* the water at 30°C *to* the water at 25°C even though the internal energy of the 25°C water is much greater because there is so much more of it.

Internal Energy of an Ideal Gas

Let us calculate the internal energy of n moles of an ideal monatomic (one atom per molecule) gas. The internal energy, U, is the sum of the translational kinetic energies of all the atoms. This sum is just equal to the average kinetic energy per molecule times the total number of molecules, N:

$$U = N(\tfrac{1}{2} m\overline{v^2}).$$

Using Eq. 13–8, $\overline{KE} = \tfrac{1}{2} m\overline{v^2} = \tfrac{3}{2} kT$, we can write this as

$$U = \tfrac{3}{2} NkT$$

or (recall Section 13–9)

$$U = \tfrac{3}{2} nRT, \qquad \text{[ideal monatomic gas]} \quad \textbf{(14–1)}$$

Internal energy of ideal monatomic gas

where n is the number of moles. Thus, the internal energy of an ideal gas depends only on temperature and the number of moles of gas.

If the gas molecules contain more than one atom, then the rotational and vibrational energy of the molecules (Fig. 14–2) must also be taken into account. The internal energy will be greater at a given temperature than for a monatomic gas, but it will still be a function only of temperature for an ideal gas.

The internal energy of real gases also depends mainly on temperature, but where real gases deviate from ideal gas behavior, their internal energy depends also somewhat on pressure and volume (due to atomic potential energy).

The internal energy of liquids and solids is quite complicated, for it includes electrical potential energy associated with the forces (or "chemical" bonds) between atoms and molecules.

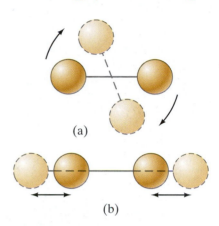

(a)

(b)

FIGURE 14–2 Besides translational kinetic energy, molecules can have (a) rotational kinetic energy, and (b) vibrational energy (both kinetic and potential).

14–3 Specific Heat

If heat flows into an object, the object's temperature rises (assuming no phase change). But how much does the temperature rise? That depends. As early as the eighteenth century, experimenters had recognized that the amount of heat Q required to change the temperature of a given material is proportional to the mass m of the material present and to the temperature change ΔT. This remarkable simplicity in nature can be expressed in the equation

$$Q = mc\,\Delta T, \qquad \textbf{(14–2)}$$

where c is a quantity characteristic of the material called its **specific heat**. Because $c = Q/m\,\Delta T$, specific heat is specified in units of $J/kg \cdot C°$ (the proper SI unit) or $kcal/kg \cdot C°$. For water at 15°C and a constant pressure of 1 atm, $c = 4.19 \times 10^3\ J/kg \cdot C°$ or $1.00\ kcal/kg \cdot C°$, since, by definition of the cal and the joule, it takes 1 kcal of heat to raise the temperature of 1 kg of water by 1 C°. Table 14–1 gives the values of specific heat for other substances at 20°C. The values of c depend to some extent on temperature (as well as slightly on pressure), but for temperature changes that are not too great, c can often be considered constant.

Relation between heat transfer and temperature change

Specific heat

TABLE 14–1 Specific Heats
(at 1 atm constant pressure and 20°C unless otherwise stated)

Substance	Specific Heat, c kcal/kg·C° (= cal/g·C°)	J/kg·C°
Aluminum	0.22	900
Alcohol (ethyl)	0.58	2400
Copper	0.093	390
Glass	0.20	840
Iron or steel	0.11	450
Lead	0.031	130
Marble	0.21	860
Mercury	0.033	140
Silver	0.056	230
Wood	0.4	1700
Water		
Ice (−5°C)	0.50	2100
Liquid (15°C)	1.00	4186
Steam (110°C)	0.48	2010
Human body (average)	0.83	3470
Protein	0.4	1700

EXAMPLE 14–2 How heat transferred depends on specific heat. (a) How much heat input is needed to raise the temperature of an empty 20-kg vat made of iron from 10°C to 90°C? (b) What if the vat is filled with 20 kg of water?

APPROACH We apply Eq. 14–2 to the different materials involved.

SOLUTION (a) Our system is the iron vat alone. From Table 14–1, the specific heat of iron is $450\ J/kg \cdot C°$. The change in temperature is $(90°C - 10°C) = 80\ C°$. Thus,

$$Q = mc\,\Delta T = (20\ \text{kg})(450\ J/kg \cdot C°)(80\ C°) = 7.2 \times 10^5\ J = 720\ kJ.$$

(b) Our system is the vat plus the water. The water alone would require

$$Q = mc\,\Delta T = (20\ \text{kg})(4186\ J/kg \cdot C°)(80\ C°) = 6.7 \times 10^6\ J = 6700\ kJ,$$

or almost 10 times what an equal mass of iron requires. The total, for the vat plus the water, is $720\ kJ + 6700\ kJ = 7400\ kJ$.

NOTE In (b), the iron vat and the water underwent the same temperature change, $\Delta T = 80\ C°$, but their specific heats are different.

If the iron vat in part (a) of Example 14–2 had been *cooled* from 90°C to 10°C, 720 kJ of heat would have flowed *out* of the iron. In other words, Eq. 14–2 is valid for heat flow either in or out, with a corresponding increase or decrease in temperature. We saw in part (b) that water requires almost 10 times as much heat as an equal mass of iron to make the same temperature change. Water has one of the highest specific heats of all substances, which makes it an ideal substance for hot-water space-heating systems and other uses that require a minimal drop in temperature for a given amount of heat transfer. It is the water content, too, that causes the apples rather than the crust in hot apple pie to burn our tongues, through heat transfer.

Practical effects of water's high specific heat

> **CONCEPTUAL EXAMPLE 14–3** | **A very hot frying pan.** You accidentally let an empty iron frying pan get very hot on the stove (200°C or even more). What happens when you dunk it into a few inches of cool water in the bottom of the sink? Will the final temperature be midway between the initial temperatures of the water and pan? Will the water start boiling? Assume the mass of water is roughly the same as the mass of the frying pan.
>
> **RESPONSE** Experience may tell you that the water warms up—perhaps by as much as 10 or 20 degrees. The water doesn't come close to boiling. The water's temperature increase is a lot less than the frying pan's temperature decrease. Why? Because the mass of water is roughly equal to that of the pan, and iron has a specific heat nearly 10 times smaller than that of water (Table 14–1). As heat leaves the frying pan and enters the water, the iron pans' temperature change will be about 10 times greater than that of the water. If, instead, you let a few drops of water fall onto the hot pan, that very small mass of water will sizzle and boil away (the pan's mass may be hundreds of times larger than that of the water).

* Specific Heats for Gases

Specific heats for gases are more complicated than for solids and liquids, which change in volume only slightly with a change in temperature (Section 13–4). Gases change strongly in volume with a change in temperature at constant pressure, as we saw in Chapter 13 with the gas laws; or, if kept at constant volume, the pressure in a gas changes strongly with temperature. The specific heat of a gas depends very much on how the process of changing its temperature is carried out. Most commonly, we deal with the specific heats of gases kept (a) at constant pressure (c_p) or (b) at constant volume (c_v). Some values are given in Table 14–2, where we see that c_p is always greater than c_v. For liquids and solids, this distinction is usually negligible. More details are given in Appendix D on molecular specific heats and the equipartition of energy.

TABLE 14–2
Specific Heats of Gases
(kcal/kg · C°)

Gas	c_p (constant pressure)	c_v (constant volume)
Steam (100°C)	0.482	0.350
Oxygen	0.218	0.155
Helium	1.15	0.75
Carbon dioxide	0.199	0.153
Nitrogen	0.248	0.177

14–4 Calorimetry—Solving Problems

In discussing heat and thermodynamics, we shall often refer to particular systems. As already mentioned in earlier Chapters, a **system** is any object or set of objects that we wish to consider. Everything else in the universe we will refer to as its "environment" or the "surroundings." There are several categories of systems. A **closed system** is one for which no mass enters or leaves (but energy may be exchanged with the environment). In an **open system**, mass may enter or leave (as may energy). Many (idealized) systems we study in physics are closed systems. But many systems, including plants and animals, are open systems since they exchange materials (food, oxygen, waste products) with the environment. A closed system is said to be **isolated** if no energy in any form passes across its boundaries; otherwise it is not isolated.

When different parts of an isolated system are at different temperatures, heat will flow (energy is transferred) from the part at higher temperature to the

Systems

part at lower temperature—that is, within the system. If the system is completely isolated, no energy is transferred into or out of it. So the *conservation of energy* again plays an important role for us: the heat lost by one part of the system is equal to the heat gained by the other part:

$$\text{heat lost} = \text{heat gained}$$

or

$$\text{energy out of one part} = \text{energy into another part}.$$

These simple relations are very useful. Let us take an Example.

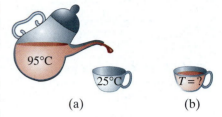
(a)   (b)

FIGURE 14–3 Example 14–4.

Energy conservation

EXAMPLE 14–4 **The cup cools the tea.** If 200 cm³ of tea at 95°C is poured into a 150-g glass cup initially at 25°C (Fig. 14–3), what will be the common final temperature T of the tea and cup when equilibrium is reached, assuming no heat flows to the surroundings?

APPROACH We apply conservation of energy to our system of tea plus cup, which we are assuming is isolated: all of the heat that leaves the tea flows into the cup. We can use the specific heat equation, Eq. 14–2, to determine how the heat flow is related to the temperature changes.

SOLUTION Since tea is mainly water, its specific heat is 4186 J/kg·C° (Table 14–1), and its mass m is its density times its volume $(V = 200\,\text{cm}^3 = 200 \times 10^{-6}\,\text{m}^3)$: $m = \rho V = (1.0 \times 10^3\,\text{kg/m}^3)(200 \times 10^{-6}\,\text{m}^3) = 0.20\,\text{kg}$. We use Eq. 14–2, apply conservation of energy, and let T be the as yet unknown final temperature:

$$\text{heat lost by tea} = \text{heat gained by cup}$$
$$m_{\text{tea}}\,c_{\text{tea}}(95°\text{C} - T) = m_{\text{cup}}\,c_{\text{cup}}(T - 25°\text{C}).$$

Putting in numbers and using Table 14–1 ($c_{\text{cup}} = 840$ J/kg·C° for glass), we solve for T, and find

$$(0.20\,\text{kg})(4186\,\text{J/kg·C°})(95°\text{C} - T) = (0.15\,\text{kg})(840\,\text{J/kg·C°})(T - 25°\text{C})$$
$$79{,}500\,\text{J} - (837\,\text{J/C°})T = (126\,\text{J/C°})T - 3150\,\text{J}$$
$$T = 86°\text{C}.$$

The tea drops in temperature by 9 C° by coming into equilibrium with the cup.

NOTE The cup increases in temperature by $86°\text{C} - 25°\text{C} = 61\,\text{C}°$. Its much greater change in temperature (compared with that of the tea water) is due to its much smaller specific heat compared to that of water.

NOTE In this calculation, the ΔT (of Eq. 14–2, $Q = mc\,\Delta T$) is a positive quantity on both sides of our conservation of energy equation. On the left is "heat lost" and ΔT is the initial minus the final temperature ($95°\text{C} - T$), whereas on the right is "heat gained" and ΔT is the final minus the initial temperature. But consider this alternate approach.

Alternate Solution We can set up this Example (and others) by a different approach. We can write that the total heat transferred into or out of the isolated system is zero:

$$\Sigma Q = 0.$$

Then each term is written as $Q = mc(T_{\text{f}} - T_{\text{i}})$, and $\Delta T = T_{\text{f}} - T_{\text{i}}$ is always the final minus the initial temperature, and each ΔT can be positive or negative. In the present Example:

$$\Sigma Q = m_{\text{cup}}\,c_{\text{cup}}(T - 25°\text{C}) + m_{\text{tea}}\,c_{\text{tea}}(T - 95°\text{C}) = 0.$$

The second term is negative because T will be less than 95°C. Solving the algebra gives the same result.

⚠️ **CAUTION**
When using
heat lost = heat gained,
ΔT is positive on both sides

➡ **PROBLEM SOLVING**
Alternate approach: $\Sigma Q = 0$

The exchange of energy, as exemplified in Example 14–4, is the basis for a technique known as **calorimetry**, which is the quantitative measurement of heat

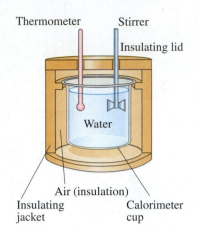

Thermometer   Stirrer

Insulating lid

Water

Air (insulation)

Insulating   Calorimeter
jacket   cup

FIGURE 14–4 Simple water calorimeter.

exchange. To make such measurements, a **calorimeter** is used; a simple water calorimeter is shown in Fig. 14–4. It is very important that the calorimeter be well insulated so that almost no heat is exchanged with the surroundings. One important use of the calorimeter is in the determination of specific heats of substances. In the technique known as the "method of mixtures," a sample of a substance is heated to a high temperature, which is accurately measured, and then quickly placed in the cool water of the calorimeter. The heat lost by the sample will be gained by the water and the calorimeter cup. By measuring the final temperature of the mixture, the specific heat can be calculated, as illustrated in the following Example.

EXAMPLE 14–5 **Unknown specific heat determined by calorimetry.**
An engineer wishes to determine the specific heat of a new metal alloy. A 0.150-kg sample of the alloy is heated to 540°C. It is then quickly placed in 400 g of water at 10.0°C, which is contained in a 200-g aluminum calorimeter cup. (We do not need to know the mass of the insulating jacket since we assume the air space between it and the cup insulates it well, so that its temperature does not change significantly.) The final temperature of the system is 30.5°C. Calculate the specific heat of the alloy.

APPROACH We apply conservation of energy to our system, which we take to be the alloy sample, the water, and the calorimeter cup. We assume this system is isolated, so the energy lost by the hot alloy equals the energy gained by the water and calorimeter cup.

SOLUTION The heat lost equals the heat gained:

$$\begin{pmatrix} \text{heat lost} \\ \text{by alloy} \end{pmatrix} = \begin{pmatrix} \text{heat gained} \\ \text{by water} \end{pmatrix} + \begin{pmatrix} \text{heat gained by} \\ \text{calorimeter cup} \end{pmatrix}$$

$$m_a c_a \, \Delta T_a = m_w c_w \, \Delta T_w + m_{cal} c_{cal} \, \Delta T_{cal}$$

where the subscripts a, w, and cal refer to the alloy, water, and calorimeter, respectively, and each $\Delta T > 0$. When we put in values and use Table 14–1, this equation becomes

$$(0.150 \, \text{kg})(c_a)(540°C - 30.5°C) = (0.40 \, \text{kg})(4186 \, \text{J/kg} \cdot \text{C°})(30.5°C - 10.0°C)$$
$$+ (0.20 \, \text{kg})(900 \, \text{J/kg} \cdot \text{C°})(30.5°C - 10.0°C)$$
$$76.4 \, c_a = (34{,}300 + 3700) \, \text{J/kg} \cdot \text{C°}$$
$$c_a = 500 \, \text{J/kg} \cdot \text{C°}.$$

In making this calculation, we have ignored any heat transferred to the thermometer and the stirrer (which is used to quicken the heat transfer process and thus reduce heat loss to the outside). It can be taken into account by adding additional terms to the right side of the above equation and will result in a slight correction to the value of c_a (see Problem 14).

In all Examples and Problems of this sort, be sure to include *all* objects that gain or lose heat (within reason). On the "heat loss" side here, it is only the hot metal alloy. On the "heat gain" side, it is both the water and the aluminum calorimeter cup. For simplicity, we have ignored very small masses, such as the thermometer and the stirrer, which will affect the energy balance only very slightly.

A **bomb calorimeter** is used to measure the thermal energy released when a substance burns. Important applications are the burning of foods to determine their Calorie content, and the burning of seeds and other substances to determine their "energy content," or heat of combustion. A carefully weighed sample of the substance, together with an excess amount of oxygen at high pressure, is placed in a sealed container (the "bomb"). The bomb is placed in the water of the calorimeter and a fine wire passing into the bomb is then heated briefly, which causes the mixture to ignite. The energy released in the burning process is gained by the water and the bomb.

EXAMPLE 14–6 **Measuring the energy content of a cookie.** Determine the energy content of 100 g of Fahlgren's fudge cookies from the following measurements. A 10-g sample of a cookie is allowed to dry before putting it in a bomb calorimeter. The aluminum bomb has a mass of 0.615 kg and is placed in 2.00 kg of water contained in an aluminum calorimeter cup of mass 0.524 kg. The initial temperature of the system is 15.0°C, and its temperature after ignition is 36.0°C.

APPROACH We apply energy conservation to our system, which we assume is isolated and consists of the cookie sample, the bomb, the calorimeter cup, and the water.

SOLUTION In this case, the heat Q released in the burning of the cookie is absorbed by the system of bomb, calorimeter, and water:

$$Q = \left(m_w c_w + m_{cal} c_{cal} + m_{bomb} c_{bomb}\right) \Delta T$$
$$= \left[(2.00\,\text{kg})(1.0\,\text{kcal/kg·C°}) + (0.524\,\text{kg})(0.22\,\text{kcal/kg·C°}) \right.$$
$$\left. + (0.615\,\text{kg})(0.22\,\text{kcal/kg·C°})\right][36.0°\text{C} - 15.0°\text{C}] = 47\,\text{kcal}.$$

In joules, $Q = (47\,\text{kcal})(4186\,\text{J/kcal}) = 197\,\text{kJ}$. Since 47 kcal is released in the burning of 10 g of cookie, a 100-g portion would contain 470 food Calories, or 1970 kJ.

14–5 Latent Heat

When a material changes phase from solid to liquid, or from liquid to gas (see also Section 13–12), a certain amount of energy is involved in this **change of phase**. For example, let us trace what happens when a 1.0-kg block of ice at −40°C is heated at a slow steady rate until all the ice has changed to water, then the (liquid) water is heated to 100°C and changed to steam above 100°C, all at 1 atm pressure. As shown in the graph of Fig. 14–5, as the ice is heated, its temperature rises at a rate of about 2 C°/kcal of heat added (since for ice, $c \approx 0.50\,\text{kcal/kg·C°}$). However, when 0°C is reached, the temperature stops increasing even though heat is still being added. The ice gradually changes to water in the liquid state, with no change in temperature. After about 40 kcal has been added at 0°C, half the ice remains and half has changed to water. After about 80 kcal, or 330 kJ, has been added, all the ice has changed to water, still at 0°C. Continued addition of heat causes the water's temperature to again increase, now at a rate of 1 C°/kcal. When 100°C is reached, the temperature again remains constant as the heat added changes the liquid water to vapor (steam). About 540 kcal (2260 kJ) is required to change the 1.0 kg of water completely to steam, after which the graph rises again, indicating that the temperature of the steam rises as heat is added.

FIGURE 14–5 Temperature as a function of the heat added to bring 1.0 kg of ice at −40°C to steam above 100°C.

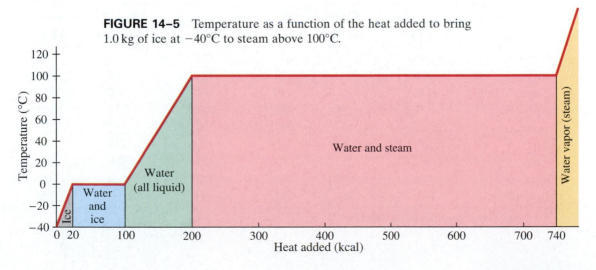

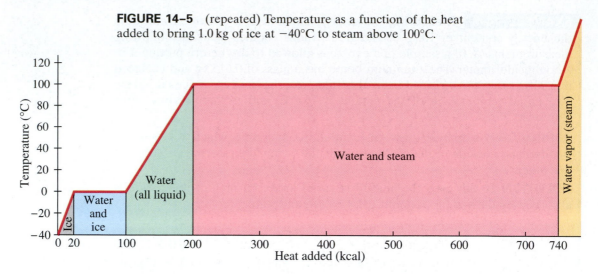

FIGURE 14–5 (repeated) Temperature as a function of the heat added to bring 1.0 kg of ice at −40°C to steam above 100°C.

Heat of fusion   The heat required to change 1.0 kg of a substance from the solid to the liquid state is called the **heat of fusion**; it is denoted by L_F. The heat of fusion of water is 79.7 kcal/kg or, in proper SI units, 333 kJ/kg $(= 3.33 \times 10^5 \text{ J/kg})$. The heat required to change a substance from the liquid to the vapor phase is called *Heat of vaporization*   the **heat of vaporization**, L_V. For water it is 539 kcal/kg or 2260 kJ/kg. Other substances follow graphs similar to Fig. 14–5, although the melting-point and boiling-point temperatures are different, as are the specific heats and heats of fusion and vaporization. Values for the heats of fusion and vaporization, which *Latent heat*   are also called the **latent heats**, are given in Table 14–3 for a number of substances.

The heats of vaporization and fusion also refer to the amount of heat *released* by a substance when it changes from a gas to a liquid, or from a liquid to a solid. Thus, steam releases 2260 kJ/kg when it changes to water, and water releases 333 kJ/kg when it becomes ice.

The heat involved in a change of phase depends not only on the latent heat but also on the total mass of the substance. That is,

Change of phase
$$Q = mL, \qquad\qquad (14\text{–}3)$$

where L is the latent heat of the particular process and substance, m is the mass of the substance, and Q is the heat added or released during the phase change. For example, when 5.00 kg of water freezes at 0°C, $(5.00 \text{ kg})(3.33 \times 10^5 \text{ J/kg}) = 1.67 \times 10^6 \text{ J}$ of energy is released.

TABLE 14–3 Latent Heats (at 1 atm)

Substance	Melting Point (°C)	Heat of Fusion kcal/kg[†]	kJ/kg	Boiling Point (°C)	Heat of Vaporization kcal/kg[†]	kJ/kg
Oxygen	−218.8	3.3	14	−183	51	210
Nitrogen	−210.0	6.1	26	−195.8	48	200
Ethyl alcohol	−114	25	104	78	204	850
Ammonia	−77.8	8.0	33	−33.4	33	137
Water	0	79.7	333	100	539	2260
Lead	327	5.9	25	1750	208	870
Silver	961	21	88	2193	558	2300
Iron	1808	69.1	289	3023	1520	6340
Tungsten	3410	44	184	5900	1150	4800

[†] Numerical values in kcal/kg are the same in cal/g.

Calorimetry sometimes involves a change of state, as the following Examples show. Indeed, latent heats are often measured using calorimetry.

EXAMPLE 14–7 **Making ice.** How much energy does a freezer have to remove from 1.5 kg of water at 20°C to make ice at −12°C?

APPROACH We need to calculate the total energy removed by adding the heat outflow (1) to reduce the water from 20°C to 0°C, (2) to change it to ice at 0°C, and (3) to lower the ice from 0°C to −12°C.

SOLUTION The heat Q that needs to be removed from the 1.5 kg of water is

$$Q = mc_w(20°C − 0°C) + mL_F + mc_{ice}[0° − (−12°C)]$$
$$= (1.5\,kg)(4186\,J/kg \cdot C°)(20\,C°) + (1.5\,kg)(3.33 × 10^5\,J/kg)$$
$$+ (1.5\,kg)(2100\,J/kg \cdot C°)(12\,C°)$$
$$= 6.6 × 10^5\,J = 660\,kJ.$$

EXAMPLE 14–8 **ESTIMATE** **Will all the ice melt?** At a reception, a 0.50-kg chunk of ice at −10°C is placed in 3.0 kg of "iced" tea at 20°C. At what temperature and in what phase will the final mixture be? The tea can be considered as water. Ignore any heat flow to the surroundings, including the container.

APPROACH Before we can write down an equation applying conservation of energy, we must first check to see if the final state will be all ice, a mixture of ice and water at 0°C, or all water. To bring the 3.0 kg of water at 20°C down to 0°C would require an energy release of

➡ **PROBLEM SOLVING**

First determine (or estimate) the final state

$$m_w c_w(20°C − 0°C) = (3.0\,kg)(4186\,J/kg \cdot C°)(20\,C°) = 250\,kJ.$$

On the other hand, to raise the ice from −10°C to 0°C would require

$$m_{ice} c_{ice}[0°C − (−10°C)] = (0.50\,kg)(2100\,J/kg \cdot C°)(10\,C°) = 10.5\,kJ,$$

and to change the ice to water at 0°C would require

$$m_{ice} L_F = (0.50\,kg)(333\,kJ/kg) = 167\,kJ,$$

for a total of 10.5 kJ + 167 kJ = 177 kJ. This is not enough energy to bring the 3.0 kg of water at 20°C down to 0°C, so we know that the mixture must end up all water, somewhere between 0°C and 20°C.

SOLUTION To determine the final temperature T, we apply conservation of energy and write

Then determine the final temperature

$$\text{heat gain} = \text{heat loss}$$

$$\begin{pmatrix} \text{heat to raise} \\ \text{0.50 kg of ice} \\ \text{from } −10°C \\ \text{to } 0°C \end{pmatrix} + \begin{pmatrix} \text{heat to change} \\ \text{0.50 kg} \\ \text{of ice} \\ \text{to water} \end{pmatrix} + \begin{pmatrix} \text{heat to raise} \\ \text{0.50 kg of water} \\ \text{from } 0°C \\ \text{to } T \end{pmatrix} = \begin{pmatrix} \text{heat lost by} \\ \text{3.0 kg of} \\ \text{water cooling} \\ \text{from } 20°C \text{ to } T \end{pmatrix}.$$

Using some of the results from above, we obtain

$$10.5\,kJ + 167\,kJ + (0.50\,kg)(4186\,J/kg \cdot C°)(T − 0°C)$$
$$= (3.0\,kg)(4186\,J/kg \cdot C°)(20°C − T).$$

Solving for T we obtain

$$T = 5.0°C.$$

EXERCISE A How much more ice at −10°C would be needed in Example 14–8 to bring the tea down to 0°C, while just melting all the ice?

1. Be sure you have sufficient information to apply energy conservation. Ask yourself: **is the system isolated** (or very nearly so, enough to get a good estimate)? Do we know or can we calculate all significant sources of energy transfer?

2. Apply **conservation of energy**:

    heat gained = heat lost.

    For each substance in the system, a heat (energy) term will appear on either the left or right side of this equation. [Alternatively, use $\Sigma Q = 0$.]

3. If **no phase changes** occur, each term in the energy conservation equation (above) will have the form

    $$Q(\text{gain}) = mc(T_f - T_i)$$

    or

    $$Q(\text{lost}) = mc(T_i - T_f)$$

    where T_i and T_f are the initial and final temperatures

of the substance, and m and c are its mass and specific heat, respectively.

4. If **phase changes** do or might occur, there may be terms in the energy conservation equation of the form $Q = mL$, where L is the latent heat. But *before* applying energy conservation, determine (or estimate) in which phase the final state will be, as we did in Example 14–8 by calculating the different contributing values for heat Q.

5. Be sure each term appears on the correct side of the **energy equation** (heat gained or heat lost) and that each ΔT is positive.

6. Note that when the system reaches thermal **equilibrium**, the final **temperature** of each substance will have the *same* value. There is only one T_f.

7. **Solve** your energy equation for the unknown.

EXAMPLE 14–9 **Determining a latent heat.** The specific heat of liquid mercury is $140\,\text{J/kg}\cdot\text{C}°$. When 1.0 kg of solid mercury at its melting point of $-39°\text{C}$ is placed in a 0.50-kg aluminum calorimeter filled with 1.2 kg of water at $20.0°\text{C}$, the final temperature of the combination is found to be $16.5°\text{C}$. What is the heat of fusion of mercury in J/kg?

APPROACH We follow the Problem Solving Box explicitly.

SOLUTION

1. **Is the system isolated?** The mercury is placed in a calorimeter, which, by definition, is well insulated. Our isolated system is the calorimeter, the water, and the mercury.

2. **Conservation of energy.** The heat gained by the mercury = the heat lost by the water and calorimeter.

3 and 4. **Phase changes.** There is a phase change, plus we use specific heat equations. The heat gained by the mercury (Hg) includes a term representing the melting of the Hg:

$$Q(\text{melt solid Hg}) = m_{Hg}L_{Hg},$$

plus a term representing the heating of the liquid Hg from $-39°\text{C}$ to $+16.5°\text{C}$:

$$Q(\text{heat liquid Hg}) = m_{Hg}c_{Hg}\big[16.5°\text{C} - (-39°\text{C})\big]$$
$$= (1.0\,\text{kg})(140\,\text{J/kg}\cdot\text{C}°)(55.5\,\text{C}°) = 7770\,\text{J}.$$

All of this heat gained by the mercury is obtained from the water and calorimeter, which cool down:

$$Q_{\text{cal}} + Q_{H_2O} = m_{\text{cal}}c_{\text{cal}}(20.0°\text{C} - 16.5°\text{C}) + m_{H_2O}c_{H_2O}(20.0°\text{C} - 16.5°\text{C})$$
$$= (0.50\,\text{kg})(900\,\text{J/kg}\cdot\text{C}°)(3.5\,\text{C}°) + (1.2\,\text{kg})(4186\,\text{J/kg}\cdot\text{C}°)(3.5\,\text{C}°)$$
$$= 19{,}200\,\text{J}.$$

5. **Energy equation.** The conservation of energy tells us the heat lost by the water and calorimeter cup must equal the heat gained by the mercury:

$$Q_{\text{cal}} + Q_{H_2O} = Q(\text{melt solid Hg}) + Q(\text{heat liquid Hg})$$

or

$$19{,}200\,\text{J} = m_{Hg}L_{Hg} + 7770\,\text{J}.$$

6. **Equilibrium temperature.** It is given as $16.5°\text{C}$, and we already used it.

7. Solve. The only unknown in our energy equation (point 5) is L_{Hg}, the latent heat of fusion (or melting) of mercury. We solve for it, putting in $m_{Hg} = 1.0\,kg$:

$$L_{Hg} = \frac{19{,}200\,J - 7770\,J}{1.0\,kg} = 11{,}400\,J/kg \approx 11\,kJ/kg,$$

where we rounded off to 2 significant figures.

Evaporation

The latent heat to change a liquid to a gas is needed not only at the boiling point. Water can change from the liquid to the gas phase even at room temperature. This process is called **evaporation** (see also Section 13–13). The value of the heat of vaporization of water increases slightly with a decrease in temperature: at 20°C, for example, it is 2450 kJ/kg (585 kcal/kg) compared to 2260 kJ/kg (= 539 kcal/kg) at 100°C. When water evaporates, the remaining liquid cools, because the energy required (the latent heat of vaporization) comes from the water itself; so its internal energy, and therefore its temperature, must drop.[†]

Evaporation of water from the skin is one of the most important methods the body uses to control its temperature. When the temperature of the blood rises slightly above normal, the hypothalamus region of the brain detects this temperature increase and sends a signal to the sweat glands to increase their production. The energy (latent heat) required to vaporize this water comes from the body, and hence the body cools.

PHYSICS APPLIED
Body temperature

Kinetic Theory of Latent Heats

We can make use of kinetic theory to see why energy is needed to melt or vaporize a substance. At the melting point, the latent heat of fusion does not act to increase the average kinetic energy (and the temperature) of the molecules in the solid, but instead is used to overcome the potential energy associated with the forces between the molecules. That is, work must be done against these attractive forces to break the molecules loose from their relatively fixed positions in the solid so they can freely roll over one another in the liquid phase. Similarly, energy is required for molecules held close together in the liquid phase to escape into the gaseous phase. This process is a more violent reorganization of the molecules than is melting (the average distance between the molecules is greatly increased), and hence the heat of vaporization is generally much greater than the heat of fusion for a given substance.

14–6 Heat Transfer: Conduction

Three methods of heat transfer

Heat transfer from one place or object to another occurs in three different ways: by *conduction, convection,* and *radiation.* We now discuss each of these in turn; but in practical situations, any two or all three may be operating at the same time. This Section deals with conduction.

When a metal poker is put in a hot fire, or a silver spoon is placed in a hot bowl of soup, the end that you hold soon becomes hot as well, even though it is not directly in contact with the source of heat. We say that heat has been *conducted* from the hot end to the cold end.

Heat **conduction** in many materials can be visualized as being carried out via molecular collisions. As one end of an object is heated, the molecules there move faster and faster. As they collide with their slower-moving neighbors, they transfer some of their kinetic energy to these molecules, whose speeds thus increase. These in turn transfer some of their energy by collision with molecules still farther along the object. Thus the kinetic energy of thermal motion is transferred by molecular collision along the object. In metals, according to modern theory, it is collisions of free electrons within the metal that are visualized as being mainly responsible for conduction.

[†] According to kinetic theory, evaporation is a cooling process because it is the fastest-moving molecules that escape from the surface (Section 13–13). Hence the average speed of the remaining molecules is less, so by Eq. 13–8 the temperature is less.

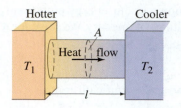

FIGURE 14–6 Heat conduction between areas at temperatures T_1 and T_2. If T_1 is greater than T_2, the heat flows to the right; the rate is given by Eq. 14–4.

Heat conduction from one point to another takes place only if there is a difference in temperature between the two points. Indeed, it is found experimentally that the rate of heat flow through a substance is proportional to the difference in temperature between its ends. The rate of heat flow also depends on the size and shape of the object. To investigate this quantitatively, let us consider the heat flow through a uniform cylinder, as illustrated in Fig. 14–6. It is found experimentally that the heat flow Q over a time interval t is given by the relation

$$\frac{Q}{t} = kA\frac{T_1 - T_2}{l} \tag{14–4}$$

where A is the cross-sectional area of the object, l is the distance between the two ends, which are at temperatures T_1 and T_2, and k is a proportionality constant called the **thermal conductivity** which is characteristic of the material. From Eq. 14–4, we see that the rate of heat flow (units of J/s) is directly proportional to the cross-sectional area and to the temperature gradient[†] $(T_1 - T_2)/l$.

The thermal conductivities, k, for a variety of substances are given in Table 14–4. Substances for which k is large conduct heat rapidly and are said to be good **conductors**. Most metals fall in this category, although there is a wide range even among them, as you may observe by holding the ends of a silver spoon and a stainless-steel spoon immersed in the same hot cup of soup. Substances for which k is small, such as wool, fiberglass, polyurethane, and goose down, are poor conductors of heat and are therefore good **insulators**. The relative magnitudes of k can explain simple phenomena such as why a tile floor is much colder on the feet than a rug-covered floor at the same temperature. Tile is a better conductor of heat than the rug; heat that flows from your foot to the rug is not conducted away rapidly, so the rug's surface quickly warms up to the temperature of your foot and feels good. But the tile conducts the heat away rapidly and thus can take more heat from your foot quickly, so your foot's surface temperature drops.

Rate of heat flow by conduction

TABLE 14–4
Thermal Conductivities

Substance	Thermal Conductivity, k	
	kcal $(s \cdot m \cdot C°)$	J $(s \cdot m \cdot C°)$
Silver	10×10^{-2}	420
Copper	9.2×10^{-2}	380
Aluminum	5.0×10^{-2}	200
Steel	1.1×10^{-2}	40
Ice	5×10^{-4}	2
Glass	2.0×10^{-4}	0.84
Brick	2.0×10^{-4}	0.84
Concrete	2.0×10^{-4}	0.84
Water	1.4×10^{-4}	0.56
Human tissue	0.5×10^{-4}	0.2
Wood	0.3×10^{-4}	0.1
Fiberglass	0.12×10^{-4}	0.048
Cork	0.1×10^{-4}	0.042
Wool	0.1×10^{-4}	0.040
Goose down	0.06×10^{-4}	0.025
Polyurethane	0.06×10^{-4}	0.024
Air	0.055×10^{-4}	0.023

Why rugs feel warmer than tile

PHYSICS APPLIED
Heat loss through windows

EXAMPLE 14–10 Heat loss through windows. A major source of heat loss from a house is through the windows. Calculate the rate of heat flow through a glass window 2.0 m × 1.5 m in area and 3.2 mm thick, if the temperatures at the inner and outer surfaces are 15.0°C and 14.0°C, respectively (Fig. 14–7).

APPROACH Heat flows by conduction through the 3.2-mm thickness of glass from the higher inside temperature to the lower outside temperature. We use the heat conduction equation, Eq. 14–4.

SOLUTION Here $A = (2.0\,\text{m})(1.5\,\text{m}) = 3.0\,\text{m}^2$ and $l = 3.2 \times 10^{-3}\,\text{m}$. Using Table 14–4 to get k, we have

$$\frac{Q}{t} = kA\frac{T_1 - T_2}{l} = \frac{(0.84\,\text{J/s}\cdot\text{m}\cdot\text{C°})(3.0\,\text{m}^2)(15.0°\text{C} - 14.0°\text{C})}{(3.2 \times 10^{-3}\,\text{m})}$$

$$= 790\,\text{J/s}.$$

NOTE This rate of heat flow is equivalent to $(790\,\text{J/s})/(4.19 \times 10^3\,\text{J/kcal}) = 0.19\,\text{kcal/s}$, or $(0.19\,\text{kcal/s}) \times (3600\,\text{s/h}) = 680\,\text{kcal/h}$.

FIGURE 14–7 Example 14–10.

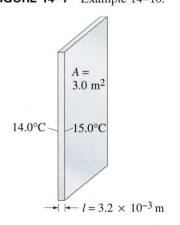

$A = 3.0\,\text{m}^2$

14.0°C 15.0°C

$l = 3.2 \times 10^{-3}\,\text{m}$

[†]Equation 14–4 is quite similar to the relations describing diffusion (Section 13–14) and the flow of fluids through a pipe (Section 10–12). In those cases, the flow of matter was found to be proportional to the concentration gradient $(C_1 - C_2)/l$, or to the pressure gradient $(P_1 - P_2)/l$. This close similarity is one reason we speak of the "flow" of heat. Yet we must keep in mind that no substance is flowing in this case—it is energy that is being transferred.

You might notice in Example 14–10 that 15°C is not very warm for the living room of a house. The room itself may indeed be much warmer, and the outside might be colder than 14°C. But the temperatures of 15°C and 14°C were specified as those at the window surfaces, and there is usually a considerable drop in temperature of the air in the vicinity of the window both on the inside and the outside. That is, the layer of air on either side of the window acts as an insulator, and normally the major part of the temperature drop between the inside and outside of the house takes place across the air layer. If there is a heavy wind, the air outside a window will constantly be replaced with cold air; the temperature gradient across the glass will be greater and there will be a much greater rate of heat loss. Increasing the width of the air layer, such as using two panes of glass separated by an air gap, will reduce the heat loss more than simply increasing the glass thickness, since the thermal conductivity of air is much less than that for glass.

Wind can cause much greater heat loss

The insulating properties of clothing come from the insulating properties of air. Without clothes, our bodies would heat the air in contact with the skin and would soon become reasonably comfortable because air is a very good insulator. But since air moves—there are breezes and drafts, and people move about—the warm air would be replaced by cold air, thus increasing the temperature difference and the heat loss from the body. Clothes keep us warm by trapping air so it cannot move readily. It is not the cloth that insulates us, but the air that the cloth traps. Goose down is a very good insulator because even a small amount of it fluffs up and traps a great amount of air.

| **EXERCISE B** Explain why drapes in front of a window reduce heat loss from a house.

R-values for Building Materials

For practical purposes the thermal properties of building materials, particularly when considered as insulation, are usually specified by *R*-values (or "thermal resistance"), defined for a given thickness *l* of material as:

$$R = \frac{l}{k}.$$

The *R*-value of a given piece of material combines the thickness *l* and the thermal conductivity *k* in one number. In the United States, *R*-values are given in British units as $\text{ft}^2 \cdot \text{h} \cdot \text{F}° / \text{Btu}$ (for example, *R*-19 means $R = 19 \, \text{ft}^2 \cdot \text{h} \cdot \text{F}° / \text{Btu}$). Table 14–5 gives *R*-values for some common building materials: note that *R*-values increase directly with material thickness. For example, 2 inches of fiberglass is *R*-6, half that for 4 inches (= *R*-12; see Table 14–5).

TABLE 14–5 *R*-values

Material	Thickness	*R*-value ($\text{ft}^2 \cdot \text{h} \cdot \text{F}°/\text{Btu}$)
Glass	$\frac{1}{8}$ inch	1
Brick	$3\frac{1}{2}$ inches	0.6–1
Plywood	$\frac{1}{2}$ inch	0.6
Fiberglass insulation	4 inches	12

14–7 Heat Transfer: Convection

Although liquids and gases are generally not very good conductors of heat, they can transfer heat quite rapidly by convection. **Convection** is the process whereby heat flows by the mass movement of molecules from one place to another. Whereas conduction involves molecules (and/or electrons) moving only over small distances and colliding, convection involves the movement of large numbers of molecules over large distances.

A forced-air furnace, in which air is heated and then blown by a fan into a room, is an example of *forced convection. Natural convection* occurs as well, and one familiar example is that hot air rises. For instance, the air above a radiator (or other type of heater) expands as it is heated (Chapter 13), and hence its density decreases. Because its density is less than that of the surrounding cooler air, it rises, just as a log submerged in water floats upward because its density is less than that of water. Warm or cold ocean currents, such as the balmy Gulf Stream, represent natural convection on a global scale. Wind is another example of convection, and weather in general is a result of convective air currents.

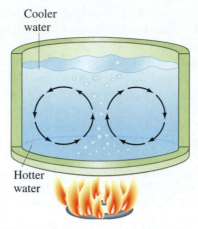

FIGURE 14–8 Convection currents in a pot of water being heated on a stove.

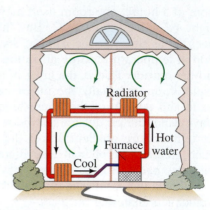

FIGURE 14–9 Convection plays a role in heating a house. The circular arrows show the convective air currents in the rooms.

PHYSICS APPLIED

House heating via convection

When a pot of water is heated (Fig. 14–8), convection currents are set up as the heated water at the bottom of the pot rises because of its reduced density. That heated water is replaced by cooler water from above. This principle is used in many heating systems, such as the hot-water radiator system shown in Fig. 14–9. Water is heated in the furnace, and as its temperature increases, it expands and rises as shown. This causes the water to circulate in the heating system. Hot water then enters the radiators, heat is transferred by conduction to the air, and the cooled water returns to the furnace. Thus, the water circulates because of convection; pumps are sometimes used to improve circulation. The air throughout the room also becomes heated as a result of convection. The air heated by the radiators rises and is replaced by cooler air, resulting in convective air currents, as shown by the green arrows in Fig. 14–9.

Other types of furnaces also depend on convection. Hot-air furnaces with registers (openings) near the floor often do not have fans but depend on natural convection, which can be appreciable. In other systems, a fan is used. In either case, it is important that cold air can return to the furnace so that convective currents circulate throughout the room if the room is to be uniformly heated.

PHYSICS APPLIED

Convection on steep hiking trails

FIGURE 14–10 Convection on a hiking trail: (a) upward movement of air in the morning because it is heated; (b) downward movement in the evening because the air is cooled.

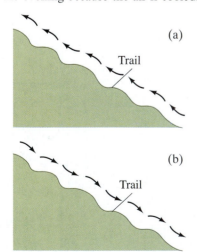

Another example of convection and its effects is given in the following excerpt from "The Winds of Yosemite Valley"[†] by the early environmentalist François Matthes:

It happens to be so ordained in nature that the sun shall heat the ground more rapidly than the air. And so it comes that every slope or hillside basking in the morning sun soon becomes itself a source of heat. It gradually warms the air immediately over it, and the latter, becoming lighter, begins to rise. But not vertically upward, for above it is still the cool air pressing down. Up along the warm slope it ascends, much as shown by the arrows in the accompanying diagram [Fig. 14–10a]. Few visitors to the valley but will remember toiling up some never ending zigzags on a hot and breathless day, with the sun on their backs and their own dust floating upward with them in an exasperating, choking cloud. Perhaps they thought it was simply their misfortune that the dust should happen to rise on that particular day. It always does on a sun-warmed slope.

But again, memories may arise of another occasion when, on coming down a certain trail the dust ever descended with the travelers, wafting down upon them from zigzag to zigzag as if with malicious pleasure. That, however, undoubtedly happened on the shady side of the valley. For there the conditions are exactly reversed. When the sun leaves a slope the latter begins at once to lose its heat by radiation, and in a short time is colder than the air. The layer next to the ground then gradually chills by contact, and, becoming heavier as it condenses, begins to creep down along the slope [Fig. 14–10b]. There is, thus,

[†]Reprinted from the *Sierra Club Bulletin*, June 1911, pp. 91–92.

normally a warm updraft on a sunlit slope and a cold downdraft on a shaded slope—and that rule one may depend on almost any day in a windless region like the Yosemite. Indeed, one might readily take advantage of it and plan his trips so as to have a dust-free journey.

The human body produces a great deal of thermal energy. Of the food energy transformed within the body, at best 20% is used to do work, so over 80% appears as thermal energy. During light activity, for example, if this thermal energy were not dissipated, the body temperature would rise about 3 C° per hour. Clearly, the heat generated by the body must be transferred to the outside. Is the heat transferred by conduction? The temperature of the skin in a comfortable environment is 33 to 35°C, whereas the interior of the body is at 37°C. A simple calculation (see Problem 55) shows that, because of this small temperature difference, plus the low thermal conductivity of tissue, direct conduction is responsible for very little of the heat that must be dissipated. Instead, the heat is carried to the surface by the blood. In addition to all its other important responsibilities, blood acts as a convective fluid to transfer heat to just beneath the surface of the skin. It is then conducted (over a very short distance) to the surface. Once at the surface, the heat is transferred to the environment by convection, evaporation, and radiation (see Section 14–8).

PHYSICS APPLIED
Body heat:
convection by blood

14–8 Heat Transfer: Radiation

Convection and conduction require the presence of matter as a medium to carry the heat from the hotter to the colder region. But a third type of heat transfer occurs without any medium at all. All life on Earth depends on the transfer of energy from the Sun, and this energy is transferred to the Earth over empty (or nearly empty) space. This form of energy transfer is heat—since the Sun's surface temperature is much higher (6000 K) than Earth's—and is referred to as **radiation** (Fig. 14–11). The warmth we receive from a fire is mainly radiant energy. (Most of the air heated by a fire in a fireplace rises by convection up the chimney and does not reach us.)

As we shall see in later Chapters, radiation consists essentially of electromagnetic waves. Suffice it to say for now that radiation from the Sun consists of visible light plus many other wavelengths that the eye is not sensitive to, including infrared (IR) radiation, which is mainly responsible for heating the Earth.

The rate at which an object radiates energy has been found to be proportional to the fourth power of the Kelvin temperature, T. That is, a body at 2000 K, as compared to one at 1000 K, radiates energy at a rate $2^4 = 16$ times as much. The rate of radiation is also proportional to the area A of the emitting object, so the rate at which energy leaves the object, $\Delta Q / \Delta t$, is

$$\frac{\Delta Q}{\Delta t} = e\sigma A T^4. \tag{14–5}$$

Radiation $\propto T^4$

This is called the **Stefan-Boltzmann equation**, and σ is a universal constant called the **Stefan-Boltzmann constant** which has the value

$$\sigma = 5.67 \times 10^{-8} \, \text{W/m}^2 \cdot \text{K}^4.$$

Stefan-Boltzmann constant

The factor e, called the **emissivity**, is a number between 0 and 1 that is characteristic of the surface of the radiating material. Very black surfaces, such as charcoal, have emissivity close to 1, whereas shiny metal surfaces have e close to zero and thus emit correspondingly less radiation. The value of e depends somewhat on the temperature of the body.

Emissivity

Not only do shiny surfaces emit less radiation, but they absorb little of the radiation that falls upon them (most is reflected). Black and very dark objects, on the other hand, absorb nearly all the radiation that falls on them—which is why light-colored clothing is usually preferable to dark clothing on a hot day. Thus, **a good absorber is also a good emitter**.

FIGURE 14–11 The Sun's surface radiates at 6000 K—much higher than the Earth's surface.

PHYSICS APPLIED
Dark vs. light clothing

Good absorber is good emitter

Any object not only emits energy by radiation but also absorbs energy radiated by other bodies. If an object of emissivity e and area A is at a temperature T_1, it radiates energy at a rate $e\sigma AT_1^4$. If the object is surrounded by an environment at temperature T_2, the rate at which the surroundings radiate energy is proportional to T_2^4, and the rate that energy is absorbed by the object is proportional to T_2^4. The *net* rate of radiant heat flow from the object is given by the equation

Net flow rate of heat radiation

$$\frac{\Delta Q}{\Delta t} = e\sigma A(T_1^4 - T_2^4), \tag{14–6}$$

where A is the surface area of the object, T_1 its temperature and e its emissivity (at temperature T_1), and T_2 is the temperature of the surroundings. Notice in this equation that the rate of heat absorption by an object was taken to be $e\sigma AT_2^4$; that is, the proportionality constant is the same for both emission and absorption. This must be true to correspond with the experimental fact that equilibrium between the object and its surroundings is reached when they come to the same temperature. That is, $\Delta Q/\Delta t$ must equal zero when $T_1 = T_2$, so the coefficients of emission and absorption terms must be the same. This confirms the idea that a good emitter is a good absorber.

Because both the object and its surroundings radiate energy, there is a net transfer of energy from one to the other unless everything is at the same temperature. From Eq. 14–6 it is clear that if $T_1 > T_2$, the net flow of heat is from the object to the surroundings, so the object cools. But if $T_1 < T_2$, the net heat flow is from the surroundings into the object, and its temperature rises. If different parts of the surroundings are at different temperatures, Eq. 14–6 becomes more complicated.

PHYSICS APPLIED

The body's radiative heat loss

EXAMPLE 14–11 ESTIMATE Cooling by radiation. An athlete is sitting unclothed in a locker room whose dark walls are at a temperature of 15°C. Estimate the rate of heat loss by radiation, assuming a skin temperature of 34°C and $e = 0.70$. Take the surface area of the body not in contact with the chair to be $1.5\,\text{m}^2$.

APPROACH We can make a rough estimate using the given assumptions and Eq. 14–6, for which we must use Kelvin temperatures.

SOLUTION We have

PROBLEM SOLVING

Must use the Kelvin temperature

$$\frac{\Delta Q}{\Delta t} = e\sigma A(T_1^4 - T_2^4)$$
$$= (0.70)(5.67 \times 10^{-8}\,\text{W/m}^2\cdot\text{K}^4)(1.5\,\text{m}^2)\big[(307\,\text{K})^4 - (288\,\text{K})^4\big]$$
$$= 120\,\text{W}.$$

NOTE The "output" of this resting person is a bit more than what a 100-W lightbulb uses.

PHYSICS APPLIED

Room comfort

A resting person naturally produces heat internally at a rate of about 100 W (Chapter 15), less than the heat loss by radiation as calculated in this Example. Hence, the person's temperature would drop, causing considerable discomfort. The body responds to excessive heat loss by increasing its metabolic rate (Section 15–3), and shivering is one method by which the body increases its metabolism. Naturally, clothes help a lot. Example 14–11 illustrates that a person may be uncomfortable even if the temperature of the air is, say, 25°C, which is quite a warm room. If the walls or floor are cold, radiation to them occurs no matter how warm the air is. Indeed, it is estimated that radiation accounts for about 50% of the heat loss from a sedentary person in a normal room. Rooms are most comfortable when the walls and floor are warm and the air is not so warm. Floors and walls can be heated by means of hot-water conduits or electric heating elements. Such first-rate heating systems are becoming more common today, and it is interesting to note that 2000 years ago

Temperature of walls and surroundings, not only the air, affects comfort

the Romans, even in houses in the remote province of Great Britain, made use of hot-water and steam conduits in the floor to heat their houses.

EXAMPLE 14–12 **ESTIMATE** **Two teapots.** A ceramic teapot ($e = 0.70$) and a shiny one ($e = 0.10$) each hold 0.75 L of tea at 95°C. (a) Estimate the rate of heat loss from each, and (b) estimate the temperature drop after 30 min for each. Consider only radiation, and assume the surroundings are at 20°C.

APPROACH We are given all the information necessary to calculate the heat loss due to radiation, except for the area. The teapot holds 0.75 L, and we can approximate it as a cube 10 cm on a side (volume = 1.0 L), with five sides exposed. To estimate the temperature drop in (b), we use the concept of specific heat and ignore the contribution of the pots compared to that of the water.

SOLUTION (a) The teapot, approximated by a cube 10 cm on a side with five sides exposed, has a surface area of about $5 \times (0.1\,\text{m})^2 = 5 \times 10^{-2}\,\text{m}^2$. The rate of heat loss would be about

$$\frac{\Delta Q}{\Delta t} = e\sigma A(T_1^4 - T_2^4)$$
$$= e(5.67 \times 10^{-8}\,\text{W/m}^2\cdot\text{K}^4)(5 \times 10^{-2}\,\text{m}^2)[(368\,\text{K})^4 - (293\,\text{K})^4]$$
$$\approx e(30)\ \text{W},$$

or about 20 W for the ceramic pot ($e = 0.70$) and 3 W for the shiny one ($e = 0.10$).

(b) To estimate the temperature drop, we use the specific heat of water and ignore the contribution of the pots. The mass of 0.75 L of water is 0.75 kg. (Recall that $1.0\,\text{L} = 1000\,\text{cm}^3 = 1 \times 10^{-3}\,\text{m}^3$ and $\rho = 1000\,\text{kg/m}^3$.) Using Eq. 14–2 and Table 14–1, we get

$$\frac{\Delta Q}{\Delta t} = mc\frac{\Delta T}{\Delta t}.$$

Then

$$\frac{\Delta T}{\Delta t} = \frac{\Delta Q/\Delta t}{mc} \approx \frac{e(30)\ \text{J/s}}{(0.75\,\text{kg})(4.186 \times 10^3\,\text{J/kg}\cdot\text{C}°)} = e(0.01)\ \text{C}°/\text{s}.$$

After 30 min (1800 s), $\Delta T = e(0.01\,\text{C}°/\text{s})\Delta t = e(0.01\,\text{C}°/\text{s})(1800\,\text{s}) = 18e\ \text{C}°$, or about 12 C° for the ceramic pot $e = 0.70$ and about 2 C° for the shiny one ($e = 0.10$). The shiny one clearly has an advantage, at least as far as radiation is concerned.

NOTE Convection and conduction could play a greater role than radiation.

Heating of an object by radiation from the Sun cannot be calculated using Eq. 14–6 since this equation assumes a uniform temperature, T_2, of the environment surrounding the object, whereas the Sun is essentially a point source. Hence the Sun must be treated as a separate source of energy. Heating by the Sun is calculated using the fact that about 1350 J of energy strikes the atmosphere of the Earth from the Sun per second per square meter of area at right angles to the Sun's rays. This number, 1350 W/m², is called the **solar constant**. The atmosphere may absorb as much as 70% of this energy before it reaches the ground, depending on the cloud cover. On a clear day, about 1000 W/m² reaches the Earth's surface. An object of emissivity e with area A facing the Sun absorbs energy from the Sun at a rate, in watts, of about

$$\frac{\Delta Q}{\Delta t} = (1000\,\text{W/m}^2)eA\cos\theta, \tag{14–7}$$

where θ is the angle between the Sun's rays and a line perpendicular to the area A (Fig. 14–12). That is, $A\cos\theta$ is the "effective" area, at right angles to the Sun's rays.

PHYSICS APPLIED
Radiation from the Sun

Solar constant

FIGURE 14–12 Radiant energy striking a body at an angle θ.

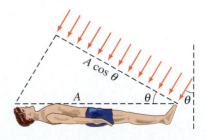

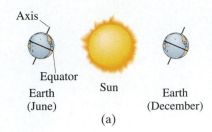

Axis

Equator

Earth   Sun   Earth
(June)   (December)

(a)

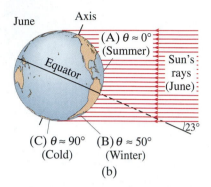

June   Axis

(A) θ ≈ 0°
(Summer)

Equator   Sun's
                           rays
                           (June)

                           23°

(C) θ ≈ 90°   (B) θ ≈ 50°
(Cold)   (Winter)

(b)

FIGURE 14–13 (a) Earth's seasons arise from the $23\frac{1}{2}°$ angle Earth's axis makes with its orbit around the Sun. (b) June sunlight makes an angle of about 23° with the equator. Thus θ in the southern United States (A) is near 0° (direct summer sunlight), whereas in the Southern Hemisphere (B), θ is 50° or 60°, and less heat can be absorbed—hence it is winter. Near the poles (C), there is never strong direct sunlight; cos θ varies from about $\frac{1}{2}$ in summer to 0 in winter; so with little heating, ice can form.

FIGURE 14–14 Thermograms of a healthy person's arms and hands (a) before and (b) after smoking a cigarette, showing a temperature decrease due to impaired blood circulation associated with smoking. The thermograms have been color-coded according to temperature; the scale on the right goes from blue (cold) to white (hot).

The explanation for the **seasons** and the polar ice caps (see Fig. 14–13) depends on this cos θ factor in Eq. 14–7. The seasons are *not* a result of how close the Earth is to the Sun—in fact, in the Northern Hemisphere, summer occurs when the Earth is farthest from the Sun. It is the angle (i.e., cos θ) that really matters. Furthermore, the reason the Sun heats the Earth more at midday than at sunrise or sunset is also related to this cos θ factor.

EXAMPLE 14–13 ESTIMATE Getting a tan—energy absorption. What is the rate of energy absorption from the Sun by a person lying flat on the beach on a clear day if the Sun makes a 30° angle with the vertical? Assume that $e = 0.70$ and that 1000 W/m^2 reaches the Earth's surface.

APPROACH We use Eq. 14–7 and estimate a typical human to be roughly 2 m tall by 0.4 m wide, so $A \approx (2 \text{ m})(0.4 \text{ m}) = 0.8 \text{ m}^2$.

SOLUTION Since $\cos 30° = 0.866$, we have

$$\frac{\Delta Q}{\Delta t} = (1000 \text{ W/m}^2) e A \cos \theta$$

$$= (1000 \text{ W/m}^2)(0.70)(0.8 \text{ m}^2)(0.866) = 500 \text{ W}.$$

NOTE If a person wears light-colored clothing, e is much smaller, so the energy absorbed is less.

An interesting application of thermal radiation to diagnostic medicine is **thermography**. A special instrument, the thermograph, scans the body, measuring the intensity of radiation from many points and forming a picture that resembles an X-ray (Fig. 14–14). Areas where metabolic activity is high, such as in tumors, can often be detected on a thermogram as a result of their higher temperature and consequent increased radiation.

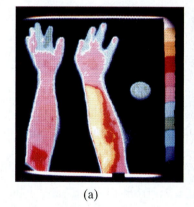
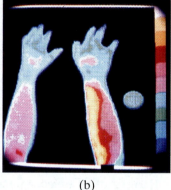

(a)   (b)

EXAMPLE 14–14 ESTIMATE Star radius. The giant star Betelgeuse emits radiant energy at a rate 10^4 times greater than our Sun, whereas its surface temperature is only half (2900 K) that of our Sun. Estimate the radius of Betelgeuse, assuming $e = 1$. The Sun's radius is $r_S = 7 \times 10^8$ m.

APPROACH We assume both Betelgeuse and the Sun are spherical, with surface area $4\pi r^2$.

SOLUTION We solve Eq. 14–5 for A:

$$4\pi r^2 = A = \frac{(\Delta Q/\Delta t)}{e\sigma T^4}.$$

Then

$$\frac{r_B^2}{r_S^2} = \frac{(\Delta Q/\Delta t)_B}{(\Delta Q/\Delta t)_S} \cdot \frac{T_S^4}{T_B^4} = (10^4)(2^4) = 16 \times 10^4.$$

Hence $r_B = \sqrt{16 \times 10^4} \, r_S = (400)(7 \times 10^8 \text{ m}) \approx 3 \times 10^{11}$ m. If Betelgeuse were our Sun, it would envelop us (Earth is 1.5×10^{11} m from the Sun).

Summary

Internal energy, U, refers to the total energy of all the molecules in an object. For an ideal monatomic gas,

$$U = \tfrac{3}{2} NkT = \tfrac{3}{2} nRT \qquad \text{(14–1)}$$

where N is the number of molecules or n is the number of moles.

Heat refers to the transfer of energy from one object to another because of a difference of temperature. Heat is thus measured in energy units, such as joules.

Heat and internal energy are also sometimes specified in calories or kilocalories (kcal), where

$$1 \text{ kcal} = 4.186 \text{ kJ}$$

is the amount of heat needed to raise the temperature of 1 kg of water by 1 C°.

The **specific heat**, c, of a substance is defined as the energy (or heat) required to change the temperature of unit mass of substance by 1 degree; as an equation,

$$Q = mc\,\Delta T, \qquad \text{(14–2)}$$

where Q is the heat absorbed or given off, ΔT is the temperature increase or decrease, and m is the mass of the substance.

When heat flows between parts of an isolated system, conservation of energy tells us that the heat gained by one part of the system is equal to the heat lost by the other part of the system. This is the basis of **calorimetry**, which is the quantitative measurement of heat exchange.

Exchange of energy occurs, without a change in temperature, whenever a substance changes phase. The **heat of fusion** is the heat required to melt 1 kg of a solid into the liquid phase; it is also equal to the heat given off when the substance changes from liquid to solid. The **heat of vaporization** is the energy required to change 1 kg of a substance from the liquid to the vapor phase; it is also the energy given off when the substance changes from vapor to liquid.

Heat is transferred from one place (or object) to another in three different ways: conduction, convection, and radiation.

In **conduction**, energy is transferred from molecules or electrons with higher kinetic energy to lower-KE neighbors when they collide.

Convection is the transfer of energy by the mass movement of molecules over considerable distances.

Radiation, which does not require the presence of matter, is energy transfer by electromagnetic waves, such as from the Sun. All objects radiate energy in an amount that is proportional to the fourth power of their Kelvin temperature (T^4) and to their surface area. The energy radiated (or absorbed) also depends on the nature of the surface (dark surfaces absorb and radiate more than do bright shiny ones), which is characterized by the emissivity, e.

Radiation from the Sun arrives at the surface of the Earth on a clear day at a rate of about 1000 W/m².

Questions

1. What happens to the work done when a jar of orange juice is vigorously shaken?

2. When a hot object warms a cooler object, does temperature flow between them? Are the temperature changes of the two objects equal?

3. (a) If two objects of different temperatures are placed in contact, will heat naturally flow from the object with higher internal energy to the object with lower internal energy? (b) Is it possible for heat to flow even if the internal energies of the two objects are the same? Explain.

4. In warm regions where tropical plants grow but the temperature may drop below freezing a few times in the winter, the destruction of sensitive plants due to freezing can be reduced by watering them in the evening. Explain.

5. The specific heat of water is quite large. Explain why this fact makes water particularly good for heating systems (that is, hot-water radiators).

6. Why does water in a metal canteen stay cooler if the cloth jacket surrounding the canteen is kept moist?

7. Explain why burns caused by steam on the skin are often more severe than burns caused by water at 100°C.

8. Explain why water cools (its temperature drops) when it evaporates, using the concepts of latent heat and internal energy.

9. Will potatoes cook faster if the water is boiling faster?

10. Does an ordinary electric fan cool the air? Why or why not? If not, why use it?

11. Very high in the Earth's atmosphere, the temperature can be 700°C. Yet an animal there would freeze to death rather than roast. Explain.

12. Explorers on failed Arctic expeditions have survived by covering themselves with snow. Why would they do that?

13. Why is wet sand at the beach cooler to walk on than dry sand?

14. If you hear that an object has "high heat content," does that mean that its temperature is high? Explain.

15. When hot-air furnaces are used to heat a house, why is it important that there be a vent for air to return to the furnace? What happens if this vent is blocked by a bookcase?

16. Ceiling fans are sometimes reversible, so that they drive the air down in one season and pull it up in another season. Which way should you set the fan for summer? For winter?

17. Down sleeping bags and parkas are often specified as so many inches or centimeters of *loft*, the actual thickness of the garment when it is fluffed up. Explain.

18. Microprocessor chips have a "heat sink" glued on top that looks like a series of fins. Why is it shaped like that?

19. Sea breezes are often encountered on sunny days at the shore of a large body of water. Explain in light of the fact that the temperature of the land rises more rapidly than that of the nearby water.

20. The floor of a house on a foundation under which the air can flow is often cooler than a floor that rests directly on the ground (such as a concrete slab foundation). Explain.

21. A 22°C day is warm, while a swimming pool at 22°C feels cool. Why?

22. Explain why air temperature readings are always taken with the thermometer in the shade.

23. A premature baby in an incubator can be dangerously cooled even when the air temperature in the incubator is warm. Explain.

24. Why is the liner of a thermos bottle silvered (Fig. 14–15), and why does it have a vacuum between its two walls?

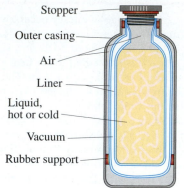

Stopper

Outer casing

Air

Liner

Liquid, hot or cold

Vacuum

Rubber support

FIGURE 14–15
Question 24.

25. Imagine you have a wall that is very well insulated—it has a very high thermal resistance, R_1. Now you place a window in the wall that has a relatively low R-value, R_2. What has happened to the overall R-value of the wall plus window, compared to R_1 and R_2? [*Hint*: The temperature *difference* across the wall is still the same everywhere.]

26. Heat loss occurs through windows by the following processes: (1) ventilation around edges; (2) through the frame, particularly if it is metal; (3) through the glass panes; and (4) radiation. (*a*) For the first three, what is (are) the mechanism(s): conduction, convection, or radiation? (*b*) Heavy curtains reduce which of these heat losses? Explain in detail.

27. A piece of wood lying in the Sun absorbs more heat than a piece of shiny metal. Yet the wood feels less hot than the metal when you pick it up. Explain.

28. The Earth cools off at night much more quickly when the weather is clear than when cloudy. Why?

29. An "emergency blanket" is a thin shiny (metal coated) plastic foil. Explain how it can help to keep an immobile person warm.

30. Explain why cities situated by the ocean tend to have less extreme temperatures than inland cities at the same latitude.

Problems

14–1 Heat as Energy Transfer

1. (I) How much heat (in joules) is required to raise the temperature of 30.0 kg of water from 15°C to 95°C?

2. (I) To what temperature will 7700 J of heat raise 3.0 kg of water that is initially at 10.0°C?

3. (II) An average active person consumes about 2500 Cal a day. (*a*) What is this in joules? (*b*) What is this in kilowatt-hours? (*c*) Your power company charges about a dime per kilowatt-hour. How much would your energy cost per day if you bought it from the power company? Could you feed yourself on this much money per day?

4. (II) A British thermal unit (Btu) is a unit of heat in the British system of units. One Btu is defined as the heat needed to raise 1 lb of water by 1 F°. Show that

$$1\,\text{Btu} = 0.252\,\text{kcal} = 1055\,\text{J}.$$

5. (II) A water heater can generate 32,000 kJ/h. How much water can it heat from 15°C to 50°C per hour?

6. (II) A small immersion heater is rated at 350 W. Estimate how long it will take to heat a cup of soup (assume this is 250 mL of water) from 20°C to 60°C.

7. (II) How many kilocalories are generated when the brakes are used to bring a 1200-kg car to rest from a speed of 95 km/h?

14–3 and 14–4 Specific Heat; Calorimetry

8. (I) An automobile cooling system holds 16 L of water. How much heat does it absorb if its temperature rises from 20°C to 90°C?

9. (I) What is the specific heat of a metal substance if 135 kJ of heat is needed to raise 5.1 kg of the metal from 18.0°C to 31.5°C?

10. (II) Samples of copper, aluminum, and water experience the same temperature rise when they absorb the same amount of heat. What is the ratio of their masses? [*Hint*: See Table 14–1.]

11. (II) A 35-g glass thermometer reads 21.6°C before it is placed in 135 mL of water. When the water and thermometer come to equilibrium, the thermometer reads 39.2°C. What was the original temperature of the water?

12. (II) What will be the equilibrium temperature when a 245-g block of copper at 285°C is placed in a 145-g aluminum calorimeter cup containing 825 g of water at 12.0°C?

13. (II) A hot iron horseshoe (mass = 0.40 kg), just forged (Fig. 14–16), is dropped into 1.35 L of water in a 0.30-kg iron pot initially at 20.0°C. If the final equilibrium temperature is 25.0°C, estimate the initial temperature of the hot horseshoe.

FIGURE 14–16
Problem 13.

14. (II) A 215-g sample of a substance is heated to 330°C and then plunged into a 105-g aluminum calorimeter cup containing 165 g of water and a 17-g glass thermometer at 12.5°C. The final temperature is 35.0°C. What is the specific heat of the substance? (Assume no water boils away.)

15. (II) How long does it take a 750-W coffeepot to bring to a boil 0.75 L of water initially at 8.0°C? Assume that the part of the pot which is heated with the water is made of 360 g of aluminum, and that no water boils away.

16. (II) Estimate the Calorie content of 75 g of candy from the following measurements. A 15-g sample of the candy is allowed to dry before putting it in a bomb calorimeter. The aluminum bomb has a mass of 0.725 kg and is placed in 2.00 kg of water contained in an aluminum calorimeter cup of mass 0.624 kg. The initial temperature of the mixture is 15.0°C, and its temperature after ignition is 53.5°C.

17. (II) When a 290-g piece of iron at 180°C is placed in a 95-g aluminum calorimeter cup containing 250 g of glycerin at 10°C, the final temperature is observed to be 38°C. Estimate the specific heat of glycerin.

18. (II) The 1.20-kg head of a hammer has a speed of 6.5 m/s just before it strikes a nail (Fig. 14–17) and is brought to rest. Estimate the temperature rise of a 14-g iron nail generated by 10 such hammer blows done in quick succession. Assume the nail absorbs all the energy.

FIGURE 14–17
Problem 18.

19. (II) A 0.095-kg aluminium sphere is dropped from the roof of a 45-m-high building. If 65% of the thermal energy produced when it hits the ground is absorbed by the sphere, what is its temperature increase?

20. (II) The *heat capacity, C,* of an object is defined as the amount of heat needed to raise its temperature by 1 C°. Thus, to raise the temperature by ΔT requires heat Q given by
$$Q = C \, \Delta T.$$
(a) Write the heat capacity C in terms of the specific heat, c, of the material. (b) What is the heat capacity of 1.0 kg of water? (c) Of 25 kg of water?

14–5 Latent Heat

21. (I) How much heat is needed to melt 16.50 kg of silver that is initially at 20°C?

22. (I) During exercise, a person may give off 180 kcal of heat in 30 min by evaporation of water from the skin. How much water has been lost?

23. (I) If 2.80×10^5 J of energy is supplied to a flask of liquid oxygen at −183°C, how much oxygen can evaporate?

24. (II) A 30-g ice cube at its melting point is dropped into an insulated container of liquid nitrogen. How much nitrogen evaporates if it is at its boiling point of 77 K and has a latent heat of vaporization of 200 kJ/kg? Assume for simplicity that the specific heat of ice is a constant and is equal to its value near its melting point.

25. (II) A cube of ice is taken from the freezer at −8.5°C and placed in a 95-g aluminum calorimeter filled with 310 g of water at room temperature of 20.0°C. The final situation is observed to be all water at 17.0°C. What was the mass of the ice cube?

26. (II) An iron boiler of mass 230 kg contains 830 kg of water at 18°C. A heater supplies energy at the rate of 52,000 kJ/h. How long does it take for the water (a) to reach the boiling point, and (b) to all have changed to steam?

27. (II) In a hot day's race, a bicyclist consumes 8.0 L of water over the span of four hours. Making the approximation that all of the cyclist's energy goes into evaporating this water as sweat, how much energy in kcal did the rider use during the ride? (Since the efficiency of the rider is only about 20%, most of the energy consumed does go to heat, so our approximation is not far off.)

28. (II) What mass of steam at 100°C must be added to 1.00 kg of ice at 0°C to yield liquid water at 20°C?

29. (II) The specific heat of mercury is 138 J/kg·C°. Determine the latent heat of fusion of mercury using the following calorimeter data: 1.00 kg of solid Hg at its melting point of −39.0°C is placed in a 0.620-kg aluminum calorimeter with 0.400 kg of water at 12.80°C; the resulting equilibrium temperature is 5.06°C.

30. (II) A 70-g bullet traveling at 250 m/s penetrates a block of ice at 0°C and comes to rest within the ice. Assuming that the temperature of the bullet doesn't change appreciably, how much ice is melted as a result of the collision?

31. (II) A 54.0-kg ice-skater moving at 6.4 m/s glides to a stop. Assuming the ice is at 0°C and that 50% of the heat generated by friction is absorbed by the ice, how much ice melts?

32. (II) At a crime scene, the forensic investigator notes that the 8.2-g lead bullet that was stopped in a doorframe apparently melted completely on impact. Assuming the bullet was fired at room temperature (20°C), what does the investigator calculate as the *minimum* muzzle velocity of the gun?

14–6 to 14–8 Conduction, Convection, Radiation

33. (I) One end of a 33-cm-long aluminum rod with a diameter of 2.0 cm is kept at 460°C, and the other is immersed in water at 22°C. Calculate the heat conduction rate along the rod.

34. (I) Calculate the rate of heat flow by conduction in Example 14–10, assuming that there are strong gusty winds and the external temperature is −5°C.

35. (I) (a) How much power is radiated by a tungsten sphere (emissivity $e = 0.35$) of radius 22 cm at a temperature of 25°C? (b) If the sphere is enclosed in a room whose walls are kept at −5°C, what is the *net* flow rate of energy out of the sphere?

36. (II) *Heat conduction to skin.* Suppose 200 W of heat flows by conduction from the blood capillaries beneath the skin to the body's surface area of 1.5 m². If the temperature difference is 0.50 C°, estimate the average distance of capillaries below the skin surface.

37. (II) Two rooms, each a cube 4.0 m per side, share a 12-cm-thick brick wall. Because of a number of 100-W lightbulbs in one room, the air is at 30°C, while in the other room it is at 10°C. How many of the 100-W bulbs are needed to maintain the temperature difference across the wall?

38. (II) How long does it take the Sun to melt a block of ice at 0°C with a flat horizontal area 1.0 m² and thickness 1.0 cm? Assume that the Sun's rays make an angle of 30° with the vertical and that the emissivity of ice is 0.050.

39. (II) A copper rod and an aluminum rod of the same length and cross-sectional area are attached end to end (Fig. 14–18). The copper end is placed in a furnace maintained at a constant temperature of 250°C. The aluminum end is placed in an ice bath held at constant temperature of 0.0°C. Calculate the temperature at the point where the two rods are joined.

FIGURE 14–18 Problem 39.

40. (II) (a) Using the solar constant, estimate the rate at which the whole Earth receives energy from the Sun. (b) Assume the Earth radiates an equal amount back into space (that is, the Earth is in equilibrium). Then, assuming the Earth is a perfect emitter ($e = 1.0$), estimate its average surface temperature.

41. (II) A 100-W lightbulb generates 95 W of heat, which is dissipated through a glass bulb that has a radius of 3.0 cm and is 1.0 mm thick. What is the difference in temperature between the inner and outer surfaces of the glass?

42. (III) Suppose the insulating qualities of the wall of a house come mainly from a 4.0-in. layer of brick and an R-19 layer of insulation, as shown in Fig. 14–19. What is the total rate of heat loss through such a wall, if its total area is 240 ft^2 and the temperature difference across it is 12 F°?

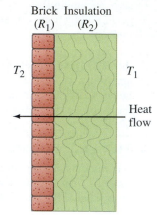

FIGURE 14–19 Two layers insulating a wall. Problem 42.

43. (III) A double-glazed window has two panes of glass separated by an air space, Fig. 14–20. (a) Show that the rate of heat flow through such a window by conduction is given by

$$\frac{Q}{t} = \frac{A(T_2 - T_1)}{l_1/k_1 + l_2/k_2 + l_3/k_3},$$

where k_1, k_2, and k_3 are the thermal conductivities for glass, air, and glass, respectively. (b) Generalize this expression for any number of materials placed next to one another.

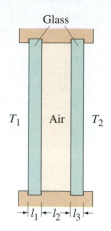

FIGURE 14–20
Problem 43.

44. (III) Approximately how long should it take 11.0 kg of ice at 0°C to melt when it is placed in a carefully sealed Styrofoam ice chest of dimensions 25 cm × 35 cm × 55 cm whose walls are 1.5 cm thick? Assume that the conductivity of Styrofoam is double that of air and that the outside temperature is 32°C.

General Problems

45. A soft-drink can contains about 0.20 kg of liquid at 5°C. Drinking this liquid can actually consume some of the fat in the body, since energy is needed to warm the water to body temperature (37°C). How many food Calories should the drink have so that it is in perfect balance with the heat needed to warm the liquid?

46. If coal gives off 30 MJ/kg when it is burned, how much coal would be needed to heat a house that requires 2.0×10^5 MJ for the whole winter? Assume that 30% of the heat is lost up the chimney.

47. To get an idea of how much thermal energy is contained in the world's oceans, estimate the heat liberated when a cube of ocean water, 1 km on each side, is cooled by 1 K. (Approximate the ocean water as pure water for this estimate.)

48. A 15-g lead bullet is tested by firing it into a fixed block of wood with a mass of 1.05 kg. The block and imbedded bullet together absorb all the heat generated. After thermal equilibrium has been reached, the system has a temperature rise measured as 0.020 C°. Estimate the entering speed of the bullet.

49. (a) Find the total power radiated into space by the Sun, assuming it to be a perfect emitter at $T = 5500$ K. The Sun's radius is 7.0×10^8 m. (b) From this, determine the power per unit area arriving at the Earth, 1.5×10^{11} m away (Fig. 14–21).

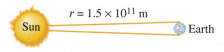

FIGURE 14–21 Problem 49.

50. During light activity, a 70-kg person may generate 200 kcal/h. Assuming that 20% of this goes into useful work and the other 80% is converted to heat, calculate the temperature rise of the body after 1.00 h if none of this heat were transferred to the environment.

51. A 340-kg marble boulder rolls off the top of a cliff and falls a vertical height of 140 m before striking the ground. Estimate the temperature rise of the rock if 50% of the heat generated remains in the rock.

52. A 2.3-kg lead ball is dropped into a 2.5-L insulated pail of water initially at 20.0°C. If the final temperature of the water–lead combination is 28.0°C, what was the initial temperature of the lead ball?

53. A mountain climber wears a goose down jacket 3.5 cm thick with total surface area 1.2 m^2. The temperature at the surface of the clothing is -20°C and at the skin is 34°C. Determine the rate of heat flow by conduction through the jacket (a) assuming it is dry and the thermal conductivity k is that of down, and (b) assuming the jacket is wet, so k is that of water and the jacket has matted to 0.50 cm thickness.

54. A marathon runner has an average metabolism rate of about 950 kcal/h during a race. If the runner has a mass of 55 kg, estimate how much water she would lose to evaporation from the skin for a race that lasts 2.5 h.

55. Estimate the rate at which heat can be conducted from the interior of the body to the surface. Assume that the thickness of tissue is 4.0 cm, that the skin is at 34°C and the interior at 37°C, and that the surface area is 1.5 m^2. Compare this to the measured value of about 230 W that must be dissipated by a person working lightly. This clearly shows the necessity of convective cooling by the blood.

56. A house has well-insulated walls 17.5 cm thick (assume conductivity of air) and area 410 m², a roof of wood 6.5 cm thick and area 280 m², and uncovered windows 0.65 cm thick and total area 33 m². (*a*) Assuming that heat is lost only by conduction, calculate the rate at which heat must be supplied to this house to maintain its inside temperature at 23°C if the outside temperature is −10°C. (*b*) If the house is initially at 10°C, estimate how much heat must be supplied to raise the temperature to 23°C within 30 min. Assume that only the air needs to be heated and that its volume is 750 m³. (*c*) If natural gas costs $0.080 per kilogram and its heat of combustion is 5.4 × 10⁷ J/kg, how much is the monthly cost to maintain the house as in part (*a*) for 24 h each day, assuming 90% of the heat produced is used to heat the house? Take the specific heat of air to be 0.24 kcal/kg·C°.

57. A 15-g lead bullet traveling at 220 m/s passes through a thin wall and emerges at a speed of 160 m/s. If the bullet absorbs 50% of the heat generated, (*a*) what will be the temperature rise of the bullet? (*b*) If the bullet's initial temperature was 20°C, will any of the bullet melt, and if so, how much?

58. A leaf of area 40 cm² and mass 4.5 × 10⁻⁴ kg directly faces the Sun on a clear day. The leaf has an emissivity of 0.85 and a specific heat of 0.80 kcal/kg·K. (*a*) Estimate the rate of rise of the leaf's temperature. (*b*) Calculate the temperature the leaf would reach if it lost all its heat by radiation to the surroundings at 20°C. (*c*) In what other ways can the heat be dissipated by the leaf?

59. Using the result of part (*a*) in Problem 58, take into account radiation from the leaf to calculate how much water must be transpired (evaporated) by the leaf per hour to maintain a temperature of 35°C.

60. An iron meteorite melts when it enters the Earth's atmosphere. If its initial temperature was −125°C outside of Earth's atmosphere, calculate the minimum velocity the meteorite must have had before it entered Earth's atmosphere.

61. The temperature within the Earth's crust increases about 1.0 C° for each 30 m of depth. The thermal conductivity of the crust is 0.80 W/C°·m. (*a*) Determine the heat transferred from the interior to the surface for the entire Earth in 1 day. (*b*) Compare this heat to the amount of energy incident on the Earth in 1 day due to radiation from the Sun.

62. In a typical game of squash (Fig. 14–22), two people hit a soft rubber ball at a wall until they are about to drop due to dehydration and exhaustion. Assume that the ball hits the wall at a velocity of 22 m/s and bounces back with a velocity of 12 m/s, and that the kinetic energy lost in the process heats the ball. What will be the temperature increase of the ball after one bounce? (The specific heat of rubber is about 1200 J/kg·C°.)

FIGURE 14–22 Problem 62.

63. What will be the final result when equal masses of ice at 0°C and steam at 100°C are mixed together?

64. In a cold environment, a person can lose heat by conduction and radiation at a rate of about 200 W. Estimate how long it would take for the body temperature to drop from 36.6°C to 35.6°C if metabolism were nearly to stop. Assume a mass of 70 kg. (See Table 14–1.)

65. After a hot shower and dishwashing, there is "no hot water" left in the 50-gal (185-L) water heater. This suggests that the tank has emptied and refilled with water at roughly 10°C. (*a*) How much energy does it take to reheat the water to 50°C? (*b*) How long would it take if the heater output is 9500 W?

66. The temperature of the glass surface of a 60-W lightbulb is 65°C when the room temperature is 18°C. Estimate the temperature of a 150-W lightbulb with a glass bulb the same size. Consider only radiation, and assume that 90% of the energy is emitted as heat.

Answers to Exercises

A: 0.21 kg.

B: The drapes trap a layer of air between the outside wall and the room, which acts as an excellent insulator.

Thermodynamics is the study of heat and work. Heat is a transfer of energy due to a difference of temperature; work is a transfer of energy by mechanical means, not due to a temperature difference. The first law of thermodynamics is a general statement of energy conservation: the heat Q added to a system minus the net work W done by the system equals the change in internal energy ΔU of the system: $\Delta U = Q - W$. The photos show two uses for a heat engine: a modern coal-burning power plant, and an old steam locomotive. Both produce steam which does work—on turbines to generate electricity, and on a piston that moves linkage to turn locomotive wheels. The efficiency of any engine is limited by nature as described in the second law of thermodynamics. This great law is best stated in terms of a quantity called entropy, which is *not* conserved, but instead is constrained always to increase in any real process. Entropy is a measure of disorder. The second law of thermodynamics tells us that as time moves forward, the disorder in the universe increases.

CHAPTER **15**

The Laws of Thermodynamics

Thermodynamics is the name we give to the study of processes in which energy is transferred as heat and as work.

In Chapter 6 we saw that work is done when energy is transferred from one object to another by mechanical means. In Chapter 14 we saw that heat is a transfer of energy from one object to a second one at a lower temperature. Thus, heat is much like work. To distinguish them, *heat* is defined as a *transfer of energy due to a difference in temperature*, whereas work is a transfer of energy that is not due to a temperature difference.

Heat distinguished from work

In discussing thermodynamics, we often refer to particular systems. A **system** is any object or set of objects that we wish to consider (see Section 14–4). Everything else in the universe will be referred to as the "environment" or the "surroundings."

In this Chapter, we examine the two great laws of thermodynamics. The first law of thermodynamics relates work and heat transfers to the change in internal energy of a system, and is a general statement of the conservation of energy. The second law of thermodynamics expresses limits on the ability to do useful work, and is often stated in terms of *entropy*, which is a measure of disorder. Besides these two great laws, we also discuss some important related practical devices: heat engines, refrigerators, heat pumps, and air conditioners.

15–1 The First Law of Thermodynamics

In Section 14–2, we defined the internal energy of a system as the sum total of all the energy of the molecules of the system. We would expect that the internal energy of a system would be increased if work was done on the system, or if heat were added to it. Similarly the internal energy would be decreased if heat flowed out of the system or if work were done by the system on something in the surroundings.

Thus it is reasonable to extend the work-energy principle and propose an important law: the change in internal energy of a closed system, ΔU, will be equal to the energy added to the system by heating minus the work done by the system on the surroundings. In equation form we write

$$\Delta U = Q - W \qquad\qquad (15\text{–}1)$$

where Q is the net heat *added* to the system and W is the net work done *by* the system. We must be careful and consistent in following the sign conventions for Q and W. Because W in Eq. 15–1 is the work done *by* the system, then if work is done *on* the system, W will be negative and U will increase. Similarly, Q is positive for heat added to the system, so if heat leaves the system, Q is negative.

Equation 15–1 is known as the **first law of thermodynamics**. It is one of the great laws of physics, and its validity rests on experiments (such as Joule's) to which no exceptions have been seen. Since Q and W represent energy transferred into or out of the system, the internal energy changes accordingly. Thus, the first law of thermodynamics is a great and broad statement of the *law of conservation of energy*.

First law of thermodynamics is conservation of energy

It is worth noting that the conservation of energy law was not formulated until the nineteenth century, for it depended on the interpretation of heat as a transfer of energy.

A given system at any moment is in a particular state and can be said to have a certain amount of internal energy, U. But a system does not "have" a certain amount of heat or work. Rather, when work is done on a system (such as compressing a gas), or when heat is added or removed from a system, the state of the system *changes*. Thus, work and heat are involved in *thermodynamic processes* that can change the system from one state to another; they are not characteristic of the state itself. Quantities which describe the state of a system, such as internal energy U, pressure P, volume V, temperature T, and mass m or number of moles n, are called **state variables**. Q and W are *not* state variables.

Internal energy is a property of the system; work and heat are not

EXAMPLE 15–1 **Using the first law.** 2500 J of heat is added to a system, and 1800 J of work is done on the system. What is the change in internal energy of the system?

APPROACH We apply the first law of thermodynamics, Eq. 15–1, to our system.

SOLUTION The heat added to the system is $Q = 2500$ J. The work W done *by* the system is -1800 J. Why the minus sign? Because 1800 J done *on* the system (as given) equals -1800 J done *by* the system, and it is the latter we need to put in Eq. 15–1 by the sign conventions given above. Hence

$$\Delta U = 2500\,\text{J} - (-1800\,\text{J}) = 2500\,\text{J} + 1800\,\text{J} = 4300\,\text{J}.$$

You may have intuitively thought that the 2500 J and the 1800 J would need to be added together, since both refer to energy added to the system. You would have been right.

NOTE We did this calculation in detail to emphasize the importance of keeping careful track of signs.

EXERCISE A What would be the internal energy change in Example 15–1 if 2500 J of heat is added to the system and 1800 J of work is done *by* the system (i.e., as output)?

The First Law of Thermodynamics Extended

To be really complete about the first law, consider a system that is moving, so it has kinetic energy KE, and suppose there is also potential energy PE. Then the first law of thermodynamics would have to include these terms and would be written as

$$\Delta \text{KE} + \Delta \text{PE} + \Delta U = Q - W. \qquad (15\text{–}2)$$

EXAMPLE 15–2 **Kinetic energy transformed to thermal energy.** A 3.0-g bullet traveling at a speed of 400 m/s enters a tree and exits the other side with a speed of 200 m/s. Where did the bullet's lost KE go, and what was the energy transferred?

APPROACH Take the bullet and tree as our system. No potential energy is involved. No work is done on (or by) the system by outside forces, nor is any heat added because no energy was transferred to or from the system due to a temperature difference. Thus the kinetic energy gets transformed into internal energy of the bullet and tree.

SOLUTION From the first law of thermodynamics as given in Eq. 15–2, we are given $Q = W = \Delta \text{PE} = 0$, so we have

$$\Delta \text{KE} + \Delta U = 0$$

or

$$\Delta U = -\Delta \text{KE} = -\left(\text{KE}_f - \text{KE}_i \right) = \tfrac{1}{2} m \left(v_i^2 - v_f^2 \right)$$
$$= \tfrac{1}{2} (3.0 \times 10^{-3} \text{ kg}) \left[(400 \text{ m/s})^2 - (200 \text{ m/s})^2 \right] = 180 \text{ J}.$$

NOTE The internal energy of the bullet and tree both increase, as both experience a rise in temperature. If we had chosen the bullet alone as our system, work would be done on it and heat transfer would occur.

15–2 Thermodynamic Processes and the First Law

Let us analyze some thermodynamic processes in light of the first law of thermodynamics. To begin, we choose a very simple system: a fixed mass of an ideal gas enclosed in a container fitted with a movable piston as shown in Fig. 15–1.

First we consider an idealized process that is carried out at constant temperature. Such a process is called an **isothermal** process (from the Greek meaning "same temperature"). If an isothermal process is carried out on our ideal gas, then $PV = nRT$ (Eq. 13–3) becomes $PV = \text{constant}$. Thus the process follows a curve like AB on the PV diagram shown in Fig. 15–2, which is a curve for $PV = \text{constant}$ (as in Fig. 13–12). Each point on the curve, such as point A, represents a state of the system—that is, its pressure P and volume V at a given moment. At a lower temperature, another isothermal process would be represented by a curve like A′B′ in Fig. 15–2 (the product $PV = nRT = \text{constant}$ is less when T is less). The curves shown in Fig. 15–2 are referred to as *isotherms*.

We assume that the gas is in contact with a **heat reservoir** (a body whose mass is so large that, ideally, its temperature does not change significantly when heat is exchanged with our system). We also assume that the process of compression (volume decrease) or expansion (volume increase) is done very slowly to make certain that all of the gas stays in equilibrium at the same constant temperature. If the gas is initially in a state represented by point A in Fig. 15–2, and an amount of heat Q is added to the system, the pressure and volume will change and the state of the system will be represented by another point, B, on the diagram. If the temperature is to remain constant, the gas must expand and do an amount of work W on the environment (it exerts a force on the piston in Fig. 15–1 and moves it through a distance). The temperature is kept constant so, from Eq. 14–1, the internal energy does not change: $\Delta U = \tfrac{3}{2} nR \, \Delta T = 0$. Hence, by the first law of thermodynamics (Eq. 15–1), $\Delta U = Q - W = 0$, so $W = Q$: the work done by the gas in an isothermal process equals the heat added to the gas.

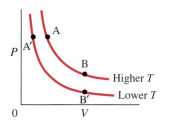

FIGURE 15–1 An ideal gas in a cylinder fitted with a movable piston.

Isothermal process ($\Delta T = 0$)

Movable piston

Ideal gas

Heat reservoir

FIGURE 15–2 *PV* diagram for an ideal gas undergoing isothermal processes at two different temperatures.

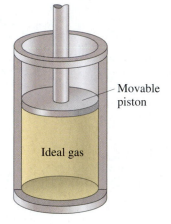

Isothermal process (ideal gas):
$T = constant, \ \Delta U = 0, \ Q = W$

An **adiabatic** process is one in which no heat is allowed to flow into or out of the system: $Q = 0$. This situation can occur if the system is extremely well insulated, or the process happens so quickly that heat—which flows slowly— has no time to flow in or out. The very rapid expansion of gases in an internal combustion engine is one example of a process that is very nearly adiabatic. A slow adiabatic expansion of an ideal gas follows a curve like that labeled AC in Fig. 15–3. Since $Q = 0$, we have from Eq. 15–1 that $\Delta U = -W$. That is, the internal energy decreases if the gas expands; hence the temperature decreases as well (because $\Delta U = \frac{3}{2} nR\, \Delta T$). This is evident in Fig. 15–3 where the product $PV (= nRT)$ is less at point C than at point B (curve AB is for an isothermal process, for which $\Delta U = 0$ and $\Delta T = 0$). In the reverse operation, an adiabatic compression (going from C to A, for example), work is done *on* the gas, and hence the internal energy increases and the temperature rises. In a diesel engine, the fuel–air mixture is rapidly compressed adiabatically by a factor of 15 or more; the temperature rise is so great that the mixture ignites spontaneously.

Isothermal and adiabatic processes are just two possible processes that can occur. Two other simple thermodynamic processes are illustrated on the *PV* diagrams of Fig. 15–4: (*a*) an **isobaric** process is one in which the pressure is kept constant, so the process is represented by a straight horizontal line on the *PV* diagram (Fig. 15–4a); (*b*) an **isovolumetric** or *isochoric* process is one in which the volume does not change (Fig. 15–4b). In these, and in all other processes, the first law of thermodynamics holds.

Adiabatic process $(Q = 0)$

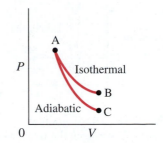

FIGURE 15–3 *PV* diagram for adiabatic (AC) and isothermal (AB) processes on an ideal gas.

Isobaric process:
$P = constant$, $W = P\,\Delta V$

Isovolumetric process:
$V = constant$, $W = 0$

FIGURE 15–4 (a) Isobaric ("same pressure") process. (b) Isovolumetric ("same volume") process.

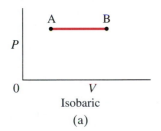

Isobaric
(a)

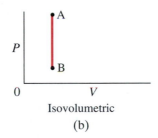

Isovolumetric
(b)

FIGURE 15–5 Work is done on the piston when the gas expands, moving the piston a distance *d*.

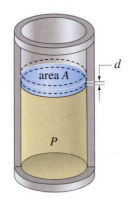

It is often valuable to calculate the work done in a process. If the pressure is kept constant during a process (isobaric), the work done is easily calculated. For example, if the gas in Fig. 15–5 expands slowly against the piston, the work done by the gas to raise the piston is the force *F* times the distance *d*. But the force is just the pressure *P* of the gas times the area *A* of the piston, $F = PA$. Thus,

$$W = Fd = PAd.$$

Because $Ad = \Delta V$, the change in volume of the gas, then

$$W = P\,\Delta V. \qquad \text{[constant pressure]} \quad (15\text{–}3)$$

Work done in volume changes

Equation 15–3 also holds if the gas is *compressed* at constant pressure, in which case ΔV is negative (since *V* decreases); *W* is then negative, which indicates that work is done *on* the gas. Equation 15–3 is also valid for liquids and solids, as long as the pressure is constant during the process.

In an isovolumetric process (Fig. 15–4b) the volume does not change, so no work is done, $W = 0$.

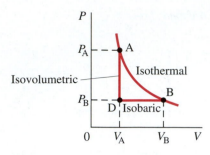

FIGURE 15–6 *PV* diagram for different processes (see the text), where the system changes from A to B.

Work = area under PV curve

FIGURE 15–7 Work done by a gas is equal to the area under the *PV* curve.

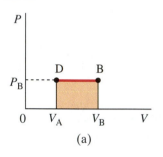

(a)

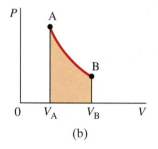

(b)

Figure 15–6 shows the isotherm AB we saw in Fig. 15–2 as well as another possible process represented by the path ADB. In going from A to D, the gas does no work since the volume does not change. But in going from D to B, the gas does work equal to $P_B(V_B - V_A)$, and this is the total work done in the process ADB.

If the pressure varies during a process, such as for the isothermal process AB in Fig. 15–2, Eq. 15–3 cannot be used directly to determine the work. A rough estimate can be obtained, however, by using an "average" value for P in Eq. 15–3. More accurately, the work done is equal to the area under the *PV* curve. This is obvious when the pressure is constant: as Fig. 15–7a shows, the shaded area is just $P_B(V_B - V_A)$, and this is the work done. Similarly, the work done during an isothermal process is equal to the shaded area shown in Fig. 15–7b. The calculation of work done in this case can be carried out using calculus, or by estimating the area on graph paper.

CONCEPTUAL EXAMPLE 15–3 **Work in isothermal and adiabatic processes.** In Fig. 15–3 we saw the *PV* diagrams for a gas expanding in two ways, isothermally and adiabatically. The initial volume V_A was the same in each case, and the final volumes were the same $(V_B = V_C)$. In which process was more work done by the gas?

RESPONSE Our system is the gas. More work was done by the gas in the isothermal process, which we can see in two simple ways by looking at Fig. 15–3. First, the "average" pressure was higher during the isothermal process AB, so $W = P_{av} \Delta V$ was greater (ΔV is the same for both processes). Second, we can look at the area under each curve: the area under curve AB, which represents the work done, was greater (since curve AB is higher) than that under AC.

EXERCISE B Is the work done by the gas in process ADB of Fig. 15–6 greater than, less than, or equal to the work done in the isothermal process AB?

CONCEPTUAL EXAMPLE 15–4 **Simple adiabatic process.** Here is an example of an adiabatic process that you can do with just a rubber band. Hold a thin rubber band loosely with two hands and gauge its temperature with your lips. Stretch the rubber band suddenly and again touch it lightly to your lips. You should notice an increase in temperature. Explain clearly why the temperature increases.

RESPONSE Stretching the rubber band *suddenly* makes the process adiabatic because there is no time for heat to enter or leave the system (the rubber band), so $Q = 0$. You do work on the system, representing an energy input, so W is negative in Eq. 15–1 $(\Delta U = Q - W)$. Hence ΔU must be positive. An increase in internal energy corresponds to an increase in temperature (for an ideal gas it is given by Eq. 14–1).

Table 15–1 gives a brief summary of the processes we have discussed.

TABLE 15–1 Simple Thermodynamic Processes and the First Law

Process	What is constant:	The first law predicts:
Isothermal	T = constant	$\Delta T = 0$ makes $\Delta U = 0$, so $Q = W$
Isobaric	P = constant	$Q = \Delta U + W = \Delta U + P \Delta V$
Isovolumetric	V = constant	$\Delta V = 0$ makes $W = 0$, so $Q = \Delta U$
Adiabatic	$Q = 0$	$\Delta U = -W$

EXAMPLE 15–5 **First law in isobaric and isovolumetric processes.** An ideal gas is slowly compressed at a constant pressure of 2.0 atm from 10.0 L to 2.0 L. This process is represented in Fig. 15–8 as the path B to D. (In this process, some heat flows out of the gas and the temperature drops.) Heat is then added to the gas, holding the volume constant, and the pressure and temperature are allowed to rise (line DA) until the temperature reaches its original value $(T_A = T_B)$. Calculate (a) the total work done by the gas in the process BDA, and (b) the total heat flow into the gas.

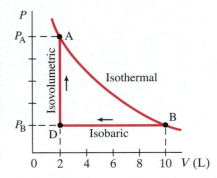

FIGURE 15–8 Example 15–5.

APPROACH (a) Work is done only in the compression process BD. In process DA, the volume is constant so $\Delta V = 0$ and no work is done (Eq. 15–3). (b) We use the first law of thermodynamics, Eq. 15–1.

SOLUTION (a) During the compression BD, the pressure is $2.0 \text{ atm} = 2(1.01 \times 10^5 \text{ N/m}^2)$ and the change in volume is

$$\Delta V = (2.0 \times 10^{-3} \text{ m}^3) - (10.0 \times 10^{-3} \text{ m}^3) = -8.0 \times 10^{-3} \text{ m}^3.$$

Then the work done is

$$W = P \Delta V = (2.02 \times 10^5 \text{ N/m}^2)(-8.0 \times 10^{-3} \text{ m}^3) = -1.6 \times 10^3 \text{ J}.$$

The total work done by the gas is -1.6×10^3 J, where the minus sign means that $+1.6 \times 10^3$ J of work is done *on* the gas.

(b) Because the temperature at the beginning and at the end of process BDA is the same, there is no change in internal energy: $\Delta U = 0$. From the first law of thermodynamics we have

$$0 = \Delta U = Q - W,$$

so

$$Q = W = -1.6 \times 10^3 \text{ J}.$$

Since Q is negative, 1600 J of heat flows out of the gas for the whole process, BDA.

EXERCISE C In Example 15–5, if the heat lost from the gas in the process BD is 8.4×10^3 J, what is the change in internal energy of the gas during process BD?

Additional Examples

EXAMPLE 15–6 **Work done in an engine.** In an engine, 0.25 moles of an ideal monatomic gas in the cylinder expands rapidly and adiabatically against the piston. In the process, the temperature of the gas drops from 1150 K to 400 K. How much work does the gas do?

APPROACH We take the gas as our system (the piston is part of the surroundings). The pressure is not constant, so we can't use Eq. 15–3. Instead, we can use the first law of thermodynamics because we can determine ΔU given $Q = 0$ (the process is adiabatic).

SOLUTION We determine ΔU from Eq. 14–1 for the internal energy of an ideal monatomic gas:

$$\Delta U = U_f - U_i = \tfrac{3}{2} nR(T_f - T_i)$$
$$= \tfrac{3}{2}(0.25 \text{ mol})(8.314 \text{ J/mol} \cdot \text{K})(400 \text{ K} - 1150 \text{ K})$$
$$= -2300 \text{ J}.$$

Then, from the first law of thermodynamics, Eq. 15–1,

$$W = Q - \Delta U = 0 - (-2300 \text{ J}) = 2300 \text{ J}.$$

EXAMPLE 15–7 **ΔU for boiling water to steam.** Determine the change in internal energy of 1.00 liter of water (mass 1.00 kg) at 100°C when it is fully boiled from liquid to gas, which results in 1671 liters of steam at 100°C. Assume the process is done at atmospheric pressure.

APPROACH Our system is the water. The heat required here does not result in a temperature change; rather, a change in phase occurs. We can determine the heat Q required using the latent heat of water, as in Section 14–5. Work too will be done: $W = P\,\Delta V$. The first law of thermodynamics will then give us ΔU.

SOLUTION The latent heat of vaporization of water (Table 14–3) is $L_V = 22.6 \times 10^5\ \mathrm{J/kg}$. So the heat input required for this process is

$$\begin{aligned} Q = mL &= (1.00\ \mathrm{kg})(22.6 \times 10^5\ \mathrm{J/kg}) \\ &= 22.6 \times 10^5\ \mathrm{J}. \end{aligned}$$

The work done by the water is (Eq. 15–3)

$$\begin{aligned} W = P\,\Delta V &= (1.01 \times 10^5\ \mathrm{N/m^2})\left[(1671 \times 10^{-3}\ \mathrm{m^3}) - (1 \times 10^{-3}\ \mathrm{m^3})\right] \\ &= 1.69 \times 10^5\ \mathrm{J}, \end{aligned}$$

where we used $1\ \mathrm{atm} = 1.01 \times 10^5\ \mathrm{N/m^2}$ and $1\ \mathrm{L} = 10^3\ \mathrm{cm^3} = 10^{-3}\ \mathrm{m^3}$. Then

$$\begin{aligned} \Delta U = Q - W &= (22.6 \times 10^5\ \mathrm{J}) - (1.7 \times 10^5\ \mathrm{J}) \\ &= 20.9 \times 10^5\ \mathrm{J}. \end{aligned}$$

NOTE Most of the heat added goes to increasing the internal energy of the water (increasing molecular energy to overcome the attraction that held the molecules close together in the liquid state). Only a small part ($< 10\%$) goes into doing work.

EXERCISE D Equation 14–1, $U = \frac{3}{2}nRT$, tells us that $\Delta U = 0$ in Example 15–7 because $\Delta T = 0$. Yet we determined that $\Delta U = 21 \times 10^5\ \mathrm{J}$. What is wrong?

*15–3 Human Metabolism and the First Law

PHYSICS APPLIED
Energy in the human body

Human beings and other animals do work. Work is done when a person walks or runs, or lifts a heavy object. Work requires energy. Energy is also needed for growth—to make new cells, and to replace old cells that have died. A great many energy-transforming processes occur within an organism, and they are referred to as *metabolism*.

We can apply the first law of thermodynamics,

$$\Delta U = Q - W,$$

to an organism: say, the human body. Work W is done by the body in its various activities; if this is not to result in a decrease in the body's internal energy (and temperature), energy must somehow be added to compensate. The body's internal energy is not maintained by a flow of heat Q into the body, however. Normally, the body is at a higher temperature than its surroundings, so heat usually flows *out* of the body. Even on a very hot day when heat is absorbed, the body has no way of utilizing this heat to support its vital processes. What then is the source of energy that allows us to do work? It is the internal energy (chemical potential energy) stored in foods (Fig. 15–9). In a closed system, the internal energy changes only as a result of heat flow or work done. In an open system, such as a human, internal energy itself can flow into or out of the system. When we eat food, we are bringing internal energy into our bodies directly, which thus increases the total internal energy U in our bodies. This energy eventually goes into work and heat flow from the body according to the first law.

FIGURE 15–9 Bike rider getting an input of energy.

The metabolic rate is the rate at which internal energy is transformed within the body. It is usually specified in kcal/h or in watts. Typical metabolic rates for a variety of human activities are given in Table 15–2 for an "average" 65-kg adult.

EXAMPLE 15–8 **Energy transformation in the body.** How much energy is transformed in 24 h by a 65-kg person who spends 8.0 h sleeping, 1.0 h at moderate physical labor, 4.0 h in light activity, and 11.0 h working at a desk or relaxing?

APPROACH The energy transformed during each activity equals the metabolic rate (Table 15–2) multiplied by the time.

SOLUTION Table 15–2 gives the metabolic rate in watts (J/s). Since there are 3600 s in an hour, the total energy transformed is

$$\left[\begin{array}{l}(8.0\,\text{h})(70\,\text{J/s}) + (1.0\,\text{h})(460\,\text{J/s}) \\ + (4.0\,\text{h})(230\,\text{J/s}) + (11.0\,\text{h})(115\,\text{J/s})\end{array}\right](3600\,\text{s/h}) = 1.15 \times 10^7\,\text{J}.$$

NOTE Since $4.186 \times 10^3\,\text{J} = 1\,\text{kcal}$, this is equivalent to 2800 kcal; a food intake of 2800 Cal would compensate for this energy output. A 65-kg person who wanted to lose weight would have to eat less than 2800 Cal a day, or increase his or her level of activity.

TABLE 15–2
Metabolic Rates (65-kg human)

Activity	Metabolic Rate (approximate)	
	kcal/h	watts
Sleeping	60	70
Sitting upright	100	115
Light activity (eating, dressing, household chores)	200	230
Moderate work (tennis, walking)	400	460
Running (15 km/h)	1000	1150
Bicycling (race)	1100	1270

15–4 The Second Law of Thermodynamics— Introduction

The first law of thermodynamics states that energy is conserved. There are, however, many processes we can imagine that conserve energy but are not observed to occur in nature. For example, when a hot object is placed in contact with a cold object, heat flows from the hotter one to the colder one, never spontaneously the reverse. If heat were to leave the colder object and pass to the hotter one, energy could still be conserved. Yet it doesn't happen spontaneously.[†] As a second example, consider what happens when you drop a rock and it hits the ground. The initial potential energy of the rock changes to kinetic energy as the rock falls. When the rock hits the ground, this energy in turn is transformed into internal energy of the rock and the ground in the vicinity of the impact; the molecules move faster and the temperature rises slightly. But have you seen the reverse happen—a rock at rest on the ground suddenly rise up in the air because the thermal energy of molecules is transformed into kinetic energy of the rock as a whole? Energy could be conserved in this process, yet we never see it happen.

There are many other examples of processes that occur in nature but whose reverse does not. Here are two more. (1) If you put a layer of salt in a jar and cover it with a layer of similar-sized grains of pepper, when you shake it you get a thorough mixture. But no matter how long you shake it, the mixture does not separate into two layers again. (2) Coffee cups and glasses break spontaneously if you drop them. But they don't go back together spontaneously (Fig. 15–10).

The first law of thermodynamics (conservation of energy) would not be violated if any of these processes occurred in reverse. To explain this lack of reversibility, scientists in the latter half of the nineteenth century formulated a new principle known as the second law of thermodynamics.

[†]By spontaneously, we mean by itself without input of work of some sort. (A refrigerator does move heat from a cold environment to a warmer one, but only by doing work.)

(a) Initial state.　　(b) Later: cup reassembles and rises up.　　(c) Later still: cup lands on table.

FIGURE 15–10 Have you ever observed this process, a broken cup spontaneously reassembling and rising up onto a table?

The **second law of thermodynamics** is a statement about which processes occur in nature and which do not. It can be stated in a variety of ways, all of which are equivalent. One statement, due to R. J. E. Clausius (1822–1888), is that

SECOND LAW OF THERMODYNAMICS
(Clausius statement)

heat can flow spontaneously from a hot object to a cold object; heat will not flow spontaneously from a cold object to a hot object.

Since this statement applies to one particular process, it is not obvious how it applies to other processes. A more general statement is needed that will include other possible processes in a more obvious way.

Heat engine

The development of a general statement of the second law of thermodynamics was based partly on the study of heat engines. A **heat engine** is any device that changes thermal energy into mechanical work, such as steam engines and automobile engines. We now examine heat engines, both from a practical point of view and to show their importance in developing the second law of thermodynamics.

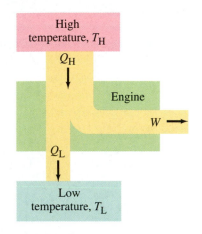

FIGURE 15–11 Schematic diagram of energy transfers for a heat engine.

15–5 Heat Engines

It is easy to produce thermal energy by doing work—for example, by simply rubbing your hands together briskly, or indeed by any frictional process. But to get work from thermal energy is more difficult, and a practical device to do this was invented only about 1700 with the development of the steam engine.

The basic idea behind any heat engine is that mechanical energy can be obtained from thermal energy only when heat is allowed to flow from a high temperature to a low temperature. In the process, some of the heat can then be transformed to mechanical work, as diagrammed schematically in Fig. 15–11. We will be interested only in engines that run in a repeating *cycle* (that is, the system returns repeatedly to its starting point) and thus can run continuously. In each cycle the change in internal energy of the system is $\Delta U = 0$ because it returns to the starting state. Thus a heat input Q_H at a high temperature T_H is partly transformed into work W and partly exhausted as heat Q_L at a lower temperature T_L (Fig. 15–11). By conservation of energy, $Q_H = W + Q_L$. The high and low temperatures, T_H and T_L, are called the **operating temperatures** of the engine. Note carefully that we are now using a new sign convention: we take Q_H, Q_L, and W as always positive. The direction of each energy transfer is found from the applicable diagram, such as Fig. 15–11.

⚠ **C A U T I O N**

New sign convention:
$Q_H > 0, Q_L > 0, W > 0$

👤 **P H Y S I C S   A P P L I E D**

Engines

Steam Engine and Internal Combustion Engine

The operation of a steam engine is illustrated in Fig. 15–12. Steam engines are of two main types, each making use of steam heated by combustion of coal, oil,

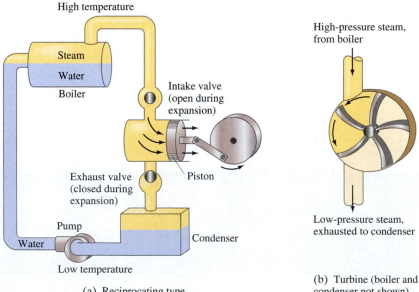

FIGURE 15–12 Steam engines.

(a) Reciprocating type

(b) Turbine (boiler and condenser not shown)

gas, or nuclear energy. In the so-called reciprocating type, Fig. 15–12a, the heated steam passes through the intake valve and expands against a piston, forcing it to move. As the piston returns to its original position, it forces the gases out the exhaust valve. In a steam turbine, Fig. 15–12b, everything is essentially the same, except that the reciprocating piston is replaced by a rotating turbine that resembles a paddlewheel with many sets of blades. Most of our electricity today is generated using steam turbines.[†] The material that is heated and cooled, steam in this case, is called the **working substance**. In a steam engine, the high temperature is obtained by burning coal, oil, or other fuel to heat the steam.

In an internal combustion engine (used in most automobiles), the high temperature is achieved by burning the gasoline–air mixture in the cylinder itself (ignited by the spark plug), as described in Fig. 15–13.

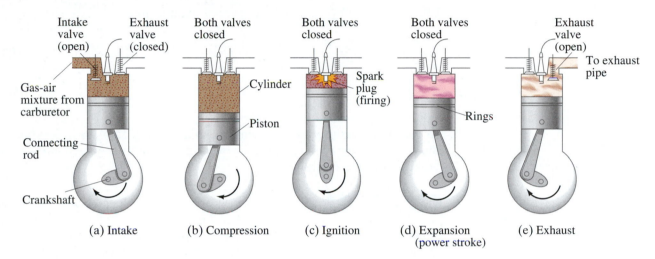

(a) Intake   (b) Compression   (c) Ignition   (d) Expansion (power stroke)   (e) Exhaust

FIGURE 15–13 Four-stroke-cycle internal combustion engine: (a) the gasoline–air mixture flows into the cylinder as the piston moves down; (b) the piston moves upward and compresses the gas; (c) the brief instant when firing of the spark plug ignites the highly compressed gasoline–air mixture, raising it to a high temperature; (d) the gases, now at high temperature and pressure, expand against the piston in this, the power stroke; (e) the burned gases are pushed out to the exhaust pipe; when the piston reaches the top, the exhaust valve closes and the intake valve opens, and the whole cycle repeats. (a), (b), (d), and (e) are the four strokes of the cycle.

Why a ΔT Is Needed to Drive a Heat Engine

To see why a *temperature difference* is required to run an engine, let us examine the steam engine. In the reciprocating engine, for example, suppose there were no condenser or pump (Fig. 15–12a), and that the steam was at the same temperature throughout the system. This would mean that the pressure of the gas being exhausted would be the same as that on intake. Thus, although work would be done by the gas *on* the piston when it expanded, an equal amount of work would have to be done *by* the piston to force the steam out the exhaust; hence, no net work would be done. In a real engine, the exhausted gas is cooled to a lower temperature and condensed so that the exhaust pressure is less than the intake pressure. Thus, although the piston must do work on the gas to expel it on the exhaust stroke, it is less than the work done by the gas on the piston during the intake. So a net amount of work can be obtained—but only if there is a difference of temperature. Similarly, in the gas turbine if the gas isn't cooled, the pressure on each side of the blades would be the same. By cooling the gas on the exhaust side, the pressure on the back side of the blade is less and hence the turbine turns.

[†]Even nuclear power plants utilize steam turbines; the nuclear fuel—uranium—merely serves as fuel to heat the steam.

Efficiency

The **efficiency**, e, of any heat engine can be defined as the ratio of the work it does, W, to the heat input at the high temperature, Q_H (Fig. 15–11):

$$e = \frac{W}{Q_H}.$$

This is a sensible definition since W is the output (what you get from the engine), whereas Q_H is what you put in and pay for in burned fuel. Since energy is conserved, the heat input Q_H must equal the work done plus the heat that flows out at the low temperature (Q_L):

$$Q_H = W + Q_L.$$

Thus $W = Q_H - Q_L$, and the efficiency of an engine is

Efficiency of any heat engine

$$e = \frac{W}{Q_H} \tag{15–4a}$$

$$= \frac{Q_H - Q_L}{Q_H} = 1 - \frac{Q_L}{Q_H}. \tag{15–4b}$$

To give the efficiency as a percent, we multiply Eq. 15–4 by 100. Note that e could be 1.0 (or 100%) only if Q_L were zero—that is, only if no heat were exhausted to the environment.

EXAMPLE 15–9 **Car efficiency.** An automobile engine has an efficiency of 20% and produces an average of 23,000 J of mechanical work per second during operation. (*a*) How much heat input is required, and (*b*) how much heat is discharged as waste heat from this engine, per second?

APPROACH We want to find the heat input Q_H as well as the heat output Q_L, given $W = 23,000$ J each second and an efficiency $e = 0.20$. We can use the definition of efficiency, Eq. 15–4 in its various forms, to find first Q_H and then Q_L.

SOLUTION (*a*) From Eq. 15–4, $e = W/Q_H$, we solve for Q_H:

$$Q_H = \frac{W}{e} = \frac{23,000\ \text{J}}{0.20}$$

$$= 1.15 \times 10^5\ \text{J} = 115\ \text{kJ}.$$

The engine requires 115 kJ/s = 115 kW of heat input.

(*b*) We now use the last part of Eq. 15–4 $\left(e = 1 - Q_L/Q_H\right)$ to solve for Q_L:

$$\frac{Q_L}{Q_H} = 1 - e$$

so

$$Q_L = (1 - e)Q_H = (0.80)115\ \text{kJ}$$

$$= 92\ \text{kJ}.$$

The engine discharges heat to the environment at a rate of 92 kJ/s = 92 kW.

NOTE Of the 115 kJ that enters the engine per second, only 23 kJ does useful work whereas 92 kJ is wasted as heat output.

NOTE The problem was stated in terms of energy per unit time. We could just as well have stated it in terms of power, since 1 J/s = 1 W.

Carnot Engine

To see how to increase efficiency, the French scientist Sadi Carnot (1796–1832) examined the characteristics of an ideal engine (now called a **Carnot engine**). No Carnot engine actually exists, but as a theoretical idea it played an important role in the development of thermodynamics.

Carnot (ideal) engine

The idealized Carnot engine consisted of four processes done in a cycle, two of which are adiabatic ($Q = 0$) and two are isothermal ($\Delta T = 0$). This idealized cycle is shown in Fig. 15–14. Each of the processes was considered to be done **reversibly**. That is, each of the processes (say, during expansion of the gases against a piston) was done so slowly that the process could be considered a series of equilibrium states, and the whole process could be done in reverse with no change in the magnitude of work done or heat exchanged. A real process, on the other hand, would occur more quickly; there would be turbulence in the gas, friction would be present, and so on. Because of these factors, a real process cannot be done precisely in reverse—the turbulence would be different and the heat lost to friction would not reverse itself. Thus, real processes are **irreversible**.

FIGURE 15–14 The Carnot cycle. Heat engines work in a cycle, and the cycle for the Carnot engine begins at point a on this PV diagram. (1) The gas is first expanded isothermally, with the addition of heat Q_H, along the path ab at temperature T_H. (2) Next the gas expands adiabatically from b to c—no heat is exchanged, but the temperature drops to T_L. (3) The gas is then compressed at constant temperature T_L, path cd, and heat Q_L flows out. (4) Finally, the gas is compressed adiabatically, path da, back to its original state.

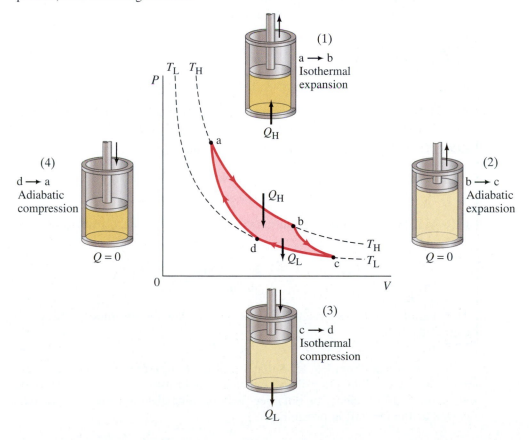

Carnot showed that for an ideal reversible engine, the heats Q_H and Q_L are proportional to the operating temperatures T_H and T_L (in kelvins), so the efficiency can be written as

Carnot (ideal) efficiency

$$e_{ideal} = \frac{T_H - T_L}{T_H} = 1 - \frac{T_L}{T_H}. \qquad \left[\begin{array}{c}\text{Carnot (ideal)} \\ \text{efficiency}\end{array}\right] \quad \textbf{(15–5)}$$

Equation 15–5 expresses the fundamental upper limit to the efficiency. Real engines always have an efficiency lower than this because of losses due to friction and the like. Real engines that are well designed reach 60 to 80% of the Carnot efficiency.

EXAMPLE 15–10 **Steam engine efficiency.** A steam engine operates between 500°C and 270°C. What is the maximum possible efficiency of this engine?

APPROACH The maximum possible efficiency is the idealized Carnot efficiency, Eq. 15–5. We must use kelvin temperatures.

SOLUTION We first change the temperature to kelvins by adding 273 to the given Celsius temperatures: $T_H = 773$ K and $T_L = 543$ K. Then

$$e_{ideal} = 1 - \frac{543}{773} = 0.30.$$

To get the efficiency in percent, we multiply by 100. Thus, the maximum (or Carnot) efficiency is 30%. Realistically, an engine might attain 0.70 of this value, or 21%.

NOTE In this Example the exhaust temperature is still rather high, 270°C. Steam engines are often arranged in series so that the exhaust of one engine is used as intake by a second or third engine.

EXAMPLE 15–11 **A phony claim?** An engine manufacturer makes the following claims: An engine's heat input per second is 9.0 kJ at 435 K. The heat output per second is 4.0 kJ at 285 K. Do you believe these claims?

APPROACH The engine's efficiency can be calculated from the definition, Eq. 15–4. It must be less than the maximum possible, Eq. 15–5.

SOLUTION The claimed efficiency of the engine is

$$e = \frac{Q_H - Q_L}{Q_H} = \frac{9.0\,\text{kJ} - 4.0\,\text{kJ}}{9.0\,\text{kJ}} = 0.56.$$

However, the maximum possible efficiency is given by the Carnot efficiency, Eq. 15–5:

$$e_{ideal} = \frac{T_H - T_L}{T_H} = \frac{435\,\text{K} - 285\,\text{K}}{435\,\text{K}} = 0.34.$$

The manufacturer's claims violate the second law of thermodynamics and cannot be believed.

It is quite clear from Eq. 15–5 that at normal temperatures, a 100% efficient engine is not possible. Only if the exhaust temperature, T_L, were at absolute zero could 100% efficiency be obtained. But reaching absolute zero is a practical (as well as theoretical) impossibility.[†]

[†]Careful experimentation suggests that absolute zero is unattainable. This result is known as the **third law of thermodynamics**.

Because no engine can be 100% efficient, we can say that

no device is possible whose sole effect is to transform a given amount of heat completely into work.

This is known as the **Kelvin-Planck statement of the second law of thermodynamics**. Figure 15–15 diagrams the ideal perfect heat engine, which does not exist.

If the second law were not true, so that a perfect engine could be built, rather remarkable things could happen. For example, if the engine of a ship did not need a low-temperature reservoir to exhaust heat into, the ship could sail across the ocean using the vast resources of the internal energy of the ocean water. Indeed, we would have no fuel problems at all!

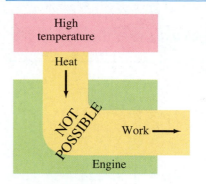

FIGURE 15–15 Diagram of an impossible perfect heat engine in which all heat input is used to do work.

15–6 | Refrigerators, Air Conditioners, and Heat Pumps

The operating principle of refrigerators, air conditioners, and heat pumps is just the reverse of a heat engine. Each operates to transfer heat *out* of a cool environment into a warm environment. As diagrammed in Fig. 15–16, by doing work W, heat is taken from a low-temperature region, T_L (such as inside a refrigerator), and a greater amount of heat is exhausted at a high temperature, T_H (the room). You can often feel this heat blowing out beneath a refrigerator. The work W is usually done by an electric compressor motor which compresses a fluid, as illustrated in Fig. 15–17.

FIGURE 15–16 Schematic diagram of energy transfers for a refrigerator or air conditioner.

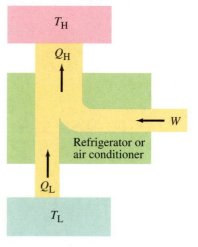

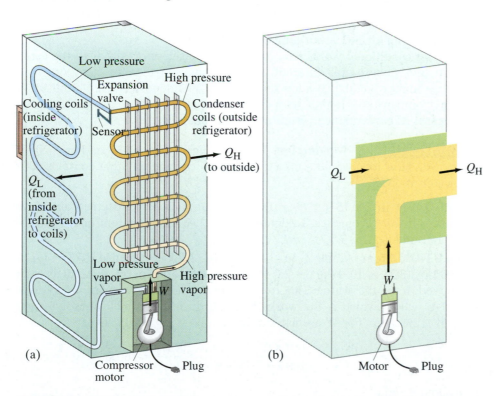

FIGURE 15–17 (a) Typical refrigerator system. The electric compressor motor forces a gas at high pressure through a heat exchanger (condenser) on the rear outside wall of the refrigerator, where Q_H is given off, and the gas cools to become liquid. The liquid passes from a high-pressure region, via a valve, to low-pressure tubes on the inside walls of the refrigerator; the liquid evaporates at this lower pressure and thus absorbs heat (Q_L) from the inside of the refrigerator. The fluid returns to the compressor, where the cycle begins again. (b) Schematic diagram, like Fig. 15–16.

A perfect **refrigerator**—one in which no work is required to take heat from the low-temperature region to the high-temperature region—is not possible. This is the **Clausius statement of the second law of thermodynamics**, already mentioned in Section 15–4: it can be stated formally as

| SECOND LAW OF THERMODYNAMICS |
| (Clausius statement) |

no device is possible whose sole effect is to transfer heat from one system at a temperature T_L into a second system at a higher temperature T_H.

To make heat flow from a low-temperature object (or system) to one at a higher temperature, work must be done. Thus, *there can be no perfect refrigerator.*

The **coefficient of performance** (COP) of a refrigerator is defined as the heat Q_L removed from the low-temperature area (inside a refrigerator) divided by the work W done to remove the heat (Fig. 15–16):

$$\text{COP} = \frac{Q_L}{W}. \qquad \left[\begin{array}{l}\text{refrigerator and}\\ \text{air conditioner}\end{array}\right] \quad \textbf{(15–6a)}$$

This makes sense since the more heat, Q_L, that can be removed from inside the refrigerator for a given amount of work, the better (more efficient) the refrigerator is. Energy is conserved, so from the first law of thermodynamics we can write $Q_L + W = Q_H$, or $W = Q_H - Q_L$ (see Fig. 15–16). Then Eq. 15–6a becomes

$$\text{COP} = \frac{Q_L}{W} = \frac{Q_L}{Q_H - Q_L}. \qquad \left[\begin{array}{l}\text{refrigerator and}\\ \text{air conditioner}\end{array}\right] \quad \textbf{(15–6b)}$$

For an ideal refrigerator (not a perfect one, which is impossible), the best one could do would be

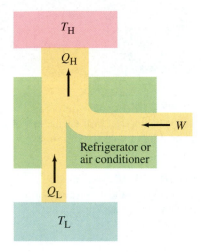

FIGURE 15–16 (repeated) Schematic diagram of energy transfers for a refrigerator or air conditioner.

$$\text{COP}_{\text{ideal}} = \frac{T_L}{T_H - T_L}, \qquad \left[\begin{array}{l}\text{refrigerator and}\\ \text{air conditioner}\end{array}\right] \quad \textbf{(15–6c)}$$

analogous to an ideal (Carnot) engine (Eq. 15–5).

An **air conditioner** works very much like a refrigerator, although the actual construction details are different: an air conditioner takes heat Q_L from inside a room or building at a low temperature, and deposits heat Q_H outside to the environment at a higher temperature. Equations 15–6 also describe the coefficient of performance for an air conditioner.

EXAMPLE 15–12 **Making ice.** A freezer has a COP of 3.8 and uses 200 W of power. How long would it take to freeze an ice-cube tray that contains 600 g of water at 0°C?

APPROACH In Eq. 15–6b, Q_L is the heat that must be transferred out of the water so it will become ice. To determine Q_L, we use the latent heat of fusion of water and Eq. 14–3, $Q = mL$.

SOLUTION From Table 14–3, $L = 333 \text{ kJ/kg}$. Hence $Q = mL = (0.600 \text{ kg})(3.33 \times 10^5 \text{ J/kg}) = 2.0 \times 10^5 \text{ J}$ is the total energy that needs to be removed from the water. The freezer does work at the rate of $200 \text{ W} = 200 \text{ J/s} = W/t$, which is the work W it can do in t seconds. We solve for t: $t = W/(200 \text{ J/s})$. For W, we use Eq. 15–6b: $W = Q_L/\text{COP}$. Thus

$$t = \frac{W}{200 \text{ J/s}} = \frac{Q_L/\text{COP}}{200 \text{ J/s}} = \frac{2.0 \times 10^5 \text{ J}}{(3.8)(200 \text{ J/s})} = 260 \text{ s},$$

or about $4\frac{1}{2}$ min.

Heat naturally flows from high temperature to low temperature. Refrigerators and air conditioners do work to accomplish the opposite: to make heat flow from cold to hot. We might say they "pump" heat from cold areas to hotter areas, against the natural tendency of heat to flow from hot to cold, just as water can be pumped uphill, against the natural tendency to flow downhill. The term

heat pump is usually reserved for a device that can heat a house in winter by using an electric motor that does work W to take heat Q_L from the outside at low temperature and delivers heat Q_H to the warmer inside of the house; see Fig. 15–18. As in a refrigerator, there is an indoor and an outdoor heat exchanger (coils of the refrigerator) and an electric compressor motor. The operating principle is like that for a refrigerator or air conditioner; but the objective of a heat pump is to heat (deliver Q_H), rather than to cool (remove Q_L). Thus, the coefficient of performance of a heat pump is defined differently than for an air conditioner because it is the heat Q_H delivered to the inside of the house that is important now:

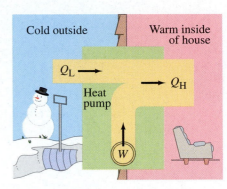

FIGURE 15–18 A heat pump uses an electric motor to "pump" heat from the cold outside to the warm inside of a house.

$$\text{COP} = \frac{Q_H}{W}. \qquad \text{[Heat pump]} \quad \textbf{(15–7)}$$

The COP is necessarily greater than 1. Most heat pumps can be "turned around" and used as air conditioners in the summer.

EXAMPLE 15–13 **Heat pump.** A heat pump has a coefficient of performance of 3.0 and is rated to do work at 1500 W. (*a*) How much heat can it add to a room per second? (*b*) If the heat pump were turned around to act as an air conditioner in the summer, what would you expect its coefficient of performance to be, assuming all else stays the same?

APPROACH We use the definitions of coefficient of performance, which are different for the two devices in (*a*) and (*b*).

SOLUTION (*a*) We use Eq. 15–7 for the heat pump, and, since our device does 1500 J of work per second, it can pour heat into the room at a rate of

$$Q_H = \text{COP} \times W = 3.0 \times 1500\,\text{J} = 4500\,\text{J}$$

per second, or at a rate of 4500 W.

(*b*) If our device is turned around in summer, it can take heat Q_L from inside the house, doing 1500 J of work per second to then dump $Q_H = 4500$ J per second to the hot outside. Energy is conserved, so $Q_L + W = Q_H$ (see Fig. 15–18, but reverse the inside and outside of the house). Then

$$Q_L = Q_H - W = 4500\,\text{J} - 1500\,\text{J} = 3000\,\text{J}.$$

The coefficient of performance as an air conditioner would thus be (Eq. 15–6a)

$$\text{COP} = \frac{Q_L}{W} = \frac{3000\,\text{J}}{1500\,\text{J}} = 2.0.$$

NOTE The coefficients of performance are defined differently for heat pumps and air conditioners.

CAUTION
Heat pumps and air conditioners have different COP definitions

A good heat pump can sometimes be a money saver and an energy saver, depending on the cost of the unit and installation, etc. Compare, for example, our heat pump in Example 15–13 to, say, a 1500-W electric heater. We plug the latter into the wall, it draws 1500 W of electricity and delivers 1500 W of heat to the room. Our heat pump when plugged into the wall also draws 1500 W of electricity (which is what we pay for), but it delivers 4500 W of heat!

* SEER Rating

Cooling devices such as refrigerators and air conditioners are often given a rating known as SEER (Seasonal Energy Efficiency Ratio), which is defined as

$$\text{SEER} = \frac{(\text{heat removed in Btu})}{(\text{electrical input in watt-hours})},$$

as measured by averaging over varying (seasonal) conditions. The definition of the SEER is basically the same as the COP except for the (unfortunate) mixed units. Given that 1 Btu = 1055 J (see Section 14–1 and Problem 4 in Chapter 14), then a SEER = 1 is a COP equal to $(1\,\text{Btu}/1\,\text{W}\cdot\text{h}) = (1055\,\text{J})/(1\,\text{J/s} \times 3600\,\text{s}) = 0.29$. A COP = 1 is a SEER = 1/0.29 = 3.4.

15-7 Entropy and the Second Law of Thermodynamics

We have seen several aspects of the second law of thermodynamics; and the different statements of it that we have discussed can be shown to be completely equivalent. But what we really need is a general statement of the second law of thermodynamics. It was not until the latter half of the nineteenth century that the *second law of thermodynamics* was finally stated in a general way—

Entropy

namely, in terms of a quantity called **entropy**, introduced by Clausius in the 1860s. Entropy, unlike heat, is a function of the state of a system. That is, a system in a given state has a temperature, a volume, a pressure, and so on, and also has a particular value of entropy. In the next Section, we will see that entropy can be interpreted as a measure of the order or disorder of a system.

When we deal with entropy—as with potential energy—it is the *change* in entropy during a process that is important, not the absolute amount. According to Clausius, the change in entropy S of a system, when an amount of heat Q is *added* to it by a reversible[†] process at constant temperature, is given by

Entropy change

$$\Delta S = \frac{Q}{T}, \tag{15–8}$$

where T is the kelvin temperature.

EXAMPLE 15–14 **Entropy change in melting.** An ice cube of mass 56 g is taken from a storage compartment at 0°C and placed in a paper cup. After a few minutes, exactly half of the mass of the ice cube has melted, becoming water at 0°C. Find the change in entropy of the ice/water.

APPROACH We consider the 56 g of water, initially in the form of ice, as our system. To determine the entropy change, we first must find the heat needed to melt the ice, which we do using the latent heat of fusion of water, $L = 333\ \text{kJ/kg}$ (Section 14–5).

SOLUTION The heat required to melt 28 g of ice (half of the 56-g ice cube) is

$$Q = mL = (0.028\ \text{kg})(333\ \text{kJ/kg}) = 9.3\ \text{kJ}.$$

The temperature remains constant in our process, so we can find the change in entropy from Eq. 15–8:

$$\Delta S = \frac{Q}{T} = \frac{9.3\ \text{kJ}}{273\ \text{K}} = 34\ \text{J/K}.$$

NOTE The change in entropy of the surroundings (cup, air) has not been computed.

The temperature in Example 15–14 was constant, so the calculation was easy. If the temperature varies during a process, a summation of the heat flow over the changing temperature can often be calculated using calculus or a computer. However, if the temperature change is not too great, a reasonable approximation can be made using the average value of the temperature, as indicated in the next Example.

EXAMPLE 15–15 **ESTIMATE** **Entropy change when mixing water.** A sample of 50.0 kg of water at 20.00°C is mixed with 50.0 kg of water at 24.00°C. Estimate the change in entropy.

APPROACH The final temperature of the mixture will be 22.00°C, since we started with equal amounts of water. We use the specific heat of water and the methods of calorimetry (Sections 14–3 and 14–4) to determine the heat transferred. Then we use the average temperature of each sample of water to estimate the entropy change ($\Delta Q/T$).

[†]Real processes are irreversible. Because entropy is a state variable, the change in entropy ΔS for an irreversible process can be determined by calculating ΔS for a reversible process between the same two states.

SOLUTION A quantity of heat,

$$Q = mc\,\Delta T = (50.0\,\text{kg})(4186\,\text{J/kg}\cdot\text{C}°)(2.00\,\text{C}°) = 4.186 \times 10^5\,\text{J},$$

flows out of the hot water as it cools down from 24°C to 22°C, and this heat flows into the cold water as it warms from 20°C to 22°C. The total change in entropy, ΔS, will be the sum of the changes in entropy of the hot water, ΔS_H, and that of the cold water, ΔS_C:

$$\Delta S = \Delta S_\text{H} + \Delta S_\text{C}.$$

We estimate entropy changes by writing $\Delta S = Q/T_\text{av}$, where T_av is an "average" temperature for each process, which ought to give a reasonable estimate since the temperature change is small. For the hot water we use an average temperature of 23°C (296 K), and for the cold water an average temperature of 21°C (294 K). Thus

$$\Delta S_\text{H} \approx -\frac{4.186 \times 10^5\,\text{J}}{296\,\text{K}} = -1414\,\text{J/K}$$

which is negative because this heat flows out, whereas heat is added to the cold water:

$$\Delta S_\text{C} \approx \frac{4.186 \times 10^5\,\text{J}}{294\,\text{K}} = 1424\,\text{J/K}.$$

Note that the entropy of the hot water (S_H) decreases since heat flows out of the hot water. But the entropy of the cold water (S_C) increases by a greater amount. The total change in entropy is

$$\Delta S = \Delta S_\text{H} + \Delta S_\text{C} \approx -1414\,\text{J/K} + 1424\,\text{J/K} \approx 10\,\text{J/K}.$$

In Example 15–15, we saw that although the entropy of one part of the system decreased, the entropy of the other part increased by a greater amount; the net change in entropy of the whole system was positive. This result, which we have calculated for a specific case in Example 15–15, has been found to hold in all other cases tested. That is, the total entropy of an isolated system is found to increase in all natural processes. The second law of thermodynamics can be stated in terms of entropy as follows: *The entropy of an isolated system never decreases. It can only stay the same or increase.* Entropy can remain the same only for an idealized (reversible) process. For any real process, the change in entropy ΔS is greater than zero:

Entropy of an isolated system never decreases

$$\Delta S > 0. \tag{15–9}$$

If the system is not isolated, then the change in entropy of the system, ΔS_s, plus the change in entropy of the environment, ΔS_env, must be greater than or equal to zero:

$$\Delta S = \Delta S_\text{s} + \Delta S_\text{env} \geq 0. \tag{15–10}$$

Only idealized processes have $\Delta S = 0$. Real processes have $\Delta S > 0$. This, then, is the *general statement of the second law of thermodynamics*:

the total entropy of any system plus that of its environment increases as a result of any natural process.

SECOND LAW OF THERMODYNAMICS (general statement)

Although the entropy of one part of the universe may decrease in any process (see Example 15–15), the entropy of some other part of the universe always increases by a greater amount, so the total entropy always increases.

Now that we finally have a quantitative general statement of the second law of thermodynamics, we can see that it is an unusual law. It differs considerably from other laws of physics, which are typically equalities (such as $F = ma$) or conservation laws (such as for energy and momentum). The second law of thermodynamics introduces a new quantity, the entropy S, but does not tell us it is conserved. Quite the opposite. Entropy is *not* conserved in natural processes; it always increases in time.

15-8 Order to Disorder

The concept of entropy, as we have discussed it so far, may seem rather abstract. To get a feel for the concept of entropy, we can relate it to the more ordinary concepts of *order* and *disorder*. In fact, the entropy of a system can be considered a *measure of the disorder of the system*. Then the second law of thermodynamics can be stated simply as:

Natural processes tend to move toward a state of greater disorder.

Exactly what we mean by disorder may not always be clear, so we now consider a few examples. Some of these will show us how this very general statement of the second law applies beyond what we usually consider as thermodynamics.

Let us look at the simple processes mentioned in Section 15–4. First, a jar containing separate layers of salt and pepper is more orderly than a jar in which the salt and pepper are all mixed up. Shaking a jar containing separate layers results in a mixture, and no amount of shaking brings the orderly layers back again. The natural process is from a state of relative order (layers) to one of relative disorder (a mixture), not the reverse. That is, disorder increases. Second, a solid coffee cup is a more "orderly" and useful object than the pieces of a broken cup. Cups break when they fall, but they do not spontaneously mend themselves (as faked in Fig 15–10). Again, the normal course of events is an increase of disorder.

When a hot object is put in contact with a cold object, heat flows from the high temperature to the low until the two objects reach the same intermediate temperature. At the beginning of the process we can distinguish two classes of molecules: those with a high average kinetic energy (the hot object), and those with a low average kinetic energy (the cooler object). After the process in which heat flows, all the molecules are in one class with the same average kinetic energy; we no longer have the more orderly arrangement of molecules in two classes. Order has gone to disorder. Furthermore, the separate hot and cold objects could serve as the hot- and cold-temperature regions of a heat engine, and thus could be used to obtain useful work. But once the two objects are put in contact and reach the same temperature, no work can be obtained. Disorder has increased, since a system that has the ability to perform work must surely be considered to have a higher order than a system no longer able to do work.

When a stone falls to the ground, its kinetic energy is transformed to thermal energy. (We noted earlier that the reverse never happens: a stone never absorbs thermal energy and rises into the air of its own accord.) This is another example of order changing to disorder. Thermal energy is associated with the disorderly random motion of molecules, but the molecules in the falling stone all have the same velocity downward in addition to their own random velocities. Thus, the more orderly kinetic energy of the stone is changed to disordered thermal energy when the stone strikes the ground. Disorder increases in this process, as it does in all processes that occur in nature.

15-9 Unavailability of Energy; Heat Death

In the process of heat conduction from a hot object to a cold one, we have seen that entropy increases and that order goes to disorder. The separate hot and cold objects could serve as the high- and low-temperature regions for a heat engine and thus could be used to obtain useful work. But after the two objects are put in contact with each other and reach the same uniform temperature, no work can be obtained from them. With regard to being able to do useful work, order has gone to disorder in this process.

The same can be said about a falling rock that comes to rest upon striking the ground. Before hitting the ground, all the kinetic energy of the rock could have been used to do useful work. But once the rock's mechanical kinetic energy becomes thermal energy, doing useful work is no longer possible.

Both these examples illustrate another important aspect of the second law of thermodynamics:

in any natural process, some energy becomes unavailable to do useful work.

In any process, no energy is ever lost (it is always conserved). Rather, energy becomes less useful—it can do less useful work. As time goes on, **energy is degraded**, in a sense; it goes from more orderly forms (such as mechanical) eventually to the least orderly form, internal, or thermal, energy. Entropy is a factor here because the amount of energy that becomes unavailable to do work is proportional to the change in entropy during any process.

Energy degradation

A natural outcome of this degradation of energy is the prediction that as time goes on, the universe will approach a state of maximum disorder. Matter will become a uniform mixture, and heat will have flowed from high-temperature regions to low-temperature regions until the whole universe is at one temperature. No work can then be done. All the energy of the universe will have become degraded to thermal energy. All change will cease. This prediction, called the **heat death** of the universe, has been much discussed by philosophers. The tendency toward this final state would seem an inevitable consequence of the second law of thermodynamics, although it would lie very far in the future.

"Heat death"

*15–10 Evolution and Growth; "Time's Arrow"

An interesting example of the increase in entropy relates to biological evolution and to growth of organisms. Clearly, a human being is a highly ordered organism. The theory of evolution describes the process from the early macromolecules and simple forms of life to *Homo sapiens*, which is a process of increasing order. So, too, the development of an individual from a single cell to a grown person is a process of increasing order. Do these processes violate the second law of thermodynamics? No, they do not. In the processes of evolution and growth, and even during the mature life of an individual, waste products are eliminated. These small molecules that remain as a result of metabolism are simple molecules without much order. Thus they represent relatively higher disorder or entropy. Indeed, the total entropy of the molecules cast aside by organisms during the processes of evolution and growth is greater than the decrease in entropy associated with the order of the growing individual or evolving species.

PHYSICS APPLIED
Biological evolution

Another aspect of the second law of thermodynamics is that it tells us in which *direction* processes go. If you were to see a film being run backward, you would undoubtedly be able to tell that it *was* run backward. For you would see odd occurrences, such as a broken coffee cup rising from the floor and reassembling on a table, or a torn balloon suddenly becoming whole again and filled with air. We know these things don't happen in real life; they are processes in which order increases—or entropy decreases. They violate the second law of thermodynamics. When watching a movie (or imagining that time could go backward), we are tipped off to a reversal of time by observing whether entropy (and disorder) is increasing or decreasing. Hence, entropy has been called **time's arrow**, for it can tell us in which direction time is going.

The ideas of entropy and disorder are made clearer with the use of a statistical or probabilistic analysis of the molecular state of a system. This statistical approach, which was first applied toward the end of the nineteenth century by Ludwig Boltzmann (1844–1906), makes a clear distinction between the "macrostate" and the "microstate" of a system. The **microstate** of a system would be specified in giving the position and velocity of every particle (or molecule). The **macrostate** of a system is specified by giving the macroscopic properties of the system—the temperature, pressure, number of moles, and so on. In reality, we can know only the macrostate of a system. There are generally far too many molecules in a system to be able to know the velocity and position of every one at a given moment. Nonetheless, it is important to recognize that a great many different microstates can correspond to the *same* macrostate.

Let us take a very simple example. Suppose you repeatedly shake four coins in your hand and drop them on a table. Specifying the number of heads and the number of tails that appear on a given throw is the macrostate of this system. Specifying each coin as being a head or a tail is the microstate of the system. In the following Table we see how many microstates correspond to each macrostate:

Macrostate	Possible Microstates (H = heads, T = tails)	Number of Microstates
4 heads	H H H H	1
3 heads, 1 tail	H H H T, H H T H, H T H H, T H H H	4
2 heads, 2 tails	H H T T, H T H T, T H H T, H T T H, T H T H, T T H H	6
1 head, 3 tails	T T T H, T T H T, T H T T, H T T T	4
4 tails	T T T T	1

Probabilities

A basic assumption behind the statistical approach is that *each microstate is equally probable.* Thus the number of microstates that give the same macrostate corresponds to the relative probability of that macrostate occurring. The macrostate of two heads and two tails is the most probable one in our case of tossing four coins; out of the total of 16 possible microstates, six correspond to two heads and two tails, so the probability of throwing two heads and two tails is 6 out of 16, or 38%. The probability of throwing one head and three tails is 4 out of 16, or 25%. The probability of four heads is only 1 in 16, or 6%. If you threw the coins 16 times, you might not find that two heads and two tails appear exactly 6 times, or four tails exactly once. These are only probabilities or averages. But if you made 1600 throws, very nearly 38% of them would be two heads and two tails. The greater the number of tries, the closer the percentages are to the calculated probabilities.

If we toss more coins—say, 100 all at the same time—the relative probability of throwing all heads (or all tails) is greatly reduced. There is only one microstate corresponding to all heads. For 99 heads and 1 tail, there are 100 microstates since each of the coins could be the one tail. The relative probabilities for other macrostates are given in Table 15–3. About 10^{30} microstates are possible.[†] Thus the relative probability of finding all heads is 1 in 10^{30}, an incredibly unlikely event! The probability of obtaining 50 heads and 50 tails (see Table 15–3) is $(1.0 \times 10^{29})/10^{30} = 0.10$, or 10%. The probability of obtaining anything between 45 and 55 heads is 90%.

Thus we see that as the number of coins increases, the probability of obtaining the most orderly arrangement (all heads or all tails) becomes extremely unlikely. The least orderly arrangement (half heads, half tails) is the most probable, and the probability of being within, say, 5% of the most probable

[†]Each coin has two possibilities, heads or tails. Then the possible number of microstates is $2 \times 2 \times 2 \times \cdots = 2^{100} = 1.27 \times 10^{30}$ (using a calculator or logarithms).

TABLE 15–3
Probabilities of Various Macrostates for 100 Coin Tosses

| Macrostate | | Number of | |
heads	tails	microstates	Probability
100	0	1	8.0×10^{-31}
99	1	1.0×10^2	8.0×10^{-29}
90	10	1.7×10^{13}	1.0×10^{-17}
80	20	5.4×10^{20}	4.0×10^{-10}
60	40	1.4×10^{28}	0.01
55	45	6.1×10^{28}	0.05
50	50	1.0×10^{29}	0.08
45	55	6.1×10^{28}	0.05
40	60	1.4×10^{28}	0.01
20	80	5.4×10^{20}	4.0×10^{-10}
10	90	1.7×10^{13}	1.0×10^{-17}
1	99	1.0×10^2	8.0×10^{-29}
0	100	1	8.0×10^{-31}

arrangement greatly increases as the number of coins increases. These same ideas can be applied to the molecules of a system. For example, the most probable state of a gas (say, the air in a room) is one in which the molecules take up the whole space and move about randomly; this corresponds to the Maxwellian distribution, Fig. 15–19a (and see Chapter 13). On the other hand, the very orderly arrangement of all the molecules located in one corner of the room and all moving with the same velocity (Fig. 15–19b) is extremely unlikely.

From these examples, it is clear that probability is directly related to disorder and hence to entropy. That is, the most probable state is the one with greatest entropy, or greatest disorder and randomness.

In terms of probability, the second law of thermodynamics—which tells us that entropy increases in any process—reduces to the statement that those processes occur which are most probable. The second law thus becomes a trivial statement. However, there is an additional element now. The second law in terms of probability does not *forbid* a decrease in entropy. Rather, it says the probability is extremely low. It is not impossible that salt and pepper should separate spontaneously into layers, or that a broken teacup should mend itself. It is even possible that a lake should freeze over on a hot summer day (that is, for heat to flow out of the cold lake into the warmer surroundings). But the probability for such events occurring is miniscule. In our coin examples, we saw that increasing the number of coins from 4 to 100 drastically reduced the probability of large deviations from the average, or most probable, arrangement. In ordinary systems, we are dealing not with 100 molecules, but with incredibly large numbers of molecules: in 1 mole alone there are 6×10^{23} molecules. Hence the probability of deviation far from the average is incredibly tiny. For example, it has been calculated that the probability that a stone resting on the ground could transform 1 cal of thermal energy into mechanical energy and rise up into the air is much less likely than the probability that a group of monkeys typing randomly would by chance produce the complete works of Shakespeare.

Entropy in terms of probability

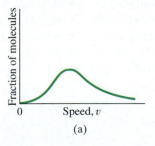
(a)

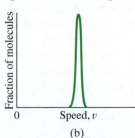
(b)

FIGURE 15–19 (a) Most probable distribution of molecular speeds in a gas (Maxwellian, or random); (b) orderly, but highly unlikely, distribution of speeds in which all molecules have nearly the same speed.

(a)

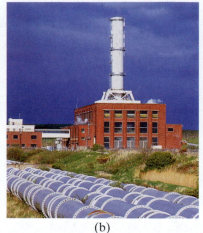
(b)

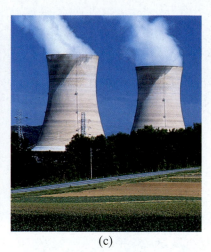
(c)

FIGURE 15–20 (a) An array of mirrors focuses sunlight on a boiler to produce steam at a solar energy installation. (b) A fossil-fuel steam plant. (c) Large cooling towers at an electric generating plant.

FIGURE 15–21 Mechanical or heat energy is transformed to electric energy with a turbine and generator.

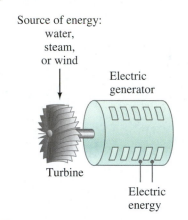

Source of energy:
water,
steam,
or wind

Electric generator

Turbine

Electric energy

PHYSICS APPLIED

Heat engines and thermal pollution

Much of the energy we utilize in everyday life—from motor vehicles to most of the electricity produced by power plants—makes use of a heat engine. Electricity produced by falling water at dams, by windmills, or by solar cells (Fig. 15–20a) does not involve a heat engine. But over 90% of the electric energy produced in the U.S. is generated at fossil-fuel steam plants (coal, oil, or gas—see Fig. 15–20b), and they make use of a heat engine (essentially steam engines). In electric power plants, the steam drives the turbines and generators (Fig. 15–21) whose output is electric energy. The various means to turn the turbine are discussed briefly in Table 15–4, along with some of the advantages and disadvantages of each. Even nuclear power plants use nuclear fuel to run a steam engine.

The heat output Q_L from every heat engine, from power plants to cars, is referred to as **thermal pollution** because this heat (Q_L) must be absorbed by the environment—such as by water from rivers or lakes, or by the air using large cooling towers (Fig. 15–20c). This heat raises the temperature of the cooling water, altering the natural ecology of aquatic life (largely because warmer water holds less oxygen). In the case of air cooling towers, the output heat Q_L raises the temperature of the atmosphere, which affects the weather.

Air pollution—by which we mean the chemicals released in the burning of fossil fuels in cars, power plants, and industrial furnaces—gives rise to smog and other problems. One big problem is the buildup of CO_2 in the Earth's atmosphere due to the burning of fossil fuels. This CO_2 absorbs some of the infrared radiation that the Earth naturally emits (Section 14–8), causing **global warming**, a serious problem that can be addressed by limiting the burning of fossil fuels.

Thermal pollution, however, is unavoidable. Engineers can try to design and build engines that are more efficient, but they cannot surpass the Carnot efficiency and must live with T_L being at best the ambient temperature of water or air. The second law of thermodynamics tells us the limit imposed by nature. What we can do, in the light of the second law of thermodynamics, is use less energy and conserve our fuel resources.

TABLE 15–4 Electric Energy Resources

Form of Electric Energy Production	% of Production (approx.)		Advantages	Disadvantages
	U.S.	World		
Fossil-fuel steam plants: burn coal, oil, or natural gas to boil water, producing high-pressure steam that turns a turbine of a generator (Figs. 15–12b, 15–21); uses heat engine.	87	86	We know how to build them; for now relatively inexpensive.	Air pollution; thermal pollution; limited efficiency; land devastation from extraction of raw materials (mining); global warming; accidents such as oil spills at sea; limited fuel supply (estimates range from a couple of decades to a few centuries).
Nuclear energy:				
Fission: nuclei of uranium or plutonium atoms split ("fission") with release of energy (Chapter 31) that heats steam; uses heat engine.	8	6	Normally almost no air pollution; less contribution to global warming; relatively inexpensive.	Thermal pollution; accidents can release damaging radioactivity; difficult disposal of radioactive by-products; possible diversion of nuclear material by terrorists; limited fuel supply.
Fusion: energy released when isotopes of hydrogen (or other small nuclei) combine or "fuse" (Chapter 31).	0	0	Relatively "clean"; vast fuel supply (hydrogen in water molecules in oceans); less contribution to global warming.	Not yet workable.
Hydroelectric: Falling water turns turbines at the base of a dam.	4	7	No heat engine needed; no air, water, or thermal pollution; relatively inexpensive; high efficiency; dams can control flooding.	Reservoirs behind dams inundate scenic land or canyons; dams block upstream migration of salmon and other fish for reproduction; few locations remain for new dams; drought.
Geothermal: natural steam from inside the Earth comes to the surface (hot springs, geysers, steam vents); or cold water passed down into contact with hot, dry rock is heated to steam.	<1	<1	No heat engine needed; little air pollution; good efficiency; relatively inexpensive and "clean."	Few appropriate sites; small production; mineral content of spent hot water can pollute.
Wind power: 3-kW to 5-MW windmills (vanes up to 50 m wide) turn a generator.	<1	<1	No heat engine; no air, water or thermal pollution; relatively inexpensive.	Large array of big windmills might affect weather and be eyesores; hazardous to migratory birds; winds not always strong.
Solar energy:	<0.1	<1		
Active solar heating: rooftop solar panels absorb the Sun's rays, which heat water in tubes for space heating and hot water supply.			No heat engine needed; no air or thermal pollution; unlimited fuel supply.	Space limitations; may require back-up; relatively expensive; less effective when cloudy.
Passive solar heating: architectural devices—windows along southern exposure, sunshade over windows to keep Sun's rays out in summer.			No heat engine needed; no air or thermal pollution; relatively inexpensive.	Almost none, but other methods needed too.
Solar cells (photovoltaic cells): convert sunlight directly into electricity without use of heat engine.			No heat engine; thermal, air, and water pollution very low; good efficiency (>30% and improving).	Expensive; chemical pollution at manufacture; large land area needed as Sun's energy not concentrated.

1. Define the **system** you are dealing with; distinguish the system under study from its surroundings.

2. When applying the first law of thermodynamics, be careful of **signs** associated with **work** and **heat**. In the first law, work done *by* the system is positive; work done *on* the system is negative. Heat *added* to the system is positive, but heat *removed* from it is negative. With heat engines, we usually consider the heat intake, the heat exhausted, and the work done as positive.

3. Watch the **units** used for work and heat; work is most often expressed in joules, and heat can be in calories, kilocalories, or joules. Be consistent: choose only one unit for use throughout a given problem.

4. **Temperatures** must generally be expressed in kelvins; temperature *differences* may be expressed in C° or K.

5. **Efficiency** (or coefficient of performance) is a ratio of two energy transfers: useful output divided by required input. Efficiency (but *not* coefficient of performance) is always less than 1 in value, and hence is often stated as a percentage.

6. The **entropy** of a system increases when heat is added to the system, and decreases when heat is removed. If heat is transferred from system A to system B, the change in entropy of A is negative and the change in entropy of B is positive.

Summary

The **first law of thermodynamics** states that the change in internal energy ΔU of a system is equal to the heat *added* to the system, Q, minus the work done *by* the system, W:

$$\Delta U = Q - W. \qquad (15\text{–}1)$$

This is a statement of the conservation of energy, and is found to hold for all types of processes.

An **isothermal** process is a process carried out at constant temperature.

In an **adiabatic** process, no heat is exchanged ($Q = 0$).

The work W done by a gas at constant pressure P is given by

$$W = P\,\Delta V, \qquad (15\text{–}3)$$

where ΔV is the change in volume of the gas.

A **heat engine** is a device for changing thermal energy, by means of heat flow between two temperatures, into useful work.

The **efficiency** e of a heat engine is defined as the ratio of the work W done by the engine to the heat input Q_H. Because of conservation of energy, the work output equals $Q_H - Q_L$, where Q_L is the heat exhausted at low temperature to the environment; hence

$$e = \frac{W}{Q_H} = 1 - \frac{Q_L}{Q_H}. \qquad (15\text{–}4)$$

The *upper limit* on the efficiency (the *Carnot efficiency*) can be written in terms of the higher and lower operating temperatures (in kelvins) of the engine, T_H and T_L, as

$$e_{ideal} = 1 - \frac{T_L}{T_H}. \qquad (15\text{–}5)$$

The operation of **refrigerators** and **air conditioners** is the reverse of that of a heat engine: work is done to extract heat from a cool region and exhaust it to a region at a higher temperature. The coefficient of performance (COP) for either is

$$COP = \frac{Q_L}{W}, \qquad \begin{bmatrix} \text{refrigerator or} \\ \text{air conditioner} \end{bmatrix} \;(15\text{–}6a)$$

where W is the work needed to remove heat Q_L from the area with the low temperature.

A **heat pump** does work W to bring heat Q_L from the cold outside and deliver heat Q_H to warm the interior. The coefficient of performance of a heat pump is

$$COP = \frac{Q_H}{W}. \qquad \text{[heat pump] }(15\text{–}7)$$

The **second law of thermodynamics** can be stated in several equivalent ways:

(a) heat flows spontaneously from a hot object to a cold one, but not the reverse;

(b) there can be no 100% efficient heat engine—that is, one that can change a given amount of heat completely into work;

(c) natural processes tend to move toward a state of greater disorder or greater **entropy**.

Statement (c) is the most general statement of the second law of thermodynamics, and can be restated as: the total entropy, S, of any system plus that of its environment increases as a result of any natural process:

$$\Delta S > 0. \qquad (15\text{–}9)$$

The change in entropy in a process that transfers heat Q at a constant temperature T is

$$\Delta S = \frac{Q}{T}. \qquad (15\text{–}8)$$

Entropy is a quantitative measure of the disorder of a system.

As time goes on, energy is degraded to less useful forms—that is, it is less available to do useful work.

[*The second law of thermodynamics tells us in which direction processes tend to go, so entropy is called "time's arrow."]

[*All heat engines give rise to **thermal pollution** because they exhaust heat to the environment.]

Questions

1. What happens to the internal energy of water vapor in the air that condenses on the outside of a cold glass of water? Is work done or heat exchanged? Explain.

2. Use the conservation of energy to explain why the temperature of a gas increases when it is quickly compressed, whereas the temperature decreases when the gas expands.

3. In an isothermal process, 3700 J of work is done by an ideal gas. Is this enough information to tell how much heat has been added to the system? If so, how much?

4. Is it possible for the temperature of a system to remain constant even though heat flows into or out of it? If so, give one or two examples.

5. Explain why the temperature of a gas increases when it is adiabatically compressed.

6. Can mechanical energy ever be transformed completely into heat or internal energy? Can the reverse happen? In each case, if your answer is no, explain why not; if yes, give one or two examples.

7. Can you warm a kitchen in winter by leaving the oven door open? Can you cool the kitchen on a hot summer day by leaving the refrigerator door open? Explain.

8. Would a definition of heat engine efficiency as $e = W/Q_L$ be useful? Explain.

9. What plays the role of high-temperature and low-temperature areas in (a) an internal combustion engine, and (b) a steam engine?

10. Which will give the greater improvement in the efficiency of a Carnot engine, a 10 C° increase in the high-temperature reservoir, or a 10 C° decrease in the low-temperature reservoir? Explain.

11. The oceans contain a tremendous amount of thermal (internal) energy. Why, in general, is it not possible to put this energy to useful work?

12. A gas is allowed to expand (a) adiabatically and (b) isothermally. In each process, does the entropy increase, decrease, or stay the same? Explain.

13. A gas can expand to twice its original volume either adiabatically or isothermally. Which process would result in a greater change in entropy? Explain.

14. Give three examples, other than those mentioned in this Chapter, of naturally occurring processes in which order goes to disorder. Discuss the observability of the reverse process.

15. Which do you think has the greater entropy, 1 kg of solid iron or 1 kg of liquid iron? Why?

16. (a) What happens if you remove the lid of a bottle containing chlorine gas? (b) Does the reverse process ever happen? Why or why not? (c) Can you think of two other examples of irreversibility?

17. You are asked to test a machine that the inventor calls an "in-room air conditioner": a big box, standing in the middle of the room, with a cable that plugs into a power outlet. When the machine is switched on, you feel a stream of cold air coming out of it. How do you know that this machine cannot cool the room?

18. Think up several processes (other than those already mentioned) that would obey the first law of thermodynamics, but, if they actually occurred, would violate the second law.

19. Suppose a lot of papers are strewn all over the floor; then you stack them neatly. Does this violate the second law of thermodynamics? Explain.

20. The first law of thermodynamics is sometimes whimsically stated as, "You can't get something for nothing," and the second law as, "You can't even break even." Explain how these statements could be equivalent to the formal statements.

* 21. Entropy is often called "time's arrow" because it tells us in which direction natural processes occur. If a movie were run backward, name some processes that you might see that would tell you that time was "running backward."

* 22. Living organisms, as they grow, convert relatively simple food molecules into a complex structure. Is this a violation of the second law of thermodynamics?

Problems

15–1 and 15–2 First Law of Thermodynamics

1. (I) An ideal gas expands isothermally, performing 3.40×10^3 J of work in the process. Calculate (a) the change in internal energy of the gas, and (b) the heat absorbed during this expansion.

2. (I) A gas is enclosed in a cylinder fitted with a light frictionless piston and maintained at atmospheric pressure. When 1400 kcal of heat is added to the gas, the volume is observed to increase slowly from $12.0 \, \text{m}^3$ to $18.2 \, \text{m}^3$. Calculate (a) the work done by the gas and (b) the change in internal energy of the gas.

3. (I) One liter of air is cooled at constant pressure until its volume is halved, and then it is allowed to expand isothermally back to its original volume. Draw the process on a PV diagram.

4. (I) Sketch a PV diagram of the following process: 2.0 L of ideal gas at atmospheric pressure are cooled at constant pressure to a volume of 1.0 L, and then expanded isothermally back to 2.0 L, whereupon the pressure is increased at constant volume until the original pressure is reached.

5. (II) A 1.0-L volume of air initially at 4.5 atm of (absolute) pressure is allowed to expand isothermally until the pressure is 1.0 atm. It is then compressed at constant pressure to its initial volume, and lastly is brought back to its original pressure by heating at constant volume. Draw the process on a PV diagram, including numbers and labels for the axes.

6. (II) The pressure in an ideal gas is cut in half slowly, while being kept in a container with rigid walls. In the process, 265 kJ of heat left the gas. (a) How much work was done during this process? (b) What was the change in internal energy of the gas during this process?

7. (II) In an engine, an almost ideal gas is compressed adiabatically to half its volume. In doing so, 1850 J of work is done on the gas. (a) How much heat flows into or out of the gas? (b) What is the change in internal energy of the gas? (c) Does its temperature rise or fall?

8. (II) An ideal gas expands at a constant total pressure of 3.0 atm from 400 mL to 660 mL. Heat then flows out of the gas at constant volume, and the pressure and temperature are allowed to drop until the temperature reaches its original value. Calculate (a) the total work done by the gas in the process, and (b) the total heat flow into the gas.

9. (II) One and one-half moles of an ideal monatomic gas expand adiabatically, performing 7500 J of work in the process. What is the change in temperature of the gas during this expansion?

10. (II) Consider the following two-step process. Heat is allowed to flow out of an ideal gas at constant volume so that its pressure drops from 2.2 atm to 1.4 atm. Then the gas expands at constant pressure, from a volume of 6.8 L to 9.3 L, where the temperature reaches its original value. See Fig. 15–22. Calculate (a) the total work done by the gas in the process, (b) the change in internal energy of the gas in the process, and (c) the total heat flow into or out of the gas.

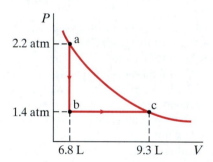

FIGURE 15–22 Problem 10.

11. (II) The PV diagram in Fig. 15–23 shows two possible states of a system containing 1.35 moles of a monatomic ideal gas. $(P_1 = P_2 = 455 \, \text{N/m}^2, \; V_1 = 2.00 \, \text{m}^3, \; V_2 = 8.00 \, \text{m}^3.)$ (a) Draw the process which depicts an isobaric expansion from state 1 to state 2, and label this process A. (b) Find the work done by the gas and the change in internal energy of the gas in process A. (c) Draw the two-step process which depicts an isothermal expansion from state 1 to the volume V_2, followed by an isovolumetric increase in temperature to state 2, and label this process B. (d) Find the change in internal energy of the gas for the two-step process B.

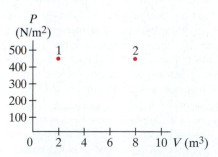

FIGURE 15–23 Problem 11.

12. (III) When a gas is taken from a to c along the curved path in Fig. 15–24, the work done by the gas is $W = -35 \, \text{J}$ and the heat added to the gas is $Q = -63 \, \text{J}$. Along path abc, the work done is $W = -48 \, \text{J}$. (a) What is Q for path abc? (b) If $P_c = \frac{1}{2}P_b$, what is W for path cda? (c) What is Q for path cda? (d) What is $U_a - U_c$? (e) If $U_d - U_c = 5 \, \text{J}$, what is Q for path da?

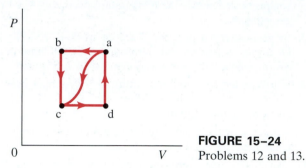

FIGURE 15–24
Problems 12 and 13.

13. (III) In the process of taking a gas from state a to state c along the curved path shown in Fig. 15–24, 80 J of heat leaves the system and 55 J of work is done *on* the system. (a) Determine the change in internal energy, $U_a - U_c$. (b) When the gas is taken along the path cda, the work done by the gas is $W = 38 \, \text{J}$. How much heat Q is added to the gas in the process cda? (c) If $P_a = 2.5P_d$, how much work is done by the gas in the process abc? (d) What is Q for path abc? (e) If $U_a - U_b = 10 \, \text{J}$, what is Q for the process bc? Here is a summary of what is given:

$$Q_{a \to c} = -80 \, \text{J}$$
$$W_{a \to c} = -55 \, \text{J}$$
$$W_{cda} = 38 \, \text{J}$$
$$U_a - U_b = 10 \, \text{J}$$
$$P_a = 2.5P_d.$$

* **15–3 Human Metabolism**

* 14. (I) How much energy would the person of Example 15–8 transform if instead of working 11.0 h she took a noontime break and ran for 1.0 h?

* 15. (I) Calculate the average metabolic rate of a person who sleeps 8.0 h, sits at a desk 8.0 h, engages in light activity 4.0 h, watches television 2.0 h, plays tennis 1.5 h, and runs 0.5 h daily.

* 16. (II) A person decides to lose weight by sleeping one hour less per day, using the time for light activity. How much weight (or mass) can this person expect to lose in 1 year, assuming no change in food intake? Assume that 1 kg of fat stores about 40,000 kJ of energy.

15–5 Heat Engines

17. (I) A heat engine exhausts 8200 J of heat while performing 3200 J of useful work. What is the efficiency of this engine?

18. (I) A heat engine does 9200 J of work per cycle while absorbing 22.0 kcal of heat from a high-temperature reservoir. What is the efficiency of this engine?

19. (I) What is the maximum efficiency of a heat engine whose operating temperatures are 580°C and 380°C?

20. (I) The exhaust temperature of a heat engine is 230°C. What must be the high temperature if the Carnot efficiency is to be 28%?

21. (II) A nuclear power plant operates at 75% of its maximum theoretical (Carnot) efficiency between temperatures of 625°C and 350°C. If the plant produces electric energy at the rate of 1.3 GW, how much exhaust heat is discharged per hour?

22. (II) It is not necessary that a heat engine's hot environment be hotter than ambient temperature. Liquid nitrogen (77 K) is about as cheap as bottled water. What would be the efficiency of an engine that made use of heat transferred from air at room temperature (293 K) to the liquid nitrogen "fuel" (Fig. 15–25)?

FIGURE 15–25 Problem 22.

23. (II) A Carnot engine performs work at the rate of 440 kW while using 680 kcal of heat per second. If the temperature of the heat source is 570°C, at what temperature is the waste heat exhausted?

24. (II) A Carnot engine's operating temperatures are 210°C and 45°C. The engine's power output is 950 W. Calculate the rate of heat output.

25. (II) A certain power plant puts out 550 MW of electric power. Estimate the heat discharged per second, assuming that the plant has an efficiency of 38%.

26. (II) A heat engine utilizes a heat source at 550°C and has an ideal (Carnot) efficiency of 28%. To increase the ideal efficiency to 35%, what must be the temperature of the heat source?

27. (II) A heat engine exhausts its heat at 350°C and has a Carnot efficiency of 39%. What exhaust temperature would enable it to achieve a Carnot efficiency of 49%?

28. (III) At a steam power plant, steam engines work in pairs, the output of heat from one being the approximate heat input of the second. The operating temperatures of the first are 670°C and 440°C, and of the second 430°C and 290°C. If the heat of combustion of coal is 2.8×10^7 J/kg, at what rate must coal be burned if the plant is to put out 1100 MW of power? Assume the efficiency of the engines is 60% of the ideal (Carnot) efficiency.

15–6 Refrigerators, Air Conditioners, Heat Pumps

29. (I) The low temperature of a freezer cooling coil is −15°C, and the discharge temperature is 30°C. What is the maximum theoretical coefficient of performance?

30. (II) An ideal refrigerator-freezer operates with a COP = 7.0 in a 24°C room. What is the temperature inside the freezer?

31. (II) A restaurant refrigerator has a coefficient of performance of 5.0. If the temperature in the kitchen outside the refrigerator is 29°C, what is the lowest temperature that could be obtained inside the refrigerator if it were ideal?

32. (II) A heat pump is used to keep a house warm at 22°C. How much work is required of the pump to deliver 2800 J of heat into the house if the outdoor temperature is (a) 0°C, (b) −15°C? Assume ideal (Carnot) behavior.

33. (II) What volume of water at 0°C can a freezer make into ice cubes in 1.0 hour, if the coefficient of performance of the cooling unit is 7.0 and the power input is 1.0 kilowatt?

34. (II) An ideal (Carnot) engine has an efficiency of 35%. If it were possible to run it backward as a heat pump, what would be its coefficient of performance?

15–7 Entropy

35. (I) What is the change in entropy of 250 g of steam at 100°C when it is condensed to water at 100°C?

36. (I) One kilogram of water is heated from 0°C to 100°C. Estimate the change in entropy of the water.

37. (I) What is the change in entropy of 1.00 m³ of water at 0°C when it is frozen to ice at 0°C?

38. (II) If 1.00 m³ of water at 0°C is frozen and cooled to −10°C by being in contact with a great deal of ice at −10°C, what would be the total change in entropy of the process?

39. (II) A 10.0-kg box having an initial speed of 3.0 m/s slides along a rough table and comes to rest. Estimate the total change in entropy of the universe. Assume all objects are at room temperature (293 K).

40. (II) A falling rock has kinetic energy KE just before striking the ground and coming to rest. What is the total change in entropy of the rock plus environment as a result of this collision?

41. (II) An aluminum rod conducts 7.50 cal/s from a heat source maintained at 240°C to a large body of water at 27°C. Calculate the rate entropy increases per unit time in this process.

42. (II) 1.0 kg of water at 30°C is mixed with 1.0 kg of water at 60°C in a well-insulated container. Estimate the net change in entropy of the system.

43. (II) A 3.8-kg piece of aluminum at 30°C is placed in 1.0 kg of water in a Styrofoam container at room temperature (20°C). Calculate the approximate net change in entropy of the system.

44. (III) A real heat engine working between heat reservoirs at 970 K and 650 K produces 550 J of work per cycle for a heat input of 2200 J. (a) Compare the efficiency of this real engine to that of an ideal (Carnot) engine. (b) Calculate the total entropy change of the universe per cycle of the real engine. (c) Calculate the total entropy change of the universe per cycle of a Carnot engine operating between the same two temperatures.

* **45.** (II) Calculate the probabilities, when you throw two dice, of obtaining (*a*) a 5, and (*b*) an 11.

* **46.** (II) Rank the following five-card hands in order of increasing probability: (*a*) four aces and a king; (*b*) six of hearts, eight of diamonds, queen of clubs, three of hearts, jack of spades; (*c*) two jacks, two queens, and an ace; and (*d*) any hand having no two equal-value cards. Discuss your ranking in terms of microstates and macrostates.

* **47.** (II) Suppose that you repeatedly shake six coins in your hand and drop them on the floor. Construct a table showing the number of microstates that correspond to each macrostate. What is the probability of obtaining (*a*) three heads and three tails, and (*b*) six heads?

* **48.** (I) Solar cells (Fig. 15–26) can produce about 40 W of electricity per square meter of surface area if directly facing the Sun. How large an area is required to supply the needs of a house that requires 22 kWh/day? Would this fit on the roof of an average house? (Assume the Sun shines about 9 h/day.)

FIGURE 15–26 Problem 48.

* **49.** (II) Energy may be stored for use during peak demand by pumping water to a high reservoir when demand is low and then releasing it to drive turbines when needed. Suppose water is pumped to a lake 135 m above the turbines at a rate of 1.00×10^5 kg/s for 10.0 h at night. (*a*) How much energy (kWh) is needed to do this each night? (*b*) If all this energy is released during a 14-h day, at 75% efficiency, what is the average power output?

* **50.** (II) Water is stored in an artificial lake created by a dam (Fig. 15–27). The water depth is 45 m at the dam, and a steady flow rate of 35 m³/s is maintained through hydroelectric turbines installed near the base of the dam. How much electrical power can be produced?

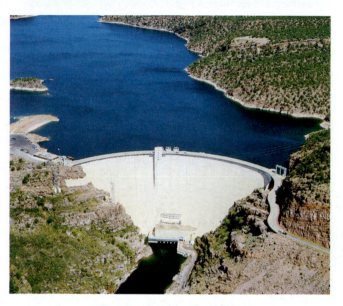

FIGURE 15–27 Problem 50.

General Problems

51. An inventor claims to have designed and built an engine that produces 1.50 MW of usable work while taking in 3.00 MW of thermal energy at 425 K, and rejecting 1.50 MW of thermal energy at 215 K. Is there anything fishy about his claim? Explain.

52. When 5.30×10^5 J of heat is added to a gas enclosed in a cylinder fitted with a light frictionless piston maintained at atmospheric pressure, the volume is observed to increase from 1.9 m³ to 4.1 m³. Calculate (*a*) the work done by the gas, and (*b*) the change in internal energy of the gas. (*c*) Graph this process on a *PV* diagram.

53. A 4-cylinder gasoline engine has an efficiency of 0.25 and delivers 220 J of work per cycle per cylinder. When the engine fires at 45 cycles per second, (*a*) what is the work done per second? (*b*) What is the total heat input per second from the fuel? (*c*) If the energy content of gasoline is 35 MJ per liter, how long does one liter last?

54. A "Carnot" refrigerator (the reverse of a Carnot engine) absorbs heat from the freezer compartment at a temperature of −17°C and exhausts it into the room at 25°C. (*a*) How much work must be done by the refrigerator to change 0.50 kg of water at 25°C into ice at −17°C? (*b*) If the compressor output is 210 W, what minimum time is needed to accomplish this?

55. It has been suggested that a heat engine could be developed that made use of the temperature difference between water at the surface of the ocean and that several hundred meters deep. In the tropics, the temperatures may be 27°C and 4°C, respectively. (*a*) What is the maximum efficiency such an engine could have? (*b*) Why might such an engine be feasible in spite of the low efficiency? (*c*) Can you imagine any adverse environmental effects that might occur?

56. Two 1100-kg cars are traveling 95 km/h in opposite directions when they collide and are brought to rest. Estimate the change in entropy of the universe as a result of this collision. Assume $T = 20°C$.

57. A 120-g insulated aluminum cup at 15°C is filled with 140 g of water at 50°C. After a few minutes, equilibrium is reached. (a) Determine the final temperature, and (b) estimate the total change in entropy.

* 58. (a) What is the coefficient of performance of an ideal heat pump that extracts heat from 6°C air outside and deposits heat inside your house at 24°C? (b) If this heat pump operates on 1200 W of electrical power, what is the maximum heat it can deliver into your house each hour?

59. The burning of gasoline in a car releases about 3.0×10^4 kcal/gal. If a car averages 41 km/gal when driving 90 km/h, which requires 25 hp, what is the efficiency of the engine under those conditions?

60. A Carnot engine has a lower operating temperature $T_L = 20°C$ and an efficiency of 30%. By how many kelvins should the high operating temperature T_H be increased to achieve an efficiency of 40%?

61. Calculate the work done by an ideal gas in going from state A to state C in Fig. 15–28 for each of the following processes: (a) ADC, (b) ABC, and (c) AC directly.

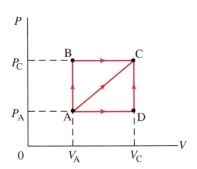

FIGURE 15–28 Problem 61.

62. A 33% efficient power plant puts out 850 MW of electrical power. Cooling towers are used to take away the exhaust heat. (a) If the air temperature is allowed to rise 7.0 C°, estimate what volume of air (km^3) is heated per day. Will the local climate be heated significantly? (b) If the heated air were to form a layer 200 m thick, estimate how large an area it would cover for 24 h of operation. Assume the air has density 1.2 kg/m³ and that its specific heat is about 1.0 kJ/kg·C° at constant pressure.

63. Suppose a power plant delivers energy at 980 MW using steam turbines. The steam goes into the turbines superheated at 625 K and deposits its unused heat in river water at 285 K. Assume that the turbine operates as an ideal Carnot engine. (a) If the river flow rate is 37 m³/s, estimate the average temperature increase of the river water immediately downstream from the power plant. (b) What is the entropy increase per kilogram of the downstream river water in J/kg·K?

64. A 100-hp car engine operates at about 15% efficiency. Assume the engine's water temperature of 85°C is its cold-temperature (exhaust) reservoir and 495°C is its thermal "intake" temperature (the temperature of the exploding gas–air mixture). (a) What is the ratio of its efficiency relative to its maximum possible (Carnot) efficiency? (b) Estimate how much power (in watts) goes into moving the car, and how much heat, in joules and in kcal, is exhausted to the air in 1.0 h.

65. An ideal gas is placed in a tall cylindrical jar of cross-sectional area 0.080 m². A frictionless 0.10-kg movable piston is placed vertically into the jar such that the piston's weight is supported by the gas pressure in the jar. When the gas is heated (at constant pressure) from 25°C to 55°C, the piston rises 1.0 cm. How much heat was required for this process? Assume atmospheric pressure outside.

66. Metabolizing 1.0 kg of fat results in about 3.7×10^7 J of internal energy in the body. (a) In one day, how much fat does the body burn to maintain the body temperature of a person staying in bed and metabolizing at an average rate of 95 W? (b) How long would it take to burn 1.0-kg of fat this way assuming there is no food intake?

67. An ideal air conditioner keeps the temperature inside a room at 21°C when the outside temperature is 32°C. If 5.3 kW of power enters a room through the windows in the form of direct radiation from the Sun, how much electrical power would be saved if the windows were shaded so that the amount of radiation were reduced to 500 W?

68. A dehumidifier is essentially a "refrigerator with an open door." The humid air is pulled in by a fan and guided to a cold coil, where the temperature is less than the dew point, and some of the air's water condenses. After this water is extracted, the air is warmed back to its original temperature and sent into the room. In a well-designed dehumidifier, the heat is exchanged between the incoming and outgoing air. This way the heat that is removed by the refrigerator coil mostly comes from the condensation of water vapor to liquid. Estimate how much water is removed in 1.0 h by an ideal dehumidifier, if the temperature of the room is 25°C, the water condenses at 8°C, and the dehumidifier does work at the rate of 600 W of electrical power.

Answers to Exercises

A: 700 J.
B: Less.
C: -6.8×10^3 J.

D: Equation 14–1 applies only to an ideal monatomic gas, not to liquid water.

Mathematical Review

A–1 Relationships, Proportionality, and Equations

One of the important aspects of physics is the search for relationships between different quantities—that is, determining how one quantity affects another. For example, how does temperature affect the air pressure in a tire? Or how does the net force on an object affect its acceleration? Sometimes a given quantity is affected by two or more quantities; for instance, the acceleration of an object is related to both its mass and the applied force. If you suspect that a relationship exists between two or more quantities, you can try to determine the precise nature of this relationship. This is done by varying one of the quantities and measuring how the other varies as a result. If it is likely that a particular quantity will be affected by more than one factor or quantity, only one quantity is varied at a time, while the others are held constant.[†]

As a simple example, the ancients found that if one circle has twice the diameter of a second circle, the first also has twice the circumference. If the diameter is three times as large, the circumference is also three times as large. In other words, an increase in the diameter results in a proportional increase in the circumference. We say that the circumference is *directly proportional to* the diameter. This can be written in symbols as $C \propto D$, where "$\propto$" means "is proportional to," and C and D refer to the circumference and diameter of a circle, respectively. The next step is to change this proportionality to an equation, which will make it possible to link the two quantities numerically. This merely entails inserting a proportionality constant, which in many cases is determined by measurement. (In some cases it can be chosen arbitrarily, if it involves only the definition of a new unit.) The ancients found that the ratio of the circumference to the diameter of any circle was 3.1416 (to keep only the first few decimal places). This number is designated by the Greek letter π. It is the constant of proportionality for the relationship $C \propto D$. To obtain an equation, we insert π into the proportion and change the $\propto$ to $=$. Thus, $C = \pi D$.

Direct proportion

Other kinds of proportionality occur as well. For example, the area of a circle is proportional to the *square* of its radius. That is, if the radius is doubled, the area becomes four times as large; and so on. In this case we can write $A \propto r^2$, where A stands for the area and r for the radius of the circle.

Sometimes two quantities are related in such a way that an increase in one leads to a proportional *decrease* in the other. This is called *inverse proportion*. For example, the time required to travel a given distance is inversely proportional to the speed of travel. The greater the speed, the less time it takes. We can write this inverse proportion as time $\propto 1/$speed. The larger the denominator of a fraction, the lower the value of the fraction is as a whole. For example, $\frac{1}{4}$ is less than $\frac{1}{2}$. Thus, if the speed is doubled, the time is halved, which is what we want to express by this inverse proportionality relationship.

Inverse proportion

[†]When one quantity affects another, we often use the expression "is a function of" to indicate this dependence; for example, we say that the pressure in a tire is a function of the temperature.

Whatever kind of proportion is found to hold, it can be changed to an equality by insertion of the proper proportionality constant. Quantitative statements or predictions about the physical world can then be made with the equation.

A–2 Exponents

When we write 10^4, we mean that you multiply 10 by itself four times: $10^4 = 10 \times 10 \times 10 \times 10 = 10{,}000$. The superscript 4 is called an *exponent*, and 10 is said to be raised to the fourth power. Any number or symbol can be raised to a power; special names are used when the exponent is 2 (a^2 is "*a* squared") or 3 (a^3 is "*a* cubed"). For any other power, we say a^n is "*a* to the *n*th power." If the exponent is 1, it is usually dropped: $a^1 = a$, since no multiplication is involved.

The rules for multiplying numbers expressed as powers are as follows:

$$(a^n)(a^m) = a^{n+m}. \tag{A–1}$$

That is, the exponents are added. To see why, consider the result of the multiplication of 3^3 by 3^4:

$$(3^3)(3^4) = (3)(3)(3) \times (3)(3)(3)(3) = (3)^7.$$

Here the sum of the exponents is $3 + 4 = 7$, so rule A–1 works. Notice that this rule works only if the base numbers (*a* in Eq. A–1) are the same. Thus we *cannot* use the rule of summing exponents for $(6^3)(5^2)$; these numbers would have to be written out. However, if the base numbers are different but the exponents are the same, we can write a second rule:

$$(a^n)(b^n) = (ab)^n. \tag{A–2}$$

For example, $(5^3)(6^3) = (30)^3$ since

$$(5)(5)(5)(6)(6)(6) = (30)(30)(30).$$

The third rule involves a power raised to another power: $(a^3)^2$ means $(a^3)(a^3)$, which is equal to $a^{3+3} = a^6$. The general rule is then

$$(a^n)^m = a^{nm}. \tag{A–3}$$

In this case, the exponents are multiplied.

Negative exponents are used for reciprocals. Thus,

$$\frac{1}{a} = a^{-1}, \qquad \frac{1}{a^3} = a^{-3},$$

and so on. The reason for using negative exponents is to allow us to use the multiplication rules given above. For example, $(a^5)(a^{-3})$ means

$$\frac{(a)(a)(a)(a)(a)}{(a)(a)(a)} = a^2.$$

Rule A–1 gives us the same result:

$$(a^5)(a^{-3}) = a^{5-3} = a^2.$$

What does an exponent of zero mean? That is, what is a^0? Any number raised to the zeroth power is defined as being equal to 1:

$$a^0 = 1.$$

This definition is used because it follows from the rules for adding exponents. For example,

$$a^3 a^{-3} = a^{3-3} = a^0 = 1.$$

But *does* $a^3 a^{-3}$ actually equal 1? Yes, because

$$a^3 a^{-3} = \frac{a^3}{a^3} = 1.$$

Fractional exponents are used to represent *roots*. For example, $a^{\frac{1}{2}}$ means the square root of *a*; that is, $a^{\frac{1}{2}} = \sqrt{a}$. Similarly, $a^{\frac{1}{3}}$ means the cube root of *a*, and so on. The fourth root of a means that if you multiply the fourth root of a by itself four times, you again get a:

$$\left(a^{\frac{1}{4}}\right)^4 = a.$$

This is consistent with rule A–3 since $\left(a^{\frac{1}{4}}\right)^4 = a^{\frac{4}{4}} = a^1 = a$.

A–3 Powers of 10, or Exponential Notation

Writing out very large and very small numbers such as the distance of Neptune from the Sun, 4,500,000,000 km, or the diameter of a typical atom, 0.00000001 cm, is inconvenient and prone to error. It also leaves in question (see Section 1–4) the number of significant figures. (How many of the zeros are significant in the number 4,500,000,000 km?) We therefore make use of the "powers of 10," or exponential notation. The distance from Neptune to the Sun is then expressed as 4.50×10^9 km (assuming that the value is significant to three digits), and the diameter of an atom 1.0×10^{-8} cm. This way of writing numbers is based on the use of exponents, where a^n signifies a multiplied by itself n times. For example, $10^4 = 10 \times 10 \times 10 \times 10 = 10,000$. Thus, $4.50 \times 10^9 = 4.50 \times 1,000,000,000 = 4,500,000,000$. Notice that the exponent (9 in this case) is just the number of places the decimal point is moved to the right to obtain the fully written-out number (4.500,000,000.)

When two numbers are multiplied (or divided), you first multiply (or divide) the simple parts and then the powers of 10. Thus, 2.0×10^3 multiplied by 5.5×10^4 equals $(2.0 \times 5.5) \times (10^3 \times 10^4) = 11 \times 10^7$, where we have used the rule for adding exponents (Appendix A–2). Similarly, 8.2×10^5 divided by 2.0×10^2 equals

$$\frac{8.2 \times 10^5}{2.0 \times 10^2} = \frac{8.2}{2.0} \times \frac{10^5}{10^2} = 4.1 \times 10^3.$$

For numbers less than 1, say 0.01, the exponent power of 10 is written with a negative sign: $0.01 = 1/100 = 1/10^2 = 1 \times 10^{-2}$. Similarly, $0.002 = 2 \times 10^{-3}$. The decimal point has again been moved the number of places expressed in the exponent. Thus, $0.020 \times 3600 = 72$; in exponential notation $(2.0 \times 10^{-2}) \times (3.6 \times 10^3) = 7.2 \times 10^1 = 72$.

Notice also that $10^1 \times 10^{-1} = 10 \times 0.1 = 1$, and by the law of exponents, $10^1 \times 10^{-1} = 10^0$. Therefore, $10^0 = 1$.

When writing a number in exponential notation, it is usual to make the simple number be between 1 and 10. Thus it is conventional to write 4.5×10^9 rather than 45×10^8, although they are the same number.[†] This notation also allows the number of *significant figures* to be clearly expressed. We write 4.50×10^9 if this value is accurate to three significant figures, but 4.5×10^9 if it is accurate to only two.

A–4 Algebra

Physical relationships between quantities can be represented as equations involving symbols (usually letters of the alphabet) that represent the quantities. The manipulation of such equations is the field of algebra, and it is used a great deal in physics. An equation involves an equals sign, which tells us that the quantities on either side of the equals sign have the same value. Examples of equations are

$$3 + 8 = 11$$
$$2x + 7 = 15$$
$$a^2b + c = 6.$$

The first equation involves only numbers, so is called an arithmetic equation. The other two equations are algebraic since they involve symbols. In the third equation, the quantity a^2b means the product of a times a times b: $a^2b = a \times a \times b$.

[†]Another convention used, particularly with computers, is that the simple number be between 0.1 and 1. Thus we could write 4,500,000,000 as 0.450×10^{10}.

Solving for an Unknown

Often we wish to solve for one (or more) symbols, and we treat it as an *unknown*. For example, in the equation $2x + 7 = 15$, x is the unknown; this equation is true, however, only when $x = 4$. Determining what value (or values) the unknown(s) can have to satisfy the equation(s) is called *solving the equation*. To solve an equation, the following rule can be used:

An equation will remain true if any operation performed on one side is also performed on the other side: for example, (*a*) addition or subtraction of a number or symbol; (*b*) multiplication or division by a number or symbol; (*c*) raising each side of the equation to the same power, or taking the same root (such as square root).

EXAMPLE A–1 Solve for x in the equation

$$2x + 7 = 15.$$

APPROACH We perform the same operations on both sides of the equation to isolate x as the only variable on the left side of the equals sign.

SOLUTION We first subtract 7 from both sides:

$$2x + 7 - 7 = 15 - 7$$

or

$$2x = 8.$$

Then we divide both sides by 2 to get

$$\frac{2x}{2} = \frac{8}{2},$$

or, carrying out the divisions,

$$x = 4,$$

and this solves the equation.

EXAMPLE A–2 (*a*) Solve the equation

$$a^2 b + c = 24$$

for the unknown a in terms of b and c. (*b*) Solve for a assuming that $b = 2$ and $c = 6$.

APPROACH We perform operations to isolate a as the only variable on the left side of the equals sign.

SOLUTION (*a*) We are trying to solve for a, so we first subtract c from both sides:

$$a^2 b = 24 - c,$$

then divide by b:

$$a^2 = \frac{24 - c}{b},$$

and finally take square roots:

$$a = \sqrt{\frac{24 - c}{b}}.$$

(*b*) If we are given that $b = 2$ and $c = 6$, then

$$a = \sqrt{\frac{24 - 6}{2}} = 3.$$

NOTE Whenever we take a square root, the number can be either positive or negative. Thus $a = -3$ is also a solution. Why? Because $(-3)^2 = 9$, just as $(+3)^2 = 9$. So we actually get two solutions: $a = +3$ and $a = -3$.

To check a solution, we put it back into the original equation (this is really a check that we did all the manipulations correctly). In the equation

$$a^2b + c = 24,$$

we put in $a = 3$, $b = 2$, $c = 6$ and find

$$(3)^2(2) + (6) \stackrel{?}{=} 24$$
$$24 = 24,$$

which checks.

EXERCISE A Put $a = -3$ into the equation of Example A–2 and show that it works too.

Two or More Unknowns

If we have two or more unknowns, one equation is not sufficient to find them. In general, if there are n unknowns, n independent equations are needed. For example, if there are two unknowns, we need two equations. If the unknowns are called x and y, a typical procedure is to solve one equation for x in terms of y, and substitute this into the second equation.

EXAMPLE A–3 Solve the following pair of equations for x and y.

$$3x - 2y = 19$$
$$x + 4y = -3.$$

APPROACH We have two unknowns and two equations; we can start by solving the second equation for x in terms of y. Then we substitute this result for x into the first equation.

SOLUTION We subtract $4y$ from both sides of the second equation:

$$x = -3 - 4y.$$

We substitute this expression for x into the first equation, and simplify:

$$
\begin{aligned}
3(-3 - 4y) - 2y &= 19 \\
-9 - 12y - 2y &= 19 \quad \text{(carried out the multiplication by 3)} \\
-14y &= 28 \quad \text{(added 9 to both sides)} \\
y &= -2. \quad \text{(divided both sides by } -14\text{)}
\end{aligned}
$$

Now that we know $y = -2$, we substitute this into the expression for x:

$$
\begin{aligned}
x &= -3 - 4y \\
&= -3 - 4(-2) = -3 + 8 = 5.
\end{aligned}
$$

Our solution is $x = 5$, $y = -2$. We check this solution by putting these values back into the original equations:

$$
\begin{aligned}
3x - 2y &\stackrel{?}{=} 19 \\
3(5) - 2(-2) &\stackrel{?}{=} 19 \\
15 + 4 &\stackrel{?}{=} 19 \\
19 &= 19 \quad \text{(it checks)}
\end{aligned}
$$

and

$$
\begin{aligned}
x + 4y &\stackrel{?}{=} -3 \\
5 + 4(-2) &\stackrel{?}{=} -3 \\
-3 &= -3. \quad \text{(it checks)}
\end{aligned}
$$

Other methods for solving two or more equations, such as the method of determinants, can be found in an algebra textbook.

The Quadratic Formula

We sometimes encounter equations that involve an unknown, say x, that appears not only to the first power, but squared as well. Such a *quadratic equation* can be written in the form

$$ax^2 + bx + c = 0.$$

The quantities a, b, and c are typically numbers or constants that are given.[†] The general solutions to such an equation are given by the *quadratic formula*:

Quadratic formula

$$x = \frac{-b \pm \sqrt{b^2 - 4ac}}{2a}. \qquad \text{(A–4)}$$

The $\pm$ sign indicates that there are two solutions for x: one where the plus sign is used, the other where the minus sign is used.

EXAMPLE A–4 Find the solutions for x in the equation

$$3x^2 - 5x = 2.$$

APPROACH Here x appears both to the first power and squared, so we use the quadratic equation.

SOLUTION First we write this equation in the standard form

$$ax^2 + bx + c = 0$$

by subtracting 2 from both sides:

$$3x^2 - 5x - 2 = 0.$$

In this case, a, b, and c in the standard formula take the values $a = 3$, $b = -5$, and $c = -2$. The two solutions for x are

$$x = \frac{+5 + \sqrt{25 - (4)(3)(-2)}}{(2)(3)} = \frac{5 + 7}{6} = 2$$

and

$$x = \frac{+5 - \sqrt{25 - (4)(3)(-2)}}{(2)(3)} = \frac{5 - 7}{6} = -\frac{1}{3}.$$

In this Example, the two solutions are $x = 2$ and $x = -\frac{1}{3}$. In physics problems, it sometimes happens that only one of the solutions corresponds to a real-life situation; in this case, the other solution is discarded. In other cases, both solutions may correspond to physical reality.

Notice, incidentally, that b^2 must be greater than $4ac$, so that $\sqrt{b^2 - 4ac}$ yields a real number. If $(b^2 - 4ac)$ is less than zero (negative), there is no real solution. The square root of a negative number is called *imaginary*.

A second-order equation—one in which the highest power of x is 2—has two solutions; a third-order equation—involving x^3—has three solutions; and so on.

A–5 The Binomial Expansion

Sometimes we end up with a quantity of the form $(1 + x)^n$. That is, the quantity $(1 + x)$ is raised to the nth power. This can be written as an infinite sum of terms, known as a *series expansion*, as follows:

$$(1 + x)^n = 1 + nx + \frac{n(n - 1)}{2!}x^2 + \cdots. \qquad \text{(A–5)}$$

This formula is useful for us mainly when x is very small compared to one ($x \ll 1$). In this case, each successive term is much smaller than the preceding

[†]Or one or more of them could be variables, in which case additional equations are needed.

term. For example, if $x = 0.01$, and $n = 2$, say, then whereas the first term equals 1, the second term is $nx = (2)(0.01) = 0.02$, and the third term is $[(2)(1)/2](0.01)^2 = 0.0001$, and so on. Thus, when x is small, we can ignore all but the first two (or three) terms and can write

$$(1 + x)^n \approx 1 + nx. \qquad \textbf{(A-6)}$$

This approximation often allows us to solve an equation easily that otherwise might be very difficult. Some examples are

$$(1 + x)^2 \approx 1 + 2x,$$

$$\frac{1}{1 + x} = (1 + x)^{-1} \approx 1 - x,$$

$$\sqrt{1 + x} = (1 + x)^{\frac{1}{2}} \approx 1 + \tfrac{1}{2}x,$$

$$\frac{1}{\sqrt{1 + x}} = (1 + x)^{-\frac{1}{2}} \approx 1 - \tfrac{1}{2}x,$$

where $x \ll 1$.

As a numerical example, let us evaluate $\sqrt{1.02}$ using the binomial expansion since $x = 0.02$ is much smaller than 1:

$$\sqrt{1.02} = (1.02)^{\frac{1}{2}} = (1 + 0.02)^{\frac{1}{2}} \approx 1 + \tfrac{1}{2}(0.02) = 1.01.$$

You can check with a calculator (and maybe not even more quickly) that $\sqrt{1.02} \approx 1.01$.

A-6 Plane Geometry

We review here a number of theorems involving angles and triangles that are useful in physics.

1. *Equal angles.* Two angles are equal if any of the following conditions are true:
   (a) They are vertical angles (Fig. A–1); *or*
   (b) the left side of one is parallel to the left side of the other, and the right side of one is parallel to the right side of the other (the left and right sides are as seen from the vertex, where the two sides meet; Fig. A–2); *or*
   (c) the left side of one is perpendicular to the left side of the other, and the right sides are likewise perpendicular (Fig. A–3).

2. *The sum of the angles* in any plane triangle is 180°.

3. *Similar triangles.* Two triangles are said to be similar if all three of their angles are equal (in Fig. A–4, $\theta_1 = \phi_1$, $\theta_2 = \phi_2$, and $\theta_3 = \phi_3$). Similar triangles thus have the same basic shape but may be different sizes and have different orientations. Two useful theorems about similar triangles are:
   (a) Two triangles are similar if any two of their angles are equal. (This follows because the third angles must also be equal since the sum of the angles of a triangle is 180°.)
   (b) The ratios of corresponding sides of two similar triangles are equal. That is (Fig. A–4),

$$\frac{a_1}{b_1} = \frac{a_2}{b_2} = \frac{a_3}{b_3}.$$

4. *Congruent triangles.* Two triangles are congruent if one can be placed precisely on top of the other. That is, they are similar triangles and they have the same size. Two triangles are congruent if any of the following holds:
   (a) The three corresponding sides are equal.
   (b) Two sides and the enclosed angle are equal ("side-angle-side").
   (c) Two angles and the enclosed side are equal ("angle-side-angle").

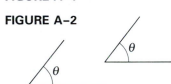

FIGURE A–1

FIGURE A–2

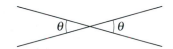

FIGURE A–3

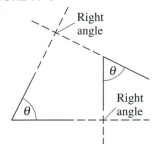

FIGURE A–4

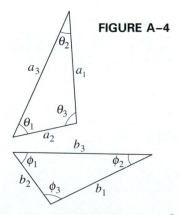

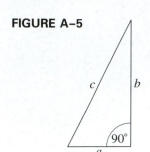

5. *Right triangles.* A right triangle has one angle that is 90° (a *right angle*); that is, the two sides that meet at the right angle are perpendicular (Fig. A–5). The two other (acute) angles in the right triangle add up to 90°.

6. *Pythagorean theorem.* In any right triangle, the square of the length of the hypotenuse (the side opposite the right angle) is equal to the sum of the squares of the lengths of the other two sides. In Fig. A–5,

$$c^2 = a^2 + b^2.$$

A-7 Trigonometric Functions and Identities

FIGURE A-6

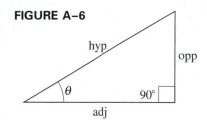

Trigonometric functions for any angle θ are defined by constructing a right triangle about that angle as shown in Fig. A–6; opp and adj are the lengths of the sides opposite and adjacent to the angle θ, and hyp is the length of the hypotenuse:

$$\sin \theta = \frac{\text{opp}}{\text{hyp}} \qquad\qquad \csc \theta = \frac{1}{\sin \theta} = \frac{\text{hyp}}{\text{opp}}$$

$$\cos \theta = \frac{\text{adj}}{\text{hyp}} \qquad\qquad \sec \theta = \frac{1}{\cos \theta} = \frac{\text{hyp}}{\text{adj}}$$

$$\tan \theta = \frac{\text{opp}}{\text{adj}} = \frac{\sin \theta}{\cos \theta} \qquad \cot \theta = \frac{1}{\tan \theta} = \frac{\text{adj}}{\text{opp}}$$

$$\text{adj}^2 + \text{opp}^2 = \text{hyp}^2 \qquad\qquad \text{(Pythagorean theorem).}$$

FIGURE A-7

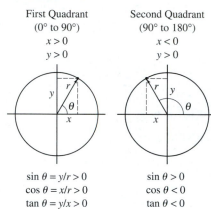

Figure A–7 shows the signs (+ or −) that cosine, sine, and tangent take on for angles θ in the four quadrants (0° to 360°). Note that angles are measured counterclockwise from the x axis as shown; negative angles are measured from *below* the x axis, clockwise: for example, −30° = +330°, and so on.

The following are some useful identities among the trigonometric functions:

$$\sin^2 \theta + \cos^2 \theta = 1$$

$$\sin 2\theta = 2 \sin \theta \cos \theta$$

$$\cos 2\theta = \cos^2 \theta - \sin^2 \theta = 2 \cos^2 \theta - 1 = 1 - 2 \sin^2 \theta$$

$$\tan 2\theta = \frac{2 \tan \theta}{1 - \tan^2 \theta}$$

$$\sin(A \pm B) = \sin A \cos B \pm \cos A \sin B$$

$$\cos(A \pm B) = \cos A \cos B \mp \sin A \sin B$$

$$\tan(A \pm B) = \frac{\tan A \pm \tan B}{1 \mp \tan A \tan B}$$

$$\sin(180° - \theta) = \sin \theta$$

$$\cos(180° - \theta) = -\cos \theta$$

$$\sin(90° - \theta) = \cos \theta$$

$$\cos(90° - \theta) = \sin \theta$$

$$\sin \tfrac{1}{2}\theta = \sqrt{\frac{1 - \cos \theta}{2}}$$

$$\cos \tfrac{1}{2}\theta = \sqrt{\frac{1 + \cos \theta}{2}}$$

$$\tan \tfrac{1}{2}\theta = \sqrt{\frac{1 - \cos \theta}{1 + \cos \theta}}$$

$$\sin A \pm \sin B = 2 \sin\left(\frac{A \pm B}{2}\right) \cos\left(\frac{A \mp B}{2}\right).$$

FIGURE A-8

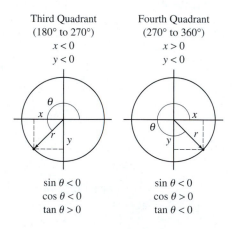

For any triangle (see Fig. A–8):

$$\frac{\sin \alpha}{a} = \frac{\sin \beta}{b} = \frac{\sin \gamma}{c} \qquad \text{(law of sines)}$$

$$c^2 = a^2 + b^2 - 2ab \cos \gamma. \qquad \text{(law of cosines)}$$

Trigonometric Table: Numerical Values of Sin, Cos, Tan

Angle in Degrees	Angle in Radians	Sine	Cosine	Tangent	Angle in Degrees	Angle in Radians	Sine	Cosine	Tangent
0°	0.000	0.000	1.000	0.000					
1°	0.017	0.017	1.000	0.017	46°	0.803	0.719	0.695	1.036
2°	0.035	0.035	0.999	0.035	47°	0.820	0.731	0.682	1.072
3°	0.052	0.052	0.999	0.052	48°	0.838	0.743	0.669	1.111
4°	0.070	0.070	0.998	0.070	49°	0.855	0.755	0.656	1.150
5°	0.087	0.087	0.996	0.087	50°	0.873	0.766	0.643	1.192
6°	0.105	0.105	0.995	0.105	51°	0.890	0.777	0.629	1.235
7°	0.122	0.122	0.993	0.123	52°	0.908	0.788	0.616	1.280
8°	0.140	0.139	0.990	0.141	53°	0.925	0.799	0.602	1.327
9°	0.157	0.156	0.988	0.158	54°	0.942	0.809	0.588	1.376
10°	0.175	0.174	0.985	0.176	55°	0.960	0.819	0.574	1.428
11°	0.192	0.191	0.982	0.194	56°	0.977	0.829	0.559	1.483
12°	0.209	0.208	0.978	0.213	57°	0.995	0.839	0.545	1.540
13°	0.227	0.225	0.974	0.231	58°	1.012	0.848	0.530	1.600
14°	0.244	0.242	0.970	0.249	59°	1.030	0.857	0.515	1.664
15°	0.262	0.259	0.966	0.268	60°	1.047	0.866	0.500	1.732
16°	0.279	0.276	0.961	0.287	61°	1.065	0.875	0.485	1.804
17°	0.297	0.292	0.956	0.306	62°	1.082	0.883	0.469	1.881
18°	0.314	0.309	0.951	0.325	63°	1.100	0.891	0.454	1.963
19°	0.332	0.326	0.946	0.344	64°	1.117	0.899	0.438	2.050
20°	0.349	0.342	0.940	0.364	65°	1.134	0.906	0.423	2.145
21°	0.367	0.358	0.934	0.384	66°	1.152	0.914	0.407	2.246
22°	0.384	0.375	0.927	0.404	67°	1.169	0.921	0.391	2.356
23°	0.401	0.391	0.921	0.424	68°	1.187	0.927	0.375	2.475
24°	0.419	0.407	0.914	0.445	69°	1.204	0.934	0.358	2.605
25°	0.436	0.423	0.906	0.466	70°	1.222	0.940	0.342	2.747
26°	0.454	0.438	0.899	0.488	71°	1.239	0.946	0.326	2.904
27°	0.471	0.454	0.891	0.510	72°	1.257	0.951	0.309	3.078
28°	0.489	0.469	0.883	0.532	73°	1.274	0.956	0.292	3.271
29°	0.506	0.485	0.875	0.554	74°	1.292	0.961	0.276	3.487
30°	0.524	0.500	0.866	0.577	75°	1.309	0.966	0.259	3.732
31°	0.541	0.515	0.857	0.601	76°	1.326	0.970	0.242	4.011
32°	0.559	0.530	0.848	0.625	77°	1.344	0.974	0.225	4.331
33°	0.576	0.545	0.839	0.649	78°	1.361	0.978	0.208	4.705
34°	0.593	0.559	0.829	0.675	79°	1.379	0.982	0.191	5.145
35°	0.611	0.574	0.819	0.700	80°	1.396	0.985	0.174	5.671
36°	0.628	0.588	0.809	0.727	81°	1.414	0.988	0.156	6.314
37°	0.646	0.602	0.799	0.754	82°	1.431	0.990	0.139	7.115
38°	0.663	0.616	0.788	0.781	83°	1.449	0.993	0.122	8.144
39°	0.681	0.629	0.777	0.810	84°	1.466	0.995	0.105	9.514
40°	0.698	0.643	0.766	0.839	85°	1.484	0.996	0.087	11.43
41°	0.716	0.656	0.755	0.869	86°	1.501	0.998	0.070	14.301
42°	0.733	0.669	0.743	0.900	87°	1.518	0.999	0.052	19.081
43°	0.750	0.682	0.731	0.933	88°	1.536	0.999	0.035	28.636
44°	0.768	0.695	0.719	0.966	89°	1.553	1.000	0.017	57.290
45°	0.785	0.707	0.707	1.000	90°	1.571	1.000	0.000	∞

A–8 Logarithms

Logarithms are defined in the following way:

$$\text{if } y = A^x, \qquad \text{then } x = \log_A y.$$

That is, the logarithm of a number y to the base A is that number which, as the exponent of A, gives back the number y. For *common logarithms*, the base is 10, so

Common logs

$$\text{if } y = 10^x, \qquad \text{then } x = \log y.$$

The subscript 10 on $\log_{10}$ is usually omitted when dealing with common logs. Another base sometimes used is the exponential base $e = 2.718\cdots$, a natural number.[†] Such logarithms are called *natural logarithms* and are written ln. Thus,

Natural logs

$$\text{if } y = e^x, \qquad \text{then } x = \ln y.$$

For any number y, the two types of logarithm are related by

$$\ln y = 2.3026 \log y.$$

Some simple rules for logarithms are as follows:

$$\log (ab) = \log a + \log b. \tag{A–7}$$

This is true because if $a = 10^n$ and $b = 10^m$, then $ab = 10^{n+m}$. From the definition of logarithm, $\log a = n$, $\log b = m$, and $\log (ab) = n + m$; hence, $\log (ab) = n + m = \log a + \log b$. In a similar way, we can show that

$$\log\left(\frac{a}{b}\right) = \log a - \log b \tag{A–8}$$

and

$$\log a^n = n \log a. \tag{A–9}$$

These three rules apply not only to common logs but to natural or any other kind of logarithm.

Logs were once used as a technique for simplifying certain types of calculation. Because of the advent of electronic calculators and computers, they are not often used any more for this purpose. However, logs do appear in certain physical equations, so it is helpful to know how to deal with them. If you do not have a calculator that calculates logs, you can easily use a *log table*, such as the small one shown here (Table A–1). The number N is given to two digits (some tables give N to three or more digits); the first digit is in the vertical column to the left, the second digit is in the horizontal row across the top. For example, the Table tells us that $\log 1.0 = 0.000$, $\log 1.1 = 0.041$, and $\log 4.1 = 0.613$. Table A–1 does not include the decimal point—it is understood. The Table gives logs for numbers between 1.0 and 9.9; for larger or smaller numbers, we use rule A–7:

$$\log (ab) = \log a + \log b.$$

For example,

$$\log (380) = \log (3.8 \times 10^2) = \log (3.8) + \log (10^2).$$

From the Table, $\log 3.8 = 0.580$; and from rule A–9,

$$\log (10^2) = 2 \log (10) = 2,$$

since $\log (10) = 1$. [This follows from the definition of the logarithm: if

---

[†]The exponential base e can be written as an infinite series:

$$e = 1 + \frac{1}{1} + \frac{1}{1\cdot 2} + \frac{1}{1\cdot 2\cdot 3} + \frac{1}{1\cdot 2\cdot 3\cdot 4} + \cdots.$$

TABLE A–1 Short Table of Common Logarithms

N	0.0	0.1	0.2	0.3	0.4	0.5	0.6	0.7	0.8	0.9
1	000	041	079	114	146	176	204	230	255	279
2	301	322	342	362	380	398	415	431	447	462
3	477	491	505	519	531	544	556	568	580	591
4	602	613	623	633	643	653	663	672	681	690
5	699	708	716	724	732	740	748	756	763	771
6	778	785	792	799	806	813	820	826	833	839
7	845	851	857	863	869	875	881	886	892	898
8	903	908	914	919	924	929	935	940	944	949
9	954	959	964	968	973	978	982	987	991	996

$10 = 10^1$, then $1 = \log(10)$.] Thus,

$$\log(380) = \log(3.8) + \log(10^2)$$
$$= 0.580 + 2$$
$$= 2.580.$$

Similarly,

$$\log(0.081) = \log(8.1) + \log(10^{-2})$$
$$= 0.908 - 2 = -1.092.$$

Sometimes we need to do the reverse process: find the number N whose log is, say, 2.670. This is called "taking the antilogarithm." To do so, we separate our number 2.670 into two parts, making the separation at the decimal point:

Antilogs

$$\log N = 2.670 = 2 + 0.670$$
$$= \log 10^2 + 0.670.$$

We now look at Table A–1 to see what number has its log equal to 0.670; none does, so we must *interpolate*: we see that $\log 4.6 = 0.663$ and $\log 4.7 = 0.672$. So the number we want is between 4.6 and 4.7, and closer to the latter by $\frac{7}{9}$. Approximately we can say that $\log 4.68 = 0.670$. Thus

Interpolation

$$\log N = 2 + 0.670$$
$$= \log(10^2) + \log(4.68) = \log(4.68 \times 10^2),$$

so $N = 4.68 \times 10^2 = 468$.

If the given logarithm is negative, say, -2.180, we proceed as follows:

$$\log N = -2.180 = -3 + 0.820$$
$$= \log 10^{-3} + \log 6.6 = \log 6.6 \times 10^{-3},$$

so $N = 6.6 \times 10^{-3}$. Notice that we added to our given logarithm the next largest integer (3 in this case) so that we have an integer, plus a decimal number between 0 and 1.0 whose antilogarithm can be looked up in the Table.

Selected Isotopes

(1) Atomic Number Z	(2) Element	(3) Symbol	(4) Mass Number A	(5) Atomic Mass†	(6) % Abundance (or Radioactive Decay‡ Mode)	(7) Half-life (if radioactive)
0	(Neutron)	n	1	1.008665	β^-	10.24 min
1	Hydrogen	H	1	1.007825	99.9885%	
	Deuterium	d or D	2	2.014102	0.0115%	
	Tritium	t or T	3	3.016049	β^-	12.33 yr
2	Helium	He	3	3.016029	0.000137%	
			4	4.002603	99.999863%	
3	Lithium	Li	6	6.015122	7.59%	
			7	7.016004	92.41%	
4	Beryllium	Be	7	7.016929	EC, γ	53.29 days
			9	9.012182	100%	
5	Boron	B	10	10.012937	19.9%	
			11	11.009306	80.1%	
6	Carbon	C	11	11.011434	β^+, EC	20.39 min
			12	12.000000	98.93%	
			13	13.003355	1.07%	
			14	14.003242	β^-	5730 yr
7	Nitrogen	N	13	13.005739	β^+, EC	9.965 min
			14	14.003074	99.632%	
			15	15.000109	0.368%	
8	Oxygen	O	15	15.003065	β^+, EC	122.24 s
			16	15.994915	99.757%	
			18	17.999160	0.205%	
9	Fluorine	F	19	18.998403	100%	
10	Neon	Ne	20	19.992440	90.48%	
			22	21.991386	9.25%	
11	Sodium	Na	22	21.994437	β^+, EC, γ	2.6019 yr
			23	22.989770	100%	
			24	23.990963	β^-, γ	14.951 h
12	Magnesium	Mg	24	23.985042	78.99%	
13	Aluminum	Al	27	26.981538	100%	
14	Silicon	Si	28	27.976927	92.2297%	
			31	30.975363	β^-, γ	157.3 min

† The masses given in column (5) are those for the neutral atom, including the Z electrons.
‡ Chapter 30; EC = electron capture.

(1) Atomic Number Z	(2) Element	(3) Symbol	(4) Mass Number A	(5) Atomic Mass	(6) % Abundance (or Radioactive Decay Mode)	(7) Half-life (if radioactive)
15	Phosphorus	P	31	30.973762	100%	
			32	31.973907	β^-	14.262 days
16	Sulfur	S	32	31.972071	94.9%	
			35	34.969032	β^-	87.38 days
17	Chlorine	Cl	35	34.968853	75.78%	
			37	36.965903	24.22%	
18	Argon	Ar	40	39.962383	99.600%	
19	Potassium	K	39	38.963707	93.258%	
			40	39.963999	0.0117%	
					β^-, EC, γ, β^+	1.277×10^9 yr
20	Calcium	Ca	40	39.962591	96.94%	
21	Scandium	Sc	45	44.955910	100%	
22	Titanium	Ti	48	47.947947	73.72%	
23	Vanadium	V	51	50.943964	99.750%	
24	Chromium	Cr	52	51.940512	83.789%	
25	Manganese	Mn	55	54.940363	100%	
26	Iron	Fe	56	55.934942	91.75%	
27	Cobalt	Co	59	58.933200	100%	
			60	59.933822	β^-, γ	5.2708 yr
28	Nickel	Ni	58	57.935348	68.077%	
			60	59.930791	26.223%	
29	Copper	Cu	63	62.929601	69.17%	
			65	64.927794	30.83%	
30	Zinc	Zn	64	63.929147	48.6%	
			66	65.926037	27.9%	
31	Gallium	Ga	69	68.925581	60.108%	
32	Germanium	Ge	72	71.922076	27.5%	
			74	73.921178	36.3%	
33	Arsenic	As	75	74.921596	100%	
34	Selenium	Se	80	79.916522	49.6%	
35	Bromine	Br	79	78.918338	50.69%	
36	Krypton	Kr	84	83.911507	57.00%	
37	Rubidium	Rb	85	84.911789	72.17%	
38	Strontium	Sr	86	85.909262	9.86%	
			88	87.905614	82.58%	
			90	89.907738	β^-	28.79 yr
39	Yttrium	Y	89	88.905848	100%	
40	Zirconium	Zr	90	89.904704	51.4%	
41	Niobium	Nb	93	92.906378	100%	
42	Molybdenum	Mo	98	97.905408	24.1%	
43	Technetium	Tc	98	97.907216	β^-, γ	4.2×10^6 yr
44	Ruthenium	Ru	102	101.904350	31.55%	
45	Rhodium	Rh	103	102.905504	100%	
46	Palladium	Pd	106	105.903483	27.33%	
47	Silver	Ag	107	106.905093	51.839%	
			109	108.904756	48.161%	

(1) Atomic Number Z	(2) Element	(3) Symbol	(4) Mass Number A	(5) Atomic Mass	(6) % Abundance (or Radioactive Decay Mode)	(7) Half-life (if radioactive)
48	Cadmium	Cd	114	113.903358	28.7%	
49	Indium	In	115	114.903878	95.71%; β^-	4.41×10^{14} yr
50	Tin	Sn	120	119.902197	32.58%	
51	Antimony	Sb	121	120.903818	57.21%	
52	Tellurium	Te	130	129.906223	34.1%; $\beta^-\beta^-$	$> 5.6 \times 10^{22}$ yr
53	Iodine	I	127	126.904468	100%	
			131	130.906124	β^-, γ	8.0207 days
54	Xenon	Xe	132	131.904155	26.89%	
			136	135.907220	8.87%; $\beta^-\beta^-$	$> 3.6 \times 10^{20}$ yr
55	Cesium	Cs	133	132.905447	100%	
56	Barium	Ba	137	136.905821	11.232%	
			138	137.905241	71.70%	
57	Lanthanum	La	139	138.906348	99.910%	
58	Cerium	Ce	140	139.905434	88.45%	
59	Praseodymium	Pr	141	140.907648	100%	
60	Neodymium	Nd	142	141.907719	27.2%	
61	Promethium	Pm	145	144.912744	EC, α	17.7 yr
62	Samarium	Sm	152	151.919728	26.75%	
63	Europium	Eu	153	152.921226	52.19%	
64	Gadolinium	Gd	158	157.924101	24.84%	
65	Terbium	Tb	159	158.925343	100%	
66	Dysprosium	Dy	164	163.929171	28.2%	
67	Holmium	Ho	165	164.930319	100%	
68	Erbium	Er	166	165.930290	33.6%	
69	Thulium	Tm	169	168.934211	100%	
70	Ytterbium	Yb	174	173.938858	31.8%	
71	Lutetium	Lu	175	174.940768	97.41%	
72	Hafnium	Hf	180	179.946549	35.08%	
73	Tantalum	Ta	181	180.947996	99.988%	
74	Tungsten (wolfram)	W	184	183.950933	30.64%; α	$> 4 \times 10^{18}$ yr
75	Rhenium	Re	187	186.955751	62.60%; β^-	4.35×10^{10} yr
76	Osmium	Os	191	190.960928	β^-, γ	15.4 days
			192	191.961479	40.78%	
77	Iridium	Ir	191	190.960591	37.3%	
			193	192.962924	62.7%	
78	Platinum	Pt	195	194.964774	33.832%	
79	Gold	Au	197	196.966552	100%	
80	Mercury	Hg	199	198.968262	16.87%	
			202	201.970626	29.9%	
81	Thallium	Tl	205	204.974412	70.476%	
82	Lead	Pb	206	205.974449	24.1%	
			207	206.975881	22.1%	
			208	207.976636	52.4%	
			210	209.984173	β^-, γ, α	22.3 yr
			211	210.988731	β^-, γ	36.1 min
			212	211.991887	β^-, γ	10.64 h
			214	213.999798	β^-, γ	26.8 min

(1) Atomic Number Z	(2) Element	(3) Symbol	(4) Mass Number A	(5) Atomic Mass	(6) % Abundance (or Radioactive Decay Mode)	(7) Half-life (if radioactive)
83	Bismuth	Bi	209	208.980383	100%	
			211	210.987258	α, γ, β^-	2.14 min
84	Polonium	Po	210	209.982416	α, γ, EC	138.376 days
			214	213.995186	α, γ	164.3 μs
85	Astatine	At	218	218.008681	α, β^-	1.5 s
86	Radon	Rn	222	222.017570	α, γ	3.8235 days
87	Francium	Fr	223	223.019731	β^-, γ, α	22.00 min
88	Radium	Ra	226	226.025403	α, γ	1600 yr
89	Actinium	Ac	227	227.027747	β^-, γ, α	21.773 yr
90	Thorium	Th	228	228.028731	α, γ	1.9116 yr
			232	232.038050	100%; α, γ	1.405×10^{10} yr
91	Protactinium	Pa	231	231.035879	α, γ	3.276×10^4 yr
92	Uranium	U	232	232.037146	α, γ	68.9 yr
			233	233.039628	α, γ	1.592×10^5 yr
			235	235.043923	0.720%; α, γ	7.038×10^8 yr
			236	236.045562	α, γ	2.342×10^7 yr
			238	238.050783	99.274%; α, γ	4.468×10^9 yr
			239	239.054288	β^-, γ	23.45 min
93	Neptunium	Np	237	237.048167	α, γ	2.144×10^6 yr
			239	239.052931	β^-, γ	2.3565 days
94	Plutonium	Pu	239	239.052157	α, γ	24,110 yr
			244	244.064198	α	8.00×10^7 yr
95	Americium	Am	243	243.061373	α, γ	7370 yr
96	Curium	Cm	247	247.070347	α, γ	1.56×10^7 yr
97	Berkelium	Bk	247	247.070299	α, γ	1380 yr
98	Californium	Cf	251	251.079580	α, γ	898 yr
99	Einsteinium	Es	252	252.082970	α, EC, γ	471.7 days
100	Fermium	Fm	257	257.095099	α, γ	100.5 days
101	Mendelevium	Md	258	258.098425	α, γ	51.5 days
102	Nobelium	No	259	259.10102	α, EC	58 min
103	Lawrencium	Lr	262	262.1097	α, EC, fission	3.6 h
104	Rutherfordium	Rf	263	263.11831	fission	10 min
105	Dubnium	Db	262	262.11415	α, fission, EC	34 s
106	Seaborgium	Sg	266	266.1219	α, fission	21 s
107	Bohrium	Bh	264	264.1247	α	0.44 s
108	Hassium	Hs	269	269.1341	α	9 s
109	Meitnerium	Mt	268	268.1388	α	0.07 s
110	Darmstadtium	Ds	271	271.14608	α	0.06 ms
111		Uuu	272	272.1535	α	1.5 ms
112		Uub	277	277	α	0.24 ms

Rotating Frames of Reference; Inertial Forces; Coriolis Effect

Inertial and Noninertial Reference Frames

In Chapters 5 and 8 we examined the motion of objects, including circular and rotational motion, from the outside, as observers fixed on the Earth. Sometimes it is convenient to place ourselves (in theory, if not physically) into a reference frame that is rotating. Let us examine the motion of objects from the point of view, or frame of reference, of persons seated on a rotating platform such as a merry-go-round. It looks to them as if the rest of the world is going around *them*. But let us focus on what they observe when they place a tennis ball on the floor of the rotating platform, which we assume is frictionless. If they put the ball down gently, without giving it any push, they will observe that it accelerates from rest and moves outward as shown in Fig. C–1a. According to Newton's first law, an object initially at rest should stay at rest if no force

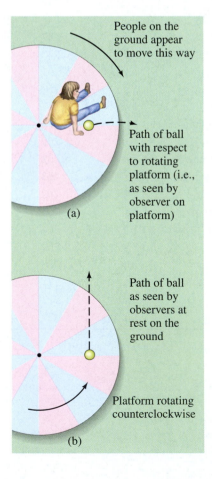

People on the
ground appear
to move this way

Path of ball
with respect
to rotating
platform (i.e.,
as seen by
observer on
platform)

(a)

FIGURE C–1 Path of a ball released on a rotating merry-go-round as seen (a) in the reference frame of the merry-go-round, and (b) in a reference frame fixed on the ground.

Path of ball
as seen by
observers at
rest on the
ground

Platform rotating
counterclockwise

(b)

acts on it. But, according to the observers on the rotating platform, the ball starts moving even though there is no force applied to it. To observers on the ground, this is all very clear: the ball has an initial velocity when it is released (because the platform is moving), and it simply continues moving in a straight-line path as shown in Fig. C–1b, in accordance with Newton's first law.

But what shall we do about the frame of reference of the observers on the rotating platform? Clearly, Newton's first law, the law of inertia, does not hold in this rotating frame of reference. For this reason, such a frame is called a **noninertial reference frame**. An **inertial reference frame** (as discussed in Chapter 4) is one in which the law of inertia—Newton's first law—does hold, and so do Newton's second and third laws. In a noninertial reference frame, such as our rotating platform, Newton's second law also does not hold. For instance in the situation described above, there is no net force on the ball; yet, with respect to the rotating platform, the ball accelerates.

Fictitious (Inertial) Forces

Because Newton's laws do not hold when observations are made with respect to a rotating frame of reference, calculation of motion can be complicated. However, we can still apply Newton's laws in such a reference frame if we make use of a trick. The ball on the rotating platform of Fig. C–1a flies outward when released (as if a force were acting on it—though as we saw above, no force actually does act on it); so the trick we use is to write down the equation $\Sigma F = ma$ as if a force equal to mv^2/r (or $m\omega^2 r$) were acting radially outward on the object in addition to any other forces that may be acting. This extra force, which might be designated as "centrifugal force" since it *seems* to act outward, is called a **fictitious force** or **pseudoforce**. It is a pseudoforce ("pseudo" means "false") because there is no object that exerts this force. Furthermore, when viewed from an inertial reference frame, the effect doesn't exist at all. We have made up this pseudoforce so that we can make calculations in a noninertial frame using Newton's second law, $\Sigma F = ma$. Thus the observer in the noninertial frame of Fig. C–1a uses Newton's second law for the ball's outward motion by assuming that a force equal to mv^2/r acts on it. Such pseudoforces are also called **inertial forces** since they arise only because the reference frame is not an inertial one.

Fictitious force (pseudoforce)

Inertial force

We can examine the motion of a particle in a centrifuge (Section 5–5) from the frame of reference of the rotating test tube. In this frame of reference, the particles move in a more-or-less straight path down the tube. (From the reference frame of the Earth, the particles go round and round.) The acceleration of a particle with respect to the rotating tube can then be calculated using $F = ma$ if we include a pseudoforce, "F," equal to $m\omega^2 r = m(v^2/r)$ acting down the tube, in addition to the drag force F_D exerted by the fluid on the particle (Fig. C–2) up the tube.

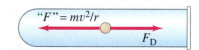

FIGURE C–2 The forces on a particle in a test tube rotating in a centrifuge, seen in the reference frame of the test tube.

In Section 5–3 we discussed the forces on a person in a car going around a curve (Fig. 5–11) from the point of view of an inertial frame. The car, on the other hand, is not an inertial frame. Passengers in such a car could interpret this being pressed outward as the effect of a "centrifugal" force. But they need to recognize that it is a pseudoforce because there is no identifiable object exerting it. It is an effect of being in a noninertial frame of reference.

The Earth itself is rotating on its axis. Thus, strictly speaking, Newton's laws are not valid on the Earth. However, the effect of the Earth's rotation is usually so small that it can be ignored, although it does influence the movement of large air masses and ocean currents. Because of the Earth's rotation, the material of the Earth is concentrated slightly more at the equator. The Earth is thus not a perfect sphere but is slightly fatter at the equator than at the poles.

Coriolis Effect

In a reference frame that rotates at a constant angular speed ω (relative to an inertial frame), there exists another pseudoforce known as the *Coriolis force*. It appears to act on a body in a rotating reference frame only if the body is moving relative to that reference frame, and it acts to deflect the body sideways. It, too, is an effect of the reference frame being noninertial and hence is referred to as an *inertial force*. To see how the Coriolis force arises, consider two people, A and B, at rest on a platform rotating with angular speed ω, as shown in Fig. C–3a. They are situated at distances r_A and r_B, respectively, from the axis of rotation (at O). The woman at A throws a ball with a horizontal velocity $\vec{v}$ (in her reference frame) radially outward toward the man at B on the outer edge of the platform. In Fig. C–3a, we view the situation from an inertial reference frame. The ball initially has not only the velocity $\vec{v}$ radially outward, but also a tangential velocity $\vec{v}_A$ due to the rotation of the platform. Now Eq. 8–4 tells us that $v_A = r_A\omega$, where r_A is the woman's radial distance from the axis of rotation at O. If the man at B had this same velocity v_A, the ball would reach him perfectly. But his speed is greater than v_A (Fig. C–3a) since he is farther from the axis of rotation. His speed is $v_B = r_B\omega$, which is greater than v_A because $r_B > r_A$. Thus, when the ball reaches the outer edge of the platform, it passes a point that the man at B has already passed because his speed in that direction is greater than the ball's. So the ball passes behind him.

Figure C–3b shows the situation as seen from the rotating platform as frame of reference. Both A and B are at rest, and the ball is thrown with velocity $\vec{v}$ toward B, but the ball deflects to the right as shown and passes behind B as previously described. This is not a centrifugal-force effect, for the latter acts radially outward. Instead, this effect acts sideways, perpendicular to $\vec{v}$, and is called a **Coriolis acceleration**; it is said to be due to the Coriolis force, which is a fictitious inertial force. Its explanation as seen from an inertial system was given above: it is an effect of being in a rotating system, wherein points that are farther from the rotation axis have higher linear speeds. On the other hand, when viewed from the rotating system, we can describe the motion using Newton's second law, $\Sigma\vec{F} = m\vec{a}$, if we add a "pseudoforce" term corresponding to this Coriolis effect.

Let us determine the magnitude of the Coriolis acceleration for the simple case described above. (We assume v is large and distances are short, so we can ignore

FIGURE C–3 The origin of the Coriolis effect. Looking down on a rotating platform, (a) as seen from a nonrotating inertial system, and (b) as seen from the rotating platform as frame of reference.

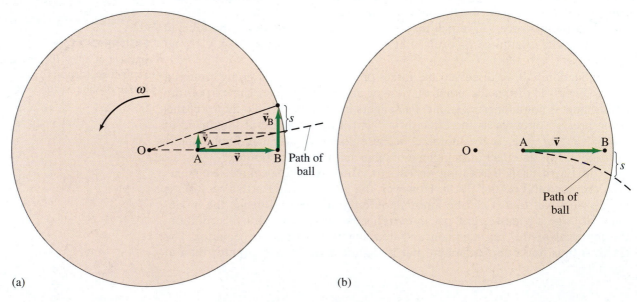

(a)   (b)

gravity.) We do the calculation from the inertial reference frame (Fig. C–3a). The ball moves radially outward a distance $r_B - r_A$ at speed v in a time t given by

$$r_B - r_A = vt.$$

During this time, the ball moves to the side a distance s_A given by

$$s_A = v_A t.$$

The man at B, in this time t, moves a distance

$$s_B = v_B t.$$

The ball therefore passes behind him a distance s (Fig. C–3a) given by

$$s = s_B - s_A = (v_B - v_A)t.$$

We saw earlier that $v_A = r_A \omega$ and $v_B = r_B \omega$, so

$$s = (r_B - r_A)\omega t.$$

We substitute $r_B - r_A = vt$ (see above) and get

$$s = \omega v t^2. \tag{C–1}$$

This same s equals the sideways displacement as seen from the noninertial rotating system (Fig. C–3b).

We see immediately that Eq. C–1 corresponds to motion at constant acceleration. For as we saw in Chapter 2 (see Eq. 2–11b), $y = \frac{1}{2}at^2$ for a constant acceleration (with zero initial velocity in the y direction). Thus, if we write Eq. C–1 in the form $s = \frac{1}{2}a_{Cor}t^2$, we see that the Coriolis acceleration a_{Cor} is

$$a_{Cor} = 2\omega v. \tag{C–2}$$

This relation is valid for any velocity in the plane of rotation—that is, in the plane perpendicular to the axis of rotation (in Fig. C–3, the axis through point O perpendicular to the page).

Because the Earth rotates, the Coriolis effect has some interesting manifestations on the Earth. It affects the movement of air masses and thus has an influence on weather. In the absence of the Coriolis effect, air would rush directly into a region of low pressure, as shown in Fig. C–4a. But because of the Coriolis effect, the winds are deflected to the right in the Northern Hemisphere (Fig. C–4b), since the Earth rotates from west to east. So there tends to be a counterclockwise wind pattern around a low-pressure area. The reverse is true in the Southern Hemisphere. Thus cyclones rotate counterclockwise in the Northern Hemisphere and clockwise in the Southern Hemisphere. The same effect explains the easterly trade winds near the equator: any winds heading south toward the equator will be deflected toward the west (that is, as if coming from the east).

The Coriolis effect also acts on a falling body. A body released from the top of a high tower will not hit the ground directly below the release point, but will be deflected slightly to the east. Viewed from an inertial frame, this is because the top of the tower revolves with a slightly higher speed than does the bottom of the tower.

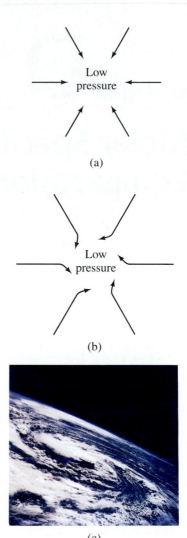

FIGURE C–4 (a) Winds (moving air masses) would flow directly toward a low-pressure area if the Earth did not rotate; (b) and (c): because of the Earth's rotation, the winds are deflected to the right in the Northern Hemisphere (as in Fig. C–3) as if a fictitious (Coriolis) force were acting.

Molar Specific Heats for Gases, and the Equipartition of Energy

Molar Specific Heats for Gases

The values of the specific heats for gases depend on how the thermodynamic process is carried out. Two important processes are those in which either the volume or the pressure is kept constant, and Table D–1 shows how different they can be.

The difference in specific heats for gases is nicely explained in terms of the first law of thermodynamics and kinetic theory. For gases we usually use **molar specific heats**, C_V and C_P, which are defined as the heat required to raise 1 mol of a gas by $1\,C°$ at constant volume and at constant pressure, respectively. In analogy to Eq. 14–2, the heat Q needed to raise the temperature of n moles of gas by ΔT is

Molar specific heats

$$Q = nC_V\,\Delta T \qquad \text{[volume constant]} \quad \textbf{(D–1a)}$$
$$Q = nC_P\,\Delta T. \qquad \text{[pressure constant]} \quad \textbf{(D–1b)}$$

It is clear from the definition of molar specific heat (compare Eqs. 14–2 and D–1) that

$$C_V = Mc_V \quad \text{and} \quad C_P = Mc_P,$$

where M is the molecular mass of the gas ($M = m/n$ in grams/mol). The values for molar specific heats are included in Table D–1. These values are nearly the same for different gases that have the same number of atoms per molecule.

Now we use kinetic theory of gases to see, first, why the specific heats of gases are higher for constant-pressure processes than for constant-volume processes.

TABLE D–1 Specific Heats of Gases at 15°C

Gas	Specific Heats (kcal/kg · K)		Molar Specific Heats (cal/mol · K)		$C_P - C_V$ (cal/mol · K)
	c_V	c_P	C_V	C_P	
Monatomic					
He	0.75	1.15	2.98	4.97	1.99
Ne	0.148	0.246	2.98	4.97	1.99
Diatomic					
N_2	0.177	0.248	4.96	6.95	1.99
O_2	0.155	0.218	5.03	7.03	2.00
Triatomic					
CO_2	0.153	0.199	6.80	8.83	2.03
H_2O (100°C)	0.350	0.482	6.20	8.20	2.00
Polyatomic					
C_2H_6	0.343	0.412	10.30	12.35	2.05

Imagine that an ideal gas is slowly heated via these two processes—first at constant volume, and then at constant pressure. In both processes, we let the temperature increase by the same amount, ΔT. In the constant-volume process, no work is done since $\Delta V = 0$. Thus, according to the first law of thermodynamics, the heat added (denoted by Q_V) all goes into increasing the internal energy of the gas:

$$Q_V = \Delta U.$$

In the constant-pressure process, work *is* done. Hence the heat added, Q_P, must not only increase the internal energy but also is used to do work $W = P\,\Delta V$. Thus, for the same ΔT, more heat must be added in the process at constant pressure than at constant volume. For the process at constant pressure, the first law of thermodynamics gives

$$Q_P = \Delta U + P\,\Delta V.$$

Since ΔU is the same in the two processes (we chose ΔT to be the same), we can combine the two above equations:

$$Q_P - Q_V = P\,\Delta V.$$

From the ideal gas law, $V = nRT/P$, so for a process at constant pressure $\Delta V = nR\,\Delta T/P$. Putting this into the above equation and using Eqs. D–1, we find

$$nC_P\,\Delta T - nC_V\,\Delta T = P\left(\frac{nR\,\Delta T}{P}\right)$$

or, after cancellations,

$$C_P - C_V = R. \tag{D–2}$$

Since the gas constant $R = 8.315\,\text{J/mol·K} = 1.99\,\text{cal/mol·K}$, we predict that C_P will be larger than C_V by about $1.99\,\text{cal/mol·K}$. Indeed, this is very close to what is obtained experimentally, as the last column in Table D–1 shows.

Now we calculate the molar specific heat of a monatomic gas using the kinetic theory. For a process carried out at constant volume, no work is done, so the first law of thermodynamics tells us that

$$\Delta U = Q_V.$$

For an ideal monatomic gas, the internal energy, U, is the total kinetic energy of all the molecules,

$$U = N\left(\tfrac{1}{2}m\overline{v^2}\right) = \tfrac{3}{2}nRT$$

as we saw in Section 14–2. Then, using Eq. D–1a, we write $\Delta U = Q_V$ as

$$\Delta U = \tfrac{3}{2}nR\,\Delta T = nC_V\,\Delta T \tag{D–3}$$

or

$$C_V = \tfrac{3}{2}R. \tag{D–4}$$

Since $R = 8.315\,\text{J/mol·K} = 1.99\,\text{cal/mol·K}$, kinetic theory predicts that $C_V = 2.98\,\text{cal/mol·K}$ for an ideal monatomic gas. This is very close to the experimental values for monatomic gases such as helium and neon (Table D–1). From Eq. D–2, C_P is predicted to be about $4.97\,\text{cal/mol·K}$, also in agreement with experiment (Table D–1).

Equipartition of Energy

The measured molar specific heats for more complex gases (Table D–1), such as diatomic (two-atom) and triatomic (three-atom) gases, increase with the increased number of atoms per molecule. We can explain this by assuming that the internal energy includes not only translational kinetic energy but other forms of energy as well. For example, in a diatomic gas (Fig. D–1), the two atoms can rotate about two different axes (but rotation about a third axis passing through the two atoms would give rise to neglible energy, since the moment of inertia is so small). The molecules can have rotational as well as translational kinetic energy.

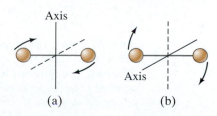

FIGURE D–1 A diatomic molecule can rotate about two different axes.

(a) (b)

It is useful to introduce the idea of **degrees of freedom**, by which we mean the number of independent ways molecules can possess energy. For example, a monatomic gas has three degrees of freedom, because an atom can have velocity along the x, y, and z axes. These are considered to be three independent motions because a change in any one of the components would not affect the others. A diatomic molecule has the same three degrees of freedom associated with translational kinetic energy plus two more degrees of freedom associated with rotational kinetic energy (Fig. D–1), for a total of five degrees of freedom.

Table D–1 indicates that the C_V for diatomic gases is about $\frac{5}{3}$ times as great as for a monatomic gas—that is, in the same ratio as their degrees of freedom.

This led nineteenth-century physicists to the **principle of equipartition of energy**. This principle states that energy is shared equally among the active degrees of freedom, and each active degree of freedom of a molecule has on the average an energy equal to $\frac{1}{2}kT$. Thus, the average energy for a molecule of a monatomic gas would be $\frac{3}{2}kT$ (which we already knew) and of a diatomic gas $\frac{5}{2}kT$. Hence the internal energy of a diatomic gas would be $U = N(\frac{5}{2}kT) = \frac{5}{2}nRT$, where n is the number of moles. Using the same argument we did for monatomic gases, we see that for diatomic gases the molar specific heat at constant volume would be $\frac{5}{2}R = 4.97\ \text{cal/mol·K}$, in accordance with measured values. More complex molecules have even more degrees of freedom and thus greater molar specific heats.

However, measurements showed that for diatomic gases at very low temperatures, C_V has a value of only $\frac{3}{2}R$, as if there were only three degrees of freedom. And at very high temperatures, C_V was about $\frac{7}{2}R$, as if there were seven degrees of freedom. The explanation is that at low temperatures, nearly all molecules have only translational kinetic energy, so, no energy goes into rotational energy and only three degrees of freedom are "active." At very high temperatures, all five degrees of freedom are active plus two additional ones. We interpret the two new degrees of freedom as being associated with the two atoms vibrating, as if they were connected by a spring (Fig. D–2). One degree of freedom comes from the kinetic energy of the vibrational motion, and the second from the potential energy of vibrational motion $(\frac{1}{2}kx^2)$. At room temperature, these two degrees of freedom are apparently not active. Why fewer degrees of freedom are "active" at lower temperatures was eventually explained by Einstein using the quantum theory.

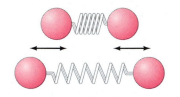

FIGURE D–2 A diatomic molecule can vibrate, as if the two atoms were connected by a spring. Of course they are not, but rather they exert forces on each other that are electrical in nature—of a form that resembles a spring force.

Solids

The principle of equipartition of energy can be applied to solids as well. The molar specific heat of any solid at high temperature is close to $3R$ ($6.0\ \text{cal/mol·K}$), Fig. D–3. This is called the *Dulong and Petit value* after the scientists who first measured it in 1819. (Note that Table 14–1 gave the specific heats per kilogram, not per mole.) At high temperatures, each atom apparently has six degrees of freedom, although some are not active at low temperatures. Each atom in a crystalline solid can vibrate about its equilibrium position as if it were connected by springs to each of its neighbors (Fig. D–4). Thus it can have three degrees of freedom for kinetic energy and three more associated with potential energy of vibration in each of the x, y, and z directions, which is in accord with measured values.

FIGURE D–3 Molar specific heats of solids as a function of temperature.

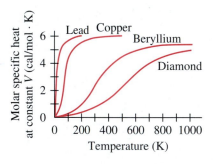

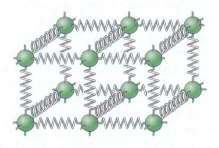

FIGURE D–4 The atoms in a crystalline solid can vibrate about their equilibrium positions as if they were connected to their neighbors by springs. (The forces between atoms are actually electrical in nature.)

Galilean and Lorentz Transformations

W e now examine in detail the mathematics of relating quantities in one inertial reference frame to the equivalent quantities in another. In particular, we will see how positions and velocities *transform* (that is, change) as we go from one frame of reference to another.

We begin with the classical, or Galilean, viewpoint. Consider two reference frames S and S′ which are each characterized by a set of coordinate axes, Fig. E–1. The axes x and y (z is not shown) refer to S, and x' and y' refer to S′. The x' and x axes overlap one another, and we assume that frame S′ moves to the right (in the x direction) at speed v with respect to S. For simplicity let us assume the origins O and O′ of the two reference frames are superimposed at time $t = 0$.

Now consider an event that occurs at some point P (Fig. E–1) represented by the coordinates x', y', z' in reference frame S′ at the time t'. What will be the coordinates of P in S? Since S and S′ overlap precisely initially, after a time t, S′ will have moved a distance vt'. Therefore, at time t', $x = x' + vt'$. The y and z coordinates, on the other hand, are not altered by motion along the x axis; thus $y = y'$ and $z = z'$. Finally, since time is assumed to be absolute in Galilean–Newtonian physics, clocks in the two frames will agree with each other; so $t = t'$. We summarize these in the following **Galilean transformation** equations:

$$x = x' + vt'$$
$$y = y'$$
$$z = z' \qquad \text{(E–1)}$$
$$t = t'.$$

Galilean transformations

These equations give the coordinates of an event in the S frame when those in the S′ frame are known. If those in the S system are known, then the S′ coordinates are obtained from

$$x' = x - vt, \qquad y' = y, \qquad z' = z, \qquad t' = t.$$

These four equations are the "inverse" transformation and are very easily obtained from Eqs. E–1. Notice that the effect is merely to exchange primed and unprimed quantities and replace v by $-v$. This makes sense because from the S′ frame, S moves to the left (negative x direction) with speed v.

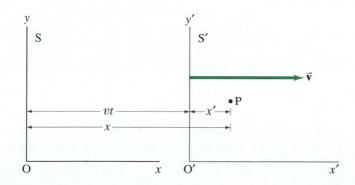

FIGURE E–1 Inertial reference frame S′ moves to the right at speed v with respect to inertial frame S.

Now suppose that the point P in Fig. E–1 represents an object that is moving. Let the components of its velocity vector in S' be u'_x, u'_y, and u'_z (we use u to distinguish it from the relative velocity of the two frames, v). Now $u'_x = \Delta x'/\Delta t'$, $u'_y = \Delta y'/\Delta t'$, and $u'_z = \Delta z'/\Delta t'$, where all quantities are as measured in the S' frame. For example, if at time t'_1 the particle is at x'_1 and a short time later, t'_2, it is at x'_2, then

$$u'_x = \frac{x'_2 - x'_1}{t'_2 - t'_1} = \frac{\Delta x'}{\Delta t'}.$$

Now the velocity of P as seen from S will have components u_x, u_y, and u_z. We can show how these are related to the velocity components in S' by using Eqs. E–1. For example,

$$u_x = \frac{\Delta x}{\Delta t} = \frac{x_2 - x_1}{t_2 - t_1} = \frac{(x'_2 + vt'_2) - (x'_1 + vt'_1)}{t'_2 - t'_1}$$

$$= \frac{(x'_2 - x'_1) + v(t'_2 - t'_1)}{t'_2 - t'_1}$$

$$= \frac{\Delta x'}{\Delta t'} + v = u'_x + v.$$

For the other components, $u'_y = u_y$ and $u'_z = u_z$, so we have

Galilean

velocity

transformations

$$u_x = u'_x + v,$$
$$u_y = u'_y, \qquad\qquad \text{(E–2)}$$
$$u_z = u'_z.$$

These are known as the **Galilean velocity transformation** equations. We see that the y and z components of velocity are unchanged, but the x components differ by v. This is just what we have used before when dealing with relative velocity. For example, if S' is a train and S the Earth, and the train moves with speed v with respect to Earth, a person walking toward the front of the train with speed u'_x will have a speed with respect to the Earth of $u_x = u'_x + v$.

Relativity theory   The Galilean transformations, Eqs. E–1 and E–2, are valid only when the velocities involved are not relativistic (Chapter 26)—that is, much less than the speed of light, c. We can see, for example, that the first of Eqs. E–2 will not work for the speed of light, c, which is the same in all inertial reference frames (a basic postulate in the theory of relativity). That is, light traveling in S' with speed $u'_x = c$ will have speed $c + v$ in S, according to Eq. E–2, whereas the theory of relativity insists it must be c in S. Clearly, then, a new set of transformation equations is needed to deal with relativistic velocities.

We will derive the required equations in a simple way, again looking at Fig. E–1. We assume the transformation is linear and of the form

$$x = \gamma(x' + vt'), \qquad y = y', \qquad z = z'.$$

That is, we modify the first of Eqs. E–1 by multiplying by a factor γ which is yet to be determined. But we assume the y and z equations are unchanged because we expect no length contraction in these directions. We won't assume a form for t, but will derive it. The inverse equations must have the same form with v replaced by $-v$. (The principle of relativity demands it, since S' moving to the right with respect to S is equivalent to S moving to the left with respect to S'.) Therefore

$$x' = \gamma(x - vt).$$

Now if a light pulse leaves the common origin of S and S' at time $t = t' = 0$, after a time t it will have traveled along the x axis a distance

$x = ct$ (in S), or $x' = ct'$ (in S'). Therefore, from the equations for x and x' above,

$$ct = \gamma(ct' + vt') = \gamma(c + v)t',$$
$$ct' = \gamma(ct - vt) = \gamma(c - v)t.$$

We substitute t' from the second equation into the first and find $ct = \gamma(c + v)\gamma(c - v)(t/c) = \gamma^2(c^2 - v^2)t/c$. We cancel out the t on each side and solve for γ to find

$$\gamma = \frac{1}{\sqrt{1 - v^2/c^2}}.$$

Now that we have found γ, we need only find the relation between t and t'. To do so, we combine $x' = \gamma(x - vt)$ with $x = \gamma(x' + vt')$:

$$x' = \gamma(x - vt) = \gamma[\gamma(x' + vt') - vt].$$

We solve for t and find $t = \gamma(t' + vx'/c^2)$. In summary,

$$x = \frac{1}{\sqrt{1 - v^2/c^2}}(x' + vt'),$$
$$y = y',$$
$$z = z',$$
$$t = \frac{1}{\sqrt{1 - v^2/c^2}}\left(t' + \frac{vx'}{c^2}\right).$$

(E–3) *Lorentz transformations*

These are called the **Lorentz transformation** equations. They were first proposed, in a slightly different form, by Lorentz in 1904 to explain the null result of the Michelson–Morley experiment and to make Maxwell's equations take the same form in all inertial systems. A year later, Einstein derived them independently based on his theory of relativity. Notice that not only is the x equation modified as compared to the Galilean transformation but so is the t equation. Indeed, we see directly in this last equation, as well as in the first, how the space and time coordinates mix.

The relativistically correct velocity equations are readily obtained. For example, using Eqs. E–3 (we let $\gamma = 1/\sqrt{1 - v^2/c^2}$),

$$u_x = \frac{\Delta x}{\Delta t} = \frac{\gamma(\Delta x' + v\,\Delta t')}{\gamma(\Delta t' + v\,\Delta x'/c^2)} = \frac{(\Delta x'/\Delta t') + v}{1 + (v/c^2)(\Delta x'/\Delta t')}$$
$$= \frac{u_x' + v}{1 + vu_x'/c^2}.$$

The others are obtained in the same way, and we collect them here:

$$u_x = \frac{u_x' + v}{1 + vu_x'/c^2},$$
$$u_y = \frac{u_y'\sqrt{1 - v^2/c^2}}{1 + vu_x'/c^2},$$
$$u_z = \frac{u_z'\sqrt{1 - v^2/c^2}}{1 + vu_x'/c^2}.$$

(E–4) *Relativistic velocity transformations*

The first of these equations is Eq. 26–9, which we used in Section 26–11 where we discussed how velocities do not add in our commonsense (Galilean) way, because of the denominator $(1 + vu_x'/c^2)$. We can now also see that the y and z components of velocity are also altered and that they depend on the x' component of velocity.

EXAMPLE E–1 **Length contraction.** Derive the length contraction formula, Eq. 26–2, from the Lorentz transformation equations.

SOLUTION Let an object of length L_0 be at rest on the x axis in S. The coordinates of its two end points are x_1 and x_2, so that $x_2 - x_1 = L_0$. At any instant in S′, the end points will be at x_1' and x_2' as given by the Lorentz transformation equations. The length measured in S′ is $L = x_2' - x_1'$. An observer in S′ measures this length by measuring x_2' and x_1' at the same time (in the S′ frame), so $t_2' = t_1'$. Then, from the first of Eqs. E–3,

$$L_0 = x_2 - x_1 = \frac{1}{\sqrt{1 - v^2/c^2}}(x_2' + vt_2' - x_1' - vt_1').$$

Since $t_2' = t_1'$, we have

$$L_0 = \frac{1}{\sqrt{1 - v^2/c^2}}(x_2' - x_1') = \frac{L}{\sqrt{1 - v^2/c^2}},$$

or

$$L = L_0\sqrt{1 - \frac{v^2}{c^2}},$$

which is Eq. 26–2.

EXAMPLE E–2 **Time dilation.** Derive the time dilation formula, Eq. 26–1, from the Lorentz transformation equations.

SOLUTION The time Δt_0 between two events that occur at the same place $(x_2' = x_1')$ in S′ is measured to be $\Delta t_0 = t_2' - t_1'$. Since $x_2' = x_1'$, then from the last of Eqs. E–3, the time Δt between the events as measured in S is

$$\Delta t = t_2 - t_1 = \frac{1}{\sqrt{1 - v^2/c^2}}\left(t_2' + \frac{vx_2'}{c^2} - t_1' - \frac{vx_1'}{c^2}\right)$$

$$= \frac{1}{\sqrt{1 - v^2/c^2}}(t_2' - t_1')$$

$$= \frac{\Delta t_0}{\sqrt{1 - v^2/c^2}},$$

which is Eq. 26–1. Notice that we chose S′ to be the frame in which the two events occur at the same place, so that $x_1' = x_2'$, and the terms containing x_1' and x_2' cancel out.

Answers to Odd-Numbered Problems

CHAPTER 1

1. (a) 1.4×10^{10} years;
   (b) 4.4×10^{17} s.
3. (a) 1.156×10^0;
   (b) 2.18×10^1;
   (c) 6.8×10^{-3};
   (d) 2.7635×10^1;
   (e) 2.19×10^{-1};
   (f) 4.44×10^2.
5. 1%.
7. (a) 4%;
   (b) 0.4%;
   (c) 0.07%.
9. 1.7 m.
11. 9%.
13. (a) 1 megavolt;
    (b) 2 micrometers;
    (c) 6 kilodays;
    (d) 18 hectobucks;
    (e) 8 nanopieces.
15. (a) 1.5×10^{11} m;
    (b) 150 gigameters.
17. 3.8 s.
19. 3.76 m.
21. 7.3%.
23. (a) 3.80×10^{13} m²;
    (b) 13.4.
25. $\approx 7 \times 10^5$ books.
27. ≈ 11 hr.
29. 8×10^4 cm³.
31. 4×10^8 kg/yr.
33. (a) Cannot be correct;
    (b) can be correct;
    (c) can be correct.
35. 50,000 chips.
37. 2×10^{-4} m.
39. (a) 10^{12} protons or neutrons;
    (b) 10^{10} protons or neutrons;
    (c) 10^{29} protons or neutrons;
    (d) 10^{68} protons or neutrons.
41. 1500 gumballs.
43. ≈ 3 ft.
45. ≈ 3500 km.
47. 150 m long, 25 m wide, 15 m high;
    6×10^4 m³.
49. 210 yd, 190 m.

51. 2.21×10^{19} m³, 49.3 Moons.
53. (a) 3%, 3%;
    (b) 0.7%, 0.2%.

CHAPTER 2

1. 72.3 km/h.
3. 61 m.
5. -2.5 cm/s.
7. (a) 2.6×10^2 km;
   (b) 77 km/h.
9. (a) 4.3 m/s;
   (b) 0 m/s.
11. 2.7 min.
13. 6.8 h, 8.7×10^2 km/h.
15. 6.73 m/s.
17. (a) 7.41 m/s²;
    (b) 9.60×10^4 km/h².
19. -5.5 m/s², -0.56 g's.
21. 2.0 m/s², 114 m.
23. 1.8×10^2 m.
25. 63.0 m.
27. -36 g's.
31. 3.1 s.
33. 51.8 m.
35. (a) 8.8 s;
    (b) 86 m/s.
37. 15 m/s, 11 m.
39. 5.61 s.
43. 4.1×10^{-2} s.
45. 46 m.
47. (a) 5.20 s;
    (b) 38.9 m/s;
    (c) 84.7 m.
49. (a) 48 s;
    (b) 90 s to 108 s;
    (c) 0 s to 38 s, 65 s to 83 s, 90 s to 108 s;
    (d) 65 s to 83 s.
51. (a) 0 s to 18 s;
    (b) 27 s;
    (c) 38 s;
    (d) both directions.
53. (a) 4 m/s²;
    (b) 3 m/s²;
    (c) 0.35 m/s²;
    (d) 1.6 m/s².

55.

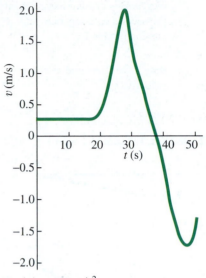

57. (a) -150 m/s²;
    (b) loosen.
59. 1.3 m.
61. (b) 14 m;
    (c) 39.4 m.
63. 31 m/s.
65. (a) 8.8 min;
    (b) 7.5 min.
67. 4.9 m/s to 5.7 m/s, 6.0 m/s to 6.9 m/s, smaller range of initial velocities.
69. 29.0 m.
71. 5.1×10^{-2} m/s².
73. 3.3 min; 5.2 km; 23.3 s, 0.61 km.
75. (a) 88 m/s;
    (b) 27 s;
    (c) 1590 m;
    (d) 36 s;
    (e) -177 m/s;
    (f) 54 s.

77.
   (a)

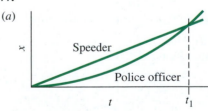

   (b) 23 s;
   (c) 3.0 m/s²;
   (d) 67 m/s.
79. 18 m/s.
81. 0.44 m/min, 2.9 burgers/min.
83. 12 m/s.

85. (a) Near the midpoint of the time interval;
(b) A;
(c) at the times when the two graphs cross; at the first crossing, bicycle B is passing bicycle A; at the second crossing, bicycle A is passing bicycle B;
(d) A;
(e) they have the same average velocity.

CHAPTER 3

1. 282 km, 12° south of west.

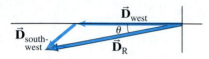

3. $\vec{V}_2 - \vec{V}_1$.

5. 58 m, 48°.

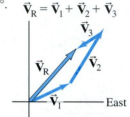

7. (a)

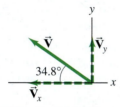

(b) -11.7 units, 8.16 units;
(c) 14.3 units, 34.8° above the $-x$ axis.

9. (a) 550 km/h, 487 km/h;
(b) 1650 km, 1460 km.

11. 64.6, 53.1°.

13. (a) 62.6, 329°;
(b) 77.5, 71.9°;
(c) 77.5, 251.9°.

15. -2450 m, 3870 m, 2450 m; 5190 m.

17. 4.0 m.

19. 13° and 77°.

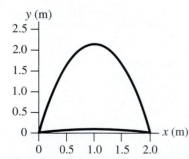

21. 7.92 m/s.

23. 12.9 m.

25. 6 times farther.

27. 5.71 s.

29. The football will not clear the bar. It is 0.76 m too low when it reaches the goal post.

31. (a) 10.4 s;
(b) 541 m;
(c) 51.9 m/s, -63.1 m/s;
(d) 81.7 m/s;
(e) 50.6° below the horizon;
(f) 78.1 m.

33. 76°.

35. (a) 481 m;
(b) 8.37 m/s down;
(c) 97.4 m/s.

37. 1.8 m/s, 19° to the river bank.

39. (a) 2.59 m/s, 62° from the shore;
(b) 3.60 m downstream, 6.90 m across the river.

41. (a) 543 km/h, 7.61° east of south;
(b) 17 km.

43. 1.41 m/s.

45. (a) 1.24 m/s;
(b) 2.28 m/s.

47. (a) 67 m;
(b) 170 s.

49. 42.2° north of east.

51. 114 km/h.

53. 6.2°.

55. 4.7 m/s² left (opposite to the truck's motion), 2.8 m/s² down.

57. $v_T/\tan\theta$.

59. 180 s, 4.8 km; 21.2 s, 0.56 km.

61. 1.9 m/s².

63. 1.9 m/s, 2.7 s.

65. 49.6°.

67. 63 m/s, 66° above the horizontal.

69. 10.8 m/s to 11.0 m/s.

71. (a) 36 m/s;
(b) 20 m/s.

73. 7.0 m/s, 97°.

75. 39 m.

CHAPTER 4

1. 75.0 N.

3. 1.15×10^3 N.

5. (a) 196 N, 196 N;
(b) 294 N, 98.0 N.

7. 68.4 N.

9. 780 N, backward.

11. 2.00 g's, 9.51×10^3 N.

13. 5.08×10^4 N, 4.43×10^4 N.

15. 2.5 m/s², down.

17. (a) 7.4 m/s², down;
(b) 1.29×10^3 N.

19. (a) 47.0 N;
(b) 17.0 N;
(c) 0 N.

21.

(a)   (b)

23. 1.41×10^3 N.

25. (a) 63 N, 31 N;
(b) 73 N, 36 N.

27. 6.9×10^3 N, 8.9×10^3 N.

29. (a) 320 N;
(b) 1.5 m/s².

31. (a)

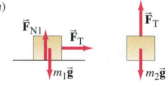

(b) $a = g\, \dfrac{m_2}{m_1 + m_2}$,

$$F_T = m_1 a = g\, \frac{m_1 m_2}{m_1 + m_2}.$$

33. (a)

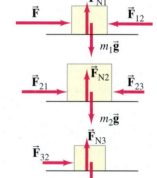

(b) $a = \dfrac{F}{m_1 + m_2 + m_3}$;

(c) $F_{1\,net} = \dfrac{m_1 F}{m_1 + m_2 + m_3}$,

$F_{2\,net} = \dfrac{m_2 F}{m_1 + m_2 + m_3}$,

$F_{3\,net} = \dfrac{m_3 F}{m_1 + m_2 + m_3}$;

(d) $F_{12} = F_{21} = \dfrac{(m_2 + m_3)F}{m_1 + m_2 + m_3}$,

$F_{23} = F_{32} = \dfrac{m_3 F}{m_1 + m_2 + m_3}$;

(e) 2.67 m/s²; 32.0 N, 64.0 N, 32.0 N.

35. 1.74 m/s^2, 22.6 N, 20.9 N.

37. (a) 0.98;
   (b) 0.91.

39. 7.8 m/s^2.

41. 73 N, 0.59.

43. (a)

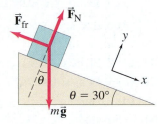

   (b) No change;
   (c) friction force direction would
      be reversed.

45. 40 N.

47. 4.1 m.

49. -7.4 m/s^2.

51. 0.40.

53. (a) 1.2 m;
   (b) 1.6 s.

55. 101 N, 0.719.

57. (a) 0.58;
   (b) 5.7 m/s;
   (c) 15 m/s.

59. 0.36.

61. 5.3×10^2 N.

63. (a) $g \dfrac{(m_1 \sin\theta - m_2)}{(m_1 + m_2)}$;
   (b) $m_1 \sin\theta > m_2$ (down the plane),
      $m_1 \sin\theta < m_2$ (up the plane).

65. 1.3×10^2 N.

67. 1.3 m.

69. 1.54×10^3 N.

71. (a) 16 m/s;
   (b) 13 m/s.

73. Yes, 3.8 m/s.

75. 82 m/s.

77. 5.9°.

79. 940 N, 79° above the horizontal.

81. (a) 9.43×10^4 N;
   (b) 1.33×10^4 N;
   (c) 1.33×10^4 N.

83. 12 m/s.

85. (a) 45 N (10 lb);
   (b) 37 N (8.4 lb);
   (c) not when pulled vertically.

87. (a) 4.1 m/s^2, 3.2 m/s^2;
   (b) 4.1 m/s^2, 3.2 m/s^2;
   (c) 3.5 m/s^2.

89. 5.3×10^2 N, 2.6×10^2 N.

CHAPTER 5

1. (a) 1.42 m/s^2;
   (b) 35.5 N.

3. $5.97 \times 10^{-3} \text{ m/s}^2$, 3.56×10^{22} N, the
   Sun.

5. 0.9 g's.

7. (a) 3.73 N;
   (b) 9.61 N.

9. 25 m/s, yes.

11. 30.4 m/s, 0.403 rev/s.

13. 8.5 m/s.

15. 11 rpm.

17. 3.38×10^4 rpm.

21. 0.22.

23. $4\pi^2 f^2(m_1 r_1 + m_2 r_2)$, $4\pi^2 m_2 r_2 f^2$.

25. 3.5×10^3 N, 5.0×10^2 N.

27. (a) 1.27 m/s;
   (b) 3.05 m/s.

29. (a) 21.0 kg, 21.0 kg;
   (b) 206 N, 252 N.

31. 4.4 m/s^2.

33. 3.9 kg, 0.1 kg.

35. 2.02×10^7 m.

37. $4.38 \times 10^7 \text{ m/s}^2$.

39. 3.2×10^{-8} N toward center of
   square.

41. 6.4×10^{23} kg.

43. 6.32×10^3 m/s.

45. 10 s/rev.

47. 7.90×10^3 m/s.

49. 2.0×10^4 s, 7.1×10^4 s.

51. (a) 21 N, toward the Moon;
   (b) 2.0×10^2 N, away from Moon.

53. (a) 5.4×10^2 N;
   (b) 5.4×10^2 N;
   (c) 7.2×10^2 N;
   (d) 3.6×10^2 N;
   (e) 0 N.

55. (b) $5.4 \times 10^3 \text{ kg/m}^3$.

57. 1.62×10^{11} m.

59. 2690×10^6 km, yes, Pluto.

61. (a) 1.90×10^{27} kg;
   (b) 1.90×10^{27} kg, 1.89×10^{27} kg,
      1.90×10^{27} kg, yes.

63. 671×10^3 km, 1070×10^3 km,
   1880×10^3 km.

65. 9.0 d.

67. 2.64×10^6 m.

69. 0.344%.

71. 2.6 m/s^2 upward.

73. (a) 2.2×10^3 m;
   (b) 5.4×10^3 N;
   (c) 3.8×10^3 N.

75. (a) $\theta = \tan^{-1} m_M R_{\text{Earth}}^2 / M_{\text{Earth}} D_M^2$;
   (b) 5×10^{13} kg;
   (c) $(8 \times 10^{-4})°$.

77. 5.07×10^3 s.

79. 26.9 m/s.

81. 5.2×10^{39} kg, 2.6×10^9 solar
   masses.

83. (a) 3.86×10^3 m/s;
   (b) 4.36×10^4 s.

85. (a) ≈ 12 h;
   (b) 1.8×10^3 m.

87. $5 \times 10^{-5} \text{ N·m}^2/\text{kg}^2$.

89. 3.8×10^{-10} N, upward.

91. $1.6 \times 10^{-4} \text{ m/s}^2$.

93. $v_{\min} = v_0 \sqrt{\dfrac{(1 - \mu_s Rg/v_0^2)}{(1 + \mu_s v_0^2/Rg)}}$,

   $v_{\max} = v_0 \sqrt{\dfrac{(1 + Rg\mu_s/v_0^2)}{(1 - \mu_s v_0^2/Rg)}}$.

CHAPTER 6

1. 7.27×10^3 J.

3. (a) 9.2×10^2 J;
   (b) 5.2×10^3 J.

5. 4.9×10^2 J.

9. (a) $1.10Mg$;
   (b) $1.10Mgh$.

11. 5.0×10^3 J.

13. 8.4×10^{-2} J.

15. 484 m/s.

17. -1.64×10^{-18} J.

19. 44 m/s.

21. 2.25.

23. 1.1 N.

25. (a) 3.24×10^3 N;
   (b) 9.83×10^3 J;
   (c) 7.13×10^4 J;
   (d) -6.14×10^4 J;
   (e) 8.31 m/s.

27. 82 J.

29. 8.1×10^4 N/m.

31. (a) 9.2×10^5 J;
   (b) 9.2×10^5 J;
   (c) yes.

33. 1.4 m, no unless length <0.7 m.

35. 5.14 m/s.

37. (a) 9.2 m/s;
   (b) −0.31 m.

39. (a) 8.3 m/s;
　　(b) 3.64 m.
41. $\frac{1}{2}mv^2 + \frac{1}{2}kx^2 = \frac{1}{2}kx_0^2$.
43. 26 m/s, 12 m/s, 20 m/s.
45. 12 Mg/h.
47. 5.3×10^6 J.
49. (a) 21 m/s;
　　(b) 2.4×10^2 m.
51. (a) 25%;
　　(b) 5.4 m/s;
　　(c) heat, sound, non-elastic
　　　deformation.
53. 23 m/s.
55. 0.40.
57. (a) 1.1×10^3 km/h;
　　(b) 2×10^3 N.
59. 5.5×10^2 N.
61. (b) 0.10 hp.
63. 2.2×10^4 W, 3.0×10^1 hp.
65. 480 W.
67. 1.0×10^3 W.
69. 18°.
71. 9.0×10^2 W.
73. 1.5×10^3 J.
75. (a) $2.5r$;
　　(b) $11mg$;
　　(c) $5mg$;
　　(d) mg.
77. (a) $\sqrt{2gL}$;
　　(b) $\sqrt{1.2\,gL}$.
79. (a) 2.5×10^5 J;
　　(b) 23 m/s;
　　(c) -1.56 m.
81. (a) 4.0×10^1 m/s;
　　(b) 3.0×10^5 W.
83. (a) 1.4×10^3 m;
　　(b) 1.6×10^2 m/s.
85. 4.2×10^4 N.
87. 3.9×10^2 W.
89. $2k$.
91. 4.6 s.
93. (a) 1×10^2 m/s;
　　(b) 4×10^7 W.

CHAPTER 7

1. 0.24 kg·m/s.
3. 4.40×10^3 N toward the pitcher.
5. 6.0×10^7 N upward.
7. 12.6 m/s.
9. 8×10^2 N, $F_{\text{wind}} > F_{\text{fr}} \approx 7 \times 10^2$ N.
11. 4.2×10^3 m/s.

13. (a) 6.9×10^3 m/s away from Earth,
　　　4.7×10^3 m/s away from Earth;
　　(b) 5.9×10^8 J.
15. (a) 2.0 kg·m/s;
　　(b) 5.8×10^2 N.
17. 2.1 kg·m/s to the left.
19. (a) 3.8×10^2 kg·m/s;
　　(b) -3.8×10^2 kg·m/s;
　　(c) 3.8×10^2 kg·m/s;
　　(d) 5.1×10^2 N.
21. 69 m.
23. 1.00 m/s west, 2.00 m/s east.
25. 0.88 m/s and 2.23 m/s, both in direc-
　　tion of tennis ball's initial motion.
27. (a) 3.62 m/s, 4.42 m/s;
　　(b) -4.0×10^2 kg·m/s,
　　　4.0×10^2 kg·m/s.
29. 0.35 m, 1.4 m.
31. $v_2 = \sqrt{2}\,v_1$.
33. (a) $-M/(m + M)$;
　　(b) -0.96.
35. 23 m/s.
37. (b) $e = \sqrt{h'/h}$.
39. (a) 1.7 m/s for both;
　　(b) -2.1 m/s, 7.4 m/s;
　　(c) 0, 4.3 m/s, reasonable;
　　(d) 2.8 m/s, 0, not reasonable;
　　(e) -4.0 m/s, 10.3 m/s, not
　　　reasonable.
41. 60° relative to eagle A, 6.7 m/s.
43. 141°.
45. 39.9 u.
47. 6.5×10^{-11} m.
49. (1.04 m, -1.04 m) relative to raft's
　　center.
51. ($1.2l, 0.9l$) relative to back left
　　corner.
53. 17% of the whole body mass.
55. 21.7 cm horizontal, 7.6 cm vertical.
57. (a) 4.66×10^6 m from center of
　　　Earth.
59. 24.8 cm.
61. $vm/(m + M)$ upward, the balloon
　　stops.
63. $v'_x = \frac{3}{2}v_0$, $v'_y = -v_0$.
65. (a) 0.194 m/s;
　　(b) 8.8×10^2 N.
67. $m_B = \frac{5}{3}m$.
69. 4.00 m.
71. 3.8×10^2 m/s.
73. (a) 2.5×10^{-13} m/s;
　　(b) 1.7×10^{-17};
　　(c) 0.19 J.

75. (a)

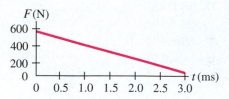

　　(b) 0.93 N·s;
　　(c) 4.2×10^{-3} kg.
77. 6.7×10^3 m/s.
79. (a) -4.4 m/s, 4.0 m/s;
　　(b) 2.0 m.
81. -29.6 km/s.

CHAPTER 8

1. (a) 0.52 rad, $\pi/6$ rad;
　　(b) 0.99 rad, $19\pi/60$ rad;
　　(c) 1.57 rad, $\pi/2$ rad;
　　(d) 6.28 rad, 2π rad;
　　(e) 7.33 rad, $7\pi/3$ rad.
3. 5.3×10^3 m.
5. 7.4×10^{-2} m.
7. (a) 2.6×10^2 rad/s;
　　(b) 46 m/s, 1.2×10^4 m/s².
9. (a) 1.99×10^{-7} rad/s;
　　(b) 7.27×10^{-5} rad/s.
11. 3.6×10^4 rpm.
13. $\omega_1/\omega_2 = R_2/R_1$.
15. 2.8×10^4 rev.
17. (a) 4.0×10^1 rev/min²;
　　(b) 4.0×10^1 rpm.
19. (a) -0.42 rad/s²;
　　(b) 210 s.
21. (a) -4.1 rad/s²;
　　(b) 7.6 s.
23. (a) 41 m·N;
　　(b) 29 m·N.
25. $mg(L_2 - L_1)$, clockwise.
27. 1.81 kg·m².
29. (a) 0.94 kg·m²;
　　(b) 2.4×10^{-2} m·N.
31. (a) 6.1 kg·m²;
　　(b) 0.61 kg·m²;
　　(c) vertical axis.
33. 20 N.
35. 62 m·N.
37. 993 rev, 10.9 s.
39. (a) 92 rad/s²;
　　(b) 7.9×10^2 N.

41. $a = \dfrac{(m_2 - m_1)}{(m_1 + m_2 + I/r^2)} g < a_{I=0}$,

$\quad a_{I=0} = \dfrac{(m_2 - m_1)}{(m_1 + m_2)} g.$

43. 1.40×10^4 J.

45. 56 J.

47. 1.42×10^4 J.

49. 3.22 m/s.

51. 2.64 kg·m²/s.

53. (a) His rotational inertia increases;
  (b) 1.6.

55. 0.77 kg·m², by pulling her arms in toward the center of her body.

57. (a) 14 kg·m²/s;
  (b) −2.7 m·N.

59. $\omega/2$.

61. (a) 1.2 rad/s;
  (b) 1.8×10^3 J, 1.1×10^3 J.

63. 5×10^{-2} rad/s, 2×10^4 KE$_i$.

65. $(2.7 \times 10^{-16})\%$.

67. −0.30 rad/s.

69. 8.21×10^{-6}.

71. 53 m·N.

73. (a) $\omega_R/\omega_F = N_F/N_R$.
  (b) 4.0;
  (c) 1.5.

75. (b) 2.2×10^3 rad/s; (c) 25 min.

77. (a) 4.3 m;
  (b) 5.2 s.

79. $Mg\sqrt{2Rh - h^2}/(R - h)$.

81. 2.8 m·N, from his arm muscles.

83. (a) 7.8 kg·m²/s;
  (b) 3.9 m·N;
  (c) 2.9 rad/s.

85. $2.7(R - r)$.

CHAPTER 9

1. 430 N, 112° clockwise from $\vec{F}_A$.

3. 6.52 kg.

5. 1.1×10^3 N.

7. 5.8×10^3 N, 8.1×10^3 N.

9. (a) 2.3 m from adult;
  (b) 2.5 m from adult.

11. 2.6×10^3 N, 3.1×10^3 N.

13. 0.32 m.

15. 6.1×10^3 N, 5.9×10^3 N.

17. 34.6 N.

19. 9.05×10^{-1} m.

21. (a) 4.25×10^2 N;
  (b) 4.25×10^2 N, 3.28×10^2 N.

23. (a) 0.78 N;
  (b) 0.98 N.

25. 55.2 N, 63.7 N.

27. 0.50.

29. 1.0×10^2 N.

31. 9.9×10^2 N.

33. 2.7×10^3 N.

35. $2.4w$.

37. (b) Yes, by 1/24 of a brick length;
  (c) $D = \displaystyle\sum_{i=1}^{n} \frac{L}{2i}$;
  (d) 35 bricks.

39. (a) 2.0×10^5 N/m²;
  (b) 4.1×10^{-6}.

41. (a) 1.4×10^5 N/m²;
  (b) 6.9×10^{-7};
  (c) 6.5×10^{-6} m.

43. 9.6×10^6 N/m².

45. $(-2 \times 10^{-2})\%$.

47. (a) 1.1×10^2 m·N, clockwise;
  (b) the wall.

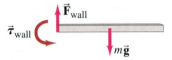

49. (a) 393 N;
  (b) thicker.

51. (a) 4.4×10^{-5} m²;
  (b) 2.7×10^{-3} m.

53. 1.2×10^{-2} m.

55. 12 m.

57. 2.94×10^{-1} kg, 2.29×10^{-1} kg, 6.56×10^{-2} kg.

59. (a) $Mg\sqrt{h/(2R - h)}$;
  (b) $Mg\sqrt{h(2R - h)}/(R - h)$.

61. (a)

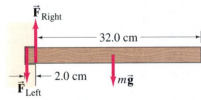

  (b) $F_{\text{Left}} = 3.7 \times 10^2$ N,
   $F_{\text{Right}} = 4.2 \times 10^2$ N,
   $mg = 49$ N;
  (c) 8.3 m·N.

63. 29°.

65. 7.7×10^{-6} m.

67. (a) $0.29mg$;
  (b) $0.58mg$;
  (c) horizontal at lowest point, 60° above the horizontal at points of attachment.

69. (a) $\mu_s < l/2h$;
  (b) $\mu_s > l/2h$.

71. (a) $F_{\text{Left}} = 3.3 \times 10^2$ N up,
   $F_{\text{Right}} = 2.3 \times 10^2$ N down;
  (b) 0.65 m;
  (c) 1.2 m.

73. $F_{\text{Left}} = 1.0 \times 10^2$ N,
   $F_{\text{Right}} = 1.9 \times 10^2$ N.

75. Average force per area = 4.5×10^5 N/m².

77. (a) 3.5×10^8 N/m²;
  (b) the bone will break;
  (c) 8.2×10^6 N/m², the bone will not break.

79. 2.34 m.

CHAPTER 10

1. 3×10^{11} kg.

3. 5.8×10^2 kg.

5. 0.8477.

7. (a) 7×10^7 N/m²;
  (b) 2×10^5 N/m².

9. (a) 4.7×10^5 N;
  (b) 4.7×10^5 N.

11. 2.2×10^3 kg.

13. 13 m.

15. 1.60×10^4 m.

17. (a) 9.6×10^5 N/m²;
  (b) 98 m.

19. (a) 1.41×10^5 Pa;
  (b) 9.8×10^4 Pa.

21. 1.06×10^3 kg/m³, 3% higher.

23. 0.199.

25. 920 kg.

27. Iron or steel.

29. (a) 7.4×10^5 N;
  (b) 1.0×10^4 N.

31. (a) 1.03×10^3 kg/m³;
  (b) $\rho_{\text{liquid}} = \rho_{\text{object}}(m_{\text{object}} - m_{\text{apparent}})/m_{\text{object}}.$

33. 0.105.

35. 0.90 m/s.

39. 4.4×10^5 s (5.1 days).

41. 5.6×10^{-3} m³/s.

43. 1.9×10^5 N.

45. 9.7×10^4 Pa (≈ 0.96 atm).

47. (b) 0.24 m/s.

49. (a) $2\sqrt{h_1(h_2 - h_1)}$;
  (b) $h_1' = h_2 - h_1$.

51. new time = 0.13 (previous time).

53. 9.9×10^2 Pa.

55. 0.9 Pa/cm.

57. (a) $Re = 2500$, so turbulent;
(b) $Re = 5000$, so turbulent.

59. 3.6×10^{-2} N/m.

61. No, 8.3×10^{-6} kg is the maximum mass that could be supported.

63. (a) 0.75 m;
(b) 0.65 m;
(c) 0.24 m.

65. 150 N to 220 N.

67. 0.047 atm.

69. 0.6 atm.

71. 0.142 m.

73. 1.3×10^2 N.

75. 1.1 m.

77. 0.33 kg.

79. 1.1 W.

81. 4.6 m.

83. (a) 9.1 m/s;
(b) 0.26 L/s;
(c) 0.91 m/s.

85. 4.0×10^{-4} m³/s.

87. 4.2×10^{-3} Pa·s.

CHAPTER 11

1. 0.72 m.

3. 1.5 Hz.

5. 3.8 Hz.

7. (a) 0.16 N/m;
(b) 2.8 Hz.

9. (a) 2.5 m/s;
(b) ±1.6 m/s;
(c) 1.8 J;
(d) $x = (0.13 \text{ m}) \cos(6.0\pi t)$.

11. $\pm\frac{1}{2}x_0$.

13. (a) 6.0×10^{-2} m;
(b) 0.58 m/s.

15. (a) 4.2×10^2 N/m;
(b) 3.3 kg.

17. ±0.707A.

19. (a) $y = (0.18 \text{ m}) \cos(2\pi t/0.65 \text{ s})$;
(b) 0.16 s;
(c) 1.7 m/s;
(d) 17 m/s², at the release point.

21. (a) 0.38 m;
(b) 1.03 Hz;
(c) 0.967 s;
(d) 0.92 J;
(e) 5.1×10^{-2} J, 0.86 J.

(f)

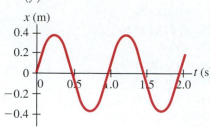

23. (a) 0.490 s, 2.04 Hz;
(b) 0.231 m;
(c) 37.9 m/s²;
(d) $y = (0.231 \text{ m}) \sin(4.08\pi t)$;
(e) 3.31 J.

27. 114 N/m, 19.4 m.

29. 0.99 m.

31. (a) 1.8 s;
(b) the pendulum will not oscillate.

33. Shorten the pendulum by 0.7 mm.

35. (a) −11°;
(b) 15°;
(c) 15°.

37. 1.31 m.

39. (a) 1.4×10^3 m/s;
(b) 4.1×10^3 m/s;
(c) 5.1×10^3 m/s.

41. 0.35 s.

43. 2.1×10^3 m.

45. 0.99 m.

47. (a) 4.6×10^9 W/m²;
(b) 2.3×10^{10} W.

49. 1.73.

51. (a)

(b)

(c) All the energy is kinetic energy.

53. 441 Hz.

55. 9.7×10^{-2} m.

57. 290 Hz, 580 Hz, 870 Hz.

59. (a) 1.3 kg;
(b) 0.32 kg;
(c) 5.2×10^{-2} kg.

61. 1.1 m/s.

63. 25°.

65. 44°.

67. 10 min.

69. (a) 3.2×10^{-2} m;
(b) 1.5 m.

71. (a) 1.8×10^4 N/m;
(b) 0.71 s.

73. 220 Hz.

75. (a) 1.22f;
(b) 0.71f.

77. (a) G: 784 Hz, 1180 Hz; A: 880 Hz, 1320 Hz;
(b) 1.26;
(c) 1.12;
(d) 0.794.

79. (a) 3.0 m/s;
(b) 5.0×10^3 m/s².

81. $\lambda = 4L/(2n - 1), n = 1, 2, 3, \cdots$.

83. Horizontal period is longer by a factor of $\sqrt{1 + l_0 k/mg}$.

85. 6.44 m from the origin of the first pulse.

87. 0.40 s.

CHAPTER 12

1. 3.4×10^2 m.

3. (a) 17 cm to 17 m;
(b) 3.4×10^{-5} m.

5. 55 m.

7. (a) 8%;
(b) 4%.

9. 63 dB.

11. 114 dB.

13. (a) 9×10^{-6} W;
(b) 1×10^7 people.

15. (a) 122 dB, 114 dB;
(b) no.

17. 1.3.

19. 4.

21. 25 dB.

23. (a) 10^9;
(b) 10^{12}.

25. (a) 76.6 Hz, 230 Hz, 383 Hz, 536 Hz;
(b) 153 Hz, 306 Hz, 459 Hz, 613 Hz.

27. 8.6 mm to 8.6 m.

29. (a) 0.18 m;
(b) 1.1 m;
(c) 440 Hz, 0.78 m.

31. −2.6%.

33. (a) 0.583 m;
(b) 862 Hz.

35. (a) 55 Hz;
(b) 2.0×10^2 m/s.

37. (a) 248 overtones;
(b) 249 overtones.

39. ± 0.50 Hz.

41. 28.5 kHz.

43. 3.0 Hz.

45. $f_A = 438$ Hz or 444 Hz,
$f_C = 437$ Hz or 445 Hz,
$f_{beat} = 1$ Hz or 7 Hz.

47. (a) 130.5 Hz, 133.5 Hz;
(b) increase by 2.3%, decrease by 2.2%.

49. (a) 1690 Hz;
(b) 1410 Hz.

51. (a) 2091 Hz and 2087 Hz;
(b) 3550 Hz and 2870 Hz;
(c) 16,000 Hz and 3750 Hz.

53. 4.32×10^4 Hz.

55. 2 Hz.

57. 0.171 m/s.

59. (a) 110 m/s;
(b) 260 m/s.

61. (a) 120;
(b) 0.48°.

65. 0.3 s.

67. (a) 57 Hz, 69 Hz, 86 Hz, 110 Hz, 170 Hz.

69. 88 dB.

71. 15 W.

73. 50 dB.

75. (a) 2.8×10^2 m/s, 48 N;
(b) 0.195 m;
(c) 880 Hz, 1320 Hz.

77. 7.4×10^2 N.

79. 504 Hz.

81. 17 m/s.

83. 2.84 m.

85. 2.29×10^3 Hz.

87. 11.5 m.

89. 34 Hz, 43 Hz, 61 Hz.

91. 10^6.

93. 17 km/h.

CHAPTER 13

1. 3.3×10^{22} atoms.

3. (a) 20°C;
(b) 3300°F.

5. (a) 5°F;
(b) −26°C.

7. 4.3×10^{-3} m.

9. 8×10^{-2} m.

11. 981 kg/m³.

13. 5.12 mL.

15. (a) −140°C;
(b) 180°C.

17. (b) 5.7×10^{-3} (0.57% increase).

21. (a) 6.1 cm;
(b) $\delta L = \dfrac{V_{0\,bulb}}{\pi r_0^2} (\beta_{Hg} - \beta_{glass}) \Delta T$.

23. 3.5×10^7 N/m².

25. (a) 27°C;
(b) 4.3×10^3 N.

27. −459.67°F.

29. 1.07 m³.

31. 1.43 kg/m³.

33. (a) 14.8 m³;
(b) 1.83×10^5 Pa.

35. 2.40×10^8 Pa.

37. 37°C.

39. 3.43 atm.

41. 2.69×10^{25} molecules/m³.

43. (a) 7×10^{22} moles;
(b) 4×10^{46} molecules.

45. 19 molecules/breath.

47. 6×10^3 m/s.

49. 899°C.

51. 25.9°C.

55. 3.9×10^2 m/s.

57. 3.34×10^{-9} m.

61. (a) solid or vapor;
(b) 5.11 atm $\leq P \leq$ 73 atm,
−56.6°C $\leq T \leq$ 31°C.

63. 14°C.

65. 91°C.

67. 1.1×10^3 Pa.

69. 3.1 kg.

71. 0.28 s, $v_{diffuse} = 5.4 \times 10^{-5}$ m/s,
$v_{rms} = 3.1 \times 10^2$ m/s,
$v_{diffuse}/v_{rms} = 1.7 \times 10^{-7}$.

73. (a) low;
(b) (1.7×10^{-2})%.

75. 0.21.

77. 260 m/s, 4×10^{-22} atm.

79. 11 L, not advisable.

81. 1.65, 1.29.

83. 1.1×10^{44} molecules.

85. 15 hours.

87. 0.66×10^3 kg/m³, −3.5%.

89. 1.6×10^{-3} cm.

91. (a) 2.20×10^3 L;
(b) 92 min;
(c) 30 min.

93. 6.8 balls/s.

95. (a) 1.7×10^3 Pa;
(b) 7.0×10^2 Pa.

97. 6% decrease.

99. 3.0 kg.

CHAPTER 14

1. 1.0×10^7 J.

3. (a) 1.0×10^7 J;
(b) 2.9 kWh;
(c) $0.29 per day, no.

5. 220 kg/h.

7. 100 kcal.

9. 2.0×10^3 J/kg·C°.

11. 40.1°C.

13. (1.9×10^2)°C.

15. 425 s.

17. 2.3×10^3 J/kg·C°.

19. 0.32 C°.

21. 5.0×10^6 J.

23. 1.3 kg.

25. 9.90×10^{-3} kg.

27. 4.7×10^3 kcal.

29. 1.12×10^4 J/kg.

31. 1.7 g.

33. 83 W.

35. (a) 95 W;
(b) 33 W.

37. 23 bulbs.

39. (1.6×10^2)°C.

41. 10 C°.

43. (b) $\dfrac{Q}{t} = A \dfrac{(T_1 - T_2)}{\sum\limits_{i=1}^{n} l_i/k_i}$.

45. 6.4 Calories.

47. 4×10^{15} J.

49. (a) 3.2×10^{26} W;
(b) 1.1×10^3 W/m².

51. 0.80 C°.

53. (a) 46 W;
(b) 7.3×10^3 W.

55. 20 W, only about 9% of the required heat loss rate.

57. (a) 44 C°;
(b) none of the bullet will melt.

59. 4.1 g/h.

61. (a) 1.2×10^{18} J;
(b) $Q_{Sun} = 1.3 \times 10^4 \, Q_{interior}$.

63. A mixture of liquid water and steam at 100°C, with the mass of liquid water equal to twice the mass of the steam.

65. (*a*) 3.1×10^7 J;
(*b*) 3.3×10^3 s.

CHAPTER 15

1. (*a*) 0 J;
(*b*) 3.40×10^3 J.

3.

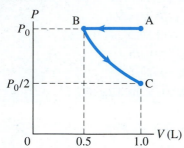

5.

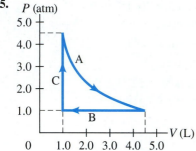

7. (*a*) 0 J;
(*b*) 1850 J;
(*c*) rise.

9. -4.0×10^2 K.

11. (*a, c*)

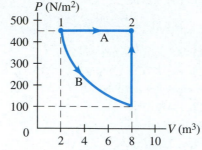

(*b*) 2.73×10^3 J, 4.10×10^3 J;
(*d*) 4.10×10^3 J.

13. (*a*) 25 J;
(*b*) 63 J;
(*c*) −95 J;
(*d*) −120 J;
(*e*) −15 J.

15. 162 W.

17. 0.28.

19. 0.23.

21. 1.6×10^{13} J/h.

23. 440°C.

25. 9.0×10^2 MW (MJ/s).

27. 250°C.

29. 5.7.

31. −21°C.

33. 76 L.

35. -1.5×10^3 J/K.

37. -1.22×10^6 J/K.

39. 0.15 J/K.

41. $4.35 \times 10^{-2} \dfrac{\text{J/K}}{\text{s}}$.

43. 1.1 J/K.

45. (*a*) 1/9;
(*b*) 1/18.

47. (*a*) 5/16;
(*b*) 1/64.

49. (*a*) 1.32×10^6 kWh;
(*b*) 7.09×10^4 kW.

51. Yes, the proposed engine operates at a higher than ideal efficiency.

53. (*a*) 4.0×10^4 J/s;
(*b*) 1.6×10^5 J/s;
(*c*) 220 s.

55. (*a*) 0.077.

57. (*a*) 45°C;
(*b*) 0.58 J/K.

59. 0.24.

61. (*a*) $P_\text{A}(V_\text{C} - V_\text{A})$;
(*b*) $P_\text{C}(V_\text{C} - V_\text{A})$;
(*c*) $\frac{1}{2}(P_\text{C} + P_\text{A})(V_\text{C} - V_\text{A})$.

63. (*a*) 5.3C°;
(*b*) 77 J/kg·K.

65. 200 J.

67. 180 W.

INDEX

Note: the abbreviation *defn* means the page cited gives the definition of the term; *fn* means the reference is in a footnote; *pr* means it is found in a Problem or Question; *ff* means also the following pages.

PHOTO CREDITS

Periodic Table of the Elements§

Legend: Symbol — **Cl** 17 — Atomic Number; Atomic Mass§ 35.4527; Electron Configuration (outer shells only) $3p^5$

Transition Elements

Group I	Group II												Group III	Group IV	Group V	Group VI	Group VII	Group VIII
H 1 1.00794 $1s^1$																		**He** 2 4.002602 $1s^2$
Li 3 6.941 $2s^1$	**Be** 4 9.012182 $2s^2$												**B** 5 10.811 $2p^1$	**C** 6 12.0107 $2p^2$	**N** 7 14.00674 $2p^3$	**O** 8 15.9994 $2p^4$	**F** 9 18.9984032 $2p^5$	**Ne** 10 20.1797 $2p^6$
Na 11 22.989770 $3s^1$	**Mg** 12 24.3050 $3s^2$												**Al** 13 26.981538 $3p^1$	**Si** 14 28.0855 $3p^2$	**P** 15 30.973761 $3p^3$	**S** 16 32.066 $3p^4$	**Cl** 17 35.4527 $3p^5$	**Ar** 18 39.948 $3p^6$
K 19 39.0983 $4s^1$	**Ca** 20 40.078 $4s^2$	**Sc** 21 44.955910 $3d^1 4s^2$	**Ti** 22 47.867 $3d^2 4s^2$	**V** 23 50.9415 $3d^3 4s^2$	**Cr** 24 51.9961 $3d^5 4s^1$	**Mn** 25 54.938049 $3d^5 4s^2$	**Fe** 26 55.845 $3d^6 4s^2$	**Co** 27 58.933200 $3d^7 4s^2$	**Ni** 28 58.6934 $3d^8 4s^2$	**Cu** 29 63.546 $3d^{10} 4s^1$	**Zn** 30 65.39 $3d^{10} 4s^2$		**Ga** 31 69.723 $4p^1$	**Ge** 32 72.61 $4p^2$	**As** 33 74.92160 $4p^3$	**Se** 34 78.96 $4p^4$	**Br** 35 79.904 $4p^5$	**Kr** 36 83.80 $4p^6$
Rb 37 85.4678 $5s^1$	**Sr** 38 87.62 $5s^2$	**Y** 39 88.90585 $4d^1 5s^2$	**Zr** 40 91.224 $4d^2 5s^2$	**Nb** 41 92.90638 $4d^4 5s^1$	**Mo** 42 95.94 $4d^5 5s^1$	**Tc** 43 (98) $4d^5 5s^2$	**Ru** 44 101.07 $4d^7 5s^1$	**Rh** 45 102.90550 $4d^8 5s^1$	**Pd** 46 106.42 $4d^{10} 5s^0$	**Ag** 47 107.8682 $4d^{10} 5s^1$	**Cd** 48 112.411 $4d^{10} 5s^2$		**In** 49 114.818 $5p^1$	**Sn** 50 118.710 $5p^2$	**Sb** 51 121.760 $5p^3$	**Te** 52 127.60 $5p^4$	**I** 53 126.90447 $5p^5$	**Xe** 54 131.29 $5p^6$
Cs 55 132.90545 $6s^1$	**Ba** 56 137.327 $6s^2$	57–71†	**Hf** 72 178.49 $5d^2 6s^2$	**Ta** 73 180.9479 $5d^3 6s^2$	**W** 74 183.84 $5d^4 6s^2$	**Re** 75 186.207 $5d^5 6s^2$	**Os** 76 190.23 $5d^6 6s^2$	**Ir** 77 192.217 $5d^7 6s^2$	**Pt** 78 195.078 $5d^9 6s^1$	**Au** 79 196.96655 $5d^{10} 6s^1$	**Hg** 80 200.59 $5d^{10} 6s^2$		**Tl** 81 204.3833 $6p^1$	**Pb** 82 207.2 $6p^2$	**Bi** 83 208.98038 $6p^3$	**Po** 84 (209) $6p^4$	**At** 85 (210) $6p^5$	**Rn** 86 (222) $6p^6$
Fr 87 (223) $7s^1$	**Ra** 88 (226) $7s^2$	89–103‡	**Rf** 104 (261) $6d^2 7s^2$	**Db** 105 (262) $6d^3 7s^2$	**Sg** 106 (266) $6d^4 7s^2$	**Bh** 107 (264) $6d^5 7s^2$	**Hs** 108 (269) $6d^6 7s^2$	**Mt** 109 (268) $6d^7 7s^2$	**Ds** 110 (271) $6d^9 7s^1$	111 (272) $6d^{10} 7s^1$	112 (277) $6d^{10} 7s^2$							

†Lanthanide Series

La 57 138.9055 $5d^1 6s^2$	**Ce** 58 140.115 $4f^1 5d^1 6s^2$	**Pr** 59 140.90765 $4f^3 5d^0 6s^2$	**Nd** 60 144.24 $4f^4 5d^0 6s^2$	**Pm** 61 (145) $4f^5 5d^0 6s^2$	**Sm** 62 150.36 $4f^6 5d^0 6s^2$	**Eu** 63 151.964 $4f^7 5d^0 6s^2$	**Gd** 64 157.25 $4f^7 5d^1 6s^2$	**Tb** 65 158.92534 $4f^9 5d^0 6s^2$	**Dy** 66 162.50 $4f^{10} 5d^0 6s^2$	**Ho** 67 164.93032 $4f^{11} 5d^0 6s^2$	**Er** 68 167.26 $4f^{12} 5d^0 6s^2$	**Tm** 69 168.93421 $4f^{13} 5d^0 6s^2$	**Yb** 70 173.04 $4f^{14} 5d^0 6s^2$	**Lu** 71 174.967 $4f^{14} 5d^1 6s^2$

‡Actinide Series

Ac 89 (227.02775) $6d^1 7s^2$	**Th** 90 232.0381 $6d^2 7s^2$	**Pa** 91 (231) $5f^2 6d^1 7s^2$	**U** 92 238.0289 $5f^3 6d^1 7s^2$	**Np** 93 (237) $5f^4 6d^1 7s^2$	**Pu** 94 (244) $5f^6 6d^0 7s^2$	**Am** 95 (243) $5f^7 6d^0 7s^2$	**Cm** 96 (247) $5f^7 6d^1 7s^2$	**Bk** 97 (247) $5f^9 6d^0 7s^2$	**Cf** 98 (251) $5f^{10} 6d^0 7s^2$	**Es** 99 (252) $5f^{11} 6d^0 7s^2$	**Fm** 100 (257) $5f^{12} 6d^0 7s^2$	**Md** 101 (258) $5f^{13} 6d^0 7s^2$	**No** 102 (259) $5f^{14} 6d^0 7s^2$	**Lr** 103 (262) $5f^{14} 6d^1 7s^2$

§ Atomic mass values averaged over isotopes in the percentages they occur on Earth's surface. For unstable elements, mass of the longest-lived known isotope is given in parentheses. 2003 revisions. (See also Appendix B.)

Useful Geometry Formulas—Areas, Volumes

Circumference of circle $\quad C = \pi d = 2\pi r$

Area of circle $\qquad\qquad A = \pi r^2 = \dfrac{\pi d^2}{4}$

Area of rectangle $\qquad A = lw$

Area of parallelogram $\quad A = bh$

Area of triangle $\qquad A = \frac{1}{2} hb$

Right triangle
    (Pythagoras) $\quad c^2 = a^2 + b^2$

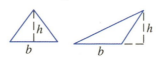

Sphere: surface area $\quad A = 4\pi r^2$
       volume $\qquad\; V = \frac{4}{3}\pi r^3$

Rectangular solid:
    volume $\qquad\qquad V = lwh$

Cylinder (right):
    surface area $\qquad A = 2\pi rl + 2\pi r^2$
    volume $\qquad\qquad V = \pi r^2 l$

Right circular cone:
    surface area $\qquad A = \pi r^2 + \pi r \sqrt{r^2 + h^2}$
    volume $\qquad\qquad V = \frac{1}{3}\pi r^2 h$

Exponents [See Appendix A–2 for details]

$(a^n)(a^m) = a^{n+m}$ [Example: $(a^3)(a^2) = a^5$]
$(a^n)(b^n) = (ab)^n$ [Example: $(a^3)(b^3) = (ab)^3$]
$(a^n)^m = a^{nm}$ $\qquad \begin{bmatrix}\text{Example: } (a^3)^2 = a^6 \\ \text{Example: } (a^{\frac{1}{4}})^4 = a \end{bmatrix}$

$a^{-1} = \dfrac{1}{a} \qquad a^{-n} = \dfrac{1}{a^n} \qquad a^0 = 1$

$a^{\frac{1}{2}} = \sqrt{a} \qquad a^{\frac{1}{4}} = \sqrt{\sqrt{a}}$

$(a^n)(a^{-m}) = \dfrac{a^n}{a^m} = a^{n-m}$ [Ex.: $(a^5)(a^{-2}) = a^3$]

$\dfrac{a^n}{b^n} = \left(\dfrac{a}{b}\right)^n$

Quadratic Formula [Appendix A–4]

Equation with unknown x, in the form
$$ax^2 + bx + c = 0,$$
has solutions
$$x = \frac{-b \pm \sqrt{b^2 - 4ac}}{2a}.$$

Logarithms [Appendix A–8; Table p. A–11]

If $y = 10^x$,   then $\quad x = \log_{10} y = \log y$.
If $y = e^x$,   then $\quad x = \log_e y = \ln y$.

$\log(ab) = \log a + \log b$

$\log\left(\dfrac{a}{b}\right) = \log a - \log b$

$\log a^n = n \log a$

Binomial Expansion [Appendix A–5]

$(1 + x)^n = 1 + nx + \dfrac{n(n-1)}{2\cdot1}x^2 + \dfrac{n(n-1)(n-2)}{3\cdot2\cdot1}x^3 + \cdots$ [for $x^2 < 1$]

$\qquad \approx 1 + nx \quad$ if $x \ll 1$

$\qquad\qquad$ [Example: $(1 + 0.01)^3 \approx 1.03$]

$\qquad\qquad$ [Example: $\dfrac{1}{\sqrt{0.99}} = \dfrac{1}{\sqrt{1-0.01}} = (1 - 0.01)^{-\frac{1}{2}} \approx 1 - (-\frac{1}{2})(0.01) \approx 1.005$]

Fractions

$\dfrac{a}{b} = \dfrac{c}{d}$ is the same as $ad = bc$

$\dfrac{\left(\dfrac{a}{b}\right)}{\left(\dfrac{c}{d}\right)} = \dfrac{ad}{bc}$

Trigonometric Formulas [Appendix A–7]

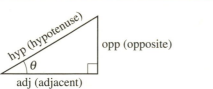

$\sin\theta = \dfrac{\text{opp}}{\text{hyp}}$

$\cos\theta = \dfrac{\text{adj}}{\text{hyp}}$

$\tan\theta = \dfrac{\text{opp}}{\text{adj}}$

$\text{adj}^2 + \text{opp}^2 = \text{hyp}^2$ (Pythagorean theorem)

$\tan\theta = \dfrac{\sin\theta}{\cos\theta}$

$\sin^2\theta + \cos^2\theta = 1$

$\sin 2\theta = 2\sin\theta\cos\theta$

$\cos 2\theta = (\cos^2\theta - \sin^2\theta) = (1 - 2\sin^2\theta) = (2\cos^2\theta - 1)$

$\sin(180° - \theta) = \sin\theta \qquad\qquad \cos(180° - \theta) = -\cos\theta$

$\left.\begin{array}{l}\sin(90° - \theta) = \cos\theta \\ \cos(90° - \theta) = \sin\theta\end{array}\right\}$ $[0 < \theta < 90°]$

$\sin\frac{1}{2}\theta = \sqrt{(1 - \cos\theta)/2} \qquad \cos\frac{1}{2}\theta = \sqrt{(1 + \cos\theta)/2}$

$\sin\theta \approx \theta$ [for small $\theta \lesssim 0.2$ rad]

$\cos\theta \approx 1 - \dfrac{\theta^2}{2}$ [for small $\theta \lesssim 0.2$ rad]

$\sin(A \pm B) = \sin A \cos B \pm \cos A \sin B$

$\cos(A \pm B) = \cos A \cos B \mp \sin A \sin B$

For any triangle:

$c^2 = a^2 + b^2 - 2ab\cos\gamma$ (law of cosines)

$\dfrac{\sin\alpha}{a} = \dfrac{\sin\beta}{b} = \dfrac{\sin\gamma}{c}$ (law of sines)

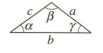